Principles of
Polymerization Engineering

Principles of Polymerization Engineering

JOSEPH A. BIESENBERGER
DONALD H. SEBASTIAN
Stevens Institute of Technology

A Wiley-Interscience Publication

JOHN WILEY & SONS

New York   Chichester   Brisbane   Toronto   Singapore

Library of Congress Cataloging in Publication Data:

Biesenberger, Joseph A., 1935–
  Principles of polymerization engineering.

  "A Wiley-Interscience publication."
  Includes index.
  1. Polymers and polymerization. I. Sebastian, Donald
H., 1952–   II. Title.

TP156.P6B48 1983   668.9   82-23746
ISBN 0-471-08616-9

Printed in the United States of America

10 9 8 7 6 5 4 3 2 1

*This book is dedicated
to the memory of Joseph Biesenberger,
father of the senior author,
whose legacy transcends
the boundaries of these pages.*

Preface

Science searches for order among diverse phenomena (natural and biological), which are often seemingly unrelated. It utilizes observation and the construction of models.

Engineering, on the other hand, applies scientific principles for the benefit of mankind. Its products often require the design, construction, and operation of complex systems (processes). These systems exhibit their own kind of order. The task of engineering science in general (and chemical engineering science in particular) is to seek and delineate order among such systems (chemical processes).

In this book, we deal specifically with processes aimed at the manufacture of polymeric materials. Consistent with the task of engineering science, we emphasize concepts and principles. Our goal is the formulation of generalizations that will be useful in the design, scaling, and modification of polymerization processes. Our strategy consists of constructing and analyzing the behavior of relevant model processes and is therefore largely heuristic. Our results, whenever possible, take the form of dimensionless algorithms or criteria, with clear physical significance to facilitate their application to actual processes. Actual processes are used occasionally as examples, where sufficient knowledge and data are available.

No claim is made that this book is an exhaustive treatise on "polymerization reaction engineering," since no attempt has been made to review the entire literature dealing with this subject. Quite the contrary, much of it is based on work done in The Chemistry and Chemical Engineering Department at Stevens Institute of Technology. We accept the blame for the inevitable errors thereby introduced and for inadvertently "reinventing the wheel" wherever that is done.

This book is directed primarily toward the industrial engineer or scientist in research or development with a firm technical background who is attempting to apply that background to polymerization processes. It should also be suitable as a text for courses dealing with chemical engineering aspects of polymer processes, since it evolved from a graduate course on polymerization engineering taught at Stevens Institute over a period of 20 years. The writing style is intentionally compact for the sake of economy, but many examples are included to assist the reader in understanding the concepts presented.

A knowledge of transport theory and chemical reaction engineering (kinetics) is presumed. The reader who requires a review of these topics as they relate to the text will find a concise treatment in Appendixes E and F, along with a brief review of polymerization chemistry (Appendix A), thermodynamics (Appendix D), and relevant mathematical methods (Appendixes B and C). Appendixes D–F, in addition to serving the immediate needs of the text, contain background material for Appendix G, which is devoted to a general discussion of characteristic times (CT's). This "text within a text" was felt to be justified, in view of the heavy use of CT's throughout the book, to facilitate their physical interpretation and demonstrate their general utility.

Chapter 1 consists of a collection of fundamental concepts on which subsequent chapters are based. Chapter 2 discusses phenomena that are inherent in polymerization reactions. Chapters 3–5 deal with physical factors that affect, or are affected by, polymerizations on an engineering scale. Chapter 6 treats postreactor separation of unreacted monomer and/or diluents.

The reader will note that Chapters 3–5 are not organized in the traditional way according to reactor type (batch, CSTR, continuous tubular, etc.). Agitated reactors, such as CSTR's, are included in Chapters 3 and 4 as part of a more general class of reactors, termed ideal mixers (IM's), and laminar-flow tubular reactors are included in Chapters 3–5 as streamline reactors. This was done to allow for the treatment of nontraditional reactors peculiar to polymerizations, especially those carried out in bulk, which may be structurally different from the traditional types and yet dynamically similar, and also to encourage an open-minded approach to the design of new polymerization reactors.

Most of the graphs were prepared by computer digitization of rough copy and then plotted in india ink, using a high-speed drum plotter interfaced with a DEC System 10 computer. The line styles and character set fonts were developed and coded by the junior author.

To our friends and colleagues who supported and assisted us throughout the preparation of this book, we wish to express our sincere appreciation. Specifically, we would like to thank: Professors Zehev Tadmor, Costas Gogos, Nick Peppas, Kenneth O'Driscoll, and Dr. Pradip Mehta for reading portions of the manuscript and offering many helpful suggestions; Drs. Ilan Duvdevani and Robert Capinpin for their (previously unpublished) contributions; Farrel Machinery Group (Emhart), especially Dr. Peter Hold, for generously supporting the research that led to some of the original contents; and, of course, the Department of Chemistry and Chemical Engineering at Stevens Institute of Technology for the use of its facilities and services. Last, but certainly not least, our heartfelt thanks go to Miss Ann Randy for typing the entire manuscript—equations, tables, and all. Her unequaled skill, dedication, and good humor helped make this formidable task a pleasure.

JOSEPH A. BIESENBERGER
DONALD H. SEBASTIAN

Hoboken, New Jersey
May 1983

Contents

2. REACTION PHENOMENA

Symbols

A	addition-type monomer
A	Helmholtz free energy or dimensionless ratio λ_{ad}/λ_v
AA	condensation-type monomer
AB	condensation-type monomer
a	ratio of CT's λ_{ad}/Λ_R
a	number of repeat units A in copolymer chain, activity or ratio of CT's λ_{ad}/λ_R.
a_K	ratio of reaction rate constants k_i/k_p
B	addition type monomer
BB	condensation-type monomer
b	number of repeat units B in copolymer chain, number of branches in polymer chain, or ratio of CT's λ_r/λ_{ad} (dimensionless)
b_k	ratio of CT's λ_k/λ_{ad} (dimensionless)
b'	number of branches per repeat unit
c	molar concentration
c_p	specific heat
D	molecular diffusivity or denominator (defined in text)
Da	Damköhler number of the first kind (dimensionless)
Da_{II}	Damköhler number of the second kind (dimensionless)
D_A	CCD dispersion index
De	devolatilization number (dimensionless)
D_L	longitudinal dispersion coefficient
D_N	number dispersion index
D_W	weight dispersion index
d	diameter

E	total energy, activation energy, or external age distribution (RTD)
$\dot{E}$	evaporation rate
E_F	process efficiency
E_f	stage efficiency
Ex	extraction number (dimensionless)
e	specific total energy, density function for E (RTD), or amount evaporated per unit mass
f	function (general) or initiator efficiency
G	Gibbs free energy, G-E function, shear modulus or ratio of pressure flow-rate to drag flow-rate
ΔG_M	free energy of mixing
g	function (general), G-E function or specific Gibbs free energy
g_e	temperature dependence of generation function
H	enthalpy, vessel dimension transverse to flow or holdback
H_k	partial molar enthalpy
H_k'	partial enthalpy
ΔH_M	heat of mixing
ΔH_r	heat of reaction
h	specific enthalpy or film thickness
$\hat{h}_A, \hat{h}_B$	dimensionless intermediate concentrations for copolymerization
I	internal age distribution
i	density function for I
i_{seg}	intensity of segregation
j_k	mass transport flux of substance k by diffusion
j_Q	molecular transport flux of property Q (general)
K	equilibrium constant
K_W, K_c, K_c'	Henry's law constants
k	reaction rate constant, mass transfer coefficient
k	thermal conductivity
k_{ap}	apparent (lumped) rate constant for polymerization
k_{ax}	apparent (lumped) rate constant for DP
k_{pt}	lumped rate constant for propagation and termination
L	longitudinal vessel dimension or streak length
l	characteristic length
l_{seg}	length of segregation
M	molecular weight
M_0	molecular weight of monomer or repeat unit
M_{A0}, M_{B0}	molecular weight of co-repeat units
m	mass

m	monomer
m_0	initiator or by-product
m_1	monomer or 1-mer
m_x	x-mer
m'_x	x-mer with a terminal double bond
N	number of moles or number of stages
N_p	number of particles or number of moles of polymer
N_s	number of samples
n	mole fraction, termination stoichiometry, reaction order (general), or power-law flow index
$\boldsymbol{n}_k$	total mass transport flux of substance k
$\boldsymbol{n}_Q$	total transport flux of property Q (general)
$\dot{O}$	output rate
$\dot{o}$	specific output rate
P	intensive property (general), or probability
p	pressure
Pe_L	longitudinal Peclet number (dimensionless)
Po	Poiseuille number (dimensionless)
P_S	equilibrium partial pressure of substance S in solution
Q	extensive property (general)
Q_c	extensive property "vector" (column matrix)
Q_k	partial molar extensive property (general)
Q'_k	partial extensive property (general)
q	specific extensive property (general)
$\boldsymbol{q}$	heat flux vector
R	ratio (general) or k_{tc}/k_t
R'	fraction of polymers terminated by combination
R_A, R_B	reactivity ratios for comonomers A and B
$\mathscr{R}_A, \mathscr{R}_B$	parameters for penultimate effect termination model
Ra	radius
R_B	bubble radius
Re	Reynolds number (dimensionless)
RD	relative dispersion
R_f	chain transfer constant
R_g	universal gas constant
R_R	recycle ratio
r	volume-specific rate (general), polymerization rate or radial position
r_B	rate of bubble nucleation per unit volume
r_e	removal function
r_e	evaporation rate per unit volume

r_k	rate of formation of substance k per unit volume
r_G	generation rate per unit volume
S	entropy, surface area, or dimensionless segregation number
S	by-product, solvent, or impurity
SD	standard deviation
SK	skewness
S_k	partial molar entropy
S_k'	partial entropy
ΔS_M	entropy of mixing
s	specific entropy, surface area, initiator stoichiometry, or sign (± 1)
s_R	stream flow ratio $\dot{V}_f/\dot{V}$
s_V	surface-to-volume ratio
T	temperature (absolute)
T	total stress tensor (dynamic and static)
t	time
U	internal energy or overall heat transfer coefficient
u	specific internal energy or unit step function
V	volume
$\dot{V}$	volumetric flow (throughput) rate
v	velocity vector
v	specific volume or velocity magnitude
v_0	velocity magnitude at the axis
W	weight or width
w	weight fraction
X	volumetric flow-rate per unit area for a process
x	number of repeat units in polymer chain (DP), position coordinate in space, or volumetric flow-rate per unit area for a stage
x'	number of equivalent volume segments in polymer chain
x_0	ratio of monomer to initiator in feed
x_c	position "vector" (column matrix)
y	fraction of comonomer A in copolymer chain (CC) or position coordinate
z	transform variable, position coordinate, or eigenvalue

GREEK

α	age
α_K	ratio of CT's λ_m/λ_i (dimensionless)
α_T	thermal diffusivity
β_K	ratio of CT's Λ_A/Λ_m (dimensionless)

γ	activity coefficient, shear strain
$\dot{\gamma}$	shear strain rate
γ_k	ratio of CT's λ_k/λ_p (dimensionless)
γ_i	ratio of CT's λ_i/λ_p (dimensionless)
γ_m	ratio of CT's λ_m/λ_p (dimensionless)
γ_r	ratio of CT's λ_r/λ_p (dimensionless)
γ_T	ratio of CT's λ_G/λ_R (dimensionless)
γ_v	ratio of CT's λ_v/λ_p (dimensionless)
Δ	final (effluent) minus initial (feed) value or ratio of CT's λ_{ad}/λ_H (dimensionless)
δ	impulse (Dirac) function, small variation, or feedback distance
δ_H	ratio of CT's λ_H/λ_{ad}
δ_K	dimensionless group describing temperature drift in CC
δ_p	penetration depth
ε	small number, shear strain, or reciprocal dimensionless activation energy
ζ_K	dimensionless group describing composition drift in ΔH_r for copolymerizations
η	viscosity
θ	dimensionless temperature or variable angle
Λ	composite CT
Λ_v	space (residence) time for complex process
λ	CT for simple process (stage)
λ_v	space (residence) time for simple process (stage)
μ_k	chemical potential of substance k
μ^k	kth moment
ν	kinematic viscosity
ν_A	kinetic chain composition
ν_b	kinetic degree of branching
ν_k	stoichiometric coefficient for substance k
ν_{jk}	stoichiometric coefficient for substance k in reaction j
ν_N	kinetic chain length
ξ	extent of reaction (general)
ρ	particle property (general) or density
σ	standard deviation or surface tension
σ^k	kth moment
σ^2	variance
τ	dynamic stress tensor
τ	residence time
τ_0	shortest RT
τ_K	ratio of CT's λ_m/λ_t (dimensionless)
Φ	conversion (general) or monomer conversion

Φ_{jk}	yield of product j from reactant k
ϕ	volume fraction, probability of propagation or copolymerization termination parameter
χ	molecular interaction parameter
Ψ_{jk}	selectivity of reactant k for product j
Ω	intensity function

SUBSCRIPTS

A	monomer of type A
a	number of repeat units A
ad	adiabatic
B	monomer of type B or bubble
b	number of repeat units B or number of branches
c	concentration or ceiling
c	column matrix
ce	chemical equilibrium
cr	critical
D	disproportionation or diffusion
d	initiator decomposition
e	equilibrium or evaporation
f	final or chain-transfer
f	film or forward
G	generation or gas
i	initiation (as in r_i)
i	initiator (as in λ_i, Φ_i, and b_i) or, occasionally, substance i (general)
j	substance j or stage j
k	substance k
L	liquid or longitudinal
l	substance l
M	mixing
m	monomer
m	mean
mac	macroscopic
me	mechanical equilibrium
mic	microscopic
N	number or Nth stage
n	normal space component
0	feed (initial) or initiator
p	propagation, pool, or pseudo

p	polymer
R	reservoir or removal
RU	repeat unit
r	reaction, reverse, or reference state (condition)
s	substance S (general), solvent, sample, or square matrix
s	steady state or stable
T	transverse
t	termination
V	volume
W	weight
x	*x*-mer
y	CC

SUPERSCRIPTS

M	mixing
0	pure component
T	transpose

MISCELLANEOUS

{ }	area-average
⟨ ⟩	volume-average
[ ]	flow-average
| |	magnitude or absolute value
‾ (single overbar)	mean quantity
＝ (double overbar)	mean sample property
ˆ (hat)	dimensionless quantity
· (overdot)	rate quantity
′ (prime)	quantity that differs in some way from unprimed quantity (general), volume-specific quantity, partial property or quantity pertaining to double-bond containing molecule
*	active intermediate or standard state
∇	"del" vector operator

Abbreviations

BM	backmixing
BR	batch reactor
CC	copolymer composition
CCD	copolymer composition distribution
CDFR	continuous drag-flow reactor
CEPR	continuous externally pressurized reactor
CSTR	continuous stirred tank reactor
CT	characteristic time
CTR	continuous tubular reactor
DB	degree of branching
DBD	degree of branching distribution
D-E	dead-ending
DF	diffusing film
DP	degree of polymerization
DPD	degree of polymerization distribution
DV	devolatilization
ERA	early runaway approximation
EXP	exponential
FB	feedback
G-E	gel-effect
IM	ideal mixer
LCA	long-chain approximation
MM	maximally micromixed
MS	microsegregated
MW	molecular weight

MWD molecular-weight distribution
PF   plug-flow
PFR plug-flow reactor
QSSA quasi steady-state approximation
R-A runaway
RC   recycle
REX reactive extrusion
RIM reactive injection molding
RSSA reactor steady-state approximation
RT   residence time
RTD residence time distribution
RX   reaction
SIMS series of IM's
SR   surface renewal
SS   stream-splitter

CHAPTER ONE

Fundamentals

We shall begin with an introduction to some concepts of a general nature and principles specific to polymers that will be used throughout the book as needed. Polymeric materials and polymerization reactions will be discussed in terms of generalized symbols, which are defined in separate tables throughout this chapter.

1.1. REACTOR DYNAMICS AND REACTION PATH

Polymeric materials consist of long-chain molecules. The composition, molecular weight (MW), and structure of these macromolecules determine their bulk physical properties (processing as well as end-use). Some of their molecular properties are in turn direct consequences of the reaction path experienced by the growing macromolecules during formation.

The term "reaction path" has at least two connotations. To the chemist, it suggests a particular chemical reaction mechanism. Since there are often several alternative chemical reaction paths leading to a given polymer, a choice must be made; that choice frequently affects the molecular structure of the product. Choice of catalyst (free radical versus ionic) is an obvious example.

To the chemical engineer, on the other hand, the term reaction path signifies the physical history experienced by a reaction as dictated by prevailing reactor dynamics. Physical reaction path is a consequence of coupling between reaction and transport processes, which are especially important in reactions conducted on an engineering scale. Such physical factors can strongly influence the outcome of polymerization reactions, and in many cases significantly alter certain molecular properties of the polymeric products.

The science of chemical kinetics deals primarily with the quantitative characterization of chemical reactions in progress, devoid of associated transport processes. It utilizes mathematical models and experimental rate studies for the elucidation of reaction mechanisms and the evaluation of specific rate constants. On the other hand, chemical engineering kinetics, recently renamed "chemical reaction engineering," deals with the interaction between the chemical and physical processes, and attempts to formulate mathematical models for the purpose of reactor design and control by applying chemical kinetics coupled with mass, momentum, and energy transport processes, as required.

It may be said that polymerization reactions possess certain characteristics that offer the chemical engineer unique challenges. Several distinguishing features in particular render these reactions especially susceptible to physical effects. One is their potential for dramatic increases in reaction viscosity; another is their exothermic nature. Virtually all polymerization reactions exhibit low diffusivities and thermal conductivities in addition to high viscosities. Reaction viscosity is sensitive to the concentration and molecular weight of dissolved polymer formed and to reaction temperature, frequently in that order. Increases from fractions of a centipoise to thousands of poises are typical for bulk polymerizations, notwithstanding the mitigating effects of rising temperatures when isothermal conditions are not maintained. Autoacceleration due to the coupling between reaction and transport processes is not uncommon; examples include chain polymerization with hindered termination and thermal runaway.

Thus "hot spots" and "flow occlusions" are likely to occur within reactors, and their occurrence can affect reaction efficiency and alter the physical properties of polymer product. An example is a runaway polymerization whose temperature approaches the ceiling temperature or is sufficiently high to support degradative reactions. Possible consequences are a significant reduction in molecular weight and an altered distribution. This touches on a most important objective in polymerization engineering: product quality. Quality of reactor output normally refers to the selectivity of a reaction with respect to desired versus undesired products. In polymerization engineering it could manifest itself as a molecular property or property distribution (molecular weight, copolymer composition, degree of branching, etc.). Such properties will certainly be affected by differences in physical reaction path. Conversely, however, it also appears possible, at least in principle, to utilize these differences to tailor polymer properties to desired specifications by means of proper manipulation of the appropriate engineering parameters. An example is the production of a copolymer with uniform composition in continuous, well-mixed vessels by taking advantage of backmixing and the steady state to eliminate "drift dispersion."

Transport processes interact with polymerizations reactions on various levels. On the molecular level, for instance, diffusion effects can hinder bimolecular chain termination, as previously mentioned, and enhance primary radical recombination, thus giving rise to autoacceleration and reduced catalyst efficiency, respectively. The former has been termed the "gel-effect" and the latter the "cage-affect." In both cases conventional kinetics are no longer obeyed and polymer properties are affected. On a larger scale, due to shear forces or incomplete mixing, it is possible for reaction zones to remain microscopically segregated and thus to experience a variety of different concentration histories, depending on the degree to which they interact by means of transport processes. This is certainly true for polymerizations in bulk and is often true for polymerizations with diluents as well. In fact, suspension, emulsion, and interfacial polymerizations are deliberately aimed at achieving microsegregation of reaction zones. Transport rates to and from these reaction zones are obviously of major importance in such systems.

The effects of chemical reaction path on polymer properties have been discussed in numerous books on polymer chemistry. The emphasis in this book will be on the effects of physical reaction path, as determined by mixing, heat-transfer, and flow conditions, on certain properties such as molecular weight, copolymer composition, degree of branching, and their distributions. In fact, the reader will recognize this list of effects as the basis for organizing the chapters. Every attempt will be made to generalize results and to present them in dimensionless form in order to facilitate their application to the widest possible class of polymerizations and reactors.

1.1.1. *Polymerization Modes*

The chemical reaction paths leading to various types of polymers have been summarized in Appendix A. Here we shall review briefly some common modes for carrying out these reactions, in particular the nature of the phases involved. The mode that is chosen for a given polymerization can affect the physical reaction path (mixing, heat transfer, flow) and, in some cases, even the mechanism of the reaction (emulsion versus suspension, solution, or bulk).

Table 1.1-1 lists some common industrial modes for polymerizations. Bulk polymerization is the simplest in principle because it involves the fewest inert materials, which could later become impurities. It is presumed that the polymer is soluble in its own monomer. If so, high reaction viscosities are possible with concomitant effects such as autoacceleration (gel-effect, thermal runaway) or incomplete mixing, as previously mentioned. If not, then the polymer will precipitate and invite autoacceleratory behavior of a different kind.

Solution polymerization employs a diluent in which the monomer, catalyst, and polymer are each presumed to be soluble. The reaction mechanism is identical to that of bulk reaction unless the diluent is not entirely inert (chain transfer, polar effects, etc.). The diluent generally mitigates the deleterious aspects of the physical

Table 1.1-1. Polymerization Modes

MODE	INITIAL STATE	REACTION SITE	FINAL STATE
BULK	HOMOGENEOUS	LIQUID PHASE	DISSOLVED POLYMER WITH POSSIBLE GEL-EFFECT OR POLYMER PRECIPITATE
SOLUTION	HOMOGENEOUS	LIQUID PHASE	
SUSPENSION	HETEROGENEOUS	DISPERSED LIQUID PHASE	SOLID POLYMER
EMULSION	HETEROGENEOUS	DISPERSED LIQUID PHASE	SOLID POLYMER
INTERFACIAL	HETEROGENEOUS	INTERFACE BETWEEN LIQUIDS	SOLID POLYMER
CATALYTIC	HETEROGENEOUS	SOLID CATALYST SURFACE	SOLID POLYMER

reaction path in bulk reactions cited earlier, but it must be removed subsequently. This necessitates the addition of a separation process that could be energy-consuming and costly.

Suspension polymerization employs a diluent (water) in which neither the monomer, catalyst, nor polymer is soluble. It has the same mitigating effects on physical reaction path (lower viscosity, better heat transfer, etc.) just described for solution polymerization and generally requires a separation step as well. The chemical reaction path in the dispersed organic phase is presumed to be identical to that of bulk reaction. Dispersion is effected through mechanical agitation although inert material is sometimes added as a stabilizing agent. Thus impurities are introduced.

Emulsion polymerization also employs water as the diluent, but the organic phase is stable and more finely dispersed. The catalyst is usually water-soluble and the polymerization cites are provided by soap micelles; the reaction mechanism is thereby altered. Emulsion polymerization will be treated in some detail in Section 2.9 along with gel-effect and precipitous polymerizations.

Interfacial polymerizations involve two immiscible liquids, one containing monomer A and the other monomer B. Reaction takes place at the liquid-liquid interface where a solid polymer film is produced. The polymerization type is random propagation (Section 1.6).

Finally, by catalytic polymerizations (Table 1.1-1), we mean reactions that involve Ziegler–Natta type heterogeneous catalysts, which contain a transition metal halide and an aluminum alkyl and produce stereospecific polymers (Section 1.5). In such reactions, the continuous phase may be a liquid or a gas, in which the solid catalyst particles are dispersed by mechanical agitation and the polymer product is generally insoluble.

1.2. REACTOR TYPES

All chemical reactors, whether industrial or experimental in scale, fall into one of three broad categories: batch, semibatch (or semicontinuous) and continuous. The batch reactor (BR) is a closed system; no material enters or leaves during reaction and therefore it proceeds toward a thermodynamic equilibrium, which is the only steady state available to it. In the semibatch reactor, either some reactants are introduced or some products are removed. The system is thus forced to remain in a perturbed state and is prevented from reaching equilibrium. It is therefore also an inherently transient reactor in which no true steady state is possible. Certain polymerizations, for instance, require continuous feeding of catalyst or monomer or continuous withdrawal of volatile by-product such as water vapor. In continuous reactors, both feed and effluent streams flow continuously; consequently these reactors generally seek a dynamic steady state in which to operate.

The chief reason for the increasing influence of transport processes on the behavior of chemical reactions during reactor scale-up is the declining surface-to-volume ratio. Since the quantities being transported, namely, mass and thermal energy, are extensive properties, their generation (or depletion) rates increase in proportion to the volume of the reacting system. However, their transport rates, being proportional to surface area, tend to increase less rapidly until they play an important, if not dominant, role in determining the overall rate and even the character of the chemical reaction.

In any chemical reactor, owing to the depletion of some substances, the production of others, and/or the absorption or release of heat, reaction variables such as concentration and temperature can, and generally do, vary with position in space as well as with time. Consequently, local neighborhoods containing reacting material (reaction elements) may experience different concentration and thermal histories (1). The nature and extent of these variations are, of course, affected by such factors as rate of mixing and rate of mass and heat transport.

While in a batch reactor, the final product is a time-composite of the reaction elements, the effluent from a continuous reactor is a flow-composite. Reaction time is therefore an average quantity, determined by the space-time of the reaction vessel

$$\lambda_v = \frac{V}{\dot{V}} \tag{1.2-1}$$

where V is reaction volume and $\dot{V}$ is the volumetric throughput rate, generally at the temperature and pressure of the feed. Throughput rate reflects the scale of the process and is generally dictated by economic factors, whereas λ_v is determined by the time required to achieve a desired conversion. Thus specifying $\dot{V}$ and λ_v fixes the reactor size V.

An additional geometric parameter that plays a significant role in determining the performance of a continuous reactor is its shape. Since batch reactors are generally the most efficient in terms of specific output rate (Section 1.3), it seems reasonable to attempt to design continuous reactors, with their inherent advantages over batch operation, that emulate the essential features of the batch reactor. Such an attempt leads naturally to the streamline reactor, whose longitudinal dimension (in the direction of flow) is large and whose transverse dimensions (orthogonal to flow) are small. Mixing, especially longitudinal, is avoided and the reactor may be pictured for convenience as consisting of a system of tiny batch reactor elements being conveyed in parallel along streamlines, in which reaction time is replaced by L/v, where L is reactor length and v is the velocity along a streamline. An equivalent definition for space time in such reactors is

$$\lambda_c = \frac{L}{\langle v \rangle} \tag{1.2-2}$$

where $\langle v \rangle$ is mean longitudinal velocity.

When laminar flow prevails, as it does in most polymerization reactors (especially for bulk polymerization), the transverse dimension H of streamline reactors is frequently limited by the rate of transport processes occurring in the transverse direction, which are of necessity molecular and consequently very slow. The most common example involves the removal of reaction heat by transverse conduction.

More specifically, we conclude from the following expression

$$\dot{V}\lambda_v \propto H^2 L \tag{1.2-3}$$

that after specifying $\dot{V}$, λ_v, and H, the only remaining geometry parameter is L. A combination of small H and large λ_v, which is not atypical for many polymerizations of commercial interest whose characteristic reaction times are of order of minutes or hours, can lead to a demand for long reactors requiring very large driving forces for flow. This requirement will be exacerbated by high reaction viscosities if the driving force is externally applied pressure, as it commonly is in such streamline reactors. An example is the continuous tubular reactor (CTR).

An alternative to the streamline reactor is a reactor in which mixing is purposely induced via mechanical agitation to assist the transport processes described earlier and to eliminate stagnancy (dead spaces), which frequently occurs when mixing is absent. Such vessels, whose dimensions are generally similar in all directions, are termed agitated reactors. They are virtually always used for batch and semibatch operation. A common example is the continuous stirred tank reactor (CSTR). Although agitation certainly facilitates transport, as intended, it simultaneously introduces a phenomenon known as backmixing (BM), which may be detrimental to reactor output.

The streamline reactor and the agitated reactor are illustrated in Fig. 1.2-1. Idealizations of these reactors are represented by the extreme conditions of plug flow (PF) and ideal mixing, respectively. In the former mixing is totally absent, and in the latter it is complete. For purposes of analysis and design, it is customary practice in chemical reaction engineering to treat actual continuous reactors as one of these two fundamentally distinct types, allowing, of course, for appropriate variations that are specific to the reactor in question. PF vessels and ideal mixers (IM's) may be coupled in series and in parallel configurations.

It is also customary practice in the design of reactors to rely on agitation to induce mixing and externally applied pressure to induce flow. When dealing with polymerizations, however, where reaction viscosities can span several orders of magnitude, as previously mentioned, high viscosities accompanying advanced reaction can cause serious mixing and conveying problems in conventional types of

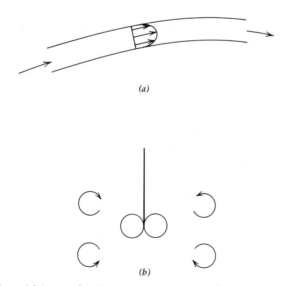

(a)

(b)

Figure 1.2-1.   (a) Continuous streamline reactor. (b) Agitated reactor.

reactors. Polymer products having viscosities of hundreds or even thousands of poises can result from monomeric liquids whose viscosities are only fractions of 1 poise, or even from monomeric gases. Large requisite values for λ_v would in general favor the agitated vessel until waxing viscosities prevent adequate mixing for the removal of reaction exotherm, at which point a streamline reactor might be preferred. Such considerations would suggest a reactor sequence consisting of a CSTR followed by a CTR, wherein the material passes from the former to the latter after a suitable reaction viscosity has been achieved. A major difficulty with this arrangement, of course, is conveying the high-viscosity material in the CTR via externally applied pressure without experiencing inordinately large pressure demands and ultimately even occlusion (plugging).

An alternative that is available to the chemical engineer faced with high reaction viscosities, but not always recognized, is the drag flow principle. By utilizing vessels with "moving walls," one can pump viscous polymeric liquids via internal pressurization (2) and mix them by the controlled action of shear strain (laminar mixing). Internal pressurization is actually enhanced by high viscosities. These methods have been successfully employed by the polymer processing engineer in such machines as the extruder. We shall name the class of continuous reactors that are based on the drag flow principle in general (illustrated in Fig. 1.2-2), the continuous drag-flow reactor (CDFR). Thus, to solve the problem alluded to earlier, it would seem reasonable to use a reactor sequence consisting of a CSTR and CDFR in series.

In Chapter 3, we shall identify three model continuous reactors from among the agitated and streamline classes, based on their mixing characteristics. They will be

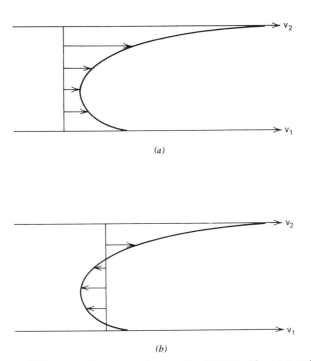

(a)

(b)

Figure 1.2-2. The parallel-plate, continuous drag flow reactor (CDFR) without (*a*) and with (*b*) reverse flow.

analyzed in detail computationally as polymerizers, but every attempt will be made to treat them abstractly, without tying them to specific reactors. Toward this end, the continuous reactor with complete backmixing will be labeled IM rather than CSTR, because agitating a tank is not the only way to achieve good mixing. The continuous plug-flow reactor will not be equated with the CTR, because forcing a fluid through a circular tube with external pressure is not the best way to achieve pluglike flow. Backmixing can be achieved in laminar, drag-flow vessels by inducing excessive backflow, and plug-flow can be approached by avoiding it. Thus, by viewing all existing and potential reaction vessels in terms of these abstract models, we hope to promote versatility and innovation in the future design of polymerization reactors.

1.3.   PROCESS OUTPUT, EFFICIENCY, AND SELECTIVITY

Some chemical processes are simple, some are complex. For our purposes a simple process is one consisting of a single element. Furthermore, we shall consider only reaction (RX) and devolatilizing (DV) elements. Examples of the former were introduced in Section 1.2 and will be discussed further in Chapters 3–5. Examples of the latter will be discussed in Chapter 6, which deals with postreactor separation processes.

Several general configurations for complex processes have been sketched in Fig. 1.3-1. The elements comprising these processes may be RX or DV and may be coupled in series or parallel as shown. The series configuration represents the important phenomenon of staging (cascading), and the parallel configuration includes bypassing (short-circuit) as a special case. The equally important phenomenon of recycle (RC), or feedback (FB), is also included. It should be noted that stream separation and recombination are represented by idealized elements labeled the stream-splitter (SS) and ideal mixer (IM), respectively.

Whether processes are simple or complex, criteria for judging their performance are required for at least two reasons. One concerns the choice of optimum design configuration from among alternative candidates; the other concerns the optimum operating mode for any given configuration. The goal of this section is to lay the foundations for the establishment of such criteria.

1.3.1.   Definitions

Process efficiency in general may be defined as the actual outcome divided by the theoretical (best) outcome. In terms of the removal of arbitrary substance k (e.g., monomer) by reaction or separation, the following alternative definitions may be used:

$$E_f \equiv \left\{ \begin{array}{c} -\dfrac{\Delta N_k}{(N_k)_0} \\[2ex] -\dfrac{\Delta N_k}{(N_k)_0 - (N_k)_e} \end{array} \right\} \qquad (1.3\text{-}1)$$

where ΔN_k signifies final minus initial number of moles of k. An example of the first

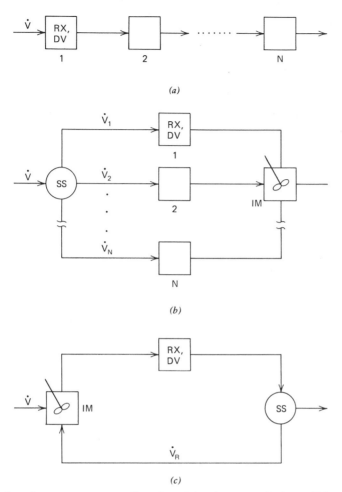

Figure 1.3-1. Several common process configurations: (*a*) series or staged, (*b*) parallel, and (*c*) recycle.

definition is the chemical conversion of reactant k, which is frequently expressed in terms of molar concentrations as

$$\Phi_k \equiv -\frac{\Delta c_k}{(c_k)_0} \qquad (1.3\text{-}2)$$

by neglecting density changes in the process. The second definition utilizes thermodynamic equilibrium as the "best" outcome and will be used in Chapter 6 to represent the efficiency of removing substance k in a separation process. Obviously, the two definitions are equivalent when $(N_k)_e \ll (N_k)_0$.

Frequently the success of a process is determined not by its efficiency alone, but by its rate as well. Therefore, we shall consider a quantity termed the output rate which we define for any continuous process as the amount of arbitrary substance k processed in unit time. In terms of substance k removed in a steady-state process, the molar output rate is $-\Delta \dot{N}_k$, where $\dot{N}_k$ represents a transport rate (Appendix F). If we neglect density changes and molecular transport across the vessel entrance and exit, we may write the molar output rate as follows for vessels with a single inlet and outlet

$$\dot{O}_k \equiv -\Delta[c_k]\dot{V} \tag{1.3-3}$$

where the square brackets signify flow-average concentration and $\dot{V}$ is volumetric throughput rate. Irrespective of which of the definitions (1.3-1) is used, it is evident that for continuous processes

$$\dot{O}_k \propto E_f \dot{V} \tag{1.3-4}$$

since

$$E_f \propto -\Delta[c_k] \tag{1.3-5}$$

Besides the obvious goal of achieving the highest possible efficiency for a specified throughput rate, another objective of good process design is to do it with the smallest possible equipment, since equipment size is reflected in capital cost. Therefore, we should seek to maximize volume-specific output rate $\dot{O}_k/V$ for processes whose rates are volume-dependent, such as chemical reaction (polymerization), and area-specific output rate $\dot{O}_k/S$ for processes whose rates are surface-area-dependent, such as transport (devolatilization), assuming that equipment costs for such processes are proportional to volume and area, respectively. These definitions may be summarized as follows:

$$\dot{o}_k = \begin{cases} -\dfrac{\Delta[c_k]\dot{V}}{V} \propto \dfrac{E_f}{\lambda_v} \\[3mm] -\dfrac{\Delta[c_k]\dot{V}}{s} \propto xE_f \end{cases} \tag{1.3-6}$$

where $\dot{o}_k$ represents specific output rate and V is the volume occupied by the materials being processed.

The quantities just defined are listed in Table 1.3-1, where s_V represents surface-to-volume ratio and x is defined as volumetric throughput rate per unit surface area,

Table 1.3-1. Summary of Definitions and Computational Methods for Process and Equipment Efficiency

E_f	$\propto$	$-\Delta[c_k]$	1st Method
	$\propto$	$\langle r_k \rangle \lambda_v$ or $\{n_k\} s_v \lambda_v$	2nd Method
$\dot{o}_k$	$\propto$	E_f/λ_v or $E_f x$	1st Method
	$=$	$\langle r_k \rangle$ or $\{n_k\}$	2nd Method

$\dot{V}/s$. Heretofore, for convenience, we have used symbols that correspond to simple processes, such as E_f, λ_v, and x. All expressions may be applied to composite processes as well by replacing these symbols with E_F, Λ_v, and X, respectively.

As noted earlier, the success of a process is frequently determined by its rate and not by its overall efficiency. This is especially true for complex chemical reactions with multiple products, some of which are more desirable than others. For instance, it may be more efficient or desirable in a global sense to "freeze" a reaction kinetically by stopping it at low conversions, or far from equilibrium, than to allow it to proceed if it has attained an optimum composition with respect to a desired product.

Considerations of product quality invite us to refine our earlier definitions of conversion, equations 1.3-1 and 1.3-2, as measures of reaction efficiency. Two suitable efficiency criteria for concurrent and consecutive reactions are selectivity and yield (1), which measure the proportion of a reaction that produces a desired product k in contrast with an undesired product l. The selectivity of reactant k for product j is defined as

$$\Psi_{jk} \equiv \frac{(\Delta N_j/\nu_j)}{(\Delta N_k/\nu_k)} = \frac{\Phi_{jk}}{\Phi_k} \qquad (1.3\text{-}7)$$

where Φ_{jk} is the yield of product j from reactant k,

$$\Phi_{jk} \equiv \frac{(\Delta N_j/\nu_j)}{(N_k)_0/\nu_k} \qquad (1.3\text{-}8)$$

and ν_j and ν_k are stoichiometric coefficients (Appendix D). Thus selectivity is a measure of the relative conversion of reactant k to product j, or the relative output. It is obviously possible for Ψ_{jk} to be high while Φ_{jk} is low, which signifies high selectivity but low conversion efficiency. For simple reactions, $\Psi_{jk} = 1$ and $\Phi_{jk} = \Phi_k$. Some of these concepts are illustrated in Example 1.3-3. Yield and selectivity in polymerizations take the form of degree-of-polymerization distribution (DPD), copolymer composition distribution (CCD), degree-of-branching distribution (DBD), and so on.

1.3.2. Computational Methods

As noted, an important objective in the analysis of continuous process performance is the calculation of quantities such as efficiency and specific output. We will use two alternative methods for such calculations, which are equivalent under steady-state conditions. The first requires computation of the exit rate of a key component k. By assuming that convection dominates over molecular transport at the exit, we can compute the flow-average concentration of the effluent (Appendix F)

$$[c_k(t)] = \frac{\displaystyle\int_{S_f} c_k(x_c, t) v_n(x_c, t)\, ds}{\displaystyle\int_{S_f} v_n(x_c, t)\, ds}, \qquad (1.3\text{-}9)$$

where v_n is the velocity component normal to the exit surface area S_f, x_c is the position "vector" (actually, column matrix), and the denominator is equal to $\dot{V}$ evaluated at exit conditions. It is evident that velocity and concentration distributions at the vessel exit are required. The time variable t signifies the validity of this equation under transient conditions as well; it is unnecessary, of course, at steady state.

The second method is based on a direct computation of average process rate. For a chemical reactor, this quantity is the volume-average reaction rate of a key reactant k; thus (Appendix F),

$$\langle r_k(t) \rangle = \int_V \frac{r_k(x_c, t)\, dV}{V} \tag{1.3-10}$$

Knowledge of the spatial distribution of the reaction rate function r_k is thus required. A steady-state material balance for component k (Appendix F) easily reveals the following relationship between the two methods for a reaction vessel with a single inlet and outlet:

$$-\langle r_k \rangle V = -\Delta[c_k]\dot{V} \equiv \dot{O}_k \tag{1.3-11}$$

This substantiates the interpretation of $\dot{o}_k$ as an average reaction rate, as mentioned earlier. For a separation process, the expression that corresponds to equation 1.3-10 is area-average transport flux of component k (Appendix F):

$$\{n_k(t)\} = \int_S \frac{n_{kn}(x_c, t)\, ds}{S} \tag{1.3-12}$$

A steady-state material balance thus yields

$$\{n_k\}S = -\Delta[c_k]\dot{V} \equiv \dot{O}_k \tag{1.3-13}$$

1.3.3. Some General Observations

To summarize, output rate is in essence a measure of quantity, whereas specific output rate and efficiency, which are intensive, are measures of quality. Furthermore, achieving a given specific output rate and achieving a given efficiency are distinctly different process objectives. In practive, both efficiency and throughput rate $\dot{V}$ are frequently specified as design constraints. The process that is capable of achieving these with the smallest equipment size (largest specific output rate) is considered to be superior. On the other hand, although large values of specific output rate appear virtuous, achieving them at the expense of reduced efficiency is often unacceptable and vice versa.

Specific output rate $\dot{o}_k$ serves as a yardstick against which to measure the performance of equipment under different operating conditions or various pieces of equipment performing the same task. For continuous chemical reaction, specific output rate takes the form $\Phi_k[c_k]_0/\lambda_v$. Thus, if we interpret Φ_k as a measure of reaction efficiency, we may interpret $\dot{o}_k$ as measure of reactor efficiency. As we have shown, specific output rate is identical to the volume-average reaction rate $\langle r_k \rangle$, where r_k is the rate function for the formation of substance k. Its use is common in chemical reaction engineering (1) to reflect such things as the adverse effects of backmixing upon reactor size (Examples 1.3-1 and 1.3-2).

For separation processes, such as devolatilization (DV) (Chapter 6), we shall use an area-specific output rate, which is proportional to the quantity $E_f \dot{V}/s = E_f x$. In the latter, which will facilitate comparison among various DV processes, s is the transport surface area. In this case $\dot{o}_k$ is identical to the area-average mass transfer flux $\langle n_k \rangle$.

Process efficiency and specific output rate depend on many parameters such as residence time (RT), temperature history, mixing history, number of stages (N), and recycle ratio (R_R), if recycle is used. The last four items mentioned all influence residence time distribution (RTD), which will be discussed in Chapter 3. Temperature effects will be discussed in Chapter 4. The effects of residence time (λ_v) will be examined briefly here.

We can increase λ_v systematically in two ways: either by reducing $\dot{V}$ and holding V fixed, or by increasing V with $\dot{V}$ fixed. The latter will generally increase process output (e.g., $-\Delta[c_k]\dot{V}$ in equation 1.3-11) at the expense of greater equipment volume, whereas the former will generally reduce it. These procedures are summarized in Table 1.3-2, together with a third one in which λ_v is fixed. The objective there is to test the effects, especially in complex processes, of varying "internal" parameters such as number of stages N, recycle ratio R_R, and flow velocity. Some effects of varying parameters are illustrated in Examples 1.3-2–1.3-4, and will be discussed further in Chapters 3 and 6.

In general, process efficiency increases with λ_v and specific output rate $\dot{o}_k$ declines. The latter is a direct reflection of the inherent behavior of such functions as reaction rate r_k (nonautoacceleratory) and transport rate n_k (interphase), both of which are monotonically decreasing, convex functions of time in batch systems, as shown in Fig. 1.3-2. Thus, as λ_v is reduced, we expect specific output rates $\langle r_k \rangle$ and

Table 1.3-2. Generalized Tests for Performance of Simple and Complex Processes

TARGETS:

     Efficiency (E_f or E_F)

     Specific output rate ($\dot{o}_k$)

PROCEDURES:

     Vary λ_v (or Λ_v)

         via V with V constant

         via V with $\dot{V}$ constant

     Vary internal parameters (e.g., N, R_R, velocity etc.)

         Constant λ_v (Λ_v) with variable $\dot{V}$ or constant $\dot{V}$

$\langle n_k \rangle$ to increase (Examples 1.3-1 and 1.3-4). The corresponding process efficiencies however, which are proportional to $\langle r_k \rangle \lambda_v$ and $\langle n_k \rangle s_v \lambda_v$, respectively, will normally decline in response to the dominance of λ_v over specific output rate in these products, unless, for example, s_v is concomitantly increased in the latter and the combined effects of rising $\langle n_k \rangle$ and s_v prevail (Example 1.3-4).

To achieve desired monomer conversion levels and polymer molecular weight in continuous polymerizations, λ_v must at least match λ_r, the characteristic time (CT) of reaction. This requirement may be expressed in terms of the dimensionless Damköhler number, $Da \equiv \lambda_v / \lambda_r$, and it justifies the choice of $Da = 1$ for the comparisons made in Example 1.3-3. As mentioned in Section 1.2, achieving a satisfactory value for Da could be difficult. This is especially true for reactions conducted in processing equipment such as extruders ($\lambda_v \sim$ minutes) or injection molding machines ($\lambda_v \sim$ seconds). One remedy is to shorten λ_r by using fast catalysts or high temperatures.

Up to this point we have considered primarily the "normal" responses of systems to changes in RT, which include an increase in E_f and a decline in $\dot{o}_k$ as λ_v is increased. Exceptions are autoacceleratory reactions (Chapter 2) and processes that are sensitive to mixing (Chapters 3 and 6). In the former, specific output rate actually rises as λ_v is increased, and in the latter, efficiency could actually increase as λ_v is reduced (Example 6.5-2). In general, we shall regard an increase in efficiency in response to a decline in residence time as anomalous.

Chemical reactions and phase changes that are diffusion-controlled are commonplace in chemical engineering. Examples abound among reactions with heterogeneous catalysts and separations with vapor-liquid interfaces, such as the devolatilization of polymer melts (Chapter 6). Inadequate mixing is frequently the root cause of diffusion control. Thus we shall classify such phenomena collectively under mixing effects.

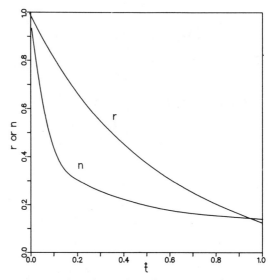

Figure 1.3-2. Typical time dependence of reaction rate (r) and mass transport flux (n).

A technique that is sometimes used to identify mixing effects in continuous-flow rate processes is to increase flow velocity v, while maintaining total residence time (λ_v or Λ_v) constant (Table 1.3-2) and to examine the effect on specific output rate ($\langle r_k \rangle$ or $\langle n_k \rangle$). An increase should be a clear indication that mixing has enhanced the rate process, since the RT has not been allowed to decline. The usual reason is that the higher flow velocity reduces the diffusion length in the rate-controlling transport process. An example is the continuous, packed-bed reactor, in which extraparticle diffusion is rate-controlling. Specific reaction rate can be enhanced by increasing the flow velocity and maintaining RT constant. Such increases in specific output rate are generally manifested as an increase in efficiency as well, by virtue of the expressions $\langle r_k \rangle \lambda_v$ and $\langle n_k \rangle s_v \lambda_v$, unless a decrease in s_v is required in the latter to maintain RT constant.

In many flow processes, however, an increase in v cannot be readily achieved without a concomitant decrease in λ_v. This is true in particular when v is accomplished via, or accompanied by, an increase in throughput rate $\dot{V}$. For instance, in a catalytic reactor, velocity is generally increased by increasing $\dot{V}$; in an extruder, by increasing the RPM (N_R), which produces a concomitant increase in $\dot{V}$. Thus maintaining λ_v constant would require enlarging the occupied volume V by increasing the longitudinal dimension L, for example. Failure to increase V would reduce λ_v and consequently tend to offset, or even reverse, any increase in efficiency that might be anticipated due to higher v. Even if an increase in $\dot{o}_k$ is experienced, it could be due to the larger $\dot{V}$ combined with an efficiency that is actually lower.

An alternative method for increasing v is to reduce a vessel dimension in a direction transverse to flow for a given $\dot{V}$ and simultaneously to increase the longitudinal dimension L as required to maintain λ_v constant. Failure to increase L would again cause λ_v to decline, which could produce the competitive effect on process efficiency cited earlier.

Should process efficiency increase anyway, in the face of a decline in λ_v, we would be compelled to ascribe its cause to a mixing effect. Such "anomalous" behavior has been observed in DV processes (Section 6.5) and rationalized in terms of diffusion theory as a surface-renewal phenomenon (Example 6.5-2) or simply a reduction in diffusion length (Example 1.3-4).

From these observations, we conclude that manipulation of $\lambda_v (\Lambda_v)$ can be useful in process analysis as well as in process design.

Example 1.3-1.   Efficiency and Specific Output of Reactors.   Polymerization rate r for the addition type is generally expressed as monomer conversion rate, which is the negative value of monomer formation rate r_m. It is frequently pseudo first-order. Thus,

$$-r_m = kc_m \equiv r$$

The steady-state monomer concentration distribution in the longitudinal direction (x) in a plug flow reactor (PFR) (Chapter 3) is then

$$c_m(x) = (c_m)_0 \exp\left(\frac{-k\lambda_v x}{L} \right)$$

where $\lambda_v = \lambda_c = L/v$, and L and v are tube length and longitudinal velocity, respectively. The corresponding distribution in an ideal mixer (IM) at steady state is (uniform)

$$c_m = \frac{(c_m)_0}{(1 + k\lambda_v)}$$

Compute $[c_m]$, monomer conversion $\Phi_m \equiv 1 - [c_m]_f/[c_m]_0$, $\langle r \rangle$, and specific output rate (Φ_m/λ_v) via the two alternative methods discussed. Then compare Φ_m and Φ_m/λ_v for the PFR and the IM at various values of λ_v using $k = 10^{-3}$ s^{-1}.

Solution. From equation 1.3-9, we obtain for the PFR

$$[c_m] = (c_m)_0 \exp(-k\lambda_v)$$

and for the IM

$$[c_m] = \frac{(c_m)_0}{(1 + k\lambda_v)}$$

From equation 1.3-10, the corresponding average rate functions are

$$\langle r \rangle = \frac{\int k c_m(x_c)\pi Ra^2\, dx}{\pi Ra^2 L}$$
$$= \frac{(c_m)_0(1 - \exp - k\lambda_v)}{\lambda_v}$$

for the PFR, where Ra is the tube radius, and

$$\langle r \rangle = \frac{k(c_m)_0}{(1 + k\lambda_v)}$$

for the IM. A comparison of these four results clearly shows that

$$\dot{o}_m = \begin{cases} ((c_m)_0 - [c_m])/\lambda_v \\ \langle r \rangle \end{cases}$$

which demonstrates the equivalence of the two computational methods.

The following chart gives several comparative values for Φ_m and Φ_m/λ_v, from which we conclude that process efficiency rises as λ_v is increased, whereas specific output rate declines. We also conclude that the PFR is a superior reactor to the IM in terms of both criteria, efficiency, and specific output. This conclusion is well known and is a consequence of backmixing in the IM (see Chapter 3).

λ_v	Φ_m		$\Phi_m/\lambda_v = o_m/[c_m]_0$	
(s)	PFR	IM	PFR	IM
10^2	9.5×10^{-2}	9.1×10^{-2}	9.5×10^{-4}	9.1×10^{-4}
10^3	6.3×10^{-1}	5.0×10^{-1}	6.3×10^{-4}	5.0×10^{-4}
10^4	~ 1.0	9.1×10^{-1}	10^{-4}	9.1×10^{-5}

Example 1.3-2. Staging. Polymerization rate r for random kinetics is generally expressed as consumption rate of a functional group, which is also equal to depletion rate of the total number of polymer molecules. It is frequently second order. Thus,

$$-r_p = kc_p^2 \equiv r$$

with initial condition $(c_p)_0 = (c_1)_0$. Compare three alternative reactor configurations using this reaction: a single IM, two equal-volume IM's in series, and three equal-volume IM's in series. Specifically, contrast their conversions (reaction efficiencies) at the same value of $Da \equiv \Lambda_v/\lambda_r = 20$ and their specific output rates (reactor efficiencies) at the same conversion $\Phi = 0.80$. Da is the Damköhler number of the first kind, λ_r is the characteristic time (CT) of the reaction (see Appendix G), and Λ_v is the total mean space time.

Solution. Assuming molecular uniformity in each IM permits us to write $\langle r_p \rangle = r_p$, or $\langle kc_p^2 \rangle = kc_p^2$, where c_p is also equal to the concentration of polymer in the effluent stream. A component balance on the jth vessel in a cascade consisting of N identical vessels ($\Lambda_v/N = \lambda_v$)

$$\frac{c_{pj} - c_{pj-1}}{\lambda_v} = -kc_{pj}^2$$

leads to the following first-order nonlinear difference equation in dimensionless form

$$\hat{c}_{pj} = \frac{N}{2Da}\left[\left(1 + \frac{4Da\,\hat{c}_{pj-1}}{N}\right)^{1/2} - 1\right]$$

where $\hat{c}_{pj} \equiv c_{pj}/(c_1)_0$ and the CT is $\lambda_r \equiv [k(c_1)_0]^{-1}$ since the reaction is second order. Conversion is defined as $\Phi_N = 1 - \hat{c}_{pN}$. Thus, for $Da = 20$ and $N = 1, 2$, and 3, respectively, we obtain from successive solutions:

$$\Phi_1 = 0.8, \qquad \Phi_2 = 0.878, \qquad \Phi_3 = 0.904$$

When $\Phi = 0.80$, a trial-and-error solution yields for $N = 1, 2$, and 3, respectively,

$$Da_1 = 20, \qquad Da_2 = 4.4, \qquad Da_3 = 2.25$$

These results are summarized in the following chart in terms of Φ/Da, which in essence is a dimensionless specific output rate.

N	Φ/Da		
	1	2	3
Da = 20	0.04	0.044	0.045
Φ = 0.8	0.04	0.182	0.356

CONCLUSIONS

Note how specific output rate increases with N when Φ is fixed. From these results, we conclude that although staging increases conversion only slightly for a fixed value of Λ_v, it dramatically reduces the value of $\Lambda_v(\mathrm{Da})$, and therefore equipment size, required to achieve a specified (high) conversion.

Example 1.3-3. Recycle. Consider the PFR with recycle shown in the sketch and the first-order polymerization of Example 1.3-1. Assuming that reaction volume V is fixed, examine the effects of recycle ratio $R_R \equiv \dot{V}_2/\dot{V}_1$ on process efficiency $E_F \equiv 1 - [c_m]/(c_m)_0$ and specific output rate (E_F/Λ_v) under the following conditions (see Table 1.3-1):

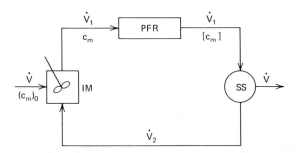

1. $\dot{V}_1$ is held constant but $\dot{V}$ is allowed to decrease as R_R increases, so that the net effect is simply an increase in $\Lambda_v \equiv V/\dot{V}$.
2. $\dot{V}$ is held constant so that the effect of raising R_R is merely to increase the internal circulation rate $\dot{V}_1$, thereby reducing λ_v. Λ_v, however, remains constant.

Solution. From Example 1.3-1, we define reactor efficiency as

$$E_f \equiv 1 - \frac{[c_m]}{c_m} = \Phi_m = 1 - \exp(-\mathrm{Da}),$$

where $\mathrm{Da} \equiv \lambda_v/\lambda_r$ and λ_v is now defined as $V/\dot{V}_1$. For first-order reactions, the CT is $\lambda_r = k^{-1}$, and it remains constant throughout this example. Since λ_v will be allowed to vary in part 2, it is convenient to define a modified Damköhler number, $\mathrm{Da}' \equiv \Lambda_v/\lambda_r$.

From a total material balance,

$$\dot{V} = \dot{V}_1 - \dot{V}_2 = \dot{V}_1(1 - R_R)$$

and from a monomer component balance,

$$(c_m)_0 \dot{V} = c_m \dot{V}_1 - [c_m] \dot{V}_2$$

Thus it can be shown that $\mathrm{Da} = \mathrm{Da}'(1 - R_R)$ and

$$E_F = \frac{E_f}{1 - R_R(1 - E_f)}$$

(a) The results in the first chart were obtained by varying R_R and thus Da' for a fixed, arbitrary value of $\mathrm{Da} = 1$. It should be noted that specific output rate has been tabulated in dimensionless form (E_F/Da') as in Example 1.3-2.

R_R	$E_f = \Phi_m$	E_F	E_F/Da'	$\dot{V}/\dot{V}_1$ (fixed $\dot{V}_1$)
0	0.632	0.632	0.632	1
0.2	0.632	0.682	0.506	0.8
0.4	0.632	0.741	0.445	0.6
0.6	0.632	0.811	0.324	0.4
0.8	0.632	0.896	0.179	0.3
1.0	0.632	1.0	0	0 (no output)

(b) The results in the second chart were obtained by varying R_R and thus Da for a fixed, arbitrary value of $\mathrm{Da}' = 1$. Again, specific output rate has been tabulated in dimensionless form (E_f/Da).

R_R	$E_f = \Phi_m$	E_F	E_f/D_a	$\dot{V}_1/\dot{V}$ (fixed $\dot{V}$)
0	0.632	0.632	0.632	1.00
0.2	0.551	0.605	0.689	1.25
0.4	0.451	0.578	0.752	1.66
0.6	0.330	0.552	0.825	2.50
0.8	0.181	0.525	0.905	5.00
0.9	0.0952	0.513	0.952	10.0
0.99	0.00995	0.501	0.995	100
1.0	0	0.500	1.0	∞ (infinitely rapid circulation)

CONCLUSIONS

Increasing residence time by recycling (part a) raises reaction efficiency (E_F), which is well known, but it reduces reactor efficiency (E_F/Λ_v) and output ($E_F\dot{V}$). On the other hand, increasing internal circulation velocity with constant residence time Λ_v and throughput rate $\dot{V}$ (part b) significantly enhances reactor efficiency (E_f/Da) while slightly reducing overall efficiency (E_F) and therefore output.

Example 1.3-4.   Efficiency and Specific Output of Film Evaporators. The process efficiency E_F of an N-stage cascade of identical film evaporators in series with N intermittant surface-renewal (mixing) operations is given by the following expression (Chapter 6)

$$E_F = 1 - \left(1 - E_f\right)^N$$

where the stage efficiency E_f is related to the mass transfer coefficient from penetration theory, $k_f = 2(D/\pi\lambda_f)^{1/2}$, under appropriate conditions by $E_f = \lambda_f k_f s_v$. Show how an increase in the film velocity, which represents a decrease in the total residence time in the evaporators ($N\lambda_f$), can lead to not only a larger specific output rate $\langle n_k \rangle$, but also a higher process efficiency, when throughput rate $\dot{V}$ is fixed.

Solution.   Clearly, k_f increases as λ_f is reduced. This is reflected by a higher mean transport flux $\langle n_n \rangle$, which is related to E_F by virtue of the proportionality, $\langle n_n \rangle N s_f \propto \dot{V} E_F$, where s_f is the surface area of each film. Thus, if we maintain a constant $\dot{V} = s_f h/\lambda_f$ as film velocity increases, by reducing film thickness $h = s_v^{-1}$ instead of area s_f (constant), it follows from $E_f = (2s_f/\dot{V})(D/\pi\lambda_f)^{1/2}$ that stage efficiency increases and, consequently, so does E_F.

1.4.   THE CONCEPT OF RELATIVE DISPERSION

It is well known from statistical analysis that variance σ^2 [or standard deviation $\mathrm{SD} = (\sigma^2)^{1/2}$] characterizes absolute dispersion of any distributed variable about its mean value $\hat{\mu}^1$ (Appendix B). However, absolute dispersion of a physical parameter or property is generally not the best measure of the effect of dispersion on the performance of an associated process or material. The role of residence time and its distribution (due to backmixing) in separation processes has been discussed elsewhere in this context (3, 4). We begin with this example and include chemical reaction at the same time.

Backmixing (BM) in continuous flow vessels is the intermingling of molecules of different ages (Chapter 3). In streamline vessels (Fig. 1.2-1), it is frequently described in terms of the so-called dispersion model. This model takes the following form for arbitrary substance k

$$\frac{\partial c_k}{\partial t} = -\langle v \rangle \frac{\partial c_k}{\partial x} + D_L \frac{\partial^2 c_k}{\partial x^2} + r_{Gk} \tag{1.4-1}$$

where D_L is a longitudinal dispersion coefficient and r_{Gk} is the rate of generation of substance k per unit volume. When applied to chemical reaction (RX) $r_{Gk} = r_k$, the rate function for formation of k; when applied to devolatilization (DV) $r_{Gk} = r_e$, the rate of evaporation per unit volume.

A noteworthy example of BM in a cylindrical tube containing a Newtonian fluid in steady laminar flow is "Taylor axial diffusion," for which (5)

$$D_L = \frac{4Ra^2\{v\}^2}{192D} \tag{1.4-2}$$

Young molecules are mixed with old ones and vice versa, not by axial diffusion as implied by the name, but by the combined action of a radial velocity distribution and radial diffusion. The resulting phenomenon has the appearance of axial dispersion. Incidentally, the same model is also used to represent turbulent axial disperison and axial dispersion in packed beds; only D_L in these cases is an "eddy diffusivity." To demonstrate the effects of BM, we examine two first-order processes

$$r_{Gk} = \begin{cases} r_k = -kc_k & \text{RX} \\ r_e = -ks_v(c_k - c_e) & \text{DV} \end{cases} \tag{1.4-3}$$

and with our differential balance in dimensionless form

$$\frac{\partial \hat{c}}{\partial \hat{t}} = -\frac{\partial \hat{c}}{\partial \hat{x}} + Pe_L^{-1}\frac{\partial^2 \hat{c}}{\partial \hat{x}^2} - \begin{cases} Da\hat{c} \\ Ex\hat{c} \end{cases} \tag{1.4-4}$$

where $\hat{t} \equiv t/\lambda_c$, $\hat{x} \equiv x/L$, and the subscript on dimensionless concentration has been dropped for convenience

$$\hat{c} \equiv \begin{cases} \dfrac{c_k}{c_0 - c_k} & \text{RX} \\ \dfrac{(c_k - c_e)}{c_0 - c_e} & \text{DV} \end{cases} \tag{1.4-5}$$

Three dimensionless groups that will be encountered again in later chapters have been introduced. They are, in terms of characteristic times, the longitudinal Peclet number, $Pe_L \equiv L^2/D_L\lambda_c = \lambda_D/\lambda_c$, the Damköhler number, $Da \equiv k\lambda_c = \lambda_c/\lambda_r$ and the extraction number (Chapter 6), $E_x \equiv ks_v\lambda_c = \lambda_c E_f/\lambda_f$.

The appropriate boundary conditions for steady-state processes $\left(\dfrac{\partial \hat{c}}{\partial \hat{t}} = 0\right)$ described by equation 1.4-4 are (6):

$$\hat{c} - Pe_L^{-1}\frac{d\hat{c}}{d\hat{x}} = 1 \qquad \text{when } \hat{x} = 0 \tag{1.4-6}$$

and

$$\frac{d\hat{c}}{d\hat{x}} = 0 \qquad \text{when } \hat{x} = 1. \tag{1.4-7}$$

Since the first one produces a discontinuity in $\hat{c}$ at $x = 0$, it is tempting to use the following boundary conditions (BC) in lieu of equation 1.4-6:

$$\hat{c} = 1 \qquad \text{when } \hat{x} = 0. \tag{1.4-8}$$

The corresponding solutions are

$$\hat{c} = \frac{A}{\left[\dfrac{(1 + A)}{2}\right]^2 \exp\left[-\dfrac{\mathrm{Pe}_L(1 - A)}{2}\right] - \left[\dfrac{(1 - A)}{2}\right]^2 \exp\left[\dfrac{-\mathrm{Pe}_L(1 + A)}{2}\right]}$$

(1.4-9)

for the BC's given in equations 1.4-6 and 1.4-7, and

$$\hat{c} = \frac{A}{\left[\dfrac{(1 + A)}{2}\right] \exp\left[\dfrac{-\mathrm{Pe}_L(1 - A)}{2}\right] - \left[\dfrac{(1 - A)}{2}\right] \exp\left[\dfrac{-\mathrm{Pe}_L(1 + A)}{2}\right]}$$

(1.4-10)

for the BC's in equations 1.4-6 and 1.4-8, where A is defined as $(1 + 4\,\mathrm{Da}/\mathrm{Pe}_L)^{1/2}$ or $(1 + 4\,E_x/\mathrm{Pe}_L)^{1/2}$ as appropriate. Equations 1.4-9 and 1.4-10 have been plotted in Fig. 1.4-1 as $E_f \equiv 1 - \hat{c}$ versus Da (or E_x) for various values of Pe_L. The reader will note that the two solutions converge for large values of Pe_L (> 10).

Two conclusions are evident from Fig. 1.4-1. First, E_F rises with Da (or Ex) and approaches unity when Da > 10. Second, for given process characteristics (fixed Da or Ex), E_F rises with rising values of Pe_L. In fact, the curves for $\mathrm{Pe}_L \geqslant 100$ virtually coincide with the PF solution (Example 1.3-1)

$$\hat{c} = \begin{cases} \exp(-\mathrm{Da}) \\ \exp(-\mathrm{Ex}) \end{cases}$$

(1.4-11)

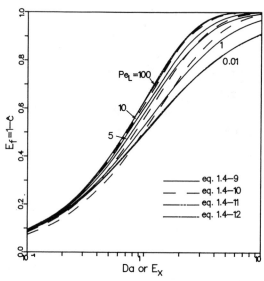

Figure 1.4-1. Process efficiency (E_F) versus Damköhler number (Da) or extraction number (Ex) for the dispersion model with longitudinal Peclet number (Pe_L) as parameter, and for the plug flow (PF) model and the ideal mixer (IM) for comparison.

and the curves for $Pe_L \leqslant 0.01$ (equation 1.4-9) virtually coincide with the solution for the IM with complete BM (cf. Example 1.3-1).

$$\hat{c} = \begin{cases} (1 + Da)^{-1} \\ (1 + Ex)^{-1} \end{cases} \tag{1.4-12}$$

Thus we conclude that Pe_L^{-1} is a measure of the importance of BM. It is also evident that equation 1.4-10 is not a valid model when BM is appreciable ($Pe_L < 10$), owing to BC 1.4-8, which does not permit a discontinuity at the entrance.

To understand the concept of relative dispersion, we inquire further into the nature of Pe_L. From the definition of Pe_L, it is clear that BM increases with space time, and from the definitions of Da and Ex it is equally clear that so does process output rate in the absence of BM. Thus we have competing effects. On the one hand, a large value of λ_c is beneficial to the processes; on the other, it simultaneously enhances the importance of BM. Therefore, we seek a measure of the relative importance of BM.

Let us examine BM in our streamline vessel by subjecting it to a transient, inert tracer analysis (Red Dye Experiment in Chapter 3) in the absence of RX and DV. We drop the last terms in equations 1.4-1 and 1.4-4 and retain the first, and utilize the following "classical" initial and boundary conditions: at $\hat{t} = 0$,

$$\hat{c} = \begin{cases} 1 & \text{when } \hat{x} < 0 \\ 0 & \text{when } \hat{x} > 0 \end{cases} \tag{1.4-13}$$

and at $\hat{t} > 0$,

$$\hat{c} = \begin{cases} 1 & \text{when } \hat{x} = -\infty \\ 0 & \text{when } \hat{x} = \infty \end{cases} \tag{1.4-14}$$

The result, which is well known (7), evaluated at the vessel exit $\hat{x} = 1$ is

$$\hat{c} = \frac{1}{2}\left\{ 1 - \text{erf}\left[\left(\frac{Pe_L}{4\hat{t}} \right)^{1/2} (1 - \hat{t}) \right] \right\}. \tag{1.4-15}$$

It represents the fraction of tracer in the effluent at time $\hat{t}$. Its derivative, which is the residence time distribution of the vessel (Chapter 3),

$$e(\hat{t}) = \left(\frac{Pe_L}{4\pi\hat{t}} \right)^{1/2} \exp\left[-\left(\frac{Pe_L}{4\hat{t}} \right)(1 - \hat{t})^2 \right] \tag{1.4-16}$$

may be approximated by the Gaussian distribution

$$e(\hat{t}) = \left(\frac{Pe_L}{4\pi} \right)^{1/2} \exp\left\{ -\frac{Pe_L}{4}(1 - \hat{t})^2 \right\} \tag{1.4-17}$$

when dispersion is relatively small, as it must be to qualify for representation by the dispersion model. The variance of this dimensionless distribution is (Appendix B) var $\hat{t} = 2/\mathrm{Pe}_L$. On the other hand, the corresponding variance in absolute time t is var $t = 2\lambda_c^2/\mathrm{Pe}_L$. Thus we see that Pe_L^{-1} actually measures the relative dispersion (RD) of residence times, that is, the absolute dispersion $(\mathrm{var}\, t)^{1/2}$ relative to the mean residence time λ_c. More precisely,

$$\left(2\,\mathrm{Pe}_L^{-1}\right)^{1/2} = \frac{\mathrm{SD}}{\lambda_c} \equiv \mathrm{RD}. \tag{1.4-18}$$

Frequently the characteristics of a process or property can be enhanced by increasing the mean value of a certain parameter, and simultaneously diminished by broadening the distribution among the values of the same parameter. In such cases it is most appropriate to characterize dispersion in terms of RD. The parameter residence time is a good example. In the foregoing example the efficiency of reaction and separation was enhanced by raising λ_c, but dispersion was broadened at the same time. It is noteworthy that absolute dispersion increased as $\lambda_c^{3/2}$ (var $t \propto \lambda_c^3$), whereas relative dispersion only increased as $\lambda_c^{1/2}$. To complete the analysis, of course, one must know the functional dependence of efficiency on λ_c. It is well known in chromatography, for instance, that increasing RT improves resolution by promoting peak separation to a greater degree than peak broadening.

Another example of the appropriateness of relative dispersion arises in connection with the effects of molecular weight, or degree of polymerization (DP), and its distribution on polymer properties. Here again a large average value is desirable since high molecular weight is the property of primary importance in polymeric materials. Concerning its dispersion, what the polymer scientist or engineer actually seeks is a measure of the relative importance of DPD on the bulk physical, or processing, properties of polymers. To illustrate this, we compare two samples of the same polymer having identical values of variance σ_N^2 (Appendix B). Only their average $\bar{x}_N$ differs. Intuitively, we would expect the effect of dispersion on properties to be more pronounced for the polymer with small $\bar{x}_N$ than for the one with large $\bar{x}_N$. Therefore, as with our first example, we are more interested in relative dispersion

$$\mathrm{RD}_N \equiv \frac{\mathrm{SD}_N}{\bar{x}_N} \tag{1.4-19}$$

than with variance σ_N^2. This conclusion is not surprising in view of the following relationship (8),

$$\mathrm{RD}_N = \left(D_N - 1\right)^{1/2} \tag{1.4-20}$$

where D_N is the well-known dispersion index (Section 1.5)

$$D_N = \frac{\bar{x}_W}{\bar{x}_N} = \frac{\overline{M}_W}{\overline{M}_N} \geqslant 1 \tag{1.5-11}$$

Indeed, D_N is used universally by polymer scientists and engineers as a measure of polydispersity. The quantity $\bar{x}_W$ is weight-average DP and the proof of equation 1.4-20 is left as an exercise for the reader.

By analogy we may define a second RD (8)

$$RD_W \equiv \frac{SD_W}{\bar{x}_W} = \left(\frac{\bar{x}_z}{\bar{x}_W} - 1 \right)^{1/2} \tag{1.4-21}$$

which in turn leads to another dispersion index (Section 1.5)

$$D_W \equiv \frac{\bar{x}_z}{\bar{x}_W} = \frac{\overline{M}_z}{\overline{M}_W} \geqslant 1 \tag{1.5-14}$$

Thus D_N and D_W are actually measures of relative dispersion of number and weight distributions, respectively, but their functional relationship to RD is nonlinear. It is apparent from the graph of RD versus D shown in Fig. 1.4-2, that D in general is disproportionately sensitive to RD for large values of the former. The opposite is true for values near unity. D_N for many polymers lies between 2 and 5.

A simple example utilizing a bimodal molecular-weight distribution (MWD), synthesized from a mixture of two monodisperse polymers with chain lengths x and y, where $y > x$, will serve to shed light on the physical interpretation of D_N and D_W. If N_x and N_y represent moles of x-mer and y-mer in the blend, respectively, it can be shown (9) that for fixed values of x and y a maximum in D_N will occur when $N_x/N_y = y/x$ and a maximum in D_W when $N_x/N_y = (y/x)^2$. This implies that a high value of D_N is caused by either a large number of low ends or a small number of high ends, or both. On the other hand, a high value of D_W requires more low ends

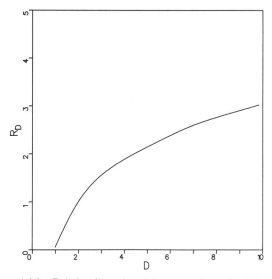

Figure 1.4-2.   Relative dispersion (RD) versus dispersion index (D).

and fewer high ends than D_N. Hence D_W is more sensitive to high-molecular-weight components than D_N, and the latter is more sensitive to low molecular weights.

Example 1.4-1. Relative Dispersion.   With absolute dispersion fixed at $\sigma_N^2 =$ 10,000 (SD = 100), compute RD_N, D_N, and $\bar{x}_W$ for arbitrary values of $\bar{x}_N$. Then, with a fixed value of $\bar{x}_W = 100,000$, compute RD_N and SD_N for arbitrary values of $\bar{x}_N$.

The results, obtained from the equations in the text, have been tabulated:

$\bar{x}_N$	100	1,000	10,000
RD_N	1	0.1	0.01
D_N	2	1.01	1.0001
$\bar{x}_W$	200	1,010	10,001

$\bar{x}_N$	D_N	RD_N	SD_N
66,000	1.5	0.707	47,133
50,000	2.0	1.00	50,000 (maximum)
40,000	2.5	1.23	49,200
33,000	3.0	1.41	47,000
25,000	4.0	1.73	43,250
20,000	5.0	2.00	40,000

The application of relative dispersion to polymer property characterization is demonstrated in Example 1.4-1. The nonuniformity of the D_N scale relative to that of RD_N is evident in the first table. More importantly, this table also shows that the effect of a fixed absolute dispersion is to diminish the gap between $\bar{x}_W$ and $\bar{x}_N$ as the latter increases. Since $\bar{x}_W$ is believed to be the primary molecular size parameter affecting the flow characteristics of polymer melts, this example suggests that the influence of absolute breadth of a DPD on melt behavior also diminishes at high DP levels (as measured by $\bar{x}_N$). This trend is reflected by the behavior of both RD_N and D_N and is consistent with the experimental observation that the presence of low ends (large D_N) strongly influences melt processing at various DP levels.

The contention that the effects of MWD breadth on melt flow behavior should be measured by RD_N, or D_N, is better demonstrated in the second table. Here the primary flow parameter $\bar{x}_W$ remains constant and $\bar{x}_N$ decreases. Instead of increasing in response to the growing presence of low ends, SD_N passes through a maximum at $D_N = 2$ and then decreases throughout the region of practical interest. On the other hand, RD_N and D_N rise monotonically as desired. This is easily confirmed with the aid of the following relationship

$$\left(\frac{\partial RD_N}{\partial \bar{x}_N} \right)_{\bar{x}_W} = - \frac{D_N}{2\bar{x}_N \sqrt{D_N - 1}} \tag{1.4-22}$$

which is negative for all permissible values of $D_N > 1$, while

$$\left(\frac{\partial SD_N}{\partial \bar{x}_N} \right)_{\bar{x}_w} = \frac{D_N - 2}{2\sqrt{D_N - 1}} \qquad (1.4\text{-}23)$$

is positive when $D_N > 2$ and negative only when $D_N < 2$.

1.5. SOME POLYMER PROPERTIES AFFECTED BY REACTION PATH

The property of paramount importance to polymeric materials is molecular weight, or degree of polymerization (DP). A large value of this property is the primary factor responsible for the unique and desirable bulk properties of these materials. Other important properties of polymer molecules, perhaps second only to their size, include their size distribution, their composition, and their structure. Most of these properties are determined by chemical reaction path, for instance, choice of catalyst. We will focus on those that are affected by physical reaction path. To clarify the distinction, we begin by reviewing briefly certain structural aspects of polymer molecules.

Close inspection of homopolymer chains reveals a structural periodicity, which is depicted in Appendix A with the chemical formula A_x. Each period is termed a repeat unit (A), or mer, and x is the DP. Sometimes these chains are linear (nonnetwork), and intermolecular cohesion derives from secondary forces. Sometimes the chains are cross-linked to one another with primary chemical bonds to form huge macromolecular networks. The former polymers are traditionally called thermoplastic because they melt, dissolve, and flow in response to applied mechanical forces, and the latter are called thermosetting because they do not. Since "linear" (nonnetwork) polymers are amenable to manufacturing processes familiar to the chemical engineer, we will focus our attention on these.

Synthetic polymers are always produced as an admixture of macromolecules with a spectrum of lengths (DP's). Their distribution (DPD) is an important property that can be affected by physical as well as chemical reaction path. In addition to being nonuniform in length, non-network polymer molecules are not always strictly linear. Depending on their chemical reaction path, they can be branched, and sometimes even multi-chain in structure. We will include degree of branching (DB) and its distribution (DBD) for consideration because these properties can also be affected by physical reaction path.

Closer examination of macromolecular structure reveals the possibility of configurational order among repeat units. For instance, they may be joined in head-to-tail or head-to-head sequence, or randomly. Examples are found among vinyl-type addition polymers. Another kind of configurational order that is common among olefin-type addition polymers is optical isomerism. Repeat units may be joined in identical, alternating, or random optical sequence, resulting in isotactic, syndiotactic, or atactic structures, respectively. Still another kind of configurational order among repeat units, which is found among polydienes for example, is geometric isomerism, leading to *cis* and *trans* configurations. Properties associated with configurational order in general will be omitted from consideration, partly because they derive from chemical reaction path by definition and partly because little or no evidence exists linking them to physical reaction path.

Increasing emphasis in polymer manufacture is being placed on polymer alloys, including chemical mixtures (copolymers, terpolymers, etc.) as well as physical blends. Copolymer chains, having the chemical formula A_aB_b (Appendix A), can vary in composition $y = a/x$ and sequence structure, which may be random or ordered. Ordered sequences include perfectly alternating and block copolymers. Copolymer molecules rarely consist of molecules having identical compositions. Of the composition and sequence properties cited, emphasis will be placed on copolymer composition (CC) and its distribution (CCD), both of which are certainly amenable to the influence of physical reaction path.

Some general property variables for hompolymers and copolymers, and their interrelationships, are defined in Tables 1.5-1–1.5-3. Although molecular weight M

Table 1.5-1. Homopolymer Properties

MW	$M = xM_o$ where real number $M > 0$
DP	x where integer $x \geq 1$
Number MWD	N_M
Weight MWD	$W_M = MN_M$
Number average MW	$\bar{M}_N = \sum_M MN_M / \sum_M N_M$
Weight average MW	$\bar{M}_W = \sum_M MW_M / \sum_M W_M$
Number DPD	N_x
Weight DPD	$W_x = x M_o N_x$
Number average DP	$\bar{x}_N = \sum_x xN_x / \sum_x N_x$
Weight average DP	$\bar{x}_W = \sum_x x W_x / \sum_x W_x$
Bivariate branching distribution	$N_{x,b}$
Number DBD	$N_b = \sum_x N_{x,b}$
Number DPD	$N_x = \sum_b N_{x,b}$
Average branch density	$\bar{b}' = \sum_x \sum_b bN_{x,b} / \sum_x \sum_b xN_{x,b}$
Number average DB	$\bar{b}_N = \sum_x \sum_b bN_{x,b} / \sum_x \sum_b N_{x,b}$
Weight average DB	$\bar{b}_W = \sum_x \sum_b bxN_{x,b} / \sum_x \sum_b xN_{x,b}$

Table 1.5-2.   Moments

k^{th} number moment	$\mu_N^k = \Sigma\, x^k\, N_x$ where k is an integer
k^{th} weight moment	$\mu_W^k = \Sigma\, x^k\, W_x$
Some identities:	$\bar{x}_N = \mu_N^1/\mu_N^0 = \mu_W^0/\mu_W^{-1} = \bar{M}_N/M_0$
	$\bar{x}_W = \mu_N^2/\mu_N^1 = \mu_W^1/\mu_W^0 = \bar{M}_W/M_0$
	$\bar{x}_z = \mu_N^3/\mu_N^2 = \mu_W^2/\mu_W^1 = \bar{M}_z/M_0$
	$\sigma_N^2 = \bar{x}_W\,\bar{x}_N - \bar{x}_N^2$
	$\sigma_W^2 = \bar{x}_z\,\bar{x}_W - \bar{x}_W^2$
	$SK_N = (\bar{x}_z\bar{x}_W - 3\bar{x}_W\,\bar{x}_N + 2\bar{x}_N^2)/\bar{x}_N^2\,(D_N-1)^{3/2}$
	$SK_W = (\bar{x}_{z+1}\bar{x}_z - 3\bar{x}_z\bar{x}_W + 2\bar{x}_W^2)/\bar{x}_W^2\,(D_W-1)^{3/2}.$
Some inequalities	$\bar{x}_N \leq \bar{x}_W \leq \bar{x}_z \leq \cdots\cdots$

is sometimes used as the distributed variable, we will find degree of polymerization x more convenient, especially for summing. Both are discrete variables, but x is always an integer whereas M is generally not. The spacing between consecutive values of M is fixed by the molecular weight of the repeat unit M_0.

Four DPD's frequently encountered in practice are listed in Table 1.5-4, together with some of their characteristics. They have been plotted in Fig. 1.5-1 (weight DPD's) with a common value of $\bar{x}_W$ (approximately 1000) for comparison; their remaining properties are listed in Table 1.5-5. The fundamental differences among these distributions are self-evident. A more complete list of distributions in general form appears in Appendix B (Table B-1), and a compilation of useful sums in Appendix C (Table C-3) to aid the interested reader in performing computations.

In summary, we aim to examine the effects of physical reaction path (viz., mass and heat transfer and flow phenomena) on such properties as DP, DPD, CC, CCD, DB, and DBD. Quantitative analysis of such effects requires a system of symbols. We shall signify degree of polymerization, composition and number of branches per molecule with x, y, and b, respectively, and their mean values with an overbar ($\bar{x}$, $\bar{y}$, $\bar{z}$). Numerous variations among the latter quantities exist, owing primarily to the existence of two distinct weighting factors: number (mole) N and weight W.

1.5.1.   DP and DPD

Two distributions associated with x are thus number and weight DPD's, N_x and w_x. Corresponding average values are number average DP

$$\bar{x}_N \equiv \sum_x xw_x \qquad (1.5\text{-}1)$$

Table 1.5-3. Copolymer Properties

MW	$M = a\,M_{Ao} + bM_{Bo} + [yM_{Ao} + (1-y)M_{Bo}]x$
DP	$x = a + b$ where integers $o \leq a,b \leq x$, $x \geq 1$
CC	$y = a/x$ where $o \leq y \leq 1$
Bivariate number distribution	$N_{a,b}$, $N_{x,y}$
Bivariate weight distribution	$W_{a,b} = (a\,M_{Ao} + b\,M_{Bo})N_{a,b}$ $W_{x,y} = [\,y\,M_{Ao} + (1-y)M_{Bo}]xN_{x,y}$
Number CCD	$N_a = \sum_b N_{a,b}$; $N_y = \sum_x N_{x,y}$
Weight CCD	$W_a = \sum_b W_{a,b}$; $W_y = \sum_x W_{x,y}$
Number average DP	$\bar{x}_N \equiv \sum_x \sum_y xN_{x,y} / \sum_x \sum_y N_{x,y}$
Weight average DP	$\bar{x}_W \equiv \sum_x \sum_y xW_{x,y} / \sum_x \sum_y W_{x,y}$
Average number CC	$y \equiv \sum_a \sum_b aN_{a,b} / \sum_a \sum_b (a+b)N_{a,b}$
Number average CC	$\bar{y}_N \equiv \sum_x \sum_y yN_{x,y} / \sum_x \sum_y N_{x,y}$
Weight average CC	$\bar{y}_W \equiv \sum_x \sum_y yW_{x,y} / \sum_x \sum_y W_{x,y}$

and weight average DP

$$\bar{x}_W \equiv \sum_x xw_x \tag{1.5-2}$$

where n_x and w_x are normalized (fractional) DPD's. A useful relationship between these DPD's, indicating their interdependence, is

$$w_x = \frac{xn_x}{\bar{x}_N} \tag{1.5-3}$$

A consequence is that w_x and $\bar{x}_W$ may be defined alternatively in terms of the

Table 1.5-4.   Some Common DPD's and Their Characteristics

Most Probable ($\phi < 1$)

$x \geq 1$

$$n_x = \phi^{x-1} (1 - \phi) \qquad\qquad w_x = x\,\phi^{x-1} (1 - \phi)^2$$

$$\bar{x}_N = 1/(1 - \phi) \qquad\qquad \bar{x}_W = (1 + \phi)/(1 - \phi)$$

$$D_N = 1 + \phi \qquad\qquad D_W = (1 + 4\phi + \phi^2)/(1 + \phi)^2$$

$$SK_N = (1 + \phi)/2\phi^{1/2} \qquad\qquad SK_W = (1 + \phi)/(2\phi)^{1/2}$$

Poisson ($z > 0$)

$x \geq 1$

$$n_x = z^{x-1} \exp(-z)/(x-1)! \qquad w_x = xz^{x-1} \exp(-z)/(1+z)(x-1)!$$

$$\bar{x}_N = 1 + z \qquad\qquad \bar{x}_W = 1 + z + z/(1 + z)$$

$$D_N = 1 + z/(1+z)^2 \qquad\qquad D_W = (z^4 + 7z^3 + 13z^2\ 8z + 1)/(z^2+3z+1)^2$$

$$SK_N = 1/z^{1/2} \qquad\qquad SK_W = \frac{z^4 + 3z3 + 2z^2 + 2z}{(z^3 + 2z^2 + 2z)^{3/2}}$$

Lansing Log Normal (10)

$$n_x\ (\alpha^{\frac{1}{2}}/x\pi^{\frac{1}{2}})\exp[-\alpha(\ln x - \ln\beta)^2] \qquad\qquad \underline{\qquad}$$

$$\bar{x}_N = \beta \exp(\tfrac{1}{4}\,\alpha) \qquad\qquad \bar{x}_w = \beta \exp(\tfrac{3}{4}\,\alpha)$$

$$D_N = \exp(\tfrac{1}{2}\,\alpha) \qquad\qquad D_W = \exp(\tfrac{1}{2}\,\alpha)$$

$$SK_N = \frac{\bar{x}_N^6 - 3x_N^2 + 2}{(\bar{x}_N^2 - 1)^{3/2}} \qquad\qquad SK_W = SK_N$$

Wesslau Log Normal (11)

$$\underline{\qquad} \qquad\qquad w_x = (\alpha^{\frac{1}{2}}/x\pi^{\frac{1}{2}}) \exp[-\alpha(\ln x - \ln\beta)^2]$$

$$\bar{x}_N = \beta \exp(-\tfrac{1}{4}\alpha) \qquad\qquad \bar{x}_W = \beta\exp(\tfrac{1}{4}\,\alpha)$$

$$D_N = \exp(\tfrac{1}{2}\,\alpha) \qquad\qquad D_W = \exp(\tfrac{1}{2}\,\alpha)$$

$$SK_N = SK_W \qquad\qquad SK_W = \frac{\bar{x}_w^6 - 3\bar{x}_w^2 + 2}{(\bar{x}_w^2 - 1)^{3/2}}$$

number DPD:

$$w_x = \frac{xN_x}{\displaystyle\sum_{\bar{x}} xN_x} \tag{1.5-4}$$

$$\bar{x}_W = \frac{\displaystyle\sum_x x^2 N_x}{\displaystyle\sum_x xN_x} \tag{1.5-5}$$

Equation 1.5-4 will subsequently be used as a model for defining a series of useful

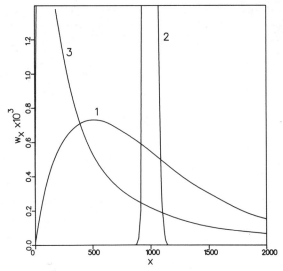

Figure 1.5-1. Three common weight-fraction distributions: the most probable (1), the Poisson (2), and the Wesslau log normal distribution (3). Each has a common $\bar{x}_W$ of approximately 1000.

pseudo properties of copolymers on which several important analyses appearing in the literature are based. Equation 1.5-5 follows directly from the method of moments, which will be used to simplify our computations later on. Appropriate moments and identities are listed in Table 1.5-2.

Two distinct sets of moments may be generated (Appendix B). The kth moments are designated as μ_N^k and σ_N^k for the number DPD, and μ_W^k and σ_W^k for the weight DPD. The following relationship between them

$$\mu_N^k \mu_W^{k-2} = \mu_N^{k-1} \mu_W^{k-1} \qquad (1.5-6)$$

Table 1.5-5. **Properties of DPD's in Figure 1.5-1**

CURVE	1	2	3
DPD	MOST PROBABLE	POISSON	WESSLAU LOG NORMAL ($\alpha=4$, $\beta=350$)
$\bar{x}_N$	500	999	128
$\bar{x}_W$	999	1000	951
D_N	1.998	1.001	7.4
D_W	1.499	1.001	7.4
SK_W	1.41	-0.009	23.73

is left as an exercise for the reader. Obviously, the larger k is, the higher the moment is for either DPD. To characterize polymers, it is customary to use an array of average MW's or DP's instead, which are related to each other by the molecular weight of the repeat unit M_0,

$$\overline{M}_k \equiv M_0 \overline{x}_k \tag{1.5-7}$$

and to the corresponding moments by

$$\overline{x}_k \equiv \frac{\mu_N^k}{\mu_N^{k-1}} \tag{1.5-8}$$

$$\overline{x}_k = \frac{\mu_W^{k-1}}{\mu_W^{k-2}} \tag{1.5-9}$$

It is evident that equation 1.5-5 is a special case. We note that equation 1.5-9 follows from equation 1.5-8 with the aid of equation 1.5-6 and that in general integer k may be either positive or negative. Furthermore, it can be shown that

$$\overline{x}_1 \leqslant \overline{x}_2 \leqslant \cdots \leqslant \overline{x}_{k-1} \leqslant \overline{x}_k \tag{1.5-10}$$

The identities and inequalities listed in Table 1.5-2 follow directly from the preceding relationships if we define $\overline{x}_1 \equiv \overline{x}_N$, $\overline{x}_2 \equiv \overline{x}_W$, and $\overline{x}_3 \equiv \overline{x}_Z$. The equalities in expression 1.5-10 apply only to monodisperse polymers, for which the so-called dispersion index,

$$D_N \equiv \frac{\overline{x}_W}{\overline{x}_N} = \frac{\overline{M}_W}{\overline{M}_N} \geqslant 1 \tag{1.5-11}$$

takes on the value of unity. More generally,

$$D_N > 1 \tag{1.5-12}$$

To appreciate the utility of these averages, we observe that high averages (large k) are more sensitive to the presence of high-molecular-weight components than low averages. Therefore, $\overline{x}_N$ is most sensitive to low ends, and $\overline{x}_Z$ to high ends.

All three averages, $\overline{x}_N$, $\overline{x}_W$, and $\overline{x}_Z$, can be measured experimentally. Gel permeation chromatography (GPC) is a convenient and reliable method for determining an entire weight DPD, from which the number DPD and all corresponding moments and averages may be computed. The somewhat less convenient methods of osmometry, light scattering, and ultracentrifugation are capable of yielding independent values of number-, weight- and z-averages directly. Another useful average, which arises from flow measurements on dilute polymer solutions, is

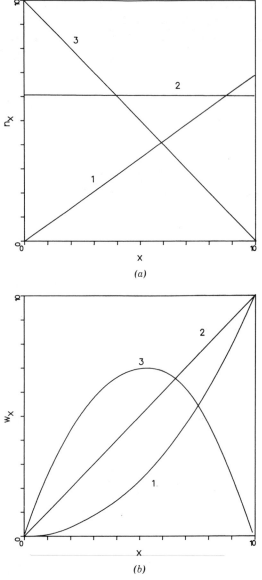

Figure 1.5-2. (*a*) Three idealized DPD's: the ramp (1), the rectangle (2), and the triangle (3) distribution. (*b*) Corresponding weight-fraction DPD's.

the viscosity-average

$$\bar{x}_v \equiv \left[\sum x^a w_x \right]^{1/a} = \left[\mu_W^a / \mu_W^o \right]^{1/a} = \overline{M}_v / M_0 \qquad (1.5\text{-}13)$$

where a is the so-called Mark–Houwink exponent.

Absolute breadth, or dispersion, of distributions are generally characterized by a parallel family of moments σ^k, defined in Appendix B. Two examples pertaining to number and weight distributions are variances σ_N^2 and σ_W^2, defined in Table 1.5-2 in terms of the three averages. These quantities, however, are seldom, if ever, used in the form shown.

Asymmetry, or skewness, of number and weight DPD's may be characterized using the expressions in Table 1.5-2 for SK_N and SK_W, where the quantity D_W appearing in the latter

$$D_W \equiv \frac{\bar{x}_z}{\bar{x}_W} = \frac{\bar{M}_z}{\bar{M}_W} \tag{1.5-14}$$

is a measure of the relative dispersion of the weight distribution.

To illustrate the relationships between shape and dispersion, we consider three simple mole-fraction DPD's: the ramp, the rectangle and the triangle (Appendix B). Their corresponding weight-fraction DPD's are the parabolic, the ramp, and the quadratic distributions, respectively. These distributions and their characteristics have been listed in Table 1.5-6 and plotted in Fig. 1.5-2. The dispersion indices for these distributions approach the following values in the limit $a \rightarrow \infty$: ramp, $D_N = 1.125$, $D_W = 1.066$; rectangle, $D_N = 4/3 = 1.333$, $D_W = 1.125$; and triangle $D_N = 1.5$, $D_W = 1.2$. From these results, we confirm that as the DPD progresses from negative (ramp) to positive (triangle) skewness, the importance of low-molecular-weight polymer grows. This is reflected by the values of D_N and D_W. Both increase as relative dispersion grows, but D_N increases more rapidly than D_W, as expected. In Section 2.3, we will encounter DPD's arising from certain mechanisms that can be approximated by these fictitious distributions.

1.5.2. CC and CCD

Simultaneous specification of both DPD and CCD obviously requires a bivariate distribution. A choice between number (mole) distribution $N_{x,y}$ and weight distribution $W_{x,y}$ is again available. Furthermore, copolymer composition itself can be weighted by number (mole) or weight fraction. We label these fractions y_N and y_W, respectively. Because copolymer composition can be weighted in more than one way, numerous alternative bivariate distributions are possible; four of these are listed in Table 1.5-3.

Computation of $\bar{x}_N$ and $\bar{x}_W$ for copolymers via the formulas in Table 1.5-1 requires expressions for copolymer DPD's. These may be obtained from the bivariate distributions by summing over composition. Thus

$$N_x \equiv \sum_y N_{x,y} \tag{1.5-15}$$

$$W_x \equiv \sum_y W_{x,y} \tag{1.5-16}$$

Normalized bivariate distributions $(n_{x,y}, w_{x,y})$ and DPD's (n_x, w_x) may be defined by analogy. For the weight pair, we obtain

$$w_x \equiv \sum_y w_{x,y} = \frac{\sum_y W_{x,y}}{\sum_x \sum_y W_{x,y}} \tag{1.5-17}$$

where

$$w_{x,y} = \frac{\left[yM_{A0} + (1 - y)M_{B0}\right]xN_{x,y}}{\sum_{x}\sum_{y}\left[yM_{A0} + (1 - y)M_{B0}\right]xN_{x,y}} \tag{1.5-18}$$

The complexity of the last expression and the attendant difficulties associated with its evaluation has led to the use of pseudo weight distributions by several researchers, whose results will be discussed later. Therefore, it is desirable to introduce the concept at this point and to state how it may be used to approximate some of the distributions listed in Table 1.5-3. Thus we introduce a pseudo bivariate weight

Table 1.5-6. Some Idealized DPD's and Their Characteristics

RAMP	NUMBER DPD		CORRESPONDING WEIGHT DPD	
$1 \leq x \leq a$	n_x	$= cx$	w_x	$= 3cx^2/(2a + 1) = c'x^2$
	$\bar{x}_N$	$= (2a + 1)/3$	$\bar{x}_w$	$= \dfrac{3a(a + 1)}{2(2a + 1)}$
	D_N	$= \dfrac{9a(a + 1)}{2(2a + 1)^2}$	D_w	$= \dfrac{2(2a + 1)^2(3a^2 + 3a + 1)}{45a^2(a + 1)^2}$
RECTANGLE				
$1 \leq x \leq a$	n_x	$= c$	w_x	$= \dfrac{2cx}{a + 1} = c'x$
	$\bar{x}_N$	$= \dfrac{(a + 1)}{2}$	$\bar{x}_w$	$= \dfrac{2a + 1}{3}$
	D_N	$= \dfrac{2(2a + 1)}{3(a + 1)}$	D_w	$= \dfrac{9a(a + 1)}{2(2a + 1)^2}$
TRIANGLE				
$1 \leq x \leq a$	n_x	$= b - cx$ where $c = b/a$	w_x	$= 3(bx - cx^2)/(a + 1)$
	$\bar{x}_N$	$= (a + 1)/3$	$\bar{x}_w$	$= a/2$
	D_N	$= \dfrac{3a}{2(a + 1)}$	D_w	$= \dfrac{2(3a^2 - 2)}{5a^2}$

fraction distribution

$$w_{px,y} \equiv \frac{xN_{x,y}}{\sum_x \sum_y xN_{x,y}} \qquad (1.5\text{-}19)$$

which is precisely equal to $w_{x,y}$ only when both repeat units have identical molecular weights $M_{A0} = M_{B0} \equiv M_0$.

Two copolymer properties of major interest are obviously CC and CCD. Alternative definitions of number-average CC, $\bar{y}$ and $\bar{y}_N$, are listed in Table 1.5-3. We may write number-average and weight-average CC's as

$$\bar{y}_N = \sum_y yn_y \qquad (1.5\text{-}20)$$

and

$$\bar{y}_W = \sum_y yw_y \qquad (1.5\text{-}21)$$

where n_y and w_y are normalized number and weight CCD's. Evaluation of these CCD's requires summation of bivariate distributions over DP. Thus,

$$w_y = \sum_x w_{x,y} = \frac{\sum_x W_{x,y}}{\sum_x \sum_y W_{x,y}} \qquad (1.5\text{-}22)$$

where $w_{x,y}$ is given by equation 1.5-18. In lieu of this complex expression, we introduce a pseudo weight fraction CCD (w_{py}), alluded to earlier, in terms of $w_{px,y}$ defined by equation 1.5-19. The result takes the relatively simple form

$$w_{py} \equiv \sum_x w_{px,y} = \frac{\sum_x xN_{x,y}}{\sum_x \sum_y xN_{x,y}} \qquad (1.5\text{-}23)$$

and leads to a corresponding pseudo average composition

$$\bar{y}_{pw} = \sum_y yw_{py} = \frac{\sum_x \sum_y xyN_{x,y}}{\sum_x \sum_y xN_{x,y}}, \qquad (1.5\text{-}24)$$

While w_{py} and $\bar{y}_{pw}$ are precisely equal to w_y and $\bar{y}_w$, respectively, only when repeat units A and B have identical molecular weights, as previously noted, the former nevertheless enjoy widespread use as approximations, owing primarily to their simple forms.

1.5.3. DB and DBD

Simultaneous specification of both DPD and DBD also requires a bivariate distribution, which gives the number of branch points per molecule (b) as well as the DP (x). Computation of $\bar{x}_N$ and $\bar{x}_W$ for branched polymers via the formulas in Table 1.5-1 requires expressions for DPD's. Such expressions have been listed in this table in terms of the bivariate number distribution $N_{b,x}$. Also included is the number degree-of-branching distribution (DBD) and the number-average and weight-average degree of branching (DB). Thus, for example, the number fraction and weight fraction DBD's are

$$n_b = \frac{\sum\limits_{x} N_{b,x}}{\sum\limits_{x}\sum\limits_{b} N_{b,x}} \tag{1.5-25}$$

and

$$w_b = \frac{\sum\limits_{x} x N_{b,x}}{\sum\limits_{x}\sum\limits_{b} x N_{b,x}}, \tag{1.5-26}$$

and the corresponding average DB's are

$$\bar{b}_N = \sum\limits_{b} b n_b \tag{1.5-27}$$

and

$$\bar{b}_W = \sum bw_b \tag{1.5-28}$$

Graessley et al. (12) have remarked that $\bar{b}_W$ is more meaningful than $\bar{b}_N$ in assessing the effects of branching on solution and melt viscosities of polymers due to the sensitivity of flow properties in general to the structure of large molecules.

Finally, we point to a third way of characterizing the extent of branching in polymer molecules, that is, the number of branches per repeat unit b', which we shall call the branch density (Table 1.5-1). It is noteworthy that b' is a fraction; therefore, it corresponds to copolymer composition y in the sense that both are intensive properties, whereas b corresponds to copolymer composition a since both are extensive.

Example 1.5-1. Compute $\bar{y}$ and $\bar{y}_N$ in Table 1.5-3 for the following double distribution matrix $N_{a,b}$:

$$N_{1,3} = 1, \qquad N_{3,5} = 2, \qquad N_{5,11} = 1, \qquad N_{7,1} = 2$$

Solution.

$$\bar{y} = \frac{1(1) + 3(2) + 5(1) + 7(2)}{4(1) + 8(2) + 16(1) + 8(2)} = \frac{1}{2}$$

$$\bar{y}_N = \frac{1}{4}\frac{1}{6} + \frac{3}{8}\frac{2}{6} + \frac{5}{16}\frac{1}{6} + \frac{7}{8}\frac{2}{6}$$

$$= \frac{49}{96}$$

Their equivalence is apparent even for small populations.

1.6. POLYMERIZATION TYPES

We begin by identifying three basic reaction types (model polymerizations) that represent the essential features of most polymerizations. For now let us focus our attention on their growth sequences and omit all concomitant steps, with the exception of termination. These sequences are listed in Table 1.6-1. Termination, as written, does not necessarily represent an elementary step (Appendix E).

1.6.1. *Random Versus Addition Propagation*

The use of m in the addition sequences in place of m_1 is customary, owing to the need to distinguish between monomer m and 1-mer m_1. Frequently the latter is different from the former chemically. It may, for instance, contain a fragment of

Table 1.6-1. Basic Propagation Sequences

RANDOM PROPAGATION

$x \geq 1, \ y \geq 1$ 　　　　　 $m_x + m_y \longrightarrow m_{x+y}$

ADDITION PROPAGATION
WITHOUT TERMINATION

$x \geq 1$ 　　　　　 $m_x + m \longrightarrow m_{x+1}$

ADDITION PROPAGATION
WITH TERMINATION

$x \geq 1$ 　　　　　 $m_x^* + m \longrightarrow m_{x+1}^*$

　　　　　 $m_x^* \longrightarrow m_x$

initiator molecule (Appendix A). This distinction is consistent with the fact that addition polymers generally attain large values of average DP quickly and contain relatively small amounts of oligomer, even when large quantities of monomer remain unreacted. Physical separation of monomer from polymer is thus facilitated. In random polymers, on the other hand, monomer and 1-mer are usually identical chemically. Furthermore, the average DP of the polymer remains relatively low until quite late in the reaction. Thus much oligomer is present and the spectrum of molecular sizes from monomer to polymer is smooth. This inhibits physical separation of residual monomer.

The fundamental difference between random and addition propagation is the manner in which polymer chains grow. In random propagation, polymer chains of all sizes combine with one another. In addition propagation, polymer chains grow one repeat unit at a time.

1.6.2.   Step Versus Chain Polymerizations

The essential distinction between addition sequences with termination and those without is that the latter are "step polymerizations" and the former are not. In the former, the polymer molecules that grow early during the reaction are not the same ones that grow later on. Only a fraction of the total number grow at any time. They grow to full size, terminate, and are succeeded by new generations, which are formed in a separate initiation reaction. In step polymerizations, the same polymer molecules (ideally) that grow at the start of reaction continue to grow until the end. This is true for random propagation as well as addition propagation without termination.

A ubiquitous class of addition polymerizations is the chain reaction, in which polymer chains grow in a very short period of time (seconds). These chains propagate via active intermediates ($*$), such as free radicals, ions, or catalyst complexes, rather than by reaction between stable molecules. The intermediates are formed (initiated) as the result of a catalyst or initiator decomposition reaction and are subject to destruction (termination) owing to their high degree of reactivity. It should be noted that chain addition sequences in which termination is averted, such as the so-called living ionic polymerizations, are step polymerizations by our definition, notwithstanding the presence of active intermediates.

1.6.3.   Classification of Reactions

The characteristic differences in polymer chain growth just described are primary factors in determining the diverse responses of various polymerizations to physical reaction path and its effect on the properties of the polymer products. To assist us in analyzing the underlying phenomena responsible, we classify all polymerizations in terms of various characteristic times (CT's) associated with their reaction sequences. This classification scheme is summarized in Table 1.6-2. Quantitative definitions will follow later.

Polymerizations may be monomer-limited (conventional) or initiator-limited (dead-end) if initiator is employed, depending on which reactant is consumed first. The CT associated with the overall reaction λ_r must therefore be equal to either λ_m or λ_i, whichever is smaller. It should be noted that λ_m is the CT for monomer

Table 1.6-2. Classification of Polymerizations in Terms of Characteristic Times

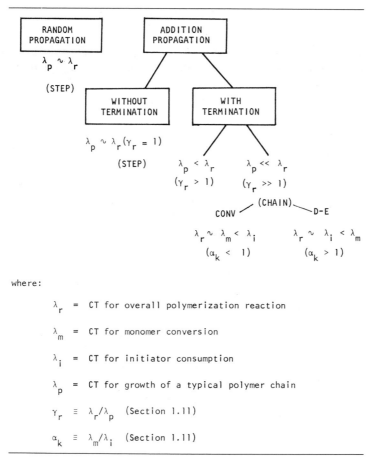

where:

λ_r = CT for overall polymerization reaction

λ_m = CT for monomer conversion

λ_i = CT for initiator consumption

λ_p = CT for growth of a typical polymer chain

$\gamma_r \equiv \lambda_r/\lambda_p$ (Section 1.11)

$\alpha_k \equiv \lambda_m/\lambda_i$ (Section 1.11)

conversion, and λ_i is the CT for initiator depletion, not initiation. The need to distinguish between the latter two processes will become clear subsequently.

It is self-evident that in all step polymerizations, as we have defined them, the CT associated with the growth of polymer chains λ_p must be of the same order of magnitude as λ_r. Clearly, then, when termination of polymer chains occurs it follows that $\lambda_p < \lambda_r$. Chain reactions are extreme cases in which $\lambda_p \ll \lambda_r$.

1.6.4. Initiation and Termination

Addition polymerizations with termination, if they are to sustain their production of polymer throughout reaction, must obviously be accompanied by initiation reactions. Such reactions generally consist of two distinct sequences, decomposition of catalyst or initiator and initiation of polymer chains. The reader is referred to Appendix A and Section 1.10 for details.

Suffice it to say at this point that initiation and termination sequences for ionic and free-radical polymerizations bear fundamental differences. Initiator stoichiometries

$$\frac{1}{s} m_0 + m = m_1^*$$ (1.6-1)

differ by a factor of 2, since one initiator molecule (m_0) can initiate two polymer chains ($s = 2$) in the free-radical case. The stoichiometry of termination

$$n m^* = \text{polymer product}$$ (1.6-2)

by combination is different from all others; it permits two residual initiator fragments to remain with each terminated polymer molecule, one at each end. Termination sequences for ionic ($n = 1$) and free-radical ($n = 2$) polymerizations are also kinetically distinct, being bimolecular and second order for free-radical polymerizations. The reader will note the formal similarity between the termination sequence for combination of free radicals and the propagation sequence for random polymerization.

1.7. STOICHIOMETRY AND CONSTRAINT EQUATIONS

When formulating mathematical rate models for reaction sequences in general, it is important to know which kinetic variables are experimentally measurable and which ones are not. It is equally important to know how many variables are independent in order to facilitate the computation of those that are not measurable in terms of those that are. When dealing with polymerizations and copolymerizations, in addition to obvious kinetic variables such as monomer and initiator consumption, we have certain polymer properties that are experimentally accessible. They include DPD, average DP, average CC, and so on. The computational aspects of these properties were discussed in Section 1.5.

At this point it is appropriate to examine how stoichiometric considerations can lead to relationships among kinetic variables. We apply the law of definite proportions, equation D-42 with $\nu \equiv -x$, to random and addition polymerizations. The results are called constraint equations because they constrain the independence of rate equations. Some details are given in Example 1.7-1.

For heterodisperse random polymers

$$1 < x < \infty \qquad\qquad x m_1 = m_x$$ (1.7-1)

we obtain the following constraint equation among end groups (terminal units at either end of an x-mer): Counting monomer as 1-mer.

$$(N_1)_0 = \mu_N^0 + \sum_{x \geqslant 1} (x - 1) N_x$$ (1.7-2)

The zeroth moment represents the number of unreacted end groups at any time, and the second term on the right-hand side (RHS) represents the number of reacted ones.

Conversion, which is defined as the fraction of reacted end groups,

$$\Phi \equiv \frac{(N_1)_0 - \sum\limits_{x \geqslant 1} N_x}{(N_1)_0} \equiv 1 - \frac{\mu_N^0}{(N_1)_0} \tag{1.7-3}$$

is therefore related to the zeroth moment of the DPD. Assuming that the feed contains pure monomer, we can also interpret equation 1.7-2 as a conservation equation for repeat units at any time

$$\sum\limits_{x \geqslant 1} xN_x \equiv \mu_N^1 = (N_1)_0. \tag{1.7-4}$$

For addition polymers, on the other hand (equations A-11 and A-12),

$$1 \leqslant x < \infty \qquad (1 + R')s^{-1}m_0 + xm = m_x, \tag{1.7-5}$$

the law of definite proportions yields for monomer

$$-\Delta N_m \equiv (N_m)_0 - N_m = \sum\limits_{x \geqslant 1} xN_x = \mu_N^1 \tag{1.7-6}$$

and for initiating species m_0, in the absence of chain transfer,

$$-\Delta N_0 \equiv (N_0)_0 - N_0 = (1 + R')s^{-1} \sum\limits_{x \geqslant 1} N_x = (1 + R')s^{-1}\mu_N^0, \tag{1.7-7}$$

where we have ignored both monomer and initiator fragments contained within the intermediates $(\sum_{x \geqslant 1} N_x^*)$. It should be noted that equations 1.7-5 and 1.7-7 include free-radical polymerizations $(s = 2)$ with termination by combination $(R' = 1)$ and disproportionation $(R' = 0)$, as well as catalytic and ionic $(s = 1, R' = 0)$ polymerizations, as special cases. We conclude from equations 1.7-6 and 1.7-7 that, for addition polymerizations, monomer conversion

$$\Phi \equiv \frac{-\Delta N_m}{(N_m)_0} \tag{1.7-8}$$

is proportional to the first moment of the DPD (N_x) and catalyst (initiator) consumption is proportional to the zeroth moment. It is also noteworthy that the molar conversion of monomer is proportional to the weight of polymer formed

$$W_p = -M_0 \Delta N_m = M_0(N_m)_0 \Phi, \tag{1.7-9}$$

where M_0 represents the molecular weight of the repeat unit.

In free-radical polymerizations, not all primary intermediates m_0^* (radicals) are effective in initiating polymer chains. A significant fraction of them recombine to form non-initiating species. It is common practice to use a constant efficiency factor f (Chapter 2) to represent the fraction that is effective. Thus, we may set $s = 2f$ in equation 1.7-7 and obtain the following useful expression:

$$\mu_N^0 = \frac{2f(-\Delta N_0)}{1 + R'} \tag{1.7-10}$$

Another useful expression for addition polymerizations in general, which provides an estimate of $\bar{x}_N$ from a knowledge of monomer conversion and initiator depletion, may be obtained by substituting the preceding expressions for μ_N^1 and μ_N^0 into the equation $\bar{x}_N = \mu_N^1/\mu_0^1$ (cf. Table 1.5-1). The result is

$$\bar{x}_N = \frac{(1 + R') \Delta N_m}{s \Delta N_0} \tag{1.7-11}$$

The special case $R' = 0$ reflects the intuitively satisfying definition of number average DP as the number of monomer molecules converted to polymer $(-\Delta N_m)$ divided by the number of polymer molecules initiated $(-s \Delta N_0)$. It should be remembered that s contains f.

For addition copolymerizations (equation A-17), monomer balances lead to the following equations if the content of intermediates $(\Sigma_{a \geqslant 0} a N_a^* + \Sigma_{b \geqslant 0} b N_b^*)$ is again ignored:

$$(N_A)_0 = N_A + \sum_{a \geqslant 0} a N_a, \tag{1.7-12}$$

$$(N_B)_0 = N_B + \sum_{b \geqslant 0} b N_b, \tag{1.7-13}$$

$$(N_m)_0 = N_m + \sum_{a \geqslant 0} \sum_{b \geqslant 0} (a + b) N_{a,b}, \tag{1.7-14}$$

where $N_m \equiv N_A + N_B$. From these equations together with the definition of y in Table 1.5-3, we obtain the simple expression for the CC (cf. Table 1.5-3)

$$y = \frac{\Delta N_A}{\Delta N_m} \tag{1.7-15}$$

which, like its counterpart for DP (equation 1.7-11), facilitates determination of the mean value of the property without knowledge of the distribution. Finally, it is also apparent that the weight CCD w_y is proportional to the weight of copolymer formed of a given composition y. This interpretation will be assigned to pseudo CCD w_{py} in Chapter 3 as a convenient approximation.

Example 1.7-1. **Constraint Equations for Random and Addition Polymerizations.** Starting with equation D-42, derive constraint equations 1.7-2, 6 and 7.

Solution. Application of equation D-42 to the stoichiometric equation for random polymerization

$$x m_1 = m_x$$

leads to the following

$$x > 1 \qquad\qquad N_x = (N_x)_0 + \xi,$$

$$x \geq 1 \qquad\qquad N_1 = (N_1)_0 - x\xi.$$

By combining these results to eliminate ξ and assuming that $(N_x)_0 = 0$ for $x > 1$, we obtain for monodisperse product

$$x > 1 \qquad\qquad (N_1)_0 = N_1 + xN_x$$

and thus for polydisperse product

$$(N_1)_0 = N_1 + \sum_{x=2}^{\infty} xN_x = \sum_{x=1}^{\infty} xN_x.$$

Similarly, application of equation D-42 to the stoichiometric equation for addition polymerization (1.7-5) leads to

$$x \geq 1 \qquad\qquad N_x = (N_x)_0 + \xi,$$

$$N_0 = (N_0)_0 - (1 + R')s^{-1}\xi,$$

$$x \geq 1 \qquad\qquad N_m = (N_m)_0 - x\xi.$$

After eliminating ξ and assuming $(N_x)_0 = 0$, we obtain for polydisperse product

$$(N_0)_0 = N_0 + (1 + R')s^{-1} \sum_{x=1}^{\infty} N_x,$$

$$(N_m)_0 = N_m + \sum_{x=1}^{\infty} xN_x.$$

1.8. EQUILIBRIUM AND RATE CONSTANTS

In most analyses of polymerization reactions, a single equilibrium constant and a single rate constant are applied to all propagation reactions, despite the fact that each reaction actually involves different molecules. Since these differences are primarily due to size (DP), it is generally assumed that K_p and k_p are independent of DP. Moreover, molar concentrations are used to express equilibrium composition and rate functions notwithstanding the vast differences that exist among the molecular weights of the species involved.

The alternative is to employ a different equilibrium constant and a different rate constant for each propagation step. This would require a very large number of constants and would enormously complicate our reaction models. The benefits of assuming DP independence are obvious. What is remarkable is that the assumption appears to be valid, as does the use of the molar concentrations. In this section, we shall examine these simplifications and some of their ramifications.

1.8.1. The Equilibrium Constant

We begin by showing that the equilibrium constants for polymerization reactions are independent of DP when expressed in the usual way, that is, in terms of molar concentrations. From the criterion for chemical reaction equilibrium (Appendix D)

$$\sum_k \nu_k \mu_k = 0, \tag{1.8-1}$$

which requires substitution of an expression for chemical potential μ_k of each substance k in terms of its activity a_k. The result is an equilibrium constant

$$K = \prod_k a_k^{\nu_k}, \tag{1.8-2}$$

which is independent of composition, being a function of temperature (and pressure) only:

$$K = \exp\left(\frac{-\Delta G_r^*}{R_g T}\right), \tag{1.8-3}$$

where ν_k is the stoichiometric coefficient for substance k and ΔG_r^* is the standard free energy of reaction

$$\Delta G_r^* \equiv \sum_k \nu_k \mu_k^*. \tag{1.8-4}$$

For reactions among small molecules, it is common practice to express compositions as molar concentrations

$$a_k = \gamma_k c_k \tag{1.8-5}$$

and let activity coefficients γ_k account for solution nonidealities. For ideal solutions, $\gamma_k \to 1$ for all k, and we obtain the well-known result

$$K = \prod_k c_k^{\nu_k} \tag{1.8-6}$$

The reader is referred to Appendix D for a brief review of chemical thermodynamics.

For polymer solutions, on the other hand, Flory and Huggins (13–16) have shown that volume fraction ϕ_k is the proper composition variable. This is, in essence, a direct consequence of the contrast in size between the large polymer solute molecules and the small solvent molecules. More will be said about polymer solutions in Chapter 6.

Volume fraction of x'-mer is defined as

$$\phi_{x'} \equiv \frac{N_{x'} v_{x'}}{N_s v_s + \sum_{x'} N_{x'} v_{x'}} \tag{1.8-7}$$

where v_s and $v_{x'}$ are molar volumes occupied by solvent (monomer) and x'-mer molecules in solution, respectively, and x' represents the number of equivalent segments of volume v_s occupied by the polymer.

$$v_{x'} = x'v_s \tag{1.8-8}$$

Obviously, x' is not in general equal to the number of structural mers, x, nor is it necessarily an integer. Volume fraction can be related to molar concentration by combining equations 1.8-7 and 1.8-8:

$$\phi_{x'} = v_s x' c_{x'} \tag{1.8-9}$$

Frequently the approximation $x' \cong x$ can be used without serious error, especially when the solvent is monomer.

To obtain the required expressions for chemical potential μ_k, where k is x-mer, we use Flory's pseudo lattice model for polydisperse polymer solutions (16), which is based on a reference (standard) state of unmixed (pure) solvent and ordered polymer. For such systems, Flory has shown that the entropy of mixing is

$$\Delta S_M = -R_g \left\{ \sum_{x'} N_{x'} \ln\left(\frac{x'}{\sigma}\right) - [\ln(\gamma - 1) - 1]\sum_{x'}(x' - 1)N_x \right.$$

disorienting ordered polymer

$$\left. - N_s \ln\phi_s + \sum_{x'} N_{x'}\ln\phi_{x'} \right\} \tag{1.8-10}$$

mixing polymer and solvent

where γ is the lattice "coordination number" and σ is a symmetry number for polymer molecules, both being constant parameters. For the special case of monodisperse polymer, this result reduces to (Chapter 6)

$$\Delta S_M = -R_g(N_s \ln\phi_s + N_{x'}\ln\phi_{x'}). \tag{1.8-11}$$

When $x = 1$, it reduces further to the well-known expression for entropy of mixing for simple binary mixtures of small molecules:

$$\Delta S_M = -R_g(N_s \ln n_s + N_1 \ln n_1). \tag{1.8-12}$$

Following Flory, we add a Van Laar type heat of mixing term to obtain the corresponding Gibbs free energy of mixing ΔG^M, and differentiate with respect to $N_{x'}$ to deduce the required chemical potential:

$$\mu_{x'} = \mu_{x'}^0 + \underbrace{R_g T\left[\chi x'(1 - \phi_p)^2\right]}_{H_{x'}^M}$$

$$+ T\underbrace{\left\{ R_g\left[\ln\left(\frac{\phi_{x'}}{x'}\right) - (x' - 1)\ln(\gamma - 1) + x'\phi_p\left(1 - \frac{1}{\bar{x}_N'}\right) + \ln\sigma\right]\right\}}_{-S_{x'}^M}$$

$$\tag{1.8-13}$$

where χ is an interaction parameter, $\mu_{x'}^0$ refers to the standard (reference) state of pure polymer cited earlier, and

$$\phi_p \equiv \sum_{x'} \phi_{x'} \tag{1.8-14}$$

For processes in which the DPD of dissolved polymer can change, as with chemical reaction and interphase transport of polymer, Flory (16) has asserted that equation 1.8-13 is required, rather than the following one in which the standard (reference) state is unmixed (pure) solvent and polydisperse, disordered polymer:

$$\mu_{x'} = \mu_{x'}^0 + R_gT\left[\chi x'(1 - \phi_p)^2 + \ln\phi_{x'} - (x' - 1) + x'\phi_p\left(1 - \frac{1}{\bar{x}_N'}\right)\right] \tag{1.8-15}$$

The corresponding chemical potential for solvent will be used in Chapter 6 (equation 6.1-19) in connection with phase equilibrium in which solvent and not polymer is transported between phases. The major difference between equations 1.8-13 and 1.8-15, as Flory has pointed out, is the term $\ln(\phi_{x'}/x')$ in the former, as contrasted with $\ln\phi_{x'}$ in the latter. It is precisely this difference, as we shall see, that gives rise to equilibrium constants of the form sought.

We now apply equation 1.8-1 to equilibrium random polymerization

$$x, y \geq 1 \qquad\qquad m_x + m_y \overset{K}{\rightleftharpoons} m_{x+y} + m_0 \tag{1.8-16}$$

and equilibrium addition polymerization

$$m_0 + m \overset{K_i}{\rightleftharpoons} m_1 \tag{1.8-17}$$

$$x \geq 1 \qquad\qquad m_x + m \overset{K_p}{\rightleftharpoons} m_{x+1} \tag{1.8-18}$$

to obtain

$$\mu_{x+y} - \mu_x - \mu_y = 0 \tag{1.8-19}$$

if m_0 is ignored for the moment, and

$$\mu_{x+1} - \mu_x - \mu_1 = 0 \tag{1.8-20}$$

respectively, from which we deduce the following results after substitution of equation 1.8-13:

$$\frac{(\phi_{x+y}/\phi_x\phi_y)xy}{(x + y)} = (\gamma - 1)\sigma\exp(-\Delta G_r^0/R_gT) \tag{1.8-21}$$

and

$$\frac{(\phi_{x+1}/\phi_x\phi_1)x}{(x + 1)} = (\gamma - 1)\sigma\exp\left(\frac{-\Delta G_r^0}{R_gT}\right) \tag{1.8-22}$$

where

$$\Delta G_r^0 = \begin{cases} \mu_{x+y}^0 - \mu_x^0 - \mu_y^0 & \text{for random} & (1.8\text{-}23) \\ \mu_{x+1}^0 - \mu_x^0 - \mu_1^0 & \text{for addition} & (1.8\text{-}24) \end{cases}$$

It should be noted that we have dropped the prime on x, thus tacitly assuming that $x' \cong x$. Finally, the left-hand side (LHS) of equations 1.8-21 and 1.8-22 may be transformed into the following equilibrium constants for random and addition polymerization, respectively:

$$x, y \geqslant 0 \qquad\qquad K = \frac{(c_{x+y})_e}{(c_x)_e (c_y)_e} \qquad\qquad (1.8\text{-}25)$$

and

$$K_p = \frac{(c_{x+1})_e}{(c_x)_e (c_m)_e} \qquad\qquad (1.8\text{-}26)$$

by substituting equation 1.8-9 for ϕ_x, and by defining their temperature dependence as follows:

$$\left.\begin{array}{c} K \\ K_p \end{array}\right\} = v_s \sigma (\gamma - 1) \exp\left(\frac{-\Delta G_r^0}{R_g T} \right) \qquad\qquad (1.8\text{-}27)$$

The validity of the following equation,

$$K_i = \frac{(c_1)_e}{(c_0)_e (c_m)_e} \qquad\qquad (1.8\text{-}28)$$

in which all components are small molecules, follows as well. We note that v_s does not appear in equation 1.8-27 for random polymerizations when by-product m_0 is included.

A final word concerning the quantity ΔG_r^0 appearing in equations 1.8-21 and 1.8-22 is in order. It is well known that the standard free-energy change of any reaction can be expressed in terms of the standard free energies of formation of its participants. For the formation of pure x-mer from pure monomer,

$$x m_1 = m_x \qquad\qquad (1.8\text{-}29)$$

we write

$$\mu_x^0 = x\mu_1^0 + (x - 1)\Delta G^0 \qquad\qquad (1.8\text{-}30)$$

where ΔG^0 is the standard free-energy change per mole of interunit bonds formed. After substituting this result into equations 1.8-23 and 1.8-24, we obtain $\Delta G_r^0 = \Delta G^0$ for both.

To summarize, we conclude that polymerization equilibrium constants are independent of DP when expressed in terms of molar concentrations. The counterpart of this conclusion for rate constants will be examined next. It is termed the principle of equal reactivity.

1.8.2. The Principle of Equal Reactivity

The irreducible molecular events that determine the mechanism of any chemical reaction are called elementary steps (Appendix E). They generally consist of complex sequences of coupled reactions. Since the kinetic rate law can only be applied to elementary steps, their identities are requisite to the *a priori* mathematical formulation of kinetic rate functions.

The rate function in its general form comprises the product of molar concentrations of the reacting species with a rate constant, which represents the specific reactivity of the reaction. In all polymerizations, such species include small molecules and macromolecules reacting with each other and among themselves. The reactivity of the macromolecules lies in their chain ends, which contain functional groups or active intermediates such as free radicals, ions, or catalyst complexes. It is therefore reasonable to adopt.the principle of equal reactivity (PER), which asserts that the specific reactivity of chemically similar polymer chains is determined only by the nature of their ends and is independent of DP. This hypothesis has been tested experimentally (17).

The implication of the PER is that a single rate constant may be assigned to all propagation steps. For example, we can represent irreversible random propagation with the following sequence of elementary steps having a single rate constant

$$\mathrm{m}_x + \mathrm{m}_y \overset{k}{\to} \mathrm{m}_{x+y} \tag{1.8-31}$$

and, similarly, irreversible addition propagation with

$$\mathrm{m}_x + m \overset{k_p}{\to} \mathrm{m}_{x+1} \tag{1.8-32}$$

However, when applying the PER to formulate rate functions for these simple reactions, we encounter a subtle aspect of this principle which should be clarified before proceeding with more complex reactions. Thus, although the rate of each molecular event as written in equation 1.8-32 can be expressed by

$$r_{px} = k_p c_m c_x \tag{1.8-33}$$

for all x, the corresponding rate function for equation 1.8-31 can be expressed as follows only when x and y are unequal:

$$x \neq y \qquad\qquad r_{pxy} = k c_x c_y \tag{1.8-34}$$

For equal x and y, we must write

$$x = y \qquad\qquad r_{pxx} = \frac{k}{2}c_x^2 \qquad\qquad (1.8\text{-}35)$$

if the same specific reactivity k is to be used.

The reason for the difference between functions 1.8-34 and 1.8-35 may be deduced from the rate law itself. We interpret the product of molar concentrations as the collision probability (number of favorable combinations) of reacting species, and the rate constant as their reaction probability (chemical affinity), once collision has occurred. We let k represent the specific reactivity of functional group pairs of the kind involved in reactions 1.8-31. Then, all rates must be directly proportional to the number of ways of combining pairs of reactant molecules, since all pairs have equal probability of reacting (same k), once combined. Thus N_x molecules of type m_x may be paired with N_y of type m_y in $N_x N_y$ ways, not counting sequential order within each pair. On the other hand, molecules of type m_x may be paired with one another in the following number of ways:

$$\binom{N_x}{2} \equiv \frac{N_x!}{(N_x - 2)!\,2!} \qquad\qquad (1.8\text{-}36)$$

However, since $N_x \gg 1$ the binomial coefficient $\binom{N_x}{2}$ (Appendix C) may be approximated by $N_x^2/2$ since $N_x \gg 1$. The origin of the 2, which distinguishes equation 1.8-35 from 1.8-34 should now be evident.

More specifically, for AB-type condensation polymerizations, we shall write

$$r_{pxy} = \begin{cases} 2kc_x c_y & \text{when } x \neq y \\ kc_x^2 & \text{when } x = y \end{cases} \qquad\qquad (1.8\text{-}37)$$

in lieu of functions 1.8-34 and 1.8-35. The reason is that we must in fact count sequential order within AB pairs, because m_x and m_y are bifunctional. Thus, although no distinction will be made between conformations $[AB]_x[AB]_y$ and $[BA]_y[BA]_x$, we will distinguish between configurations $[AB]_x[AB]_y$ and $[AB]_y[AB]_x$. This distinction raises the number of combinations uniformly by a factor of 2. Our motive for inserting the 2 is the establishment of a consistent nomenclature throughout. Thus, for bimolecular termination of free-radical intermediates, whose elementary steps are similar to 1.8-31:

$$m_x^* + m_y^* \xrightarrow{k_t} \text{products} \qquad\qquad (1.8\text{-}38)$$

we shall write

$$r_{txy} = \begin{cases} k_t c_x^* c_y^* & \text{when } x \neq y \\ \dfrac{k_t}{2}(c_x^*)^2 & \text{when } x = y \end{cases} \qquad\qquad (1.8\text{-}39)$$

because in this case only one chain end on each molecule is reactive.

Until now we have been concerned with formulating reaction rate functions consistent with the PER. A word of explanation is also in order concerning the matter of expressing formation and consumption rates for specific components participating in a reaction from the reaction rate function by applying stoichiometry. For example, the rate of each molecular event (loss of a pair of functional groups) represented by reaction 1.8-31 is expressed by function 1.8-37. It follows that the rates of formation and consumption of species m_x are $2kc_yc_{x-y}$ and $-2kc_xc_y$, respectively, when $y \neq x/2$ in the former case and $x \neq y$ in the latter, because each molecular event involves only one molecule m_x. When $y = x/2$ in the formation reaction, its rate function must be kc_y^2 by the argument advanced in the preceding paragraphs. However, when $y = x$ in the consumption reaction, we must again write $2kc_x^2$ by stoichiometry, which requires restoration of the factor 2 to reflect the fact that two m_x molecules disappear during each molecular event.

These considerations will be amplified in subsequent sections of this chapter.

Example 1.8-1. *Effect of Stoichiometry on Rate Functions.* We write $r_t = k_t(c^*)^2$ for the overall rate of termination, defined as the loss rate of active intermediates. In the literature, however, it is frequently written as $r_t = 2k_t(c^*)^2$ to reflect the fact that two intermediates are lost in each termination reaction:

$$m_x^* + m_y^* \xrightarrow{k_t} \text{product}$$

From our previous discussions, the loss rate of active x-mer in this reaction is $k_t c_x^* c_y^*$, whether x and y are different or equal. Since the total loss rate of active intermediates in each reaction is twice this, it would appear that the coefficient of 2 is required in our expression as well for it to be consistent. Resolve this apparent dilemma.

Solution. The loss rate of active x-mer is $k_t c_x^* c_y^*$ when $x \neq y$ and $k_t(c_x^*)^2$ when $y = x$. Thus we can argue that since the loss rate of active x-mer in all reactions is $k_t c_x c^*$, the total loss rate of active intermediates must be $k_t \sum_{x=1}^{\infty} c_x c^*$, which yields $k_t(c^*)^2$.

1.9. RANDOM POLYMERIZATION

The following reaction sequence may be used to represent a variety of random polymerizations.

$$x, y \geqslant 1 \qquad\qquad m_x + m_y \underset{k_r}{\overset{k_f}{\rightleftarrows}} m_{x+y} + m_0 \text{ (or S)} \qquad\qquad (1.9\text{-}1)$$

Thus, for AB-type polycondensation m_x is $[AB]_x$ (equation A-7) and S could represent water; for interchange propagation m_x is $BB[AABB]_x$ and by-product m_0 is BB (equation A-4). Interchange polymerizations are generally preceded by a distinct and separate first step (equation A-3) which may be viewed as a sort of "initiation" reaction:

$$I + 2m_0 \rightarrow m_1 + 2S \qquad\qquad (1.9\text{-}2)$$

Interchange reactions of the rearrangement type (equation A-10) may be written generally as

$$x, y \geqslant 1$$
$$x > w, \qquad y > z \qquad\qquad \mathrm{m}_x + \mathrm{m}_y \underset{k}{\overset{k}{\rightleftarrows}} \mathrm{m}_{x-w+z} + \mathrm{m}_{y+w-z} \qquad\qquad (1.9\text{-}3)$$

An important special case (equation A-5) occurs when w (or z) = 0

$$x, y \geqslant 1$$
$$y > z \qquad\qquad \mathrm{m}_x + \mathrm{m}_y \underset{k}{\overset{k}{\rightleftarrows}} \mathrm{m}_{x+z} + \mathrm{m}_{y-z} \qquad\qquad (1.9\text{-}4)$$

and another when $w = 0$, $z = y$. It should be noted that the latter represents a chain-building process in the forward direction, that is, propagation, which is identical to reaction 1.9-1. The difference is that in equation 1.9-1 we have used separate rate constants to distinguish specific reactivity for interchange involving exclusively ultimate interunit linkages (k_f) from that involving exclusively end groups on $\mathrm{m}_0(k_r)$.

1.9.1. Simple Random Propagation with Degradation

The simplest sequence for random polymerization is the one shown in Table 1.9-1. Reverse propagation, which is equivalent to random degradation, has been included. By introducing a single rate constant for each direction, it has been tacitly assumed that the PER applies to degradation as well as to propagation, that is, that the specific reactivity of both reactions is independent of DP.

By applying the rate law to this sequence, assuming that it comprises elementary kinetic steps, we obtain rate functions r_x for the net formation of x-mer that are similar to those reported by Blatz and Tobolsky (18). It should be noted that infinite series appear in the rate functions. Their use is common practice in polymerization kinetics and will be continued throughout this book. The intent is to indicate that summation should be carried out to sufficiently large values of x to include all species m_x present. Obviously, the largest value of x must always be finite on physical grounds, since x represents a DP and is therefore bounded by the number of repeat units available from the reactant molecules. The sequence c_x must therefore converge to zero when x becomes sufficiently large.

It should also be noted that our choice of coefficients reflects the probability factors associated with AB-type polymerization discussed in Section 1.8. Thus, for example, the term $2k_f$ in the forward reaction signifies that two combinations of molecules, whose functional groups each have specific reactivity k_f, will yield the same product molecule. The factor 2 appears in the reverse rates for the same reason, that is, because x-mer can be generated from larger chains by scission of either one of two appropriate interunit linkages occupying symmetrical positions at opposite ends of the chain. The coefficient $x - 1$ signifies that x-mer can be destroyed by cleavage of any one of its $x - 1$ interunit linkages.

For random polymerizations in general, we define the rate of polymerization r as the depletion rate of functional group pairs. Each time such a pair disappears, a

Table 1.9-1.   Simple Random Propagation with Degradation

$$x,y \geq 1 \qquad m_x + m_y \underset{k_r}{\overset{k_f}{\rightleftharpoons}} m_{x+y}$$

$$r_1 = -2k_f c_1 \sum_{j=1}^{\infty} c_j + 2k_r \sum_{j=2}^{\infty} c_j$$

$$x > 1, \qquad r_x = -2k_f c_x \sum_{j=1}^{\infty} c_j + k_f \sum_{j=1}^{x-1} c_j c_{x-j}$$

$$- k_r(x-1)c_x + 2k_r \sum_{j=x+1}^{\infty} c_j$$

Alternatively, if $c_0 \equiv 0$

$$x \geq 1, \qquad r_x = -2k_f c_x \sum_{j=0}^{\infty} c_j + k_f \sum_{j=0}^{x} c_j c_{x-j}$$

$$- k_r(x-1)c_x + 2k_r \sum_{j=x+1}^{\infty} c_j$$

OVERALL $\qquad r \equiv - \sum_{x=1}^{\infty} r_x = k_f c_p^2 - k_r(c_{RU} - c_p)$

where $\qquad c_p \equiv \sum_{x=1}^{\infty} c_x = \mu_c^0$

$$c_{RU} \equiv \sum_{x=1}^{\infty} x c_x = \mu_c^1$$

single polymer molecule is formed. Consequently

$$r = - \sum_{x=1}^{\infty} r_x, \tag{1.9-5}$$

where the summation represents the net molar rate of polymer formation. The minus sign reflects the decline in the molar concentration of polymer

$$c_p \equiv \sum_{x=1}^{\infty} c_x = \mu_c^0 \text{ (zeroth moment of } c_x). \tag{1.9-6}$$

Substitution of the rate functions r_x into equation 1.9-5 followed by summation leads to the simple overall rate expression in Table 1.9-1. We expect this result intuitively since $c_{RU} - c_p$ represents the concentration of interunit linkages, which act as reactants in reverse polymerization. Of course, the total concentration of repeat units

$$c_{RU} \equiv \sum_{x=1}^{\infty} x c_x = \mu_c^1 \tag{1.9-7}$$

is conserved during polymerization.

As an aid to carrying out the kind of summation shown in the preceding equations, it is convenient to use the following matrix:

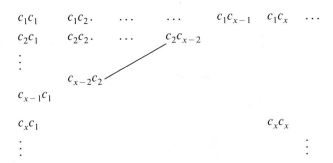

Either half of the matrix above or below the diagonal from top left to bottom right, including the diagonal, includes all possible pairs of molecules participating in propagation, since we need not count the order within each pair. The double summation $\sum_{x=1}^{\infty} c_x \sum_{j=1}^{\infty} c_j$ may be viewed as the sum of all terms in the matrix, row by row. Summation of the convolution sum $\sum_{x=1}^{\infty} \sum_{j=1}^{x-1} c_j c_{x-j}$ (Table C-1) may again be viewed as the sum of all terms in the matrix along the diagonals from lower left to upper right. Consequently, the sum of both forward rates yields c_p^2. The double summation in the last reverse rate term may be obtained by regrouping terms:

$$\sum_{x=1}^{\infty} \sum_{j=x+1}^{\infty} c_j = \sum_{x=1}^{\infty} x c_x - \sum_{x=1}^{\infty} c_x = c_{RU} - c_p. \tag{1.9-8}$$

When the forward reaction in Table 1.9-1 generates a by-product, say S, we simply replace k_r everywhere with $k_r c_0$, where c_0 is the concentration of S. Thus we obtain

$$r = k_f c_p^2 - k_r (c_{RU} - c_p) c_0 \tag{1.9-9}$$

Irreversible propagation is obviously included as the special case $k_r = 0$.

1.9.2. Interchange Polymerization

A general sequence for interchange polymerization is shown in Table 1.9-2 together with the appropriate rate functions. It should be noted that a distinction has been made between pure propagation (k_f) and rearrangement by scission (k); and that scission via by-product $m_0(k_r)$ has been segregated from other scission reactions

Table 1.9-2.   Interchange Polymerization

$i, j \geq 1$ $\qquad\qquad m_i + m_j \underset{k_r}{\overset{k_f}{\rightleftharpoons}} m_{i+j} + m_o$

$i, j \geq 1, k > 0$ $\qquad m_i + m_j \underset{k}{\overset{k}{\rightleftharpoons}} m_{i+j-k} + m_k$

$$r_o = k_f \underset{①}{\sum_{i=1}^{\infty} \sum_{j=1}^{\infty} 2c_i\, 2c_j} - k_r \underset{②}{\sum_{j=2}^{\infty} c_o(2j - 2)c_j}$$

$$r_1 = -2k_f \underset{③}{\sum_{j=1}^{\infty} 2c_1 2c_j} + k \underset{④}{\sum_{i=2}^{\infty} \sum_{j=2}^{\infty} 2c_i\, 2c_j}$$

$$- k \underset{⑤}{\sum_{j=3}^{\infty} 2c_1(2j-4)c_j} + 2k_r \underset{⑥}{\sum_{j=2}^{\infty} 2c_o 2c_j}$$

$x > 1 \qquad$
$$r_x = k_f \underset{⑦}{\sum_{i=1}^{x-1} 2c_i\, 2c_{x-i}} - 2k_f \underset{⑧}{\sum_{j=1}^{\infty} 2c_x\, 2c_j}$$

$$+ 2k_r \underset{⑨}{\sum_{j=x+1}^{\infty} 2c_o\, 2c_j} - k_r \underset{⑩}{2c_o(2x - 2)\, c_x}$$

$$+ k \underset{⑪}{\sum_{i=1}^{x-1} \sum_{k=1}^{\infty} 2c_i\, 2c_{x-i+k}} + k \underset{⑫}{\sum_{i=1}^{\infty} \sum_{j=x+1}^{\infty} 2c_i\, 2c_j}$$

$$- k \underset{⑬}{\sum_{j=1}^{\infty} 2c_x(2j-2)c_j} - k \underset{⑭}{\sum_{j=1}^{\infty} (2x - 2)c_x\, 2c_j}$$

OVERALL $\qquad \displaystyle\sum_{x=o}^{\infty} r_x = 0$

(k). The reason is that the specific reactivity for interchange between ultimate interunit linkages and m_0 could differ from that for interchange among internal linkages and end groups on m_x where $x > 0$, as mentioned earlier. Use of the same k for all rearrangement-type scission reactions obviously implies application of the PER and suggests that forward and reverse reactions are identical, which seems reasonable.

By assuming the structure $BB[AABB]_x$ for m_x and BB for m_0, we have $2x$ interunit linkages available in each molecule for scission and the net rate of formation of all species in the reaction mixture is given by the rate functions listed in Table 1.9-2. The circled number under each term identifies a corresponding statement in Table 1.9-3 which explains its rationale. Inspection of the rearrangement reaction in Table 1.9-2 reveals that, in general, any x-mer is formed in either of two ways: by combination of i-mer and j-mer such that $i + j = x$, or by splitting away a fragment x from j-mer where $j > x$. In each case, either end of each molecule may react with either one of two appropriate interunit linkages. Similarly, any x-mer disappears in either of two ways: by combining with a fragment split from another molecule of any size or by being split itself at any one of its interunit linkages.

It should be noted that for each $i \leqslant x - 1$ in ⑪, we have purposely included a term for the case $x - i + k = x$, in which no net formation of m_x actually occurs. However, this error was offset by including in ⑭ precisely the same term for each $j \leqslant x - 1$, only with the opposite sign. Similarly, all the terms with $i = x$ in ⑫, which need not have been included, are exactly canceled by equivalent terms in ⑬ with opposite signs, one for each $j \geqslant x + 1$.

The validity of the rate functions in Table 1.9-2 may be tested in several ways. First, rate functions r_1 and r_x in Table 1.9-2 reduce to those in Table 1.9-1 after setting $k = 0$ and $k_r c_0 = k_r$ in the former and dividing by 4. The latter sequence is a special case of the former without rearrangement (k). The factor 4 signifies that eight combinations are required for interchange as opposed to two for AB-type condensation. A second test is provided by setting $k_f = 0 = k_r$ in rate functions r_x in Table 1.9-2 and again dividing by 4. The result

$$x \geqslant 1 \qquad r_x = k \sum_{i=1}^{x-1} \sum_{j=x-1+k}^{\infty} c_i c_j + k \sum_{i=1}^{\infty} c_i \sum_{j=x+1}^{\infty} c_j$$

$$- k c_x \sum_{j=1}^{\infty} (j - 1) c_j - k(x - 1) c_x \sum_{j=1}^{\infty} c_j \qquad (1.9\text{-}10)$$

is in agreement with that derived by Hermans (19) specifically for rearrangement interchange reactions 1.9-14. As a final check, we expect the summation $\sum_{x=0}^{\infty} r_x$ to yield zero since the total number of molecules in the system must be conserved. This exercise is the subject of Example 1.9-1.

The interchange equations developed in this section are applied to polyethylene terephthalate PET production in Example 1.9-2. Further reference to this important polymerization will be made in several subsequent examples.

1.9.3. More General Random Propagation

Random polymerization of two different comonomers, as in AA-BB type polycondensation (Appendix A), leads to far more complex rate expressions than those previously discussed. First, we have three distinguishable x-mers: $m_x^{AB} = [AABB]_x = m_x^{BA}$, $m_x^{AA} = [AABB]_x AA$ and $m_x^{BB} = [BBAA]_x BB$. Second, we must thus distinguish among four possible propagation reactions (cf. equations A-2)

$$x, y \geqslant 0$$

$$m_x^{AA} + m_y^{BB} \rightarrow m_{x+y}^{AB}$$

$$m_x^{AB} + m_y^{AB} \rightarrow m_{x+y}^{AB} \qquad (1.9\text{-}11)$$

$$m_x^{AA} + m_y^{BA} \rightarrow m_{x+y}^{AA}$$

$$m_x^{BB} + m_y^{AB} \rightarrow m_{x+y}^{BB}$$

in lieu of just one, equation 1.9-1. By-product m_0 (or S) and reverse reactions have been omitted to simplify matters somewhat. Monomers are defined as $m_0^{AA} = AA$ and $m_0^{BB} = BB$.

The rate functions for irreversible propagation, which have previously been reported by Kilkson (20), are listed in Table 1.9-4. Coefficients 1, 2, and 4 reflect the number of ways in which molecules can pair their end groups, whose specific reactivity is represented by k, consistent with the probability factors discussed in Section 1.8. For example, m_x^{AA} and m_x^{BB} may combine in the following ways and yield the same product molecule: $(_1AA_2)_x(_1BB_2)_y$, $(_2AA_1)_x(_1BB_2)_y$, $(_1BB_2)_y(_1AA_2)_x$ and $(_1BB_2)_y(_2AA_1)_x$, where subscripts 1 and 2 identify chain ends. It should be noted that the coefficient 4 appears in the last term of function r_x, although this term is a convolution sum. Yet the ostensibly appropriate coefficient 2 in the convolution sum just preceding it was omitted. The reason is that AA and BB molecules are distinguishable from one another, whereas AB are not. For the same reason, the corresponding convolution sum is of the general type, $\sum_{j=0}^{x} f_j g_{x-j}$ (Appendix C), as compared with the special case $\sum_{j=0}^{x} f_j f_{x-j}$ encountered in Section 1.9.1.

As with AB-type polycondensation, the depletion rate of functional groups of either kind r_A or r_B must be proportional to the net molar rate of polymer formation because one polymer molecule is formed each time one of each of the functional groups A and B disappears. Consequently,

$$r_A = r_B \equiv r = \sum_{x=1}^{\infty} r_x = -\left(\sum_{x=0}^{\infty} r_x^{AA} + \sum_{x=0}^{\infty} r_x^{BB} + \sum_{x=1}^{\infty} r_x^{AB} \right). \qquad (1.9\text{-}12)$$

After substituting the appropriate rate functions from Table 1.9-4, we obtain

$$r = kc_A c_B, \qquad (1.9\text{-}13)$$

where

$$c_A \equiv 2 \sum_{x=0}^{\infty} c_x^{AA} + \sum_{x=1}^{\infty} c_x^{AB} \tag{1.9-14}$$

and

$$c_B \equiv 2 \sum_{x=0}^{\infty} c_x^{BB} + \sum_{x=1}^{\infty} c_x^{AB} \tag{1.9-15}$$

Example 1.9-1. Net Interchange Rate. Show that $\sum_{x=0}^{\infty} r_x = 0$, where r_x are the rate functions for interchange polymerization in Table 1.9-2.

Solution. After substituting appropriate rate functions and summing over all x, we obtain the following:

$$\sum_{x=0}^{\infty} r_x = \underset{①}{4k_f c_p^2} - \underset{②}{4k_r c_0(c_{RU} - c_p)} - \underset{③}{8k_f c_1 c_p} + \underset{④}{4k(c_p - c_1)^2}$$

$$- \underset{⑤}{4kc_1(c_{RU} - 2c_p + c_1)} + \underset{⑥}{8k_r c_0(c_p - c_1)} + \underset{⑦}{4k_f c_p^2}$$

$$- \underset{⑧}{8k_f c_p(c_p - c_1)} + \underset{⑨}{8k_r c_0(c_{RU} - c_p + c_1)}$$

$$- \underset{⑩}{4k_r c_0(c_{RU} - c_p)} + \underset{⑪}{4kc_p(c_{RU} - c_p)} + \underset{⑫}{4kc_p(c_{RU} - 2c_p + c_1)}$$

$$- \underset{⑬}{4k(c_p - c_1)(c_{RU} - c_p)} - \underset{⑭}{4kc_p(c_{RU} - c_p)} = 0$$

The circled number below each term refers to the corresponding term in Table 1.9-2 from which it was obtained. Details are left to the reader.

Example 1.9-2. Interchange Equilibrium. Consider the catalyzed ester interchange of di-hydroxy ethyl terephthalate as in the manufacture of PET, whose reaction sequence is frequently written in terms of functional groups (Table A-4) as:

$$\text{——OH} + \text{HO——} \underset{}{\overset{K'}{\rightleftharpoons}} \text{——ES——} + \text{EG}$$

where ES represents an internal ester group and EG is ethylene glycol by-product. The corresponding rate function for EG formation is

$$r_0 = k_f c_{OH}^2 - \frac{k_f}{K'} c_0 c_{ES}$$

where subscript zero denotes EG and K' is an equilibrium constant.

$$K' \equiv \frac{(c_{ES})_e (c_0)_e}{(c_{OH})_e^2}$$

In our terminology, we would write for the stoichiometry

$$x \geq 2 \qquad\qquad x m_1 = m_x + (x - 1) m_0$$

for the propagation sequence

$$x, y \geq 1 \qquad\qquad m_x + m_y \underset{k_r}{\overset{k_f}{\rightleftharpoons}} m_{x+y} + m_0$$

and for the rate function (Table 1.9-2)

$$r_0 = 4 k_f c_p^2 - 2 k_r c_0 \sum_{x=2}^{\infty} (2x - 2) c_x$$

where m_1 and m_x represent monomeric and polymeric di-hydroxy molecules, respectively, and where

$$c_p \equiv \sum_{x=1}^{\infty} c_x$$

Reconcile the preceding expressions for r_0 and show that

$$2K' = K \equiv \frac{k_f}{k_r}$$

Solution. Set $c_{OH} = 2 c_p$ because each polymer molecule contains two OH end groups, and define

$$c_{ES} \equiv 2 \sum_{x=1}^{\infty} x c_x - 2 \sum_{x=1}^{\infty} c_x$$

because ES groups are internal and there are two of them for each repeat unit (RU). After substituting these definitions into our rate function, using the identity

$$\sum_{x=2}^{\infty} (x c_x - c_x) = \sum_{x=1}^{\infty} (x c_x - c_x)$$

and setting $r_0 = 0$, we obtain the desired result:

$$\frac{k_f}{k_r} = \frac{2(c_0)_e (c_{ES})_0}{(c_{OH})_e^2} \equiv K = 2K'$$

Table 1.9-3. Explanation of Terms in Table 1.9-2

①   Either end of m_i with $i \geqslant 1$ can react with either of the two ultimate linkages of m_j with $j \geqslant 1$, or vice versa, giving eight combinations of m_i and m_j when $i \neq j$ and four when $i = j$ as required by the probability argument in Section 1.8.

②   Either end of m_0 can react when any of the $2j$ linkages in m_j except the ultimate ones, thus m_j does not include m_1 since such a reaction would lead to no net change in composition.

③   Either end of m_1 can react with either of the two ultimate linkages of m_j, or vice versa, giving eight combinations of m_1 and m_j even when $j = 1$ because this term represents a depletion rate.

④   Either end of m_i can react with either one of only two linkages in m_j which are appropriate for the formation of 1-mer, or vice versa, giving eight combinations with $i \neq j$ and four when $i = j$ as required by probability; the cases $i = 1$ and $j = 1$ are excluded because they would lead to no net formation of m_1.

⑤   Either end of m_1 can react with any of $2j$ linkages of m_j except the ultimate and penultimate ones which would lead to m_0 and m_1, respectively; the first exception is the province of the forward rearrangement interchange (k) and the second would result in no net depletion of m_1.

⑥   Either end of m_0 can react with any of four appropriate linkages of m_j to split out m_1, either of two at each end; eight combinations are therefore possible and case $j = 1$ is omitted because there would be no net formation of m_1.

⑦   Either end of m_i with $1 \leqslant i \leqslant x - 1$ can combine with either of two appropriate fragments of another molecule, m_1 up to m_{x-1}, remaining after cleavage of the ultimate linkage, or vice versa, to form the desired DP of x; eight combinations are therefore possible excepting the case $i = x - i$, in which there are only four.

⑧   The identical argument used in ③ applies if 1 is replaced by x.

⑨   The identical argument used in ⑥ applies, only here the appropriate combinations must form m_x.

⑩   Either end of m_0 can react with any of the linkages of m_x except the ultimate ones; this is obviously a special case of ② specifically for the disappearance of m_x rather than m_0.

⑪   Either end of m_i with $i \leqslant x - 1$ can react with either of two linkages of another molecule to form m_x by combination with either of its two appropriate fragments when $k < i + 1$, thus giving four combinations; alternatively, it may form m_x by splitting out either of two appropriate fragments when $k \geqslant i + 1$, giving eight possible combinations in all. These variations are all taken into account by the double summation, as is the reduced probability for the special case $i = x - i + k$.

⑫   Either end of m_i can react with any of four appropriate linkages of m_j with $j \geqslant x + 1$ when $1 \leqslant i \leqslant x - 1$, and with either of only two when $j \geqslant x$; the combinations eight and four, respectively, are properly accounted for by the double summation, as is the reduced probability when $i = j$.

⑬ & ⑭   Either end of m_x can react with any internal linkage of m_j $\left(⑬\right)$, or vice versa $\left(⑭\right)$, to destroy m_x; this gives a total of eight possible combinations when $j = x$, which is correct.

Table 1.9-4. AA-BB Type Random Polymerization (sequence 1.9-11)

$$x \geq 0 \qquad r_x^{AA} = -2kc_x^{AA} \sum_{j=0}^{\infty} c_j^{AB} - 4kc_x^{AA} \sum_{j=0}^{\infty} c_j^{BB}$$

$$+ 2k \sum_{j=0}^{x} c_j^{AA} c_{x-j}^{AB}$$

$$x \geq 0 \qquad r_x^{BB} = -2kc_x^{BB} \sum_{j=0}^{\infty} c_j^{AB} - 4kc_x^{BB} \sum_{j=0}^{\infty} c_j^{AA}$$

$$+ 2k \sum_{j=0}^{x} c_j^{BB} c_{x-j}^{AB}$$

$$x > 0 \qquad r_x^{AB} = -2kc_x^{AB} \sum_{j=0}^{\infty} c_j^{AB} - 2kc_x^{AB} \sum_{j=0}^{\infty} c_j^{AA}$$

$$- 2kc_x^{AB} \sum_{j=0}^{\infty} c_j^{BB} + k \sum_{j=0}^{x} c_j^{AB} c_{x-j}^{AB}$$

$$+ 4k \sum_{j=0}^{x-1} c_j^{AA} c_{x-j-1}^{BB}$$

$$\text{OVERALL} \qquad r_A = r_B = r \equiv -\sum_{x=1}^{\infty} r_x$$

$$= -\left(\sum_{x=0}^{\infty} r_x^{AA} + \sum_{x=0}^{\infty} r_x^{BB} + \sum_{x=1}^{\infty} r_x^{AB} \right)$$

$$r = kc_A c_B$$

1.10. ADDITION POLYMERIZATION

In Section 1.6, three basic polymerization classes were identified: random, addition without termination, and addition with termination. The first and third are representative of most industrially important processes. The second and third comprise the subject of this section. Although very few reactions belonging to the second class have been reported, they are nevertheless of theoretical interest because they can lead to so-called monodisperse polymers (very narrow DPD).

1.10.1. Addition Polymerization Without Termination

Reaction sequences for the general case in which initiation is separate and all reactions are reversible are listed in Table 1.10-1. The associated rate functions, obtained through application of the rate law assuming that all reactions represent elementary kinetic steps, are also tabulated. We define initiation as the formation of 1-mer. Reverse polymerization here represents chain unzipping, one repeat unit at a time, as contrasted with random scission in section 1.9.

Table 1.10-1. General Step Addition Sequences and Rate Functions

$$m_o + m \underset{k_{ir}}{\overset{k_{if}}{\rightleftharpoons}} m_1$$

$$x \geq 1 \qquad m_x + m \underset{k_{pr}}{\overset{k_{pf}}{\rightleftharpoons}} m_{x+1}$$

$$r = k_{if}c_m c_o + k_{pf}c_m c_p - k_{ir}c_1 - k_{pr}(c_p - c_1)$$

$$r_o = -k_{if}c_m c_o + k_{ir}c_1$$

$$r_1 = k_{if}c_m c_o - k_{pf}c_m c_1 - k_{ir}c_1 + k_{pr}c_2$$

$$x > 1 \qquad r_x = k_{pf}c_m c_{x-1} - k_{pf}c_m c_x + k_{pr}c_{x+1} - k_{pr}c_x$$

where

$$c_p \equiv \sum_{x=1}^{\infty} c_x = \mu_c^o$$

OVERALL

$$\sum_{x=1}^{\infty} r_x = k_{if}c_m c_o - k_{ir}c_1 = -r_o \neq r$$

$$r = \sum_{x=1}^{\infty} x r_x$$

Another distinction between the addition and random classes is the definition of rate of polymerization r. For addition types in general, we defined it as the rate of monomer conversion. It is noteworthy that r is not proportional to the molar rate of polymer formation $\sum_x r_x$ as it is for random polymerization. This is apparent in the last equation of Table 1.10-1, which follows after substitution of the appropriate rate functions for r_x and application of the convergence condition $\lim_{x \to \infty} c_x = 0$. For addition polymerizations, r is proportional to the weight rate of polymer formation $M_0 \sum_x x r_x$ by virtue of the relationship $W_x = x M_0 N_x$ from Section 1.5. This is verified in Example 1.10-1 for the addition sequence in Table 1.10-1, which yields

$$r = \sum_{x=1}^{\infty} x r_x \tag{1.10-1}$$

It should be noted that this expression and

$$r_0 = -\sum_{x=1}^{\infty} r_x \tag{1.10-2}$$

Table 1.10-2.   Special Step-Addition Sequences and Rate Functions

$$m_o + m \xrightarrow{k_i} m_1$$

$x \geq 1$
$$m_x + m \xrightarrow{k_p} m_{x+1}$$

$$r = k_i c_m c_o + k_p c_m c_p$$

$$r_o = - k_i c_m c_o$$

$$r_1 = k_i c_m c_o - k_p c_m c_1$$

$x > 1$
$$r_x = k_p c_m c_{x-1} - k_p c_m c_x$$

$x \geq o$
$$m_x + m \underset{k_r}{\overset{k_f}{\rightleftharpoons}} m_{x+1}$$

$$r = k_f c_m (c_{Ru} + c_p) - k_r c_p$$

$$r_o = - k_f c_m c_o + k_r c_1$$

$$r_x = k_f c_m c_{x-1} - k_f c_m c_x + k_r c_{x+1} - k_r c_x$$

are analogous to the constraint equations in Section 1.8. They demonstrate the interdependence of rate functions r, r_0, and r_x.

Two special cases of the sequence in Table 1.10-1 are listed in Table 1.10-2, together with their rate functions. They are consequences of the following limiting conditions, respectively; $k_{ir} = 0 = k_{pr}$; and $k_{if} = k_{pf} \equiv k_f$, $k_{ir} = k_{pr} \equiv k_r$. Further specialization of the last case, $k_f = k_r \equiv k$, gives us a Poisson process that produces the monodisperse polymer, as previously mentioned. This reaction is therefore of theoretical interest and will be discussed further in Chapter 2.

1.10.2.   Addition Polymerization with Termination

Chain polymerizations with intermediates and termination may be represented schematically by the following reaction series:

$$\text{reactants} \xrightarrow{\text{initiation}} \text{intermediates} \xrightarrow{\text{termination}} \text{products.} \tag{1.10-3}$$

This sequence resembles the familiar consecutive sequence

$$A \rightarrow B \rightarrow C, \tag{1.10-4}$$

but is far more complex. It includes at elast two reactants, initiator m_0 and monomer m:

$$m_0 + m \rightarrow m_1^*, \tag{1.10-5}$$

a myriad of consecutive propagation reactions among the intermediates that compete for monomer:

$$x \geqslant 1 \qquad\qquad m_x^* + m \rightarrow m_{x+1}^*, \tag{1.10-6}$$

and alternative termination reactions that affect the distribution of products.

$$x \geqslant 1 \qquad\qquad m_x^* \rightarrow products, \tag{1.10-7}$$

$$x \geqslant 1, y \geqslant 1 \qquad\qquad m_x^* + m_y^* \rightarrow products. \tag{1.10-8}$$

As before, we define initiation as the formation of 1-mer, in this case intermediate m_1^*. This reaction therefore actually represents polymer chain initiation

$$m_0^* + m \rightarrow m_1^* \tag{1.10-9}$$

and must be distinguished from kinetic chain initiation, which is a separate, decomposition reaction that provides the necessary active species m_0^* (Appendix A). Whether the initiation step and its corresponding termination step involve one or two active intermediates depends, as previously mentioned, on whether the intermediates are ionic, catalytic, or free-radical. In any case, as written reaction 1.10-5 is generally not an elementary step.

A general set of elementary reactions representing a variety of chain-addition polymerizations is listed in Table 1.10-3, and their corresponding rate functions are given in Table 1.10-4. Initiation and termination sequences of various types have been included for the sake of completeness, notwithstanding their possible chemical incompatibility. The parentheses in the initiation sequence allow for the option of uncatalyzed chain initiation, for instance, by thermal decomposition of monomer. The primed quantities (m_y', c_x', k_p') refer to polymer molecules with terminal double bonds (Table A-7). Their delineation is necessary when dealing with polymer chain branching (12). When utilizing these rate models in later chapters, we will discard those kinetic steps and rate functions that do not apply to the reaction in question.

We represent rate of initiation (reaction 1.10-5) with

$$r_i = k_i c_m^p c_0^q, \tag{1.10-10}$$

because this general expression includes a variety of industrially important rate functions as special cases. Examples are: $p = 0$, $q = 1$ for free-radical initiation; $2 \leqslant p \leqslant 3$, $q = 0$ for thermal initiation; and $p = 1$, $q = 1$ for ionic initiation (Example 1.10-2).

The rate function for polymerization r is a composite expression, resulting from a combination of the particular elementary reaction sequences that best describe the polymerization in question. Its development from the basic rate laws (Table 1.10-4)

Table 1.10-3. Chain-Addition Sequences

CATALYST DECOMPOSITION

$$m_o \longrightarrow m_o^* \qquad\qquad r_d$$

INITIATOR DECOMPOSITION

$$m_o \longrightarrow 2m_o^* \qquad\qquad r_d$$

INITIATION

$$m(\; +\; m_o^*) \xrightarrow{\; k_i'\;} m_1^* \qquad \text{primary} \qquad r_{ip}$$

$$m + s \xrightarrow{\; k_{is}\;} m_1^* \qquad \text{secondary} \qquad r_{is}$$

PROPAGATION

$$x \geq 1 \quad m_x^* + m \xrightarrow{\; k_p\;} m_{x+1}^* \qquad \text{propagation with monomer} \qquad r_p$$

$$x \geq 1 \quad m_x^* + m_y' \xrightarrow{\; k_p'\;} m_{x+y}^* \qquad \text{propagation with polymer} \qquad r_p'$$

CHAIN TRANSFER

$$x \geq 1 \quad m_x^* + m \xrightarrow{\; k_{fm}\;} \left\{ \begin{array}{l} m_x + m_1'^{*} \\ m_x' + m_1^* \end{array} \right\} \quad \text{to monomer} \qquad r_{fm}$$

$$x \geq 1 \quad m_x^* + m_y \xrightarrow{\; k_{fp}\;} m_x + m_y^* \qquad \text{to polymer} \qquad r_{fp}$$

$$x \geq 1 \quad m_x^* + s \xrightarrow{\; k_{fs}\;} m_x + s^* \qquad \text{to foreign substance} \qquad r_{fs}$$

TERMINATION

$$x \geq 1 \quad m_x^* \xrightarrow{\; k_{ts}\;} m_x \qquad \text{spontaneous (unimolecular)} \qquad r_{ts}$$

$$m_x^* + m_y^* \; \begin{array}{c} \nearrow^{\; k_{tc}} \; m_{x+y} \\[4pt] \searrow_{\; k_{tD}} \; m_x + m_y' \end{array} \qquad \begin{array}{l} \text{combination} \\ \text{(bimolecular)} \end{array} \quad r_{tc} \\ \begin{array}{l} \text{disproportionation} \\ \text{(bimolecular)} \end{array} \quad r_{tD}$$

$$x \geq 1 \quad m_x^* + s^* \xrightarrow{\; k_{ta}\;} m_x$$
$$\qquad\qquad\qquad\qquad\qquad \text{secondary (bimolecular)}$$
$$s^* + s^* \xrightarrow{\; k_{tb}\;} s_2$$

generally makes use of two kinetic principles, which greatly simplify the analysis and which will be discussed in detail in Section 1.11. They are the quasi-steady-state approximation (QSSA) and the long-chain approximation (LCA). In general terms they may be written as follows if we ignore chain transfer reactions:

$$r_i \cong r_t \qquad \text{QSSA} \qquad\qquad (1.10\text{-}11)$$

$$r_i \ll r_p \qquad \text{LCA} \qquad\qquad (1.10\text{-}12)$$

where r_t is the rate of polymer chain termination:

$$r_t = k_t (c^*)^n \qquad\qquad (1.10\text{-}13)$$

Table 1.10-4. Chain-Addition Rate Functions

INITIATION

$$r_i = \underset{r_{ip}}{\underbrace{k_i c_m^p c_o^q}} + \underset{r_{is}}{\underbrace{k_{is} c_m c_s^*}}$$

PROPAGATION

$$r_p = k_p c_m c^* + k_p c_p c^* \quad \text{where} \quad c^* \equiv \sum_x c_x^* \quad \text{and} \quad c_p' \equiv \sum_y c_y'$$

TERMINATION

$$r_t = k_{ts} c^* + k_t (c^*)^2$$

$$\text{where} \quad k_t \equiv k_{tc} + k_{tD}$$

TRANSFER

$$r_f = k_{fm} c_m c^* + k_{fs} c_s c^* + k_{fp} c^* \sum_j j c_j$$

MONOMER

$$r = r_i + r_p + r_{fm} \equiv \text{rate of polymerization}$$

FOREIGN SUBSTANCE

$$r_s = -k_{fs} c_s c^*$$

INTERMEDIATES

$$r_1^* = r_i + k_{fm} c_m c^* - k_p c_m c_1^* - k_p' c_p' c_1^*$$

$$\qquad - k_{fm} c_m c_1^* - k_{fs} c_s c_1^* - k_t c_1^* c^*$$

$$\qquad - k_{ts} c_1^* - k_{ta} c_1^* c_s^*$$

Table 1.10-4. (*Continued*)

INTERMEDIATES (continued)

$x > 1$

$$r_x^* = k_p c_m c_{x-1}^* + xk_{fp} c_x c^* + k_p' \sum_{j=1}^{x-1} c_j' c_{x-j}^* - k_p c_m c_x^* - k_p' c_p' c_x^* - k_{fm} c_m c_x^*$$

$$- k_{fs} c_s c_x^* - k_{fp} c_x^* \sum_{j=1}^{\infty} j c_j - k_t c_x^* c^* - k_{ts} c_x^*$$

$$- k_{ta} c_x^* c_s^*$$

$$r^* \equiv \sum_{x=1}^{\infty} r_x^* = r_i - k_{fs} c_s c^*$$

$$- k_t (c^*)^2 - k_{ts} c^* - k_{ta} c^* c_s$$

$$r_s^* = k_{fs} c_s c^* - k_{is} c_m c_s^* - k_{ta} c^* c_s^* - k_{tb} (c_s^*)^2$$

POLYMERIC PRODUCTS

$x \geq 1$

$$r_x = (k_{fm} c_m + k_{fs} c_s + k_{fp} \sum_j j c_j + k_{tD} c^* + k_{ta} c_s^* + k_{ts}) c_x^*$$

$$- xk_{fp} c_x c^* - k_p' c_x' c^* + (k_{tc}/2) \sum_{j=0}^{\infty} c_j^* c_{x-j}^* \quad \text{where} \quad c_o^* \equiv 0$$

$$\sum_{x=1}^{\infty} r_x = (k_{fm} c_m + k_{fs} c_s + k_{fp} \mu_c^1 c^* + k_{tD} c^* + k_{ta} c_s^* + k_{ts}) c^*$$

$$+ (k_{tc}/2) (c^*)^2 - k_{fp} \mu_c^1 c^* - k_p' c_p' c^*$$

---

r_p is the rate of propagation:

$$r_p = k_p c_m c^* \tag{1.10-14}$$

and

$$c^* \equiv \sum_{x=1}^{\infty} c_x^* \tag{1.10-15}$$

We note again that termination may be first or second order depending on whether the intermediates are ions or free radicals, respectively. Both the LCA and the QSSA will be examined in detail in later sections.

Applying the QSSA, we equate rate functions 1.10-10 and 1.10-13 and obtain an approximate expression for c^*,

$$c^* \cong \left(\frac{k_i}{k_t}\right)^{1/n} c_m^{p/n} c_0^{q/n} \tag{1.10-16}$$

which yields the following function for rate of polymerization after substitution into expression 1.10-14 and application of the LCA:

$$r \cong r_p = k_{ap} c_m^{1+p/n} c_0^{q/n} \tag{1.10-17}$$

The lumped ("apparent") rate constant

$$k_{ap} \equiv k_p \left(\frac{k_i}{k_t} \right)^{1/n} \tag{1.10-18}$$

is the one that would be measured experimentally in kinetic studies of the dependence of polymerization rate on the concentrations of reactants m and m_0. A most important special case is chemically initiated free-radical polymerization. Its rate function is

$$r = k_p \left(\frac{2fk_d}{k_t} \right)^{1/2} c_m c_0^{1/2} \tag{1.10-19}$$

where k_d is the rate constant for initiator decomposition and f is initiator efficiency (Example 1.10-2). For further discussion of initiator efficiency, the reader is referred to Chapter 2.

Example 1.10-1. *Rate of Step-Addition Polymerization.* Show that $\sum_{x=0}^{\infty} x r_x = r$, where r_x and r are defined in Table 1.10-1.

Solution. After substituting the appropriate rate functions and summing over all x, we obtain the following:

$$\sum_{x=1}^{\infty} x r_x = k_{if} c_m c_0 + k_{pf} c_m \sum_{x=2}^{\infty} x c_{x-1} + k_{pr} \sum_{x=2}^{\infty} (x-1) c_x$$

$$- k_{pf} c_m \sum_{x=1}^{\infty} x c_x - k_{ir} c_1 - k_{pr} \sum_{x=2}^{\infty} x c_x$$

$$= k_{if} c_m c_0 + k_{pf} c_m c_{RU} + k_{pf} c_m c_p + k_{pr} c_{RU}$$

$$- k_{pr} c_p - k_{pf} c_m c_{RU} - k_{ir} c_1 - k_{pr} c_{RU} + k_{pr} c_1 = r,$$

where

$$c_p \equiv \sum_{x=1}^{\infty} c_x = \mu_c^0$$

and

$$c_{RU} \equiv \sum_{x=1}^{\infty} x c_x \equiv \mu_c^1.$$

Example 1.10-2. *Chain Initiation Rates.* Using the elementary step for polymer chain initiation

$$m_0^* + m \xrightarrow{k_i'} m_1^*$$

together with the "local" QSSA applied to intermediate m_0^* only, derive the expressions for lumped constant k_i in equation 1.10-10 in terms of the appropriate elementary rate constants for: ionic polymerization

$$m_0 \overset{K_d}{\rightleftharpoons} m_0^*$$

and free-radical polymerization

$$m_0 \overset{k_d}{\to} 2m_0^*.$$

Solution. Applying the rate law to the initiation reaction yields $r_i = k_i' c_0^* c_m$. For ionic initiation (equilibrium) $c_0^* = K_d c_0$ so that $r_i = k_i' K_d c_m c_0$ and $k_i \equiv k_i' K_d$. For free-radical initiation (local QSSA) $k_i' c_0^* c_m \cong 2k_d c_0$ so that $r_i = 2k_d c_0$ and $k_i \equiv 2k_d$. When species c_0^* is not 100 percent efficient in initiating polymer chains, due to the "cage effect," it is customary to use the constant efficiency factor $f \leqslant 1$; thus $k_i c_0^* c_m \cong 2fk_d c_0$, so that $k_i \equiv 2fk_d$.

1.11. SOME KINETIC APPROXIMATIONS

It is appropriate at this point to review several basic kinetic approximations that are routinely applied to chain-addition polymerizations to facilitate their mathematical tractability. They are the long-chain approximation (LCA), the quasi-steady-state approximation (QSSA), and the reactor steady-state approximation (RSSA).

1.11.1. The Long-Chain Approximation

The LCA is based on the requirement that most of the monomer being consumed is added to existing polymer chains in the propagation step, as opposed to participating in non-chain-building reactions such as starting new polymer chains by reaction with initiating species m_0^*. Its validity may be easily confirmed by observing the DP of the product, for if it were not valid then the polymerization in question would not produce high polymer.

A property of chain reactions in general, well known to kineticists, is the kinetic chain length. It is defined as the ratio of kinetic chain propagation rate to kinetic chain initiation rate. When applying it to polymerizations, we modify its definition as follows

$$\nu_N \equiv \frac{r_p}{r_i} \tag{1.11-1}$$

so that it reflects the polymer chain length more closely. Here r_p and r_i are rates of polymer chain propagation and polymer chain initiation, respectively. The distinction between kinetic chains and polymer chains is delineated in Table A-11. To illustrate the effect of this distinction on ν_N, we note that the traditional definition would require inclusion of polymer chain initiation rate r_i as well as all chain transfer rates r_f in the numerator, in addition to r_p, since they all represent kinetic chain propagation rates. On the other hand, r_i in the denominator of equation 1.11-1 should actually include chain transfer rates (cf. Table 1.10-3), since such reactions can contribute to polymer chain initiation and termination.

At this point, the reader will doubtless anticipate that a relationship exists between ν_N and the LCA. To formalize this relationship, we write a monomer balance for addition polymerizations, step or chain, ignoring transfer reactions for the moment (cf. Tables 1.10-1 and 1.10-4):

$$-\frac{dc_m}{dt} = r_i + r_p \qquad (1.11-2)$$

By following the method outlined in Appendix G and using dimensionless concentration $\hat{c}_m \equiv c_m/(c_m)_0$, equation 1.11-2 may be written in semidimensionless form

$$-\frac{d\hat{c}_m}{dt} = s(x_0\lambda_i)^{-1}\hat{r}_i + \lambda_m^{-1}\hat{r}_p \qquad (1.11-3)$$

This immediately identifies two characteristic times, one for monomer conversion $\lambda_m \equiv (c_m)_0/(r_p)_0$ and the other for initiator (catalyst) consumption $\lambda_i \equiv (c_0)_0/(r_d)_0$, where r_i and r_d are the rates of polymer chain initiation and rate of initiator (catalyst) decomposition, respectively. The latter rates are related by stoichiometry, $r_i = sr_d$ (cf. equation 1.6-1). The following relationship between the CT's and ν_N can now be deduced by substituting $(r_p)_0$ and $(r_i)_0$ into equation 1.11-1:

$$(\nu_N)_0 = \frac{x_0}{s\alpha_K} \qquad (1.11-4)$$

where $x_0 \equiv (c_m)_0/(c_0)_0$ is the monomer-initiator feed ratio and

$$\alpha_K \equiv \frac{\lambda_m}{\lambda_i} \qquad (1.11-5)$$

The LCA requires that the propagation reaction dominate the consumption of monomer (Section 1.10) (i.e., $\lambda_m \ll x_0\lambda_i/s$ from equation 1.11-3). Thus we obtain the following criterion for the LCA:

$$(\nu_N)_0 \gg 1 \qquad (1.11-6)$$

This inequality may also be used as an indicator for the production of high DP, which may be effected by either a large monomer-to-initiator consumption rate (small α_K) or a large monomer-to-initiator feed ratio (large x_0). Both parameters are adjustable. Example 1.11-1 illustrates an application of the LCA to chain-addition kinetics. Table 1.11-1 demonstrates the role of $(\nu_N)_0$ as a DP indicator in the free-radical polymerization of styrene with AIBN (21) when $\alpha_K < 1$ and when $\alpha_K > 1$. Final $\bar{x}_N$ should be compared with $(\bar{x}_N)_0$, which is twice $(\nu_N)_0$ because termination was assumed to occur exclusively by combination.

Table 1.11-1. Computer Simulation of Isothermal Styrene Polymerization with AIBN (21)

x_0	α_k	$\left(v_N\right)_0$	$\left(\overline{x}_N\right)_0 = 2\left(v_N\right)_0$	Φ	$\overline{x}_N$	D_N
			INITIAL CONDITIONS		FINAL CONDITIONS	
544	0.48	567	1134	0.80	907	1.51
4840	1.44	1680	3360	0.48	4708	2.22

It should be noted that criterion 1.11-6 for large DP applies to step-addition polymerizations as well, since definition 1.11-1 is not inherently limited to chain polymerizations, notwithstanding our earlier analogy with the kinetic chain length.

Example 1.11-1. The LCA Applied to Free-Radical Polymerizations. Use the LCA to simplify the integration of monomer balance 1.11-2 for free-radical polymerization $r_p = k_{ap}c_mc_0^{1/2}$. Solve this equation simultaneously with equation 1.11-17 to obtain an analytical expression for monomer conversion $\Phi_m = 1 - \hat{c}_m$ with time $\hat{t} \equiv t/\lambda_m$. Estimate the value of $\hat{t}$ required to achieve 80% monomer conversion when $\alpha_K = 1$.

Solution. We reduce all concentrations with their corresponding feed values and write our balances in dimensionless form:

$$-\frac{d\hat{c}_0}{d\hat{t}} = \alpha_K\hat{c}_0,$$

$$-\frac{d\hat{c}_m}{d\hat{t}} = \hat{c}_m\hat{c}_0^{1/2} + \left(v_N\right)_0^{-1}\hat{c}_0 \overset{LCA}{\cong} \hat{c}_m\hat{c}_0^{1/2}.$$

The solution of the first subject to IC $(\hat{c}_0)_0 = 1$ is

$$\hat{c}_0 = \exp\left(-\alpha_K\hat{t}\right).$$

Substitution of this result into the second equation, following application of the LCA, yields the desired solution, subject to IC $(\hat{c}_m)_0 = 1$:

$$\Phi_m = 1 - \hat{c}_m = 1 - \exp\left\{\frac{2}{\alpha_K}\left[\exp\left(\frac{-\alpha_K\hat{t}}{2}\right) - 1\right]\right\}.$$

Using this result we obtain $\Phi_m = 0.8$ when $\hat{t} = 3.2$. The corresponding initiator conversion is $\Phi_i = 1 - \hat{c}_0 = 0.96$.

Example 1.11-2. Limiting Rate Equations for Free-Radical Polymerizations. Using the results of Example 1.11-1, show that free-radical polymerizations become pseudo first-order and pseudo half-order reactions, respectively, under the following limiting conditions:

$$\alpha_K \ll 1 \quad \text{very slow initiator decomposition (CONV)}$$
$$\alpha_K \gg \quad \text{very rapid initiator decomposition (D-E)}$$

Solution. When $\alpha_K \ll 1$, it follows from the initiator balance that $c_0 \cong$ constant $\cong (c_0)_0$. By substituting $\hat{c}_0 = 1$ into the monomer balance, we obtain the first-order solution

$$\Phi \cong 1 - \exp(-\hat{t}),$$

which could have been deduced directly from the solution for conversion in Example 1.11-1 by using the approximation (Appendix C)

$$\exp\left(\frac{-\alpha_K \hat{t}}{2}\right) \cong 1 - \frac{\alpha_K \hat{t}}{2}.$$

Similarly, when $\alpha_K \gg 1$, it follows that $c_m \cong$ constant $= (c_m)_0$. Therefore, the rate equation for monomer is half-order with respect to the initiator:

$$-\frac{d\hat{c}_m}{d\hat{t}} \cong \hat{c}_0^{1/2}.$$

1.11.2. The Quasi Steady-State Approximation

Many reaction sequences like 1.10-3 involve active intermediates (free-radical, ionic, or catalytic). Such sequences are called chain reactions. One manifestation of chain reactions is that the total concentration of active intermediates c^* at all times is very small. A typical value of c^* for free-radical polymerizations is 10^{-8} mol/l. Another manifestation is the existence of a quasi steady-state, in which initiation and termination rates of intermediates are virtually equal. The QSSA is remarkably accurate under most conditions (Example 1.11-3) and will be used whenever possible. Most kinetic analyses of chain reactions rely on the QSSA for computation of intermediate concentrations (Example 1.11-4), which are necessary to compute product compositions that are difficult to obtain otherwise.

A typical population balance for active intermediates is

$$\frac{dc^*}{dt} = sr_d(t) - k_t(c^*)^n, \tag{1.11-7}$$

where sr_d may now be interpreted as the birth rate of intermediates, and coefficient s the birth-to-decomposition ratio (cf. equation 1.6-1). Exponent n is the order of the destruction reaction (equation 1.6-2). We let the function $c^* \equiv f(t)$ represent the solution of the balance equation and define the QSSA as another function (cf. equation 1.10-16)

$$(c^*)_s = \left[\frac{sr_d(t)}{k_t}\right]^{1/n} \equiv g(t), \tag{1.11-8}$$

which is obtained by equating the RHS of the balance equation to zero and solving the resulting algebraic equation for c^*. It is important to note that $g(t)$ is not a solution of equation 1.11-7 and that c^* need not remain constant for the QSSA to be valid. Replacing the time derivative with zero merely recognizes that its value is small compared to the remaining terms on the RHS of the balance equation. The QSSA is considered to be valid when $g(t)$ is a good approximation of $f(t)$, that is, when it tracks $f(t)$ with sufficient accuracy. A consequence of the QSSA is that the total concentration of intermediates $(c^*)_s$ is extremely small, thus qualifying them as active intermediates (22). We seek a simple criterion for the validity of the QSSA.

Following our usual procedure, we write the balance equation in semidimensionless form, after reducing concentration with a reference value $\hat{c}^* \equiv c^*/(c^*)_0$. It should be noted that $(c^*)_0$ refers to a fictitious reference state based on $(c_0)_0$; the true initial condition should, of course, be zero. The result

$$\frac{d\hat{c}^*}{dt} = \lambda_1^{-1}\hat{r}_d - \lambda_2^{-1}(\hat{c}^*)^n \tag{1.11-9}$$

yields two characteristic times. One is associated with the forcing function (disturbance)

$$\lambda_1 = \frac{(c^*)_0}{s(r_d)_0} \tag{1.11-10}$$

and the other with the response of c^* to the disturbance

$$\lambda_2 \equiv \frac{1}{k_t(c^*)_0^{n-1}} \tag{1.11-11}$$

Validity of the QSSA would certainly suggest that these CT's are of the same order of magnitude. Thus, after equating 1.11-10 and 1.11-11, solving for $(c^*)_0$, and substituting the result into either equation, we obtain a definition for the "relaxation time" (22) of the intermediates:

$$\lambda_p = \frac{[s(r_d)_0]^{1/n-1}}{k_t^{1/n}} \tag{1.11-12}$$

Next, we suppose that λ_r is the CT for the overall reaction (23), reactants $\rightarrow$ products, and use it to make equation 1.11-9 dimensionless:

$$\gamma_r^{-1}\frac{d\hat{c}^*}{d\hat{t}} = \frac{\lambda_p}{\lambda_1}\bar{r}_d - \frac{\lambda_p}{\lambda_2}(\hat{c}^*)^n \tag{1.11-13}$$

where $\hat{t} \equiv t/\lambda_r$ and $\gamma_r \equiv \lambda_r/\lambda_p$. Applicability of the QSSA during the time scale of this reaction should require that $\gamma_r \gg 1$.

These conclusions are easily tested for two simple cases for which the analytical solutions of equation 1.11-1 are known. In both, the initiator decomposition rate (sr_d) is constant and the IC is $c^* = 0$. When $n = 1$,

$$c^* = \frac{sr_d}{k_t}[1 - \exp(-k_t t)], \tag{1.11-14}$$

and when $n = 2$,

$$c* = \left(\frac{sr_d}{k_t}\right)^{1/2}\left[\tanh(sr_dk_t)^{1/2}t\right]. \qquad (1.11\text{-}15)$$

The definition of λ_p, equation 1.11-12, is confirmed by the argument of each function t/λ_p, and in each case $c*$ converges to the QSSA value predicted by equation 1.11-8 for large values of t/λ_p. We also conclude that, as anticipated, the smaller the value of λ_p, the quicker in time t is the response of the system in approaching the QSSA.

For chain-addition polymerizations in particular, $\lambda_p \equiv (c*)_0/(r_i)_0$ represents the lifetime of growing polymer chains. From the foregoing analysis, we expect the QSSA to be valid at all times $t > \lambda_p$ during the polymerization following an initial transient period λ_p, provided that the inequality $\gamma_r \equiv \lambda_r/\lambda_p \gg 1$ is satisfied. Furthermore, since polymerizations in general involve at least two reactants, monomer and initiator, λ_r is given by either λ_m for conventional polymerizations ($\alpha_K < 1$) or λ_i for dead-end polymerizations ($\alpha_K > 1$), whichever is smaller. Thus we modify our QSSA criterion as follows and require that both inequalities be satisfied (23):

$$\left.\begin{array}{ll} \text{CONV} & \gamma_m \equiv \lambda_m/\lambda_p \\ \text{D-E} & \gamma_i \equiv \lambda_i/\lambda_p \end{array}\right\} \gg 1. \qquad (1.11\text{-}16)$$

The reader will recall that these inequalities were previously used to classify addition polymerizations (Section 1.6). Table 1.11-2 summarizes the general definitions for CT's and dimensionless groups and lists those for the important special case of chemically initiated free-radical polymerizations as well.

It should be noted that our entire analysis has been based on CT's evaluated under reference (feed) conditions and our only test so far involved constant forcing.

Table 1.11-2. Definitions

	GENERAL	FREE-RADICAL ($s = 2f$, $n = 2$)
CHARACTERISTIC TIME		
λ_p	$(c*)_0/(r_i)_0$	$1/(2fk_dk_t)^{1/2}(c_o)_o^{1/2}$
λ_i	$(c_o)_o/(r_d)_0$	$1/k_d$
λ_m	$(c_m)_0/(r_p)_0$	$k_t^{1/2}/k_p(2fk_d)^{1/2}(c_o)_o^{1/2}$
DIMENSIONLESS GROUP		
γ_i	λ_i/λ_p	$(2fk_t)^{1/2}(c_o)_o^{1/2}/k_d^{1/2}$
γ_m	λ_m/λ_p	k_t/k_p

The important question is whether or not the QSSA and our criteria for it will be valid in situations of engineering interest, such as polymerizations with rapid initiator decay (large k_d), hindered termination (small k_t), and nonuniform temperature (strongly variable kinetic "constants"). Such conditions present a strong challenge and will be examined more closely in Chapter 2 with the aid of detailed computer models for specific polymerizations.

Examination of the parameters listed in Table 1.11-2 suggests that the QSSA might be violated when k_d increases, as it does with D-E prone initiators and rising reaction temperatures, or when k_t decreases, as it does when termination is hindered. Our criteria predict that, whereas small values of k_t will adversely affect both relaxation time (λ_p) and range of applicability (λ_r), large values of k_d should produce opposing effects by simultaneously shortening both the relaxation time and the range of applicability.

Figures 1.11-1–1.11-3 substantiate these claims. The graphs apply to chemically initiated free-radical polymerizations (equation 1.10-17 with $f = 1$) subject to the following balance equation for initiator.

$$-\frac{dc_0}{dt} = r_d = k_d c_0 \qquad (1.11\text{-}17)$$

Variable $\varepsilon \equiv 1 - c^*/(c^*)_s$ represents the error incurred by using the QSSA. The results were obtained (23) by substituting the solution of equation 1.11-17 into equation 1.11-8 with $s = 2$ (or, equivalently, equation 1.10-16 with $p = 0$, $q = 1$, $k_i = 2k_d$) to determine $(c^*)_s$, and into equation 1.11-7 to determine c^*. The latter computation was done numerically with the aid of a computer. Since all were initiator-limited polymerizations, time was reduced to dimensionless form with λ_i.

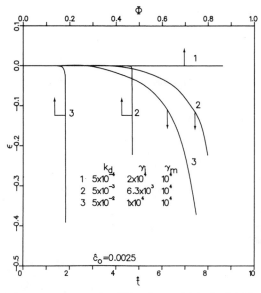

Figure 1.11-1. Computed monomer conversion (Φ) and error (ε) in the total free-radical concentration (c^*), due to the QSSA, versus dimensionless time ($\hat{t} \equiv t/\lambda_i$) for various values of $k_d(\mathrm{s}^{-1})$, γ_i, and γ_m.

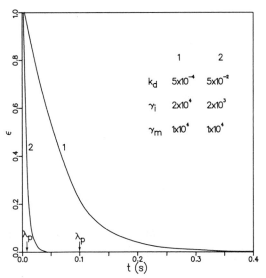

Figure 1.11-2. Computed error (ε) in the total free-radical concentration (c^*), due to the QSSA, versus time t in seconds for various values of $k_d(\mathrm{s}^{-1})$, γ_i, and γ_m.

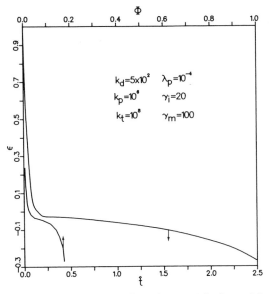

Figure 1.11-3. Computed monomer conversion (Φ) and error (ε) in the total free-radical concentration (c^*), due to the QSSA, versus dimensionless time ($\hat{t} \equiv t/\lambda_i$) for various values of $k_d(\mathrm{s}^{-1})$, $k_p(\mathrm{l/mol\ s})$, $k_t(\mathrm{l/mol\ s})$, and $\lambda_p(\mathrm{s})$.

Table 1.11-3. Typical Kinetic Properties for Free-Radical Polymerizations

---

$k_d \sim 10^{-5} \text{ s}^{-1}$

$k_p \sim 10^4 \text{ }\ell/\text{mol} \cdot \text{s}$

$k_t \sim 10^8 \text{ }\ell/\text{mol} \cdot \text{s}$

$\lambda_p \sim 10^{-1} - 10 \text{ s}$

Kinetic parameters were varied systematically. For comparison, typical values for free-radical polymerizations are listed in Table 1.11-3.

In Figs. 1.11-1 and 1.11-2, all cases satisfy our criteria, and indeed the QSSA is seen to be valid. Figure 1.11-1 shows that deviations begin only at the point of D-E, where total depletion of initiator occurs, and that the range of applicability of the QSSA is shortened as k_d declines, as predicted. Figure 1.11-2, in which the time scale has been expanded, shows that ε remains small after time $t \cong \lambda_p$ and that relaxation to the QSSA occurs at shorter times with rising k_d, as predicted.

In Fig. 1.11-3, unreasonably large values of k_d were required to violate our criteria. This further necessitated the use of exceptionally large values of k_p to

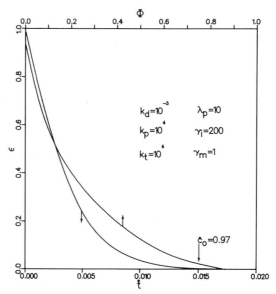

Figure 1.11-4. Computed monomer conversion (Φ) and error (ε) in the total free-radical concentration (c^*), due to the QSSA, versus dimensionless time ($\hat{t} \equiv t/\lambda_i$) for various values of $k_d(\text{s}^{-1})$, $k_p(\text{l/mol s})$, $k_t(\text{l/mol s})$, and $\lambda_p(\text{s})$.

achieve discernible monomer conversion at the time of D-E. A by-product of this condition is an unusually small value for λ_p. The kinetic constants used in Fig. 1.11-4 are much more reasonable, since values as high as 2×10^{-3} s^{-1} for k_d and as low as 10^4 1/mol s for k_t have been reported. Even with such small values of k_t, c^* does not rise far in excess of 10^{-5} mol/l, which is probably acceptable as an upper limit. The results shown in both Figs. 1.11-3 and 1.11-4 show significant deviations in ε over large reaction periods. It also appears that the QSSA is more sensitive to diminishing k_t than it is to increasing k_d.

Example 1.11-3. **The QSSA Applied to Free-Radical Polymerizations.** Verify the applicability of the QSSA to the benzoyl peroxide initiated free-radical polymerization of styrene at 50°C

$$fk_d = 6.378 \times 10^{13}\exp\left(\frac{-29{,}700}{R_gT}\right) \text{s}^{-1}$$

$$k_p = 1.058 \times 10^7\exp\left(\frac{-7068}{R_gT}\right) \text{1/mol s}$$

$$k_t = 1.255 \times 10^9\exp\left(\frac{-1675}{R_gT}\right) \text{1/mol s}$$

Solution. From Table 1.11-2, the ratio

$$\frac{\lambda_m}{\lambda_p} = \frac{k_t}{k_p}$$

is independent of the choice of initiator. Thus, at the specified temperature

$$\gamma_m \equiv \frac{\lambda_m}{\lambda_p} = 5.26 \times 10^5 \gg 1$$

The choice of λ_m over λ_i as the CT for the reaction can be verified by evaluating α_K,

$$\alpha_K \equiv \frac{\lambda_m}{\lambda_i} = 0.0278(c_0)_0^{-1/2}$$

which implies that $\lambda_m < \lambda_i$ provided that $(c_0)_0 > 7.7 \times 10^{-4}$ mol/l and thus monomer is the rate-limiting component.

Example 1.11-4. **The QSSA Applied to Heterogeneously Catalyzed Polymerizations.** Consider the following sequence for polymerization of an α-olefin by Ziegler–Natta type heterogeneous catalysts (Table A-10)

$$m + c \xrightarrow{k_i} m_1^* \qquad \text{adsorption}$$

$$m_x^* + m \xrightarrow{k_p} m_{x+1}^* \qquad \text{Rideal propagation}$$

$$m_x^* \xrightarrow{k_t} m_x + c \qquad \text{desorption}$$

where c signifies an empty site on the catalyst surface and m_x^* are adsorbed catalytic intermediates. Using the LCA and the QSSA together with the constraint equation

$$c_0 = c_0^* + c_p^*$$

show that the overall rate of polymerization is

$$r = \frac{k_p K c_0 c_m^2}{(1 + K c_m)}$$

where c_0 represents the total catalyst (site) concentration (constant if unpoisoned), $K \equiv k_i/k_t$ is an adsorption isotherm, and all rate constants have units that correspond to the units chosen for concentration (Appendix E).

Solution.   The rate of monomer consumption is

$$r = k_i c_m c_0^* + k_p c_m c_p^* \underset{\text{LCA}}{\cong} k_p c_m c_p^*$$

By combining the QSSA, $k_i c_m c_0^* = k_t c_p^*$, with the constraint equation, we obtain an expression for c_p^*:

$$c_p^* = \frac{K c_m c_0}{1 + K c_m}$$

Substitution into the rate function yields the desired result.

1.11.3.   *The Reactor Steady-State Approximation*

It is frequently taken for granted that the kinetic relationships arrived at through application of the QSSA to reactions conducted in closed systems will be equally valid in describing reactions in flow reactors. Specifically, the QSSA is used to determine the concentration of active intermediates in terms of reactant concentrations and kinetic constants (cf. equation 1.10-16). These results are frequently applied to flow reactors, even though they were derived for batch systems.

The problem arises because exit from the reactor represents an additional means for consuming intermediates, over and above termination reactions.

Denbigh (24) postulated that the QSSA could be applied without significant error provided that the lifetime of the active intermediates is considerably shorter than the space time of the reactor, $\lambda_p \ll \lambda_v$. Ray (25) examined its applicability to free-radical polymerizations in IM's in particular and arrived at the same conclusion. We shall briefly review this approximation in the context of continuous chain-addition polymerizations in general.

A population balance for intermediates in continuous reactors must contain the rate of exit. Examples are

$$\frac{dc^*}{dt} = sr_d(t) - k_t(c^*)^n - \frac{\dot{V}}{V}c^* \tag{1.11-18}$$

for the IM and

$$\frac{\partial c^*}{\partial t} = sr_d(t) - k_t(c^*)^n - v\frac{\partial c^*}{\partial x} \tag{1.11-19}$$

for the PFR (Appendix F). Since these reactors are open systems, they are capable of achieving a true steady-state, for which the transient term on the LHS of each equation is zero. The resulting equations could thus be solved for c^*. If the QSSA were applied, we would drop the last term on the RHS in addition and then solve for c^*. The latter result is obviously identical to that of Section 1.11.2 (cf. equation 1.11-8).

For unsteady states, for example, during start-up, the first procedure described earlier, in which the transient terms are set equal to zero, leads to approximate expressions for c^*. Ray (25) termed this procedure the RSSA and demonstrated that errors in c^* incurred through its use relax to zero more rapidly in general than do those stemming from the QSSA. The expressions from the RSSA may, however, be more difficult to solve than those from the QSSA. As noted earlier, Ray also concluded that the QSSA is valid for polymerization in which the chain lifetime λ_p is sufficiently short, such as free-radical polymerizations.

Equations 1.11-18 and 19 may be written in combined, dimensionless form by defining $\hat{r}_{exit}$ as $\hat{c}^*$ for the IM and $L\dfrac{\partial \hat{c}^*}{\partial x}$ for the PFR, and by multiplying both sides by λ_p. Thus

$$\left.\begin{array}{c} \dfrac{\lambda_p}{\lambda_v}\dfrac{dc^*}{d\hat{t}} \\[2em] \dfrac{\lambda_p}{\lambda_v}\dfrac{\partial \hat{c}^*}{\partial \hat{t}} \end{array}\right\} = \dfrac{\lambda_p}{\lambda_1}\hat{r}_d - \dfrac{\lambda_p}{\lambda_2}(\hat{c}^*)^n - \dfrac{\lambda_p}{\lambda_v}\hat{r}_{exit} \qquad (1.11\text{-}20)$$

where $\hat{t} \equiv t/\lambda_v$ and λ_v is $V/\dot{V}$ (IM) or L/v (PFR). By analogous reasoning to that used earlier, we deduce the following criterion for the validity of the QSSA in continuous reactors.

$$\gamma_v \equiv \frac{\lambda_v}{\lambda_p} \gg 1 \qquad (1.11\text{-}21)$$

This result represents a formal statement of Denbigh's conclusion and is the counterpart of inequalities 1.11-16.

Example 1.11-5.   The RSSA Applied to Free-Radical Polymerizations. Consider a free-radical polymerization conducted in an IM. The rate of radical formation is given by

$$r_i = 2fk_d c_0$$

while termination rate is second-order in radical concentration. Compare the QSSA and RSSA expressions for $c^*(t)$ and show that for radical lifetimes much less than the reactor residence time both solutions converge.

Solution. The QSSA is satisfied when

$$\frac{dc^*}{dt} = 2fk_d c_0 - k_t(c^*)^2 \cong 0$$

thus

$$(c^*)_{\mathrm{QSSA}} = \sqrt{\frac{2fk_d c_0}{k_t}}$$

The RSSA for an IM yields

$$\frac{dc^*}{dt} = 2fk_d c_0 - k_t(c^*)^2 - \frac{\dot{V}}{V}c^* \cong 0$$

so

$$(c^*)_{\mathrm{RSSA}} = \frac{(\dot{V}/V)}{2k_t}\left[-1 + \left(1 + \frac{8fk_d k_t c_0}{[\dot{V}/V]^2} \right)^{1/2} \right]$$

Note that

$$\lambda_{\mathrm{v}} = \frac{V}{\dot{V}}, \qquad \lambda_{\mathrm{p}} = 1/\sqrt{2fk_d k_t c_0}$$

so

$$(c^*)_{\mathrm{RSSA}} = \frac{1}{2k_t \lambda_{\mathrm{v}}}\left\{ -1 + \left[1 + 4\left(\frac{\lambda_{\mathrm{v}}}{\lambda_{\mathrm{p}}} \right)^2 \right]^{1/2} \right\}$$

If $\lambda_{\mathrm{p}} \ll \lambda_{\mathrm{v}}$, then

$$-1 + \left[1 + 4\left(\frac{\lambda_{\mathrm{v}}}{\lambda_{\mathrm{p}}} \right)^2 \right]^{1/2} \cong 2\left(\frac{\lambda_{\mathrm{v}}}{\lambda_{\mathrm{p}}} \right)$$

so

$$(c^*)_{\mathrm{RSSA}} = \frac{1}{k_t \lambda_{\mathrm{p}}} = \sqrt{\frac{2fk_d c_0}{k_t}} = (c^*)_{\mathrm{QSSA}}$$

1.11.4.   *The Constant-Density Approximation*

In the formulation of kinetic rate models for liquid-phase reactions, it is customary to neglect density variations. For polymerizations in particular, this practice occasionally comes under attack owing to the volume shrinkage that can accompany such a chemical aggregation process. Since we shall make heavy use of the constant-density approximation, it is worthwhile to at least estimate the potential errors that might be incurred through its use.

We shall use simple homogeneous kinetics ($r = k_{ap}c_m c_0^{1/2}$) and assume, as a first approximation, that volumes are additive. Thus

$$\rho^{-1} = \rho_m^{-1} w_m + \rho_p^{-1} w_p \tag{1.11-22}$$

where w_m and w_p are weight fractions of monomer and polymer, respectively:

$$w_m = \frac{M_0 N_m}{W} \tag{1.11-23}$$

$$w_p = \frac{M_0\left[(N_m)_0 - N_m\right]}{W} \tag{1.11-24}$$

and W represents the total weight of the system. Thus from the rigorous definition of monomer conversion

$$\Phi \equiv 1 - \frac{N_m}{(N_m)_0} \tag{1.11-25}$$

we obtain, using $W \cong M_0(N_m)_0$,

$$\rho^{-1} = \rho_m^{-1}(1 - \Phi) + \rho_p^{-1}\Phi \tag{1.11-26}$$

or equivalently, after multiplying both sides of W,

$$V = V_0(1 - B\Phi) \tag{1.11-27}$$

where V is total volume, $V_0 = W/\rho_m$, and

$$B \equiv \frac{(\rho_p - \rho_m)}{\rho_p} \tag{1.11-28}$$

The rigorous batch rate equations are (Appendix E) for monomer consumption

$$-V^{-1}\frac{dN_m}{dt} = k_{ap}c_m c_0^{1/2} = (N_m)_0 V^{-1}\frac{d\Phi}{dt} \tag{1.11-29}$$

and for initiator consumption

$$-V^{-1}\frac{dN_0}{dt} = k_d c_0 = k_d N_0 V^{-1} \tag{1.11-30}$$

The solution of the last equation in

$$N_0 = (N_0)_0 \exp(-k_d t) \tag{1.11-31}$$

Substitution of N_m/V for c_m and N_0/V for c_0 into equation 1.11-29 yields the following rate equation for monomer conversion

$$\frac{d\Phi}{dt} = k_{ap}(c_0)_0^{1/2}\frac{(1 - \Phi)}{(1 - B\Phi)^{1/2}}\exp\left(\frac{-k_d t}{2}\right) \tag{1.11-32}$$

or, in dimensionless form,

$$\frac{d\Phi}{d\hat{t}} = \frac{(1 - \Phi)}{(1 - B\Phi)^{1/2}\exp\left(\dfrac{-\alpha_K \hat{t}}{2}\right)} \tag{1.11-33}$$

where $\hat{t} \equiv t/\lambda_m$ and $\alpha_K \equiv \lambda_m/\lambda_i$. The solution of this equation for the special case of no density variation ($B = 0$) appears in Example 1.11-1. The effect of various degrees of density variation (values of B) on monomer conversion profiles can be seen in Fig. 1.11-5, which is a graph of the numerical solution of equation 1.11-33 for a conventional polymerization ($\alpha_K < 1$). As expected, the effect worsens with conversion, but appears to be surprisingly small in general. A 10-percent change in density ($B = 0.1$) has very little effect. Even a variation of 50 percent is not as detrimental as one might expect.

1.12. COPOLYMERIZATION

Simultaneous addition polymerization of two comonomers, say m_A and m_B, rarely results in a perfectly alternating sequence of A and B groups in the copolymer chain. In other words, AB cannot be regarded as the repeat unit, as it can in AB-type condensation polymerizations. Another subscript, in addition to x, is therefore required to identify each polymer chain. In Section 1.5 double subscript notation (a, b) was introduced, which simultaneously specifies the DP ($x \equiv a + b$) and composition ($y \equiv a/x$) of molecule $m_{a, b}$.

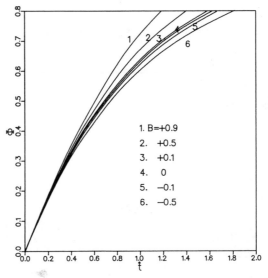

Figure 1.11-5. A graph of the solution of equation 1.11-33 showing the effect of volume shrinkage ($B > 1$) as well as growth ($B < 1$) on the monomer conversion profile of a conventional ($\alpha_K = 0.01$) chain-addition polymerization.

Table 1.12-1. Copolymerization Sequences

INITIATION

$$m_o \xrightarrow{k_d} m_o^*$$

$$j = A, B \qquad m_o^* + m_j \xrightarrow{k_{ij}} m_1^{j*}$$ primary

$$s^* + m_j \xrightarrow{k_{isj}} m_1^{j*}$$ secondary

Propagation and Transfer to Momomer

$$x \geq 1 \qquad k = A, B \qquad m_x^{j*} + m_k \begin{array}{c} \xrightarrow{k_{pjk}} m_{x+1}^{k*} \\ \xrightarrow{k_{fjk}} m_x + m_1^{k*} \end{array}$$

Transfer to Foreign Substance

$$x \geq 1 \qquad m_x^j + S \xrightarrow{k_{fsj}} m_x + S^*$$

Termination (Bimolecular)

$$x \geq 1 \qquad y \geq 1 \qquad m_x^{j*} + m_y^{k*} \begin{array}{c} \xrightarrow{k_{tDjk}} m_x + m_y \\ \xrightarrow{k_{tcjk}} m_{x+y} \end{array}$$ GM, PF

$$i = A, B \qquad \ell = A, B \qquad m_x^{ij*} + m_y^{k\ell} \begin{array}{c} \xrightarrow{k_{tDijk\ell}} m_x + m_y \\ \xrightarrow{k_{tcijk\ell}} m_{x+y} \end{array}$$ PE

For intermediates, a further expansion of notation is required. Since their relative reactivity with respect to each monomer, as well as with respect to each other, is determined primarily by the chemical nature of their terminal mer units, and even their penultimate ones to a lesser extent, our notation must be able to distinguish among such intermediates. To accomplish this, superscripts will be used. Thus $m_{a,b}^{A*}$ and $m_{a,b}^{B*}$ will represent intermediates with terminal units A and B, and $m_{a,b}^{AA*}$, $m_{a,b}^{AB*}$, $m_{a,b}^{BA*}$, and $m_{a,b}^{BB*}$ will represent intermediates with terminal pairs AA, AB, and so on, as indicated.

Table 1.12-2.   Copolymerization Rate Functions—Steps

INITIATION

$j = A,B$
$$r_{ij} = k_{ij}c_j^p c_o^q + k_{isj}c_j c_s^*$$

$$r_i = \sum_{j=A}^{B} r_{ij}$$

PROPAGATION

$k = A,B$
$$r_{pjk} = k_{pjk}c^{j*}c_k \qquad\qquad c^{j*} \equiv \sum_x c_x^{j*}$$

$$r_{pk} = \sum_{j=A}^{B} r_{pjk}$$

$$r_p = \sum_{k=A}^{B} r_{pk} = \sum_j \sum_k k_{pjk}c^{j*}c_k$$

TERMINATION

    GM, PF MODELS

$$r_{tjk} = k_{tjk}c^{j*}c^{k*}$$

$$r_t = \sum_j \sum_k r_{tjk} = \sum_j \sum_k k_{tjk}c^{j*}c^{k*}$$

$$= \begin{cases} \left(\sum_j k_{tjj}^{1/2} c^{j*} \right)^2 & \text{GM} \\[2ex] \sum_j \sum_k \phi^{(1-\delta_j^k)} k_{tjj}^{1/2} k_{tkk}^{1/2} c^{j*}c^{k*} & \text{PF} \end{cases}$$

where δ_j^k in the exponent is the Kronecker

delta:
$$\delta_j^k = \begin{cases} 1 \text{ when } j = k \\ 0 \text{ when } j \neq k \end{cases}$$

Table 1.12-2 (*Continued*)

TERMINATION (continued)

PE MODEL

$$i = A, B$$
$$\ell = A, B$$

$$r_{tijk\ell} = k_{tijk\ell} c^{ij*} c^{\ell k*}$$

$$r_{tjk} = \sum_{i=A}^{B} \sum_{\ell=A}^{B} r_{tijk\ell}$$

$$r_t = \sum_{j=A}^{B} \sum_{k=A}^{B} r_{tjk} = \sum_i \sum_j \sum_k \sum_\ell k_{tijk\ell} c^{ij*} c^{\ell k*}$$

$$= \left(\sum_j \sum_k k_{tjkkj}^{1/2} c^{jk*} \right)^2$$

TRANSFER

$$r_{fjk} = k_{fjk} c^{j*} c_k$$

$$r_{fsj} = k_{fsj} c^{j*} c_s$$

$$r_{fk} = \sum_j k_{fjk} c^{j*} c_k$$

$$r_f = \sum_k r_{fk} + \sum_j r_{fsj}$$

Sequences for chain-addition copolymerizations are listed in Table 1.12-1 and their rate functions in Tables 1.12-2–1.12-4. The rate functions for the two distinguishable initiation reactions (equations A-16)

$$j = A, B \qquad\qquad m_0 + m_j \rightarrow m_1^{j*} \qquad\qquad (1.12\text{-}1)$$

may be represented by a format similar to that of equation 1.10-10 for homopolymerizations:

$$r_{ij} = k_{ij} c_j^p c_0^q \qquad\qquad (1.12\text{-}2)$$

where the overall initiation rate is the sum of r_{ij} over j. Some simplifications are noteworthy. When the initiator is a free-radical type, overall initiation rate can be

Table 1.12-3. Copolymerization Rate Functions—Components

MONOMER

$$r = \sum_k r_k = r_i + r_p + r_f = \sum_j \sum_k (k_{pjk} + k_{fjk}) c^{j*} c_k + \sum_j k_{fsj} c^{j*} c_s$$

$$+ \sum_j \sum_k k_{tDjk} c^{j*} c^{k*} + \sum_j \sum_k k_{tcjk} c^{j*} c^{k*}$$

$$r_k = r_{ik} + r_{pk} + r_{fk}$$

FOREIGN SUBSTANCE

$$r_s = -\sum_j r_{fsj}$$

INTERMEDIATES

$$r_1^{j*} = r_{ij} + \sum_k k_{fkj} c^{k*} c_j - \sum_k k_{pjk} c_1^{j*} c_k - \sum_k k_{fjk} c_1^{j*} c_k$$

$$- k_{fsj} c_1^{j*} c_s - \sum_k k_{tjk} c_1^{j*} c^{k*}$$

$$x \geq 1 \qquad r_x^{j*} = \sum_k k_{pkj} c_{x-1}^{k*} c_j - \sum_k k_{pjk} c_x^{j*} c_k - \sum_k k_{fjk} c_x^{j*} c_k$$

$$- k_{fsj} c_x^{j*} c_s - \sum_k k_{tjk} c_x^{j*} c^{k*}$$

$$\begin{aligned} a > 1 \\ b > 1 \end{aligned} \qquad r_{a,b}^{j*} = \sum_k k_{pkj} c_{a-1,b}^{k*} c_j - \sum_k k_{pjk} c_{a,b}^{j*} c_k - \sum_k k_{fjk} c_{a,b}^{j*} c_k$$

$$- k_{fsj} c_{a,b}^{j*} c_s - \sum_k k_{tjk} c_{a,b}^{j*} c^{k*}$$

$$\ell \neq j \qquad r^{j*} = \sum_x r_x^{j*} = \sum_a \sum_b r_{a,b}^{j*}$$

$$= r_{ij} + (k_{p\ell j} + k_{f\ell j}) c^{\ell *} c_j - (k_{pj\ell} + k_{fj\ell}) c^{j*} c_\ell$$

$$- k_{fsj} c^{j*} c_s - \sum_k k_{tjk} c^{j*} c^{k*}$$

Table 1.12-3 (*Continued*)

INTERMEDIATES (cont'd)

$$r^* = \sum_j r^{j*} = r_i + r_s + r_t$$

$$r_s^* = -r_s - \sum_k k_{isk} c_s^* c_k$$

PRODUCTS

$$r_{a,b} = \sum_j \sum_k k_{fjk} c_{a,b}^{j*} c_k + \sum_j k_{fsj} c_{a,b}^{j*} c_s + \sum_j \sum_k k_{tDjk} c_{a,b}^{j*} c^{k*}$$

$$+ \sum_j \sum_k (k_{tcjk}/2) \sum_{i=0}^{a} \sum_{\ell=0}^{b} c_{i,\ell}^{j*} c_{a-i,b-\ell}^{k*}$$

$$r_x = \sum_j \sum_k k_{fjk} c_x^{j*} c_k + \sum_j k_{fsj} c_x^{j*} c_s + \sum_j \sum_k k_{tDjk} c_x^{j*} c^{k*}$$

$$+ \sum_j \sum_k (k_{tcjk}/2) \sum_{i=1}^{x-1} c_i^{j*} c_{x-i}^{k*}$$

expressed as usual

$$r_i = f k_d c_0 \tag{1.12-3}$$

and when composition drift is minimal, individual rates can be expressed in terms of equation 1.12-2 with $p = 0$, $q = 1$ and

$$j = A, B \qquad\qquad k_{ij} = f_j k_d \tag{1.12-4}$$

Equation 1.12-2 may also be applied to ionic initiation with $p = 1 = q$, but only for the case $k_{i1} = k_{i2} \equiv k_i$ may one assume that

$$r_i = k_i c_m c_0 \tag{1.12-5}$$

When ultimate mers alone determine reactivity, there are four distinguishable propagation steps (equations A-17),

$$\begin{array}{l} k = A, B \\ x \geqslant 1 \end{array} \qquad\qquad m_x^{j*} + m_k \xrightarrow{k_{pjk}} m_{x+1}^{k*} \tag{1.12-6}$$

and when the penultimate mer influences reactivity as well, the number increases to eight

$$\begin{array}{l} j, k, l = A, B \\ x \geqslant 1 \end{array} \qquad\qquad m_x^{jk*} + m_l \xrightarrow{k_{pjkl}} m_{x+1}^{kl*} \tag{1.12-7}$$

Table 1.12-4. Active Intermediate Concentrations for Free-Radical Copolymerization

$$\begin{array}{l} j \neq \ell \\ j = A, B \\ \ell = A, B \end{array} \qquad c^{j*} = k_{p\ell j} c_j H (r_i / k_{tjj} k_{t\ell\ell})^{1/2}$$

(GM)

$$H = \left[\frac{k_{PBA} c_A}{k_{tBB}^{1/2}} + \frac{k_{PAB} c_B}{k_{tAA}^{1/2}} \right]^{-1}$$

(PF)

$$H = \left[\left(\frac{k_{PBA} c_A}{k_{tBB}^{1/2}} \right) + 2\phi \frac{k_{PBA} k_{PAB} c_A c_B}{(k_{tAA} k_{tBB})^{1/2}} + \left(\frac{k_{PAB} c_B}{k_{tAA}^{1/2}} \right)^2 \right]^{-1/2}$$

(PE)

$$H = \left[\frac{k_{PBA} c_A}{k_{tBB}^{1/2}} \left[\frac{R_A c_A + \left(\frac{k_{tBAAB}}{k_{tAAAA}} \right)^{1/2} c_B}{R_A c_A + c_B} \right] + \frac{k_{PAB} c_B}{k_{tAA}^{1/2}} \left[\frac{R_B c_B + \left(\frac{k_{tABBA}}{k_{tBBBB}} \right)^{1/2} c_A}{R_B c_B + c_A} \right] \right]^{-1}$$

By neglecting penultimate effects, which is common practice, and by applying the long-chain and quasi-steady-state approximations (LCA and QSSA), we can deduce the symmetry relation $r_{pjk} = r_{pkj}$ among the four propagation rate functions (Example 1.13-4)

$$\begin{array}{l} j = A, B \\ k = A, B \end{array} \qquad\qquad r_{pjk} = k_{pjk} c^{j*} c_k \qquad\qquad (1.12\text{-}8)$$

As usual, the LCA applies here when propagation rate dominates initiation and termination rates.

Free-radical termination sequences and rate functions for both ultimate and penultimate reactions are listed in Tables 1.12-1 and 1.12-2. Various models have been proposed to relate termination rate constants for unlike radical ends to those for like ends. The incentive for doing this obviously stems from the experimental difficulties associated with measuring cross-termination constants. The simplest model, called the geometric mean (GM), assumes that

$$\left. \begin{array}{l} j = A, B \\ k = A, B \end{array} \right\} \qquad\qquad k_{tjk} = \left(k_{tjj} k_{tkk} \right)^{1/2} = k_{tkj} \qquad\qquad (1.12\text{-}9)$$

as the name implies. The more widely used phi factor model (PF) modifies this form with a weighting factor ϕ

$$\left. \begin{array}{l} j = A, B \\ k = A, B \end{array} \right\} \qquad\qquad k_{tjk} = \phi \left(k_{tjj} k_{tkk} \right)^{1/2} = k_{tkj} \qquad\qquad (1.12\text{-}10)$$

where $\phi > 1$ indicates that cross-termination is favored. The penultimate effect model (PE) is more complex, but may be simplified somewhat by the following symmetry relationships:

$$
\left.\begin{array}{l} i = A, B \\ j = A, B \\ k = A, B \\ l = A, B \end{array}\right\} \qquad k_{t\,ijkl} = \left(k_{t\,ijji}\,k_{t\,lkkl} \right)^{1/2} \qquad \qquad (1.12\text{-}11)
$$

where $k_{t\,ijjj} = k_{t\,jj}$ for $j = A, B$. The rate functions in Tables 1.12-2 for each termination model make use of the appropriate symmetry relations.

Propagation and termination sequences showing composition details are (cf. Equations A-17 and A-18):

$$
\mathrm{m}^{A*}_{a,b} + \mathrm{m}_A \xrightarrow{k_{pAA}} \mathrm{m}^{A*}_{a+1,b}
$$

$$
\mathrm{m}^{A*}_{a,b} + \mathrm{m}_B \xrightarrow{k_{pAB}} \mathrm{m}^{B*}_{a,b+1}
$$

$$
\mathrm{m}^{B*}_{a,b} + \mathrm{m}_A \xrightarrow{k_{pBA}} \mathrm{m}^{A*}_{a+1,b} \qquad \qquad (1.12\text{-}12)
$$

$$
\mathrm{m}^{B*}_{a,b} + \mathrm{m}_B \xrightarrow{k_{pBB}} \mathrm{m}^{B*}_{a,b+1}
$$

and

$$
\left.\begin{array}{l} j = A, B \\ k = A, B \end{array}\right\} \qquad \mathrm{m}^{j*}_{a,b} + \mathrm{m}^{k*}_{\alpha,\beta} \;\begin{array}{c} \xrightarrow{k_{tDjk}} \mathrm{m}_{a,b} + \mathrm{m}_{\alpha,\beta} \\[4pt] \xrightarrow[k_{tcjk}]{} \mathrm{m}_{a+\alpha,\,b+\beta} \end{array} \qquad (1.12\text{-}13)
$$

Rate functions $r^{j*}_{a,b}$ and $r_{a,b}$, which are included in Table 1.12-3, require consideration of such details. It should be noted that all termination expressions in that Table apply to the GM and PF models, but not to the PE model.

The QSSA can be applied to individual intermediate populations $\mathrm{m}^{j*}_{a,b}$ and m^{j*} separately, as well as to the total radical population m^*, to facilitate the evaluation of expressions for intermediate concentrations $c^{j*}_{a,b}$ and $c^{j*} = \sum_a \sum_b c^{j*}_{a,b}$, which are required for substitution into the propagation and termination rate functions. The results are shown in unified form in Table 1.12-4. The origin and applicability of the various termination models to specific systems will be discussed in Chapter 2.

Example 1.12-1.   Rate of Initiation for Copolymerization. Show that the formula for rate of initiation of A type radicals (from Section 2.8) reduces to equation equivalent to 1.12-3 under the assumption of constant composition:

$$
r_{iA} = f_A k_d c_0
$$

where

$$
f_A \equiv \frac{k'_{iA} c_A c^*_0}{k'_{iA} c_A c^*_0 + k'_{iB} c_B c^*_0 + k_0 c^{*2}_0}
$$

Solution.   We can rewrite f_A as

$$f_A = \left(\frac{k'_{iA}c_A c_0^*}{k'_{iA}c_A c_0^* + k_0 c_0^{*2}} \right) \left(\frac{k'_{iA}c_A c_0^* + k_0 c_0^{*2}}{k'_{iA}c_A c_0^* + k'_{iB}c_B c_0^* + k_0 c_0^{*2}} \right)$$

$$= \left[\frac{k'_{iA}c_A}{k'_{iA}c_A + k_0 c_0^*} \right] \Big/ \left(1 + \frac{k'_{iB}c_B}{k'_{iA}c_A} \left[\frac{k'_{iA}c_A}{k'_{iA}c_A + k_0 c_0^*} \right] \right)$$

By recognizing that each bracketed quantity is the initiator efficiency in pure monomer A, f_A^0 (equation 2.8-19), the only additional assumption needed to hold f_A constant is that c_B/c_A remain fixed.

Example 1.12-2.   Calculation of Intermediate Concentrations.   Find the expression for c_A^* using the GM termination model.

Solution.   Application of the QSSA in the absence of solvent radical self-termination leads to

$$0 = r_i - r_{t_i} \qquad \text{for } c^*.$$

From the c^* balance,

$$r_i = k_{tAA}(c^{A*})^2 + 2(k_{tAA}k_{tBB})^{1/2}c^{A*}c^{B*} + k_{tBB}(c^{B*})^2$$

$$= \left(k_{tAA}^{1/2}c^{A*} + k_{tBB}^{1/2}c^{B*} \right)^2$$

or

$$r_i^{1/2} = k_{tAA}^{1/2}c^{A*} + k_{tBB}^{1/2}c^{B*}$$

Using equation 1.12-8

$$k_{PAB}c^{A*}c_B = k_{PBA}c^{B*}c_A$$

thus

$$r_i^{1/2} = k_{tAA}^{1/2}c^{A*} + \frac{k_{PAB}}{k_{PBA}} \frac{c_B}{c_A} k_{tBB}^{1/2}c^{A*}$$

or

$$c^{A*} = \frac{r_i^{1/2}}{\left(k_{tAA}^{1/2} + \dfrac{k_{PAB}}{k_{PBA}} k_{tBB}^{1/2} \dfrac{c_A}{c_B} \right)}$$

$$= \frac{(r_i/k_{tAA})^{1/2}\left(c_A k_{PBA}/k_{tBB}^{1/2} \right)}{\dfrac{c_A k_{PBA}}{k_{tBB}^{1/2}} + \dfrac{c_B k_{PAB}}{k_{tAA}^{1/2}}}$$

1.13. INSTANTANEOUS PROPERTIES

Some polymer molecules produced in a given reaction are grown concurrently, and some are grown consecutively. Step polymerizations produce mainly the former kind, as noted in Section 1.6, whereas chain polymerizations produce mainly the latter. In chain polymerizations and copolymerizations with termination, only a small fraction of the ultimate product molecules grew concurrently, but many, many generations follow one another in succession.

Properties associated with polymeric reaction products, such as DP, DB, and CC, may be viewed as cumulative in terms of reaction time. These properties reflect the combined histories of the macromolecules that determine them. To illustrate this point, we rewrite two definitions in Section 1.5 as

$$\bar{x}_N = \frac{\sum_x x \, \Delta N_x}{\sum_x \Delta N_x} \tag{1.13-1}$$

and

$$\bar{y} = \frac{\sum_a a \, \Delta N_a}{\sum_x x \Delta N_x} \tag{1.13-2}$$

where ΔN_x and ΔN_a represent the number of molecules having properties x and a, respectively, that are formed during reaction time Δt.

In chain polymerizations, the fraction of polymer product growing at any instant, albeit small, still consists of a very large number of molecules, of all sizes, Furthermore, these molecules grow to their final size so rapidly that they may be regarded as instantaneous product without appreciable error. Thus, when Δt (λ_p) is sufficiently large to include the growth of many molecules, yet sufficiently small compared to overall polymerization time (λ_r) to be considered infinitesimal, we designate the corresponding properties as "instantaneous" and associate them with the generation of chain polymers formed instantaneously at t.

1.13.1.   DP and DPD

From Equation 1.13-1, we define instantaneous DP as

$$(\bar{x}_N)_{inst} \equiv \frac{\sum_x x r_x}{\sum_x r_x} \tag{1.13-3}$$

where the interpretation of r_x

$$r_x = \lim_{\Delta t \to 0} \frac{\Delta N_x}{\Delta t} \tag{1.13-4}$$

is analogous to that of a point property in continuum mechanics, such as density

$$\rho = \lim_{V \to 0} \frac{m}{V} \qquad (1.13\text{-}5)$$

where m represents the mass in volume V. Here ρ is assumed to be a continuous function of position in space in the same way that $r_x(t)$ is a continuous function of time. It is important to remember that the instantaneous property is meaningful only in chain polymerization for which $\lambda_p \ll \lambda_r$.

The polymer grown at any instant, like the polymer produced in any polymerization, consists of a spectrum of molecular sizes (DPD) reflecting the random nature of the molecular events that make up its reaction sequence. We shall refer to the dispersion due to molecular statistics as "statistical dispersion" and discuss it in detail later on. For the time being, it is clear that we can define a collection of instantaneous properties based entirely on rate functions r_x with the aid of Table 1.10-4. Thus

$$(n_x)_{\text{inst}} \equiv \frac{r_x}{\sum_x r_x} \qquad (1.13\text{-}6)$$

$$(w_x)_{\text{inst}} \equiv \frac{x r_x}{\sum_x x r_x} = \frac{x(n_x)_{\text{inst}}}{(\bar{x}_N)_{\text{inst}}} \qquad (1.13\text{-}7)$$

and so on. Furthermore, since the final product is an accumulation of all instantaneous products formed during the course of reaction, the cumulative properties may be computed from the corresponding instantaneous properties by integration. For example (cf. Example 1.13-3), neglecting density changes

$$\bar{x}_N(\alpha) = \frac{\int_0^\alpha \sum x r_x(\alpha')\, d\alpha'}{\int_0^\alpha \sum r_x(\alpha')\, d\alpha'} \qquad (1.13\text{-}8)$$

where α represents time (t) in a batch reactor or age (x/v) in a continuous, plug-flow reactor. The reader is cautioned not to confuse instantaneous properties with the properties of active intermediates, such as $n_x^* \equiv c_x^*/\sum c_x^*$ and $\bar{x}_N^* \equiv \sum x n_x^*$.

It should be apparent at this point that a relationship exists between $(\bar{x}_N)_{\text{inst}}$ and the kinetic chain length ν_N defined in Section 1.11. In fact, under special circumstances, they are equal. This is not, however, the case in general. The numerators $\sum x r_x$ and r_p in the defining equations, 1.11-1 and 1.13-3, respectively, are identical only when the LCA applies so that $D \cong k_p c_m$ (Example 1.13-1). For the denominators to be equal, we require in addition that chain transfer be absent. Even then, from Table 1.10-4,

$$\sum r_x = k_{ts} c^* + \left(k_{tD} + \frac{k_{tc}}{2} \right)(c^*)^2 \qquad (1.13\text{-}9)$$

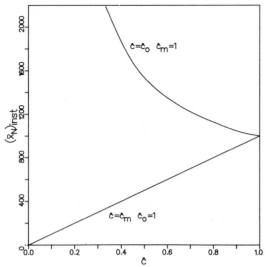

Figure 1.13-1. A graph of equation 1.13-11 without chain transfer; instantaneous DP, $(\bar{x}_N)_{inst}$, versus dimensionless monomer, $\hat{c}_m \equiv c_m/(c_m)_0$, and dimensionless initiator, $\hat{c}_0 \equiv c_0/(c_0)_0$, concentrations.

whereas

$$r_i \underset{QSSA}{\cong} r_t = k_{ts}c^* + \underbrace{(k_{tD} + k_{tc})}_{k_t}(c^*)^2 \qquad (1.13\text{-}10)$$

Thus we conclude that $(\bar{x}_N)_{inst} = \nu_N$ only when free-radical combination is absent and that $(\bar{x}_N)_{inst} = 2\nu_N$ when it is the exclusive termination mode.

A well-known expression that is used by kineticists to distinguish free-radical termination mechanisms and to determine which chain transfer reactions (excluding chain branching) are important (Example 1.13-2) is (17)

$$(\bar{x}_N)_{inst}^{-1} = R_{fm} + R_{fs}\frac{c_s}{c_m} + \left[\frac{(2-R)}{2k_{ax}}\right]c_0^{1/2}/c_m \qquad (1.13\text{-}11)$$

where $R \equiv k_{tc}/k_t = 2R'/(1 + R')$ and

$$k_{ax} \equiv k_p/(k_i k_t)^{1/2} \qquad (1.13\text{-}12)$$

The ratios $R_{fm} \equiv k_{fm}/k_p$ and $R_{fs} \equiv k_{fs}/k_p$ are called chain transfer constants and the ratio R' has been used earlier (Section 1.7). The desired information is usually obtained by conducting low-conversion experiments and measuring the dependence of DP on the feed concentration of monomer, initiator, and chain transfer agent S independently.

Equation 1.13-11 has been plotted in Fig. 1.13-1, neglecting chain transfer, in order to demonstrate the dependence of $(\bar{x}_N)_{inst}$ on monomer concentration and initiator concentration. The effect of declining concentrations during polymerization on the DP of growing polymer chains is clearly different for monomer than for initiator. The import of this difference will be discussed in detail in later chapters.

1.13.2.   CC and CCD

For copolymers, it is common practice to define instantaneous CC as a kinetic composition variable

$$\nu_A \equiv \frac{r_{pA}}{r_p} \cong (\bar{y})_{\text{inst}} \tag{1.13-13}$$

which utilizes the LCA and thereby neglects monomer introduced into the copolymer chains during the initiation steps. With the aid of the rate functions in Table 1.12-2, we can relate $(\bar{y})_{\text{inst}}$ to the composition of the reaction mixture in the same way that equation 1.13-11 relates $(\bar{x}_N)_{\text{inst}}$ to reactant compositions c_m and c_0. The result is the well known copolymer composition equation (see Example 1.13-4)

$$\nu_A = \frac{R_A n_A^2 + n_A(1 - n_A)}{R_A n_A^2 + 2n_A(1 - n_A) + R_B(1 - n_A)^2} \tag{1.13-14}$$

where n_j represents the mole fraction of unreacted comonomer j and R_j is the reactivity ratio

$$R_j \equiv k_{pjj}/k_{pjk} \tag{1.13-15}$$

Graphs of equation 1.13-14 have been plotted in Figs 1.13-2–1.13-5. We shall refer to these as CC diagrams. They are the analogues of Fig. 1.13-1 except that, unlike in the latter, reactor composition can vary in either direction.

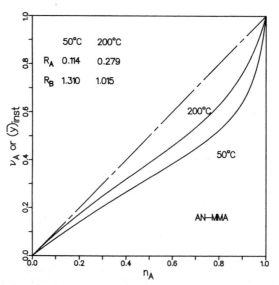

Figure 1.13-2. A graph of equation 1.13-14; the CC diagram for acrylonitrile–methyl methacrylate (AN/MMA) at different temperatures (26); A = AN, B = MMA.

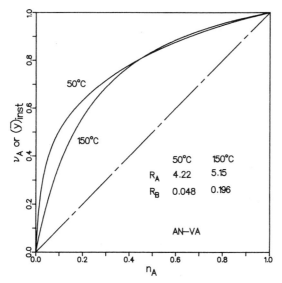

Figure 1.13-3. A graph of equation 1.13-14; the CC diagram for acrylonitrile-vinyl acetate (AN/VA) at different temperatures (26); A = AN, B = VA.

Four distinct classes of CC diagrams are possible, one for each reactivity ratio pair R_A, R_B. They have been sketched in Fig. 1.13-6. The role of R_j in determining their shape is seen from the two limiting slopes:

$$\lim_{n_A \to 0} \frac{dv_A}{dn_A} = R_B^{-1} \tag{1.13-16}$$

$$\lim_{n_A \to 1} \frac{dv_A}{dn_A} = R_A^{-1} \tag{1.13-17}$$

Thus, when $R_B < 1$ the curve near $n_A = 0$ has a slope greater than unity and therefore lies above the diagonal. On the other hand, when $R_B > 1$, the curve lies below the diagonal. Similarly, when $R_A > 1$, the curve near $n_A = 1$ lies above the diagonal, and when $R_A < 1$, it lies below. The point at which the curve intersects the diagonal is called the azeotropic composition. Its value is given by the equation

$$v_A = \frac{1 - R_B}{2 - R_A - R_B} \tag{1.13-18}$$

The analogy between the CC diagram and the vapor-liquid equilibrium diagram for binary systems is clarified by rearranging equation 1.13-14 and specializing it for the case $R_A R_B = 1$, which is known as ideal copolymerization. The result

$$\frac{v_A(1 - n_A)}{n_A(1 - v_A)} = R_A = \text{constant} \tag{1.13-19}$$

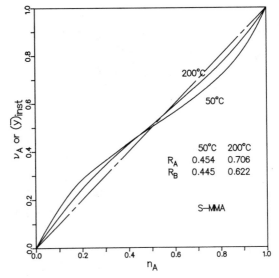

Figure 1.13-4. A graph of equation 1.13-14; the CC diagram for styrene-methyl methacrylate (S/MMA) at different temperatures (26); A = S, B = MMA.

may then be compared to its counterpart for distillation

$$\frac{y(1-x)}{x(1-y)} = \alpha_R = \text{constant} \tag{1.13-20}$$

where x and y are mole fractions of the more volatile component in the liquid and

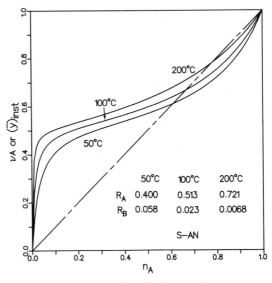

Figure 1.13-5. A graph of equation 1.13-14; the CC diagram for styrene-acrylonitrile (S/AN) at different temperatures (26); A = S, B = AN.

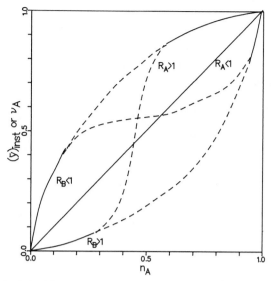

Figure 1.13-6. A sketch of the CC diagram, equation 1.13-14, showing the effect of reactivity ratios R_A and R_B on the shape.

the vapor, respectively, and α_R is the relative volatility. The implication is that copolymerization may be viewed as a two-phase separation process similar to simple batch distillation, in which vapor is removed as it is formed. Thus equation 1.13-14 may be thought of as an equation of state, like 1.13-20, and R_j a separation factor, like α_R.

Example 1.13-2.   **Instantaneous DP.**   Starting with the definition of $(\bar{x}_N)_{inst}$ and the rate functions in Table 1.10-4, derive equation 1.13-11.

Solution.   After multiplying rate function 2.2-51 by x

$$x r_x = \left(D - k_p c_m - k_{tc} c^* \right) c^* x \phi^{x-1} (1 - \phi) + \frac{k_{tc}}{2} (c^*)^2 x (x - 1) \phi^{x-2} (1 - \phi)^2$$

and evaluating the required sums with the aid of Table C-3

$$\sum_{x=1}^{\infty} x \phi^{x-1} (1 - \phi) = \frac{(1 - \phi)}{\phi} \sum_{x=1}^{\infty} x \phi^x = \frac{1}{1 - \phi}$$

$$\sum_{x=1}^{\infty} x (x - 1) \phi^{x-2} (1 - \phi)^2 = \left[\frac{(1 - \phi)}{\phi} \right]^2 \left(\sum_{x=1}^{\infty} x^2 \phi^x - \sum_{x=1}^{\infty} x \phi^x \right) = \frac{2}{(1 - \phi)}$$

we obtain

$$\sum_{x=1}^{\infty} x r_x = \frac{\left(k_{fm} c_m + k_{fs} c_s + k_t c^* + k_{ts} \right) c^*}{1 - \phi} = D c^*$$

which is equal to $r_p = k_p c_m c^*$ only when $D \cong k_p c_m$, that is, when propagation dominates monomer consumption.

Example 1.13-2.   Instantaneous DP.   Starting with the definition of $(\bar{x}_N)_{inst}$ and the rate functions in Table 1.10-4, derive equation 1.13-11.

Solution.   Omit spontaneous termination and neglect $k_{ta} c_s c^*$ relative to the other terms. After substitution into $\Sigma r_x / r_p$ and application of the QSSA (equation 1.10-11) to eliminate c^*, equation 1.13-3 yields

$$(\bar{x}_N)_{inst}^{-1} = \frac{\Sigma r_x}{r_p}$$

$$= \frac{k_{fm}}{k_p} + \frac{k_{fs} c_s}{k_p c_m} + \frac{(k_{tD} + k_{tc}/2)(k_i/k_t)^{1/2} c_0^{1/2}}{k_p c_m}$$

which reduces to equation·1.13-11 following substitution of the definitions for k_{ax} and R.

Example 1.13-3.   Cumulative Properties from Instantaneous Properties.   Show that the following, simple equations are valid for cumulative averages under conditions analogous to those specified for equation 1.13-8 as long as the chain transfer rate to foreign substance S is not excessive:

$$\bar{x}_N = \frac{\int_0^\alpha r(\alpha')\, d\alpha'}{\int_0^\alpha \frac{r(\alpha')\, d\alpha'}{(\bar{x}_N)_{inst}}}$$

$$\bar{x}_W = \frac{\int_0^\alpha r(\alpha')(\bar{x}_W)_{inst}\, d\alpha'}{\int_0^\alpha r(\alpha')\, d\alpha'}$$

where $(\bar{x}_N)_{inst}$ and $(\bar{x}_W)_{inst}$ are also functions of α.

Solution.   Using the same reasoning leading to equation 1.13-8, we may write

$$\bar{x}_W(\alpha) = \int_0^\alpha \Sigma x^2 r_x(\alpha')\, d\alpha' \Big/ \int_0^\alpha \Sigma x r_x(\alpha')\, d\alpha'$$

From the moment equations in Table 1.5-2, we can relate both the sum $\Sigma x^2 r_x$ in this result and the sum in equation 1.13-8 to the sum $\Sigma x r_x$, which is equal to r (cf. r in Table 1.10-4 and $\Sigma x r_x$ in Example 1.13-1) Thus

$$(\bar{x}_N)_{inst} = \frac{\Sigma x r_x}{\Sigma r_x} \cong \frac{r}{\Sigma r_x}$$

and

$$(\bar{x}_W)_{inst} = \sum x^2 r_x \sum x r_x \cong \frac{\sum x^2 r_x}{r}$$

Following substitution for $\sum r_x$, $\sum x r_x$, and $\sum x^2 r_x$ in terms of r in the appropriate equations for cumulative $\bar{x}_N$ and x_W, we obtain the desired results.

Example 1.13-4. *The Copolymer Composition Equation.* Starting with the rate functions for r_{pA}, r_{pB}, and r_p in Table 1.12-2, derive the CC equation (1.13-14) by applying the QSSA separately to intermediates c^{A*} and c^{B*}.

Solution. For each intermediate, the "local" QSSA implies that $r^{j*} = 0$ and $r_i^{j*} = r_t^{j*}$. Thus we obtain the symmetry relation, $r_{pjk} = r_{pkj}$, or

$$\begin{matrix} j = A, B \\ k = A, B \end{matrix} \qquad k_{pjk}c^{j*}c_k = k_{pkj}c^{k*}c_j$$

After applying these results to the quotient r_{pA}/r_p and eliminating c^{A*} and c^{B*}, we obtain

$$\nu_A = \frac{R_A c_A^2 + c_A c_B}{R_A c_A^2 + 2 c_A c_B + R_B c_B^2}$$

or, equivalently, equation 1.13-14.

1.13.3. DB and DBD

Our analysis of branching will be confined to long-chain branching in chain reactions, which produces nonlinear, non-network polymers. Long-chain branching occurs via either one of two reactions listed in Table 1.10-3: propagation with double-bond-containing polymer or chain-transfer to polymer followed by propagation with monomer. It is distinguished from short-chain branching, which is the result of intramolecular chain transfer ("backbiting"). A similar mechanism is responsible for the formation of five- and six-membered rings.

Following the procedure used for instantaneous DP and CC, we could define a kinetic degree of branching DB

$$\nu_b \equiv \frac{r_b}{r_i} = \frac{\nu_N r_0}{r_p} \tag{1.13-21}$$

where r_b is the rate of branch point generation, defined (Table 1.10-4)

$$r_b \equiv k_p' c_p' c^* + k_{fp} \mu_c^1 c^* \tag{1.13-22}$$

However, we choose instead a more rigorous quantity, that is, the instantaneous degree of branching DB

$$(\bar{b}_N)_{inst} \equiv \frac{r_b}{\sum r_x} \cong (\bar{x}_N)_{inst} \frac{r_b}{r_p} \tag{1.13-23}$$

for reasons analogous to those enumerated earlier in connection with our definition of $(\bar{x}_N)_{inst}$. This is the instantaneous property that corresponds to cumulative $\bar{b}_N$ defined in Table 1.5-1. The quotient r_p/r_b, whose reciprocal (branch density) appears in equation 1.13-23, represents the average "distance" between branch points in terms of repeat units, assuming that no more than one branch point per repeat unit is permitted. This assumption is tacit in the rate functions listed in Table 1.10-4.

REFERENCES

1. K. G. Denbigh, *Chemical Reactor Theory*, Cambridge University, London (1965).
2. Z. Tadmor and C. G. Gogos, *Principles of Polymer Processing*, Wiley, New York (1979).
3. P. Klinkenberg and F. Sjenitzer, *Chem. Eng. Sci.*, **5**, 258 (1956).
4. J. A. Biesenberger and A. Quano, *J. Appl. Polym. Sci.*, **14**, 471 (1970).
5. G. I. Taylor, *Proc. R. Soc. London Ser. A*, **219**, 186 (1953).
6. P. V. Danckwerts, *Chem. Eng. Sci.*, **2**, 1 (1953).
7. W. Jost, *Diffusion*, Academic, New York (1960).
8. J. A. Biesenberger and Z. Tadmor, *Polym. Eng. Sci.*, **6**, 299 (1966).
9. Z. Tadmor, Ph.D. thesis in chemical engineering, Stevens Institute of Technology (1966).
10. W. D. Lansing and E. O. Kraemer, *J. Am. Chem. Soc.*, **57**, 1369 (1935).
11. H. Wesslau, *Makromol. Chem.*, **20**, 111 (1956).
12. K. Nagasubramanian and W. W. Graessley, *Chem. Eng. Sci.*, **25**, 1549 (1970).
13. M. L. Huggins, *J. Phys. Chem.*, **46**, 151 (1942).
14. M. L. Huggins, *Ann. N.Y. Acad. Sci.*, **43**, 1 (1942).
15. P. J. Flory, *J. Chem. Phys.*, **10**, 51 (1942).
16. P. J. Flory, *J. Chem. Phys.*, **12**, 425 (1944).
17. P. J. Flory, *Principles of Polymer Chemistry*, Cornell University Ithaca (1953).
18. P. J. Blatz and A. V. Tobolsky, *J. Phys. Chem.*, **49**, 77 (1945).
19. J. J. Hermans, *J. Pol. Sci.*, **C12**, 345 (1966).
20. H. Kilkson, *I.E.C. Fund.*, **3**, 281 (1964).
21. J. A. Biesenberger and R. Capinpin, *Polym. Eng. Sci.*, **14**, 737 (1974).
22. M. Boudart, *Kinetics of Chemical Processes*, Prentice-Hall, Englewood Cliffs, NJ (1968).
23. J. A. Biesenberger and R. Capinpin, *J. Appl. Polym. Sci.*, **16**, 695 (1972).
24. K. G. Denbigh, *Trans. Faraday Soc.*, **43**, 648 (1947).
25. W. H. Ray, *Can. J. Chem. Eng.*, **47**, 503 (1969).
26. R. Capinpin, Ph.D. thesis in chemical engineering, Stevens Institute of Technology (1975).

CHAPTER TWO

Reaction Phenomena

Kinetic variables of importance in polymerization and copolymerization reactions, in addition to amount of monomer and initiator consumed, are polymer composition and structure. Examples discussed earlier, some of which are experimentally measurable, are DP, DPD, DB, DBD, CC, and CCD. A critical test of any proposed reaction sequence is its ability to reliably predict the evolution of such properties during reaction.

Since the main objective of this book is to demonstrate by computation the influence of reactor dynamics on polymerization behavior and polymer properties, we will use representative, yet simple, reaction sequences that embody all the essential characteristics of linear (non-network) polymerizations of industrial interest. Such sequences, identified in Section 1.6, will serve in later chapters to illustrate the effects of physical reaction path on polymerizations and polymer properties and how such effects vary depending on the basic nature of the sequence in question. We begin in this chapter with an examination of their anatomy.

2.1. BASIC DP CHARACTERISTICS

Two reaction variables previously cited that are experimentally accessible are rate of polymerization r and average DP. Evolution of the latter variable with reaction time is more revealing than the former. As we will show, it vividly depicts the contrasting characteristics of particular polymerization types.

Graphs of mole fraction of monomer or functional groups converted, Φ, versus dimensionless time $\hat{t} \equiv t/\lambda_r$ have been plotted in Fig. 2.1-1 for first- and second-order reactions, where λ_r is a relaxation time in the former and a half-life in the latter. Such reaction orders are not uncommon for random and addition polymerizations, respectively. It is evident that these curves do not differ significantly from one another, even for such diverse propagation sequences as random and addition.

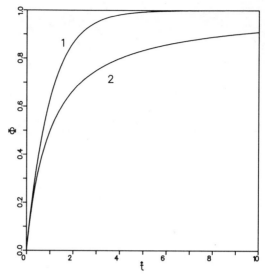

Figure 2.1-1. Conversion (Φ) versus dimensionless time for first-order (1), $\Phi = 1 - \exp(-\hat{t})$, and second-order (2), $\Phi = \hat{t}/(1 + \hat{t})$, rate functions.

Graphs of $\bar{x}_N$, on the other hand, differ markedly, as shown in Figs. 2.1-2–2.1-4. At the risk of oversimplification, these graphs were based on idealized polymerizations in which all side reactions were omitted. They are presented mainly for illustrative purposes and are only indicative of the basic features of actual polymerizations. Each pair of curves in Figs. 2.1-2 and 2.1-3 represents a different step polymerization ($\lambda_p \sim \lambda_r$). The abscissae of each pair are related by functions $\Phi(\hat{t})$,

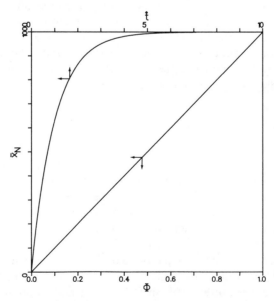

Figure 2.1-2. Degree of polymerization versus conversion, $\bar{x}_N = 1 + x_0\Phi$, and dimensionless time, $\bar{x}_N = 1 + x_0[1 - \exp(-\hat{t})]$, for step-addition polymerizations ($\lambda_p \sim \lambda_r$); Table 1.10-2 with $k_i = k_p$ or sequence 2.3-1.

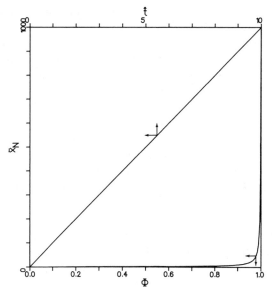

Figure 2.1-3. Degree of polymerization versus conversion, $\bar{x}_N = 1/(1 + \Phi)$, and dimensionless time, $\bar{x}_N = 1 + \hat{t}$, for random polymerizations ($\lambda_p \sim \lambda_r$); Table 1.9-1 with $k_r = 0$ or sequence 2.2-12.

which appear in Fig. 2.1-1; curve 1 corresponds to Fig. 2.1-2 and curve 2 corresponds to Fig. 2.1-3. We note that in the former case $\bar{x}_N$ is linear with conversion, while in the latter it is linear in time. This may be explained, at least qualitatively, by closer inspection of the corresponding propagation sequences.

For step-addition polymerizations without termination, it is reasonable to expect $\bar{x}_N$ to increase linearly with Φ, as shown in Fig. 2.1-2. A given incremental loss in the

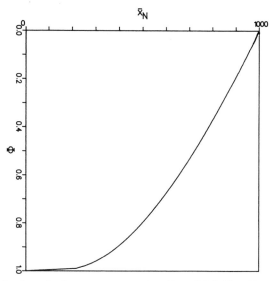

Figure 2.1-4. Degree of polymerization versus conversion, $\bar{x}_N = 100\Phi/\{1 - [1 + (1/20)\ln(1 - \Phi)]^2\}$, for chain-addition polymerizations ($\lambda_p \ll \lambda_r$); Table 1.10-3 with combination termination and without chain transfer.

number of moles of monomer at any level of conversion must produce a corresponding incremental increase in DP; that is $(d\bar{x}_N/d\Phi)$ is constant with Φ. This is a direct consequence of adding molecules to polymer chains one by one. Furthermore, since $-(dc_m/dt) \equiv r \propto c_m$ for addition polymerizations, it follows that $d\Phi/d\hat{t}$ decreases with decreasing c_m, unless the polymerization is autoacceleratory. Thus, with the aid of the identity $d\bar{x}_N/d\hat{t} = (d\bar{x}_N/d\Phi)(d\Phi/d\hat{t})$, we can also explain the similarity between the concave curve in Fig. 2.1-2 and curve 1 in Fig. 2.1-1.

In Fig. 2.1-3, however, a given increment of conversion at high levels results in a substantially greater incremental increase in $\bar{x}_N$ than at low levels. In fact, the average value of DP barely exceeds 1 until virtually 100-percent conversion, at which point it rises almost vertically to high values. Such behavior is compatible with random propagation. We associate with each increment of conversion the loss of a certain fraction of available functional groups. Since molecules of all sizes have identical functional groups, the loss of a given fraction of them when they are attached to large molecules, as is the case at high conversions, has a greater effect on raising $\bar{x}_N$ than when they are attached to smaller molecules, as at low conversions. Therefore, $d\bar{x}_N/d\Phi$ increases with Φ and becomes enormous as Φ approaches unity. Since Φ versus time is concave in shape, as shown in Fig. 2.1-1, it is reasonable, by the previous identity, to expect $\bar{x}_N$, which is convex with respect to conversion, to "straighten out" when plotted against time.

The polymerization illustrated in Fig. 2.1-4 is clearly distinct from either of those just described. Here DP is large from the outset and it may increase further with monomer conversion, or remain virtually constant or even decline, indicating that an extremum is possible. Such behavior can only be explained by postulating that those polymer chains whose length is reflected by $\bar{x}_N$ at high conversion are not the same molecules that determined $\bar{x}_N$ at low conversion, that is by admitting the existence of termination ($\lambda_p < \lambda_r$). The sudden rise in $\bar{x}_N$ to a high value, as shown in Fig. 2.1-4, is characteristic of chain reactions ($\lambda_p \ll \lambda_r$). Chain-addition polymerizations will be discussed in greater detail later in this chapter.

2.2. THE MOST PROBABLE DPD

A ubiquitous DPD among polymers of all kinds is the most probable distribution (Appendix B). To understand why, we shall explore the physical origin of this important distribution from two viewpoints, thermodynamic and kinetic. We begin by deducing that the equilibrium DPD is most probable for two diverse step polymerizations, that is, random

$$x, y \geqslant 1 \qquad\qquad m_x + m_y \overset{K}{\rightleftharpoons} m_{x+y} \qquad\qquad (1.8\text{-}16)$$

and addition

$$m_0 + m \overset{K_i}{\rightleftharpoons} m_1 \qquad\qquad (1.8\text{-}17)$$

$$x \geqslant 1 \qquad\qquad m_x + m \overset{K_p}{\rightleftharpoons} m_{x+1} \qquad\qquad (1.8\text{-}18)$$

2.2.1. Equilibrium Random Polymerization

Successive application of the equilibrium constant, equation 1.8-25, to each reaction in equation 1.8-16, ignoring m_0 for the moment, leads to the general expression for x-mer concentration, $(c_x)_e = K^{x-1}(c_1)_e^x$, from which we immediately obtain the most probable DPD (Table 1.5-4)

$$(n_x)_e \equiv \frac{(c_x)_e}{\sum\limits_{x \geqslant 1} (c_x)_e} = \phi_e^{x-1}(1 - \phi_e) \tag{2.2-1}$$

after summation, where $\phi_e \equiv K(c_1)_e$

The physical interpretation of ϕ as the fraction of reacted end groups, defined earlier as Φ, can be deduced from the definition of Φ, equation 1.7-3,

$$\Phi_e \equiv 1 - \frac{\sum\limits_x (c_x)_e}{\sum\limits_x x(c_x)_e} = 1 - \frac{(c_p)_e}{c_{RU}} \leqslant 1 \tag{2.2-2}$$

assuming that volume (density) changes are negligible and noting that the concentration of repeat units $c_{RU} \equiv \sum_{x=1}^{\infty} xc_x$, remains constant because monomer is counted as 1-mer. Verification of equation 2.2-1 and the assertion that Φ and ϕ are identical requires substitution of the equation for $(c_x)_e$ into the definitions of n_x and Φ followed by summation and is left as an exercise for the reader. Another useful result, which follows directly from equation 2.2-1, is

$$(\bar{x}_N)_e \equiv \frac{1}{1 - \Phi_e} \tag{2.2-3}$$

As previously mentioned, the requisite summations are actually finite (Table C-3), the largest value of x being, say, X, beyond which $c_x = 0$ for all $x \geqslant X$ on physical grounds. Nevertheless, we shall use infinite series $\sum_{x=1}^{\infty}$ for convenience, noting that their convergence requires that $\Phi < 1$ (Appendix C), which is guaranteed by inequality 2.2-2.

It can also be shown (Section 1.9) that

$$K = \frac{c_{RU} - (c_p)_e}{(c_p)_e^2} \tag{2.2-4}$$

By using this expression to eliminate $(c_p)_e$ from equation 2.2-2 and solving the resulting quadratic equation (2.2-37) for Φ_e, we obtain the following expression for equilibrium conversion:

$$\Phi_e = 1 - \frac{(1 + 4Kc_{RU})^{1/2} - 1}{2Kc_{RU}} \tag{2.2-5}$$

These results and the foregoing analysis also apply to random polymerizations with by-products, such as m_0 in reaction 1.8-16. In this case, we simply replace K with $K/(c_0)_e$ in the preceding equations. This is shown in Example 2.2-1.

Example 2.2-1.   Equilibrium Interchange with By-Product.   For the interchange polymerization leading to PET described in Example 1.9-2, in which the by-product m_0 is ethylene glycol (EG), derive equations that describe equilibrium conversion, Φ_e, and DP, $(\bar{x}_N)_e$, in terms of EG concentration, $(c_0)_e$; and shown how both quantities increase as EG is removed by devolatilization.

Solution.   By analogy with equation 2.2-4 (cf. Section 1.9)

$$K \equiv (c_0)_e \left[c_{RU} - (c_p)_e \right] \Big/ (c_p)_e^2$$

After combining this definition with equation 2.2-2 to eliminate c_p, and solving the resulting quadratic equation for Φ_e,

$$\frac{Kc_{RU}}{(c_0)_e} = \frac{\Phi_e}{(1 - \Phi_e)^2}$$

we obtain

$$\Phi_e = 1 - \frac{\left[1 + \dfrac{4K}{(\hat{c}_0)_e} \right]^{1/2} - 1}{2K/(\hat{c}_0)_e}$$

where dimensionless EG concentration has been defined as $\hat{c}_0 \equiv c_0/c_{RU}$. It is noteworthy that this equation follows directly from equation 2.2-5 after replacing K with $K/(c_0)_e$ as suggested in the text.

It is evident that Φ_e increases as $(c_0)_e$ is reduced with c_{RU} remaining constant. Furthermore, from equation 2.2-5, $(\bar{x}_N)_e = (1 - \Phi_e)^{-1}$, we conclude that equilibrium DP increases as well.

2.2.2.   Equilibrium Addition Polymerization

Application of the equilibrium constants for propagation and initiation, equations 1.8-26 and 28, respectively, to reactions 1.8-17 and 18 leads directly to the linear, homogeneous difference equation in x

$$x \geqslant 1 \qquad\qquad (c_{x+1})_e - K_p (c_m)_e (c_x)_e = 0 \qquad\qquad (2.2\text{-}6)$$

with constant coefficient $K_p(c_m)_e$, and its initial condition (IC)

$$(c_1)_e = K_i (c_0)_e (c_m)_e \qquad\qquad (2.2\text{-}7)$$

The solution of this equation (Appendix C)

$$(c_x)_e = K_i K_p^{x-1} (c_m)_e^x (c_0)_e \qquad\qquad (2.2\text{-}8)$$

also yields the most probable DPD,

$$(n_x)_e \equiv \frac{(c_x)_e}{\sum_x (c_x)_e} = \phi_e^{x-1}(1 - \phi_e) \qquad (2.2\text{-}9)$$

where this time

$$\phi_e \equiv K_p(c_m)_e \qquad (2.2\text{-}10)$$

Again, $\Phi_e < 1$ (Example 2.2-2) thus permitting the use of infinite series. Obviously, then, $(\bar{x}_N)_e$ is given by equation 2.2-3 for this reaction too. Furthermore, since this simple addition reaction conforms to stoichiometry 1.7-5 with $s = 1$ and $R = 0$, it follows from equation 1.7-11 ($f = 1$) that

$$(\bar{x}_N)_e = \frac{\Delta c_m}{\Delta c_0} \qquad (2.2\text{-}11)$$

when volume (density) changes are neglected. For the relation between equilibrium Φ_e and ϕ_e, see Example 2.2-3.

The most probable distribution can be shown to be the equilibrium DPD for numerous other sequences (Examples 1.2-3 and 2.2-4). As the name implies, it appears to be the equilibrium distribution for all linear polymers, assuming, of course, that sufficient time is available (Section 2.3) for the establishment of true chemical equilibrium.

Example 2.2-2. Addition Equilibrium Parameter ϕ_e. Using the rate functions in Table 1.10-1 applied to equilibrium conditions, show that $\phi_e < 1$ where $K_p \equiv k_{pf}/k_{pr}$.

Solution. From Table 1.10-1, we write for the initiation rate

$$r_i = k_{if}c_m c_0 - k_{ir}c_1$$

and for the monomer consumption (polymerization) rate

$$r = r_i + k_{pf}c_m \sum_{x=1}^{\infty} c_x - k_{pr} \sum_{x=2}^{\infty} c_x$$

At equilibrium, $r_i = 0 = r$, therefore

$$\frac{k_{pf}}{k_{pr}} \equiv K_p = \frac{\sum_{x=2}^{\infty} (c_x)_e}{\sum_{x=1}^{\infty} (c_x)_e (c_m)_e}$$

But since $\sum_{x=1}^{\infty}(c_x)_e > \sum_{x=2}^{\infty}(c_x)_e$, it follows that $K_p(c_m)_e < 1$.

Example 2.2-3.   *Addition Equilibrium Conversion* Φ_e.   Find the relationship between Φ_e and ϕ_e for the sequence

$$m_0 + m \overset{K_i}{\rightleftharpoons} m_1$$

$$m_x + m \overset{K_p}{\rightleftharpoons} m_{x+1}$$

starting with $(c_x)_e = [K_i(c_0)_e/K_p]\phi_e$, where $\phi_e \equiv K_p(c_m)_e < 1$ and

$$\Phi \equiv 1 - \frac{(c_m)_e}{(c_m)_0} = \frac{(c_{RU})_e}{(c_m)_0}$$

Solution.   Solving constraint equations (Section 1.7)

$$(c_p)_e + (c_0)_e = \sum_{x=1}^{\infty} (c_x)_e = (c_0)_0$$

and

$$(c_{R\dot{U}})_e + (c_m)_e = \sum_{x=1}^{\infty} x(c_x)_e = (c_m)_0$$

simultaneously and eliminating $(c_0)_0$, we obtain

$$\Phi_e = \frac{\dfrac{a'_K \phi_e}{(1 + a'_K \phi_e - \phi_e)}(\bar{x}_N)_e}{x_0}$$

where

$$a'_K \equiv \frac{K_i}{K_p}$$

and

$$(\bar{x}_N)_e = \frac{1}{1 - \phi_e}$$

When $a'_K = 1$ and $(\bar{x}_N)_e = x_0$, we conclude that $\Phi_e = \phi_e$.

Example 2.2-4.   *Random Propagation and Interchange Equilibrium*.   Show that the following random propagation and interchange reactions

$$m_x + m_y \overset{K}{\rightleftharpoons} m_{x+y} + m_0$$

$$m_x + m_y \overset{K'}{\rightleftharpoons} m_{x+z} + m_{y-z}$$

are not independent of reactions 1.8-17 and 1.8-18, and also that

$$K = \frac{K_p}{K_i}$$

$$K' = 1$$

In this way we will have shown that the equilibrium composition (DPD) of reactions 1.8-17 and 1.8-18 is not altered by the presence of the preceding side reactions. A polymerization in which all four reactions could occur simultaneously is the addition propagation of caprolactam ring-type monomer (m) in the presence of water (m_0) to AB-type polyamide (m_x), which could propagate randomly and participate in amide interchange as well.

Solution. It can be shown that all four reactions in question are linear combinations of the following formation reaction

$$m_0 + xm = m_x$$

The proof is left as an exercise for the reader. Thus, the preceding side reactions are not independent of reactions 1.8-17 and 1.8-18. Furthermore, from equation 2.2-8, we see that the equilibrium constant of the formation reaction is

$$\frac{(c_x)_e}{(c_m)_e^x (c_0)_e} = \frac{K_i K_p^x}{K_p}$$

After substituting this result into the definitions of the equilibrium constants for the side reactions

$$K \equiv \frac{c_{x+y} c_0}{c_x c_y}$$

$$K' \equiv \frac{c_{x+z} c_{y-z}}{c_x c_y}$$

we obtain the desired relationships between these constants and K_i and K_p.

2.2.3. Irreversible Random Polymerization

The rate equations for simple irreversible random sequence

$$x, y \geq 1 \qquad\qquad m_x + m_y \xrightarrow{k} m_{x+y} \qquad\qquad (2.2\text{-}12)$$

in a batch reactor, neglecting volume (density) changes, are (Table 1.9-1):

$$x \geq 1 \qquad\qquad \frac{dc_x}{dt} = r_x = -2k c_x c_p + k \sum_{j=0}^{x} c_j c_{x-j} \qquad\qquad (2.2\text{-}13)$$

and

$$-\frac{dc_p}{dt} = r = k c_p^2 \qquad\qquad (2.2\text{-}14)$$

and the constraint equation for all $t \geq 0$, which reflects the dependence of equation 2.2-14 on 2.2-13, is

$$\sum_{x=0}^{\infty} c_x = c_p \qquad\qquad (2.2.15)$$

A natural CT for this second-order polymerization rate equation is (Appendix G)

$$\lambda_r = \frac{1}{k(c_1)_0} \qquad\qquad (2.2\text{-}16)$$

It is convenient to reduce time with λ_r ($\hat{t} \equiv t/\lambda_r$) and to reduce all concentrations with $(c_1)_0$, $\hat{c} \equiv c/(c_1)_0$. In this way, the dimensionless version of equation 2.2-13 may be divided by the dimensionless rate of polymerization

$$-\frac{d\hat{c}_p}{d\hat{t}} = \hat{c}_p^2 \tag{2.2-17}$$

to give the following equations:

$$x \geqslant 1 \qquad \frac{d\hat{c}_x}{d\hat{c}_p} - \frac{2\hat{c}_x}{\hat{c}_p} = -\sum_{j=0}^{x} \frac{\hat{c}_j \hat{c}_{x-j}}{\hat{c}_p^2} \tag{2.2-18}$$

which may be solved with IC's

$$(\hat{c}_x)_0 = \begin{cases} 1 & \text{when } x = 1 \\ 0 & \text{when } x > 1 \end{cases} \tag{2.2-19}$$

by successive application of the integrating factor $\exp(-\int 2\, d\hat{c}_p/\hat{c}_p) = \hat{c}_p^{-2}$. The result is a family of solutions whose general form is

$$\hat{c}_x = \left(1 - \hat{c}_p\right)^{x-1} \hat{c}_p^2 \tag{2.2-20}$$

Substitution into equation 2.2-18 verifies this solution. Thus, after evaluating the convolution sum on the RHS, we obtain (1)

$$x \geqslant 1 \qquad \frac{d\hat{c}_x}{d\hat{c}_p} - \frac{2\hat{c}_x}{\hat{c}_p} = (1 - x)\left(1 - \hat{c}_p\right)^{x-2} \hat{c}_p^2 \tag{2.2-21}$$

The rest is left as an exercise for the interested reader.

Substitution of equation 2.2-20 into the definition of mole fraction and summing yields the familiar result

$$x \geqslant 1 \qquad n_x \equiv \frac{\hat{c}_x}{\sum_x \hat{c}_x} = \phi^{x-1}(1 - \phi) \tag{2.2-22}$$

where

$$\phi \equiv 1 - \hat{c}_p \tag{2.2-23}$$

from which we conclude that for random polymerizations the DPD is most probable, not only at equilibrium but during irreversible propagation as well.

Solution of the rate equation for polymerization, equation 2.2-14, leads to an expression for conversion of end groups with time

$$\Phi \equiv 1 - \hat{c}_p = \frac{\hat{t}}{1 + \hat{t}} \tag{2.2-24}$$

which has been plotted in Fig. 2.1-1. Since $\Phi = \phi$, we conclude that $\bar{x}_N$ grows with conversion and time according to the function (equation 2.2-3)

$$\bar{x}_N = \frac{1}{1 - \phi} = 1 + \hat{t} \tag{2.2-25}$$

which has been plotted in Fig. 2.1-3.

It is clear from equation 2.2-24 that λ_r represents the half-life for this reaction, as we would expect from its second-order rate equation. Supposing that $\lambda_r = 10^4$ s and $(c_1)_0 = 1$ mol/l, we obtain an order-of-magnitude value for k of 10^{-4} 1/mol s. Catalyzed condensation polymerizations with by-product m_0 are frequently far more rapid than this (Example 2.2-6). In fact, it is not uncommon for the polymerization rate to be controlled by the rate of removal of m_0 by devolatilization. In such cases, the reverse reaction is important and equilibrium is approached at conversions that are insufficient to produce high-molecular-weight product. Removal of m_0 drives the forward reaction toward higher conversions. The reader will recall that random polymers attain high molecular weights only near complete conversion (Example 2.2-5).

The expressions for weight fraction DPD, weight average DP, and dispersion indices D_N and D_W for the most probable distribution are listed in Table 1.5-4, and the expressions for the moments in Table 1.5-1. The evolution of the DPD with conversion is shown in Figs. 2.2-1 and 2.2-2. We observe that dispersion increases with conversion and that the number DPD shows no maximum. This indicates that 1-mer is the most abundant component and thereby contributes to dispersion. These observations are consistent with the computations of Example 2.2-3, from which we conclude that both average DP and DPD increase steadily with conversion, and that D_N and D_W approach values of 2 and 1.5, respectively, near complete conversion. The higher value of D_N reflects the presence of low-DP chains.

Example 2.2-5. Property Evolution of Random Polymers. Compute $\bar{x}_N$, $\bar{x}_W$, D_N, and D_W at various conversions for the most probable distribution, as it applies to random polymerization.

The results tabulated below were obtained from the equations in Table 1.5-4 for arbitrary values of conversion. Corresponding values for dimensionless time pertain

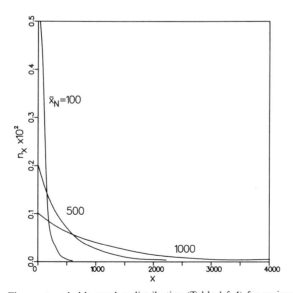

Figure 2.2-1. The most probable number distribution (Table 1.5-4) for various values of $\bar{x}_N$.

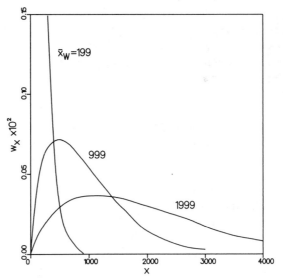

Figure 2.2-2. The corresponding weight distribution (Table 1.5-4) for various values of $\bar{x}_W$.

to irreversible polymerization only and were computed from equation 2.2-24. As expected, the DP remains low throughout most of the reaction and rises suddenly at high conversions.

	$\hat{t}$	Φ or ϕ	$\bar{x}_N$	$\bar{x}_W$	D_N	D_W
	0	0	1	1	1	1
	0.33	0.25	1.33	1.67	1.25	1.32
half-life →	1	0.50	2	3	1.50	1.44
	3	0.75	4	7	1.75	1.49
	9	0.90	10	19	1.90	1.50
	99	0.99	100	199	1.99	1.50
	999	0.999	1000	1999	1.999	1.50

2.2.4.   *Reversible Random Polymerization*

From equations 1.8-4 and 1.8-7 the rate equations for the simplest case of reversible random polymerization

$$\text{m}_x + \text{m}_y \underset{k_r}{\overset{k_f}{\rightleftarrows}} \text{m}_{x+y} \tag{1.8-1}$$

are

$$x \geqslant 1 \quad \frac{dc_x}{dt} = -2k_f c_x c_p + k_f \sum_{j=0}^{x} c_j c_{x-j} - k_r(x-1)c_x + 2k_r \sum_{j=x+1}^{\infty} c_j \tag{2.2-26}$$

and

$$-\frac{dc_{\mathrm{p}}}{dt} = k_f c_{\mathrm{p}}^2 - k_r\big[(c_1)_0 - c_{\mathrm{p}}\big] \tag{2.2-27}$$

Blatz and Tobolsky (2) have shown that these equations are satisfied by the solutions of the irreversible sequence

$$\hat{c}_x = \phi^{x-1}(1-\phi)^2 \tag{2.2-28}$$

and

$$\hat{c}_{\mathrm{p}} = 1 - \phi \tag{2.2-29}$$

subject to the same IC's (equations 2.2-19), only now ϕ depends on time according to the following function (Example 2.2-7) in lieu of 2.2-24

$$\phi = \frac{2\hat{K}}{(2\hat{K}+1) + (4\hat{K}+1)^{1/2}\coth\left[\dfrac{(4\hat{K}+1)^{1/2}\hat{t}}{2\hat{K}}\right]} \tag{2.2-30}$$

where $\hat{K}$ is a dimensionless constant, defined as

$$\hat{K} \equiv \frac{k_f(c_1)_0}{k_r} \tag{2.2-31}$$

Thus we conclude that polymer resulting from random polymerization of pure monomer has a most probable DPD throughout its formation, whether it is far from equilibrium or at equilibrium.

Since equation 2.2-30 claims to represent the general solution for random polymerization, it must reduce to the irreversible solution when $k_r \to 0$ and the equilibrium solution when $t \to \infty$. To demonstrate the former, we observe that when $\hat{K} \to \infty$, equation 2.2-30 may be approximated by the expression

$$\phi \approx \frac{1}{1 + \hat{K}^{-1/2}\coth\left(\dfrac{\hat{t}}{\hat{K}^{1/2}}\right)} \tag{2.2-32}$$

which approaches the function 2.2-24 because for small values of $\hat{t}/\hat{K}^{1/2}$

$$\coth\frac{\hat{t}}{\hat{K}^{1/2}} \approx \frac{\hat{K}^{1/2}}{\hat{t}} \tag{2.2-33}$$

To verify the equilibrium limit of equation 2.2-30

$$\phi_\infty \equiv \phi(\hat{t} \to \infty) = \frac{2\hat{K}}{(2\hat{K}+1) + (4\hat{K}+1)^{1/2}} \tag{2.2-34}$$

we need an expression for ϕ_e in terms of the equilibrium constant with which to compare it. Such an expression may be obtained from equations 2.2-1 and 2.2-2

$$\phi_e = 1 - \frac{(c_p)_e}{(c_1)_0} \tag{2.2-35}$$

and

$$(n_1)_e = \frac{(c_1)_e}{(c_p)_e} = 1 - \phi_e \tag{2.2-36}$$

where $\phi_e = K(c_1)_e$, by combining them to eliminate the quantities $(c_1)_e$ and $(c_p)_e$. The result is a quadratic equation in ϕ_e

$$\hat{K}\phi_e^2 - (2\hat{K} + 1)\phi_e + \hat{K} = 0 \tag{2.2-37}$$

where $\hat{K}$ is a dimensionless equilibrium constant, $\hat{K} \equiv K(c_1)_0$, which corresponds to the one defined in equation 2.2-31. The solution of equation 2.2-37 (cf. equation 2.2-5)

$$\phi_e = \frac{(2\hat{K} + 1) - (4\hat{K} + 1)^{1/2}}{2\hat{K}} \tag{2.2-38}$$

may easily be shown to be identical to equation 2.2-34 by proving that the product $\phi_e^{-1}\phi_\infty$ is equal to unity. The corresponding expression for final DP is

$$(\bar{x}_N)_\infty = \frac{2\hat{K}}{(4\hat{K} + 1)^{1/2} - 1} \tag{2.2-39}$$

When the forward reaction is favored ($\hat{K} \gg 1$), this becomes $(\bar{x}_N)_\infty \cong \hat{K}^{1/2}$, and when degradation is favored ($\hat{K} \ll 1$), $(\bar{x}_N)_\infty \to 1$, as expected.

Graphs of ϕ and $\bar{x}_N = 1/(1 - \phi)$ versus $\log \hat{t}$ with $\hat{K}$ as a parameter have been plotted in Figs. 2.2-3 and 4. We see that equilibrium conditions are achieved in one to ten time constants. Even when the CT for the reverse reaction ($1/k_r$) is fifty times greater than that for the forward reaction $[1/k_f(c_1)_0]$, the effect on $\bar{x}_N$ is significant.

From the standpoint of DP growth with time for various values of $\hat{K}$, it is again confirmed in the figures that values of $\hat{K} > 10^4$ are required if DP's > 100 are to be achieved. A small amount of reversibility, say $\hat{K} = 50$, limits $\bar{x}_N$ to a value less than 10, which underscores the importance of suppressing the reverse reaction. Although our idealized sequence contains no by-product m_0 and therefore does not facilitate its suppression, most real reactions would. Continuous removal of this by-product is essential to ensure large DP's. Examples 2.2-1 and 2.2-6 consider the catalyzed interchange leading to PET, in which the reverse reaction and by-product m_0 are factors.

The method of moments (Appendix C) is convenient for obtaining expressions for average properties from component balances. The moment balances for sequence 1.8-1 can be found by summing the product of x^k with equation 2.2-26 and using the relations between various summations and the moments of the DPD given in

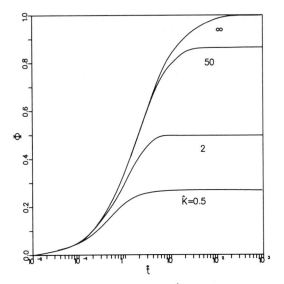

Figure 2.2-3. Conversion (Φ) versus dimensionless time, $\hat{t} = t/k_f (c_1)_0$ for simple reversible random polymerization (sequence 1.8-1) with various degree of reversibility (2).

Appendix B. The results are listed in Table 2.2-1 and applied in Example 2.2-7. Note that the balances for second and higher moments each depend on the next higher moment, and the latter appears as a result of the reverse reaction.

Example 2.2-6. Rate of Interchange with By-Product. Starting with the appropriate rate function for the formation of ethylene glycol in the catalyzed interchange formation of PET at 270°C (Example 1.9-2), and using the values of

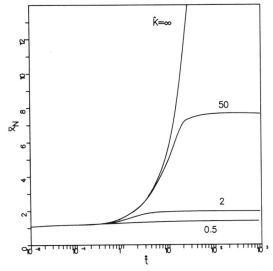

Figure 2.2-4. Degree of polymerization ($\bar{x}_N$) versus dimensionless time, $\hat{t} = t/k_f(c_1)_0$ for simple reversible random polymerization (sequence 1.8-1) with various degrees of reversibility (2).

Table 2.2-1. Moment Balances for Simple Reversible Random Polymerization (Table 1.9-1)

$$\frac{d\mu_c^0}{dt} = -k_f(\mu_c^0)^2 + k_r(\mu_c^1 - \mu_c^0)$$

$$\frac{d\mu_c^1}{dt} = 0$$

$$\frac{d\mu_c^2}{dt} = 2k_f(\mu_c^1)^2 - \frac{1}{3}k_r(\mu_c^3 - \mu_c^1)$$

$$\frac{d\mu_c^3}{dt} = 6k_f\mu_c^1\mu_c^2 - \frac{1}{2}k_r(\mu_c^4 - 4\mu_c^3 + \mu_c^2)$$

where

$$\mu_c^k \equiv \sum_{x=1}^{\infty} x^k c_x$$

$$c_p = \mu_c^0$$

and

$$c_{RU} = \mu_c^1$$

$k_f' \equiv k_f(c_1)_0 = k_f c_{RU} = 0.05 \text{ s}^{-1}$ and $\rho = 1.16$ g/cm^3 reported by Pell and Davis (3) and Stevenson and Schumann (4), respectively, compute k_f in 1/mol RU s and estimate the CT for polymerization, λ_r, in seconds. How does k_f compare with the values 0.01 (5) and 0.0011 (6) reported elsewhere in the literature?

Solution. Noting that for PET, $(C_{10}H_8O_4)_x$,

$$M_0 = 192 \text{ g RU/mol RU}$$

and

$$\rho = 1.16 \text{ g RU/cm}^3$$

we can immediately write

$$c_{RU} = \frac{(1.16)(1000)}{192} = 6.04 \text{ moles RU/1}$$

and therefore

$$k_f = \frac{0.05}{6.04} = 0.083 \text{ 1/mol s}$$

Concerning λ_r, we follow the procedure for finding CT's described in Appendix G and write the rate equation (BR) for m_0 in semidimensionless form

$$\frac{d\hat{c}_0}{dt} = \frac{\hat{r}_0}{\lambda_r}$$

where $\hat{c}_0 \equiv c_0/c_{RU}$, $\hat{r}_0 \equiv r_0/(r_0)_0$, and (Example 1.9-2)

$$r_0 = 4k_f c_p^2 - 4k_r c_0(c_{RU} - c_p)$$

Thus,

$$\lambda_r = \frac{c_{RU}}{(r_0)_0} = \frac{1}{4k_f(c_1)_0} = 5 \text{ s}$$

which indicates a very rapid reaction.

Example 2.2-7. Application of the Method of Moments. Using the method of moments (Appendix C), solve the kinetic equations for reversible random polymerization to verify the expression for $\bar{x}_N$ given by Blatz and Tobolsky (2).

Solution. Equation 2.2-27 is found by summing 2.2-26 over all x (recall $c_p = \mu_c^0$). Multiplying 2.2-26 by x and summing over x would yield $d\mu_c^1/dt$ and confirm that it is zero; thus the first moment remains constant throughout the reaction, $\mu_c^1 \equiv (c_1)_0$, and

$$\frac{d\mu_c^0}{dt} = k_f(\mu_c^0)^2 - k_r(\mu_c^1 - \mu_c^0)$$

Let

$$\hat{t} \equiv \frac{t}{k_f(c_1)_0} \qquad \hat{\mu}^0 \equiv \frac{\mu_c^0}{(c_1)_0} = \bar{x}_N^{-1}$$

and

$$\hat{K} = \frac{k_f(c_1)_0}{k_r}$$

Then

$$\frac{d\hat{\mu}^0}{d\hat{t}} = (\hat{\mu}^0)^2 - \frac{1}{\hat{K}}(1 - \hat{\mu}^0)$$

Thus

$$\int_1^{\bar{x}_N^{-1}} \frac{d\hat{\mu}^0}{(\hat{\mu}^0)^2 + \hat{\mu}^0 \hat{K}^{-1} - \hat{K}^{-1}} = \hat{t}$$

By expanding the denominator of the integrand via partial fractions and then integrating, we find:

$$\ln\left[\left(\frac{\bar{x}_N^{-1} - R_2}{\bar{x}_N^{-1} - R_1}\right)\left(\frac{1 - R_1}{1 - R_2}\right)\right] = (R_2 - R_1)\hat{t}$$

where R_1, R_2 are the roots of the quadratic in the denominator of the integrand. Solving for $\bar{x}_N$ yields

$$\bar{x}_N = \frac{(1 - R_1) - (1 - R_2)\exp(R_2 - R_1)\hat{t}}{R_2(1 - R_1) - R_1(1 - R_2)\exp(R_2 - R_1)\hat{t}}$$

and after simplification

$$\bar{x}_N = \frac{1 + 2\hat{K} + (4\hat{K} + 1)^{1/2}\coth\left\{\dfrac{(4\hat{K} + 1)^{1/2}\hat{t}}{2\hat{K}}\right\}}{1 + (4\hat{K} + 1)^{1/2}\coth\left\{\dfrac{(4\hat{K} + 1)^{1/2}\hat{t}}{2\hat{K}}\right\}}$$

which is consistent with equation 2.2-30 and $\bar{x}_N = (1 - \phi)^{-1}$

2.2.5. *Effect of Feed Composition*

Heretofore we have considered primarily the random polymerization of pure monomer. We shall now examine the more general sequence

$$x, y \geqslant 1 \qquad\qquad m_x + m_y \underset{k_r}{\overset{k_f}{\rightleftarrows}} m_{x+y} + m_0 \qquad\qquad (1.9\text{-}1)$$

$$x, y \geqslant 1, y > z \qquad\qquad m_x + m_y \underset{k}{\overset{k}{\rightleftarrows}} m_{x+z} + m_{y-z} \qquad\qquad (1.9\text{-}4)$$

in piecemeal fashion, giving emphasis to its action on feeds other than pure monomer, for example, polymers having DPDs other than those that are most probable and comonomers such as AA and BB with unequivalent stoichiometric proportions.

Kilkson (7) has shown that simple irreversible sequences such as in equation 2.2-12 (1.9-1 with $k_r = 0 = m_0$), will always yield the most probable DPD throughout reaction, providing that the feed is either pure monomer or polymer whose DPD is already most probable. If the feed DPD is not most probable, then the product DPD will become most probable only when conversion is complete. More specifically, if average DP and dispersion index of the feed are $(\bar{x}_N)_0$ and $(D_N)_0$, respectively, then the corresponding quantities at any conversion level ϕ are given by the first two equations in Table 2.2-2. As expected, these results reduce to our earlier expressions for the special case of pure monomer feed $(\bar{x}_N)_0 = 1 = (D_N)_0$. Finding an analytical expression for the entire DPD during polymerization starting with feed of arbitrary DPD is difficult, if not impossible. However, it is possible via the method of moments (Appendix C) to relate the kth moment of the product distribution to the kth and lower moments of the feed distribution. The results in Table 2.2-2 were obtained in this manner. For further details, the reader is referred to the literature.

Kilkson (7) also concluded that polymerization in the presence of small amounts of chain stopper, m_{cs}, would yield a lower average DP and a DPD essentially the same as before. Expressions for conversion with time ($\hat{t}$ defined as before) and ultimate average DP's appear in Table 2.2-2. As an example of their application, when the initial chain stopper–monomer ratio is $(\hat{c}_{cs})_0 = 0.005$, we obtain after reaction is complete a polymer having a most probable DPD with $(\bar{x}_N)_\infty = 200$ and $(\bar{x}_W)_\infty = 400$.

Random degradation

$$m_0 + m_{x+y} \xrightarrow{k_r} m_x + m_y \tag{2.2-40}$$

Table 2.2-2. Effect of Feed Composition upon Irreversible Random Polymerization (sequence 2.2-12)

WITH POLYMERIC FEED

$$\bar{x}_N = (\bar{x}_N)_0 / (1 - \phi)$$

$$D_N = (D_N)_0 (1 - \phi) + 2\phi$$

WITH CHAIN STOPPER

$$\Phi = \frac{1}{1 + (\hat{c}_{cs})_0} + \frac{(c_{cs})_0}{[1 + (\hat{c}_{cs})_0]^2 \exp[(\hat{c}_{cs})_0 \hat{t}] - (\hat{c}_{cs})_0 - 1]}$$

where:

$$\Phi \equiv \frac{(c_1)_0 - c_p}{(c_1)_0 + (c_{cs})_0} = \frac{1 - \hat{c}_p}{1 + (\hat{c}_{cs})_0}$$

$$(\phi)_\infty = \frac{1}{1 + (\hat{c}_{cs})_0}$$

$$(\bar{x}_N)_\infty = \frac{1 + (\hat{c}_{cs})_0}{(\hat{c}_{cs})_0}$$

$$(\bar{x}_W)_\infty = \frac{2 + (\hat{c}_{cs})_0}{(\hat{c}_{cs})_0}$$

is the reverse reaction of sequence 1.9-1. Tobolsky (8) derived the expressions in Table 2.2-3 for DPD at any time in terms of initial DPD $(n_x)_0$. Montroll and Simha (9) reported similar results for degradation of monodisperse polymer using statistical arguments. Simha (10) later analyzed this special case kinetically.

Tobolsky concluded that infinite chains degrade randomly to the most probable distribution and retain it, irrespective of their initial distribution. On the other hand, if the initial distribution is already most probable, even if the mean chain length is low, it will remain most probable in the face of random degradation. Obviously, these results are not limited to condensation polymers, but they do require that degradation be of type 2.2-40 (random).

In random interchange, sequence 1.9-4, the total number of polymer molecules is conserved (Table 1.9-2) as well as the total number of repeat units. It follows that $\bar{x}_N$ remains constant and $\bar{x}_W$ changes until the most probable distribution is attained. Example 2.2-8 illustrates this point using an initially bimodal distribution. This sequence, whose rate function is given by equation 1.9-10, has been treated by Hermans (11), who showed that while the equilibrium DPD must always be most probable, intermediate distributions need not be. Approximate expressions for DPDs evolving with time when the feed is monodisperse polymer with a large value of $\bar{x}_N$ are listed in Table 2.2-4. One conclusion from these results is that the larger molecules reach their equilibrium concentrations more rapidly than do the smaller ones.

Until now we have only examined sequences that apply to A-B condensation and interchange polymerizations. Another important condensation polymerization is the AA-BB type, which again produces polymer having a most probable DPD throughout its formation from monomer, but only if the feed consists of precisely equivalent molar quantities of comonomers m_0^{AA} and m_0^{BB}. In this case, we can interpret ϕ as the fraction of either end group reacted. However, in the absence of such stoichiometric equivalence, it is not possible to obtain high-molecular-weight polymer. This is so because the number-average DP of the product is extremely sensitive to the ratio of

Table 2.2-3.   Random Degradation (sequence 2.2-40)

$$n_x(t) = (n_x)_0 \, \alpha_r^{x-1} + \sum_{j-x+1}^{\infty} (n_j)_0 [(j-x-1)\alpha_r^{x+1} - 2(j-x)\alpha_r^x + (j-x+1)\alpha_r^{x-1}]$$

where

$$\alpha_r = \exp(-k_r t)$$

$$\bar{x}_N(t) = 1/[1 - (1-(x_N)_0^{-1})\alpha_r]$$

$$\bar{x}_W(t) = 1 + 2 \sum_j (n_j)_0 \alpha_r^{j+1}/(1 - \alpha_r)^2 \sum_j j(n_j)_0$$

$$+ 2\alpha_r/(1 - \alpha_r) - 2\alpha_r/(\bar{x}_N)_0$$

Table 2.2-4. Simple Random Interchange (sequence 1.9-4) of Monodisperse Feed with Large DP ($\bar{x}_N$)

$x \neq \bar{x}_N$	$n_x \cong (n_x)_e [1 - \alpha^{x-1+\bar{x}_N}]$
$x = \bar{x}_N$	$n_x \cong (n_x)_e + [1 - (n_x)_e]\alpha^{2\bar{x}_N-1}$

where:

$$\alpha \equiv \exp(-kt)$$

and

$$(n_x)_e \quad \text{is most probable}$$

$$\bar{x}_w/\bar{x}_N = 1 + \gamma$$

where

$$\gamma = \tanh[2(\bar{x}_N - 1)t/3]$$

or

$$\bar{x}_w/\bar{x}_N \cong 2/\left\{1 + \exp[-4(\bar{x}_N - 1)t/3]\right\}$$

comonomers in the feed.

$$R_0 \equiv \frac{\left(N_0^{AA}\right)_0}{\left(N_0^{BB}\right)_0} \qquad (2.2\text{-}41)$$

To demonstrate this sensitivity, we invoke the basic definition of $\bar{x}_N$, that is the number of repeat units in the polymer divided by the number of polymer molecules, and we count unreacted monomer as polymer (1-mer) as is customary with condensation polymers. Thus the total number of repeat units is $[(N_0^{AA})_0 + (N_0^{BB})_0]/2$ and, if we let ϕ be the fraction of reacted A groups, then the total number of polymer molecules is $[(N_0^{AA})_0(1 - \phi) + (N_0^{BB})_0 - \phi(N_0^{AA})_0]/2$, since the term in brackets is equivalent to the number of unreacted end groups of both kinds. Consequently, we obtain

$$\bar{x}_N = \frac{1}{1 - \dfrac{2R_0\phi}{(1 + R_0)}} \qquad (2.2\text{-}42)$$

which reduces the familiar special case for stoichiometric equivalence $R_0 = 1$. It is easy to show that $d\bar{x}_N/dR_0 > 0$ when $R_0 < 1$ so that $\bar{x}_N$ grows with R_0. The sensitivity of $\bar{x}_N$ to R_0 is demonstrated in Table 2.2-5 for complete conversion of A groups.

These conclusions are general since either group may be labeled A as required. Therefore, $\bar{x}_N$ has a maximum value when $R_0 = 1$. The low value of $\bar{x}_N$ when $R_0 \neq 1$ is a manifestation of the capping of polymeric molecules with end groups of like kind by the monomer in excess. The resulting growth inhibition is magnified when the limiting monomer is near complete conversion, owing to the sudden rise in $\bar{x}_N$ at the end that is characteristic of random polymerizations.

Example 2.2-8. Change in the DPD of a Mixture of Random Polymers. Consider a mixture of (1) 100 g of polyester with $\bar{M}_N = 500$ and $\bar{M}_W = 1000$ and (2) 100 g of polyester with $\bar{M}_N = 5000$ and $\bar{M}_W = 10,000$. The mixture is heated and ester interchange occurs until equilibrium is achieved. Find $\bar{M}_N$, $\bar{M}_W$, and D_N of the mixture before and after heating.

Solution.   Before heating:

$$\bar{M}_N = \bar{M}_{N_1} n_1 + \bar{M}_{N_2} n_2$$

$$n_1 = \frac{\dfrac{100}{500}}{\dfrac{100}{500} + \dfrac{100}{5000}} = 0.909$$

$$n_2 = 1 - n_1 = 0.091$$

Thus

$$\bar{M}_N = 0.909(500) + 0.091(5000) = 909$$

Table 2.2-5. Sensitivity of Random Polymers to Feed Stoichiometry

R_0	$\bar{x}_N$
0.5	3
0.7	5.67
0.9	19
0.95	37
0.99	199
0.999	1999

but

$$\overline{M}_W = \overline{M}_{W_1} w_1 + \overline{M}_{W_2} w_2$$

and

$$w_1 = w_2 = 0.5$$

therefore

$$\overline{M}_W = 0.5(1000) + 0.5(10,000) = 5500$$

and

$$D_N = \frac{5500}{909} = 6.05$$

After heating, the number of repeat units and the number of chains has not changed, so $\overline{M}_N$ is the same and the DPD is most probable. Therefore, $D_N = 2.0$ and $\overline{M}_W = 1818$.

2.2.6.   Chain-Addition Polymerization

The aim of Section 2.2 is to examine various reaction sequences that lead to the most probable DPD, as well as certain characteristics that cause deviations from it. We have seen that equilibrium polymers in general and irreversible random polymers in particular are so distributed. We will now show that the distribution of intermediates in chain-addition polymerizations with termination are also most probable. As in Example 1.13-2, we neglect secondary termination by chain-transfer species s* $(k_{ta}c_s^*c_x^* = 0)$. After applying the "local" QSSA separately to the rate functions listed in Table 1.10-4 for each polymeric active intermediate, that is, $r_x^* = 0$ for each $x \geqslant 1$ and $r^* = 0$, we obtain the familiar difference equation in x

$$x \geqslant 1 \qquad\qquad c_{x+1}^* - \phi c_x^* = 0 \qquad\qquad (2.2\text{-}43)$$

with IC $c_1^* = (1 - \phi)c^*$, where

$$\phi \equiv \frac{k_p c_m}{D} \leqslant 1 \qquad\qquad (2.2\text{-}44)$$

and denominator D, omitting branching reactions, is defined as

$$D \equiv k_p c_m + k_{fm} c_m + k_{fs} c_s + k_t c^* + k_{ts} \qquad\qquad (2.2\text{-}45)$$

Its solution is, of course, the most probable DPD

$$\frac{c_x^*}{c^*} \equiv n_x^* = \phi^{x-1}(1 - \phi) \qquad\qquad (2.2\text{-}46)$$

The quantity ϕ, analogous to all previous ϕ's has statistical significance, which will be discussed in Section 2.6.

Two points concerning this result are noteworthy. First, it is general since we have included both termination modes, that is, first and second order, as well as all chain-transfer reactions. Second, while ϕ is a reaction variable subject to time-drift (Section 2.7), its value throughout polymerization must always be close to, but less than, unity. This characteristic follows directly from the LCA and the QSSA (Example 2.2-9) and immediately distinguishes the behavior of these intermediates from that of random polymer product, for which ϕ grows steadily in value from 0 to 1 (Example 2.2-5). It signifies that $D_N^* = 1 + \phi$ remains virtually constant throughout polymerization, and approximately equal to 2, as long as high polymer (large $\bar{x}_N^* = (1 - \phi)^{-1}$ is formed. Deviations are possible at high temperatures where dead-ending is enhanced and polymer with low DP is formed. Such phenomena will be discussed later.

Finally, in anticipation of future needs, it is appropriate at this point to discuss the exponential distribution (Appendix B) as an approximation of the most probable DPD (12). Throughout most chain-addition polymerizations, as previously noted, ϕ is a fraction whose value is very close to unity. Therefore, $\varepsilon \equiv 1 - \phi$ is a very small number and we can write the most probable DPD for active intermediates, equation 2.2-46, in the approximate, continuous form

$$0 \leqslant x \leqslant \infty \qquad\qquad n^*(x) \cong \varepsilon \exp(-\varepsilon x) \qquad\qquad (2.2\text{-}47)$$

But, since $\bar{x}_N^* = 1/(1 - \phi)$ we can write $n^*(\hat{x})\, d\hat{x}$ for the mole fraction of active intermediate whose normalized DP lies between $\hat{x}$ and $\hat{x} + d\hat{x}$, where

$$0 \leqslant \hat{x} \leqslant \infty \qquad\qquad n^*(\hat{x}) = \exp(-\hat{x}) \qquad\qquad (2.2\text{-}48)$$

and

$$\hat{x} \equiv \frac{x}{\bar{x}_N^*} \qquad\qquad (2.2\text{-}49)$$

Furthermore, from the relation $w^*(\hat{x}) = \hat{x} n^*(\hat{x})$, we can write $w^*(\hat{x})\, d\hat{x}$ for the weight fraction of active $\hat{x}$-mer, where

$$0 \leqslant \hat{x} \leqslant \infty \qquad\qquad w^*(\hat{x}) = \hat{x} \exp(-\hat{x}) \qquad\qquad (2.2\text{-}50)$$

It is easy to show that the mean value of DPD's $n^*(\hat{x})$ and $w^*(\hat{x})$ are $\bar{\hat{x}}_N^* = 1$ and $\bar{\hat{x}}_W^* = 2$, respectively (Example 2.2-10).

Example 2.2-9. The Probability of Propagation for Chain-Addition Polymerizations. Argue via the LCA that ϕ must have a value close to 1 and verify your conclusion using the following reaction data (Tables 2.7-3 and 2.7-4 and references 13 and 14) and parameters for styrene polymerization initiated with benzoyl peroxide in benzene at 75°C. Activation energies are expressed in calories.

$$k_d = 6.38 \times 10^{13}\exp\left(-\frac{29700}{R_g T}\right) s^{-1} = 1.42 \times 10^{-5} s^{-1}$$

$$k_p = 1.06 \times 10^7\exp\left(-\frac{7068}{R_g T}\right) 1/\text{mol s} = 3.85 \times 10^2\, 1/\text{mol s}$$

$$\frac{k_{fm}}{k_p} = 2.20 \times 10^{-1}\exp\left(-\frac{5603}{R_g T}\right) = 6.66 \times 10^{-5}$$

$$\frac{k_{fs}}{k_p} = 9.47 \times 10^3\exp\left(-\frac{14790}{R_g T}\right) = 4.87 \times 10^{-6}$$

$$k_t = 1.26 \times 10^9\exp\left(-\frac{1677}{R_g T}\right) 1/\text{mol s} = 1.11 \times 10^8\, 1/\text{mol s}$$

$$k_{ts} = 0$$
$$c_m = 2.4 \text{ mol}/1$$
$$c_0 = 0.015 \text{ mol}/1$$
$$c_s = 7.44 \text{ mol}/1$$

Solution. The intent in requiring that v_N be large in equation 1.11-1 for the attainment of long polymer chains is to ensure that the rate of addition of monomer to polymer chains far exceeds the formation of polymer chain ends. Thus we could have used r_t in the denominator in place of r_i. In either case, we require that r_p be much greater than any one of the following rates: r_t, r_{fm}, and r_{fs}. Therefore, $k_p c_m \gg k_{fm}c_m + k_{fs}c_s + k_t c^* + k_{ts}$ and $\phi \cong 1$.

With aid of the QSSA, we can write equation 2.2-44 as

$$\phi = c_m \Big/ \left[c_m + \left(k_{fm}/k_p\right) + \left(k_{fs}/k_p\right)\left(c_s/c_m\right) + k_{ax}^{-1}c_0^{1/2}\right]$$

where

$$k_{ax} \equiv k_p/\left(2k_d k_t\right)^{1/2} = 6.86$$

so that

$$\phi = 2.4/\left(2.4 + 6.66 \times 10^{-5} + 1.6 \times 10^{-5} + 1.79 \times 10^{-2}\right) = 0.993$$

Clearly, the term pertaining to r_p in the denominator is far greater than any other in magnitude. The second largest is the one pertaining to r_t.

Example 2.2-10. *The Exponential Approximation of the Most Probable DPD.* Verify the following identities, which must be true for normalized distributions 2.2-48 and 2.2-50:

$$\int_0^\infty n^*(\hat{x})\, d\hat{x} = 1$$

$$\int_0^\infty w^*(\hat{x})\, d\hat{x} = \int_0^\infty \hat{x}n^*(\hat{x})\, d\hat{x} = 1$$

as well as the following mean values:

$$\bar{x}_N = \int_0^\infty \hat{x} n^*(\hat{x})\, d\hat{x} = 1$$

$$\bar{x}_W^* = \int_0^\infty \hat{x} w^*(\hat{x})\, d\hat{x} = \int_0^\infty \hat{x}^2 n^*(\hat{x})\, d\hat{x} = 2$$

Solution. By simple integration, we obtain

$$\int_0^\infty \exp(-\hat{x})\, d\hat{x} = -e^{-\hat{x}}\Big|_0^\infty = 1$$

and by integration by parts, we obtain

$$\int_0^\infty \hat{x}\exp(-\hat{x})\, d\hat{x} = -\hat{x}\exp(-\hat{x})\Big|_0^\infty - \int_0^\infty \exp(-\hat{x})\, d\hat{x} = 1$$

and also

$$\int_0^\infty \hat{x}^2\exp(-\hat{x})\, d\hat{x} = -\hat{x}^2\exp(-\hat{x})\Big|_0^\infty - \int_0^\infty 2\hat{x}\exp(-\hat{x})\, d\hat{x} = 2$$

Thus, since $n^*(\hat{x}) = \exp(-\hat{x})$, the first result verifies the first identity; the second result verifies the second identity and first mean value; and the third result verifies the second mean value.

2.2.7. The Effects of Termination

Among the reactions responsible for the termination of active polymeric inter-mediates are spontaneous termination, disproportionation, combination, chain transfer to monomer, and chain transfer to foreign substances. It is self-evident that some of these will produce instantaneous polymer product whose DPD's are different from that of the intermediates (equation 2.2-46) from which they derive. To compute $(n_x)_{inst}$ from equation 1.13-6, we require r_x and $\sum_{x=1}^\infty r_x$. By substituting equation 2.2-46 for c_x^* into the rate function r_x in Table 1.10-4, omitting branching reactions,

$$r_x = \left(D - k_p c_m - k_{tc} c^*\right)\phi^{x-1}(1-\phi)c^* + \frac{k_{tc}}{2}(x-1)\phi^{x-2}(1-\phi)^2 c^{*2}$$

$$(2.2\text{-}51)$$

and summing the result we obtain

$$\sum_{x=1}^\infty r_x = \left(D - k_p c_m - \frac{k_{tc}}{2}c^*\right)c^* \qquad (2.2\text{-}52)$$

where the term $k_{ta}c_s^* c_x^*$ has again been neglected. Dividing equation 2.2-51 by 52 and rearranging yields

$$(n_x)_{inst} = P_1\phi^{x-1}(1-\phi) - P_2(x-1)\phi^{x-2}(1-\phi)^2 \qquad (2.2\text{-}53)$$

where P_1 is the fraction of instantaneous polymer product formed by spontaneous termination, disproportionation, and chain transfer

$$P_1 \equiv 1 - P_2 \qquad (2.2\text{-}54)$$

and P_2 is the fraction formed by combination

$$P_2 \equiv \frac{\dfrac{k_{tc}c^*}{2}}{D - k_p c_m - \dfrac{k_{tc}c^*}{2}} \qquad (2.2\text{-}55)$$

The denominator in equations 2.2-54 and 2.2-55 is proportional to the total rate of formation of polymer product molecules by deactivation of intermediates. The terms in the numerators are proportional to the individual rates by each deactivation mechanism.

Computational details are given in Example 2.2-11 and more characteristics of the instantaneous polymer are developed in Example 2.2-12. Like ϕ, P_1 and P_2 have statistical significance, which will be discussed further in Section 2.6. It is evident from these results that instantaneous polymers formed by spontaneous termination, disproportionation, and chain transfer are most probable, whereas those formed by combination are distributed differently. In fact, the latter instantaneous DPD is more narrowly dispersed than the most probable for a given value of ϕ (Example 2.2-12). In either case, as long as ϕ remains close to unity and therefore instantaneous DP is large, $(D_N)_{inst}$ remains virtually constant throughout polymerization.

Example 2.2-11. The Instantaneous DPD of Chain-Addition Polymers. Derive DPD 2.2-53.

Solution. After substituting $c_x^* = \phi^{x-1}(1 - \phi)c^*$ into rate function r_x and neglecting the term $k_{ta}c_s^* c_x^*$, we obtain equation 2.2-51. The origin of the first term on the RHS is obvious; the origin of the second involves the following summation:

$$\sum_{j=1}^{x-1} \phi^{j-1} \phi^{x-j-1} = \frac{\phi^x}{\phi^2} \sum_{j=1}^{x-1} \phi^j \phi^{-j} = \frac{(x-1)\phi^x}{\phi^2}$$

Finally, the summation $\sum_{x=1}^{\infty} r_x$ may be performed with the aid of Table C-3. The two key terms are:

$$\sum_{x=1}^{\infty} \phi^{x-1}(1 - \phi) = \frac{(1-\phi)}{\phi} \sum_{x=1}^{\infty} \phi^x = 1$$

and

$$\sum_{x=1}^{\infty} (x-1)\phi^{x-2}(1-\phi)^2 = \left[\frac{(1-\phi)}{\phi} \right]^2 \left(\sum_{x=1}^{\infty} x\phi^x - \sum_{x=1}^{\infty} \phi^x \right) = 1$$

Thus

$$\sum_{1}^{\infty} r_x = (D - k_p c_m - k_{tc}c^*)c^* + \frac{k_{tc}}{2}(c^*)^2$$

Example 2.2-12. Characteristics of the Instantaneous DPD. Contrast the characteristics of the instantaneous DPD's for two special cases: when no termination by combination occurs ($P_1 = 1$, $P_2 = 0$) and when termination occurs exclusively by combination ($P_1 = 0$, $P_2 = 1$).

Solution. In the first case:

$$(n_x)_{\text{inst}} = \phi^{x-1}(1 - \phi) \qquad (\bar{x}_N)_{\text{inst}} = \frac{1}{1 - \phi}$$

$$(w_x)_{\text{inst}} = x\phi^{x-1}(1 - \phi)^2 \qquad (\bar{x}_W)_{\text{inst}} = \frac{1 + \phi}{1 - \phi}$$

Thus $(D_N)_{\text{inst}} = 1 + \phi \cong 2$. In the second case:

$$(n_x)_{\text{inst}} = (x - 1)\phi^{x-2}(1 - \phi)^2 \qquad (\bar{x}_N)_{\text{inst}} = \frac{2}{1 - \phi}$$

$$(w_x)_{\text{inst}} = [x(x - 1)/2]^{x-2}(1 - \phi)^3 \qquad (\bar{x}_W)_{\text{inst}} = \frac{2 + \phi}{1 - \phi}$$

Thus $(D_N)_{\text{inst}} = (2 + \phi)/2 \cong 3/2$. The summations were obtained with the aid of Table C-3, as in Example 2.2-11 and are left as an exercise for the interested reader. From these results, we conclude that the modified instantaneous DPD in the second case has a narrower relative dispersion than the most probable DPD in the first case.

Example 2.2-13. The Instantaneous DPD of Heterogeneously Catalyzed Polymers. For the Ziegler–Natta sequence in Example 1.11-4 derive the instantaneous DPD and show that $(D_N)_{\text{inst}} \cong 2$ when high-polymer is produced.

Solution. The rate function for formation of product x-mer

$$r_x = k_t c_x^*$$

requires knowledge of the intermediate distribution c_x^*. To obtain this, we write the required rate functions

$$r_1^* = k_i c_m c_0^* - k_p c_m c_1^* - k_t c_1^*$$

$$x > 1 \qquad r_x^* = k_p c_m c_{x-1}^* - k_p c_m c_x^* - k_t c_x^*$$

After applying the QSSA in detail, $r_x^* = 0$ for all $x \geqslant 1$, and the constraint equation, $c_0^* + c_p^* = c_0 = $ constant, and combining the results we obtain the most probable DPD,

$$n_x^* \equiv \frac{c_x^*}{c_p^*} = \phi^{x-1}(1 - \phi)$$

which should not be surprising, where

$$c_p^* \equiv \sum_{x=1}^{\infty} c_x^* = \frac{Kc_0 c_m}{(1 + Kc_m)}$$

$K \equiv k_i/k_t$ (adsorption isotherm) and (probability of propagation)

$$\phi \equiv \frac{k_p c_m}{k_p c_m + k_t}$$

Substitution into r_x yields

$$r_x = \frac{\dfrac{Kk_t^2}{k_p} c_0 \phi^x}{1 + Kc_m}$$

Thus we obtain the most probable DPD again

$$(n_x)_{inst} \equiv \frac{r_x}{\sum_x r_x} = \phi^{x-1}(1 - \phi)$$

from which we conclude immediately that

$$(D_N)_{inst} = 1 + \phi \cong 2$$

2.3. THE POISSON DISTRIBUTION

A DPD that can occur and is of considerable interest because it is narrow and virtually monodisperse, is the Poisson distribution (Appendix B). In this section, we shall examine its kinetic origin, as well as certain factors that thwart its attainment in practice.

2.3.1. The Poisson Process

The simplest sequence for step addition polymerization is

$$x \geqslant 0 \qquad\qquad m_x + m \xrightarrow{k} m_{x+1} \qquad\qquad (2.3\text{-}1)$$

with a single rate constant $k_i = k_p = k$. The corresponding rate equations, neglecting volume (density) changes, are

$$x \geqslant 1 \qquad\qquad \frac{dc_x}{dt} = r_x = kc_m c_{x-1} - kc_m c_x \qquad\qquad (2.3\text{-}2)$$

$$-\frac{dc_0}{dt} = -r_0 = kc_m c_0 \qquad\qquad (2.3\text{-}3)$$

$$-\frac{dc_m}{dt} = r = kc_m c_0 + kc_m c_p \qquad\qquad (2.3\text{-}4)$$

where the rate functions on the RHS were obtained by specializing the appropriate expressions in Table 1.10-1. The constraint equations (see Section 1.7), which reflect the dependency of 2.3-3 and 2.3-4 upon 2.3-2 at any time $t \geqslant 0$, are

$$c_m + \sum_{x=1}^{\infty} x c_x = (c_m)_0 \qquad (2.3-5)$$

and

$$c_0 + c_p = \sum_{x=0}^{\infty} c_x = (c_0)_0 \qquad (2.3-6)$$

The last result signifies that the total concentration of growing chains remains constant during polymerization when m_0 is counted as 1-mer. It serves to simplify the rate equation for polymerization

$$-\frac{dc_m}{dt} = k c_m (c_0)_0 \qquad (2.3-7)$$

The natural CT for this first-order reaction is (Appendix G)

$$\lambda_m = \left[k (c_0)_0 \right]^{-1} \equiv \lambda_r \qquad (2.3-8)$$

and its solution with IC $c_m = (c_m)_0$ yields the following expression for conversion, which has been plotted in Fig. 2.1-1,

$$\Phi \equiv 1 - \frac{c_m}{(c_m)_0} = 1 - \exp(-\hat{t}) \qquad (2.3-9)$$

where

$$\hat{t} \equiv \frac{t}{\lambda_r} \qquad (2.3-10)$$

It is convenient to reduce time with λ_r ($\hat{t} \equiv t/\lambda_r$) and to reduce all concentrations with $(c_0)_0$ ($\hat{c} \equiv c/(c_0)_0$). By applying the integrating factor $\exp \int k c_m \, dt = \exp \int \hat{c}_m \, d\hat{t}$ to equations 2.3-2 and 2.3-3 in dimensionless form and integrating them successively, subject to the IC's

$$\hat{c}_x = \begin{cases} 1 & \text{when } x = 0 \\ 0 & \text{when } x > 0 \end{cases} \qquad (2.3-11)$$

we obtain the Poisson distribution,

$$x \geqslant 0 \qquad \qquad \hat{c}_x = \frac{z^x \exp(-z)}{x!} \qquad (2.3-12)$$

which may be interpreted as the DPD ($n_x = \hat{c}_{x-1}$) if we count m_0 as 1-mer, as noted earlier. The variable $z \equiv \int_0^t k c_m \, dt$ arises in the so-called eigenzeit transformation (Example 2.3-1), which may be used to simplify equations 2.3-2 and 2.3-3. Expression 2.3-12 can be confirmed as the solution of the latter equations (Example 2.3-2) by substituting it into their transformed counterparts (Example 2.3-1). Thus, following substitution into the RHS, we obtain

$$x \geqslant 1 \qquad \frac{d\hat{c}_x}{dz} + \hat{c}_x = \frac{z^{x-1}\exp(-z)}{(x-1)!} \qquad (2.3\text{-}13)$$

The rest is left as an exercise for the reader.
A physical interpretation of z:

$$z = -\frac{\Delta c_m}{(c_0)_0} = \Phi x_0 \qquad (2.3\text{-}14)$$

may be obtained from the solution of equation 2.3-4 (Example 2.3-1), subject to the IC ($z = 0$)

$$\hat{c}_m = (c_m)_0/(c_0)_0 \equiv x_0 \qquad (2.3\text{-}15)$$

Thus z represents the number of monomer molecules polymerized for each initiator molecule available. Its value therefore rises with conversion and approaches x_0. Its magnitude at any time must be of the order of $\bar{x}_N$; more precisely,

$$\bar{x}_N = 1 + z \qquad (2.3\text{-}16)$$

The last result was obtained in the customary way by counting m_0 as a repeat unit, defining $n_x \equiv \hat{c}_{x-1}$ for all $x \geqslant 1$ and performing the required summation. The time and conversion dependence of $\bar{x}_N$, that is,

$$\bar{x}_N = 1 + x_0\left[1 - \exp(-\hat{t})\right] = 1 + x_0\Phi \qquad (2.3\text{-}17)$$

has been plotted in Fig. 2.1-2.

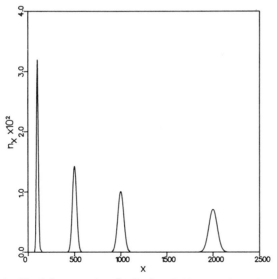

Figure 2.3-1. The Poisson number distribution (Table 1.5-4) for various values of z.

The characteristics of the Poisson distribution are listed in Table 1.5-4 and illustrated in Figs. 2.3-1 and 2.3-2. We conclude that the dispersion index

$$D_{\mathrm{N}} = \frac{1 + z}{(1 + z)^2} \tag{2.3-18}$$

approaches a value of unity (monodisperse polymer) after starting at unity and passing through a maximum value of 1.25 when $z = 1$. For practical purposes, a polymer having a value for D_{N} of 1.05 or less is generally considered to be monodisperse. Equations 2.3-17 and 2.3-18 may be verified by the reader with the aid of the appropriate definitions in Section 1.5 and the summations in Table C-3.

A more precise way to define DPD's and averages for the kinetic sequence being considered is not to count m_0 as 1-mer. The results are listed in Table 2.3-1. It is noteworthy that the weight DPD is then a Poisson distribution, instead of the number DPD as before, and the number-average DP is the quotient of monomer converted and initiator consumed, which is more acceptable intuitively than equation 2.3-16.

Example 2.3-1.   The Eigenzeit Transform Method.   Using z as a transform variable, simplify the set of equations 2.3-2–2.3-4.

Solution.   We substitute the eigenzeit transformation $kc_{\mathrm{m}}\, dt \equiv dz$ into our equations and obtain the transformed set

$$\frac{d\hat{c}_0}{dz} = -\hat{c}_0$$

$x \geqslant 1$
$$\frac{d\hat{c}_x}{dz} = \hat{c}_{x-1} - \hat{c}_x$$

$$\frac{d\hat{c}_{\mathrm{m}}}{dz} = -1$$

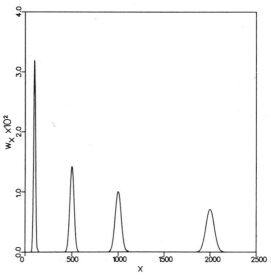

Figure 2.3-2.   The corresponding weight distribution (Table 1.5-4) for various values of z.

Table 2.3-1. Polymer Product Characteristics of Sequence 2.3-1

$x \geq 0$	$m_x + m \xrightarrow{\ k\ } m_{x+1}$ (2.3-1)

$$x \geq 1 \qquad n_x \equiv c_x \bigg/ \sum_{x=1}^{\infty} c_x = z^x \exp(-z) \bigg/ x! \, [1 - \exp(-z)]$$

$$x \geq 1 \qquad w_x = x n_x / \overline{x}_N = z^{x-1} \exp(-z)/(x-1)!$$

$$\overline{x}_N \equiv \sum_{x=1}^{\infty} x c_x \bigg/ \sum_{x=1}^{\infty} c_x = \Delta c_m / \Delta c_0 = z/[1 - \exp(-z)]$$

$$\overline{x}_w \equiv \sum_{x=1}^{\infty} x^2 c_x \bigg/ \sum_{x=1}^{\infty} x c_x = 1 + z$$

$$D_N \equiv \overline{x}_w / \overline{x}_N = (1 + z)[1 - \exp(-z)]/z$$

where all concentration have been reduced with $(c_0)_0$. Successive integration of the first two yields equation 2.3-12. Integration of the third gives equation 2.3-14.

Example 2.3-2.   The Generating Function Method. Using the generating function (Appendix C), show that the Poisson distribution (equation 2.3-12) is the DPD for step-addition sequence 2.3-1 in a batch reactor.

Solution.   The set of equations 2.3-2–2.3-4 in dimensionless form are

$$\frac{d\hat{c}_0}{d\hat{t}} = -\hat{c}_m \hat{c}_0$$

$$x > 0 \qquad \frac{d\hat{c}_x}{d\hat{t}} = \hat{c}_m \hat{c}_{x-1} - \hat{c}_m \hat{c}_x$$

$$\frac{d\hat{c}_m}{d\hat{t}} = -\hat{c}_m$$

when $\hat{t} \equiv t/\lambda_r$ and all concentrations are reduced with $(c_0)_0$. The solution of the monomer balance with initial condition $\hat{c}_m = x_0$ is

$$\hat{c}_m = x_0 \exp(-\hat{t})$$

Define the generating function (Table C-1; primes dropped for convenience)

$$F(z, \hat{t}) \equiv \sum_{x=0}^{\infty} \hat{c}_x(\hat{t}) z^x$$

with the properties

$$\frac{\partial F}{\partial \hat{t}} = \sum_{x=0}^{\infty} \frac{d\hat{c}_x}{d\hat{t}} z^x$$

and $F(z,0) = 1$, assuming the IC's: $\hat{c}_0 = 1$ and $\hat{c}_x = 0$ for $x > 0$. After substituting the polymer balances

$$\frac{\partial F}{\partial z} = -\hat{c}_m \sum_{x=0}^{\infty} \hat{c}_x z^x + \hat{c}_m z \sum_{x=0}^{\infty} \hat{c}_x z^x$$

and the solution of the monomer balance, applying the definition of F, we obtain the equation

$$\frac{\partial F}{\partial z} = (z - 1) x_0 F \exp(-\hat{t})$$

whose solution, subject to the initial condition $F = 1$, by the method of integrating factor (Appendix C) is

$$F = \exp(-\zeta) \exp(\zeta z)$$

where $\zeta \equiv x_0(1 - \exp - \hat{t})$. Finally, we expand the last term in an infinite series and compare the result

$$F = \exp(-\zeta) \sum_{x=0}^{\infty} \frac{(\zeta z)^x}{x!}$$

with the definition of f, thereby deducing that

$$\hat{c}_x = \frac{\zeta^x \exp(-\zeta)}{x!}$$

which agrees with equation 2.3-12. To complete this exercise, we note that ζ is precisely the eigenzeit transform variable (Example 2.3-1). It has been denoted here as ζ rather than z to avoid confusing it with the generating function variable. Thus the eigenzeit transform is defined as

$$\int_0^t k c_m \, dt = \int_0^{\hat{t}} \hat{c}_m \, d\hat{t}$$

which after substitution of the solution of the monomer balance yields $x_0(1 - \exp - \hat{t}) \equiv \zeta$.

2.3.2. Deviations from the Poisson Process

A more general sequence than 2.3-1 for step-addition polymerization is

$$m_0 + m \xrightarrow{k_i} m_1$$

$$x \geqslant 1 \qquad\qquad m_x + m \xrightarrow{k_p} m_{x+1} \qquad\qquad (2.3\text{-}19)$$

The corresponding rate equations (cf. Table 1.10-1)

$$x > 1 \qquad \frac{dc_x}{dt} = r_x = k_p c_m c_{x-1} - k_p c_m c_x \qquad (2.3\text{-}20)$$

$$x = 1 \qquad \frac{dc_1}{dt} = r_1 = k_i c_m c_0 - k_p c_m c_1 \qquad (2.3\text{-}21)$$

$$\frac{dc_0}{dt} = r_0 = k_i c_m c_0 \qquad (2.3\text{-}22)$$

$$-\frac{dc_m}{dt} = r = k_i c_m c_0 + k_p c_m c_p \qquad (2.3\text{-}23)$$

subject to IC's 2.3-11 and 2.3-15 were analyzed by Gold (15) and by Nanda and Jain (16). The results for product distribution are listed in Table 2.3-2, where $z \equiv \int_0^t k_p c_m \, dt$ is the eigenzeit transform variable and a_K is the ratio of kinetic constants.

$$a_K \equiv \frac{k_i}{k_p} \qquad (2.3\text{-}24)$$

The result for monomer conversion is

$$\Phi = x_0^{-1}\{z + (a_K - 1)a_K^{-1}[1 - \exp(-a_K z)]\} \qquad (2.3\text{-}25)$$

Values of a_K different from unity represent departures from the Poisson process. When $a_K = 1$, all these expressions reduce to the corresponding ones for the Poisson sequence (Table 2.3-1) as expected.

Graphs of $\bar{x}_N$ versus x_0 and D_N versus $\bar{x}_N$ with a_K as parameter are shown in Figs. 2.3-3 and 2.3-4. When $a_K > 0.01$, $\bar{x}_N$ is virtually linear with conversion (cf. step-addition behavior in Section 2.1) and the polymer product at high conversions

Table 2.3-2. Polymer Product Characteristics of Sequence 2.3-19

$$x \geq 1 \qquad n_x = \left\{ [a_k(1 - a_k)^{-x}\exp(-z)]/[1 - \exp(-a_k z)] \right\} \sum_{j=x}^{\infty} [z(1 - a_k)]^j/j!$$

$$\bar{x}_N = z/[1 - \exp(-a_k z)] - a_k^{-1}(1 - a_k)$$

$$\bar{x}_w = \frac{(a_k z)^2 + 3a_k^2 z - 2a_k z + (a_k^2 - 3a_k + 2)[1 - \exp(a_k z)]}{a_k\{a_k z - (1 - a_k)[1 - \exp(-a_k z)]\}}$$

$$D_N = \frac{\{(a_k z)^2 + 3a_k^2 z - 2a_k z + (a_k^2 - 3a_k + 2)[1-\exp(-a_k z)]\}[1-\exp(-a_k z)]}{\{a_k z - (1 - a_k)[1 - \exp(-a_k z)]\}^2}$$

is effectively monodisperse, as measured by final D_N. Graphs of $\bar{x}_N$ versus conversion acquire curvature with decreasing values of a_K, especially at low conversions, and graphs of D_N exhibit maxima at values of $\bar{x}_N$ below 10. These maxima increase and shift to higher values of $\bar{x}_N$ as a_K decreases. Nanda and Jain have shown that the largest maximum value of D_N (1.375) occurs when $a_K \to 0$ (corresponding values are $z = 6$ and $\bar{x}_N = 4$), and that the largest final value of D_N possible is $4/3$ when a_K is sufficiently small. More specifically, $D_N \to 4/3$ when $z \gg 1$ provided that $a_K z \ll 1$; and $D_N \to 1$ when $z \gg 1$ and $a_K z \gg 1$.

These conclusions may be restated in terms of a familiar kinetic parameter introduced earlier (Section 1.11), that is,

$$\alpha_K \equiv \frac{\lambda_m}{\lambda_i} \qquad (2.3\text{-}26)$$

where the CT's for this reaction scheme, defined in the usual way, are $\lambda_m \equiv (c_m)_0/(r_p)_0 = 1/k_p(c_0)_0$ and $\lambda_i \equiv (c_0)_0/(r_i)_0 = 1/k_i(c_m)_0$. The value of α_K indicates which feed component is reaction-limiting and is related to a_K by

$$\alpha_K = a_K x_0 \qquad (2.3\text{-}27)$$

Thus we see that α_K is determined by feed ratio x_0, which is a parameter, as well as by the property ratio a_K, which is fixed. Graphs of final dispersion $(D_N)_f$ versus α_K are shown in Fig. 2.3-5. From these we conclude that when $\alpha_K \ll 1$, $D_N \to 4/3$ and when $\alpha_K \gg 1$, $D_N \to 1$. Note from equation 2.3-25 that the maximum value of z is $x_0 \Phi$ when $a_K > 1$, and therefore the upper limit on z is x_0. Caution must be exercised, however, since z may not attain a larger value if the initiator is not completely consumed at the time of total monomer conversion. Nevertheless, the

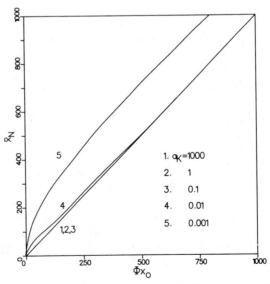

Figure 2.3-3. Evolution of $\bar{x}_N$ for step-addition sequence 2.3-19 for various values of parameter a_K

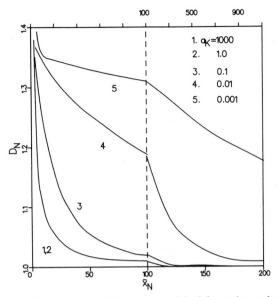

Figure 2.3-4. Evolution of D_N for step-addition sequence 2.3-19 for various values of parameter a_K.

magnitude of z is of the order of x_0 and thus the criteria of Nanda and Jain can be expressed solely in terms of α_K as defined in equation 2.3-27. The smaller α_K is, the broader the final DPD of the polymer product, because new chains will be started late during the reaction without adequate opportunity to grow. Conversely, when α_K is large, most of the chains are formed at the outset and grow simultaneously.

When the initiator runs out during the polymerizations, the value of D_N will pass through a maximum. The DPD will have negative skewness (Appendix B) at this

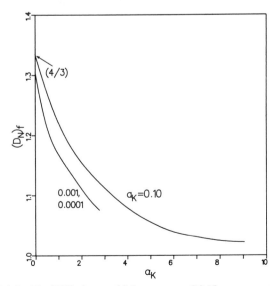

Figure 2.3-5. Final DP of step-addition sequence 2.3-19 versus parameter a_K.

time. However, subsequent conversion will permit the low-DP fractions to grow. The DPD will thus be shifted to the right, since all chains now have an equal probability of growth, and small chains could have ample time to grow in length. Although the absolute spread about the mean will remain the same, D_N will fall as $\bar{x}_N$ rises (see Section 1.4).

An idealized addition sequence that is simpler to analyze than 2.3-19 and yet exhibits the limiting behavior we seek, is the following:

$$m \overset{k_i}{\rightleftarrows} m_1$$

$$x \geqslant 1 \qquad\qquad m_x + m \overset{k_p}{\to} m_{x+1} \qquad\qquad (2.3\text{-}28)$$

Its product characteristics are examined in Examples 2.3-3 and 2.3-4. They span the same spectrum as those for sequence 2.3-19. When initiation is very rapid, $D_N \to 1$ (Example 2.3-3) and both sequences behave like Poisson processes. When initiation is very slow, $D_N \to 4/3$ for sequence 2.3-28 (Example 2.3-2) as it does for sequence 2.3-19. The equivalence of these sequences for very slow initiation is most easily seen from the rate function $r_i = k_i c_m c_0$ for the former, which approaches that for the latter when c_0 remains virtually constant, owing to slow initiation. The reason that the DPD is broadened is that new chains are initiated throughout polymerization as long as monomer is present.

From the results of Example 2.3-3, we can write the weight-fraction DPD for sequence 2.3-28 with irreversible initiation as

$$w_x = x \frac{2}{z(2+z)} \exp(-z) \sum_{j=x}^{\infty} \frac{z^j}{j!} \qquad\qquad (2.3\text{-}29)$$

This distribution has been plotted in Fig. 2.3-6 ($f_x = w_x$; curve 1) for $\bar{x}_N = 500$. Its resemblance to the ramp distribution (Section 1.5) is obvious. A ramp approximation having identical mean value and area has been plotted on the same graph (curve 2). The corresponding approximate number distribution ($f_x = n_x$; curve 3) has also been plotted at the top of this figure. The reader will note that the number distribution is approximately a rectangle, as discussed in Section 1.5.

For purposes of comparison, a Poisson number DPD has also been plotted in Fig. 2.3-6 (curve 4) together with its rectangular approximation (curve 5). Their mean values and areas are identical to rectangle 3 on the same graph. Comparing number DPD's 3 and 5 demonstrates visually how slow initiation can broaden a distribution, causing D_N to go from 1.00 to 1.33 in this case. The manifestation of continually introducing low ends in this manner is the ramplike weight DPD with negative skewness.

The DPD's resulting from sequence 2.3-19 have similar characteristics. Their evolution with monomer conversion is illustrated in Figs. 2.3-7 and 2.3-8, where weight-fraction DPD's have been plotted as functions of conversion for values of α_K an order of magnitude above and below the Poisson condition $\alpha_K = 1$. These were calculated from the equations in Table 2.3-2. When $\alpha_K = 10$, as in Fig. 2.3-7, the initiator is consumed early in the reaction, although not instantaneously as required

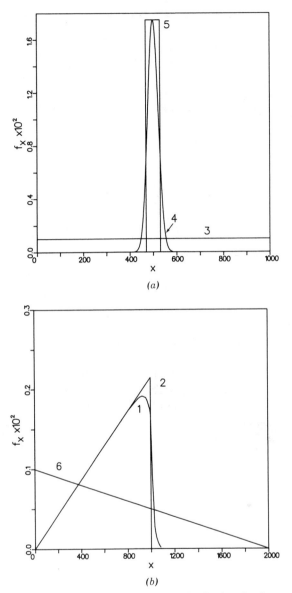

Figure 2.3-6. Graphs of: distribution 2.3-29 (1); the ramp distribution (2); the rectangular distribution (3 and 5); the Posson distribution (4); and the triangle distribution (6).

to qualify for a Poisson process. This puts a low-DP tail on the distribution, but after the initiator is consumed, all chains continue to grow at the same rate. The number of low-DP species declines since no new chain are created and the DPD shifts upwards with increasing $\bar{x}_N$. By contrast, when $\alpha_K = 0.1$ (Fig. 2.3-8) the initiation process continues throughout the reaction. New chains are always formed and a ramp DPD results. Note that $\bar{x}_N$ in Fig. 2.3-7 rises to the maximum in the DPD at high conversions (as $D_N \rightarrow 1$) while it remains at the same relative position on each DPD of Fig. 2.3-8.

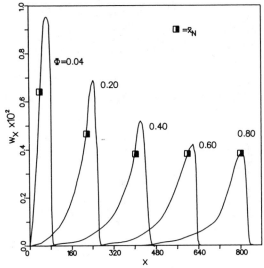

Figure 2.3-7. Weight DPD for step-addition sequence 2.3-19 with $\alpha_K = 0.1$, $x_0 = 1000$, and monomer conversion (Φ) as parameter.

It is noteworthy that slow initiation at most produces only ramp weight distributions, and corresponding rectangular number distributions. The introduction of termination, as in chain-addition polymerizations, however, can produce number DPD's with positive skewness (most probable), indicating an increased emphasis of low ends (Section 1.5). The corresponding weight distribution has a maximum. In a crude way, the triangle number distribution with its corresponding parabolic weight distribution (Section 1.5) represents such skewness. The triangle distribution has been included in Fig. 2.3-6 ($f_x = n_x$; curve 6) for comparison with the ramp. Their

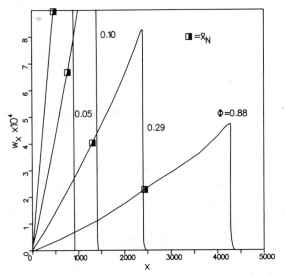

Figure 2.3-8. Weight DPD for step-addition sequence 2.3-19 with $\alpha_K = 10$, $x_0 = 1000$, and monomer conversion (Φ) as parameter.

mean values and areas are identical. The DPD's of chain-addition polymers will be discussed in more detail later on.

Example 2.3-3. *Characteristics of Addition Sequence 2.3-28.* Derive the expressions for conversion Φ, product distribution n_x, and its characteristics ($\bar{x}_N$, $\bar{x}_W$, D_N) as functions of z for sequence 2.3-28 with irreversible initiation.

Solution. We begin by writing the rate equations in dimensionless form

$$\frac{d\hat{c}_1}{dz} = \alpha_K - \hat{c}_1$$

$x > 1$

$$\frac{d\hat{c}_x}{dz} = \hat{c}_{x-1} - \hat{c}_x$$

$$\frac{d\hat{c}_m}{dz} = \alpha_K + \hat{c}_p$$

where all dimensionless concentrations have been reduced with $(c_m)_0$, $z \equiv \int_0^t k_p c_m \, dt$, and $\alpha_K = a_K/(c_m)_0$; as before, $\hat{c}_p \equiv \sum_{x=1}^{\infty} \hat{c}_x$ and $a_K \equiv k_i/k_p$. Following successive integration of the first two subject to the IC's ($z = 0$) $\hat{c}_x = 0$ for all x, we obtain the result

$$\hat{c}_x = \alpha_K \sum_{j=x}^{\infty} \frac{z^j}{j!} \exp(-z)$$

from which the desired characteristics are computed with the aid of the definitions in Section 1.5.

$$n_x = z^{-1} \exp(-z) \sum_{j=x}^{\infty} \frac{z^j}{j!}$$

$$\bar{x}_N = \frac{z + 2}{2}$$

$$\bar{x}_W = \frac{2z^2 + 9z + 6}{3z + 6}$$

$$D_N = \frac{4z^2 + 18z + 12}{3(z + 2)^2}$$

The dimensionless monomer balance, following summation over $\hat{c}_x$ (Table C-3)

$$-\frac{d\hat{c}_m}{dz} = \alpha_K(1 + z)$$

and integration subject to initial condition $\hat{c}_m = 1$, yields an expression for conversion

$$z = \left(1 + \frac{2\Phi}{\alpha_K}\right)^{1/2} - 1$$

From these results we conclude that $D_N \to 4/3$ when $\alpha_K \ll 1$ and $z \gg 1$ and that $D_N \to 1$ when $\alpha_K \gg 1$. However, in the latter case, $z \to 0$ as well, from which it follows that no polymer is produced ($\bar{x}_N \to 1$).

Example 2.3-4.   Characteristics of Step-Addition with Initiation Equilibrium.
Derive expressions for conversion $\Phi \equiv 1 - c_m/(c_m)_0$ and DPD n_x for sequence
2.3-28 at any time following the instantaneous establishment of initiation equilibrium $K_i = (c_1)_e/[(c_m)_0 - (c_1)_e]$.

Solution.   By reducing all concentrations with $(c_m)_e \equiv (c_m)_0 - (c_1)_e$, we obtain
the dimensionless equations

$$\frac{d\hat{c}_1}{dz} = -\hat{c}_1$$

$x > 1$
$$\frac{d\hat{c}_x}{dz} = \hat{c}_{x-1} - \hat{c}_x$$

$$-\frac{d\hat{c}_m}{dz} = \hat{c}_p$$

The solution of the first two subject to the IC ($z = 0$)

$$\hat{c}_x = \begin{cases} (\hat{c}_1)_e = \dfrac{(\hat{c}_m)_0 K_i}{1 + K_i} & \text{when } x = 1 \\ 0 & \text{when } x > 1 \end{cases}$$

yields a Poisson distribution

$x \geqslant 1$
$$\hat{c}_x = \frac{\dfrac{K_i(\hat{c}_m)_0}{1 + K_i} z^{x-1} \exp(-z)}{(x-1)!}$$

Solution of the last differential equation subject to IC $\hat{c}_m = 1$ gives an expression for
conversion

$$z = \frac{(1 + K_i)\Phi}{K_i}$$

From these results, we conclude that rapid initiation with equilibrium restores the
monodisperse DPD, and provides large DP's as well, if $K_i \ll 1$.

2.3.3.   The Approach to Equilibrium

In Section 2.2, we concluded that the most probable distribution represented the
equilibrium DPD for polymerizations in general. Yet in this section we have
encountered DPD's for irreversible step-addition polymerizations at high conversions that are quite different; in the case of Poisson processes, the DPD's are much
narrower. Furthermore, when we examine the product characteristics of irreversible
chain-addition polymerizations, we will encounter various other distributions, some
of which may be viewed as variation of the most probable distribution.

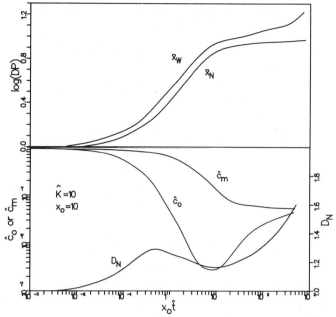

Figure 2.3-9. Evolution of composition and DP with time for reversible step-addition sequence 2.3-10 (17).

It is not uncommon for chemical reactions in general to yield product distributions that are not in chemical equilibrium when the reactions are conducted under irreversible conditions and the products are not allowed the opportunity to approach equilibrium. Such products may be viewed as being kinetically "frozen" in a metastable state. Given sufficient time and conducive circumstances, such products would seek an equilibrium state. A well-known example of such kinetic metastability is a mixture of H_2 and O_2, which will react to form water very, very slowly without a catalyst, notwithstanding the fact that the equilibrium state favors water, which would not spontaneously decompose to form H_2 and O_2.

By analogy, we would expect a DPD such as the Poisson distribution to be transformed into the most probable distribution at equilibrium if the reverse reactions were permitted to occur. To minimize analytical difficulties, we examine the simple reversible addition sequence.

$$x \geqslant 0 \qquad\qquad m_x + m \underset{k_r}{\overset{k_f}{\rightleftarrows}} m_{x+1} \qquad\qquad (2.3\text{-}30)$$

whose equilibrium characteristics were studied in Section 2.2 and whose rate functions are special cases of those listed in Table 1.10-1. Even such an ostensibly simple reaction is not mathematically tractable. Miyake and Stockmayer (17) studied the evolution of this reaction, using numerical analysis where necessary, and showed that the transformation from Poisson to most probable does indeed occur, as expected. Some of their graphs are shown in Figs. 2.3-9–2.3-11, and some of their approximate results are applied in Example 2.3-5.

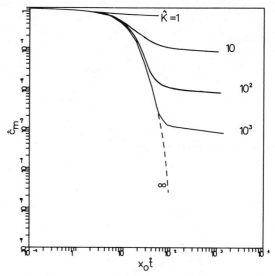

Figure 2.3-10. Monomer consumption for reversible step-addition sequence 2.3-10 with equilibrium constant $\hat{K}$ as parameter (17).

By specializing the rate functions in Table 1.10-1, we can write the monomer balance as

$$\frac{dc_m}{dt} = -k_f c_m (c_0)_0 + k_r c_p \tag{2.3-31}$$

where $c_p \equiv \sum_{x=1}^{\infty} c_x$ as always and the first term on the RHS has submitted to

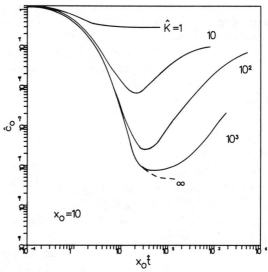

Figure 2.3-11. Evolution of initiator concentration for reversible step-addition sequence 2.3-10 with equilibrium constant $\hat{K}$ as parameter (17).

simplification by virtue of constraint equation $c_0 + c_p = (c_0)_0$. After writing this balance in semidimensionless form using $(c_m)_0$ as the reference concentration for the reverse reaction, since $(c_p)_0 = 0$, we define CT's for forward and reverse polymerization as

$$\lambda_{mf} \equiv \frac{1}{k_f(c_0)_0} \tag{2.3-32}$$

and

$$\lambda_{mr} \equiv \frac{1}{k_r} \tag{2.3-33}$$

respectively. Thus the dimensionless parameter $\hat{K}$ in the figures is

$$\hat{K} \equiv K(c_m)_0 = \frac{x_0 \lambda_{mr}}{\lambda_{mf}}$$

where $K \equiv k_f/k_r$; and the dimensionless variables are $\hat{c}_m \equiv c_m/(c_m)_0$, $\hat{c}_0 \equiv c_0/(c_0)_0$, and $\hat{t} \equiv x_0 t/\lambda_{mf}$.

The results of Miyake and Stockmayer indicate that $\hat{c}_m$ and $\bar{x}_N$ approach their final values early, in contrast with $\bar{x}_W$ and D_N, and that $(\bar{x}_N)_e \cong x_0$. This rules out both Φ and $\bar{x}_N$ as reliable indicators of the nearness of equilibrium. The graphs show a minimum value for D_N, in addition to the maximum value cited earlier. They also show that at relatively short reaction times the polymerization is virtually irreversible and the DPD is Poisson-like. At about the time $\hat{c}_m$ and $\bar{x}_N$ level out, the DPD begins to broaden and to approach the most probable distribution, as reflected by D_N. Miyake and Stockmayer estimated three CT's of interest for this reversible addition sequence. They are: the time at which the polymerization just ceases to be irreversible, $\lambda_1 \equiv \lambda_{mf} \ln(\hat{K}/x_0) = \lambda_{mf} \ln(\lambda_{mr}/\lambda_{mf})$; the time at which $\hat{c}_m$ and $\bar{x}_N$ level out, $\lambda_2 = \lambda_{mf} \ln \hat{K}$; and the time at which equilibrium is reached, $\lambda_3 \equiv (x_0/k_r)^2$.

Numerical solutions of the product balance equations have been plotted in Fig. 2.3-12. They show in detail the development of the DPD at low conversions and for small values of DP. The cause of the minimum value in D_N is clearly seen to be the evolution of the DPD from an initial triangle with positive skewness, through a fairly symmetrical distribution and, finally, to the most probable distribution with positive skewness again.

Example 2.3-5. Polystyrene Equilibration. Using the approximate results of Miyake and Stockmayer (17), estimate Φ_e, $(\bar{x}_N)_e$, λ_1, λ_2, and λ_3 for the addition sequence 2.3-30 using the following data of Szwarc (18) for anionic ("living") styrene polymerization at 25°C:

$$k_f = 10^3 \ 1/\text{mol s}$$
$$k_r = 2 \times 10^{-4} \ \text{s}^{-1}$$
$$(c_m)_0 = 2 \ \text{mol/l}$$
$$(c_0)_0 = 2 \times 10^{-3} \ \text{mol/l}$$

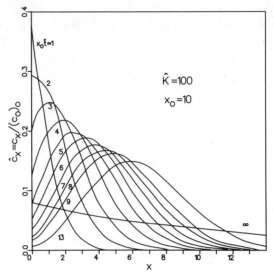

Figure 2.3-12. Evolution of DPD with time for reversible step-addition sequence 2.3-10 (17).

Solution. From the approximate results stated in the text, we conclude that

$$(\bar{x}_N)_e \cong x_0 = \frac{(c_m)_0}{(c_0)_0} = 10^3$$

Using this result and the well-known expression

$$(\bar{x}_N)_e = \frac{1}{1 - \phi_e}$$

we compute

$$\phi_e = 1 - (\bar{x}_N)_e^{-1} = 0.999$$

From the result of Example 2.3-4, that is, $\Phi_e = \phi_e (\bar{x}_N)_e/x_0$ when $K_i/K_p = 1$, we conclude that $\Phi_e \cong 0.999$. From the data given, we compute the following: $K = 5 \times 10^6 \, 1/\text{mol}$, $\hat{K} = 10^7$, $\lambda_{mf} = 0.5 \, \text{s}$, and $\lambda_{mr}/\lambda_{mf} = \hat{K}/x_0 = 10^4$. Substitution into the remaining approximate results stated in the text yields: $\lambda_1 = 4.6 \, \text{s}$, $\lambda_2 = 8.1 \, \text{s}$, and $\lambda_3 = 2.5 \times 10^{13} \, \text{s} \cong 80 \, \text{yrs}$.

From these results, we conclude that it should be possible to produce monodisperse polystyrene in a very short time, which remains as such for a very long time. This is, of course, consistent with experience.

2.4. THE TRANSITION FROM STEP- TO CHAIN-ADDITION

In our treatment of model polymerization schemes we have discarded the termination step for step-addition and tacitly assumed applicability of the QSSA for chain-addition reactions. It should be clear that it is both the presence and magnitude of chain termination that distinguishes the two different schemes. We might expect an ionic reaction with termination, via association, for example, to show characteristics somewhere between the two extremes of the model types. Similarly, with Ziegler–Natta catalysis where intermediates have been reported with lifetimes considerably longer than those of free-radical intermediates, deviations from ideal chain-addition could occur.

To examine the range of applicability of each model addition scheme, we propose to use the complete equations for chain-addition polymerization without the assumption of the QSSA. That is, the moment balances for both intermediates and polymer product will be considered along with the reactant balances. By varying the relative importance of termination, we should observe reaction behavior transform from step- to chain-addition. The equations are listed in Table 2.4-1. They were made dimensionless as follows: $\hat{c}_m \equiv c_m/(c_m)_0$, $\hat{c}_0 \equiv c_0/(c_0)_0$, $\hat{\mu}^k \equiv \mu_c^k/(c_m)_0$, $(\hat{\mu}^*)^k/(c^*)_0$, and $\hat{t} \equiv t/\lambda_m$, where $(c^*)_0$ is a "ficticious" initial intermediate concentration (Section 1.11). Three important dimensionless groups arise: $\alpha_K = \lambda_m/\lambda_i$ (equation 1.11-5), $x_0 = (c_m)_0/(c_0)_0$, and

$$\tau_K = \frac{\lambda_m}{\lambda_t} \tag{2.4-1}$$

where λ_t is the CT for termination.

The importance of α_K will be discussed in later sections (2.7 and 2.8). In both step and chain reactions, it characterizes the relative consumption rates of monomer and initiator. Although α_K for chain reactions is generally smaller than for step-addition, there could be regions of overlap. The value $\alpha_K = 1$, in particular, provides a bonus. Setting $\alpha_K = 1$ minimizes drift dispersion (Section 2.7) in chain-addition. Furthermore, for chain-addition polymers with termination by disproportionation, the initial value of $\bar{x}_N$ is $(\nu_N)_0 \equiv x_0\alpha_K$, which would then be equal to x_0. For step-addition polymers, the final value of $\bar{x}_N$ is ideally equal to x_0, so that the two schemes should ultimately produce polymer of the same degree of polymerization.

The group τ_K is formed by the ratio of CT's for monomer consumption and chain termination. Intermediates for chain-addition polymer terminate almost instantly relative to the time scale for monomer consumption. In step-addition, monomer is consumed much more rapidly than chains are terminated; indeed there is no termination at all. Therefore, we would expect τ_K to change drastically as the kinetics change in character from step to chain. Dimensionless group τ_K reduces to the ratio of termination to propagation rate constants.

$$\tau_K = \frac{\dfrac{(c_m)_0}{k_p(c_m)_0(c^*)_0}}{\dfrac{(c^*)_0}{k_t(c^*)_0^2}} = \frac{k_t}{k_p} \tag{2.4-2}$$

Consequently, we would expect small values of τ_K to signal step-addition behavior and large values to signal chain-addition behavior.

Table 2.4-1. Dimensionless Reactant, Polymer, and Intermediate Moment Balances Equations

$$\frac{d\hat{c}_o}{d\hat{t}} = -\alpha_K \hat{c}_o$$

$$\frac{d\hat{c}_m}{d\hat{t}} = -\left(\hat{c}_m \hat{\mu}^{o*} + \frac{2f\alpha_K}{x_o}\hat{c}_o\right)$$

$$\frac{d\hat{\mu}^{o*}}{d\hat{t}} = \tau_K \left(\frac{f}{(f)_o}\hat{c}_o - G(\hat{\mu}^{o*})^2\right)$$

$$\frac{d\hat{\mu}^o}{d\hat{t}} = \frac{2(f)_o \alpha_K}{x_o}\left(\frac{2-R}{2}\right) G(\hat{\mu}^{o*})^2$$

$$\frac{d\hat{\mu}^{1*}}{d\hat{t}} = \tau_K \left[\frac{f}{(f)_o}\hat{c}_o + \frac{x_o}{2(f)_o \alpha_K}\hat{c}_m \hat{\mu}^{o*} - G \hat{\mu}^{o*} \hat{\mu}^{1*}\right]$$

$$\frac{d\hat{\mu}^1}{d\hat{t}} = \frac{2(f)_o \alpha_K G}{x_o}\hat{\mu}^{o*} \hat{\mu}^{1*}$$

$$\frac{d\hat{\mu}^{2*}}{d\hat{t}} = \alpha_K \left[\frac{f}{(f)_o}\hat{c}_o + \frac{x_o}{2(f)_o \alpha_K}\hat{c}_m(\hat{\mu}^{o*} + 2\hat{\mu}^{1*}) - G \hat{\mu}^{o*} \hat{\mu}^{2*}\right]$$

$$\frac{d\hat{\mu}^2}{d\hat{t}} = \frac{2(f)_o \alpha_K}{x_o} G\left[\hat{\mu}^{o*} \hat{\mu}^{2*} + R(\hat{\mu}^{1*})^2\right]$$

$$\frac{d\hat{\mu}^{3*}}{d\hat{t}} = \tau_K \left[\frac{f}{(f)_o}\hat{c}_o + \frac{x_o}{2(f)_o \alpha_K}\hat{c}_m(\hat{\mu}^{o*} + 3\hat{\mu}^{1*} + 3\hat{\mu}^{2*}) - G \hat{\mu}^{o*} \hat{\mu}^{3*}\right]$$

$$\frac{d\hat{\mu}^3}{d\hat{t}} = \frac{2(f)_o \alpha_K}{x_o} G\left[\hat{\mu}^{o*} \hat{\mu}^{3*} + 3R \hat{\mu}^{1*} \hat{\mu}^{2*}\right]$$

$G = 1$ (no G-E); $G = \exp(-B(1 - \hat{c}_m))$ (G-E)

$f = (f)_o$ (constant efficiency); $f = f(\hat{c}_m, \hat{c}_o, (f)_o)$ (variable; Sec. 2.8)

$R \equiv k_{tc}/(k_{tc} + k_{tD})$

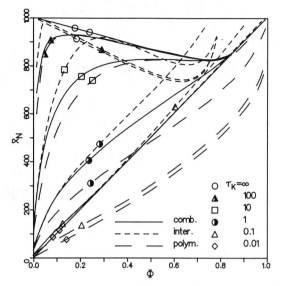

Figure 2.4-1. Effect of τ_K on number-average DP growth with conversion, $\alpha_K = 1$ and $x_0 = 2000$.

In Figs. 2.4-1–2.4-3, the results of numerical integration of the equations in Table 2.4-1 are shown for a wide range of values of τ_K. These results are shown together with the numerical solutions of the reactant and polymer product moment balances following application of the QSSA (i.e., $\tau_K \to \infty$) as in Table 2.4-2.

Consider first the variation in $\bar{x}_N$ as a function of conversion. We know that the QSSA case represents one extreme in behavior, in which the intermediates' properties are the "instantaneous" properties, whereas the cumulative product and com-

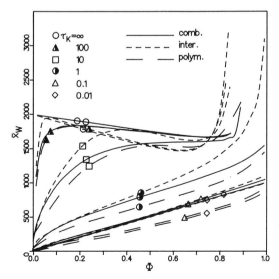

Figure 2.4-2. Effect of τ_K on weight-average DP growth with conversion, $\alpha_K = 1$ and $x_0 = 2000$.

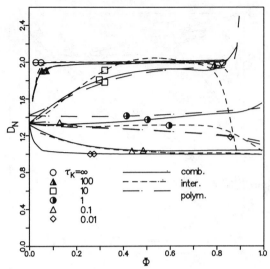

Figure 2.4-3. Effect of τ_K on dispersion index drift with conversion, $\alpha_K = 1$ and $x_0 = 2000$.

Table 2.4-2. Dimensionless Balances with the QSSA

$$\frac{d\hat{c}_o}{d\hat{t}} = -\alpha_K \hat{c}_o$$

$$\frac{d\hat{c}_m}{d\hat{t}} = -\left(\frac{f\hat{c}_o}{(f)_o G}\right)^{1/2} \hat{c}_m - \frac{2f\alpha_K}{x_o} \hat{c}_o$$

$$\frac{d\hat{\mu}^o}{d\hat{t}} = \frac{2f\alpha_K}{x_o} \left(\frac{2-R}{2}\right) \hat{c}_o$$

$$\frac{d\hat{\mu}^1}{d\hat{t}} = -\frac{d\hat{c}_m}{d\hat{\tau}}$$

$$\frac{d\hat{\mu}^2}{d\hat{t}} = (1 + R)\frac{2f\alpha_K \hat{c}_o}{x_o} + (3 + 2R)\hat{c}_m \left[\frac{f\hat{c}_o}{(f)_o G}\right]^{1/2} + (2 + R)\frac{x_o \hat{c}_m^2}{2(f)_o \alpha_K G}$$

$$\frac{d\hat{\mu}^3}{d\hat{t}} = (1 + 3R)\frac{2f\alpha_K \hat{c}_o}{x_o} + (7 + 12R)\hat{c}_m \left[\frac{f\hat{c}_o}{(f)_o G}\right]^{1/2} + (12 + 15R)\frac{x_o \hat{c}_m^2}{2(f)_o \alpha_K G} +$$

$$\frac{(6 + 6R)x_o^2 c_m^3}{(2(f)_o \alpha_K G)^2} \left[\frac{(f)_o G}{f\hat{c}_o}\right]^{1/2}$$

152

posite (cumulative plus intermediates) properties are virtually identical owing to the exceptionally low concentration of intermediates. At the other extreme, we know that the polymer formed by step-addition should show linear growth of $\bar{x}_N$ with conversion (Section 2.1) Furthermore, with no termination, the cumulative properties should be the same as those of the intermediates.

In Figs. 2.4-1 and 2.4-2, we see that the transition between extremes of behavior falls in the region:

$$0.01 < \tau_K < 100 \qquad (2.4\text{-}3)$$

Indeed, $\tau_K = 0.10$ still shows molecular weight growth characteristic of step-addition. Notice that the polymer product for small τ_K has a lower DP than the intermediates. This is due to termination prematurely stopping chain growth. Nonetheless, the concentration of terminated polymer is so low that cumulative chain length grows linearly to a final value of x_0.

When $\tau_K = 100$, we see behavior very much like that described by kinetics based on the QSSA. High-molecular-weight product $[\bar{x}_N \cong (\nu_N)_0]$ is produced early in the reaction, and the composite properties are virtually identical to those of the cumulative polymer product. The intermediate chain length tracks the value of the instantaneous chain length computed for the QSSA case. Note that even for $\tau_K = 10$, the final values of $\bar{x}_N$ for polymer, intermediate, and composite populations are very close to those obtained with the QSSA. The conclusion is that the transition in behavior occurs mainly in the range of τ_K between 0.10 and 10.

The dispersion index is another fingerprint of the reaction mechanism. The step-addition scheme forms a polymer having narrow dispersion with an asymptotic D_N of 1.0 and a maximum of 1.33 (16). On the other hand, chain-addition polymers conforming to the QSSA show a D_N of 2.0 for the intermediates, and for the product as well when termination is by disproportionation. This behavior is clearly visible in Fig. 2.4-3. The exceptional rise in D_N for $\tau_K = 10$ is caused by late dead-ending (D-E) and sudden transition from chain to step kinetics in the course of reaction. Notice how D_N for the intermediates drops from 2.0 to 1.0 at 80-percent conversion. Termination is just slow enough in this case to allow significant chain growth by the intermediates before their species become extinct as a result of D-E.

It is interesting to note that τ_K forms an effective criterion for applicability of the QSSA. Although the results for $\tau_K = 100$ are fairly close to those computed with QSSA, when τ_K is raised to 1000 the differences are virtually indistinguishable.

Example 2.4-1. Evaluation of Dimensionless Parameter τ_K. Compute values of τ_K for Polystyrene, (PS) and Polymethylmethacrylate, (PMMA) at temperatures of 30, 60, and 90°C.

	Styrene	Methyl Methacrylate
k_p (l/mol s) :	$1.057 \times 10^7 \exp(-3557/T)$	$9 \times 10^5 \exp(-2365/T)$
k_t (l/mol s) :	$1.255 \times 10^9 \exp(-844/T)$	$1.1 \times 10^8 \exp(-604/T)$

Solution.

$$\tau_K = \frac{k_t}{k_p}$$

$$\text{styrene} \quad \tau_K = 119 \exp\left(\frac{2713}{T}\right)$$

$$\text{MMA} \quad \tau_K = 122 \exp\left(\frac{1761}{T}\right)$$

τ_K	30°C	60°C	90°C
S	9.2×10^5	4×10^5	2.1×10^5
MMA	4.1×10^4	2.4×10^4	1.6×10^4

CONCLUSION

Not only are these radical polymerizations clearly chain-addition, but they also obey the QSSA since in all cases $\tau_K \gg 10^3$.

2.5. THE GAUSSIAN CCD

By analogy with our development of instantaneous DPD's for chain-addition polymers in Sections 1.13 and 2.2.6, it should be possible to find an instantaneous CCD for chain-addition copolymers. Following the method of Stockmayer (19), we will utilize the pseudo distributions discussed briefly in Section 1.5 and the exponential approximation of the most probable distribution, which was reviewed in Section 2.2.6.

We begin with the active intermediates and apply the "local" QSSA to each rate function ($r_{a,b}^{j*} = 0$) in Table 1.12-3. This leads to the following set of coupled, partial, first-order difference equations among intermediate concentrations

$$a \geqslant 1, b \geqslant 0 \qquad c_{a,b}^{A*} = \phi_{AA} c_{a-1,b}^{A*} + \frac{R_A}{R_{AB} R_B} \phi_{AB} c_{a-1,b}^{B*} \qquad (2.5\text{-}1)$$

$$a \geqslant 0, b \geqslant 1 \qquad c_{a,b}^{B*} = \frac{R_{BA} R_B}{R_A} \frac{c_B}{c_A} \phi_{BA} c_{a,b-1}^{A*} + \phi_{BB} c_{a,b-1}^{B*} \qquad (2.5\text{-}2)$$

with IC's: $c_{a,0}^{A*} = c_{1,0}^{A*} \phi_{AA}^{a-1}$ and $c_{0,b}^{B*} = c_{0,1}^{B*} \phi_{BB}^{b-1}$; where $R_{AB} \equiv k_{pAA}/k_{pBB}$ and $c_{1,0}^{A*}$, $c_{0,1}^{B*}$, and the probabilities ϕ_{jk} are defined in Table 2.5-1. It is apparent that the IC's represent the DPD's of homopolymeric intermediates and that they are most probable distributions. This is confirmed by comparison with DPD 2.2-93, rewritten as

$$c_x^* = c_1^* \phi^{x-1} \qquad (2.5\text{-}3)$$

The solutions of these equations (1.9-21), which also appear in Table 2.5-1. would give the bivariate distributions discussed in Section 1.5 for copolymer intermediates; thus

$$n^*_{a,b} = \frac{c^*_{a,b}}{\sum_a \sum_b c^*_{a,b}} \tag{2.5-4}$$

The corresponding number CCD would then be

$$n^*_a = \sum_b n^*_{a,b} = \frac{\sum_b c^*_{a,b}}{\sum_a \sum_b c^*_{a,b}} \tag{2.5-5}$$

In lieu of these unwieldy expressions, Stockmayer (19) derived approximate distributions that are far simpler to manipulate and to interpret. They are based on

Table 2.5-1.   Copolymer Compositions As Functions of Probabilities of Propagation

$$c^{A*}_{1,0} = (1 - \phi_{AA})c^{A*} - \phi_{BA}c^{B*}$$

$$c^{B*}_{0,1} = (1 - \phi_{BB})c^{B*} - \phi_{AB}c^{A*}$$

$$c^{A*}_{a,b} = \sum_{i=1}^{a} c^{A*}_{1,0} \binom{a-1}{i}\binom{b-1}{i-1} \phi^i_{AB} \phi^i_{BA} \phi^{a-1-i}_{AA} \phi^{b-i}_{BB}$$

$$+ \sum_{i-1}^{a} c^{B*}_{0,1} \binom{a-1}{i-1}\binom{b-1}{i} \phi^{i-1}_{AB} \phi^i_{BA} \phi^{a-i}_{AA} \phi^{b-i}_{BB}$$

$$c^{B*}_{a,b} = \sum c^{A*}_{1,0} \binom{a-1}{i-1}\binom{b-1}{i-1} \phi^i_{AB} \phi^{i-1}_{BA} \phi^{a-i}_{AA} \phi^{b-i}_{BB}$$

$$+ \sum_{i'=1}^{a} c^{B*}_{0,1} \binom{a-1}{i-1}\binom{b-1}{i} \phi^i_{AB} \phi^i_{BA} \phi^{a-i}_{AA} \phi^{b-1-i}_{BB}$$

where:

$$\phi_{jk} = \frac{k_{p_{jk}}c_k}{\left(k_{p_{jj}} + k_{f_{jj}}\right)c_j + \left(k_{p_{j\ell}} + k_{f_{j\ell}}\right)c_\ell + k_{fs_j}c_s + k_{t_{jj}}c^{j*} + k_{t_{j\ell}}c^{\ell*}}$$

$$j = A,B ; \quad k = A,B ; \quad \ell = A,B ; \quad \ell \neq j$$

the pseudo bivariate weight distribution introduced in Section 1.5 (equation 1.5-19). Stockmayer replaced the sums with integrals, as we did in Section 2.2-6, and evaluated the integrals with the aid of Stirling's formula. His result for the pseudo bivariate distribution for active copolymer intermediates is

$$w_p^*(\hat{x}, \hat{y}) = \left[\left(\frac{\hat{x}}{\pi}\right)^{1/2} \exp(-\hat{x}\hat{y}^2)\right]\left[\hat{x}\exp(-\hat{x})\right] \tag{2.5-6}$$

where the distributed variables are defined as

$$\hat{x} \equiv \frac{x}{\nu_N} \tag{2.5-7}$$

and

$$\hat{y} \equiv u^{1/2}(y - \nu_A) \tag{2.5-8}$$

and parameter u is

$$u \equiv \nu_N/2\nu_A(1 - \nu_A)\left[1 - 4\nu_A(1 - \nu_A)(1 - R_A R_B)\right]^{1/2} \tag{2.5-9}$$

Thus $w_p^* \, d\hat{x} \, d\hat{y}$ represents the fraction of intermediates having a reduced DP between $\hat{x}$ and $\hat{x} + d\hat{x}$ and a reduced CC between $\hat{y}$ and $\hat{y} + d\hat{y}$. The reader will note that this bivariate distribution is the product of two distributions, one Gaussian and the other most probable. Therefore, for each DP value of x, copolymer composition y is normally distributed about the mean value ν_A. However, the pseudo CCD, which is obtained by integration over $\hat{x}$

$$w_p^*(\hat{y}) = 2\int_0^\infty w_p^*(\hat{x}, \hat{y}) \, d\hat{x} = \frac{3}{4(1 + \hat{y}^2)^{5/2}} \tag{2.5-10}$$

is not Gaussian. The limits of integration extend to $\pm\infty$, although y is confined between zero and one. This is reasonable since the integrand is sufficiently narrow to render any errors incurred less than those due to other assumptions used in the derivation. Furthermore, the distribution is symmetrical about the origin ($\hat{y} = 0$ or $y = \nu_A$) so that the integral from $-\infty$ to $+\infty$ is the same as twice the integral from zero to $+\infty$. Graphs of the CCD have been plotted in reduced form in Fig. 2.5-1 for three different values of parameter $u(\nu_A)$ and a fixed value of ν_N.

The DPD, on the other hand, is given by

$$w^*(\hat{x}) = \int_{-\infty}^\infty w_p^*(\hat{x}, \hat{y}) \, d\hat{y} = \hat{x}\exp(-\hat{x}) \tag{2.5-11}$$

which is the most probable distribution (cf. equation 2.2-97).

It can be shown that the relative dispersion of the DPD is $(D_N^* - 1)^{1/2}$ (Section 1.4) and that the standard deviation (SD) of the CCD is (Example 2.5-1)

$$SD^* \equiv (2u)^{-1/2} \tag{2.5-12}$$

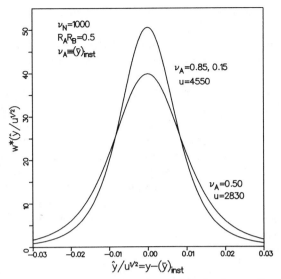

Figure 2.5-1. The copolymer composition distribution (CCD) for intermediates at three successive values of average copolymer composition (CC).

Note that this is the SD of a distribution that has been normalized, so its mean value is always $\hat{y} = 0$ or $y = \nu_A$. As such, the SD* behaves somewhat like an RD*. That is, the distribution has been forced to be centered on the same value regardless of ν_A. Therefore, the breadth of the distribution is independent of the mean. Although similar in behavior to RD*, the normalized SD* of this distribution is not the same as the RD* of the true distribution (Examples 2.5-1 and 2.5-3).

Recall from its definition (Sections 1.5 and 1.13) that ν_A is a number average CC, whereas its appearance as the mean value of 2.5-6, a pseudo weight distribution, suggests that it must also be the weight average CC. This could mislead us into concluding that statistical dispersion is nonexistent. Actually, it is an artifact due to the use of the pseudo-weight distribution. Since each repeat unit is treated as having a single molecular weight, the weight of all chains of DP $= x$ is independent of y. Therefore, averaging by weighting y with n_x^* or w_x^*, or any fraction dependent on chain length only, would give the same result. To find the dispersion, the moments of the CCD with respect to composition must be used, as illustrated in Example 2.5-3.

Equation 2.5-6 and the equations that follow from it are of a form that facilitate generalizations about CCD's, which are not apparent in more complicated forms such as the equations in Table 2.5-1. From equation 2.5-6 and Fig. 2.5-1, we conclude that chains with average CC ($\hat{y} = 0$) are the longest and also the most abundant; and from equation 2.5-12, we conclude that long chains and small values of the product $R_A R_B$ favor a narrow CCD. Notwithstanding the dependence of CCD dispersion on DP, and the fact that the instantaneous DP is fairly broadly dispersed (RD* $\cong$ 2), the normalized SD* of the instantaneous copolymer is still much less than unity for typical values of DP and $R_A R_B$ (Example 2.5-2). This conclusion forms the basis for the assumptions, commonly made, that the instantaneous CC is primarily determined by the long chains and that the instantaneous

CCD can be considered to be monodisperse for all practical purposes. We shall follow this practice when seeking the dispersive effect of drifting ν_A during polymerization on the cumulative CCD, mindful of the threat that short chains, such as those formed at high temperatures, pose to the validity of the aforementioned assumptions.

The maintenance of the narrow instantaneous CCD throughout copolymerization in the face of drift is related to the constancy of D_N^* under analogous conditions, discussed in Section 2.2.7. Both are characteristic of chain-addition reactions with termination and both require large values of DP ($\phi \cong 1$) for their validity. Another similarity between DP and CC behavior exists with respect to termination. Just as termination by any of the mechanisms previously discussed, other than combination, does not alter the DPD of the intermediates, the same is true of their CCD. Thus equations 2.6-10 and 2.6-11 also describe the final copolymer product. When a fraction P_2 of the intermediates are terminated by combination, then the instantaneous bivariate distribution of the product is (19)

$$w_p(\hat{x}, \hat{y})_{\text{inst}} = \left[\left(\frac{\hat{x}}{\pi} \right)^{1/2} \exp\left(-\hat{x}\hat{y}^2 \right) \right] \left[\hat{x} \exp(-x) \left(1 - P_2 + \frac{P_2 \hat{x}}{2} \right) \right] \quad (2.5\text{-}13)$$

Example 2.5-1.   **Standard Deviation of the Instantaneous CCD.**   Derive equation 2.5-12 from equation 2.5-10.

Solution.   By definition (Appendix C) $\sigma^2 = \mu^2 - (\mu^1)^2$ also

$$\mu_A^k = \int_{-\infty}^{\infty} \hat{y}^k w_p^*(\hat{y}) \, d\hat{y}$$

where

$$w_p^*(\hat{y}) \, d\hat{y} = \frac{3u^{1/2} \, d\hat{y}}{4\left[1 + \left(u^{1/2}\hat{y} \right)^2 \right]^{5/2}}$$

If we use the following transformation of variables

$$\tan\theta \equiv u^{1/2}(y - \nu_A) = u^{1/2}\hat{y}$$

then

$$\sec^2\theta \, d\theta = u^{1/2} \, d\hat{y}$$

and the kth moment can be expressed as:

$$\mu_A^k = \frac{3}{4} \int_{-\pi/2}^{\pi/2} \left(\frac{\tan\theta}{u^{1/2}} \right) \cos^3\theta \, d\theta$$

Integration with $k = 1$ verifies that $\mu_A^1 = 0$ as expected, since it must represent the average value of $\hat{y}$, and we know the distribution is symmetrical about the origin.

when $k = 2$, we find $\mu_A^2 = (2u)^{-1}$; thus, $\sigma^2 = (2u)^{-1}$ and

$$SD^* = \sqrt{\sigma^2} = (2u)^{-1/2}$$

Example 2.5-2.   Characteristics of the Instantaneous CCD.   For copolymerizations of industrial importance, the product $R_A R_B$ rarely exceeds a value of unity, even at high temperatures. Using $\nu_N = 1000$ and $R_A R_B = 1$, estimate the maximum value of SD*.

Solution.   Substituting definition 2.5-9 into equation 2.5-12 yields

$$SD^* = \left\{ \frac{\nu_A(1 - \nu_A)[1 - 4\nu_A(1 - \nu_A)(1 - R_A R_B)]^{1/2}}{\nu_N} \right\}^{1/2}$$

The maximum value of $\nu_A(1 - \nu_A)$ is 0.25 (22). Therefore, the maximum value of SD* is

$$SD^* = \left(\frac{0.25}{1000} \right)^{1/2} = 0.016$$

Example 2.5-3.   Relative Dispersion of the Instantaneous CCD.   Find the expression for μ_A^0, μ_A^1, μ_A^2, SD*, RD*, and D_A for the Stockmayer CCD using y rather than $\hat{y}$ as the distributed variable.

Solution.   By definition $\mu_{WA}^k = \int_0^1 y^k w(y)\, dy$ (remember this is a pseudo weight distribution), where

$$w(y)\, dy = \frac{3/4 u^{1/2}\, dy}{\left[1 + u(y - \nu_A)^2\right]^{5/2}}$$

Substitute

$$\tan \theta = u^{1/2}(y - \nu_A), \qquad \sec^2\theta\, d\theta = u^{1/2}\, dy$$

and let the limits of integration extend to $\pm \infty$ (as did Stockmayer) to obtain

$$\mu_{WA}^k = \frac{3}{4} \int_{-\infty}^{+\infty} \left(\frac{\tan \theta}{u^{1/2}} + \nu_A \right)^k \cos^3\theta\, d\theta$$

which yields

$$\mu_{WA}^0 = 1, \qquad \mu_{WA}^1 = \nu_A, \qquad \mu_{WA}^2 = \nu_A^2 + \frac{1}{2u}$$

so

$$\sigma_A^2 = \mu_{WA}^2 - \left(\mu_{WA}^2 \right)^2 = \frac{1}{2u}$$

giving

$$SD_A^* = \sqrt{\sigma^2} = (2u)^{-1/2}$$

and

$$RD_A^* = \frac{SD_A^*}{\nu_A} = (2\nu_A^2 u)^{-1/2}$$

Furthermore,

$$D_A = RD_A^2 + 1 = \frac{\mu_{WA}^2/\mu_{WA}^1}{\mu_{WA}^1/\mu_{WA}^0} = 1 + \frac{1}{2\nu_A^2 u}$$

It should be clear that since u is proportional to ν_N, it must be large. In addition, as ν_A tends to zero, u tends to infinity ensuring that $1/2\nu_A^2 u$ is small and thus the CCD is very nearly monodisperse, with $D_A \cong 1$.

2.6. STATISTICAL DISPERSION

We shall use the term statistical dispersion to describe the distribution among reaction products that results solely from molecular statistics associated with reactions. Several common product distributions were examined in Sections 2.2–2.5 from a kinetic viewpoint. In this section some of these will be reexamined using statistical arguments. This approach will be also applied to continuous reactors where appropriate in Chapter 3. The symbol ϕ, which was introduced earlier, will be used consistently to represent the probability of propagation in all statistical analyses.

2.6.1. Random Polymers

We begin with random polymerization, sequence 2.3-12, recalling that ϕ represented the fraction of end groups of either kind (A or B) that have reacted at any conversion $\Phi = \phi$. We now define ϕ as the probability that any end group has reacted. Since, by the principle of equal reactivity (PER), all groups have had the same opportunity to react at each conversion level regardless of the size of the chains to which they were attached, the probability of selecting x-mer at random is given by the combined probability of $x - 1$ successes (propagations) followed by one failure, $P_x = \phi^{x-1}(1 - \phi)$ (Appendix B). That is, if a polymer molecule is selected at random from among all those present (including m_1) at any conversion level ϕ, the probability that the penultimate functional group at either end has reacted is ϕ; the probability that the next one has reacted as well is ϕ^2, and so on. The probability that the xth one has not is $1 - \phi$. Moreover, P_x is equal to the mole fraction n_x because the selection at the outset was random. The statistical approach to the DPD for random polymerizations is due to Flory (23).

Several subtle points concerning this analysis are noteworthy. First, the use of constant probability ϕ in our statistical experiment does not imply that ϕ is constant throughout polymerization. In fact, it rises steadily, with conversion since $\phi = \Phi$. Second, the most probable DPD must fail at very high conversions where the number of polymer molecules becomes small. Third, the sum $\sum_{x=1}^{N}\phi^x$ is actually a finite geometric series, $\phi(1 - \phi^N)/(1 - \phi)$ (see Appendix C), which is closely approximated by an infinite series when $\phi < 1$, owing to the large upper limit N.

2.6.2. Step-Addition Polymers

Next, we examine ideal step addition polymerization, sequence 2.3-1. Here we define the probability of propagation as the reciprocal of the number of initiator molecules m_0 present initially, $\phi = 1/(N_0)_0$. Since, by the PER, all $(N_0)_0$ chains and potential chains have an equal opportunity to react, regardless of their size, the probability P_x that any m_0 chosen at random will grow to x-mer by the time N_m monomer molecules have reacted is given by the probability that it will experience x successes (propagations) and $N_m - x$ failures in any order $[N_m!/x!(N_m - x)!]\phi^x(1 - \phi)^{N_m-x}$. Order is unimportant because there is no termination. P_x, which is again equal to n_x because the selection was random, is called the binomial distribution (equation B-25). It may be approximated very closely by the Poisson distribution (equation B-29), since $(N_0)_0 \gg 1$ and $N_m \gg 1$ as required while their product remains a reasonable number: $N\phi = N_m/(N_0)_0 = z$. Typical values for $(N_0)_0$ and N_m are approximately 10^{20} and 10^{23}, respectively.

It is appropriate at this point to consider why sequence 2.3-19 does not follow this analysis. The reason is that the probability of propagation is not uniform. It is not the same for potential chains (initiator) as it is for actual chains.

2.6.3. Chain-Addition Polymers

The distributed properties of intermediates (Section 2.2.6) as well as the instantaneous properties (Section 2.2-7) of products in chain-addition polymerizations and copolymerizations can also be deduced using statistical arguments. We begin with intermediates, for which ϕ as defined in equations 2.2-44 and 2.2-45 signifies the probability of adding a monomer (propagating) rather than terminating or transferring activity. Thus ϕ^{x-1} represents the probability P_{1x} that any m_1^* chosen at random will grow to m_x^*. In this case, ϕ is constant because the variables (e.g., concentrations) affecting its value do not change (drift) significantly during the short period of time required for a single chain to grow. This is inherent in the nature of chain reactions. Hence the fraction of active x-mer at any time may be deduced from the fraction of active 1-mer which grows to x-mer, P_{1x}, as follows:

$$n_x^* = \frac{P_{1x}}{\sum\limits_{j=1}^{\infty} P_{1j}} = \phi^{x-1}(1 - \phi) \tag{2.6-1}$$

and the result, as expected, is the most probable DPD.

Concerning the deactivation of intermediates, $P_{tj} \equiv k_{tj}c^*/D$ represents the probability that an active chain deactivates by disproportionation ($j = D$) or combination ($j = C$), $P_{fk} = k_{fk}c_k/D$ represents the probability that it deactivates by transfer to monomer ($k = m$) or foreign substance ($k = s$), and $P_{ts} = k_{ts}/D$ represents the probability that it occurs spontaneously. Therefore, we can interpret P_2 in Section 2.2-7 as the probability that a product chain is formed by combination

$$
P_2 \equiv \frac{\dfrac{P_{tc}}{2}}{\dfrac{P_{tc}}{2} + P_{tD} + P_{ts} + P_{fm} + P_{fs}}
$$

$$
= \frac{\dfrac{k_{tc}c^*}{2}}{D - k_p c_{\mathrm{m}} - \dfrac{k_{tc}c^*}{2}}
\tag{2.6-2}
$$

because two active chains are terminated for each product chain formed (24). Clearly then, $P_1 = 1 - P_2$ is the sum of the probabilities that a product chain is formed by the remaining four deactivation mechanisms, $P_1 = (k_{tD}c^* + k_{ts} + k_{fm}c_{\mathrm{m}} + k_{fs}c_s)/(D - k_p c_{\mathrm{m}} - k_{tc}c^*/2)$. Thus $(n_x)_{\mathrm{inst}}$ may be interpreted as the probability of selecting x-mer at random from among all product molecules having just terminated.

This conclusion follows directly from combinational analysis. The probability of selecting x-mer is the sum of two probabilities. The first is the probability that the deactivation process was not combination, P_1, times the probability that the participants were active x-mer and any other polymeric intermediate $N_x^* \cdot N^*/N^{*2} = n_x^*$. The result agrees with the first term in equation 2.2-53. The second is the probability that termination occurred by combination, P_2, times the probability that it involved active y-mer and active $(x - y)$-mer for any value of y, which is given by $N_y^* N_{x-y}^* / \binom{N^*}{2} = 2n_y^* n_{x-y}^*$ when $y \neq x/2$ and $\binom{N_y^*}{2} = n_y^*$ when $y = x/2$ (cf. Section 1.8). The last term involves the convolution sum $(\sum_{j=1}^{x-1} n_j^* n_{x-j}^*)$, which has been shown to yield the second term in equation 2.2-53 excluding P_2.

2.6.4. Copolymers

Statistical methods can also be applied to the analysis of copolymer composition and sequence distributions. It is obvious from definition 1.13-15 that $R_j < 1$ implies a tendency on the part of chain end j toward heteropolymerization and $R_j > 1$ implies a tendency toward homopolymerization. Therefore, as $R_A \rightarrow 0$ and $R_B \rightarrow 0$, we should expect alternating copolymers and when $R_A > 1$ and $R_B > 1$, we should expect block copolymers. In the limit as $R_A \rightarrow \infty$ and $R_B \rightarrow \infty$, we expect no copolymer at all, but rather the concurrent formation of a mixture of homopolymers. On the other hand, for the situation $R_A \cong 1/R_B$ (where either $R_A < 1$ and $R_B > 1$, or $R_A > 1$ and $R_B < 1$), we would expect the copolymer structure to become random. Ideal (random) copolymerization occurs when $R_A R_B = 1$ and was discussed earlier. In this case $k_{pAA}/k_{pAB} = k_{pBA}/k_{pBB}$, which suggests that the relative affinity of both chain ends for the comonomers is identical. In fact, for most real polymerizations $0 < R_A R_B < 1$.

By ignoring chain transfer and termination, ϕ_{jk} in Table 2.5-1 becomes

$$j = A, B$$
$$k = B, B \qquad\qquad \phi_{jk} = k_{pjk}c_k/(k_{pjj}c_j + k_{pjl}c_l) \qquad\qquad (2.6\text{-}3)$$
$$j \neq l = A, B$$

where $\sum_j \sum_k \phi_{jk} = 1$. We interpret this quantity as the probability that an active chain ending in j propagates by adding a monomer of type k. Thus the probability that a copolymer chain grows a sequence containing a block of i consecutive repeat units of type j is

$$i = a, b$$
$$\qquad\qquad P_i = \phi_{jj}^{i-1}\phi_{jl} \qquad\qquad (2.6\text{-}4)$$
$$l \neq j$$

But, since $\sum_{i=1}^{\infty} P_i = \phi_{jl}/(1 - \phi_{jj}) = 1$ (25), we conclude that equation 2.6-4 is the most probable distribution

$$P_i = \phi_{jj}^{i-1}(1 - \phi_{jj}) \qquad\qquad (2.6\text{-}5)$$

where

$$\phi_{jj} = \frac{R_j c_j}{R_j c_j + c_l} \qquad\qquad (2.6\text{-}6)$$

The average sequence length of type A in long polymer chains is then

$$(\bar{a}_N)_{block} = 1 + \frac{R_A n_A}{1 - n_A} \qquad\qquad (2.6\text{-}7)$$

for example, where n_A is the mole fraction of monomer A in the unreacted comonomer pool. A similar result follows for type B, $(\bar{b}_N)_{block}$. As $R_j \rightarrow \infty$, $\phi_{jj} \rightarrow 1$, which is consistent with the expectations we outlined in Section 1.13.

2.6.5. Chain Branching

Long-chain branching in chain-addition polymers occurs when active intermediates propagate by adding inactive polymer product containing terminal double bonds or by transferring their activity to inactive product which subsequently propagates by adding monomer or polymer with terminal double bonds (Table 1.10-3). In addition to the resulting DPD, we are now also interested in the degree of branching distribution, DBD (Table 1.5-1), of the product molecules.

Concerning the DP and DPD of polymeric intermediates, it will be necessary to modify the definition of ϕ (equation 2.2-44 and 2.2-45) by including the appropriate rate functions that affect branching (Table 1.10-4). It should be clear, however, that n_x^* is still most probable (equation 2.2-93), notwithstanding the nonlinearity of intermediate chains. For product molecules, on the other hand, the DPD is expected to broaden as molecules of varying sizes are allowed to combine with one another via branching reactions, much in the same way that termination by combination produces broader product molecules than disproportionation does from the same intermediates. Thus, although definitions such as $(\bar{x}_N)_{inst}$ (equation 1.13-3) and $(n_x)_{inst}$ (equation 1.13-6) still apply, with appropriate modification of the expression for r_x, the product DPD's previously derived do not.

To characterize branching, a DBD was defined in Table 1.5-1 as the distribution of branches per polymer molecule b. Another distribution of interest in long-chain branching is the branch density b', that is, the number of branch points per repeat unit. The instantaneous mean branch density was defined in Section 1.13.3 as

$$\frac{r_b}{r_p} = \frac{(\bar{b}_N)_{inst}}{(\bar{x}_N)_{inst}} \tag{2.6-8}$$

Saito et al. (26) have shown that the instantaneous branch density is very narrowly distributed when chains are long relative to branch points ($x \gg b$). Assuming that branching occurs randomly along chains, the probability of finding b branches on any x-mer is given by the familiar binomial distribution

$$P_{b,x} = \binom{x}{b} \phi_b^b (1 - \phi_b)^{x-b} \tag{2.6-9}$$

where the probability that a repeat unit is branched is

$$\phi_b \equiv \frac{\sum_b b N_{b,x}}{x N_x} \tag{2.6-10}$$

This distribution may be approximated by the Poisson distribution when $\phi_b \ll 1$ and $x \gg 1$ (Appendix B)

$$P_{b,x} = \frac{(x\phi_b)^b (\exp - x\phi_b)}{b!} = \frac{N_{x,b}}{N_x} \tag{2.6-11}$$

Thus the probability ϕ_b represents the mean branch density among x-mers and the distributed variable is the number of branches per x-mer, b, with a mean value (cf. Table 1.5-1)

$$x\phi_b = \frac{\sum_b b N_{b,x}}{N_x} = \bar{b}_x$$

From equation 2.6-11, we conclude that the instantaneous values of b are narrowly distributed about their mean $\bar{b}_x$ for each x just like the instantaneous CC is narrowly distributed about its mean. However, while the branch density ϕ_b, which is a composition variable like CC, is independent of DP, the number of branches per polymer molecule b is not. Therefore, the instantaneous DBD, unlike the instantaneous CCD, is broadly dispersed, primarily because b is an extensive property rather than an intensive property like y. In fact, since b is proportional to x for long chains, the instantaneous DBD should closely resemble the instantaneous DPD. For these reasons, branch density, or its inverse, the mean distance between branch points, is often preferred over the number of branch points per molecule as a branch characteristic.

Equation 2.6-11 was propsed by Beasley (27) and it applies to instantaneous product. However, it should be clear that instantaneous DB is subject to drift, via ϕ_b, just as instantaneous DP and CC were. The molecular statistics of all chain-addition polymerizations and copolymerizations with termination are altered by drift. Therefore, we shall consider next the phenomenon called drift dispersion.

2.7. DRIFT DISPERSION

Drift dispersion occurs only in chain-addition polymerizations (including copolymerizations, terpolymerizations, etc.) with deactivation of intermediates. It is the additional dispersion experienced by cumulative product during formation above and beyond the statistical dispersion that is inherent in the instantaneous product comprising it (Section 1.13). The superposition of drift dispersion on statistical dispersion, illustrated in Fig. 2.7-1, is a direct consequence of variations in succeeding generations of instantaneous properties caused by drifting compositions and/or temperatures on which these properties depend.

If drift is allowed to occur, dispersion of cumulative properties will exceed that of instantaneous properties. The least dispersion one can hope for in the cumulative product is statistical dispersion, which has been prevented from "smearing" through the elimination of drift. This occurs when the pertinent reaction variables (composition and temperature) are fixed, as in thermostated semibatch reactors with programmed makeup feed, and in spatially uniform, continuous reactors at steady state (Chapter 3). Obviously, composition drift will occur in batch reactors even when

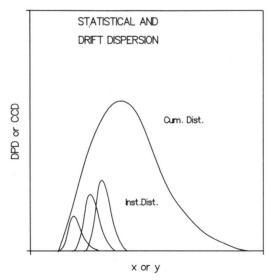

Figure 2.7-1. A sketch showing the additivity of instantaneous distributions to form a cumulative distribution.

temperature is maintained constant in time and space. The absence of drift dispersion in Example 2.7-1 represents a special case.

The separation of property dispersion into instantaneous and drift components is a useful aid to interpreting the response of polymerizations to variations in reaction parameters. Thus we expect a change in composition and/or temperature to manifest itself in two ways: an immediate change in the instantaneous product followed by a delayed change in the cumulative product (28). The instantaneous response should be more sensitive to parameter variations than the drift response is, because the latter is the result of a time-averaging process. Moreover, it is evident that conclusions about the direction and magnitude of the parametric dependence of cumulative properties that are inferred from the dependence of instantaneous properties are subject to the errors imposed by drift. An example is the use of equation 1.13-11 to predict the dependence of product DP on composition and temperature. This is commonly done, notwithstanding the fact that the equation actually applies to instantaneous DP only.

The effects of drift on polymer properties are significant. They will be examined in detail in various sections throughout this book within the context of specific phenomena causing the drift, such as initiator depletion (Section 2.8), hindered termination (Section 2.9), incomplete mixing (Chapter 3), and temperature variation (Chapter 4). For now, we will be content with a few simple computations to illustrate the concept.

2.7.1. Homopolymerization

The relationship between statistical dispersion and drift dispersion in DP has been sketched in Fig. 2.7-2. From the figure it is clear that the behavior of $(\bar{x}_N)_{inst}$ with conversion is an indicator of the direction and magnitude of drift in the ensuing DPD. Thus a steady value signals the absence of drift dispersion.

By using the definitions introduced in Section 1.11 and equation 1.13-11, we can express $(\bar{x}_N)_{inst}$ in terms of dimensionless concentrations $\hat{c}_m \equiv c_m/(c_m)_0$ and $\hat{c}_0 \equiv c_0/(c_0)_0$ as follows, if chain transfer is ignored:

$$(\bar{x}_N)_{inst} = \frac{2(\nu_N)_0}{2 - R} \frac{\hat{c}_m}{(\hat{c}_0)^{1-1/n}} \qquad (2.7\text{-}1)$$

This expression applies when termination occurs by combination ($n = 2$, $R = 1$), by disproportionation ($n = 2$, $R = 0$) or spontaneously ($n = 1$, $R = 0$). Concentration drift in batch reactors can be expressed in terms of a semidimensionless monomer balance.

$$-\frac{d\hat{c}_m}{dt} = \lambda_m^{-1}\hat{c}_m\hat{c}_0^{1/n} \qquad (2.7\text{-}2)$$

Definitions of $(\nu_N)_0$ and λ_m are listed in Table 2.7-1 in general form as well as for the special case of free-radical intermediates.

The reader will recall that λ_m characterizes the speed of polymerization and $(\nu_N)_0$ characterizes the product DP. Thus both are key parameters. From Table 2.7-1, we

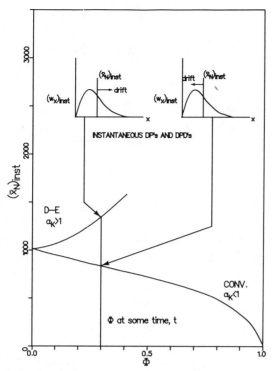

Figure 2.7-2. A sketch showing the instantaneous DPD at one level of conversion and the drift with conversion of the corresponding instantaneous average DP (28).

Table 2.7-1. Characteristic Times and Dimensionless Groups

CHARACTERISTIC TIMES	λ_m	λ_i
GENERAL	$1/k_{ap}\left(c_o\right)_o^{1/n}$	s/k_i
FREE-RADICAL	$1/k_p(2fk_d/k_t)^{1/2}\left(c_o\right)_o^{1/2}$	$1/k_d$
DIMENSIONLESS GROUPS	$\alpha_k\ (\equiv \lambda_m/\lambda_i)$	$\left(\nu_N\right)_o\ (\equiv x_o/s\alpha_K)$
GENERAL	$1/sk_{ax}\left(c_o\right)_o^{1/n}$	$x_o k_{ax}\left(c_o\right)_o^{1/n}$
FREE-RADICAL	$(2fk_dk_t)^{1/2}/2fk_p\left(c_o\right)_o^{1/2}$	$\left(c_m\right)_o k_p/(2fk_dk_t)^{1/2}\left(c_o\right)_o^{1/2}$
LUMPED RATE CONSTANTS	k_{ap}	$k_{ax}\ (\equiv k_{ap}/k_i)$
GENERAL	$k_p(k_i/k_t)^{1/n}$	$k_p/k_i^{1-1/n}k_t^{1/n}$
FREE-RADICAL	$k_p(2fk_d/k_t)^{1/2}$	$k_p/(2fk_dk_t)^{1/2}$

can relate them to laboratory parameters in the following way:

$$\lambda_m^{-1} = k_{ap}(c_0)_0^{1/n} \tag{2.7-3}$$

$$(\nu_N)_0 = x_0 k_{ax}(c_0)_0^{1/n} \tag{2.7-4}$$

where k_{ap} and k_{ax} are "lumped" kinetic constants, also listed in Table 2.7-1.

The simplest expression for cumulative DP is stoichiometric equation 1.7-11. For free-radical polymerizations ($s = 2$) without chain transfer, for instance, we obtain, neglecting density changes,

$$\bar{x}_N = \frac{2\alpha_K (\nu_N)_0}{(2 - R)} \frac{(1 - \hat{c}_m)}{(1 - \hat{c}_0)} \tag{2.7-5}$$

where the fraction of combination terminations R' has been replaced by the kinetic ratio R (Section 1.13). This expression and its instantaneous counterpart, equation 2.7-1, can both be expressed more conveniently in terms of monomer conversion, $\Phi_m \equiv 1 - \hat{c}_m$, by combining the equations in Example 1.11-1 for $\hat{c}_m$ and $\hat{c}_0$ as functions of time $\hat{t}$ to eliminate $\hat{c}_0$. The results for free-radical polymers,

$$(\bar{x}_N)_{inst} = \frac{\dfrac{2(\nu_N)_0}{(2 - R)}(1 - \Phi_m)}{1 + \dfrac{\alpha_K}{2}\ln(1 - \Phi_m)} \tag{2.7-6}$$

and

$$\bar{x}_N = \frac{2\alpha_K (\nu_N)_0 \Phi_m}{(2 - R)\left\{1 - \left[1 + \dfrac{\alpha_K}{2}\ln(1 - \Phi_m)\right]^2\right\}} \tag{2.7-7}$$

have been plotted in Fig. 2.7-3 for three values of parameters α_K and for the following constant parameters: $(\nu_N)_0 = 500$ and $R = 1 = f$. It can be shown that $\bar{x}_N \to (\bar{x}_N)_{inst}$ when $\Phi_m \to 0$, as expected. The reader will perhaps recall that the cumulative DP curve for $\alpha_K = 0.1$ and $R = 1$ was used earlier in Fig. 2.1-4 to represent chain polymerization behavior. In Fig. 2.7-4, graphs of monomer conversion and initiator conversion ($\Phi_i \equiv 1 - \hat{c}_0$) have been plotted for the same three values of α_K, together with the equation for the locus of dead-end (D-E) conversions

$$1 - \Phi_m = \exp\left(\frac{-2}{\alpha_K}\right) \tag{2.7-8}$$

obtained from the combined results of Example 1.11-1 alluded to earlier, that is,

$$(1 - \Phi_i)^{1/2} = 1 + \frac{\alpha_K}{2}\ln(1 - \Phi_m) \tag{2.7-9}$$

after setting $\Phi_i = 1$.

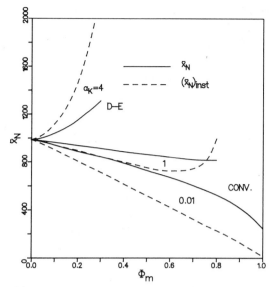

Figure 2.7-3. Computed instantaneous and cumulative DP for three values of α_K ranging from conventional (CONV) to dead-end (D-E) conditions.

From these equations (cf. equations 2.7-6 and 2.7-7) and graphs, we interpret $(\nu_N)_0$ as a "target" parameter for the desired value of DP and α_K as a drift parameter. The latter determines the direction and extent that actual DP drifts from the target value. These interpretations of $(\nu_N)_0$ and α_K are consistent with those proposed earlier, that is, they characterize the LCA and D-E, respectively. Thus, when a large DP is desired, the value of $(\nu_N)_0$ should be made large. When $\alpha_K < 1$,

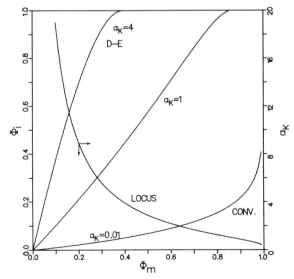

Figure 2.7-4. Computed initiator conversion (Φ_i)–monomer conversion (Φ_m) profiles for various values of α_K and locus of monomer conversions for all dead-end (D-E) polymerizations within the range $1 \leqslant \alpha_K \leqslant 20$.

Table 2.7-2. Dimensionless Criteria

TARGET PROPERTY	CRITERION	
	INSTANTANEOUS RESPONSE	DRIFT RESPONSE
Large DP/LCA	$\left(\nu_N\right)_o \gg 1$	$\alpha_K > 1$
Narrow DPD	$\phi \ll 1^*$	$\alpha_K \sim 1$
Large CC (in A)	$\left(\nu_A\right)_o \gg 1$	$\beta_K > 1$
Narrow CCD	$R_A R_B \ll 1$	$\beta_K \sim 1$

*Unrealistic because it precludes a large DP.

we can expect conventional (CONV) polymerization with a downward drift in $\bar{x}_N$; when $\alpha_K > 1$, we can expect D-E polymerization with an upward drift. These dimensionless criteria are summarized in Table 2.7-2.

It is instructive to perform some order-of-magnitude calculations for typical chain-addition polymerizations analogous to those done in Sections 2.2 and 2.3 for random and step-addition polymerizations. Some kinetic constants for vinyl monomers and free-radical initiators are listed in Tables 2.7-3 and 2.7-4, respectively. From these we compute values for k_{ap} when $f = 0.5$ and $T = 80°C$, say, that range from $1.01 \times 10^{-5} \, l^{1/2} \, mol^{-1/2} \, s^{-1}$ for the system S/DTBP to 2.66×10^{-2} for AN/AIBN. These systems are referred to as the "slowest" and "fastest," respec-

Table 2.7-3. Kinetic Properties of Some Vinyl Monomers

	STYRENE (36) (S)	METHYLMETHACRYLATE (37) (MMA)	ACRYLONITRILE* (AN)	VINYL CHLORIDE (39) (VC)	VINYL ACETATE ((VA)
$A_p \, (l.mol^{-1}s^{-1})$	1.06×10^7	7.35×10^5	4.6×10^8	3.3×10^6	3.2×10^7
$E_p \, (Kcal.mol.^{-1})$	7.07	4.8	8.1	3.7	5.0
$A_t \, (l.mol^{-1}s^{-1})$	1.26×10^9	1.8×10^8	3.5×10^9	6.0×10^{11}	3.7×10^9
$E_t \, (Kcal \, mol^{-1})$	1.68	1.3	5.4	4.2	3.2

*in 2.5M DMF (38)

tively, among those examined in Example 2.7-2. To establish a time frame, we note that when $k_{ap} = 10^{-3}\ l^{1/2}\ mol^{-1/2}\ s^{-1}$ the value of λ_m is approximately 10^4 s. When $(c_m)_0 = 1\ mol/l$ and $(c_0)_0 = 0.01\ mol/l$ in our slow and fast examples, the values for $(\nu_N)_0$ are 1760 and 1914, respectively, both of which indicate that high polymer should be formed. Furthermore, from equation 1.11-4, the corresponding values of α_K are 0.028 and 0.026, both of which predict conventional polymerizations.

It is noteworthy that the probability of propagation ϕ is relatively insensitive to drift. This can be seen by expressing it in the following way,

$$\phi = \frac{k_{ax}}{k_{ax} + \dfrac{c_0^{1/2}}{c_m}} \qquad (2.7\text{-}10)$$

where chain transfer has again been ignored for illustrative purposes, and by noting that k_{ax} takes on values of the order of $10^2\ l^{1/2}\ mol^{-1/2}$. For example, in the slowest and fastest cases cited earlier, the values for k_{ax} are 176 and 191 $l^{1/2}\ mol^{-1/2}$, respectively. As a corollary, we conclude further by using the expression for instantaneous DPD dispersion, $(D_N)_{inst} = 1 + \phi$, that $(D_N)_{inst}$ is also relatively insensitive to drift. The curves in Fig. 2.7-5, which correspond to the case $\alpha_K = 0.1$ in Fig. 2.7-3, illustrate the insensitivity predicted.

If $(D_N)_{inst}$ remains insensitive to drift, how then can cumulative D_N increase significantly and thereby reflect drift dispersion? This question is perhaps best answered with an example (Example 2.7-4). More generally, as stated earlier, a good indicator of drift dispersion is the variation in $(\bar{x}_N)_{inst}$. Such variation can be significant, as evident in the figures. Furthermore, it can be shown that in general, cumulative D_N is bounded by the extreme values of $(\bar{x}_N)_{inst}$ in accordance with the following inequality (29, 30)

$$D_N \leqslant \frac{(D_N)_{inst}(\bar{x}_N)_{inst,\,max}}{(\bar{x}_N)_{inst,\,min}} \qquad (2.7\text{-}11)$$

This relationship is predicated on the assumption that $(D_N)_{inst}$ is constant and follows directly from the definition of D_N as the quotient $\bar{x}_W/\bar{x}_N$, together with the fact that cumulative $\bar{x}_W$ cannot exceed the largest instantaneous $\bar{x}_W$ and cumulative $\bar{x}_N$ cannot be less than the smallest instantaneous $\bar{x}_N$. Thus $D_N \leqslant$

Table 2.7-4. Kinetic Properties of Some Free-Radical Initiators

	A_d (s^{-1})	E_d ($Kcal.mol^{-1}$)
Benzoyl peroxide (BP) (41)	6.38×10^{13}	29.7
Azobis-isobutyronitrile (AIBN) (42)	2.67×10^{15}	31.1
Di-t-butyl peroxide (DTBP) (43)	4.3×10^{15}	37

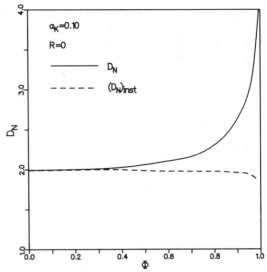

Figure 2.7-5. Instantaneous, $(D_N)_{inst}$, and cumulative, D_N, dispersion index versus monomer conversion Φ for $\alpha_K = 0.1$.

$(\bar{x}_W)_{inst, max}/(\bar{x}_N)_{inst, min}$ and equation 2.7-11 follows from the relationship $(D_N)_{inst}$ $\cong (\bar{x}_W)_{inst, max}/(\bar{x}_N)_{inst, max}$.

In summary, a narrow statistical dispersion (small $D_N)_{inst}$ requires that ϕ be small (Table 2.7-2). However, this is incompatible with large values of $(\bar{x}_N)_{inst}$. Therefore, to achieve a large DP with the least cumulative dispersion, the best one can do is reduce drift dispersion by making the value of α_K close to unity.

Several common ways to increase the DP of polymer product are: reduce initiator concentration, reduce temperature, and select a slow initiator. Before discussing the instantaneous and drift responses of chain-addition polymers to these laboratory parameters, especially temperature, it will be helpful to examine the temperature dependence of k_{ap} and k_{ax}, that is, to examine E_{ap} and E_{ax}. Using Table 2.7-1, we can write

$$E_{ap} = E_p + \frac{E_d - E_t}{n} \qquad (2.7\text{-}12)$$

and

$$E_{ax} = E_p - \frac{(n-1)E_d + E_t}{n} \qquad (2.7\text{-}13)$$

where n is the order of the termination reaction. From these equations, it is evident that negative values of E_{ap} and E_{ax} are possible, depending, of course, on the magntidues of the individual elementary activation energies. After substituting the values for free-radical polymerizations listed in Tables 2.7-3 and 2.7-4, we conclude that E_d exerts a strong influence on the lumped activation energies and that is does, in fact, cause E_{ax} to be negative. This explains why DP drops instantaneously with rising temperature. It also causes E_{ap} to be larger than E_p, thereby making the overall temperature dependence greater than that of propagation. Nevertheless, E_d is still larger than E_{ap}, indicating that initiator consumption is more temperature dependent than monomer consumption. One consequence, of course, is an increasing tendency toward D-E with rising temperature. Typical values for E_{ap} and E_{ax} are of the order of 25 and -10 kcal/mol, respectively.

For cationic polymerizations ($n = 1$), on the other hand, it is not uncommon for E_t to have a large value exceeding the sum of E_d and E_p, causing the reaction to have an overall, negative temperature dependence ($E_{ap} < 0$).

The expected responses, both instantaneous and drift, to variations in laboratory parameters are summarized in Table 2.7-5. They have been expressed in terms of the general parameters in Table 2.7-1. All primary adjustments were directed toward raising the target DP. As expected, all result in a detrimental effect on monomer conversion (longer λ_m). However, as indicated, some give rise to competing responses with respect to product DP. Thus, while reducing $(c_0)_0$ tends to increase DP initially [larger $(\nu_N)_0$], as well as ultimately (larger α_K) by favoring an upward drift (D-E) in DP, reducing T tends to make the polymerization more conventional (smaller α_K), which results in a downward drift in DP counter to the initial response. These conclusions are illustrated in Figs. 2.7-6 and 2.7-7. We observe that the initial drop in instantaneous DP in Fig. 2.7-7 in response to a higher temperature level is offset by an upward drift response that allows significant recovery by cumulative DP. Note that both DP scales are logarithmic.

Although the direction of instantaneous and drift responses of free-radical polymers to variations in temperature level is due to the large value of E_d alluded to

Table 2.7-5. Response of Free-Radical Polymerization to Parameter Variations

LAB PARAMETERS	RESPONSE PARAMETERS		
	λ_m	$\left(\nu_N\right)_o$	α_K
$\left(c_0\right)_o$ ↓	↑	↑	↑
T ↓	↑	↑*	↓*
A_d ↓	↑	↑	↓
E_d ↑	↑	↑	↓

*Assuming that $E_{ax} < 0$

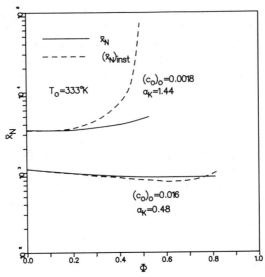

Figure 2.7-6. Molecular weight drift with monomer conversion in response to varying feed initiator concentration (28).

earlier and the concomitant negative value of E_{ax}, the existence of competition, per se, between these responses in chain-addition polymers generally is due to the opposing effects of k_{ax} on $(\nu_N)_0$ and α_K, which are evident in their definitions (Table 2.7-1). For the same reason, we expect competing responses to initiator choice, as characterized by the kinetic pair A_d, E_d and indicated in Table 2.7-5.

On a final note, the reader will find that separating responses will also prove useful in interpreting property dispersion stemming from temperature drift. For

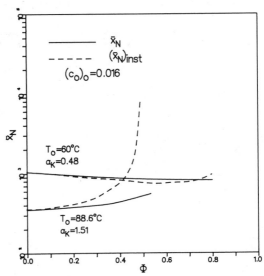

Figure 2.7-7. Molecular weight drift with monomer conversion in response to varying initial temperature (28).

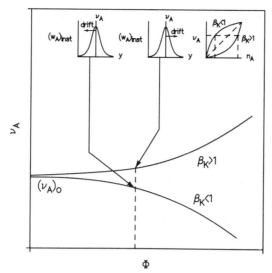

Figure 2.7-8. CC and CCD drift for different values of β_K.

instance, it is commonly believed that rising temperatures during nonisothermal, chain-addition polymerizations with inadequate heat transfer give rise to polymers and copolymers whose properties are more dispersed than those of corresponding products from isothermal reactions. In our terminology, this belief is based on the expected downward drift of instantaneous DP in response to an upward drift in temperature, with a resultant smearing of the cumulative DPD. By our reasoning, however, if the nonisothermal polymerization in question were of the D-E type, based on feed conditions ($\alpha_K < 1$), we would actually expect rising temperatures to have a beneficial effect on DPD by offsetting the tendency for DP to drift upward under isothermal conditions. In general then, a drifting temperature will not necessarily exacerbate dispersion; on the contrary, it could even mitigate dispersion in certain cases. For more on thermal effects, the reader is referred to Chapter 4.

2.7.2. Copolymerization

Composition drift during chain-addition copolymerization with deactivation is determined by the copolymer composition (CC) equation, $\nu_A = \nu_A(n_A)$ (equation 1.13-14). Thus the CC diagram (Figs. 1.13-2–1.13-6) may be viewed as being analogous to Fig. 2.7-2 in the sense that it represents instantaneous product property, in this case $(\bar{y})_{inst}$, versus the state of the reaction in terms of unreacted comonomer n_A. A more precise analogy to Fig. 2.7-2 is illustrated in Fig. 2.7-8. A corresponding CC diagram is included as an inset. Its relationship to the drift curves should be apparent. Included also is a CC drift parameter β_K, which is the counterpart of DP drift parameter α_K. Its origin and significance will be discussed shortly.

The definition of instantaneous CC in equation 1.13-13 may be written for batch reactions as

$$(\bar{y})_{\text{inst}} = \nu_A \equiv \frac{r_{pA}}{r_p} \cong \frac{d(n_A N_m)}{dN_m} \qquad (2.7\text{-}14)$$

if we assume that propagation dominates the consumption of both monomers. The last result, which is known as the Skeist equation when rearranged as follows,

$$\frac{dn_A}{\nu_A - n_A} = -\frac{d\Phi}{1 - \Phi} \qquad (2.7\text{-}15)$$

also has a counterpart in simple batch distillation, known as the Rayleigh equation

$$y = \frac{d(xL)}{dL} \qquad (2.7\text{-}16)$$

Here y represents the composition of an element of vaporized liquid dL, which has been removed under equilibrium conditions from a liquid reservoir having composition x at the time of removal, and which takes with it an amount of volatile component $d(xL)$.

When equation 2.7-15 is integrated, following substitution of the CC equation $\nu_A = \nu_A(n_A)$, one obtains a functional relationship between n_A and Φ, which can then be substituted into the CC equation to give an expression for instantaneous CC (ν_A) versus total monomer conversion (Φ). Meyer and Lowry (31) integrated equation 2.7-15. Their result is

$$\Phi = \left[\frac{1 + n_A}{1 - (n_A)_0} \right]^\alpha \left[\frac{n_A}{(n_A)_0} \right]^\beta \left[\frac{(n_A)_0 - \delta}{n_A - \delta} \right]^\gamma \qquad (2.7\text{-}17)$$

where $\alpha \equiv R_A/(1 - R_A)$, $\beta \equiv R_B/(1 - R_B)$, $\gamma \equiv (1 - R_A R_B)/(1 - R_A)(1 - R_B)$, and $\delta \equiv (1 - R_B)/(2 - R_A - R_B)$. Note that δ represents the azeotropic composition as given in equation 1.13-18. O'Driscoll and Knorr (32), on the other hand, derived an expression for n_A as a function of time in a batch reactor starting with the following rate equations (Table 1.12-2):

$$k = A, B \qquad -\frac{dc_k}{dt} \underset{\text{LCA}}{\cong} r_{pk} = \sum_{j=A}^{B} r_{pjk} = \sum_{j=A}^{B} k_{pjk} c^{j*} c_k \qquad (2.7\text{-}18)$$

Their result is

$$\ln\left\{ \left[\frac{1 - n_A}{1 - (n_A)_0} \right]^a \left[\frac{n_A}{(n_A)_0} \right]^b \left[\frac{(n_A)_0 - \delta}{n_A - \delta} \right]^c \right\} \qquad (2.7\text{-}19)$$

$$= -\left[k_{pBA} - Xk_{pBB} \right] \int_0^t c^* \, dt$$

where $a \equiv \alpha(1 - X) - X$, $b \equiv \beta(1 - X) + 1$, $c \equiv \gamma(1 - X)$, and $X \equiv (k_{pAA} - k_{pBA})/(k_{pAB} - k_{pBB})$. It should be pointed out that neither result applies to the azeotropic case $\nu_A = \nu_B$. Equation 2.7-19 is generally applicable; however, its use requires one to evaluate the integral on the RHS. This can only be done with knowledge of the termination kinetics, and realistic models do not lead to an analytically tractable expression.

From a simple material balance, equation 1.7-16, we may express cumulative average CC as follows, if density changes are neglected during reaction.

$$\bar{y} = \frac{[(n_A)_0(c_m)_0 - n_A c_m]}{[(c_m)_0 - c_m]}$$

$$= \frac{[(n_A)_0 - n_A(1 - \Phi)]}{\Phi} \qquad (2.7\text{-}20)$$

Elimination of n_A with the aid of equation 2.7-17 yields an expression for CC as a function of conversion $\bar{y}(\Phi)$. Thus $\nu_A(\Phi)$ and $\bar{y}(\Phi)$ are the instantaneous and cumulative CC property functions corresponding to equations 2.7-6 and 2.7-7, respectively, for DP.

Before we compute CC drift profiles, let us attempt to develop dimensionless target and drift parameters and criteria in a manner parallel to that used for DP. We begin by observing that Λ_m and λ_i, defined earlier, may be interpreted alternatively as reciprocal initial time gradients (33, 34)

$$\Lambda_m \equiv \left(\frac{d\hat{c}_m}{dt} \right)_0^{-1}$$

$$\lambda_i \equiv \left(\frac{d\hat{c}_0}{dt} \right)_0^{-1} \qquad (2.7\text{-}21)$$

This follows from the procedure outlined in Appendix G for deducing CT's. Similarly, we define a CT for monomer A as

$$\Lambda_A \equiv \left(\frac{d\hat{c}_A}{dt} \right)_0^{-1} \qquad (2.7\text{-}22)$$

All concentrations are reduced with their corresponding feed values, $\hat{c}_k \equiv c_k/(c_k)_0$. By analogy with α_K, we now define a CC drift parameter as

$$\beta_K \equiv \frac{\Lambda_A}{\Lambda_m} \qquad (2.7\text{-}23)$$

With the aid of equations 2.7-14, 2.7-20, 2.7-22, this definition can be rewritten as

$$(\nu_A)_0 = \frac{(n_A)_0}{\beta_K} \qquad (2.7\text{-}24)$$

This result should be compared with its counterpart for DP's, $(\nu_N)_0 = x_0/s\alpha_K$

(equation 1.11-4). Thus we interpret $(\nu_A)_0$ as the target parameter for desired CC and β_K as the drift parameter. To avert drift dispersion, one should prevent instantaneous CC from drifting away from its feed value $(\nu_A)_0$; for instance, by introducing makeup feed via a semibatch reactor or by taking advantage of the steady state in a well-mixed continuous reactor, as discussed earlier in this section.

Returning now to the CC diagram for representation of drift, we observe first that irreversible copolymerization can cause n_A to move monotonically in either direction along the abscissa, depending on whether the curve lies above or below the diagonal. If it lies above the diagonal, the initial copolymer formed, which corresponds to feed composition $(n_A)_0$, is richer in a monomer A, that is, $(\nu_A)_0 > 1$. Consequently, the pool of unreacted comonomers becomes leaner in A. This causes n_A to move toward the left on the CC diagram with a concomitant downward drift in ν_A. When the curve lies below the diagonal the opposite occurs; that is, n_A moves toward the right and ν_A drifts upward. Furthermore, the larger the displacement from the diagonal, the more pronounced the drift effect (35). The crossover situation corresponding to $R_A > 1$ and $R_B > 1$ (Fig. 1.13-6) represents an analogy to azeotropic distillation. Compositions drift toward the point of intersection of the curve and the diagonal, and copolymer of constant composition is produced thereafter. This crossover, however, is rarely encountered in practice. The other one, corresponding to $R_A < 1$ and $R_B < 1$, is not uncommon. In this case drift is away from the intersection on either side. Only for the special case of a feed composition that corresponds precisely to the intersection, or to the unlikely situation corresponding to $R_A = 1 = R_B$, does no drift occur at all.

It can be shown that the location of the CC curve relative to the diagonal is conveniently given by the value of β_K relative to unity. By substituting $(n_A)_0$ from equation 2.7-24 into the difference $(\nu_A)_0 - (n_A)_0$, which measures displacement, we obtain

$$(\nu_A)_0 - (n_A)_0 = (\nu_A)_0(1 - \beta_K) \qquad (2.7\text{-}25)$$

Thus the value of β_K is an indication of both direction and magnitude of CC drift (Example 2.7-5), analogous to α_K for DP drift. When $\beta_K > 1$ we expect drift to be upward with respect to the content of monomer A; the larger the value, the greater will be the drift. Similarly, when $\beta_K < 1$, the opposite is true. These conclusions are illustrated in Fig. 2.7-8 and summarized in Table 2.7-2. The criterion for narrow instantaneous CCD in the Table 2.7-2 follows from the discussion of Sections 1.13 and 2.5. Copolymerization drift may be analyzed by studying the results of numerical simulation. The form of the balance equations is such that the choice of termination model (Section 2.9) is irrelevant. The term pertaining to termination is common to both the total and individual monomer balances and thus drift is shaped solely by the propagation rates.

A somewhat unusual application of the QSSA is used in copolymerization kinetics. For the intermediate population of either type ultimate unit, we can say:

$$\frac{dc^{j*}}{dt} = r_{ij} + r_{pjk} - r_{pkj} - r_{tj} \cong 0 \qquad (2.7\text{-}26)$$

However, since propagation rates are far greater than those of initiation or termina-

Table 2.7-6. Vinyl Copolymerization Kinetic Constants

Co-monomers (A/B)	R_A	R_B	ϕ	R_A	R_B
S/AN	$2.56\ \exp(-599/T)$	$6.67 \times 10^{-5}\ \exp(2184/T)$	23	-	-
S/MMA	$1.83\ \exp(-450/T)$	$1.27\ \exp(-340/T)$	13	$1.615\ \exp(166/T)$	$0.38\ \exp(1610/T)$
AN/MMA	$1.94\ \exp(-916/T)$	$0.59\ \exp(260/T)$	200	-	-
AN/VA	$9.75\ \exp(-270/T)$	$18\ \exp(-1902/T)$	-	-	-

$$R_j = k_{pjj}/k_{pkj}$$

$$\phi = k_{tAB}/(k_{tAA}k_{tBB})^{1/2}$$

$$R_j = k_{tkjjk}/k_{tjjjj} \quad \text{where} \quad j = A,B \ ; \ k \neq j$$

tion (LCA), the QSSA reduces to $r_{pjk} = r_{pkj}$, or

$$k_{pkj}c_k c^{j*} = k_{pjk}c_j c^{k*} \tag{2.7-27}$$

Expressions for the radical intermediates can be deduced via the QSSA and are dependent on the termination model. Results are shown in Table 1.12-4. Substitution into the rate equations (Section 1.12) produces a means for evaluating the dimensionless drift parameters, in particular β_K. The expression proves not to be a function of termination kinetics:

$$\beta_K = \frac{R_A n_A^2 + 2n_A n_B + R_B n_B^2}{R_A n_A + n_B} \tag{2.7-28}$$

Using the kinetic constants for copolymer systems of various characteristics CC diagrams, we can examine CC drift behavior. In particular, we focus on three (Figs. 1.13-2, 1.13-4, and 1.13-5): AN/MMA, a nonazeotropic system; S/MMA, a system with CC curve close to the 45° line (and $R_A \cong R_B$); and S/AN, a system with an azeotrope, large deviations from the 45° line, and $R_A \gg R_B$. Reactivity ratios and termination parameters for these comonomer pairs are given in Table 2.7-6.

Figures 2.7-9–2.7-11 show the drift of instantaneous and cumulative CC for these systems covering the full range of feed compositions. Note that, as expected, the drift for AN/MMA is always towards AN-rich compositions since the CC lies below the 45° line for all values of $(n_A)_0$. Furthermore, the value of β_K goes through a maximum as does the magnitude of drift in ν_A. When AN composition is high ($(n_A)_0 > 0.70$), MMA is exhausted prior to completion of the reaction. This results in an accelerated rate of reaction as it changes in character from a copolymerization of AN/MMA to a homopolymerization of AN (Fig. 2.7-9). The conversion history looks much like that observed in the presence of gel-effect (Section 2.10) but is not a manifestation of hindered termination.

For the system S/MMA, which has an azeotrope near $(n_A)_0 = 0.50$, Fig. 2.7-10 shows that trends in both the magnitude and direction of CC drift correlate with the

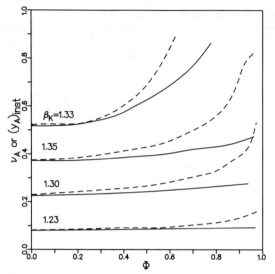

Figure 2.7-9. Instantaneous and cumulative CC drift for various values of $(n_A)_0$ for the system AN/MMA.

value of β_K. Premature consumption of either comonomer only occurs at very high levels of total monomer conversion. The system S/AN (Fig. 2.7-11) shows much more peculiar drift behavior. In spite of a large value of $\nu_A - n_A$ for $n_A < 0.70$, the CC drift at low conversions is rather mild owing to the relatively flat slope of the CC curve (Fig. 1.13-5) in this region. However, since ν_A is considerably larger than n_A, styrene is preferentially depleted and is consumed far short of full conversion of monomer. Thus, when β_K is less than 1.00 for this system, the reaction may be

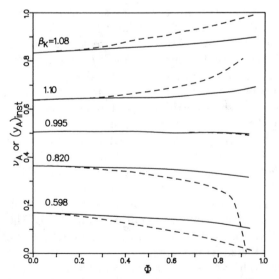

Figure 2.7-10. Instantaneous and cumulative CC drift for various values of $(n_A)_0$ for the systems S/MMA.

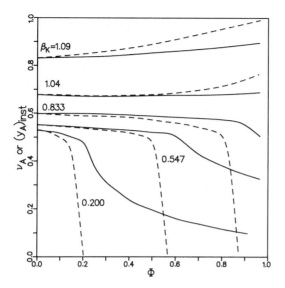

Figure 2.7-11. Instantaneous and cumulative CC drift for various values of $(n_A)_0$ for the system S/AN.

characterized as a relatively drift-free copolymerization followed in time by an AN homopolymerization. The slope effect is discussed in Example 2.7-5.

A word of caution is necessary concerning our S/AN simulations. It is well known that AN homopolymerization is precipitous while S/AN copolymerizations below the azeotrope have some degree of heterogeneity. The simulations were conducted using the AN constants for homogeneous solution polymerization since the kinetic model presumes this type of behavior. Nonetheless, the drift profiles are characteristic of the behavior to be expected from copolymer systems with CC diagrams similar to S/AN.

Just as composition drift leads to an increase in the polydispersity of the DPD beyond that imposed by statistical dispersion, so must it affect the CCD. In spite of the fact that the instantaneous CCD is narrow enough to be presumed monodispersed, as ν_A drifts in response to compositional changes, a cumulative CCD will be described (as in Fig. 2.7-1).

Calculation of moments of the CCD is directly analogous to the procedure for finding moments of the DPD. Furthermore, the dispersion index formed by

$$D_A = \frac{\mu_A^2/\mu_A^1}{\mu_A^1/\mu_A^0} \tag{2.7-29}$$

has the same relationship to the RD of the CCD as D_N has to its counterpart, the DPD. The moments of the CCD are given as:

$$\hat{\mu}_A^k = \int_0^\Phi (\nu_A)^k \, d\Phi \tag{2.7-30}$$

Table 2.7-7. Moments of the Cumulative CCD

$$(\hat{\mu}_A^0)^{'} \equiv \frac{\hat{\mu}_A^0}{\Phi_f} = 1$$

$$(\hat{\mu}_A^1)^{'} \equiv \frac{(\hat{\mu}_A^1)}{\Phi_f} = \frac{1}{\Phi_f} \int_0^\Phi \nu_A d\Phi \equiv \overline{\nu}_A$$

$$(\hat{\mu}_A^2)^{'} = \frac{(\hat{\mu}_A^2)}{\Phi_f} = \frac{1}{\Phi_f} \int_0^\Phi \nu_A^2 \, d\Phi$$

$$(\sigma_A^2)^{'} = \frac{1}{\Phi_f} \int_0^\Phi (\nu_A - \overline{\nu}_A)^2 d\Phi = (\mu_A^2)^{'} - [(\hat{\mu}_A^1)^{'}]^2$$

where

$$\hat{\mu}_A^k \equiv \frac{\mu_A^k}{(c_m)_0} \tag{2.7-31}$$

Clearly the zeroth moment is nothing more than the monomer conversion. We might readily normalize the distributions, then, by the final conversion to find the results shown in Table 2.7-7.

Just as D_N proved to be a valuable indicator of the breadth of the DPD, D_A may be used to characterize the breadth of the CCD. Since statistical dispersion is presumed to be absent from the CCD, D_A will have a minimum value of one and grow upward as spread occurs.

In Fig. 2.7-12, we see that D_A responds as expected to drift in composition. For the AN/MMA system with $(n_A)_0 = 0.70$, initial drift in n_A has little effect on ν_A, and thus the CCD remains essentially monodisperse. The value of D_A remains close to one. Beyond 50-percent monomer conversion, however, the MMA becomes depleted and rapid changes in ν_A result. The sudden rise in D_A corresponds to this period of reaction. Note that at 200°C, the rise in D_A occurs at higher conversions than at 50°C. This is consistent with the fact that at the higher temperature the CC curve lies closer to the 45° line (Fig. 1.13-2). Therefore, the disparity between ν_A and n_A is not as great, and it takes longer for the MMA to become depleted.

Example 2.7-1.   Absence of Drift in Batch Reactors.   Suppose that a free-radical polymerization is thermally initiated (no initiator) and that the initiation rate is second order with respect to monomer $r_i = k_i c_m^2$. Show that in the absence of chain transfer this special polymerization would exhibit no drift dispersion in a batch

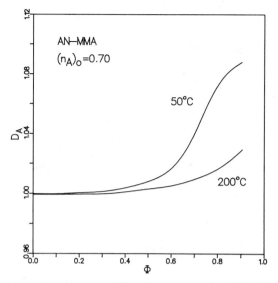

Figure 2.7-12. Growth in the breadth of the CCD with conversion for AN/MMA with $(n_A)_0 = 0.7$ at two initial temperatures.

reactor, and consequently the cumulative product DPD would be identical to the instantaneous DPD, $(n_x)_{inst}$, for any termination mechanisms.

Solution. Since the intermediates are free radicals, $r_t = k_t(c^*)^2$. Therefore, from the QSSA (Section 1.11)

$$c^* = \left(\frac{k_i}{k_t}\right)^{1/2} c_m$$

Substituting this result into equation 2.2-91 shows that ϕ is constant

$$\phi = \frac{k_p c_m}{k_p c_m + (k_i k_t)^{1/2} c_m} = \text{constant}$$

and from equation 2.6-1 that n_x^* is also constant and most probable. Furthermore, assuming that the termination mode as reflected by P_1 and P_2 remains constant, we conclude from equation 2.2-101 that $(n_x)_{inst}$ is constant in time, regardless of its specific functional form

$$(n_x)_{inst} = f_x = \text{constant}$$

Finally, after substituting the definition of instantaneous DPD, $(n_x)_{inst} \equiv r_x/\Sigma_x r_x$ (equation 1.13-6) into the definition of cumulative DPD (following equation 1.13-8)

$$n_x = \frac{\int_0^t r_x \, dt}{\int_0^t \Sigma_x r_x \, dt}$$

we obtain the desired result, that is, $n_x = (n_x)_{inst}$.

Example 2.7-2. Some CT's for Typical Free-Radical Polymerizations. It is frequently of interest to rank monomers and initiators according to their relative reaction speeds, that is, to know which are "fast" and which are "slow." Using the data in Tables 2.7-3 and 2.7-4, rank monomers S, MMA, and AN and initiators BP, AIBN, and DTBP in terms of characteristic times (CT's) at, say, T = 80°C. Estimate λ_m when $(c_0)_0 = 5 \times 10^{-3}$ mol/l.

Solution. We have concluded that λ_i and λ_m characterize initiator and monomer consumption, respectively. However, the latter quantity is not independent of initiator characteristics. To isolate its monomer dependence, the definition of λ_m (cf. Table 2.7-1) may be rearranged as follows for long, free-radical chains:

$$\lambda_m^{-1} \underset{\text{LCA}}{\cong} \frac{(r_p)_0}{(c_m)_0} = \frac{\dfrac{k_p (c_m)_0^{1/2}}{k_t^{1/2}} k_i^{1/2}}{x_0^{1/2}}$$

Noting that the first term in brackets on the RHS serves that purpose, we define a new CT for monomer reactivity alone

$$\lambda_{pt} \equiv \frac{k_t}{k_p^2 (c_m)_0}$$

and rewrite λ_m as a product of independent CT's

$$\lambda_m = \lambda_{pt}^{1/2} \lambda_i^{1/2} \left(\frac{x_0}{s} \right)^{1/2}$$

where, as usual,

$$\lambda_i \equiv k_d^{-1}$$

Substitution of the constants in Tables 2.7-3 and 2.7-4 into these definitions for λ_{pt} and λ_i leads to the following values. The former was tabulated as $\lambda_{pt}(c_m)_0$ to remove its dependence on feed concentration. From these results, we conclude that the fastest monomer and initiator are AN and AIBN, respectively, and the slowest are S and DTBP.

Some monomer and initiator CT's at 80°C are

	S	MMA	AN
k_p (1/mol s)	4.53×10^2	7.84×10^2	4.44×10^3
k_t (1/mol s)	1.15×10^8	1.99×10^7	8.41×10^8
k_t / k_p^2 (mol s/l) $\propto \lambda_{pt}$	560	460	42.7

	DTBP	BP	AIBN
$k_d (s^{-1})$	5.74×10^{-8}	2.76×10^{-5}	1.48×10^{-4}
$\lambda_i (s)$	1.74×10^7	3.62×10^4	6.76×10^3

The following values of λ_m in seconds are obtained for $(c_0)_0 = 5 \times 10^{-3}$ mol/l.

	DTBP	BP	AIBN
S	9.9×10^5	4.5×10^4	1.9×10^4
MMA	8.9×10^5	4.1×10^4	1.8×10^4
AN	2.7×10^5	1.2×10^4	5.4×10^3

Corresponding values of α_K for each monomer-initiator pair can now be calculated.

Example 2.7-3. *Typical Chain Characteristics for Free-Radical Polymerizations.* Estimate the chain lifetimes (λ_p) and free-radical concentrations (c^*) for the slowest and fastest monomer-initiator pair in Example 2.7-2 when $f = 0.5$ and $(c_0)_0 = 0.01$ mol/l.

Solution. From Section 1.11,

$$\lambda_p = \frac{1}{(2fk_dk_t)^{1/2}(c_0)_0^{1/2}}$$

and

$$(c^*)_0 = \left(\frac{2fk_d}{k_t}\right)^{1/2}(c_0)_0^{1/2}$$

The reader will note that both definitions are based on feed concentration and that $(c^*)_0$ is therefore fictitious, the true value being zero. After substituting the appropriate numbers, we obtain 3.9 S and 0.022 s. for λ_p of the slowest and fastest systems, respectively, and 2.23×10^{-9} mol/l and 3.03×10^{-8} mol/l for $(c^*)_0$. These values are consistent with those quoted in the text as being "typical" for chain-addition polymerizations.

Example 2.7-4. *A Contrast Between Statistical Dispersion and Drift Dispersion.* With a simple, approximate calculation, we can demonstrate how drifting $(\bar{x}_N)_{\text{inst}}$ increases cumulative dispersion (D_N) while $(D_N)_{\text{inst}}$ remains constant. Assuming that $(D_N)_{\text{inst}} = 2$, compute cumulative $\bar{x}_N$, $\bar{x}_W$, and D_N for three different drift modes from the following seven instantaneous polymer samples of weight fraction w_j each taken from a batch reactor.

j	$(w_p)_j$	$(\bar{x}_N)_j$ Conventional	$(\bar{x}_N)_j$ Dead-end	$(\bar{x}_N)_j$ Gel-effect (Mild)
0	0	1000	1000	1000
1	0.4	800	1100	900
2	0.2	600	1500	800
3	0.1	500	2000	850
4	0.05	400	3000	900
5	0.03	300	6000	1000
6	0.02	200	10,000	1200

Solution.   With the aid of Example 1.13-3, we obtain the relationship $dW_p/dt = M_0 r \cong M_0 \Sigma x r_x$. After substitution into the equations for cumulative averages, the latter yield the following approximate equations:

$$\bar{x}_N = \frac{\displaystyle\int_0^t \frac{dW_p}{dt} dt}{\displaystyle\int_0^t (\bar{x}_N)_{inst}^{-1} \frac{dW_p}{dt} dt} \cong \frac{\displaystyle\sum_j (w_p)_j}{\displaystyle\sum_j (\bar{x}_N)_j^{-1}(w_p)_j}$$

$$\bar{x}_W = \frac{(D_N)_{inst} \displaystyle\int_0^t (\bar{x}_N)_{inst} \frac{dW_p}{dt} dt}{\displaystyle\int_0^t \frac{dW_p}{dt} dt} \cong \frac{2\displaystyle\sum_j (\bar{x}_N)_j (w_p)_j}{\displaystyle\sum_j (w_p)_j}$$

from which the results listed in the following chart were computed. We observe that D_N increases in all cases: during conventional polymerization because $\bar{x}_N$ falls more rapidly than $\bar{x}_W$ and during dead-end polymerization because it rises less rapidly. These conclusions are consistent with those in Section 1.4 that $\bar{x}_N$ gives greater weight to the low end of the DP spectrum generated during CONV polymerization, while $\bar{x}_W$ favors the high end produced in D-E polymerization.

CONV

i	$(w_p)_j$	$(\bar{x}_N)_j$	$(\bar{x}_W)_j$	$\bar{x}_N$	$\bar{x}_W$	D_N
0	0	1000	2000	—	—	—
1	0.4	800	1600	800	1600	2.00
2	0.2	600	1200	720	1467	2.04
3	0.1	500	1000	680	1400	2.06
4	0.05	400	800	647	1360	2.10
5	0.03	300	600	619	1331	2.15
6	0.02	200	400	588	1308	2.22

D-E

0	0	1000	2000	—	—	—
1	0.4	1100	2200	1099	2200	2.00
2	0.2	1500	3000	1207	2500	2.07
3	0.1	2000	4000	1280	2714	2.12
4	0.05	3000	6000	1330	2933	2.21
5	0.03	6000	12,000	1371	3282	2.39
6	0.02	10,000	20,000	1401	3700	2.64

G-E (mild)

0	0	1000	2000	—	—	—
1	0.04	900	1800	901	1800	2.00
2	0.2	800	1600	865	1733	2.00
3	0.1	850	700	862	1729	2.01
4	0.05	900	1800	864	1733	2.01
5	0.03	1000	2000	869	1756	2.02
6	0.02	1200	2400	874	1773	2.03

Example 2.7-5. *Extent and Direction of CC Drift.* O'Driscoll (35) has suggested the magnitude of the displacement $\nu_A - n_A$ as a measure of the extent of CC drift. Capinpin (22) has pointed out that it should be the product of the displacement and the slope of the CC curve $d\nu_A/dn_A$. Reconcile these suggestions.

Solution. The Skeist equation (2.7-15) characterizes drift in the comonomer pool with conversion,

$$\frac{dn_A}{d\Phi} = - \frac{(\nu_A - n_A)}{1 - \Phi}$$

However, what we seek is a measure of CC drift with conversion $d\nu_A/d\Phi$, which is given by the product $(d\nu_A/dn_A)(dn_A/d\Phi)$. Thus,

$$\frac{d\nu_A}{d\Phi} = - \frac{d\nu_A}{dn_A} \frac{\nu_A - n_A}{1 - \Phi}$$

from which we conclude that, not only does a large slope of the CC curve, $d\nu_A/dn_A$, amplify the magnitude of drift by entering as a product with displacement, but increasing values of total conversion Φ amplify it as well. On the other hand, small slopes can offset large displacements. The reader will also note that the sign of the displacement determines the sign of $d\nu_A/d\Phi$ (direction of drift). Furthermore, the initial drift is given by

$$\left(\frac{d\nu_A}{d\Phi}\right)_0 = - \left(\frac{d\nu_A}{dn_A}\right)_0 (\nu_A)_0 (1 - \beta_K)$$

Example 2.7-6. *Some CT's and Dimensionless Parameters for Free-Radical Copolymerizations.* Evaluate time constants for the disappearance of styrene and total monomer (Λ_A, Λ_m) and then find α_K and β_K for the system S/MMA/AIBN with $(n_A)_0 = 0.70$, and $(c_0)_0 = 0.12$ at 55°C. Use phi factor kinetics.

Solution. By definition, $\Lambda_A = (c_A)_0/(r_A)_0$ and $\Lambda_m = (c_m)_0/(r)_0$. Hence

$$\Lambda_A^{-1} = \frac{(r_A)_0}{(c_A)_0} \cong \sum_{j=A}^{B} \frac{(r_{pjA})_0}{(c_A)_0} = \sum_{j=A}^{B} \lambda_{jA}^{-1}$$

$$\Lambda_m^{-1} = \frac{(r)_0}{(c_m)_0} \cong \sum_{j=A}^{B} \sum_{k=A}^{B} (n_k)_0 \frac{(r_{pjk})_0}{(c_k)_0} = \sum_{j=A}^{B} \sum_{k=A}^{B} (n_k)_0 \lambda_{jk}^{-1}$$

Using the kinetic expressions in Section 1.12 and the constants from 2.7 we find:

$$\lambda_{AA}^{-1} = 1.36 \times 10^{-5}, \qquad \lambda_{BA}^{-1} = 2.92 \times 10^{-5}, \qquad \lambda_{AB}^{-1} = 1.25 \times 10^{-5},$$

$$\lambda_{BB}^{-1} = 5.67 \times 10^{-6}$$

Hence, $\Lambda_A = 3.83 \times 10^4$ s and $\Lambda_m = 3.48 \times 10^4$ s As a result,

$$\beta_K \equiv \frac{\Lambda_A}{\Lambda_m} = 1.101$$

$$\alpha_K = \frac{\lambda_i}{\Lambda_m} = 0.211$$

We predict that the copolymerization will show a moderate drift towards styrene-rich compositions since $\beta_K > 1$. The DP drift will be downward, but moderate ($0.1 < \alpha_K < 1.0$).

2.8.   CHAIN INITIATION EFFECTS

Monomers involved in random polymerization generally contain the functional groups necessary for propagation and can therefore polymerize without additional chemical reagents. Catalysts may be added to increase the rate of reaction, but such compounds are not consumed in the course of reaction. Addition polymers that are formed via chemical intermediates generally require an added substance to transform the monomer into a reactive intermediate. This substance is called an initiator and, unlike a catalyst, is consumed during the course of reaction. Consumption of initiator, frequently ignored in kinetic models can have important consequences in shaping both the rate of reaction and the properties of the polymer produced.

2.8.1.   Dead-End Polymerization

Early kinetic studies of chain-addition polymerizations were conducted at sufficiently low temperatures that the half-life of the common initiators (AIBN and BP) were considerably greater than the length of time required for monomer conversion. The amount of initiator consumed was insignificant and the rate of initiation was presumed constant and equal to its initial value. Such polymerizations have been referred to as conventional (CONV). It should be clear that we can identify them via the inequalities (Section 1.6):

$$\lambda_i \gg \lambda_m \tag{2.8-1}$$

or

$$\alpha_K \ll 1 \tag{2.8-2}$$

which may then be used to simplify the dimensionless rate equation for monomer consumption

$$\frac{d\hat{c}_m}{d\hat{t}} = -\hat{c}_m \tag{2.8-3}$$

where $\hat{t} = t/\lambda_m$. One must ask, however, to what extremes one can press the rate of initiation, either through diminished initiator in the feed or through an enlarged constant for decay, such that condition 2.8-1 is violated and approximation 2.8-3 does not apply. Furthermore, one must ask what the consequences of such behavior would be with regard to polymer properties.

Tobolsky (44) coined the phrase "dead-end" (D-E) polymerization to describe those reactions which, in marked contrast to CONV polymerizations, ran out of initiator prior to complete conversion of monomer. Although an error in his equations, pointed out by Tadmor and Biesenberger (45), led him to overstate the effects of D-E, the behavior of such reactions is still significantly different from that of the CONV type. From our discussion of drift phenomena, we know that the instantaneous DP is shaped by the competition between propagation and initiation rates. If composition drift affects one rate more than the other, then a drift in the instantaneous DP will occur. In the event of D-E, the rate of initiation falls more rapidly than rate of propagation, owing to the higher order of dependence that r_i has upon c_0 (first order) than r_p has (half order). As a result, one would expect instantaneous $\bar{x}_N$ and $\bar{x}_W$ to drift upward as the reaction proceeds. This production of high ends should be especially apparent in $\bar{x}_W$ and the values of D_N and D_W will both rise.

Graphs of cumulative $\bar{x}_N$, $\bar{x}_W$, and D_N versus conversion Φ, based on analytical solutions of the rate equations for termination by combination only and neglecting chain transfer (Example 2.8-1), were used by Tadmor and Biesenberger (45) to demonstrate that D_N always increases with conversion when drift occurs, whether $\bar{x}_N$ falls (CONV) or rises (D-E). When $\bar{x}_N$ falls ($\alpha_K < 1$), $\bar{x}_W$ falls less rapidly and consequently D_N increases. When $\bar{x}_N$ rises ($\alpha_K > 1$), $\bar{x}_W$ rises more rapidly and, again, D_N increases.

Alternatively, we could integrate the moment balance equations in Table 2.4-2 numerically to compute $\bar{x}_N$, $\bar{x}_W$, D_N, and even D_W. This was done for termination by disproportionation ($R = 0$) and for a range of values of α_K. Initiator efficiency f and the gel-effect function G were both set equal to one.

In Figs. 2.8-1 and 2.8-2, these trends are illustrated for various values of α_K, the drift parameter described in Section 2.7. It should be immediately obvious that $\alpha_K = 1$ does indeed represent a critical value for a transition in behavior. Not only do we see that the initiator is prematurely exhausted, but the downward drift in both $\bar{x}_N$ and $(\bar{x}_N)_{inst}$, characteristic of conventional polymerization, begins to fade, and at high conversions reverses. The value of $(\bar{x}_N)_{inst}$ goes through a dramatic rise just prior to D-E.

The onset of this behavior is not continuous with increasing values of α_K. The behavior for $\alpha_K = 0.01$ and 0.1 is virtually identical while a significant change takes place when α_K is raised one decade further to a value of one.

Finally in Fig. 2.8-3, we find the most conspicuous effect of D-E. The reaction is incapable of achieving complete conversion owing to the premature consumption of all initiator. Again, increasing α_K from 0.01 to 0.1 has little consequence in altering reaction characteristics. The next order-of-magnitude increase makes quite a difference and coincides with the critical value of $\alpha_K = 1$. This is consistent with our postulate that reaction behavior should change in character when time scales for initiator and monomer consumption become equal.

Example 2.8-1. Cumulative Properties from Instantaneous Properties. Cumulative $\bar{x}_N$ for batch, free-radical polymerization with termination by combination

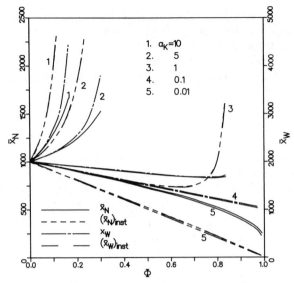

Figure 2.8-1. Drift of instantaneous and cumulative $\bar{x}_N$ and $\bar{x}_W$ with conversion with α_K as parameter.

$(R = 1)$ and initiator efficiency f is given by the expression (cf. equation 2.7-7)

$$\bar{x}_N = \frac{\dfrac{2(\nu_N)_0 \Phi \alpha_K}{f}}{1 - \left[1 + \dfrac{\alpha_K}{2f} \ln(1 - \Phi)\right]^2}$$

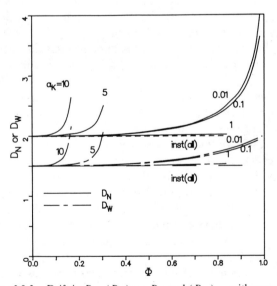

Figure 2.8-2. Drift in D_N, $(D_N)_{inst}$, D_W and $(D_W)_{inst}$ with conversion.

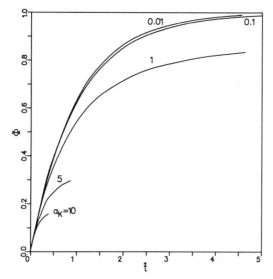

Figure 2.8-3. Monomer conversion versus dimensionless time (t/λ_m) with α_K as parameter.

Using the relationships for instantaneous properties presented in Section 1.13, deduce the corresponding expression for cumulative $\bar{x}_W$.

Solution. From Example 1.13-3,

$$\bar{x}_W = \frac{\int_0^t (\bar{x}_W)_{inst}\, dc_m}{\Delta c_m}$$

and, since $(D_N)_{inst}$ is constant (Section 2.7), we can write $(\bar{x}_W)_{inst} = 1.5\,(\bar{x}_N)_{inst}$ for termination by combination (Section 2.3-7) where $(\bar{x}_N)_{inst} \cong r/\Sigma r_x = 2r/r_i$ from Example 1.13-3 and equations 1.13-9 and 1.13-10. Following substitution into the first equation and integration with the aid of the results of Example 1.11-1, we obtain (30, 45)

$$\bar{x}_W = 6(\nu_N)_0 f\alpha_K^{-1}\Phi^{-1}\left\{ E_1\left(4f\alpha_K^{-1}\right) - E_1\left[4f\alpha_K^{-1}\left(1 + \alpha_K(2f)^{-1}\ln(1-\Phi)\right)\right]\right\}$$

where $E_1(z) \equiv \int_1^\infty \zeta^{-1}\exp(-z\zeta)\, d\zeta$ is the tabulated exponential integral (Appendix C).

2.8.2. D-E and the QSSA

In our preliminary discussions of the QSSA in Section 1.11, we concluded that validity of the approximation demanded that initiation and termination rates be roughly equivalent. One might question whether this is indeed possible under D-E conditions when rate of initiation becomes quite high. With a fixed and small

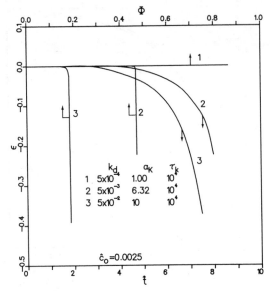

Figure 2.8-4. Error in calculation of radical concentration by use of QSSA (46).

termination rate constant, can we reasonably assume that termination keeps pace with the large rate of radical generation during D-E polymerizations?

To test the validity of the QSSA in the face of D-E, the exact and approximate equations can be integrated for large values of α_K with τ_K fixed at the very large values characteristic of chain addition ($\tau_K = 10^4$, Section 2.4). The kinetic equations with and without the QSSA were listed in Section 2.4 (Tables 2.4-1 and 2.4-2). In

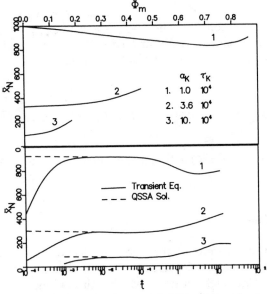

Figure 2.8-5. Comparison of exact and QSSA solutions for $\bar{x}_N$ (46).

Figs. 2.8-4 and 2.8-5, the comparison is presented. It is evident in 2.8-6 that for very large values of α_K, there may be an appreciable error in the value of c^* at the point where the limiting D-E conversion is reached (46), otherwise no error is apparent.

Of greater practical significance is the error in the computed values of the polymer properties. If the error in c^* has little effect on computed polymer properties, then the QSSA may be used with impunity. Figure 2.8-7 shows this to be the case. The true and QSSA solutions follow the same track throughout the extremes of reaction as D-E is approached. The tiny increment of conversion over which error occurs contributes only a very small amount to the cumulative polymer.

One might question whether the agreement also holds when τ_K is pressed to the lower limits of applicability of the QSSA. Figures 2.8-6 and 2.8-7 bear out the same trend observed for very large τ_K. With $\tau_K = 100$ and $\tau_K = 10$, the QSSA still correctly predicts the trajectory of chain-length growth with the exception of a noticeable error at low conversions, just as we observed in Section 2.4. These results lead us to the conclusion that even for values of kinetic constants exceeding those characteristic of radical polymerization, use of the QSSA yields no appreciable error in computed conversion or molecular weight histories in the face of severe D-E conditions.

2.8.3. Initiator Efficiency

In Section 1.10 we saw that the initiation sequence for chain-addition polymers is a multistep process involving initiator cleavage followed by attack of the initiator fragment on the monomer. It is commonly accepted that for many initiators not all fragments survive to initiate monomers. Referred to as the "cage effect" (47), this loss in radical generation is generally accounted for by introducing a proportionality constant, the initiator efficiency f (Example 1.10-2). Simply stated, the assumption is

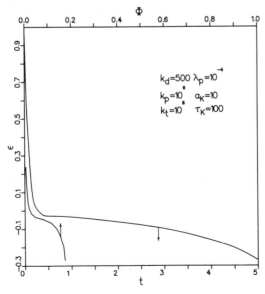

Figure 2.8-6. Error in calculation of radical concentration by use of QSSA (46).

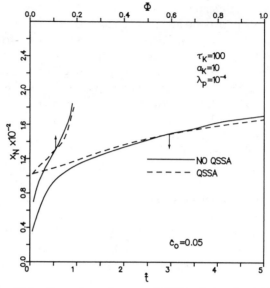

Figure 2.8-7. Comparison of exact and QSSA solutions for $\bar{x}_N$ (46).

that throughout the course of reaction a constant fraction of initiator radicals survive to initiate polymerization, and thus the rate of initiation can be expressed as (Example 1.10-2):

$$r_i = 2fk_dc_0 \tag{2.8-4}$$

Let us consider a more exact treatment of the initiation mechanism. The loss of initiator radicals must be ascribed to either the re-combination of fragments or reaction with other radical species present to form inactive products. If we accept the scheme proposed for initiators like AIBN and BP: (Table A-7):

$$I \rightarrow 2m_0^* + G \uparrow \tag{2.8-5}$$

$$m_0^* + m \rightarrow m_1^* \tag{2.8-6}$$

where G represents some gaseous by-product like N_2 or CO_2, then recombination of m_0^*

$$m_0^* + m_0^* \rightarrow P \tag{2.8-7}$$

will not regenerate the initiator, but will instead yield nonreactive product P. The first step of the preceding sequence is presumed to take place very rapidly, and the evolution of a gaseous by-product permits the process to be considered irreversible. Discussion of efficiency must then focus on the fate of radical species m_0^*. While we might allow for reaction with any and all species present in the reaction mixture, we propose that reaction with monomer or another radical of m_0^* should dominate. The former is a consequence of the large concentration of monomer relative to any other species, and the latter to the close physical proximity of the two fragment radicals formed by cleavage of a single initiator molecule.

Consider the following scheme:

$$I \xrightarrow{k_d} 2m_0^* \qquad\qquad r_1 = k_d c_0 \tag{2.8-9}$$

$$m_0^* + m \xrightarrow{k_i'} m_1^* \qquad\qquad r_2 = k_i' c_0^* c_m \tag{2.8-10}$$

$$m_0^* + m_0^* \xrightarrow{k_c} P \qquad\qquad r_3 = k_c c_0^{*2} \tag{2.8-11}$$

$$m_x^* + m \xrightarrow{k_p} m_{x+1}^* \qquad\qquad r_p = k_p c_m c^* \tag{2.8-12}$$

$$m_x^* + m_y^* \xrightarrow{k_t} m_x + m_y \qquad\qquad r_t = k_t c^{*2} \tag{2.8-13}$$

Although we have neglected other side reactions involving m_0^*, the following analysis can be extended to include these should they be of consequence to a particular reaction mechanism. The pertinent rate equations for intermediate species are:

$$r_0^* = 2r_1 - r_2 - 2r_3 \tag{2.8-14}$$
$$r^* = r_2 - r_t \tag{2.8-15}$$

and by definition the rate of initiation is r_2. Invoking the QSSA for initiator fragment radicals, we find

$$2r_1 = r_2 + 2r_3 \tag{2.8-16}$$

The initiator efficiency at any given moment is the rate at which fragments successfully initiate polymerization divided by the total rate at which these fragments are consumed:

$$f = \frac{r_2}{r_2 + 2r_3} \tag{2.8-17}$$

Thus, in accord with the result of Example 1.10-2,

$$r_i \equiv r_2 = 2fr_1 \tag{2.8-18}$$

where

$$f = \left(1 + \frac{2k_c c_0^*}{k_i c_m}\right)^{-1} \tag{2.8-19}$$

Solving equation 2.8-16 for c_0^*, we find

$$c_0^* = \frac{2fk_d c_0}{k_i c_m} \tag{2.8-20}$$

which, upon insertion in equation 2.8-19 followed by solution for f, yields

$$f = \frac{\left[\left(1 + \dfrac{16k_d k_c c_0}{k_i'^2 c_m^2}\right)^{1/2} - 1\right]}{\left(\dfrac{8k_d k_c c_0}{k_i'^2 c_m^2}\right)} \tag{2.8-21}$$

From the preceding, it should be clear that there are two distinct instances when efficiency may be assumed constant, or conversion-independent. First, when $k_c \ll k_i'$ initiator efficiency is essentially unity and it is only near complete conversion (for CONV or D-E reactions), when there is an extreme imbalance in consumption of one reactant over the other, that any drift in f might be apparent. Second, under borderline D-E conditions, when $\alpha_K \cong 1$ and both initiator and monomer are consumed at equal rates, the value of f should not change.

Use of equation 2.8-21 is rather impractical in the absence of measured values of k_c. However, we can express f in terms of its initial value, f_0, which can be determined from low-conversion rate data.

$$f = \frac{f_0^2 \hat{c}_m^2}{2(1 - f_0)\hat{c}_0}\left[\left(1 + \frac{4(1 - f_0)\hat{c}_0}{f_0^2 \hat{c}_m^2}\right)^{1/2} - 1\right] \tag{2.8-22}$$

For a preliminary examination we assume constant initiator concentration ($\hat{c}_0 = 1$) and plot f as a function of monomer conversion for various values of initial efficiency as shown in Fig. 2.8-8. The most striking result is that as the initial efficiency drops from unity, the decrease of f with conversion becomes increasingly pronounced. The commonly used initiator AIBN is generally assigned an efficiency of 0.6 and most investigators assume it to be constant throughout the course of reaction. From Fig. 2.8-8, we suspect that this might give rise to significant errors in both conversion rate and DP.

To test this hypothesis, we substitute the functional form of equation 2.8-22 into the reactant and moment balances listed in Table 2.4-2. By varying x_0 to maintain a constant value of $(\nu_N)_0$ for different values of f_0, we find that when f is assumed to be constant, regardless of f_0, the same dimensionless conversion history (Φ vs. $\hat{t}$) and DP history ($\bar{x}_N$ vs. Φ) results. In Figs. 2.8-9–2.8-11 we compare the results for conversion-dependent efficiency to this reference case for various values of α_K.

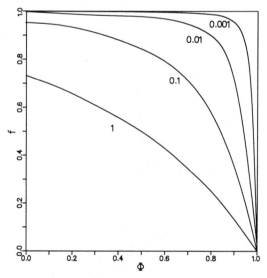

Figure 2.8-8. Variation of initiator efficiency factor with conversion. Initial value f_0 is the parameter.

Several important trends are apparent. The assumption of a constant value of f gives rise to greater drift in polymer properties regardless of α_K. For conventional polymerizations, the final value of $\bar{x}_N$ is greater for the variable f cases, while under D-E conditions it is less. That is, drift dispersion is minimized by the fact that the decay in efficiency counteracts the preferential conversion of either of the reactants. The deviation from the trajectories predicted by constant efficiency increase as f_0 decreases. From the conversion histories, it is apparent that the effect of variable f does not manifest itself until conversion reaches 10 percent for strong D-E conditions and 30–40 percent for conventional polymerization. Initial rate studies conducted in this low conversion regime would not show the effects of efficiency drift.

These observations have many ramifications. Because the use of constant initiator efficiency is common practice and its validity is thereby unquestioned, kineticists analyzing high-conversion regimes often resort to proposing complex variations in propagation and termination rates in an attempt to account for behavior that might be attributed to an initiation effect.

The initiator efficiency for copolymerization may be analyzed in a similar manner. In place of equation 2.8-10, there would be two separate monomer initiation steps

$$j = A, B \qquad\qquad m_0^* + m_j \xrightarrow{k_{ij}} m_1^{j*} \qquad\qquad (2.8\text{-}23)$$

where the rate of initiation of the jth monomer is $r_{ij} = k_{ij} c_0^* c_j$. The overall efficiency of initiation, consistent with the definition for homopolymerization, is

$$f \equiv \frac{r_{iA} + r_{iB}}{r_{1A} + r_{iB} + 2r_3} \qquad\qquad (2.8\text{-}24)$$

To express r_{iA} and r_{iB} as fractions of the rate of initiator decomposition, the

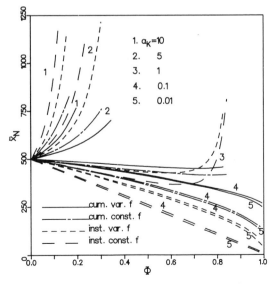

Figure 2.8-9. Comparison of DP drift with and without assumption of constant initiator efficiency, $f_0 = 0.5$.

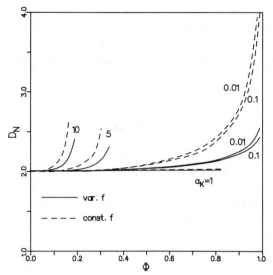

Figure 2.8-10. Comparison of D_N drift with and without assumption of constant initiator efficiency, $f_0 = 0.5$.

individual efficiencies must be:

$$f_A = \frac{r_{iA}}{r_{iA} + r_{iB} + 2r_3} \tag{2.8-25}$$

$$f_B^{\backslash} = \frac{r_{iB}}{r_{iA} + r_{iB} + 2r_3} \tag{2.8-26}$$

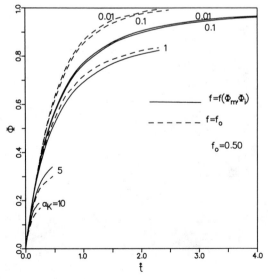

Figure 2.8-11. Comparison of conversion histories with and without assumption of constant initiator efficiency, $f_0 = 0.5$.

so that

$$j = A, B \qquad\qquad r_{ij} = 2f_j k_d c_0 \qquad\qquad (2.8\text{-}27)$$

The most useful forms for f_A and f_B are those expressed in terms of the reactant concentrations and initiator efficiencies for homopolymerization. The result is

$$f_A^{-1} = 1 + \frac{n_B}{n_A} \left[\frac{\dfrac{f_B^0}{1 - f_B^0}}{\dfrac{f_A^0}{1 - f_A^0}} \right] + n_A \left(\frac{1 - f_A^0}{f_A^0} \right) \qquad (2.8\text{-}28)$$

where f_A^0 and f_B^0 represent initiator efficiencies (equation 2.8-22) for homopolymerization of pure comonomers A and B, respectively, and n_A and n_B are mole fractions of the comonomers.

2.8.4. Thermal Initiation

Although free-radical polymers generally are initiated by a separate chemical species, several monomers are capable of self-initiation. Specifically, MMA, to a small extent, and styrene, to a very large extent, are capable of generating free radicals when heated. For styrene, the rate of thermal initiation becomes significant at temperatures greater than 100°C and probably dominates catalytic initiation at temperatures in excess of 150°C. Thermal initiation is used industrially as an alternative to chemical initiation. It should also be taken into account when chemical initiators are used at elevated temperature levels, where thermal initiation could become important or even dominant.

The mechanism by which styrene monomer forms radicals is not clearly understood. Hui and Hamielec (48) have presented a mechanism due to Pryor and Lasswell (Table 2.8-1), which has two important limiting forms. One predicts a rate of initiation that is second order in monomer and the other predicts a rate of initiation that is third order. Flory postulated a second-order mechanism and this is commonly accepted. Hamielec et al., on the other hand, conclude that third-order initiation most closely describes their experimental conversion histories. It would seem, however, that alternative explanations are possible. Thermal initiation rates are low and even trace amounts of dissolved oxygen can cause spurious results. Furthermore, at the elevated temperatures that thermal initiation rates are appreciable, many side reactions like chain transfer become significant. It should also be noted that Hamielec et al. used the third-order initiation step to determine the specific rate constants for initiation, propagation, termination, and monomer transfer by a curve-fitting technique, in which the rate constants were complex functions of conversion and temperature. If one considers the lattitude that exists when dealing with curve-fitting parameters and the ensuing functional relationships, one can show that third-order kinetics could behave like second-order kinetics.

The possible effects of initiation order (q in equation 1.10-10 with $p = 0$) on polymer properties were described by Tadmor and Biesenberger (45). They are

Table 2.8-1.   Kinetic Mechanism for Styrene Thermal Initiation (48)

1. $m + m \underset{k_{1r}}{\overset{k_{1f}}{\rightleftharpoons}} AH$

2. $m + m \underset{k_{2r}}{\overset{k_{2f}}{\rightleftharpoons}} (D^*)_2$

3. $(D^*)_2 \xrightarrow{k_{cyc}} 1,2 \text{ diphenylcyclobutane}$

4. $m + AH \xrightarrow{k_o} m_o^* + A'^*$

5. $A'^* \xrightarrow[\text{fast}]{} A^*$

6. $m + A^* \xrightarrow{k_{i1}} \cdot m_1^*$

7. $m + m_o^* \xrightarrow{k_{i2}} m_1^*$

where:

AH = (structure)   A'* = (structure)   A* = (structure)

$m_o^* = \cdot CHCH_3$ (structure)

$(D^*)_2 = \overset{\cdot}{C}HCH_2CH_2\overset{\cdot}{C}H$ (structure)

and $r_i = r_6 + r_7$

illustrated in Figs. 2.8-12–2.8-14. When $q < 2$, the reaction appears "conventional," and when $q > 2$, it resembles a D-E polymerization. Of particular interest is the second-order case. Because it results in a propagation rate that is also second-order in monomer, there is no drift in molecular properties with conversion, and the only source of dispersion is statistical (Example 2.7-1). In the third-order case, on the other hand, the initiation step is higher order than propagation with respect to monomer concentration (3 vs. 2.5). Therefore, the effect of monomer consumption is

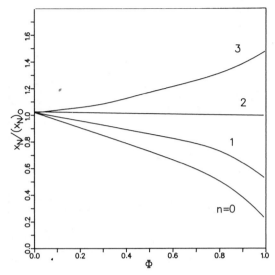

Figure 2.8-12. Drift in $\bar{x}_N$ with conversion for nth order monomer initiation.

more pronounced. As a result, $\bar{x}_N$ and $\bar{x}_W$ drift upward with conversion and D_N rises.

In light of this knowledge, we examine the experimental data of Hamielec, et al. (48) for thermally polymerized styrene. Shown in Fig. 2.8-15 are experimental cumulative $\overline{M}_N$ and $\overline{M}_W$ curves along with the authors' third-order initiation model. It is apparent that the degree of polymerization scarcely varies over the entire range of conversion. Higher-temperature data (200°C) show a slight downward drift. One would expect the author's postulate of third-order initiation with gel-effect (G-E) to

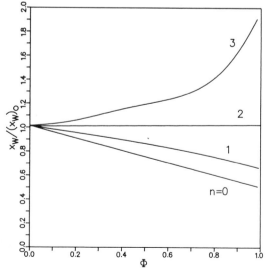

Figure 2.8-13. Drift in $\bar{x}_W$ with conversion for nth order monomer initiation.

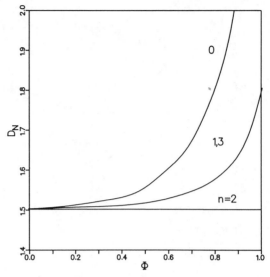

Figure 2.8-14. Drift in D_N with conversion for nth order monomer initiation.

lead to a marked upward drift in the molecular weight. Indeed, the authors claim the G-E as justification for using conversion-dependent rate constants. The data, on the other hand, seem to suggest second-order kinetics. The pronounced autoacceleration in conversion, characteristic of G-E, is absent in Fig. 2.8-16 as is the sharp rise in $\overline{M}_W$ and $\overline{M}_N$ in Fig. 2.8-15. It is quite likely that the conversion-dependence of the rate constants offsets the third-order dependence of initiation. Note that the dispersion index remains roughly 1.5, which indicates that termination by combination is the dominant mechanism, not chain transfer or disproportionation as sometimes

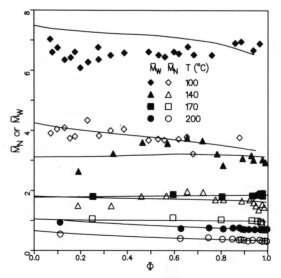

Figure 2.8-15. Experimental DP drift for thermally initiated styrene polymerization (48).

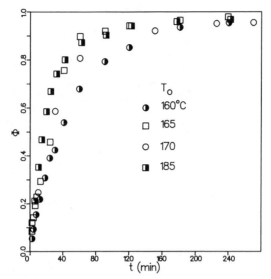

Figure 2.8-16. Experimental conversion histories for thermally initiated styrene polymerization (48).

speculated. Indeed, one is tempted to dismiss the importance of transfer as a consequence of this evidence. However, the calculation of D_N from GPC data could be somewhat inaccurate.

Example 2.8-2. *Thermally Initiated Polymerization.* Hui and Hamielec (48) report rate constants for thermal initiation of styrene assuming second- and third-order initiation:

2nd
$$k_i = 5.007 \times 10^5 \exp\left(-\frac{13,600}{T}\right) 1/\text{mol} \cdot \text{s}$$

3rd
$$k_i = 1.095 \times 10^5 \exp\left(-\frac{13,810}{T}\right) 1^2/\text{mol}^2 \cdot \text{s}$$

Using these values, calculate the time needed for 50-percent conversion, and $(x_N)_{\text{inst}}$ at that conversion when $T = 140°C$.

Note

$$k_p = 1.051 \times 10^7 \exp\left(-\frac{3557}{T}\right) 1/\text{mol} \cdot \text{s}$$

$$k_t = 1.255 \times 10^9 \exp\left(-\frac{844}{T}\right) 1/\text{mol} \cdot \text{s}$$

$$(c_m)_0 = 8.7 \text{ mol}/1$$

Solution.

2nd order
$$k_{ap} = k_p\sqrt{\frac{k_i}{k_t}} = 7.5 \times 10^{-6} \, 1/\text{mol} \cdot \text{s}$$

$$\frac{dc_m}{dt} = -k_{ap}c_m^2$$

so

$$t = \frac{1}{k_{ap}(c_m)_0}\left[\frac{\phi}{1 - \phi}\right] = 15{,}333 \text{ s}$$

$$(\bar{x}_N)_{inst} = 2\nu_N = \frac{2k_{ap}c_m^2}{k_i c_m^2} = \frac{2k_{ap}}{k_i} = 13{,}750$$

3rd order $k_{ap} = 2.718 \times 10^6 \text{ 1/mol} \cdot \text{s}$

$$\frac{dc_m}{dt} = -k_{ap}c_m^{3/2}$$

$$t = \frac{2}{3k_{ap}(c_m)_0^{3/2}}\left[\frac{1}{(1 - \phi)^{3/2}} - 1\right] = 17{,}480 \text{ s}$$

$$(\bar{x}_N)_{inst} = 2\nu_N = \frac{2k_{ap}c_m^{5/2}}{k_i c_m^3} = \frac{2k_{ap}}{k_i c_m^{1/2}} = \frac{2k_{ap}}{k_i(c_m)_0^{1/2}(1 - \phi)^{1/2}} = 7920$$

Although use of third-order kinetics predicts a slower reaction and lower molecular weight than second order, the author's predicted rates are higher than second order, indicating the influence of adjusting the kinetic constants continuously with conversion.

2.9. CHAIN TERMINATION EFFECTS

From our analysis of the instantaneous response of chain-addition polymerizations (Section 1.13), we would expect a decline in k_t to raise both the reaction rate and the instantaneous DP of the product. This is apparent from Table 2.7-1 and equation 1.13-11. As k_t declines, both k_{ap} and k_{ax} increase and, consequently, so do λ_m^{-1} and $(\bar{x}_N)_{inst}$. However, α_K decreases. From our analysis of drift behavior (Section 2.7), this indicates that instantaneous DP should drift downward (Section 2.7) as reaction progresses. Ideally, graphs of monomer conversion Φ versus reaction time should behave like curve 1 in Fig. 2.9-1. Thus the rate of polymerization (slope) should decrease with consumption of monomer and initiator.

Frequently, however, polymerizations exhibit autoacceleratory behavior, as illustrated by curve 2, wherein the rate and DP both increase with time. Autoacceleration could result from at least two distinct causes. One stems from rising temperatures and is therefore termed autothermal, and the other is termed "autocatalytic." However, in the first case, while the rate characteristic should be similar to curve 2, we might expect a downward drift in DP rather than an upward drift. Thermal drift will be analyzed in detail in Chapter 4, where it will be shown that, contrary to our expectations, upward drift in DP is sometimes possible under certain conditions.

The aforementioned "autocatalytic" behavior is due to acceleration of reaction rate by the presence of polymer product. It leads to enhancement of both rate and DP. Thus polymerization rate rises with weight of polymer produced w_p (or Φ) instead of declining as expected from $r \propto (1 - \Phi)$. Unlike true autocatalysis, however, it is caused by a physical rather than a chemical process, that is the hindered

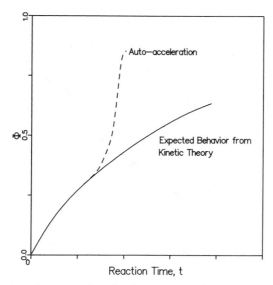

Figure 2.9-1. Typical conversion versus time plots for nonautoacceleratory (1) and autoacceleratory (2) polymerizations (28).

termination of polymeric intermediates. Termination of free-radical intermediates, which is a bimolecular reaction involving two macromolecules, can be hindered inadvertently by diffusion limitations or deliberately by isolating intermediates from one another during growth as a dispersion (oil droplets) in an immiscible (aqueous) continuous phase. An example of the latter is emulsion polymerization. A common characteristic among these processes is the simultaneous rise in rate and product DP over corresponding homogeneous reactions, owing to preferential hindrance of termination over propagation with a concomitant rise in the free-radical population.

This section deals with hindered termination in the broad sense, including emulsion polymerization as well, which we shall view as an extreme case of hindered termination. It is important to remember, however, that emulsification of reacting centers alters the reaction mechanism and causes it to differ in certain important aspects from its bulk and solution counterparts. Suspension polymerization, on the other hand, actually consists of isolated bulk reactions that are not sufficiently small to be considered emulsified. Consequently, they are generally expected to obey bulk kinetics, at least in batch reactors where additional factors such as residence time distribution (Chapter 3) are absent.

Diffusion-controlled termination of polymeric intermediates can occur in two distinct situations. In homogeneous polymerizations, where the polymer product is soluble in its own monomer or in a mutual solvent, rising reaction viscosities can impair the mutual termination of free-radical intermediates. This has been termed the gel-effect (G-E) to distinguish it from a similar phenomenon occurring in heterogeneous polymerizations, in which the polymer product is insoluble in the reaction medium. Here free-radical intermediates are trapped in a matrix of precipitating polymer product. In both cases, the intermediates, despite their reduced mobility, are believed to propagate at a sustained rate because monomer diffusion is sufficiently rapid to feed them. Termination rate, on the other hand, drops as reaction proceeds. Occasionally, propagation also ceases, owing to vitrification of the

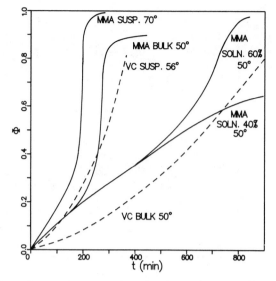

Figure 2.9-2. Experimental conversion versus time plots for: MMA polymerization in benzene solution and bulk (49); MMA suspension polymerization (50); VC bulk polymerization (51); and VC suspension polymerization (52).

polymerizing medium. This occurs when the glass transition temperature of the medium reaches the reaction temperature.

Bulk and solution polymerizations of methyl methacrylate (MMA) exhibit severe G-E, and bulk styrene polymerizations exhibit mild G-E. Vinyl chloride (VC) and acrylonitrile (AN) polymerizations are examples of precipitous reactions. All four, incidentally, are also amenable to polymerization in emulsion. Figure 2.9-2 shows typical autoacceleratory behavior of both kinds (G-E and precipitous) observed experimentally (49–52).

It is customary to model hindered termination using a time-dependent or conversion-dependent termination constant $k_t(\Phi)$. Although this practice is convenient for purposes of modelling, and thus we will use it as well, it ignores the mass transport processes that are inextricably coupled with the kinetics of hindered termination. The simplest model is

$$k_t = (k_t)_0 \exp(-B\Phi) \qquad (2.9\text{-}1)$$

where $(k_t)_0$ is the true termination constant, prior to hindered termination ($\Phi = 0$) and B is a constant parameter. Figure 2.9-3 illustrates this model for various values of B. Comparison with Fig. 2.9-2 demonstrates its ability to simulate hindered termination.

2.9.1. The Gel-Effect

The phenomenon of hindered termination in radical polymerization is well documented. Although it is commonly referred to as the "gel-effect" (G-E), it is also known as the Trommsdorff effect. As noted earlier, the decrease in termination rate

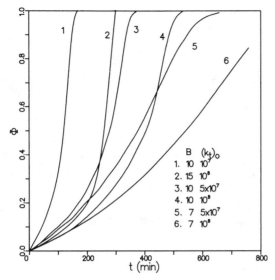

Legend in figure:

	B	$(k_t)_0$
1.	10	10^7
2.	15	10^8
3.	10	5×10^7
4.	10	10^8
5.	7	5×10^7
6.	7	10^8

Figure 2.9-3. Simulated conversion versus time plots (equation 2.8-1) with $k_d = 10^{-5}$ s^{-1} and $k_p = 10^3$ 1/mol s.

leads to the enhancement of polymer molecular weight. This serves to increase solution viscosity (Chapter 5) and thus further aggravates the conditions that first caused the G-E. Attempts to model the hindered termination process based on physicochemical models have been the focus of much research. Although many models successfully reproduce specific reaction histories, to date there is no universally accepted relationship between the kinetic constant for termination and molecular weight, polymer concentration, and temperature, all of which contribute to the G-E. Later in this section we shall critically review some current models that draw upon very different approaches in describing the G-E. Throughout much of the following discussion, however, we will employ far simpler relationships. Like the model reactions used throughout this text, these relations preserve all the major qualitative effects of the phenomenon without attempting to capture all the kinetic details.

Analysis of the G-E has been the subject of continued investigation for a number of years. Early work by North (53) showed that the termination rate was inversely related to solution viscosity even at low conversion. Thus it appeared misleading to refer to G-E as the onset of diffusion control at some finite conversion, inasmuch as diffusional effects seemed evident from the very start of reaction. There is some debate as to precisely which diffusion process is rate-limiting in the termination step. Both chain translation and chain-end segmental diffusion might be involved. North argued that only the latter could be responsible for G-E (54), while many later theories are predicated on the assumption that both chain translation and chain-end segmental diffusion produce the G-E. The important difference is that chain entanglement offers much higher resistance to chain translation than simple chain segment reorientation. In Example 2.9-1, we show that in spite of extremely low radical concentrations, growing chain radicals are actually quite closely spaced. Although it is easy to agree that the termination rate is related to the viscosity, the choice of viscosity is not so clear cut. North used intrinsic viscosity, which was appropriate for the description of dilute polymer solutions, but as higher conversion the zero shear viscosity η_0 might be more appropriate. The zero shear viscosity is a measure of chain center mobility in concentrated solutions, however, so its pertinence in describing the low-concentration regime might also be questioned. Clearly, a diffusion model to describe the broad range of conditions between dilute solution and polymer melt is needed.

We shall not discuss models that derive from curve-fitting rate constants. An example is the work of Duerksen and Hamielec (55), which is often cited. Although such models utilize data and are thus useful for the specific systems from which they derive, they cannot be generalized readily to other reaction conditions or other systems. More recent investigations have attempted to provide a theoretical basis for the functional relationships proposed.

Two approaches based on fundamentally distinct viewpoints have been advanced to describe the termination process. The work of Hamielec et al. (56) and Laurence et al. (57) is based on free-volume theory, while O'Driscoll et al. (58–61) utilize the chain entanglement model.

Free-volume theory, even as modified for polymer solutions, is generally regarded as best near the glass transition temperature (T_g) of the polymer solution. Bueche (62) extended Eyring's rate theory for diffusion in liquids (63) to describe segmental motion of polymer chains. Hamielec et al. draw on the relationship between viscosity and free volume to express the change in k_t as a function of conversion and polymer molecular weight. The earliest work, that of Balke and Hamielec (64), indicated a strong correlation between the value of k_t and the logarithm of free volume, which is the dependence predicted by the Bueche relationship. This linear dependence, shown in Fig. 2.9-4, does not hold for the entire range of conversion. In more recent work (56) it was postulated that k_p must also be a function of conversion, particularly when the polymerization temperature is below the T_g of the polymer.

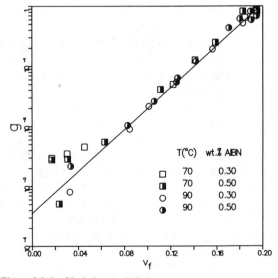

Figure 2.9-4. Variation in G-E function with free volume (64).

These authors subdivide the polymerization process into three regions: (1) low conversion with constant kinetics; (2) onset of G-E, with variable k_t; (3) high conversion with variable k_p and k_t. The rate constants are expressed as:

$$\frac{k_t}{(k_t)_0} = \left(\frac{\overline{M}_{W_{cr1}}}{\overline{M}_W}\right)\exp\left[-A\left(\frac{1}{v_f} - \frac{1}{v_{f_{cr1}}}\right)\right] \tag{2.9-2}$$

$$\frac{k_p}{(k_p)_0} = \exp\left[-B\left(\frac{1}{v_f} - \frac{1}{v_{f_{cr2}}}\right)\right] \tag{2.9-3}$$

where the two critical points (cr1, cr2) refer to the start of regions 2 and 3, respectively, and $A = 1.11$, $B = 1.0$, $n = 1.75$. These values were obtained by curve-fitting experimental data.

The entanglement theory presumes that growing chain radicals are either entangled with dissolved polymer, or not, and that each type has a distinct termination rate constant. Thus G-E will not occur until the polymer volume fraction and $\bar{x}_N$ surpass a critical value for chain entanglement given by the following relationship

$$K_{cr} = \phi_{Pcr}\bar{x}_{N_{cr}}^{1/2} \tag{2.9-4}$$

where K_{cr} is a constant characteristic of the solvent-polymer system although the authors suggest using it as a curve-fitting parameter. The resulting expression for k_t is of the form

$$\frac{k_t}{(k_t)_0} = \left[1 + (c_1g_1 - 1)\exp(-c_2g_2)\right] \tag{2.9-5}$$

where

$$c_1g_1 = \left\{\left(\frac{1+B}{\alpha_0^2 K_{cr}}\right)\left(\frac{\Phi}{1+B\Phi}\right)\left(\frac{(c_m)_0}{-2B}\right)^{1/2}\left(\frac{-\ln(1+B\Phi)}{f[(c_0)_0 - c_0]}\right)^{1/2}\right\}^{1/2} \tag{2.9-6}$$

and

$$c_2g_2 = \frac{2k_t^{1/2}}{k_p(c_m)_0}\left(\frac{K_{cr}}{1+B}\right)\left(\frac{1+B\Phi}{\Phi}\right)^2\left(\frac{1+B\Phi}{1-\Phi}\right)r_i^{1/2} \tag{2.9-7}$$

and where Φ is the monomer conversion, B is the volumetric contraction coefficient (Section 1.11.4), and α_0 is a parameter.

Although the free-volume and chain-entanglement theories are based on totally different physical models, both result in a termination constant that is an exponential function of conversion. As shown in Figs. 2.9-5 and 2.9-6, they give essentially the same fit of Balke's experimental data. The free-volume model does have one advantage in that it predicts the limiting conversion reached when T_g of the monomer-polymer mixture reaches the reaction temperature. In a subsequent section, we will show that drift of initiator efficiency as well as violation of the QSSA

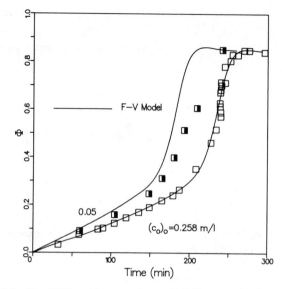

Figure 2.9-5. Fit of FV model to experimental MMA conversion histories (56).

can significantly alter computational results when the G-E is present. Neither effect was considered in the development of the free-volume and chain-entanglement models.

In preceding sections we have demonstrated the reliability of the QSSA in the face of a variety of kinetic conditions. However, the G-E places a special stress on the assumptions implicit in the QSSA. Since chain initiation is unaffected by the G-E, hindered termination must cause an accumulation of radical intermediates (see

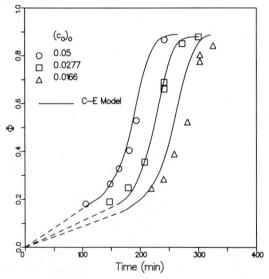

Figure 2.9-6. Fit of CE model to experimental MMA conversion histories (58).

equation 1.11-15). We are led to question the accuracy of the QSSA under such conditions.

To test the viability of the QSSA, we use a simple relationship to express the dependence of k_t on conversoin, equation 2.9-1. The resulting population balance on intermediates, which is an extension of equation 1.11-9, is written as:

$$\frac{d\hat{c}^*}{dt} = \lambda_i^{-1}\hat{c}_0 - \lambda_2^{-1}G(\hat{c}_m)\hat{c}^{*2} \qquad (2.9\text{-}8)$$

The QSSA solution for $\hat{c}^*$ would be

$$c_s^* = \left\{ \frac{\lambda_2}{\lambda_i}\hat{c}_0 G^{-1}(\hat{c}_m) \right\}^{1/2} \qquad (2.9\text{-}9)$$

where

$$G(\hat{c}_m) \equiv \frac{k_t}{(k_t)_0} = \exp\left[-B(1-\hat{c}_m)\right] \qquad (2.9\text{-}10)$$

Substitution of equation 2.9-9 into the reactant and moment balances yields the set of equations given in Table 2.4-2. Use of equation 2.9-8 requires solution of the full set of balance equations (reactants, polymer product moments, and intermediate moments) given in Table 2.4-1. In both cases, we let f remain constant at a value of one. In Fig. 2.9-7, a few sample conversion histories computed from the QSSA balances were plotted to illustrate the effectiveness of our simple model in producing realistic profiles.

In Figs. 2.9-8 and 2.9-9, we focus on the deviation in calculated properties between the solution of the balance equations of Tables 2.4-1 and 2.4-2, (exact vs.

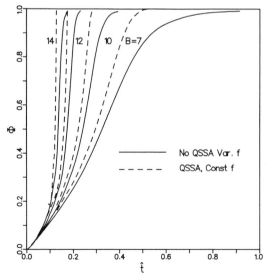

Figure 2.9-7. Comparison of conversion histories computed with fixed initiator efficiency and QSSA with exact solutions and variable f.

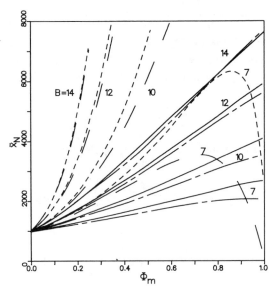

Figure 2.9-8. Comparison of $\bar{x}_N$ computed with fixed initiator efficiency and QSSA with exact solutions and variable f.

QSSA). From the figures it should be evident that significant errors may be incurred through use of the QSSA at high conversions when G-E is present. It should be noted that this is only true for a strong G-E as characterized by large values of B in the figures. However, these values are not unrealistic for monomers that exhibit a strong G-E such as MMA.

Although the QSSA predicts an ever increasing radical concentration for $B \geqslant 14$, we see that actually the concentration levels off or passes through a maximum. Thus the overall rate of reaction should follow the same trend at elevated conversions. Use of the QSSA would fail to predict this trend, and, therefore, attempts to model experimental results by manipulating k_t alone should fail. Breakdown of the QSSA acts in concert with variable initiator efficiency. Both effects are ignored in most analyses of high-conversion regions of polymerization with G-E. Yet it is clear from the figures that substantial error is incurred when a strong G-E is simulated with equations based on the QSSA and constant f. As a result of ignoring these factors, either the decrease in k_t must be mitigated or a decrease in k_p must be invoked in order to match the model to the actual behavior at high conversions.

A broad range of conditions exist for which the QSSA may be safely applied in describing free-radical polymerization with G-E. Few vinyl monomers show the excessive G-E associated with MMA. We shall now examine the influence of G-E on the polymer product formed in its presence.

From the discussions in Sections 2.2-6 and 2.7, we know that the instantaneous DPD of free-radical polymers is always most probable or narrower, and that changes in reactant composition during the course of reaction can lead to a spread in the total product distribution. Thus, if initiator is consumed more rapidly than monomer, the polymer formed in successive increments of conversion will increase in average DP. Clearly, it must be the change in the relative rates of chain initiation

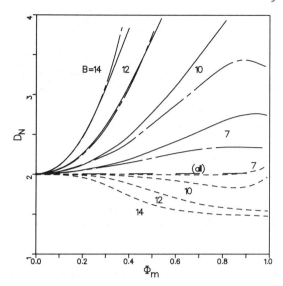

Figure 2.9-9. Comparison of D_N computed with fixed initiator efficiency and QSSA with exact solutions and variable f.

and chain propagation that cause this drift. By altering kinetic constants during the course of reaction, one might expect a similar result.

From equation 1.13-11, we note that the instantaneous DP of growing polymer is inversely proportional to the square root of k_t. The general expression for instantaneous DP including G-E is given by

$$(\bar{x}_N)_{\text{inst}} = \frac{d\mu^1}{d\mu^0} = \left(\frac{2}{2-R}\right)\left\{1 + \frac{\hat{c}_m}{(\nu_N)_0\sqrt{\hat{c}_0 G}}\right\} \qquad 2.9\text{-}11$$

where $R = 1$ for termination by combination only and $R = 0$ for termination by disproportionation. Since G-E reduces k_t, and thus G, as conversion increases, we expect $\bar{x}_N$ to rise. Qualitatively, one would suppose that G-E might have consequences similar to D-E in shaping DP drift. To test this hypothesis, we shall integrate the reactant and moment balances of Table 2.4-2 using the simple model for G-E of Sacks et al. (65).

$$G = \left(\alpha_0 + \alpha_1\Phi + \alpha_2\Phi^2 + \alpha_3\Phi^3\right)^\beta \qquad (2.9\text{-}12)$$

This model was obtained by curve-fitting experimental conversion histories for MMA and S, and used to demonstrate the effects of strong and weak G-E. The kinetic constants used are listed in Table 2.9-1 and the parameters for G in Table 2.9-2.

Table 2.9-3 summarizes the results of a series of simulations using the kinetic constants for styrene polymerization. Note that in all cases the presence of G-E increases the final DP, $\bar{x}_{N_f}$. It does not follow, however, that G-E necessarily leads to a more disperse polymer product. To aid in discussion, we refer to Figs. 2.9-10 and

Table 2.9-1. Kinetic Data for PS and PMMA Polymerizations

Polymer	$R = k_{tc}/k_t$	A_p (ℓ/mol sec)	E_p (kcal/mol)	A_t (ℓ/mol sec)	E_t (kcal/mol)	$(c_m)_o$ (mol/ℓ)	$(c_o)_o$ (mol/ℓ)
PS	1	1.05×10^7	7.06	1.26×10^9	1.68	8.7	0.348
PMMA	0	5.1×10^6	6.30	7.8×10^8	2.8	8.7	0.00725

Table 2.9-2. Gel-Effect Constants of Sacks et al. (65)

Polymer	Conversion range	α_o	α_1	α_2	α_3	β
PS	$0 \leq \Phi \leq 0.30$	1	0	0	0	1
	$0.30 < \Phi \leq 0.8$	1.522	-1.818	0	0	2
PMMA	$0 \leq \Phi \leq 0.15$	1	0	0	0	1
	$0.15 < \Phi \leq 0.60$	1	-28.72	228.2	-239.9	-2

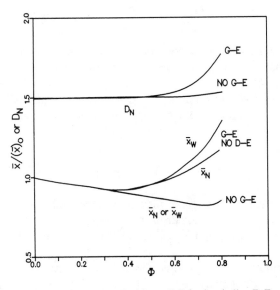

Figure 2.9-10. DP and DPD drift with and without G-E for borderline D-E polymerization (28).

2.9-11 and consider the nature of the drift associated with the polymerization in the absence of G-E. In Section 2.7, it was shown that during CONV polymerizations both $\bar{x}_N$ and $\bar{x}_W$ drift downward, but $\bar{x}_N$ drops more rapidly than $\bar{x}_W$ in response to the production of low DP polymer at high conversions, whereas during D-E polymerizations (Section 2.8), both $\bar{x}_N$ and $\bar{x}_W$ drift upward with $\bar{x}_W$ rising more rapidly owing to the creation of high-molecular-weight ends at elevated conversions. When G-E is superimposed on such drift behavior, a variety of end results may occur.

Under strong D-E conditions, as in Fig. 2.9-10 and cases 1–4 of Table 2.9-3. G-E aggravates drift by creating an increased amount of high-DP polymer. The G-E broadens the distribution as measured by larger values of D_N, but has the redeeming feature of postponing D-E by consuming monomer more rapidly than otherwise, owing to autoacceleration. Coupled with borderline D-E conditions (cases 3–6), the G-E has the effect of creating a drift profile much like strong D-E, but without limiting the ultimate attainable conversion.

When a conventional polymerization experiences G-E (cases 9–12), the strong downward drift of $\bar{x}_N$ and $\bar{x}_W$ is mitigated. The instantaneous DP's remain fairly constant with conversion. As a result, G-E produces a product with D_N much closer to the narrowest possible (statistical) dispersion. For these cases, G-E is beneficial in controlling drift dispersion.

The various effects of G-E on the breadth of the DPD are summarized in Fig. 2.9-12. Note that for low values of α_K (i.e., conventional polymerization), G-E gives a lower D_N than no G-E. At intermediate values, G-E has little added effect, while it sharply increases D_N for large α_K (D-E). The minimum in the curve without G-E at intermediate values of α_K is the result of weak D-E counteracting the "conventional" downward drift.

Concerning termination effects in copolymerizations, early research focused primarily on copolymer composition. As a result, information on termination kinetics

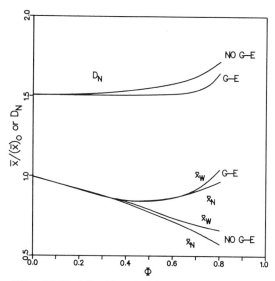

Figure 2.9-11. DP and DPD drift with and without G-E for CONV polymerization (28).

Table 2.9-3. Gel-Effect with Various Initiation Rates

Case No.	c_o (moles/ℓ)	A_d (sec^{-1})	Gel Effect (G-E)	α_k	Dead end (D-E)	x_o	$(\hat{c}_m)_f$	$(\hat{c}_o)_f$	$(\bar{x}_N)_o$	$(\bar{x}_N)_f$	$(D_N)_f$
1	0.0018	1.58×10^{15}	Yes	1.44	Yes	2420	.472	.0003	3270	5179	3.04
2	0.0018	1.58×10^{15}	No	1.44	Yes	2420	.520	.0003	3270	4708	2.22
3	0.0160	1.58×10^{15}	Yes	0.483	No	272	.20	.3063	1091	1256	1.77
4	0.0160	1.58×10^{15}	No	0.483	Yes	272	.20	.0388	1091	907	1.51
5	0.0160	7.9×10^{14}	Yes	0.351	No	272	.20	.4684	1541	1637	1.70
6	0.0160	7.9×10^{14}	No	0.351	No	272	.20	.1872	1541	1070	1.58
7	0.0160	3.45×10^{14}	Yes	0.232	No	272	.20	.6262	2331	2328	1.66
8	0.0160	3.45×10^{14}	No	0.232	No	272	.20	.3905	2331	1427	1.65
9	0.0160	1.58×10^{14}	Yes	0.157	No	272	.20	.7375	3444	3314	1.64
10	0.0160	1.58×10^{14}	No	0.157	No	272	.20	.5570	3444	1962	1.70
11	0.0160	1.58×10^{13}	Yes	0.0496	No	272	.20	.9126	10887	9958	1.62
12	0.0160	1.58×10^{13}	No	0.0496	No	272	.20	.8457	10887	5640	1.78

Note: The subscript final (f) implies $\phi = 0.80$ or dead-end, whichever occurred first.
In all tabulated cases $(c_m)_o = 8.7$ moles/ℓ, $E_d = 30.8$ kcal/mole (AIBN),
$T = 60°C$ and $1.498 \leq (D_N)_{inst} < 1.500$.

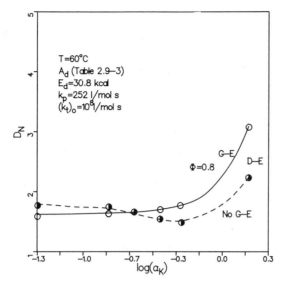

Figure 2.9-12. Variation in D_N with α_K with and without G-E.

was not necessary. When overall reaction rate data are required, however, the termination step must be taken into account. Kinetic models for the bimolecular termination of free-radical copolymers generally attempt to modify the homopolymerization rate constants to describe the coupling of chains with unlike ultimate mers. Three models are currently used (Section 1.12); they are the geometric mean (GM), phi factor (PF), and penultimate effect (PE) models.

The GM model proposes that the rate constant for hetero-termination is merely the geometric mean of the rate constants for the two homo-termination reactions (equation 1.12-9), while the latter retain their values from their respective homopolymerization reactions. A geometric mean relationship simply implies that the cross-termination rate is weighted by the probability of finding dissimilar chain ends. While this leads to simplified analysis of the rate equations, a few systems are adequately described by the GM model. It is limited to describing systems in which the comonomers are chemically similar, such as styrene and α-methyl styrene.

A direct extension of the GM model is the PF model which merely postulated that the cross-termination rate constant is proportional to the geometric mean of the homo-termination constants. The proportionality factor is called phi. This is the most frequently used technique and has particular applicability to several systems. The comonomer pair styrene–methyl methacrylate conforms quite well to the rate predicted by a constant value of phi equal to thirteen (67). Initial rate data with the predicted values are shown in Fig. 2.9-13 for styrene–methyl methacrylate (S/MMA) and styrene–methyl acrylate (S/MA).

There are many systems, however, for which the PF model fails to predict initial rates. One may regard ϕ as a function of composition, but this defeats the utility of the method. Russo et al. have recently introduced the PE model (68, 69). The author's claim that the chain-end segment is responsible for both the chemical reactivity and diffusional characteristics of the copolymer radical. The properties of

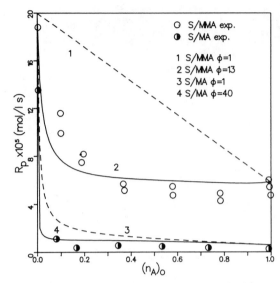

Figure 2.9-13. Copolymerization initial rate data for S/MMA and S/MA with PF prediction.

the end segment are determined by the last four carbon atoms, or the last two mer units for vinyl polymers. Thus, by considering the penultimate and ultimate repeat unit, one should be able to account for both physical and chemical hindrances to termination. This picture leads to a two-parameter model for the rate of termination. The model has been tested extensively for S/MMA both in bulk and in solvent (69). The two parameters proved to follow simple Arrhenius dependence on temperature, and excellent agreement between experimental and calculated rates was achieved. We note, however, that this comonomer pair also conforms quite well to PF kinetics.

Examination of a kinetically more demanding pair, styrene–acrylonitrile (S/AN) proved both PF and PE kinetics to be inadequate (70). Frequently, experimental rate data require negative values of the termination parameters to achieve fit over the entire composition spectrum. One must be aware that AN homopolymerizes precipitously while styrene is a homogeneous polymerization. It may be wholly unrealistic to expect models based solely on the modification of homopolymerization rate constants to adequately describe the chemical reactivity of the same species in comonomer media that are very different in nature from either pure monomer.

To examine the effect of termination kinetics, we compare computed conversion, DP, and composition histories for S/MMA using GM, PF, and PE kinetics. The moment equations of the DPD for PF and GM kinetics are given in Table 2.9-4, where the rate functions refer to those listed in Tables 1.12-2–1.12-4. Very little is known about the comparative degree to which disproportionation and combination participate in hetero-termination reactions. Experimental methods for determining copolymer DPD's lags considerably the sophistication achieved in modeling copolymerization kinetics, so evaluation of these parameters is difficult. For the purpose of simulation, we assume a single value of R.

The results pictured in Figs. 2.9-14 and 2.9-15 show that for S/MMA, choice of termination mechanism is not critical. Although the absolute value of conversion rate may vary, in dimensionless form the conversion histories are virtually identical.

Table 2.9-4. Gel-Effect with Various Initiation Rates

$$\frac{d\mu_{jc}^{o*}}{dt} = r_{ij} + (r_{pkj} - r_{pjk}) - r_{tj}$$

$$\frac{d\mu_c^{o*}}{dt} = r_i - r_t$$

$$\frac{d\mu_c^o}{dt} = \sum_{j=A}^{B}\sum_{k=A}^{B}\left(\frac{2-R_{jk}}{2}\right)r_{tjk}$$

$$\frac{d\mu_{jc}^{1*}}{dt} = r_{ij} + r_{pj} + \left[r_{pkj}(\mu_{kc}^{1*}/\mu_{jc}^{o*}) - r_{pjk}(\mu_{jc}^{1*}/\mu_{jc}^{o*})\right] - r_{tj}(\mu_{jc}^{1*}/\mu_{jc}^{o*})$$

$$\frac{d\mu_c^{1*}}{dt} = r_i + r_p - \left[r_{tA}(\mu_{AC}^{1*}/\mu_{AC}^{o*}) + r_{tB}(\mu_{BC}^{1*}/\mu_{AC}^{o*})\right]$$

$$\frac{d\mu_c^1}{dt} = \sum_{j=A}^{B}(\mu_{jc}^{1*}/\mu_{jc}^{o*})r_{tj}$$

$$\frac{d\mu_{jc}^{2*}}{dt} = r_{ij} + r_{pj} + 2\left[r_{pjj}(\mu_{jc}^{1*}/\mu_{jc}^{o*}) + r_{pjk}(\mu_{kc}^{1*}/\mu_{kc}^{o*})\right] + \left[r_{pkj}\left(\frac{\mu_{kc}^{2*}}{\mu_{kc}^{1*}}\right) - r_{pjk}\left(\frac{\mu_{kc}^{2*}}{\mu_{jc}^{o*}}\right) - r_{tj}\left(\frac{\mu_{jc}^{2*}}{\mu_{jc}^{o*}}\right)\right]$$

$$\frac{d\mu_c^{2*}}{dt} = r_i + r_p + 2\left[(r_{PAA}+r_{PAB})(\mu_{AC}^{1*}/\mu_{AC}^{o*}) + (r_{PBB}+r_{PBA})(\mu_{BC}^{1*}/\mu_{BC}^{o*})\right] - \left[r_{tA}(\mu_{AC}^{2*}/\mu_{AC}^{o*}) + r_{tB}(\mu_{BC}^{2*}/\mu_{BC}^{o*})\right]$$

$$\frac{d\mu_c^2}{dt} = \sum_{j=A}^{B}\sum_{k=A}^{B}r_{tjk}(\mu_{kc}^{2*}/\mu_{kc}^{o*}) + \sum_{j=A}^{B}\sum_{k=A}^{B}\left[r_{tjk}(\mu_{jc}^{1*}/\mu_{jc}^{o*})(\mu_{kc}^{1*}/\mu_{kc}^{o*})\right]$$

where $\mu_{jc}^{k*} = \sum_x x^k c_x^{j*}$ and $c_x^{j*} = \sum_y c_{x,y}^{j*}$

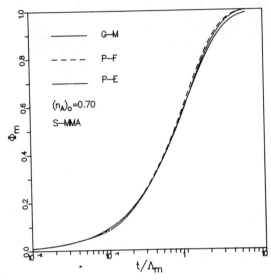

Figure 2.9-14. Effect of termination model on computed conversion histories for S/MMA with $(n_A)_0$ = 0.70.

The variations in $\bar{x}_N$ and $\bar{y}_N$ with conversion for the three termination models are so small as to be indiscernible and, therefore, have not been plotted. This is true for initial compositions on either side of the azeotrope, which is near 0.50 mole fraction styrene for S/MMA. Indeed, the termination mechanism has no influence on the copolymer composition. The propagation rate of comonomer relative to total monomer is sufficient to determine ν_A and $\bar{y}_N$.

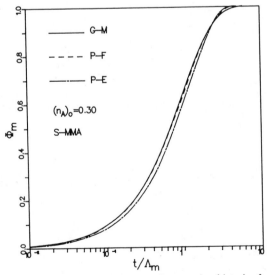

Figure 2.9-15. Effect of termination model on computed conversion histories for S/MMA with $(n_A)_0$ = 0.30.

Example 2.9-1.   Chain-End Separation of Free-Radicals. Calculate the volume per radical for a styrene polymerization initiated by 0.01 mol/l AIBN at 60°C. Estimate the chain-end separation using the results of Debye and Bueche (66)

$$[\eta] = 0.037M^{0.62} = \frac{4\pi Ra^3 Av}{3M} 1.63$$

using $M = 10^6$

Solution.

$$c^* = \left[\frac{2fk_d(c_0)_0}{k_t}\right]^{1/2} = 2.77 \times 10^{-8} \text{ mol/l}$$

$$V = \frac{10^3 Av}{c^*} = 5.99 \times 10^{-14} \text{ cm}^3/\text{radical}$$

From the Debye–Bueche relation,

$$Ra = 500 \text{ Å}$$

so the end-to-end distance $= 2Ra \cong 10^{-5}$ cm

2.9.2 Precipitous Polymerizations

As previously mentioned, AN and VC polymerizations conducted in bulk produce polymer that precipitates virtually from the outset accompanied by autoacceleration of polymerization rate. The solubility of VC in PVC has been estimated to be as low as 0.03 weight percent (71) at ambient temperature. PVC particles approximately 10^{-5} cm in size have been observed by electron microscopy (72) immediately after the start of reaction, followed by rapid coalescence into aggregates at conversions less than 1 percent.

For PAN, which has been described as even less soluble in its monomer than PVC, Bamford et al. (14) proposed a kinetic model in which buried radicals are unable to terminate; therefore, they grow in number as well as size. The absence of chain transfer to monomer precludes escape by the radicals to the monomer phase where they could terminate bimolecularly. Experiments showing near first-order dependence of polymerization rate on initiator concentration support this model.

Magat (73) suggested that VC radicals proliferate similarly and that the QSSA is never attained. This can lead to estimates for c^* as high as 10^{-4} mol/l. VC polymerization does, however, exhibit chain transfer to monomer. Furthermore, kinetic experiments show the usual half-order dependence of rate on initiator concentration ($r \propto c_0^{1/2}$), suggesting bimolecular termination. To explain this as well as the two-thirds-order dependence on polymer weight observed, ($r \propto w_p^{2/3}$), Bengough and Norrish (74) proposed a surface immobilization model in which radicals stick to the surface of (spherical) polymer particles and transfer their activity via monomer (chain-transfer) to the continuous phase where termination occurs. Mickley et al. (75) added radical occlusion with monomer diffusion in the polymer particles to account from the departure they observed from two-thirds-order dependence of the rate.

More recent descriptions of VC polymerization have been offered by Barclay (76) and Ravey et al. (71) and more recent models by Talamini et al. (77) and Hamielec et al. (78). Polymer particles swollen with monomer (solubility of VC in PVC is ca. 30 weight percent) and containing initiator precipitate immediately. Chain transfer to monomer occurs, thereby producing mobile monomer radicals in the polymer phase. Diffusion of monomer into the polymer, where most of the polymerization occurs, is sufficiently rapid to sustain propagation and maintain a constant monomer composition (at the solubility level). The primary polymer particles form aggregates that continue to grow by polymerization of imbibed monomer and by capture of newly precipitated polymer contributed by the monomer phase, but no new aggregates are formed beyond approximately 1-percent conversion. Initiator permeates both phases, and its concentration is thus assumed to be uniform in space. This mechanism is believed to apply in the conversion range 1–70 percent, after which polymerization stops short of 100 percent due to vitrification of the polymer particles (T_g of particles reaches T of reaction).

We shall briefly examine the model of Hamielec et al. (78) for batch, bulk VC polymerization. The aim of the model is to compute monomer conversion rate, $d\Phi/dt$; cumulative DPD, $w(x)$; and cumulative average DP's, $\bar{x}_N$ and $\bar{x}_W$. To facilitate the inclusion of volume change during polymerization, we use the exact definition of monomer conversion, $\Phi \equiv 1 - N_m/(N_m)_0$, and define the following weights: W_m and W_p for monomer phase and polymer phase, respectively; and weight fractions: w_m and w_p for monomer and polymer within each phase. Total weight, of course, is $W = W_m + W_p \cong (N_m)_0 M_0$. Thus, since the polymer is presumed to be insoluble in monomer, $(w_p)_m = 0$, and we may write

$$(w_p)_p W_p = \Phi W \tag{2.9-13}$$

Furthermore, the weight fraction of polymer in the polymer-rich phase $(w_p)_p$ is constant. From these relations, the total rate of polymer formation is

$$(w_p)_p \frac{dW_p}{dt} = \left(\frac{dW_p}{dt}\right)_m + \left(\frac{dW_p}{dt}\right)_p \tag{2.9-14}$$

Following Hamielec et al., we write the contributions to the formation rate by the monomer and polymer-rich phases, respectively, as

$$\left(\frac{dW_p}{dt}\right)_m = \left(\frac{d\Phi}{dt}\right)_m (W - W_p) \tag{2.9-15}$$

and

$$\left(\frac{dW_p}{dt}\right)_p = \left(\frac{d\Phi}{dt}\right)_p 1 - (w_p)_p \tag{2.9-16}$$

and account for hindered termination in the polymer phase by relating monomer conversion rates in the two phases with a constant factor, $P \geqslant 1$,

$$\left(\frac{d\Phi}{dt}\right)_{\mathrm{p}} = P\left(\frac{d\Phi}{dt}\right)_{\mathrm{m}} \qquad (2.9\text{-}17)$$

since monomer and initiator concentrations are assumed to be constant in both phases. After substituting equations 2.9-15 and 2.9-16 into 2.9-14 and replacing W_{p} and $(d\Phi/dt)_{\mathrm{p}}$ with the aid of equations 2.9-13 and 2.9-17, respectively, we obtain for total conversion rate

$$\frac{d\Phi}{dt} = \left(\frac{d\Phi}{dt}\right)_{\mathrm{m}}(1 + Q\Phi) \qquad (2.9\text{-}18)$$

where

$$Q \equiv \frac{P\left[1 - (w_{\mathrm{p}})_{\mathrm{p}}\right] - 1}{(w_{\mathrm{p}})_{\mathrm{p}}} \qquad (2.9\text{-}19)$$

For conversion rate $(d\Phi/dt)_{\mathrm{m}}$, the authors used the following rate equation from homogeneous bulk kinetics:

$$\left(\frac{d\Phi}{dt}\right)_{\mathrm{m}} = \frac{k_{\mathrm{ap}}(c_0)_0^{1/2}}{(1 - B\Phi)^{1/2}}\exp\left(\frac{-k_d t}{2}\right) \qquad (2.9\text{-}20)$$

where $B \equiv (\rho_{\mathrm{p}} - \rho_{\mathrm{m}})/\rho_{\mathrm{p}}$. This expression is based on constant monomer concentration c_{m} in the monomer phase and assumes uniform initiator concentration in both phases. It also takes volume change into account utilizing additivity of volumes, as we did in Section 1.11.4, where only a single phase was involved.

Cumulative DPD and average DP's at final conversion Φ_f were computed from the corresponding instantaneous properties. Thus, for cumulative DPD,

$$w_x = \frac{\int_0^t x r_x V\, dt}{\int_0^t \sum_x x r_x V\, dt} \qquad (2.9\text{-}21)$$

and by assuming that $\sum x r_x \cong r$ (Section 1.13), it follows that $x r_x \cong r(w_x)_{\mathrm{inst}}$ and therefore

$$w_x = \Phi_f^{-1}\int_0^{\Phi_f}(w_x)_{\mathrm{inst}}\, d\Phi \qquad (2.9\text{-}22)$$

since

$$\Phi(t) \propto \int_0^t V r\, dt \qquad (2.9\text{-}23)$$

Similarly for cumulative average DP's (Example 1.13-3)

$$\bar{x}_{N} = \frac{\int_{0}^{t} Vr\, dt}{\int_{0}^{t} (\bar{x}_{N})_{inst}^{-1} Vr\, dt} = \frac{\Phi_{f}}{\int_{0}^{\Phi_{f}} (\bar{x}_{N})_{inst}^{-1}\, d\Phi} \tag{2.9-24}$$

and

$$\bar{x}_{W} = \frac{\int_{0}^{t} Vr(\bar{x}_{W})_{inst}\, dt}{\int_{0}^{t} Vr\, dt} = \Phi_{f}^{-1} \int_{0}^{\Phi_{f}} (\bar{x}_{W})_{inst}\, d\Phi \tag{2.9-25}$$

The integrations in equations 2.9-22, 2.9-24, and 2.9-25 were performed with the aid of equation 2.9-18 and the integrands were assumed to be average values of the corresponding quantities in the monomer and polymer phases, weighted by the weight fraction of polymer produced by that phase. For example, in equation 2.9-22,

$$(w_{x})_{inst} = \frac{(r_{p})_{m} V_{m} (w_{x})_{inst, m} + (r_{p})_{p} V_{p} (w_{x})_{inst, p}}{(r_{p})_{m} V_{m} + (r_{p})_{p} V_{p}} \tag{2.9-26}$$

Simulations using this model were compared with experiments by Abdel and Hamielec (78). Some results for VC polymerization at 70°C are shown in Figs. 2.9-16–2.9-18.

Equation 2.9-18 is a key element in the foregoing model. It derives from equations 2.9-15 and 2.9-16 and is dependent on the expression for $(d\Phi/dt)_{m}$, equation 2.9-20. It should be noted that certain aspects of these expressions are open to question. For

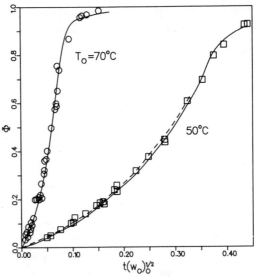

Figure 2.9-16. Experimental and computed conversion histories for VC polymerization (78).

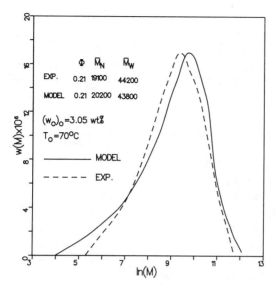

Figure 2.9-17. Experimental and computed DPD for VC polymerization at 70°C (78).

instance, while equation 2.9-20 follows from the rate equation for monomer conversion in Section 1.11.4, by substituting $(N_m)_0/V_0$ for c_m and N_0/V for c_0, we would obtain a different result, that is,

$$\left(\frac{d\Phi}{dt}\right)_m = \frac{k_{ap}(c_0)_0^{1/2}(1-\Phi)}{(1-B\Phi)^{1/2}}\exp\left(-\frac{k_d t}{2}\right) \tag{2.9-27}$$

if we additionally substitute $V_m = V_0(1-\Phi)$ for V in the remaining term

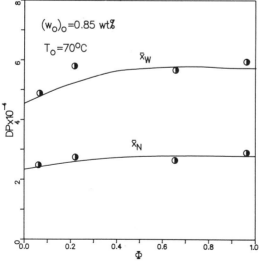

Figure 2.9-18. DP drift with conversion at 70°C with $(w_0)_0 = 0.85$ (78).

$(N_m)_0 V^{-1}(d\Phi/dt)$ of the rate equation for monomer conversion in that section, as it appears we should to properly account for volume change. Equation 2.9-27, which is identical to the equation deduced in Section 1.11-4, is clearly different from equation 2.9-20 used in the model. Total volume V is used for c_0 as opposed to V_m because the initiator pervades both phases. Its conversion-dependence is given in Section 1.11-4.

2.9.3. Emulsion Polymerization

Chain-addition polymerizations are frequently carried out in organic (oil) particles suspended in water. This procedure facilitates mixing by agitation and heat transfer, but it introduces additional material (water, stabilizers, emulsifiers) that may subsequently require separation.

In suspension (bead) polymerization, the dispersed phase is sustained by mechanical agitation. Particles are small compared to reactor dimensions but sufficient in size to contain a large number of propagating intermediates. Initiator, when used, is dissolved in the dispersed phase. Thus the reaction mechanisms and rate functions should be identical to those for bulk polymerization.

In emulsion polymerization, on the other hand, the suspension is stabilized by soap and the initiator is normally dissolved in the continuous phase. The particles are much smaller (submicron range) than in suspension polymerization. In fact, they contain less than one radical per particle on the average. Consequently, the mechanism for emulsion polymerization is distinctly different than for bulk, solution, and suspension polymerization. Two important manifestations of this difference are a more rapid rate and, simultaneously, a larger DP.

Evidently four phases play a role during free-radical emulsion polymerization: water, micelles, monomer droplets, and polymer particles. The water contains initiator and acts as a radical supplier. Most of the soap (emulsifier) is initially present in the form of micelles, which contain some of the monomer within them. Micelles are numerous and tiny. They are transformed into polymer particles the instant that a radical enters and initiates a polymer chain. Those not transformed supply soap to the growing polymer particles. Thus micelles act as particle generators and soap suppliers. The suspended monomer droplets initially contain most of the monomer and some adsorbed soap. They are larger in size and fewer in number than micelles. They supply monomer to the polymer particles, which swell with monomer and grow in size as the radical chains within them grow. The polymer particles are the principle loci of polymerization.

The entire reaction period is generally subdivided into three intervals. The first interval begins with the onset of free-radical generation and continues until micellar soap is depleted. During this interval polymer particles form and grow by adsorbing monomer, which diffuses through the water phase from the monomer droplets, and by absorbing soap from the surrounding micelles. The first interval typically ends after 1–5 percent conversion of monomer.

The second interval starts after the micelles have disappeared and continues until the monomer droplets are depleted. During this period, particles grow in volume but remain constant in number. The concentration of monomer within them apparently remains constant as well. This interval typically persists through 60–80 percent conversion.

The final interval starts with the depletion of monomer as a separate phase and continues until polymerization ceases. Polymer particles swollen with monomer grow during this period, and monomer concentration decreases. Anomalous effects associated with bulk polymerization, such as the G-E, have been attributed to this interval.

The physical picture of emulsion polymerization just described and the quantitative theory are the result of the combined efforts of numerous researchers. Only a few references are cited here (79–85), including three review articles (82–84). The original quantitative theory of Smith and Ewart (80), which still constitutes the basis for most modern theories, will be discussed briefly here.

By analogy with the rate function for homogeneous polymerization $r \cong k_p c_m c^*$, we write rate of emulsion polymerization as

$$r = k_p (c_m)_p \frac{N_p}{V} \frac{\bar{n}}{\text{Av}} \qquad (2.9\text{-}28)$$

where $(c_m)_p$ represents the monomer concentration in the polymer particles, N_p/V represents number of particles (N_p) per total reaction volume (V), $\bar{n}$ is the average number of growing radicals per particle, and Av is Avogadro's number. Obviously, N_p/V and $\bar{n}$ are the key variables. The objective of the theory is to develop analytical expressions for these variables.

Before commencing with our analysis, we shall attempt to account for the simultaneously large values for r and DP normally observed in emulsion polymerizations via a simple comparison with homogeneous polymerizations (80) using the experimental value of 10^{12}–10^{15} per cubic centimeter (or 10^{15}–10^{18} per liter) reported for N_p/V (85). We assume, as a first approximation, that a free-radical, having once entered a polymer particle, must remain there and propagate until another radical enters and terminates it, and that the time for such termination is short compared to the interval between successive radical entrances. Under these conditions, it follows that $\bar{n} \cong \frac{1}{2}$. The quantity $(N_p/V)(\bar{n}/\text{Av})$ in equation 2.9-28 could then be as large as 10^{-6} mol/l as contrasted with 10^{-8} mol/l quoted earlier as a typical value for c^* in homogeneous polymerization (Example 2.7-3). This could account for the high rate r. To account for the large DP, we contrast the characteristic time (CT) for polymer chain growth

$$\lambda_p = \frac{\dfrac{N_p}{V} \bar{n}}{\text{Av}\left(\dfrac{r_0^*}{s}\right)} \qquad (2.9\text{-}29)$$

with that for homogeneous polymerization, $\lambda_p = c^*/(r_0^*/s)$ (Section 1.11), and apply the identical argument used for r. It should be noted that equation 2.9-29 assumes that the rate of primary radical formation $V\text{Av}r_0^*$ is equal to the rate at which radicals enter polymer particles in the second interval, where r_0^* is expressed in moles per unit volume per unit time.

Returning now to one objective of the theory, which is to compute N_p at the end of the first interval ($t = t_1$), we begin with a population balance for polymer particles. Since the rate of polymer formation $\dot{N}_p$ is identical to the rate at which radicals enter micelles, this balance for a batch system may be written as

$$\frac{dN_p}{dt} = \dot{N}_p = n_n(t)S_M(t) \tag{2.9-30}$$

where $n_n(t)$ is the normal component of the radical diffusion flux and $S_M(t)$ is the total miscellar area remaining at time t. Although it is known that $n_n(t)$ is a function of radius (Chapter 6), Smith and Ewart neglected this dependence. Thus the flux into the growing particles is the same as into the micelles. In essence, this procedure assigns to the polymer particles a greater ability to capture radicals than they actually possess and should therefore lead to an underestimation of the number of micelles transformed into polymer particles. The authors also assume that the total interfacial capacity S of the soap introduced per unit reaction volume (V) is the sum of the surface areas per volume of the micelles and polymer particles at any time.

$$S = S_M(t) + S_p(t) = \text{constant} \tag{2.9-31}$$

After substituting this constraint equation into balance 2.9-30, we obtain

$$V^{-1}\frac{dN_p}{dt} = Avr_0^*(t)\left(1 - \frac{S_p(t)}{S}\right) \tag{2.9-32}$$

with the aid of the relation $VAvr_0^*(t) = Sn_n(t)$. The total surface area of the particles at any time t is given by the integral

$$S_p(t) = \int_0^t s_p(t, \alpha)\dot{N}_p(\alpha)\, d\alpha \tag{2.9-33}$$

Where $s_p(t, \alpha)$ is the area at time t per particle with age $t - \alpha$. The latter quantity may be related by simple geometry

$$s_p(t, \alpha) = (4\pi)^{1/3}3^{2/3}\left[V_p(t, \alpha)\right]^{2/3} \tag{2.9-34}$$

to the volume per particle v_p, whose rate of growth is assumed to be constant

$$\frac{dv_p}{dt} \equiv \mu = \text{constant} \tag{2.9-35}$$

because the monomer concentration is constant. Thus we may substitute

$$s_p(t, \alpha) = \theta(t - \alpha)^{2/3} \tag{2.9-36}$$

into equation 2.9-31 and the result into 2.9-32, which yields the following integral equation

$$V^{-1}\frac{dN_p}{dt} = Avr_0^*(t)\left[1 - S^{-1}\int_0^t \theta(t - \alpha)^{2/3}\frac{dN_p}{d\alpha}\, d\alpha\right] \tag{2.9-37}$$

where $\theta \equiv [3(4\pi)^{1/2}\mu]^{2/3}$. The authors solved this equation for N_p at t_1, which is the time at which $dN_p/dt = 0$. Their result for constant r_0^* is

$$\frac{N_p}{V} = 0.37\left(\frac{Avr_0^*}{\mu}\right)^{2/5} S^{3/5} \qquad (2.9\text{-}38)$$

As noted, the last result should give a low estimate for N_p/V. By contrast, if radicals absorbed by polymer particles are neglected, one should obtain a high estimate for N_p/V. Thus, by dropping the term $S_p(t)$ from equation 2.9-32, we conclude that N_p is constant when r_0^* is constant. By substituting this result together with equation 2.9-36 into 2.9-32 and setting the LHS of the latter equal to S, we obtain an equation in terms of the upper limit of integration, which gives t_1. Applying this value for t_1 to equation 2.9-32 ($N_p = VAvr_0^*t_1$) yields

$$\frac{N_p}{V} = 0.53\left(\frac{Avr_0^*}{\mu}\right)^{2/5} S^{3/5} \qquad (2.9\text{-}39)$$

The difference between the two estimates is seen to be small.

The second objective of the theory is to determine $\bar{n}$ during the second time interval for substitution into equation 2.9-28 together with the quantity N_p/V just computed. To accomplish this, we shall require a population balance for free radicals. The molar rate per unit particle volume of initiation and termination of all radicals in all particles due to entering primary radicals is $r_0^*/N_p v_p$ if we assume, as before, that primary radicals diffuse into particles at the same rate that they are formed, r_0^*. The molar rate of mutual termination of radicals within a particle per unit volume is $k_t N^*(N^* - 1)/v_p^2 Av^2$, where N^* represents the number of radicals in the particle. The reader should note that the concentration squared term $(N^*/v_p Av)^2$ normally used has been replaced by a more precise expression, owing to the small number of radicals in each particle. The probabilistic basis for this expression was discussed in detail in Section 1.8.2. To these rate functions, we add a radical escape rate $k_e s_p N^*/v_p Av$ where the molar rate constant k_e is similar to a mass transfer coefficient.

By designating N_n as the number of polymer particles containing n free radicals and applying the quasi-steady-state approximation (QSSA) (Section 1.11.2), we obtain the following population balance for each n at any instant in time:

$$\frac{v_p Avr_0^* N_{n-1}}{N_p} + \frac{k_t(n + 2)(n + 1)N_{n+2}}{v_p Av} + \frac{k_e s_p(n + 1)N_{N+1}}{v_p}$$

$$= \frac{v_p Avr_0^* N_n}{N_p} + \frac{k_t n(n - 1)N_n}{v_p Av} + \frac{k_e s_p n N_n}{v_p}$$

$$(2.9\text{-}40)$$

where the LHS represents the rate of formation of n-type particles and the RHS represents their destruction. The total number of particles is given by

$$N_p = \sum_n N_n \qquad (2.9\text{-}41)$$

and the average number of radicals per particle sought is

$$\bar{n} \equiv \frac{\sum\limits_n nN_n}{N_p} \qquad (2.9\text{-}42)$$

Difference equation 2.9-40 may be written more conveniently as

$$(n+2)(n+1)N_{n+2} + \beta(n+1)N_{n+1} + \alpha N_{n-1} = [\beta n + n(n-1) + \alpha]N_n \qquad (2.9\text{-}43)$$

where $\alpha \equiv (v_p \text{Av})^2 r_p^* / k_t$ and $\beta \equiv \text{Av} k_e s_p / k_t$.

Stockmayer (81) solved this difference equation using generating functions. His solution for negligible escape rate ($\beta = 0$) after substitution into equation 2.9-42 leads to

$$\bar{n} = \frac{I_0(a)}{I_1(a)} \qquad (2.9\text{-}44)$$

where $a \equiv (8\alpha)^{1/2}$ and I_0 and I_1 are Bessel functions of the first kind. This result is shown in Fig. 2.9-19. His general solution for $\beta \neq 0$ has been reported to be incorrect (83).

For further details on emulsion polymerization theory and experiments, the reader is referred to the research literature (85).

Example 2.9-2. Particle Size in Emulsion Polymerization.   Estimate the average particle size for which $\bar{n} = \frac{1}{2}$ using the following quantities from homogeneous polymerization kinetics: $r_0^* = 10^{-7}$ mol/l s and $k_t = 10^8$ 1/mol s.

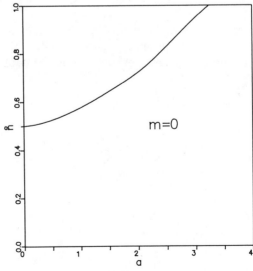

Figure 2.9-19.   Average number of radicals per particle $\bar{n}$ as a function of a (81).

Solution. From Fig. 2.9-19, $a \leqslant 1$ when $\bar{n} = \frac{1}{2}$. Therefore, $\alpha \leqslant 0.125$ and from the definition of α, we obtain

$$v_p \leqslant \left(\frac{\alpha k_t}{Av^2 r_0^*} \right)^{1/2} = 1.86 \times 10^{-14} \text{ cm}^3$$

which is equivalent to a sphere of diameter 3.3×10^{-5} cm or 0.33 micron.

2.10. CHAIN-TRANSFER EFFECTS

Reactions in which active intermediates transfer their activity, rather than propagate, were listed in Table 1.10-3. Recipients of chain transfer can be the monomer itself, the polymer product, a diluent, or a transfer agent that is added intentionally. Chain transfer to polymer will be treated separately in Section 2.10.2. We begin with chain transfer to monomer and foreign substance S.

2.10.1. *DP Regulation and Rate Retardation*

Referring to Tables 1.10-3 and 1.10-4, rate function for polymeric and foreign intermediates yields, after separate application of the QSSA,

$$r^* = r_{ip} + k_{is} c_m c_s^* - k_{fs} c_s c^* - k_t (c^*)^2 - k_{ta} c^* c_s \cong 0 \qquad (2.10\text{-}1)$$

and

$$r_s^* = \underbrace{k_{fs} c_s c^*}_{r_{fs}} - \underbrace{k_{is} c_m c_s^*}_{r_{is}} - \underbrace{k_{ta} c^* c_s^*}_{r_{ta}} - \underbrace{k_{tb} (c_s^*)^2}_{r_{tb}} \cong 0 \qquad (2.10\text{-}2)$$

respectively. In Chapter 1, we routinely neglected rate functions r_{ta} and r_{tb}, assuming that there is no buildup of c_s^*. Thus, from equation 2.10-2 we conclude that transfer and secondary initiation account for the formation and destruction of s, respectively, and that their rates are balanced, $r_{is} \cong r_{fs}$. The reader will recall that while these reactions represent kinetic chain propagation, they also represent polymer chain termination. It follows from equation 2.10-1 that the population of polymeric intermediates is determined by primary initiation and termination in the usual way, $r_{ip} = r_t$, and that the rate of polymerization is therefore unaffected. However, the additional polymer chain termination reactions result in a reduction of DP. The same is true for chain transfer to monomer. These conclusions are evident from equation 1.13-11, which states that $(\bar{x}_N)_{inst}$ decreases as chain-transfer constants k_{fm}/k_p and k_{fs}/k_p increase. From this equation, one may also infer that if $(\bar{x}_N)_{inst}$ is independent of c_0, then chain transfer is important, and if $\bar{x}_N$ is independent of c_m as well, then monomer transfer in particular is important.

When intermediate c_s^* is less active than c_0^*, S is called a rate retarder. When it is inactive ($r_{is} \cong 0$), S is an inhibitor. In such cases c_s^* increases and, as a consequence, termination rates r_{ta} and r_{tb}, especially the former, can become more important than $k_t(c^*)^2$. If r_{is}, r_t, and r_{ta} are neglected in equations 2.10-1 and 2, we conclude that the equation $r_{fs} \cong r_{tb}$ determines the population of S*, and that the equation $r_{ip} \cong r_{fs}$ determines the population of m*. After substituting the last equation into a balance for S (Table 1.10-4), and integrating the result,

$$\frac{dc_s}{dt} = -k_{fs}c_sc^* \cong r_{ip} \qquad (2.10\text{-}3)$$

subject to initial condition $c_s = (c_s)_0$, we obtain an estimate of the inhibition period (when $c_s \to 0$)

$$\lambda_{inh} \cong \frac{(c_s)_0}{(r_{ip})_0} \qquad (2.10\text{-}4)$$

The preceding relationship can be verified from an analytical treatment of the equations, detailed in Example 2.10-1, from which we find

$$\lambda_{inh} = -k_d^{-1}\ln\left[\frac{1 - (c_s)_0}{2f(c_0)_0}\right] \qquad (2.10\text{-}5)$$

For small values of the ratio of inhibitor and initiator concentrations equations 2.10-4 and 2.10-5 are identical. This is generally valid for powerful inhibitors.

2.10.2. Branching

To analyze long-chain branching in free-radical polymerizations, it is necessary to include in our reaction sequence chain transfer to polymer, which can occur with any repeat unit in the product chain, as well as propagation of polymeric intermediates by addition of product chains having terminal double bonds. Such double bonds can arise from chain transfer with monomer and from termination by disproportionation. The pertinent reactions are listed in Table 1.10-3.

The simplest mechanism for long-chain branching is the sequence consisting only of chain transfer to polymer followed by propagation of the active polymer by addition of monomer in the usual way. An important branch variable that drifts with reaction is the instantaneous branch density r_b/r_p (Section 2.6.5). Its variation with conversion can be estimated, assuming that no more than one branch point per repeat unit is permissible. By neglecting the propagation of terminal double bonds in polymer product, equation 1.13-22 reduces to

$$r_b = k_{fp}\mu_c^1 c^* \cong k_{fp}\Phi(c_m)_0 c^* \qquad (2.10\text{-}6)$$

Division by $r_p = k_p c_m c^* = k_p(1 - \Phi)(c_m)_0 c^*$ yields

$$\frac{\dfrac{k_{fp}}{k_p}\Phi}{(1 - \Phi)} = \frac{r_b}{r_p} \equiv (c_m)_0^{-1}\frac{d\left(\displaystyle\sum_x\sum_b bc_{b,x}\right)}{d\Phi} \qquad (2.10\text{-}7)$$

where the RHS may be interpreted as instantaneous branch density, b'_{inst} (Section

1.13), and the LHS thus describes its drift with conversion. Following integration, subject to the initial condition $\sum\sum bc_{b,x} = 0$ when $\Phi = 0$, and subsequently division by $\Phi(c_m)_0$, we obtain

$$\frac{\sum\limits_{x}\sum\limits_{b} bc_{b,x}}{\Phi(c_m)_0} = -\left(\frac{k_{fp}}{k_p}\right)\left[1 - \Phi^{-1}\ln(1-\Phi)\right] \tag{2.10-8}$$

which is a measure of cumulative branch density b' as a function of conversion. Instantaneous and cumulative b' have been plotted in Fig. 2.10-1 for vinyl acetate polymerization (86) with the latter quantity appearing in reciprocal form as the mean distance between branches. It is noteworthy that the quotient k_{fp}/k_p is equal to the branch density when $\Phi \cong 0.8$.

One structural consequence of branching, discussed in Section 2.6, is a broadening of the DPD. The reintroduction of terminated polymer into the chain-building process causes chain-addition polymerizations to resemble step polymerizations, and even random polymerizations, in a certain sense (Section 1.6). Each branching reaction, especially at high conversions where branch densities are large, offers a growing chain a wide spectrum of possible size increases and structural changes, as in random polymerization at high conversions. Moreover, the lifetime of a growing chain λ_p is thereby extended beyond the period from initiation to termination normally available to chain-addition polymers with termination.

Graessley et al. (86–88) have studied branching in free-radical polymerizations computationally and experimentally. They assumed that the total population of intermediates and consequently conversion rate were controlled by initiation and termination reactions, as usual, but that the distributions among the various types of

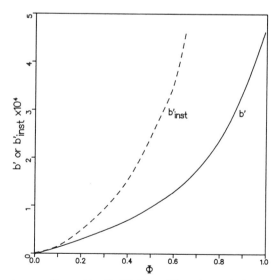

Figure 2.10-1. Instantaneous, $b'_{inst} = (k_{pf}/k_p)[\Phi/(1-\Phi)]$, and cumulative $b' = b_N/\bar{x}_N$, branch density versus monomer conversion, Φ, for vinyl acetate polymerization at 60°C with the parameters in Table 2.10-1 (86).

intermediates was determined by the propagation and transfer reactions in Table 1.10-3. This situation is somewhat analogous to the one that prevails during chain transfer with monomer and foreign substances described in this section where no rate retardation or inhibition occurs. When chain transfer reactions to foreign substances, in addition to chain initiation and termination, are relatively unimportant as polymer chain generators and terminators, then every polymer chain will have exactly one terminal double bond. Consequently, only type m^l_x polymer will be present. When solvent transfer occurs, on the other hand, molecules without terminal double bonds will be formed (Tables A-7 and 1.10).

Thus, for an undiluted (no transfer to S) free-radical polymerization in which the DPD is controlled by the propagation and transfer reactions in Table 1.10-3, the rate equations to be solved are

$$\frac{dc_x}{dt} = r_x \tag{2.10-9}$$

and

$$\frac{dc^*_x}{dt} = r^*_x \underset{QSSA}{\cong} 0 \tag{2.9-10}$$

where rate functions r_x and r^*_x are listed in Table 1.10-4. However, it is generally easier to calculate the first few moments, μ^k_{c*} and μ^k_c for $k = 0$, 1, and 2 (Appendix C) and from them the first two averages and dispersion index (Section 1.5). By differentiating the appropriate definitions,

$$\mu^k_c \equiv \sum_x x^k c_x \tag{2.9-11}$$

and

$$\mu^k_{c*} \equiv \sum_x x^k c^*_x \tag{2.10-12}$$

and exchanging the order of summation and differentiation operations, we obtain, following substitution of equations 2.10-9 and 2.10-10, coupled differential and algebraic equations for μ^k_c and μ^k_{c*}, respectively. The resulting evaluation of $\bar{x}_N$, $\bar{x}_W$, and D_N with conversion are shown in Figs. 2.10-2–2.10-4 (86, 87). These curves were obtained by choosing ratios k'_p/k_p, k_{fm}/k_p and k_{fp}/k_p to best fit experimental data on vinyl acetate bulk polymerizations at 60 and 72°C. Chain-transfer dominance was confirmed by the independence of DP from initiator concentration.

It is apparent from Fig. 2.10-2 that cumulative dispersion D_N increases, as expected. Furthermore, the growth of DP with conversion, especially at high conversions, is somewhat characteristic of step polymerizations, (Section 2.1) for which $\lambda_p \sim \lambda_r$ (Section 1.6), as we speculated earlier.

Example 2.10-1. Initiation Induction Period.   Diphenyl picrylhydrazlyn (DPPH) is a stable free radical that acts as a powerful inhibitor in vinyl polymerizations. If one assumes that one DPPH radical consumes one initiator radical, calculate the length of the induction period.

Solution.

$$(c_s)_0 - c_s = (c_0^*)_0 - c_0^*$$

but

$$\Delta c_0^* = 2f\Delta c_0$$

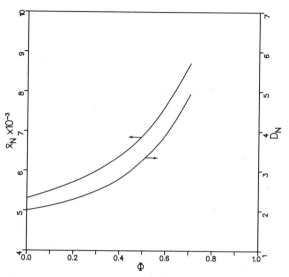

Figure 2.10-2. Degree of polymerization $\bar{x}_N$ and dispersion index D_N versus monomer conversion Φ for vinyl acetate polymerization at 60°C with the parameters in Table 2.10-1 (86).

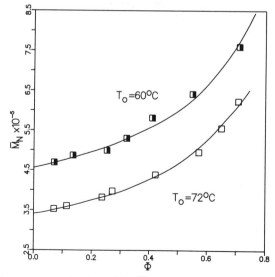

Figure 2.10-3. Number-average molecular weight $\bar{M}_N$ versus monomer conversion Φ for vinyl acetate polymerization with the parameters in Table 2.10-1 (87).

so

$$(c_s)_0 - c_s = 2f[(c_0)_0 - c_0]$$

For the initiator,

$$c_0 = (c_0)_0 \exp(-k_d t)$$

Therefore,

$$(c_s)_0 - c_s = 2f(c_0)_0[1 - \exp(-k_d t)]$$

At the end of the induction period. $t = \lambda_{inh}$ and $(c_s) = 0$. Solving the preceding equation yields:

$$\lambda_{inh} = -k_d^{-1} \ln\left[1 - \frac{(c_s)_0}{2f(c_0)_0}\right]$$

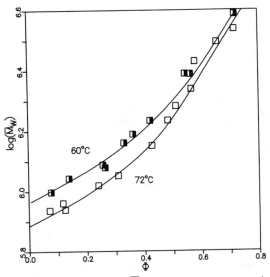

Figure 2.10-4. Weight-average molecular weight $\overline{M}_W$ versus monomer conversion Φ for vinyl acetate polymerization with the parameters in Table 2.10-1 (87).

Table 2.10-1.   Kinetic Parameters for Figures 2.10-1–2.10-4 Obtained by Curve-Fitting (86, 87)

Polymerization temperature	k_p'/k_p	$k_{fp}/k_p \times 10^4$	$k_{fm}/k_p \times 10^4$
60°C	0.66	2.36	2.43
72°C	0.82	2.41	3.00

Linearizing the ln with a Taylor series expansion about $(c_s)_0/2f(c_0)_0 = 0$ gives the approximation:

$$\lambda_{inh} \cong \frac{(c_s)_0}{2fk_d(c_0)_0}$$

which is valid if $(c_s)_0 \ll (c_0)_0$.

2.11. CUMULATIVE DISTRIBUTIONS

With the aid of modern digital computers, it is possible to compute the entire spectrum of product properties, such as DP, CC, and DB, by solving the appropriate system of differential balance equations. The resulting DPD's, CCD's and DBD's, being cumulative, include all dispersion effects stemming from the reaction per se, both statistical and drift, as well as any other dispersion effects that are due to the reactor. The latter effects are reflected in the nature of the balance equations and are the subject of discussion in Chapters 3–5. In this section, we shall restrict our examples to batch reactors.

2.11.1. *Homopolymers*

The primary balance equations to be solved, of course, are those for product x-mer

$$x \geqslant 1 \qquad\qquad \frac{dc_x}{dt} = r_x \qquad\qquad (2.11\text{-}1)$$

where the appropriate rate functions may be taken from Table 1.10-4. As shown, these expressions are functions of reactants, monomer, and initiator, and any other reactive substance (solvent, chain transfer agent, etc.). This necessitates the simultaneous solution of equation 2.11-1 with all coupled balance equations, such as

$$-\frac{dc_m}{dt} = r \qquad\qquad (2.11\text{-}2)$$

and

$$-\frac{dc_0}{dt} = \frac{r_i}{s} \qquad\qquad (2.11\text{-}3)$$

where rate functions r and r_i are also listed in Table 1.10-4. In addition to their dependence on major reaction components, rate functions r_x also depend on the concentrations of intermediates, individually (c_x^*) and collectively (c^*), as shown in Table 1.10-4. A common technique for solving this system of equations is to invoke the QSSA and thus eliminate the intermediate concentration by substituting the functional relationships for c^* and c_x^* in terms of c_0 and ϕ, respectively, developed in Sections 1.11 and 2.6, and by utilizing the equation for $\phi(c_0, c_m, c_s,$ etc.) defined in Section 2.2.6. This technique is remarkably accurate for all but the most severe cases of concentration drift. In the limiting case of very slow initiator consumption, on the other hand, the corresponding equations can be solved analytically by holding c_0 constant (30, 89).

Let us consider the case of free-radical polymerization with initiator decay. To simplify our example somewhat, we shall omit chain transfer and assume that $f = 1$. We also assume that termination occurs exclusively by combination. The balance equations to be solved are, in semi-dimensionless form:

$$x \geqslant 1 \qquad \frac{d\hat{c}_x}{dt} = 2(x-1)\lambda_i^{-1}(\hat{c}^*)^2 \phi^{x-2}(1-\phi)^2 \qquad (2.11\text{-}4)$$

$$-\frac{d\hat{c}_0}{d\tau} = \lambda_i^{-1}\hat{c}_0 \qquad (2.11\text{-}5)$$

$$-\frac{d\hat{c}_m}{dt} = 2x_0\lambda_i^{-1}\hat{c}_0 + \lambda_m^{-1}\hat{c}_m\hat{c}_0^{1/2} \qquad (2.11\text{-}6)$$

where all concentrations except $\hat{c}_x$ have been reduced with their own feed values. Using the definition of $(c^*)_0$ in Example 2.7-3 for the intermediates, we obtain

$$\hat{c}^* = \hat{c}_0^{1/2} \qquad (2.11\text{-}7)$$

and

$$\phi = \frac{\lambda_m^{-1}x_0\hat{c}_m}{\lambda_m^{-1}x_0\hat{c}_m + 2\lambda_i^{-1}\hat{c}_0^{1/2}} = \left[1 + \frac{\hat{c}_0^{1/2}}{\hat{c}_m(\nu_N)_0}\right]^{-1} \qquad (2.11\text{-}8)$$

Feed initiator concentration was chosen to reduce c_x. Either λ_m or λ_i (Table 2.7-1) could be used to render these equations completely dimensionless, depending on whether the reaction is initiator-limited or monomer-limited, respectively. The result shows that only two dimensionless groups are independent. We choose $(\nu_N)_0$ and α_K for convenience. Number and weight DPD's can be computed from the solution $\hat{c}_x(t)$ in the usual way (Table 1.5-2)

The analytical solutions for constant c_0 are most easily obtained by dividing equation 2.11-4 by equation 2.11-6 to eliminate time and integrating the quotient

$$\frac{d\hat{c}_x}{d\hat{c}_m} = -\left[\frac{(x-1)}{2}\right]\phi^{x-2}(1-\phi)^3 \qquad (2.11\text{-}9)$$

where

$$\phi \equiv \frac{(\nu_N)_0\hat{c}_m}{1 + (\nu_N)_0\hat{c}_m} \qquad (2.11\text{-}10)$$

by parts (30). The results, which apply without significant error when $\alpha_K = 0.01$ (Section 2.8), are listed in Tables 2.11-1 and 3.5-1 as functions of conversion (cf.

Example 1.11-1):

$$\Phi = \left[1 + (\nu_N)_0^{-1}\right]\left[1 - \exp\left(\frac{-t}{\lambda_m}\right)\right] \underset{LCA}{\cong} 1 - \exp\left(\frac{-t}{\lambda_m}\right) \qquad (2.11\text{-}11)$$

For values of $\alpha_K > 0.01$, numerical solutions are necessary. The weight fraction DPD found by analytical solution is shown in Fig. 2.11-1 (30). It is useful to note that when the polymer is formed from intermediates with the most probable distribution, regardless of whether termination is by combination or disproportiona- tion, there is a maximum in $(w_x)_{inst}$ versus x at a value very close to $(\bar{x}_N)_{inst}$. If drift dispersion is small, then the cumulative distribution may be approximated by the instantaneous distribution (most probable) and the peak in w_x versus x regarded as a measure of $\bar{x}_N$. (For verification, see Example 2.11-1).

The drift effects expected for a CONV polymerization are clearly evident in Fig. 2.11-1. As conversion proceeds, the peak of the distribution shifts to the left, indicative of an increasing fraction of low-molecular-weight species. Furthermore, $\bar{x}_N$ drops faster than $\bar{x}_W$, which is consistent with the fact that high conversions lead to rising values of D_N. One should be aware that the analytical results contain the LCA. As a result, the DPD for 100-percent conversion shows a maximum at $x = 0$. This is an artifact owing to the assumption that $r_i = (r_i)_0$ while $r_p = 0$ when monomer is totally consumed, and thus $(\bar{x}_N)_{inst} \propto r_p/r_i = 0$. Actually, initiation also requires monomer and, as demonstrated in Section 2.8, the rate of initiation would also approach zero as $c_m \to 0$.

The effect on polymer properties of "nonideal" kinetic behavior such as D-E, and G-E have been illustrated in preceding sections using the DP averages $\bar{x}_N$ and $\bar{x}_W$. Although this information is sufficient for most analyses, it is instructive to examine the effect of drift on the entire DPD. This necessitates numerical integration of reactant balances coupled with those for product of selected chain lengths and presumes the most probable distribution for the intermediates. Allowing for either termination mode, equation 2.11-4 is more generally written as

$$\frac{d\hat{c}_x}{dt} = \frac{f}{f_0}(\nu_N)_0^{-1}\hat{c}_0\left[(1 - R)\phi^{x-1}(1 - \phi) + xR\phi^{x-2}(1 - \phi)^2\right] \qquad (2.11\text{-}12)$$

Table 2.11-1. Expressions for DPD's of Conventional Chain-Addition Polymerizations

$$n_x = \frac{\left[\frac{(\nu_N)_0}{(\nu_N)_0 + 1}\right]^x [x^{-1} + (\nu_N)_0^{-1}] - \left[\frac{(\nu_N)_0(1 - \phi)}{(\nu_N)_0(1 - \phi) + 1}\right]^x [x^{-1} + (\nu_N)_0^{-1}(1 - \phi)^{-1}]}{\ln\frac{[(\nu_N)_0 + 1]}{(\nu_N)_0(1 - \phi) + 1}}$$

$$w_x = \frac{\left[\frac{(\nu_N)_0}{(\nu_N)_0 + 1}\right]^x [1 + x(\nu_N)_0^{-1}] - \left[\frac{(\nu_N)_0(1 - \phi)}{(\nu_N)_0(1 - \phi) + 1}\right]^x [1 + x(\nu_N)_0^{-1}(1 - \phi)^{-1}]}{2(\nu_N)_0\phi}$$

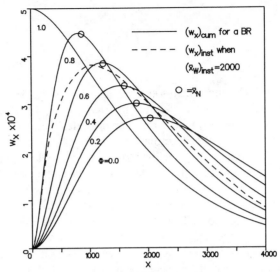

Figure 2.11-1. Drift in DPD with conversion for conventional polymerization, $\alpha_K \ll 1$.

where $\hat{t} \equiv t/\lambda_m$, while the expression for ϕ, equation 2.11-8, must be modified to account for gel-effect and initiator efficiency drift. Thus

$$\phi \equiv \left[1 + \frac{\left(\dfrac{f}{f_0}\right)^{1/2} G^{1/2} \hat{c}_0^{1/2}}{(\nu_N)_0 \hat{c}_m} \right]^{1/2} \tag{2.11-13}$$

where the functional forms of G and f have been described elsewhere (Sections 2.8 and 2.9).

Consider first the effect of D-E conditions, illustrated in Fig. 2.11-2. In Section 2.8 we showed that D-E caused an upward drift in both $\bar{x}_N$ and $\bar{x}_W$ with a more rapid rise in $\bar{x}_W$ leading to an increase in D_N with conversion. The faster rise of $\bar{x}_W$ was attributed to the production of a high-molecular-weight tail as the rate of initiation was curtailed by waning initiator concentrations. Note in the full DPD's, the increase in fraction of product composed of longer chain polymers. This is observed as the position of the high-molecular-weight tail rises with conversion. Thus the cumulative weight fraction of high ends, proportional to the area under that part of the curve, must also increase. The value of $\bar{x}_N$ approximated by the peak in the DPD curve also drifts slightly upward with conversion. This effect is smaller than the increase in $\bar{x}_W$, however, and D_N rises with conversion as a result.

Qualitatively one might expect G-E to have a similar effect. This is not only evident in Figs. 2.11-3 and 2.11-4, but the effect is even more pronounced than with D-E, since complete conversions are attained and drift may be observed over the full duration. Again it is apparent that the largest drift occurs in the magnitude of $\bar{x}_W$, and that the presence of increased concentrations of high molecular weight species gives rise to the increase in D_N with conversion.

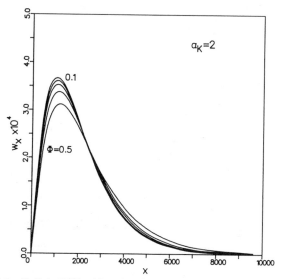

Figure 2.11-2. Drift in DPD with conversion for mild D-E polymerization, $\alpha_K = 2$.

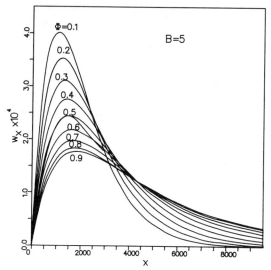

Figure 2.11-3. Drift in DPD with conversion for mild G-E polymerization, $B = 5$.

Once again the reader is reminded that drift effects may be minimized by conversion-dependent initiator efficiency. In Fig. 2.11-5, we see the conversion-dependence of the DPD for an initial efficiency $f_0 = 0.6$, when f is a function of conversion (Section 2.8-3). All parameters were chosen to make this case initially identical to the conventional polymerization in Fig. 2.11-1. However, the conversion drift in the DPD is substantially less when f is not constant at its initial value. This is consistent with the drift in average properties discussed in Section 2.8-3. One is led to speculate that some of the drastic drift effects reported in the literature, which are

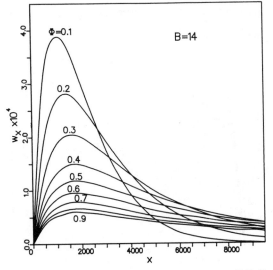

Figure 2.11-4. Drift in DPD with conversion for strong G-E, $B = 14$.

based on computation, could be due at least in part to use of the LCA and a conversion-independent value of the initiator efficiency factor.

2.11.2. Copolymers

In Section 1.5 a pseudo weight CCD was introduced in anticipation of its application in the present context. However, in lieu of the discrete distribution w_{py} defined there (equation 1.5-23) it will be more convenient to define a density function here,

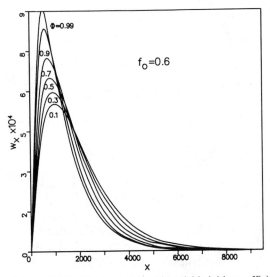

Figure 2.11-5. Drift in DPD with conversion with variable initiator efficiency, $f_0 = 0.60$.

$w_p(y)$, as the derivative of its corresponding distribution function (Appendix B)

$$W_p(y) \equiv \int_0^y w_p(y') \, dy' \qquad (2.11\text{-}14)$$

The reasons for this will become apparent presently as we seek a computational procedure for determining the CCD during copolymerization.

In Section 2.6, we concluded that the instantaneous, pseudo CCD for most copolymers was probably relatively narrowly dispersed. Consequently, drift dispersion must dominate the distribution represented by equation 2.11-14. If we assume that the instantaneous CCD is monodisperse, we can replace the distributed variable y with $(y)_{inst}$ or, equivalently, ν_A. Furthermore, in Section 2.7, it was concluded that $(y)_{inst}$, unlike $(\bar{x}_N)_{inst}$, drifts monotonically with respect to Φ (or time). Thus, assuming for now that ν_A drifts upward with Φ ($\beta_K > 1$), equation 2.11-14, which counts only molecules having compositions less in value than a specific ν_A, may be replaced with the expression

$$\hat{W}_p(\nu_A) = \Phi_f^{-1} \int_{(\nu_A)_0}^{\nu_A} \frac{d\Phi}{d\nu_A} \, d\nu_A = \frac{\Phi(\nu_A)}{\Phi_f} \qquad (2.11\text{-}15)$$

where $\Phi_f \leqslant 1$ represents the final conversion corresponding to the total cumulative copolymer product, and the conversion at any CC, $\Phi(\nu_A)$, may be obtained by combining equations 2.7-17 and 2.7-19. This result follows because molar conversion of monomer is proportional to weight of polymer formed (cf. equation 1.5-23). We have thus replaced cumulative sample in the statistical sense (equation 2.11-14) with cumulative sample in the temporal sense. The situation in which ν_A drifts downward is considered in Example 2.11-2. The cumulative CCD represented by equation 2.11-15 is simply the drift curve inverted (Fig. 2.11-8) and the CC density function $d\Phi/d\nu_A$ is just the inverse of the slope of the drift curve $\nu_A(\Phi)$. Following O'Driscoll et al. (32), this quantity may be computed from the product

$$\hat{W}_p(\nu_A) \equiv \frac{1}{\Phi_f} \frac{d\Phi}{d\nu_A} = \frac{1}{\Phi_f} \frac{d\Phi}{dn_A} \frac{dn_A}{d\nu_A} \qquad (2.11\text{-}16)$$

where the first derivative is the derivative of equation 2.7-17 and the second one is the inverse of the derivative of the CC equation, 1.13-14 (cf. Example 2.7-4). Several CCD's computed in this way are illustrated in Figs. 2.11-6–2.11-9.

Some general conclusions about the shape of the CCD ($d\Phi/d\nu_A$) are possible from the CC diagram together with the following relationship (Example 2.7-5)

$$\frac{d\nu_A}{d\Phi} = \frac{d\nu_A}{dn_A} \frac{\nu_A - n_A}{1 - \Phi} \qquad (2.11\text{-}17)$$

It is evident that the magnitude of the quantity $(d\nu_A/dn_A)/(1 - \Phi)$ must increase with conversion (Φ), while the magnitude of the displacement $\nu_A - n_A$ could pass through a maximum. Consequently, the distribution $d\Phi/d\nu_A$ either monotonically increases ($\beta_K < 1$), monotonically decreases ($\beta_K > 1$), or passes through a minimum value; a maximum value is not possible. Sample curves have been sketched in Fig. 2.11-6.

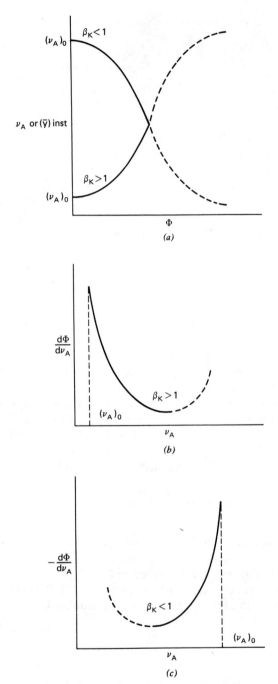

Figure 2.11-6. Inverted CC distribution functions: drift curves (a) and corresponding CC distributions (b) and (c).

Equations 3.11-15 and 2.11-16 allow one to compute cumulative and differential CCD's provided one can analytically link total monomer conversion to instantaneous CC. This can be done using the equations of Meyer and Lowry (3), previously discussed, which relate Φ to n_A as required in equation 2.7-17. By using the CC equation (1.13-14), which relates n_A to ν_A, equation 2.11-15 can be evaluated directly. This is demonstrated in Example 2.11-3. Computation of differential CCD's requires evaluation of the derivative of equation 2.11-15 with respect to ν_A, as suggested by 2.11-16.

$$\hat{w}_p(\nu_A)\, d\nu_A = \Phi_f^{-1}(1 - \Phi)\left[\frac{\alpha}{(1 - n_A)} - \frac{\beta}{n_A} + \gamma(n_A - \delta)\right]$$

$$\frac{\left[(R_A + R_B - 2)n_A^2 + 2n_A(1 - R_B) + R_B\right]^2}{R_A(1 - R_B)n_A^2 + R_A R_B + R_B(1 - R_A)(1 - n_A)^2}\, d\nu_A$$

$$(2.11\text{-}18)$$

In the preceding equation, Φ refers to the conversion at which copolymer product of composition ν_A is formed (and is the solution to equation 2.7-18). It must be less than or equal to the value of Φ_f.

The important conclusion is that the CCD for isothermal copolymerization can be evaluated analytically without numerical integration of balance equations. Although computationally tedious, evaluation of the cumulative and differential CCD's is quite straightforward. Results for several comonomer systems are presented in Figs. 2.11-7–2.11-9. In each case several initial feed compositions are used, and the corresponding CCD drift with conversion is illustrated. Note that drift is always unidirectional and that most CCD's remain rather narrow until high conversions are reached. Substantial spreading of the CCD may occur at advanced conversions, particularly when one comonomer is consumed prematurely. As one reaches such a point, the slope of the CC curve is generally quite steep, indicating large changes in

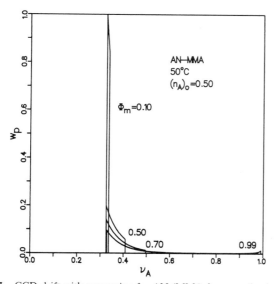

Figure 2.11-7. CCD drift with conversion for AN/MMA for several values of $(n_A)_0$.

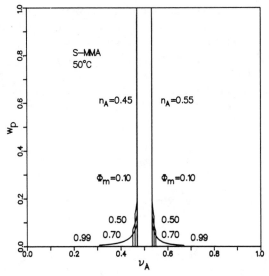

Figure 2.11-8.   CCD drift with conversion for S/MMA for several values of $(n_A)_0$.

ν_A with respect to rather small changes in n_A. S/AN with small initial values of styrene and AN/MMA with large values of AN both show bimodal CCD's at high levels of conversion owing to the sharp change in slope of the CC curve at its extreme values.

Example 2.11-1.   Interpretation of the Maximum in the DPD Curve.   Show that the maximum in the weight fraction DPD for the most probable distribution occurs at the number-average DP.

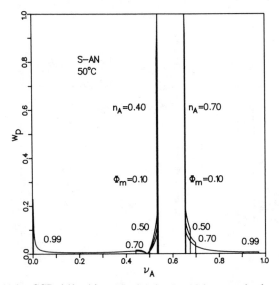

Figure 2.11-9.   CCD drift with conversion for S/AN for several values of $(n_A)_0$.

Solution. For the most probable DPD, $n_x = \phi^{x-1}(1 - \phi)$ and $\bar{x}_N = 1/(1 - \phi)$. By definition $w_x = xn_x/\bar{x}_N$. Hence $w_x = x\phi^{x-1}(1 - \phi)^2$. If there is a maximum in W_x versus x, then

$$\frac{dw_x}{dx} = 0 = (1 - \phi)^2\phi^{x-1}(1 + x \ln \phi)$$

which is satisfied when

$$x|_{\max} = -\frac{1}{\ln \phi}$$

Expanding $\ln \phi$ in a Taylor series around $\phi = 1$ and dropping terms of the order of ϕ^2 and higher (Appendix C) gives

$$\ln \phi \cong \ln(1) + (\phi - 1)$$

Thus, by substitution, the DP of the maximum weight fraction is given by:

$$x|_{\max} = \frac{1}{1 - \phi}$$

which is precisely $\bar{x}_N$ for the most probable DPD.

When termination occurs by combination,

$$n_x = (x - 1)\phi^{x-2}(1 - \phi)^2 \quad \text{and} \quad x_N = \frac{2}{1 - \phi}$$

Hence,

$$w_x = \frac{1}{2}(x^2 - x)\phi^{x-2}(1 - \phi)^3$$

and

$$\frac{dw_x}{dx} = 0 = \frac{1}{2}(1 - \phi)^3\phi^{x-2}\left[(2x - 1) + (x^2 - x)\ln \phi\right]$$

Solving for x as before yields

$$\left.\frac{x^2 - x}{2x - 1}\right|_{\max} = -\frac{1}{\ln \phi} \cong \frac{1}{1 - \phi} = \frac{\bar{x}_N}{2}$$

The LHS can be simplified somewhat if $x \gg 1$

$$\left.\frac{x^2 - x}{2x - 1}\right|_{\max} = \frac{x}{2} - \left.\frac{\dfrac{x}{2}}{2x - 1}\right|_{\max} = \left.\frac{x}{2}\right|_{\max}$$

Thus we find as before:

$$x|_{\max} = \bar{x}_N$$

Example 2.11-2. *Cumulative CCD's.* Derive the counterpart of equation 2.11-15 for cumulative CCD when ν_A drifts downward with Φ ($\beta_K < 1$).

Solution. In this case, copolymer low in composition ν_A is produced last and $d\Phi/d\nu_A < 0$ (Fig. 2.11-6). Therefore, the distribution function must begin counting product at the end of the reaction with the final (lowest) value of ν_A, $(\nu_A)_f$.

$$\hat{W}_p(\nu_A) = \Phi_f^{-1} \int_{(\nu_A)_f}^{\nu_A} \left(-\frac{d\Phi}{d\nu_A} \right) d\nu_A$$

$$= \Phi_f^{-1} \left[\int_{(\nu_A)_f}^{(\nu_A)_0} \left(-\frac{d\Phi}{d\nu_A} \right) d\nu_A - \int_{\nu_A}^{(\nu_A)_0} \left(-\frac{d\Phi}{d\nu_A} \right) d\nu_A \right] = 1 - \frac{\Phi(\nu_A)}{\Phi_f}$$

Alternatively, of course, we could perform the computations discussed in the text for monomer B, whose characteristics compliment those of A, and transform the results at the end.

Example 2.11-3. A Specific Cumulative CCD.   Use equation 2.7-17 in conjunction with equation 2.11-5 to determine the cumulative CCD, $\hat{W}_p(\nu_A)$, for a S/MMA polymerization with a feed content of styrene, $(n_A)_0 = 0.40$. The reaction is carried to completion at a temperature of 333 K. Use the kinetic constants given in Table 2.7-6.

Solution.   From Table 2.7-6, $R_A = 0.474$ and $R_B = 0.457$. Thus the parameters for 2.7-17 are: $\alpha = 0.901$, $\beta = 0.842$, $\gamma = 2.743$, $\delta = 0.508$. Substituting values into equations 1.13-14 and 2.7-17, respectively, we find:

$$\nu_A = \frac{0.474 n_A^2 + n_A(1 - n_A)}{0.474 n_A^2 + 2 n_A(1 - n_A) + 0.457(1 - n_A)^2}$$

and

$$\Phi = 1 - \left(\frac{1 - n_A}{0.60} \right)^{0.901} \left(\frac{n_A}{0.40} \right)^{0.842} \left(\frac{-0.108}{n_A - 0.508} \right)^{2.743}$$

Furthermore, since $\Phi_f = 1.0$, from equation 2.11-15 we may say that $\hat{W}_p(\nu_A) = \Phi(\nu_A)$. Choosing values of n_A, we can then construct the following chart.

n_A	ν_A	$\hat{W}_p(\nu_A)$
0.40	0.44	0.0
0.39	0.43	0.22
0.38	0.43	0.38
0.37	0.42	0.50
0.36	0.41	0.59
0.35	0.40	0.66
0.30	0.37	0.85
0.25	0.33	0.92
0.20	0.28	0.96

From the values shown in the preceding chart, we conclude that the resulting CCD is fairly narrow. More than 80 percent of the copolymer has a composition within 7 mole percent of the product formed from the feed. It is also evident that most of the smearing occurs at elevated conversions.

2.12. CHEMICAL SIMILARITY

Kinetic studies are mostly conducted at moderate temperatures (below 100°C). Production temperatures, however, are generally much higher (possibly as high as 200°C) in order to obtain the greatest specific output rate consistent with the requirements for thermal stability and product quality (e.g., no degradation). Reactive processing, in particular, demands short reaction times (minutes and even seconds) from polymerizations that might otherwise be conducted more slowly.

Of course, there are means for increasing reaction rate other than raising reaction temperatures, such as seeking more rapid catalysts. However, since chemical reaction path falls outside the scope of this book (Section 1.1), we shall concentrate on the thermal method.

The objective is to define the kinetic parameters required at some elevated temperature T to produce the identical polymer at the same conversion in a shorter time than a similar polymerization can at some reference temperature T_r. By identical polymer, we mean one having the same DP and DPD, as characterized by the dispersion index D_N. Obviously, this is a problem in scaling. Therefore, we shall employ dimensional analysis in the same way that we do in conventional scale-up problems.

For illustrative purposes, we shall focus on a chain-addition polymerization and seek a free-radical initiator that will give us the same polystyrene at T as benzoyl peroxide initiator does at 80°C. That is, we require specification of initiator rate constants A_d and E_d such that properties Φ_m, $\bar{x}_N$ and $\bar{x}_W$ are identical in both polymerizations. Both reactions will be assumed to be isothermal and simple kinetics will be used; that is, thermal initiation, chain transfer, and gel-effect will be ignored, and termination by combination ($R = 1$) will be assumed to dominate. Implicit in this analysis, of course, is the requirement that the increase in temperature sought will not alter the reaction mechanism, for example, by introducing thermal initiation or by increasing the importance of termination by disproportionation. Such changes would render the reactions chemically dissimilar.

The simple kinetic model proposed is described by the following dimensionless balance equations in a batch ($\alpha = t$) or plug-flow reactor ($\alpha = x/v$):

$$-\frac{d\hat{c}_0}{d\hat{\alpha}} = \alpha_K \hat{c}_0 \tag{2.12-1}$$

$$-\frac{d\hat{c}_m}{d\hat{\alpha}} = \hat{c}_m \hat{c}_0^{1/2} + \frac{\hat{c}_0}{(\nu_N)_0} = \frac{d\hat{\mu}^1}{d\hat{\alpha}} \tag{2.12-2}$$

$$\frac{d\hat{\mu}^0}{d\hat{\alpha}} = \left(\frac{2-R}{\dfrac{2x_0}{f}}\right)\left(-\frac{d\hat{c}_0}{d\hat{\alpha}}\right) \tag{2.12-3}$$

$$\frac{d\hat{\mu}^2}{d\hat{\alpha}} = \left(\frac{1+R}{(\nu_N)_0}\right)\hat{c}_0 + (3+2R)\hat{c}_m \hat{c}_0^{1/2} + (2+R)(\nu_N)_0 \hat{c}_m^2 \tag{2.12-4}$$

where $\hat{\alpha} \equiv \alpha/\lambda_m$, $\hat{c}_0 \equiv c_0/(c_0)_0$, $\hat{c}_m \equiv c_m(c_m)_0$, and $\hat{\mu} \equiv \mu_c^k/(c_m)_0$. Obviously, use has been made of the constraint equations (Section 1.7) to reduce the number of independent equations to those shown.

From these equations is it clear that constancy of only two dimensionless quantities, viz., α_K and $(\nu_N)_0$, guarantees chemical similarity, since R is assumed to retain a value of unity and

$$\frac{x_0}{f} = \alpha_K (\nu_N)_0 \qquad (2.12\text{-}5)$$

Thus we can expect identical values of monomer conversion ($\Phi_m = 1 - \hat{c}_m$), initiator conversion ($\Phi_i = 1 - \hat{c}_0$) and polymer properties ($\bar{x}_N = \hat{\mu}^1/\hat{\mu}^0$ and $\bar{x}_W = \hat{\mu}^2/\hat{\mu}^1$) for the same reduced age, $\hat{\alpha}$, which actually corresponds to a shorter real age α because λ_m will have been reduced by the temperature increase.

Two additional useful identities are (Table 2.7-1):

$$\lambda_m = \left(\frac{x_0}{f}\right)^{1/2} (\lambda_i \lambda_{pt})^{1/2} \qquad (2.12\text{-}6)$$

and

$$(\nu_N)_0 = \left(\frac{x_0}{f}\right)^{1/2} (\lambda_i/\lambda_{pt})^{1/2} \qquad (2.12\text{-}7)$$

The quantities λ_m and $(\nu_N)_0$ are associated with rate and target DP, respectively (Section 2.7). Their functional dependence on initiator and monomer has been separated through the use of characteristic times λ_i and λ_{pt}. The latter is a measure of the combined rate of propagation and termination, an inherent property of the monomer alone. It results from the elimination of initiator dependence from rate functions via the quotient r_p^2/r_i, and is a commonly used kinetic technique in the determination of the combined specific rate constants k_p^2/k_t from rate data.

By simple graphical analysis, we can demonstrate that it is feasible, at least in principle, to increase polymerization rate (lower λ_m) while maintaining constant molecular weight [identical $(\nu_N)_0$]. After reducing the LHS of equations 2.12-6 and 2.12-7 through division by $(x_0/f)^{1/2}$ and taking logarithms of both sides, we obtain equations for the straight lines of Fig. 2.12-1.

$$\ln\left[\frac{\lambda_m}{\left(\frac{x_0}{f}\right)^{1/2}}\right] = \frac{E_{ap}}{R_g}\frac{1}{T} - \ln A_{ap}(c_m)_0^{1/2} \qquad (2.12\text{-}8)$$

$$\ln\left[\frac{(\nu_N)_0}{\left(\frac{x_0}{f}\right)^{1/2}}\right] = \frac{-E_{ax}}{R_g}\frac{1}{T} + \ln A_{ax}(c_m)_0^{1/2} \qquad (2.12\text{-}9)$$

The lines represent the loci of chemically similar styrene polymerizations with

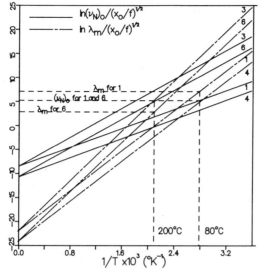

Figure 2.12-1. Temperature dependence of reduced characteristic time for polymerization and reduced chain length.

different free-radical initiators whose rate constants are listed in Table 2.12-1. Some real initiators have been included to give significance to the fictious constants used in the example. They are: benzoyl peroxide BP, azo-bisisobutyronitrile AIBN, and ditert–butyl peroxide DTBP. It is apparent from the graphs that λ_m drops more rapidly than $(\nu_N)_0$ with rising temperatures. A comparison is shown between initiators 1 and 6, the latter of which would be "slower" than BP at 80°C, but capable of giving more rapid rates at higher temperatures without a reduction in DP.

The criteria for chemical similarity at the two temperature levels are:

$$\alpha_K = (\alpha_K)_r \tag{2.12-10}$$

and

$$(\nu_N)_0 = (\nu_N)_r \tag{2.12-11}$$

which, by virtue of equations 2.12-5 and 2.12-6, give

$$\frac{x_0}{f} = \left(\frac{x_0}{f}\right)_r \tag{2.12-12}$$

and

$$\frac{\lambda_m}{(\lambda_i \lambda_{pt})^{1/2}} = \frac{(\lambda_m)_r}{(\lambda_i \lambda_{pt})_r^{1/2}} \tag{2.12-13}$$

Criterion 2.12-11 leads to

$$\ln(A_d)^{1/2} = \frac{1}{R_g} \frac{-E_{ax}}{T} + \ln C_\nu \tag{2.12-14}$$

Table 2.12-1.   Some Initiator Rate Constants

Curve Number[*]	Name	A_d (sec^{-1})	E_d (kcal)
1	BP	1×10^{13}	30
2	Fictitious	1×10^{13}	35
3	Fictitious	1×10^{13}	40
4	Fictitious	1×10^{15}	30
-	AIBN	1×10^{15}	30.5
5	Fictitious	1×10^{15}	35
-	DTBP	4.3×10^{15}	37
6	Fictitious	1×10^{15}	40

[*]Figure 2.12-1

and if we require that $\lambda_m = y(\lambda_m)_r$ such that $y < 1$, criterion, 2.12-13 yields

$$\ln(A_d)^{1/2} = \frac{1}{R_g}\frac{E_{ap}}{T} - \ln yC_\lambda \qquad (2.12\text{-}15)$$

where the constants

$$C_\nu \equiv (A_d)_r^{1/2}\exp\left(\frac{E_{ax}}{R_gT_r}\right) \qquad (2.12\text{-}16)$$

and

$$C_\lambda \equiv 1/(A_d)_r^{1/2}\exp\left(\frac{E_{ap}}{R_gT_r}\right) \qquad (2.12\text{-}17)$$

are evaluated at reference conditions. Thus all similar polymerizations must satisfy equations 2.12-14 and 2.12-15. By combining these equations we obtain a relationship between polymerization temperature T and y

$$T = \frac{E_{pt}}{R_g\ln yC_\nu C_\lambda} \qquad (2.12\text{-}18)$$

and an expression for the locus of rate constant pairs A_d, E_d

$$\ln A_d = \frac{1}{R_gT}E_d + \ln\frac{C_\nu}{yC_\lambda} \qquad (2.12\text{-}19)$$

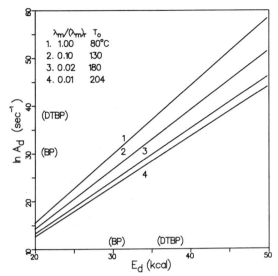

Figure 2.12-2.   Loci of initiator pairs (A_d, E_d) for chemically similar styrene polymerizations.

at any reaction temperature T. A family of graphs of equation 2.12-19 is shown in Fig. 2.12-2 using BP-initiated styrene polymerization at 80°C as the reference condition, for which $C_\nu = 12$ and $C_\lambda = 4.2 \times 10^6$. This figure contains the loci of initiator pairs A_d, E_d that lead to chemically similar styrene polymerizations at different temperature levels.

From Fig. 2.12-2, we estimate that an order-of-magnitude increase in polymerization rate seems attainable at temperatures below 200°C, where depolymerization could become significant. More specifically, by applying a similarity transformation as outlined above, it was concluded that DTBP-initiated styrene polymerization at 120°C should be chemically similar to BP-initiated styrene polymerization at 80°C and should yield identical polymer more rapidly by almost an order of magnitude (i.e., $y \cong 0.2$). The reader is reminded that we have presumed identical mechanisms at both temperatures and that this example was presented primarily to illustrate the method of chemical similarity.

REFERENCES

1.   J. A. Biesenberger, *AIChE J.*, **11**, 369 (1965)

2.   P. J. Blatz and A. V. Tobolsky, *J. Phys. Chem.*, **49**, 77 (1945).

3.   T. M. Pell and T. G. Davis, *J. Polym. Sci.*, **11**, 1671 (1973).

4.   H. D. Schumann, *Faserforsch. Textiltech.*, **22**(8), 389 (1971).

5.   R. W. Stevenson, *J. Polym. Sci. Part A*, **7**, 395 (1969).

6.   F. Kobayashi, K. Matsukura, K. Suga, and H. Olma, *Kogyo Kagaku Zasshi*, **74**, 1244 (1971).

7.   H. Kilkson, *Ind. Eng. Chem. Fund.*, **3**, 281 (1964).

8.   A. V. Tobolsky, *J. Polym. Sci.*, **26**, 247 (1957).

9.   E. W. Montroll and R. Simha, *J. Chem. Phys.*, **8**, 721 (1940).

10.   R. Simha, *J. Appl. Phys.*, **12**, 569 (1941).

11. J. J. Hermans, *J. Polym. Sci.*, **C12**, 345 (1966).

12. E. F. G. Herrington and A. Robertson, *Trans. Faraday Soc.*, **38**, 490 (1942).

13. A. V. Tobolsky and J. Ottenbach, *J. Polym. Sci.*, **16**, 311 (1955).

14. C. H. Bamford, W. G. Barb, A. D. Jenkins, and P. V. Onyon, *The Kinetics of Vinyl Polymerization by Radical Mechanisms*, Butterworths, London (1958).

15. L. Gold, *J. Chem. Phys.*, **28**, 91 (1958).

16. V. S. Nanda and R. K. Jain, *J. Chem. Phys.*, **39**, 1363 (1963).

17. A. Miyake and W. H. Stockmayer, *Polymer Prepr. Am. Chem. Soc. Div. Polym. Chem.*, **6**, 273 (1965) —ACS Meeting, Detroit (1965).

18. M. Szwarc, *Proc. R. Soc. London Ser. A*, **279**, 260 (1964).

19. W. H. Stockmayer, *J. Chem. Phys.*, **13**, 199 (1945).

20. R. Simha and H. Branson, *J. Chem. Phys.*, **12**, 253 (1944).

21. W. H. Ray, T. L. Douglas, and E. W. Godslave, *Macromolecules*, **4**, 166 (1971)

22. R. D. Capinpin, Ph.D. thesis in chemical engineering, Stevens Institute of Technology (1975).

23. P. J. Flory, *J. Am. Chem. Soc.*, **58**, 1877 (1936).

24. J. A. Biesenberger and Z. Tadmor, *J. Appl. Polym. Sci.*, **9**, 3409 (1965).

25. W. H. Ray, Polymer Reactor Engineering, Symposium at Fifty-fifth Chemical Conference and Exhibition, Chemical Institute of Canada, Quebec (1972).

26. O. Saito, K. Nagasubramanian, and W. W. Graessley, *J. Polym. Sci.*, A-2, **7**, 1937 (1969).

27. J. K. Beasly, *J. Am. Chem. Soc.*, **75**, 6123 (1953).

28. J. A. Biesenberger and R. Capinpin, *Polym. Eng. Sci.*, **14**, 737 (1974).

29. Z. Tadmor and J. A. Biesenberger, *J. Polym. Sci.*, **B3**, 753 (1965).

30. Z. Tadmor, Ph.D. thesis in chemical engineering, Stevens Institute of Technology (1966).

31. V. Meyer and G. Lowry, *J. Polym. Sci. Part A*, **3**, 2843 (1965).

32. K. F. O'Driscoll and R. Knorr, *Macromol.*, **1**, 367 (1968).

33. J. A. Biesenberger, *Polymerization Reactors and Processes*, ACS Symposium Series No. 104, American Chemical Society, Washington, D.C. (1979).

34. D. H. Sebastian, Ph.D. thesis in chemical engineering, Stevens Institute of Technology, Hoboken, NJ (1977).

35. K. F. O'Driscoll, *Polymer Reaction Engineering*, Symposium at Fifty-fifth Chemical Conference and Exhibition, Chemical Institute of Canada, Quebec (1972).

36. M. S. Matheson, E. E. Auer, E. B. Bevilacqua, and E. J. Hart, *J. Am. Chem. Soc.*, **73**, 1700 (1951).

37. G. Schulz, G. Henrici-Olive, and S. Olive, *Z. Phys. Chem.*, **27**, 1 (1961).

38. D. H. Sebastian, Master's thesis in chemical engineering, Stevens Institute of Technology, Hoboken, NJ (1975).

39. G. M. Burnett and W. W. Wright, *Proc. R. Soc. London Ser. A*, **221**, 41 (1954).

40. G. E. Ham, *Vinyl Polymerization*, Part I, Interscience, New York (1964).

41. G. Odian, *Principles of Polymerization*, McGraw-Hill, New York (1970).

42. P. F. Onyon, *J. Polym. Sci.*, **22**, 13 (1956).

43. J. A. Offenbach and A. V. Tobolsky, *J. Am. Chem. Soc.*, **83**, 1213 (1961).

44. A. V. Tobolsky, *J. Am. Chem. Soc.*, **80**, 5927 (1958).

45. Z. Tadmor and J. A. Biesenberger, *J. Polym. Sci.*, **B3**, 753 (1965).

46. J. A. Biesenberger and R. Capinpin, *J. Appl. Polym. Sci.*, **16**, 695 (1972).

47. J. Franck and E. Rabinowitch, *Trans. Faraday Soc.*, **46**, 63 (1950).

48. A. W. Hui and A. E. Hamielec, *J. Appl. Polym. Sci.*, **16**, 749 (1972).

49. G. V. Schulz and G. Harborth, *Makromol. Chem.*, **1**, 106 (1947).

50. E. Trommsdorff, H. Köhle, and P. Lagally, *Makromol. Chem.*, **1**, 169 (1948).

51. A. Schindler and J. W. Breitenbach, *Ric. Sci.*, **25A**, 34 (1955).

52. M. R. Meeks, *Polym. Eng. Sci.*, **8**, 141 (1969).

53. A. M. North and G. A. Reed, *J. Am. Chem. Soc.*, **84**, 859 (1962).

54. S. W. Benson and A. M. North, *J. Am. Chem. Soc.*, **84**, 935 (1962).

55. J. H. Duerksen and A. E. Hamielec, *AIChE J.*, **13**, 1081 (1967).

56. F. L. Martin and A. E. Hamielec, *Polymerization Reactors and Processes*, ACS Symposium, Series No. 104, American Chemical Society, Washington, D.C. (1979).

57. R. T. Ross and R. L. Laurence, *AICHE. Symp. Ser.* 160, **72**, 74 (1976).

58. K. F. O'Driscoll, J. M. Dionisio, and H. Mahabadi, *Polymerization Reactors and Processes*, ACS Symposium Series No. 104, American Chemical Society, Washington, D.C. (1979).

59. J. Cardenas and K. F. O'Driscoll, *J. Polym. Sci.*, A-1, **14**, 883 (1976).

60. J. Cardenas and K. F. O'Driscoll, *J. Polym. Sci.*, A-1, **15**, 1883 (1977).

61. J. Cardenas and K. F. O'Driscoll, *J. Polym. Sci.*, A-1, **15**, 2097 (1977).

62. F. Bueche, *Physical Properties of Polymers*, Interscience, New York (1962).

63. S. Glasstone, K. J. Laidler, and H. Eyring, *Theory of Rate Processes*, McGraw-Hill, New York (1941).

64. S. T. Balke and A. E. Hamielec, *J. Appl. Polym. Sci.*, **17**, 905 (1973).

65. M. Sacks, S-I. Lee, and J. A. Biesenberger, *Chem. Eng. Sci.*, **28**, 241 (1973).

66. S. Middleman, *The Flow of High Polymers*, Interscience, New York (1968).

67. C. Walling, *J. Am. Chem. Soc.*, **71**, 1930 (1949).

68. G. Bonta, M. G. Gallo, and S. Russo, *J. Chem. Soc. Faraday Trans. I*, **69**, 328 (1973).

69. S. Russo and S. Munari, *J. Macromol. Sci.—Chem.*, **A2(7)**, 1321 (1968).

70. D. H. Sebastian and J. A. Biesenberger, *J. Macromol. Sci.—Chem.*, **A15**, 533 (1981).

71. M. Ravey, J. A. Waterman, L. M. Shorr, and M. Kramer, *J. Polym. Sci.*, A-1, **12**, 2821 (1974).

72. J. D. Cotman, M. F. Gonzales, and G. C. Claver, *J. Polym. Sci.*, A-1, **5**, 1137 (1967).

73. M. Magat, *J. Polym. Sci.*, **16**, 491 (1955).

74. W. I. Bengough and R. G. W. Norrish, *Proc. R. Soc. London Ser. A*, **200**, 301 (1950).

75. H. S. Mickley, A. S. Michaels, and A. L. Moore, *J. Polym. Sci.*, **60**, 121 (1962).

76. L. M. Barclay, *Angew, Makromol. Chem.*, **52**, 1 (1976).

77. A. Crosato-Arnaldi, P. Grasparini, and G. Talamini, *Makromol. Chem.*, **117**, 140 (1968).

78. A. H. Abdel-Alim and A. E. Hamielec, *J. Appl. Polym. Sci.*, **16**, 783 (1972).

79. W. D. Harkins, *J. Chem. Phys.*, **13**, 381 (1945); **14**, 47 (1946); *J. Am. Chem. Soc.*, **69**, 1428 (1947).

80. W. V. Smith and R. H. Ewart, *J. Chem. Phys.*, **16**, 592 (1948).

81. W. H. Stockmayer, *J. Polym. Sci.*, **24**, 314 (1957).

82. J. L. Gardon, *Rubber Chem. Technol.*, **43**, 74 (1970).

83. J. Ugelstad and F. K. Hansen, *Rubber Chem. Technol.*, **49**, 536 (1976).

84. G. W. Poehlein and D. J. Dougherty, *Rubber Chem. Technol.*, **50**, 601 (977).

85. J. L. Gardon, *J. Polym. Sci.*, A-1, **6**, 623, 643, 665, 687, 2853, 2859 (1968).

86. W. W. Graessley, R. D. Hartung, and W. C. Uy, *J. Polym. Sci.*, A-2, **7**, 1919 (1969).

87. K. Nagasubramanian, O. Saito, and W. W. Graessley, *J. Polym. Sci.*, A-2, **7**, 1955 (1969).

88. W. W. Graessley, *Polymer Reactor Engineering*, Symposium, at Fifty-fifth Chemical Conference and Exhibition, Chemical Institute of Canada, Quebec (1972).

89. Z. Tadmor and J. A. Biesenberger, *I.E.C. Fund.*, **5**, 336 (1966).

CHAPTER THREE

Mixing Effects

Mixing requirements for chemical reactors may be classified on the basis of whether or not the reactants are premixed. We define mixing problems of the first class as those that involve reactants entering in separate feed streams, which may or may not be mutually soluble. Thus mixing is a prerequisite for reaction, and the two processes occur simultaneously. In the second class, reactants in the feed stream are either premixed, or mixing is sufficiently rapid to be regarded as virtually complete prior to reaction. Nonuniformities that require mixing as a remedy can develop subsequently as a result of differences in ages among fluid elements in the reactor or from local transport processes that occur at different rates, such as interphase diffusion and heat transfer.

It is well known that mixing processes can significantly alter the outcome of chemical reactions. As mentioned earlier, inadequate mixing can result in spatial concentration and temperature gradients, which can cause nonuniform products (e.g., DP, CC, DB), runaway reactions, and so on. On the other hand, complete mixing can reduce reactor efficiency under certain circumstances, as we shall see.

3.1. GOODNESS OF MIXING

The characterization and consequences of mixing in chemical reactors are complex and varied. We begin by examining mixtures and mixing processes in the absence of chemical reaction. Following Danckwerts (1, 2), goodness of mixing will be characterized using statistical quantities (viz., scale and intensity) analogous to those in G. I. Taylor's treatment of turbulence. It must be remembered, however, that these quantities have not been functionally related to the physical mechanisms responsible for mixing in a general way. Thus, while they provide a means for assessing the quality of mixing, both macroscopic and microscopic, they cannot be used directly in reactor calculations. We shall use them instead as a basis for classifying reactors.

256

Reactor calculations will be performed on model reactors using residence time distributions (3) or transport balances, whichever are appropriate.

3.1.1. Mixture Analysis

According to Danckwerts (1), nonuniformity in mixtures can be characterized by two quantities, the scale of segregation and the intensity of segregation. Scale is a measure of the size of unmixed regions and intensity is a measure of the magnitude of their nonuniformity. The concepts of scale and intensity, which are now widely used, are illustrated in Fig. 3.1-1. Since we are more concerned with their qualitative significance than with their quantitative evaluation, we shall define them in a somewhat unorthodox way.

Consider a region to be examined consisting of a population of N_p movable particles, termed ultimate particles, whose relative positions change as mixing

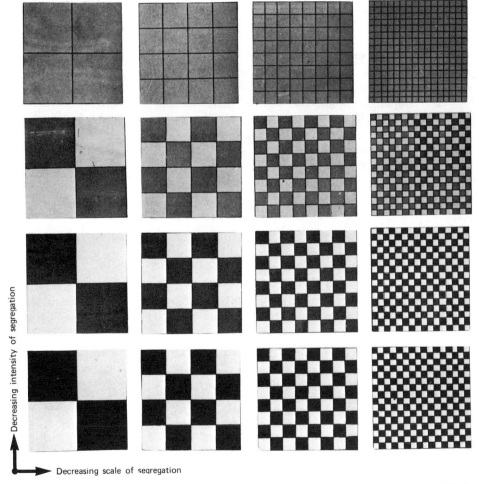

Figure 3.1-1. Scale and intensity of segregation (Reprinted with permission from Z. Tadmor and C. G. Grogos, *Principles of Polymerization*, Wiley, New York,).

progresses. Ultimate particles must be irreducible with respect to the particular mixing process in question. For instance, molecules are ultimate particles in microscopic mixing (diffusion) whereas clumps or streaks can be regarded as ultimate particles in certain macroscopic (convective) mixing processes.

Let ρ be the value of any property associated with the particles, such as composition (concentration or molecular structure), age, size, shape, color, etc. To mix our population is to increase the spatial uniformity of ρ. Yet, complete mixing is generally not associated with uniform dispersion, as illustrated in Fig. 3.1-2(c), but rather with the random dispersion, as illustrated in Fig. 3.1-2(b). Obviously the reason is that while the former state is a possible consequence of dispersion, it is an improbable one for mixing processes that are random in nature. Examples of random mixing processes are molecular and turbulent (eddy) diffusion. An exception is motionless mixing of liquids, in which uniformity is achieved through ordered subdivision and rearrangement of fluid elements. Another is laminar mixing by pure shear. Incomplete mixing results in segregation, where regions exist that consist of groups of ultimate particles possessing exclusively, or primarily, like values of ρ.

To test our population for uniformity, we must first select a region of scrutiny, which has been defined in an imprecise way as the size of a region whose state of mixedness is being examined. The region of scrutiny dictates the size of the samples to be taken from our population. Suppose we select samples containing N_p' particles each. The number of samples N_s taken depends on the sampling technique used. In inclusive sampling we would take all distinguishable samples of size N_p'. Since this is impractical, we could sample randomly by selecting the smallest number consistent with an acceptable level of sampling error.

Our population is considered to be uniform, or mixed, relative to our region of scrutiny when a suitable characteristic of property ρ for the samples is representative of the same characteristic for the entire population. One characteristic commonly used is the mean value of ρ for the sample $\bar{\rho}_s$, as compared to the mean value for the population $\bar{\rho}$. To measure deviations, it is customary to use variance of property ρ among samples. Thus,

$$\operatorname{var}\rho_s = \overline{\overline{\left(\bar{\rho}_s - \bar{\rho} \right)^2}} \qquad (3.1\text{-}1)$$

where the single bar signifies the average over ultimate particles and the double bar signifies the average over samples. Obviously, when our samples are small compared

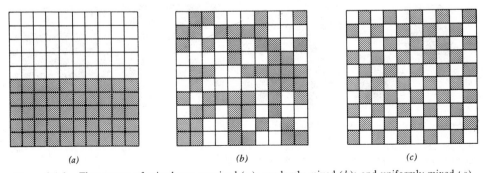

(a)   (b)   (c)

Figure 3.1-2. Three states of mixedness: unmixed (*a*); randomly mixed (*b*); and uniformly mixed (*c*).

to regions of segregation, they will exhibit large fluctuations, that is, $\text{var}\,\rho_s$ will be large. When the samples are sufficiently large compared to regions of segregation, $\text{var}\,\rho_s$ should be small. Therefore, the value of $\text{var}\,\rho_s$ will depend on the sample size. A computation of $\text{var}\,\rho_s$ for a random mixture is shown in Example 3.1-1.

When varying sample size, caution must be exercised, however. In the limit as $N'_p \to N_p$ (total number of particles), the number of distinct samples diminishes ($N_s \to 1$). Therefore, the possibility of fluctuations among samples diminishes concomitantly, and a false measure of uniformity results. By such a test, even the system in Fig. 3.1-2(a) would be judged uniform. Conversely, when scrutinized closely enough (sufficiently small sample size), all mixtures will appear nonuniform, even when well mixed. By such a test, the mixtures illustrated in Figs. 3.1-2(b) and 3.1-2(c) would be judged nonuniform in the limit as $N'_p \to 1$.

Returning now to the task at hand, that is, defining intensity and scale of segregation, we define the former as

$$i_{seg} \equiv \frac{\text{var}\,\rho_s}{\text{var}\,\rho} \qquad (3.1\text{-}2)$$

where $\text{var}\,\rho_s$ has been normalized through division by the variance of property ρ over the entire population.

$$\text{var}\,\rho \equiv \overline{(\rho - \bar{\rho})^2} \qquad (3.1\text{-}3)$$

Thus, $0 \leqslant i_{seg} \leqslant 1$. With the aid of the identity $\bar{\bar{\rho}}_s = \bar{\rho}$, we can express equation 3.1-2 in equivalent form

$$i_{seg} = \frac{\overline{\bar{\rho}_s^2} - \bar{\rho}^2}{\overline{\rho^2} - \bar{\rho}^2} \qquad (3.1\text{-}4)$$

where the single bar over ρ_s has been dropped for convenience ($\rho_s \equiv \bar{\rho}_s$). When samples contain particles of like kind predominantly, we expect their composite properties to approach the properties of the individual particles. Consequently, $\rho_s \to \rho$, or equivalently $\overline{\bar{\rho}_s^2} \to \overline{\rho^2}$, and thus $i_{seg} \to 1$. On the other hand, when the samples contain a mixture of particles of different kinds, we expect their composite properties to approach the average property of the entire population. Thus $\rho_s \to \bar{\rho}$ (or equivalently $\overline{\bar{\rho}_s^2} \to \bar{\rho}^2$) and $i_{seg} \to 0$. The dependence of i_{seg} on sample size can be demonstrated for random populations using the results of Example 3.1-1, from which we obtain $\text{var}\,\rho = \bar{\rho}(1 - \bar{\rho})$ by setting $N'_p = 1$ (7), and therefore conclude that

$$i_{seg} = \frac{1}{N'_p} \qquad (3.1\text{-}5)$$

A well-known example of i_{seg} defined by Danckwerts for reaction vessels is intensity J (2). The ultimate particles are molecules and the property scrutinized is their age in a continuous-flow vessel. By setting $\rho = \alpha$ in equation 3.1-2, we obtain

$$i_{seg} = \frac{\text{var}\,\alpha_s}{\text{var}\,\alpha} = J \qquad (3.1\text{-}6)$$

Another well-known example is Danckwerts' intensity I (1) defined for binary mixtures of uniform-size components A and B. His samples are "points" and the property scrutinized is composition, that is, volume fraction ($\rho = \phi$) of one of the components. In this way the properties of the ultimate particles (A and B) are 1 and 0 and, again, it can be shown (7) that $\mathrm{var}\,\phi = \bar{\phi}(1 - \bar{\phi})$. Thus we obtain from equation 3.1-2

$$i_{\mathrm{seg}} = \frac{\mathrm{var}\,\phi_s}{\mathrm{var}\,\phi} = \frac{\overline{\left(\phi - \bar{\phi}\,\right)^2}}{\bar{\phi}\left(1 - \bar{\phi}\,\right)} = I \tag{3.1-7}$$

To define our scale of segregation, we shall take advantage of the dependence of i_{seg} on sample size and treat N_p' as a parameter, whose value may be increased incrementally at will. As the size of the sample passes through a value that is representative of regions of segregation, we would expect the corresponding values of i_{seg} to drop abruptly. Such transition behavior facilitates the definition of a scale. A single value of N_p' (actually a corresponding length is more useful) is then chosen, such as an inflection point, as being characteristic of the size of segregation. The corresponding length is defined as the scale of segregation, l_{seg}.

Numerous computer calculations were performed on various synthesized patterns (4) to demonstrate the method described for finding l_{seg}. These patterns were synthesized from segregated square clusters whose dimensions are listed in Table 3.1-1. They are illustrated in Fig. 3.1-3 (horizontal slices) and the associated graphs of i_{seg} versus $(N_p')^{1/2}$ in Fig. 3.1-4. Indeed, inflection points appear to identify scales of segregation. The resulting values of l_{seg} are summarized in Table 3.1-1. Random sampling was also tried and, as expected, curves with "noise" resulted. All cases shown here are the product of inclusive sampling to ensure smooth curves for illustrative purposes.

We would expect that if the sizes or intensities among regions of segregation, or both, were broadly distributed the effect would be a "smearing" of graphs like those in Fig. 3.1-4. In such cases, which are closer to reality than the synthetic patterns used in the analysis, individual values of l_{seg} would be less readily discernible, and an appropriately defined central value of l_{seg} would perhaps be of more use. In fact, the scale of segregation introduced by Danckwerts' for binary mixtures (1), when applied to systems consisting of a major and a minor component arranged in a pattern of unequal-size stripes, for example, turns out to be proportional to the harmonic means of the width of the stripes (7) and is thus similar to a mean striation thickness (Section 3.1.2).

Danckwerts' scale is defined

$$l_{\mathrm{seg}} \equiv \int_0^{\zeta} R(r)\, dr \tag{3.1-8}$$

in terms of the correlation coefficient

$$R(r) = \frac{\overline{\left(\phi_1 - \bar{\phi}\,\right)\left(\phi_2 - \bar{\phi}\,\right)}}{\overline{\left(\phi - \bar{\phi}\,\right)^2}} = \mathrm{cov}\,\phi_s/\mathrm{var}\,\phi \tag{3.1-9}$$

where cov ϕ_s is the covariance among sample "points" 1 and 2 separated by a distance r and ϕ_1 and ϕ_2 are concentrations (volume fractions). The integration limit ζ is the value of r for which $R(r) \to 0$, and the double bar signifies an average over many points (samples) for each value of r. Thus r corresponds to our sample size N'_p, being a parameter that is varied at will.

Correlograms (graphs of equation 3.1-9) were plotted (4) for some of our synthesized patterns (Fig. 3.1-3) and corresponding values of Danckwerts' l_{seg} were computed. The correlograms were very similar to our intensity graphs (Fig. 3.1-4).

Incidentally, we note that pattern b in Fig. 3.1-3 may be associated with a microsegregated IM (MSIM) when property ρ is concentration (ϕ) and pattern c with a plug-flow (PF) vessel when property ρ is age (α).

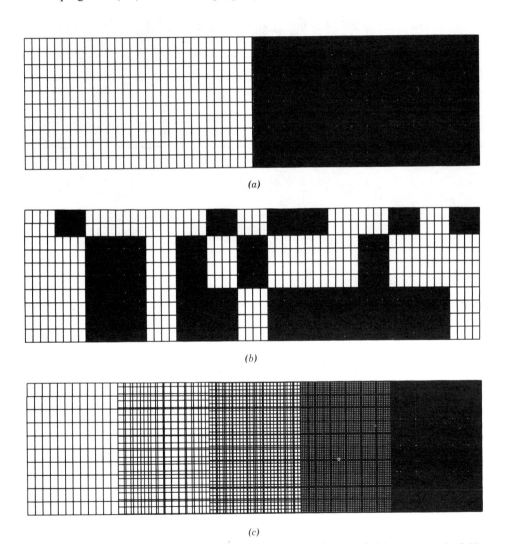

(a)

(b)

(c)

Figure 3.1-3. Synthesized mixing patterns from macrosegregated (a) and (c) to macromixed (b), including partly micromixed (d and e).

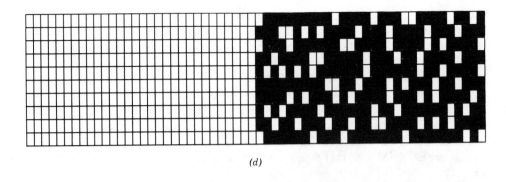

(d)

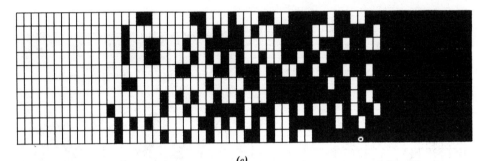

(e)

Figure 3.1-3. (*Continued*)

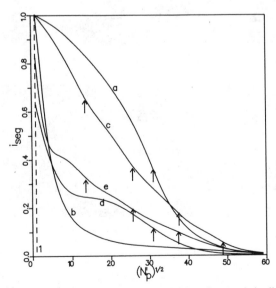

Figure 3.1-4. Intensity of segregation versus linear sample size (characteristics listed in Table 3.1-1).

Table 3.1-1. Characteristics of Patterns in Figures 3.1-3 and 3.1-4

Pattern	a	b	c	d	e
ℓ_{seg} in terms of $(N'_p)^{1/2}$	30	4	12,24 36,48	1,30	1,12,24, 36,48

Example 3.1-1. Sample Variance from a Random Population. The probability distribution associated with sampling from a randomly mixed population is the binomial distribution. Thus the fraction of randomly drawn samples of size N'_p that contain N'_1 ultimate particles of type 1 is given by (7)

$$f_{N'_p, N'_1} = \binom{N'_p}{N'_1} P^{N'_1}(1 - P)^{N'_p - N'_1}$$

where P is the fraction of type 1 particles in the population, $P = N_1/N_P$. Using this result, show that $\operatorname{var} \rho_s = P(1 - P)/N'_p$ when $\rho_s \equiv N'_1/N'_p$.

Solution. From the mean value of the binomial distribution (Appendix B) for fixed N'_p

$$\mathbf{N}'_1 = \sum_{N'_1=1}^{N'_p} N'_1 f_{N'_p, N'_1} = N'_p P$$

we conclude that

$$\bar{\rho}_s \equiv \frac{\mathbf{N}'_1}{N'_p} = P$$

and, similarly, from the variance (Appendix B)

$$\operatorname{var} N'_1 = \sum_{N'_1=1}^{N'_p} (N'_1 - \mathbf{N}'_1)^2 f_{N'_p, N1} = N'_p P(1 - P)$$

we obtain our desired relation

$$\operatorname{var} \rho_s = \frac{\operatorname{var} N'_1}{(N'_P)^2} = \frac{P(1 - P)}{N'_p}$$

with the aid of the first result. This relation has been used (7) to illustrate the narrowing effect of many ultimate particles on the sample distribution, as in true solutions consisting of molecules in contrast with blends consisting of clumps.

3.1.2. Mixing Processes

What constitutes a satisfactory state of mixedness obviously depends on the scale of scrutiny chosen relative to the size of the ultimate particles involved. In the processing of polymeric materials (5–7), for instance, mixing is frequently judged adequate when the product appears uniform to the eye, or when its mechanical properties are satisfactory. Regions of scrutiny could thus depend on the resolving power of the eye, and the ultimate particles could be quite large compared to molecules. Tolerable scales of segregation in such cases may clearly be much larger than molecular neighborhoods. The mixing processes responsible involve size reduction as well as spatial rearrangement. They are macroscopic.

By contrast, the manufacture of polymers involves molecular events, such as chemical reaction and diffusion. Here the term complete mixing implies uniformity down to the scale of molecular neighborhoods. Such mixing is called microscopic. The formulation of a point rate function r presumes complete mixedness locally, that is, within a small neighborhood δV of each point in space. Since good microscopic mixing is impossible to achieve in a reasonable time in the absence of macroscopic mixing, we must examine mixing on both levels. Examples of macroscopic mixing mechanisms for the reactor types discussed in Section 1.2 are turbulent (eddy) transport in agitated vessels (low viscosity) and laminar (shear) mixing in streamline vessels. Both are actually examples of convection (bulk flow). In microscopic mixing, on the other hand, diffusion is the sole mechanism.

Numerous criteria have been proposed to characterize "goodness" of macromixing and micromixing. We shall review some of these briefly. From our earlier discussions, we expect the main consequence of macromixing to be a reduction in scale of segregation l_{seg}, and that of micromixing to be a reduction in intensity i_{seg}. Macromixing enhances micromixing by reducing diffusion lengths, or equivalently, by increasing the surface area between regions being mixed. Following this line of reasoning, we shall utilize the method of characteristic times and assign a CT to each mixing process, labeled λ_{mac} and λ_{mic}. We define λ_{mac} as the time allotted to the macromixing process during which it reduces the scale of segregation to a value l_{seg}, which then becomes the mean diffusion length for the micromixing process. Under these conditions it is reasonable to require for the attainment of maximum micromixedness that

$$\lambda_{mic} < \lambda_{mac} \tag{3.1-10}$$

or, in terms of the dimensionless segregation number (8), that

$$S \equiv \frac{\lambda_{mic}}{\lambda_{mac}} < 1 \tag{3.1-11}$$

An example of λ_{mac} for continuous-flow vessels is the space time λ_v. For λ_{mic}, we use the CT for diffusion (λ_D), defined here as

$$\lambda_{mic} \equiv \frac{l_{seg}^2}{D} \tag{3.1-12}$$

Thus l_{seg} is some appropriate characteristic (mean) length associated with microsegretation.

The crucial quantity in these criteria is, of course, l_{seg}. It bears the same label as the statistical scale of segregation because there is a conceptual similarity. In the absence of macromixing, the value of l_{seg} can be as large as a characteristic reactor dimension (transverse or longitudinal). In the presence of macromixing, its value depends on the mechanism and degree of mixing, as well as the time available, λ_{mac}. In this context, it resembles the so-called mixing length in fluid-mechanical theories, that is, the smallest scale to which macroscopic mixing is effective.

In turbulent mixing, numerous candidates have been advanced for mixing length. We shall use Kolmogoroff's microscale for isotropic turbulence as a first approximation (9)

$$l_{seg} \equiv \left(\frac{\nu^3}{\dot{\varepsilon}} \right)^{1/4} \qquad (3.1\text{-}13)$$

where ν is kinematic viscosity, η/ρ, and $\dot{\varepsilon}$ is the power supplied to the turbulent field on a macroscopic scale, l_0, which is subsequently dissipated on the microscale. The power supplied is of the order $\dot{\varepsilon} = v^3/l_0$, where v is a characteristic large-scale (macroscopic) velocity. After $\dot{\varepsilon}$ is eliminated from these expressions, we obtain a relationship between l_{seg} and l_0

$$\frac{l_{seg}}{l_0} = \text{Re}^{-3/4} \qquad (3.1\text{-}14)$$

where the Reynolds number is defined here as $\text{Re} \equiv vl_0/\nu$. This result predicts a decline in l_{seg} with increasing mixing intensity, as measured by Re, which is consistent with our expectations. The macroscale l_0 is associated with the origin of the turbulent field. Two examples of l_0 are a pipe diameter and a characteristic mixing blade dimension.

Owing to the high viscosities frequently attained by polymerizing media and the creeping flows demanded by their long reaction times, we would expect turbulent mixing to be unimportant in such reactions. However, in premixing monomers and catalysts and during the early stages of polymerization, turbulence might indeed prevail as the dominant mixing mechanism. One example of a mixing problem of the first class is the premixing of polyfunctional alcohols and diisocyanates in reactive injection molding (RIM) of polyurethanes (Example 3.1-2). Since the comonomers enter in separate feed streams and the ensuing reaction is a random-type polymerization, good micromixing is essential for the attainment of both high conversion and high DP.

In laminar mixing, a logical candidate for l_{seg} is the striation thickness l_f (12) which is illustrated in Fig. 3.1-5 for cylindrical couette flow. It is noteworthy that this quantity, too, appears in the form of a ratio l_f/l_0, since its magnitude at any time depends on the magnitude of an initial (reference) striation thickness l_0. Thus, in a sense, l_0 is the analogue of the macroscale in turbulence. However, there is a significant distinction as well. In laminar mixing the magnitude of l_{seg} additionally depends on the orientation of l_0, as illustrated in Fig. 3.1-5. It can be shown that the optimum orientation is orthogonal to the direction of flow (7). From this observation, one can expect the quality of mixing in physical processes and the role of mixing in chemical reactions to be affected by the location and geometry of the feed stream. Sometimes it is more convenient to calculate the ratio of streak lengths than

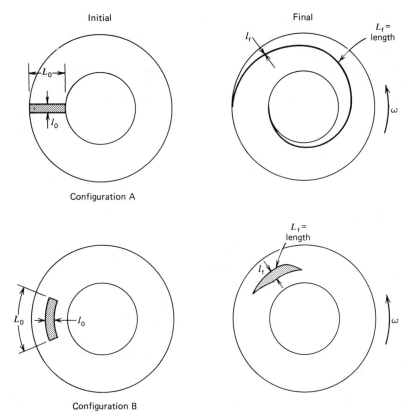

Figure 3.1-5. Effect of couette flow (outer cylinder rotating, inner cylinder stationary) on two streaks of different initial configuration.

streak widths (striation thickness). Middleman (6) uses streak length ratio, L_0/L_f (Fig. 3.1-5), which is identical to striation ratio for very thin streaks.

For simple shear flow, the striation ratio l_f/l_0 is inversely proportional to total shear strain γ (6, 7), which is also a ratio of final to initial dimensions. For complex flow patterns with nonuniform strain histories, the flow-average strain of the vessel effluent $[\gamma]$ is frequently used to measure goodness of laminar mixing. It is commonly believed that 10^4 strain units signify good mixing in polymer processing. As a laminar counterpart to equation 3.1-14, therefore, it seems reasonable to express our mean characteristic length ratio in terms of $[\gamma]^{-1}$

$$\frac{l_{\text{seg}}}{l_0} = f\left([\gamma]^{-1}\right) \qquad (3.1\text{-}15)$$

The quantity $[\gamma]$ is relatively easy to calculate for simple flow geometries.

Until now, we have focused on scale as a criterion for macromixing. At least three additional considerations are relevant when judging "goodness" of macromixing of fluid elements in continuous-flow reactors. The first concerns the uniformity of scale among elements of similar age. The second concerns the age at which their scale is reduced (i.e., how "young" the elements are), and the third, how thoroughly the members of various age groups intermingle with one another.

To characterize nonuniformity of scale among fluid elements in laminar mixing, the strain distribution (SD), $f(\gamma)$, has been proposed, where $f(\gamma)\,d\gamma$ is the fraction of fluid experiencing a strain between γ and $\gamma + d\gamma$. This addresses the first consideration above, but not the last two. It can be shown that $[\gamma]$ is equal to the mean value of the SD. Example 3.1-3 illustrates the use of mean strain and streak ratio applied to the tubular vessels.

Regardless of which ratio is used, the microscopic segregation criterion developed in this section requires that an appropriate reference length l_0 be chosen. This choice clearly depends on mixing requirements as well as prevailing flow patterns. Mixing problems of the first class, for instance, generally require that initially heterogeneous regions be homogenized (Example 3.1-2). In such cases, the choice of l_0 is clear. We would take it to be some appropriate dimension that is characteristic of the initially segregated regions, such as the inlet pipe diameter. If complete micromixing is required for chemical reaction, then mixing and reaction act in parallel, and the second consideration raised earlier is important. Therefore, the macromixing processes must reduce the macroscale to a sufficiently small value, l_{seg}, to permit diffusion to complete the task. As a first approximation, we could, as in Example 3.1-2, ignore the more subtle aspects of the flow patterns such as longitudinal mixing (backflow), which relates to the third consideration and could also affect the outcome of the reaction. These aspects will be taken up later.

In mixing problems of the second class, where the feed is initially homogeneous and microscopic mixing (backmixing) may or may not be beneficial, the choice of an appropriate value for l_0 is difficult at best. It depends on the nature of the reaction in addition to the prevailing flow patterns. In streamline vessels, for instance, transverse micromixing can be beneficial if it reduces the adverse effects of an age distribution, whereas longitudinal mixing can be detrimental. Example 3.1-4 demonstrates the application of our method to a mixing problem of the second class in a traditional, pressure-flow tubular polymerizer. Mixing of this kind is generally classified as longitudinal (backmixing), and as such may be described by the dispersion model (Section 1.4) and characterized by the longitudinal Peclet number, Pe_L. The latter quantity is another widely used measure of mixing, which is limited in applicability to relatively small degrees of dispersion (Example 3.1-4).

A streamline vessel especially suitable for polymerizations with high reaction viscosities is the continuous drag-flow reactor (CDFR) with moving walls (Section 1.2). The criterion for the existence of drag flow (longitudinal pressure increase) as opposed to pressure flow (longitudinal pressure decrease) and the criterion for the onset of reverse flow (longitudinal mixing) (Fig. 1.2-2) have been formulated in terms of the ratio of corresponding volumetric flow rates (7)

$$G \equiv -\frac{\dot{V}_p}{\dot{V}_d} \tag{3.1-16}$$

Thus G is negative for pressure flow. When its value passes through zero and becomes positive, drag flow prevails. When it attains a certain positive "critical" value G_{cr}, which depends on specific flow parameters, reverse flow begins. For steady, Newtonian flow between parallel flat plates in particular

$$G = \frac{H^2}{12\eta v_m} \frac{\Delta P}{L} \qquad (3.1\text{-}17)$$

where H represents the distance between plates, L is their length, and v_m their mean velocity. In this case G can be expressed in terms of the ratio of two CT's and our criterion for longitudinal mixing thus becomes

$$P_0 = \frac{\lambda_M}{\lambda_F} > (P_0)_{cr} \qquad (3.1\text{-}18)$$

where λ_M is the CT for momentum transport by shear stress $[(H/2)^2/\nu]$, λ_F is the CT for momentum transport by pressure $(\rho v_m/(\Delta P/L)$, and $P_0 = |G|/3$ is the dimensionless Poiseuille number (Appendix G).

One of the earliest measures of shear mixing propsed (13) was the ratio of final to initial interfacial area between two components s/s_0, to which the striation ratio is proportional for simple shear (12). It is also a function of total shear strain $f(\gamma)$. Erwin (15) demonstrated the importance of intermittant reorientation to laminar mixing in terms of s/s_0. He analyzed a collection of randomly oriented surfaces of initial area s_0, which were subjected to a large uniform shear strain followed by randomization in stages. For N equal shear-mixing stages in which each stage experiences the same fraction, γ/N, of the total shear magnitude γ, the familiar difference equation for the jth stage $s_j/s_{j-1} = f(\gamma/N)$ leads to the following solution

$$\frac{s_N}{s_0} = \left[f\left(\frac{\gamma}{N}\right) \right]^N \qquad (3.1\text{-}19)$$

Where $f(\gamma/N)$ represents the functional relationship between the shear action and the area ratio produced in each stage. For a system of randomly oriented surfaces experiencing large strains in simple shear flow, $s_N/s_0 = \gamma/2$ (7). We therefore conclude from equation 3.1-19 that for such a staged process $s_N/s_0 = (\gamma/2N)^N$ (12).

The role of staging in motionless mixers, which are actually cascades of stream-splitting and remixing operations in series, has been discussed elsewhere (6, 7). The analogy between this mixing mechanism and that in the packed bed reactor was pointed out by Middleman (6). If the stream in a motionless mixer is halved in each stage, one would anticipate that after N equal stages $l_{seg}/l_0 \sim 2^{-N}$. Thus, to reduce 10 cm to molecular dimensions (1000 Å) would require approximately 20 stages. The reader is reminded, however, that since such mixing in essence occurs transverse to flow, it acts in parallel with reaction, and therefore the appropriateness of motionless mixers as reaction vessels depends on the nature of the reaction and the class of mixing problem involved.

One manifestion of mixing, especially macroscopic, which has been mentioned thus far only in passing, is the residence time distribution (RTD). It is frequently used to characterize macromixing in continuous-flow vessels. This characteristic is important in polymerization engineering as well as in polymer processing. Knowledge of the RTD not only yields information on mixing processes, but it also provides the chemical engineer with a computational tool for performing reactor calculations.

Example 3.1-2.   Turbulent Mixing in a Jet.   Macosko et al. (10) and Suh et al. (11) have investigated impingement mixing in the RIM process and have concluded that the effectiveness of mixing indeed depends on Re, as suggested by equation 3.1-14, and that turbulence prevails when Re > 150, based on nozzle diameter, d_N (11).

For the values (10) $d_N = 0.1$ cm and Re = 200, estimate whether or not complete micromixing can be expected when the space time in the mixing chamber is of the order of 10^{-3} s. Use a value of 10^{-5} cm²/s for D.

Solution.   By setting $l_0 = d_N = 0.1$ cm, we obtain

$$\frac{l_{seg}}{l_0} = Re^{-3/4} = 1.88 \times 10^{-2}$$

so that $l_{seg} = 1.88 \times 10^{-3}$ cm. $= 18.8$ μm, as compared with the value of 13 μm computed by Macosko et al. From this we conclude that

$$\lambda_{mic} = \frac{l_{seg}^2}{D} = 0.353 \text{ s}$$

which is too large compared with the time scale for macromixing, $\lambda_{mac} = 10^{-3}$ s, to achieve good micromixing. Alternatively, we could have combined equations 3.1-11, 3.1-12, and 3.1-14 to obtain

$$S = l_0^2 / D\lambda_{mac} Re^{3/2} = 353$$

which clearly violates inequality 3.1-11.

Although this conclusion disagrees with that of Macosko et al., which was based on a far more detailed and complex analysis than ours and an even lower value for D (10^{-7} cm²/s), it appears to be in agreement with the conclusion of Suh et al.

Example 3.1-3.   Laminar Mixing in a Tube

For the laminar-flow tubular vessel shown, where $L \gg Ra$, compare the mean strain [γ] with the streak ratio L_f/L_0 at L. Take the initial (reference) streak to be $L_0 = 2\,Ra$ and evaluate L_f at time $t = L/v_0$, where v_0 is the velocity at the tube axis.

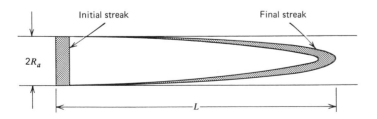

Initial streak   Final streak

$2R_a$

L

Solution. The relevant component of the strain-rate tensor (14) is $\dot{\gamma}_{rz} = dv_z/dr$. Assuming Newtonian flow, the steady-state velocity profile is $v(r) = v_0[1 - (r/\text{Ra})^2]$. Therefore, the total strain, $\gamma(r) = |\dot{\gamma}_{rz}|t$, evaluated at the tube exit, $t = L/v_z(r)$, is

$$\gamma(r) = 2Lr/\text{Ra}^2\left[1 - (r/\text{Ra})^2\right]$$

and the flow-average (Appendix F) strain is

$$[\gamma] = \int_0^{\text{Ra}} \gamma(r) \underbrace{\frac{v_z(r)(2\pi r\,dr)}{\left(\dfrac{v_0}{2}\right)(\pi\text{Ra}^2)}}_{f(\gamma)\,d\gamma} = \frac{8L}{\text{Ra}}$$

where $f(\gamma)$ is the so-called strain distribution (7).

Following Middleman (6), we compute L_f by integrating the differential form $(dL_f)^2 = dr^2 + dz^2$ where the equation of the streak after time $t = L/v_0$ is $z = L[1 - (r/\text{Ra}^2)]$. After substituting this equation into the differential form and integrating the result $(dL_f)^2 = [1 + (2Lr/\text{Ra}^2)^2](dr)^2$, we obtain

$$L_f = \int_{-\text{Ra}}^{\text{Ra}} \left[1 + \left(\frac{2Lr}{\text{Ra}^2}\right)^2\right]^{1/2} dr$$

$$= \text{Ra}\left[1 + \left(\frac{2L}{\text{Ra}}\right)^2\right]^{1/2} + \frac{\text{Ra}^2}{L \ln\left\{\dfrac{\left[1 + (2L/\text{Ra})^2\right]^{1/2} + 2L/\text{Ra}}{\left[1 + (2L/\text{Ra})^2\right]^{1/2} - 2L/\text{Ra}}\right\}}$$

When $L/\text{Ra} \gg 1$, $L_f \to 2L$ which we would expect from inspection of the figure. Consequently, $L_f/L_0 \cong L/\text{Ra}$, which compares favorably with $[\gamma]$.

Example 3.1-4. Taylor "Axial" Diffusion. Consider a laminar-flow tubular vessel with uniform cross section in which differences in concentration among streamlines can occur due to RTD coupled with reaction. Here transverse mixing, which is by diffusion alone, is desirable. G. T. Taylor's criterion (16) for effective transverse mixing requires that

$$\frac{\text{Ra}^2\{v\}}{LD} < \frac{2}{(3.8)^2}$$

where L is tube length, $\{v\}$ is mean axial velocity, and D is molecular diffusivity. The dimensionless group on the LHS has previously been named Taylor number, Ta (17).

Starting with equations 3.1-11 and 3.1-12 deduce an equivalent criterion to the inequality above and show that S is equivalent to Ta. Show also that Taylor-type longitudinal dispersion is always small (large Pe_L).

Solution. Since radial mixing is entirely by diffusion it is reasonable to define $l_{\text{seg}} = \text{Ra}$. As suggested in the text, we set $\lambda_{\text{mac}} = \lambda_v = L/\{v\} = \lambda_c$. From inequality

3.1-11 it follows directly that

$$S = \frac{Ra^2\{v\}}{DL} < 1$$

which is equivalent for practical purposes to Taylor's inequality.

The requirement for small longitudinal dispersion is that Pe_L be large (Section 1.4)

$$Pe_L \equiv \frac{L\{v\}}{D_L}$$

where, for Taylor dispersion (equation 1.4-2),

$$D_L \equiv \frac{4Ra^2\{v\}^2}{192\,D}$$

After substituting the last definition into the penultimate one and applying the earlier definition and inequality involving S, we obtain

$$Pe_L = \frac{192}{4S} > 48$$

which guarantees that longitudinal dispersion will be small.

3.2. RESIDENCE TIME DISTRIBUTION

All continuous-flow vessels, whether of the agitated or streamline type, are nonideal in the sense that neither the fluid elements within them, nor those in the effluent, will be of uniform age α at any given time. The distribution of ages among elements within the vessel at any instant is called the internal age distribution $i(\alpha)$, and that among ages in the effluent at any instant is the exit age distribution $e(\alpha)$. Since the age of a fluid element in the effluent is equal to its residence time in the vessel τ, the age distribution among elements leaving the vessel is also called its residence time distribution. Chemical engineering was introduced to the RTD by Danckwerts (3). One of its chief assets is the relative ease with which it is measured experimentally. Another is the vast amount of information it contains.

The primary cause of RTD in streamline reactors with stationary walls is the zero velocity of the fluid immediately adjacent to the walls. The idealized state of plug-flow (PF) for such reactors is therefore not appropriate unless transverse mixing is very effective. It would appear to be more appropriate for vessels with moving walls (CDFR's) of a certain kind with minimal backflow, which will be discussed in this chapter.

3.2.1. The Red Dye Experiment

The RTD of any continuous-flow vessel is readily determined in the absence of chemical reaction by using tracers. The most common technique involves the introduction of a tracer into the feed stream, in the form of either an impulse (Dirac) function $\delta(t)$ or a unit step function $u(t)$, and the measurement of the resulting output signal with time $\phi(t)$. Both methods have been described in numerous texts on chemical reaction engineering (18, 19). It is customary in such experiments to neglect density variations and to interpret the effluent tracer concentration $\phi(t)$ as a volume fraction.

We shall utilize the step input technique and call it the "red dye experiment." Into a continuous-flow vessel under steady-state conditions, we suddenly introduce at time $t = 0$ a feed stream containing a red dye of unit concentration $\phi = 1$. It follows then that the mean concentration of red dye in the vessel at time t is equal to the volume fraction of fluid elements in vessel with ages between 0 and t

$$\langle \phi(t) \rangle = \int_0^t i(\alpha) \, da \equiv I(t)$$ (3.2-1)

where $I(\alpha)$ is the distribution function corresponding to density function $i(\alpha)$ (Appendix B). Furthermore, the mean concentration of red dye in the vessel effluent at time t is equal to the volume fraction of fluid elements leaving the vessel with ages between τ_s and t

$$[\phi(t)] = \int_{\tau_s}^t e(\tau) \, d\tau \equiv E(t)$$ (3.2-2)

where $E(\tau)$ is the distribution function corresponding to $e(\tau)$ and τ_s is the smallest RT. Volume-average $\langle \phi \rangle$ and flow-average $[\phi]$ quantities are defined in Appendix F (ϕ is a volume-specific point property q').

A material balance for red dye at time t on any vessel with a single inlet and outlet, assuming constant density (Appendix F),

$$V \frac{d}{dt} \langle \phi \rangle = \dot{V}(1 - [\phi])$$ (3.2-3)

thus yields the following general relationship between internal and exit age distributions, with the aid of equations 3.2-1 and 3.2-2,

$$\lambda_v i(t) = 1 - E(t)$$ (3.2-4)

where the space time is equal to the mean RT (20)

$$\lambda_v \equiv \frac{V}{\dot{V}} = \int_{\tau_s}^\infty \tau e(\tau) \, d\tau \equiv \bar{\tau}$$ (3.2-5)

With this result, one may determine either distribution when the other is known.

3.2.2. Some Streamline Vessels and Ideal Mixers

Both distributions can be computed theoretically for vessels whose mixing processes are amenable to mathematical description. Examples are the PF vessel and IM described in Section 1.2. The former is a vessel without a RTD. It may be used as an idealized model for any streamline vessel. The latter is a vessel with complete mixing (broad RTD). It may be used as an idealized model for an agitated tank as well as a streamline vessel with extensive backflow. For PF, it is evident that $i(\alpha)$ is a uniform function (linear and horizontal) and $e(\tau)$ is an impulse function $\delta(t)$. All internal ages are uniformly represented and all elements entering together leave together as well. By contrast, the IM represents an idealized state of the opposite kind. Internal ages and exit ages are distributed identically, owing to complete mixing. Consequently, $i(\alpha)$ and $e(\tau)$ are identical and $[\phi(t)] = \langle \phi(t) \rangle$. However, if N deal mixers are connected in series and upstream (longitudinal) mixing is thus prevented, dispersion in RTD is diminished, and with it the effects of backmixing on chemical reactions. As $N \to \infty$ with total volume V fixed, the RTD approaches that of the PF vessel. In practice, their specific output rates are generally regarded as sufficiently similar when $N > 10$. Conversely, if vessels of either kind (PF of IM) with different mean RT's are connected in parallel and not permitted to mix transversely, then we can simulate two related phenomena called channeling and stagnency, which will be discussed subsequently.

The solution of equation 3.2-3 for a single IM

$$\lambda_v \frac{d\phi}{dt} = 1 - \phi \qquad (3.2\text{-}6)$$

subject to IC $\phi(0) = 0$ is

$$\phi = 1 - \exp\left(-\frac{t}{\lambda_v}\right) \qquad (3.2\text{-}7)$$

from which we find the RTD for the IM to be

$$e(\tau) = \bar{\tau}^{-1} \exp\left(\frac{-\tau}{\bar{\tau}}\right) \qquad (3.2\text{-}8)$$

A similar analysis of N equal-volume IM's in series yields the tanks-in-series (TIS) distribution (Appendix B)

$$N \geqslant 1 \qquad e(\tau) = \frac{N\bar{\tau}^{-1}(N\tau/\bar{\tau})^{N-1}}{(N-1)!} \exp(-N\tau/\bar{\tau}) \qquad (3.2\text{-}9)$$

whose relative dispersion (Section 1.4) is (19)

$$\text{RD} = \frac{\sigma}{\bar{\tau}} = \frac{1}{\sqrt{N}} \qquad (3.2\text{-}10)$$

Its limits are the single IM ($N = 1$), equation 3.2-8, for which $\text{RD} = 1$, and the PF vessel ($N \to \infty$) for which $\text{RD} \to 0$. Derivation of equation 3.2-9 from the red dye experiment may be found in texts on chemical reactor theory (18) and is left as an exercise for the interested reader. We shall deduce it using statistical arguments in order to expose its similarities and differences with respect to certain DPD's that we have encountered earlier.

Since exit of fluid elements from an IM is a random process and since successive withdrawals occur with replacement (inflow), we have Bernoulli trials (Appendix B) for which we let constant P represent the probability that any element survives a withdrawal attempt. Next, we associate the total number of attempts (trials) r with the RT, and require that the rth be a failure (exit). Thus, for one IM the probability that an entering fluid element survives $r - 1$ withdrawal attempts and exits during the rth is just the most probable distribution $P_r = P^{r-1}(1 - P)$. This result compares directly with the most probable DPD, in which $x - 1$ successes (propagations) are followed by a failure (termination) at the end. Furthermore, since fluid elements are assumed to be small in size (δV) compared to vessels (V), it follows that $P = (1 - \delta V/V)$ is a number very close to unity. We would therefore expect equation 3.2-8 to be equivalent to P_r, in the same way that the exponential DPD

$$f(x) = \bar{x}^{-1}\exp\left(\frac{-x}{\bar{x}}\right)$$

(3.2-11)

is a good approximation of the most probable DPD for chain-addition polymers, as shown in Chapter 2.

For N ideal mixers in series with identical volumes, we again have Bernoulli trails with constant P. The probability that an element entering such a cascade will survive $r - N$ withdrawal attempts and exit from the Nth vessel during the rth attempt is given by the so-called negative binomial distribution (Appendix B)

$$P_{r,\mathrm{N}} = \binom{r - 1}{N - 1} P^{r-N}(1 - P)^N$$

(3.2-12)

In this case N failures are required and only $r - 1$ of the trials are permutable, excluding of course the terminal failure. Since P is a number close to unity and $r \gg N$ for most values of r, we can apply the approximations (Appendix B) $(r - 1)!/(r - N)! \cong r^{N-1}$ and $P^{r-N} \cong \exp r(1 - P)$ to equation 3.2-12 with negligible error. Thus,

$$P_{r,\mathrm{N}} = \frac{(1 - P)(1 - P)^{N-1}}{(N - 1)!}\exp[-r(1 - P)]$$

(3.2-13)

This result may be transformed into its continuous form, equation 3.2-9, by associating r with a continuous RT, $\tau \equiv r\,\delta\tau$, and by defining the probability $1 - P \equiv \dot{V}\delta\tau/(V/N) = N\,\delta\tau/\bar{\tau}$ where V is the total occupied volume of the cascade. When these definitions are used to eliminate P and r from equation 3.2-13 and $\delta\tau$ is treated as an infinitesimal, the RHS takes the form $e(\tau)\,d\tau$ which represents the probability than an entering element leaves the cascade with a RT between τ and $\tau + d\tau$.

For lack of a better name, we have called distribution 3.2-9 the tanks-in-series (TIS) distribution. Its discrete form, equation 3.2-13, has been shown to be a good approximation of the negative binomial distribution, equation 3.2-12, under conditions similar to those that cause the Poisson distribution to be a good approximation of the binomial distribution. Furthermore, the TIS distribution bears a close resemblance to the Poisson distribution. Actually, of course, the two are quite distinct (Appendix B). In terms of parameter $N\tau/\bar{\tau}$, τ is the distributed variable in the former distribution as opposed to N in the latter, which appeared in Chapter 2 as the DPD of certain step addition polymers.

For streamline vessels, it is generally more convenient to compute $E(\tau)$ directly via $[\phi(t)]$. One example is the externally pressurized cylindrical tube (CTR). Thus the RTD for laminar flow of a non-Newtonian, power-law fluid in such a tube in the absence of diffusion is (21)

$$e(\tau) = \frac{2n(3n + 1)\tau_0^2\left(1 - \dfrac{\tau_0}{\tau}\right)^{(n-1)/(n+1)}}{(n + 1)^2\tau^3} \tag{3.2-14}$$

where n is the flow index and τ_0 is the RT at the tube axis. The Newtonian fluid, of course, is a well-known special case for which $n = 1$ (Example 3.2-1).

$$e(\tau) = \frac{2\tau_0^2}{\tau^3} \tag{3.2-15}$$

On the basis of our discussion on mixing in the preceding section, we may associate this RTD with laminar shear macromixing. When transverse mixing (molecular or eddy) occurs as well, the combined result is frequently manifested as longitudinal dispersion (actual or "Taylor" axial), and the so-called dispersion model applies (Example 3.1-4). The associated RTD was discussed in Section 1.4. Another example is drag-induced flow by one or two moving vessel walls (CDFR). The RTD for laminar flow of a Newtonian fluid between infinite plates uniformly spaced in the absence of backflow (Fig. 1.2-2) is listed in Table 3.2-1 and calculated in Example 3.2-2. Note that for the special case of single plate, pure drag-flow (no pressure-flow) the RTD is the same as for a Newtonian microsegregated continuous tubular reactor (MSCTR) (Section 3.3). The RTD for a filled, single-screw extruder, which is also devoid of backflow (5-7), was computed for a Newtonian fluid by Pinto and Tadmor (22) and for a non-Newtonian fluid by Biggs and Middleman (23).

Some common internal and external age distributions are listed in Table 3.2-1 and are illustrated in Figs. 3.2-1–3.2-7. It is apparent that broad RTD's can arise when mixing is good as well as when it is poor (stagnancy and channeling). In the latter case bimodal distributions are also possible.

In addition to the more conventional reactor schemes, RTD's for parallel-plate and moving-wall reactors have been included as representative abstractions of polymer processing equipment and will be discussed further in Chapter 5. Where possible, analytical expressions for the holdback, a measure of the breadth of the RTD (Section 3.2-3), have also been included in Table 3.2-1.

The IM is the only reactor scheme for which the internal and exit age distributions, $i(\hat{\alpha})$ and $e(\hat{\tau})$, respectively, are identical. When IM's are placed in series, however, the RTD functions change significantly. Relevant distributions are shown in Fig. 3.2-1. In fact, as the number of IM's increases, the RT behavior begins to resemble that of a PF vessel. It is common practice to represent the PFR with an infinite number of tanks in series where each differential element of axial length

Table 3.2-1. RTD Functions for Various Reactor Configurations

REACTOR	$e(\hat{\tau})$	$i(\hat{\alpha})$	H
STREAMLINE VESSELS			
PFR	$\delta(\hat{\tau} - 1)$	1	0
CEPR's			
MSCTR (power law)	$(2n/3n+1)[1-(n+1)/(3n+1)\hat{\tau}]^{\frac{n-1}{n+1}}\hat{\tau}^{-3}$	$1-[1+2n/(3n+1)\hat{\alpha}][1-(n+1)/(3n+1)\hat{\alpha}]^{\frac{2n}{n+1}}$	$0 < [1-(n+1)/(3n+1)]^{\frac{3n+1}{n+1}} < 0.296$
parallel plate (Newtonian)	$(1-2/3\hat{\tau})^{-1/2}/3\hat{\tau}^3$	$1-[1+1/3\hat{\alpha}][1-2/3\hat{\alpha}]^{1/2}$	0.289
CDFR's [*]	$(N/2\hat{\tau}^3)(4a\hat{\tau}^{-1} + d)^{-1/2}$	$1+(N/24a^2)\left\{[d+4a\hat{\tau}_o^{-1}]^{1/2}[d-2a\hat{\tau}_o^{-1}] - [d+4a\hat{\alpha}^{-1}]^{1/2}[d-2a\hat{\alpha}^{-1}]\right\}$	$(N/24a^2)[d+4a\hat{\tau}_o^{-1}]^{1/2}[d-2a\hat{\tau}_o^{-1}][\hat{\tau}_o^{-1}]$ $-(d+4a)^{3/2}\big/4a + (d+4a\hat{\tau}_o^{-1})^{3/2}\big/4a\hat{\tau}_o^{-1}$
AGITATED VESSELS [+]			
SIMS	$[N(N\hat{\tau})^{N-1}/(N-1)!]\exp(-N\hat{\tau})$	$\exp(-N\hat{\alpha})\sum_{j=o}^{N-1}((N\hat{\alpha})/j!)^j$	$0 < [N^{N-1}/(N-1)!]\exp(-N) < 0.363$
Parallel IM's	$\sum_{k=1}^{2}f_k\hat{\tau}_k^{-1}\exp(-\hat{\tau}/\hat{\tau}_k)$	$\sum_{k=1}^{2}f_k\exp(-\hat{\alpha}\,\hat{\tau}_k)$	$\sum_{k=1}^{2}f_k\hat{\tau}_k\exp(-\hat{\tau}_k^{-1})$
Parallel SIMS	$\sum_{k=1}^{2}[f_k\hat{\tau}_k^{N_k-1}/(N_k-1)!](N_k/\hat{\tau}_k)^{N_k}\exp(-N_k\hat{\tau}/\hat{\tau}_k)$	$\sum_{k=1}^{2}[f_k\exp(-N_k\hat{\alpha}/\hat{\tau}_k)]\sum_{j=o}^{N_k-1}(N_k\hat{\alpha}/\hat{\tau}_k)^j/j!$	$\sum_{k=1}^{2}(f_k\hat{\tau}_k/N_k)[\exp(-N_k/\hat{\tau}_k)]\sum_{j=1}^{N_k-1}(N_k-j)(N_k/\hat{\tau}_k)^j/j!$

[*] where: N = number of moving plates (1 or 2); $a \equiv 3G/2(1 - G)$; $b \equiv (\hat{v}_2 - \hat{v}_1)/2(1 - G)$; $d = b^2 - 4a[(1 - G)^{-1} - a]$;

and, $N = 1$, $-1/3 \leq G \leq 1/3$, for which $\hat{\tau}_o = (1 - G)/2$

$N = 2$, $\hat{v}_2 = \hat{v}_1$, $G < 0$, for which $\hat{\tau}_o = \frac{?}{}(1 - G)/(2 - 3G)$

$N = 2$, $\hat{v}_2 = \hat{v}_1$, $0 < G < 2/3$ for which $\hat{\tau}_o = 1 - G$

[+] where N_k = number of IM's in series in branch k; $f_k \equiv \dot{V}_k/\dot{V}$; and $\hat{\tau}_k \equiv \tau_k/\tau$

276

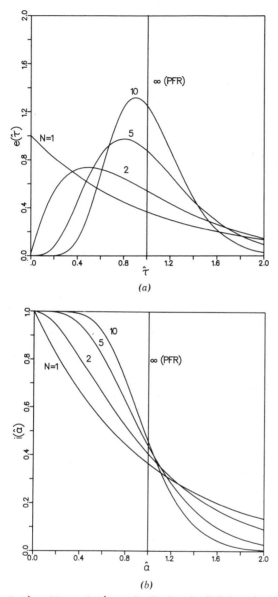

Figure 3.2-1. External $e(\hat{\tau})$ and internal $i(\hat{\alpha})$ age distributions for IM's in series (SIMS) with number of IM's (N) as parameter.

represents a single tank. We may rely on this similarity to simulate the performance of the PFR when analytical treatment of its response to a Dirac input function is inconvenient.

The results for the dispersion model actually look quite similar to those for a series of IM's (SIMS), as is apparent in Fig. 3.2-2. Recall, however, that the boundary conditions used for the former (Section 1.4) are only realistic when

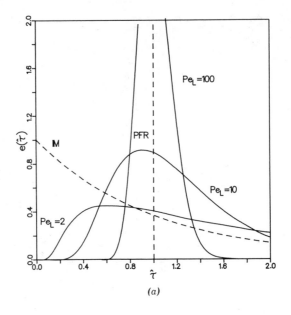

(a)

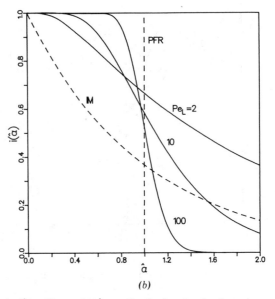

(b)

Figure 3.2-2. External $e(\hat{\tau})$ and internal $i(\hat{\alpha})$ age distributions for the dispersion model with longitudinal Peclet number (Pe_L) as parameter.

$Pe_L > 10$. Lower values were included anyway to illustrate the behavior of the functional forms of $e(\hat{\tau})$ and $i(\hat{\alpha})$. A large Pe_L can result not only from poor axial dispersion but also from extensive radial diffusion coupled with good laminar shear mixing (i.e., Taylor diffusion). This is important to bear in mind when comparing mixing capabilities of streamline flow reactors, where shear provides the macroscopic mixing and transverse diffusion produces micromixing.

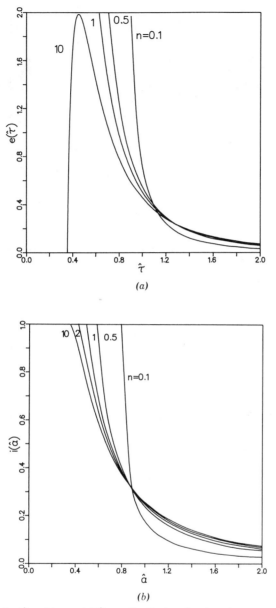

(a)

(b)

Figure 3.2-3. External $e(\hat{\tau})$ and internal $i(\hat{\alpha})$ age distributions for the microsegregated (MS) continuous tubular reactor (CTR) with power-law index n as parameter.

If we consider the MSCTR (Section 3.3) with the various power-law indices, Fig. 3.2-3, the transition from a pluglike RTD with increasing n is apparent. For $n < 1$, $e(\hat{\tau})$ asymptotically approaches infinity at the minimum value of $\hat{\tau}$, while for $n > 1$, $e(\hat{\tau})$ passes through a maximum. The maximum occurs because the volume fraction of any given age in the exit stream is the product of the streamline velocity and cross-sectional area. The differential area increases linearly with increasing radius, whereas the velocity is a declining function of the radius to a power. A power law

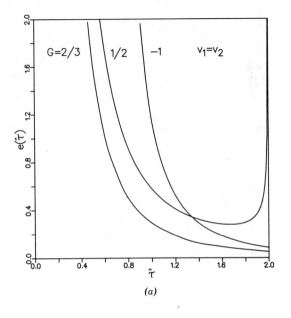

(a)

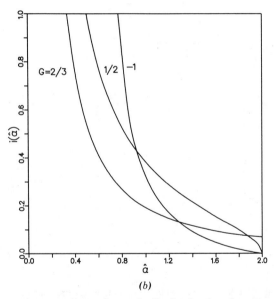

(b)

Figure 3.2-4. External $e(\hat{\tau})$ and internal $i(\hat{\alpha})$ age distribution for the microsegregated (MS) continuous drag flow reactor (CDFR) with two moving plates ($v_1 = v_2$) and with G as parameter.

index of 10 may be considered an upper bound in that further increases have little effect on the shape of the velocity profile, which may be approximated by a linear decrease from centerline to wall:

$$v(r) = v_0\left(1 - \frac{r}{\mathrm{Ra}}\right)$$

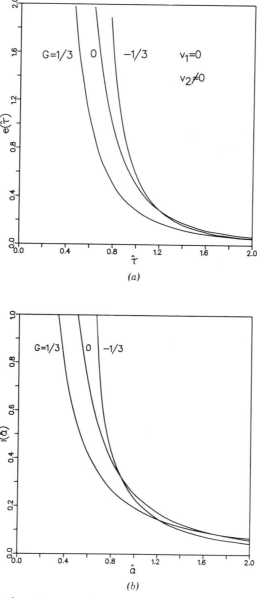

Figure 3.2-5. External $e(\hat{\tau})$ and internal $i(\hat{a})$ age distributions for the microsegregated (MS) continuous frag-flow reactor (CDFR) with one moving plate ($v_1 = 0$) and with G as parameter.

We shall categorize streamline reactors in general by the dominant driving force for flow. Thus all continuous externally pressurized streamline reactors will be labeled with the mnemonic CEPR. This includes the CTR as well as the parallel plate vessel in which the plates are stationary and flow in the gap is the result of a pressure drop only. If one or both of the plates move, we shall call this configuration a continuous drag flow reactor (CDFR), whether or not a pressure gradient exists in either longitudinal direction.

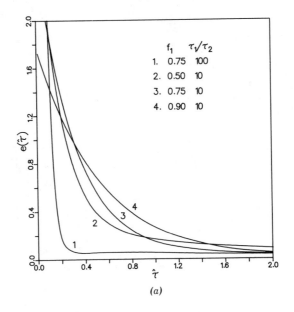

(a)

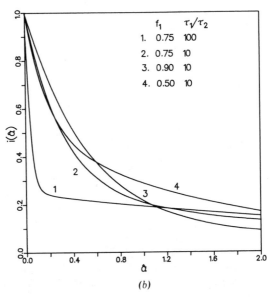

(b)

Figure 3.2-6. External $e(\hat{\tau})$ and internal $i(\hat{\alpha})$ age distributions for two ideal mixers (IM's) in parallel with flow-rate fraction f_1 and RT ratio (τ_2/τ_1) as parameters.

The shortest RT, τ_0, in streamline reactors is generally nonzero because longitudinal macromixing is not sufficient to provide "instantaneous" bypassing. Therefore $e(\hat{\tau})$ $[= \lambda_v e(\tau)]$ must be zero for all $\hat{\tau} < \hat{\tau}_0$. Equation 3.2-4 then dictates that $i(\hat{\alpha})$ $[= \lambda_v i(\alpha)]$ must be constant (unity) for all $\hat{\alpha} < \hat{\tau}_0$. This signifies that the internal age distribution among all elements younger than τ_0 is uniform and that $i(\hat{\alpha}) \, d\hat{\alpha}[= i(\alpha) \, d\alpha]$ simply represents the fraction of elements within the vessel that entered it at

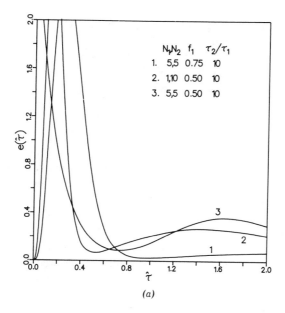

(a)

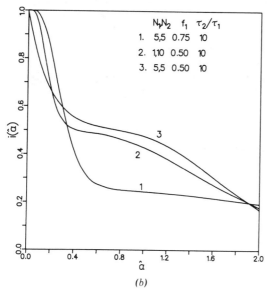

(b)

Figure 3.2-7. External $e(\hat{\tau})$ and internal $i(\hat{\alpha})$ age distributions for two SIMS (N_1 and N_3 ideal mixers) in parallel with various flow-rate fractions (f_1) mean RT ratios (τ_2/τ_1) and IM ratios (N_1/N_2), where N_k is the number of IM's in the k th branch.

a time $\hat{\alpha}$ earlier (for $\hat{\alpha} < \hat{\tau}_0$) regardless of the degree to which they may have been spatially redistributed or distorted by laminar (shear) mixing.

The CDFR with both plates moving ($\hat{v}_1 = \hat{v}_2$) presents a unique case. Not only is there a nonzero shortest RT, but there can also be a finite longest RT. If the pressure drops, the shortest RT occurs at the centerline while the longest is at the walls. If the pressure rises, the shortest RT occurs at the moving plates while the center has the longest, which is finite provided that there is no backflow (e.g., $0 \leqslant G < \frac{2}{3}$ when $\hat{v}_1 = \hat{v}_2$; Example 3.3-2). Figure 3.2-4 shows RTD functions for the CDFR with two moving plates, while 3.2-5 shows the same for one moving plate. Note that the curve $G = -1/3$ is identical to that for the CEPR.

The RTD's for CDFR's with one and two moving plates can be the same. For instance, reactors with one ($\hat{v}_1 = 0$) and two ($\hat{v}_1 = \hat{v}_2$) moving plates operating with the maximum pressure rise before the onset of backflow ($G = \frac{1}{3}$ and $\frac{2}{3}$, respectively) can have identical RTDs. Also, the parallel-plate CEPR and the CDFR with one plate moving ($G = -\frac{1}{3}$, the smallest value to just preclude a maximum in the velocity profile) can have the same $e(\hat{\tau})$ and $i(\hat{\alpha})$. In each case, the unsymmetrical reactor ($\hat{v}_1 \neq \hat{v}_2$) has the same velocity profile from wall-to-wall as the symmetrical reactor ($\hat{v}_1 = \hat{v}_2$) has from centerline to wall. Thus the RTD's are identical. As noted earlier, the RTD for pure drag flow ($G = 0$) between parallel plates with $\hat{v}_1 = 0$ is identical to that for a Newtonian MSCTR. This is of particular importance since the extruder is often erroneously conceptualized as simple parallel-plate drag flow. Actually, however, the RTD of an extruder lies between that of a PFR and a MSCTR, so one must be wary of using the simplistic visualization in rigorous applications.

The RTD functions for parallel processes (Section 1.3) have been included because they illustrate the effects of stagnancy (24). Examples are shown in Figs. 3.2-6 and 3.2-7. The functions for parallel cascades of SIMS reduce to the special case of IM's in parallel when $N_1 = N_2 = 1$ (Table 3.2-1). These functions can also be used to simulate an IM in parallel with a PFR by setting $N_1 = 1$ and $N_2 \gg 1$. Parallel SIMS represent the only reactor configuration that clearly shows a bimodal RTD. In the limit as N_1 and N_2 increase, $e(\hat{\tau})$ would take on the form of two spikes located at the mean RT of each subsystem, which is equivalent to the RTD for two PFR's in parallel. Segregated laminar flow may therefore be visualized as many parallel flow paths, each comprising many IM's in series.

Example 3.2-1.   *RTD of the MSCTR.* Compute $e(\tau)$, $E(\tau)$, $i(\alpha)$, and $I(\alpha)$ for the Newtonian laminar-flow tube with no diffusion.

Solution. The velocity profile is well-known

$$v(r) = v_0 \left[1 - \left(\frac{r}{\text{Ra}} \right)^2 \right]$$

where v_0 is the velocity at the axis. For the red dye experiment in the time interval $\tau_0 < t < \infty$

$$[\phi(t)] = \frac{\int_0^{r(t)} 2\pi(1)v(r)r\,dr + \int_{r(t)}^{\text{Ra}} 2\pi(0)v(r)r\,dr}{\pi \text{Ra}^2 \left(\dfrac{v_0}{2} \right)}$$

$$= \frac{2r(t)^2}{\text{Ra}^2} \left[1 - \frac{r(t)^2}{2\text{Ra}^2} \right]$$

where $r(t)$ locates the tracer boundary at the exit, and v is the corresponding velocity. In terms of residence times $\tau = L/v(r)$, we obtain

$$[\phi(\tau)] = E(\tau) = 1 - \left(\frac{\tau_0}{\tau}\right)^2 = 1 - (\bar{\tau}/2\tau)^2$$

where $\tau_0 = L/v_0$ is the smallest RT and $\bar{\tau} = 2\tau_0$ is the mean. Thus,

$$e(\tau) = \frac{dE}{d\tau} = \frac{2\tau_0^2}{\tau^3} = \frac{\bar{\tau}^2}{2\tau^3}$$

and

$$i(\alpha) = \frac{(1 - E)}{\bar{\tau}} = \frac{\bar{\tau}}{4\alpha^2}$$

However, $i(\alpha) = \bar{\tau}^1$ when $0 < \alpha < \tau_0 = \dfrac{\bar{\tau}}{2}$ since $E = 0$ for $\alpha < \tau_0$. Finally, then

$$I(\alpha) = \int_0^{\bar{\tau}/2} \bar{\tau}^{-1}\, d\alpha + \int_{\bar{\tau}/2}^{\alpha} \frac{\bar{\tau}\, d\alpha'}{(2\alpha')^2} = 1 - \frac{\hat{\tau}}{4\alpha}$$

Example 3.2-2. RTD of a CDFR.   Consider flow of a Newtonian, incompressible fluid in the gap of rectangular cross section WH between two parallel plates. Let the lower and upper plates move with arbitrary velocities, v_1 and v_2. As shown in Section 5.4, the velocity profile will be

$$\hat{v}_x = 1 + \frac{\hat{v}_d}{2}\hat{y} - \frac{s\mathrm{P}_0}{2}(1 - \hat{y}^2)$$

where $s\mathrm{P}_0 = 3G$, $s \equiv \Delta P/|\Delta P| = \pm 1$, and $\hat{v}_d \equiv (v_2 - v_1)/v_m$. Use the velocity profile to find the expressions for $e(\hat{\tau})\, d\hat{\tau}$ and $i(\hat{\alpha})\, d\hat{\alpha}$.

Solution.   In general, for laminar-flow problems $e(\tau)\, d\tau = \dot{V}^{-1}\, d\dot{V}$, where $d\dot{V}$ represents the increment of throughput at the exit associated with τ. The volumetric flow rate through the entire cross section is $\langle v \rangle\, WH$. If the velocity changes monotonically from plate to plate, then each lamella may be uniquely associated with a particular RT. If this is so, then $d\dot{V} = v_x W\, dy$. If, on the other hand, flow is symmetrical about the centerline, then there are two lamellae associate with a given RT, and $d\dot{V} = 2v_x W\, dy$. Recall that $\hat{y} \equiv y/(H/2)$. Thus

$$\frac{d\dot{V}}{\dot{V}} = \frac{\hat{v}_x\, d\hat{y}}{2(1 - G)}$$

where the $\frac{1}{2}$ applies to the monotonic case only. The expression for the velocity profile can be rewritten as

$$\hat{v}_x = \alpha\hat{y}^2 + \beta\hat{y} + (1 - \alpha)$$

Solving the quadratic for $\hat{y}$ yields

$$\hat{y} = -\left(\frac{\beta}{2\alpha}\right) \pm \left(\frac{\hat{v}_x}{\alpha} + \frac{\delta}{4\alpha^2}\right)^{1/2}$$

Note: $\delta \equiv \beta^2 - 4\alpha(1-\alpha)$ and since $\hat{\tau} = (1-G)/\hat{v}_x$, we can differentiate the preceding equation to get, in terms of the constants in Table 3.2-1,

$$\frac{d\hat{y}}{d\hat{\tau}} = \frac{1}{2a\hat{\tau}^2}\left(\frac{1}{a\hat{\tau}} + \frac{d}{4a^2}\right)^{-1/2}$$

where $a \equiv \alpha/(1-G)$; $b \equiv \beta/(1-G)$; $d \equiv \delta/(1-G)$, so that

$$e(\hat{\tau}) = \left(\frac{1}{2}\right)\frac{1}{\hat{\tau}^3}\left(\frac{4a}{\hat{\tau}} + d\right)^{-1/2}$$

From equation 3.2-4

$$i(\hat{\alpha}) = 1 - \int_{\hat{\tau}_0}^{\hat{\alpha}} e(\hat{\tau})\,d\hat{\tau}$$

Using the substitution $u \equiv (4a/\hat{\tau} + d)$,

$$\int e(\hat{\tau})\,d\hat{\tau} = \left(\frac{1}{2}\right)\int \frac{-1}{16a^2}u^{-1/2}(u-d)\,du$$

which yields the result

$$i(\hat{\alpha}) = 1 - \frac{1}{24a^2}\left[\left(\frac{4a}{\hat{\tau}_0} + d\right)^{1/2}\left(\frac{2a}{\hat{\tau}_0} - d\right) - \left(\frac{4a}{\hat{\alpha}} + d\right)^{1/2}\left(\frac{2a}{\hat{\alpha}} - d\right)\right]$$

3.2.3.   *Holdback and the Intensity Function*

In general, as we shall see, residence time dispersion is detrimental to reactor performance. A measure of this dispersion would therefore be desirable. Intuitively, we might expect the worst kind of dispersion to occur when a large fraction of the effluent is young and when a large fraction of the vessel contents is old. In fact, these characteristics go together. The former is the result of channeling (short-circuiting) and the latter the result of stagnancy (dead-spaces). It can be shown via equation 3.2-4 that the fraction of the vessel contents older than λ_v,

$$\int_{\lambda_v}^{\infty} i(\alpha)\,d\alpha = 1 - I(\lambda_v) \tag{3.2-16}$$

is equal to the fraction of the feed that enters and leaves again during the same period λ_v.

$$\frac{\dot{V}\int_0^{\lambda_v} E(t)\,dt}{\lambda_v \dot{V}} = \frac{\int_0^{\lambda_v} E(t)\,dt}{\lambda_v} \tag{3.2-17}$$

Danckwerts (3) called this quantity the holdback H. It may be evaluated by either definition 3.2-16 or 3.2-17. Thus the quantity defined by expression 3.2-16 could reflect stagnancy. However, it could also reflect good macromixing (longitudinal), which is quite the opposite. Similarly, the alternative definition, equation 3.2-17, reflects channeling, as well as statistical bypassing (25), which accompanies good macromixing. This dual effect on H emphasizes the uncertainty associated with using RTD dispersion to assess goodness of mixing.

Naor and Shinnar (24) showed that H is actually a measure of the relative dispersion (Section 2.4) of the RTD in the following sense (Example 3.2-5):

$$H = \frac{\sigma'}{2\lambda_v} \equiv \frac{RD'}{2} \qquad (3.2\text{-}18)$$

where the prime is used to distinguish the quantitiy in the numerator

$$\sigma' \equiv \int_0^\infty |\tau - \bar{\tau}| e(\tau)\, d\tau \qquad (3.2\text{-}19)$$

from the standard deviation σ (Appendix B) and associated relative dispersion defined in Section 1.4.

The reader will note that 0 was taken as the lower limit of integration in the preceding equations, assuming that $e(\tau)$ would be assigned the value 0 in the interval $0 \leqslant \tau \leqslant \tau_s$ if $\tau_s > 0$, and the identity $\bar{\tau} = \lambda_v$ was used. This will be done routinely from here on.

Expressions for reactor holdback were given in Table 3.2-1 and some values of H are listed in Table 3.3-2. All values lie within the range $0 < H < 1$, starting at 0 for PF, passing through e^{-1} for complete mixing and rising towards 1 for excessive stagnancy and channeling. The expressions for SIMS, parallel IM's, and the MSCTR have been plotted in Fig. 3.2-8. The holdback for SIMS decreases continuously from its largest value of e^{-1} (0.368) for a single IM to an asymptotic value of zero. This is consistent with the observation that many IM's in series approach PF behavior.

When IM's are placed in parallel the maximum holdback depends upon the ratio of flow rates through each IM. As the ratio of average RT's in each IM diverges it may be shown that

$$\lim_{\hat{\tau}_2/\hat{\tau}_1 \to \infty} H = \exp\left(f_1^{-1}\right) \qquad (3.2\text{-}20)$$

where f_1 is the fraction of flow rate through IM number one ($\dot{V}_1/\dot{V}$). The holdback is bounded by the value for a single IM as the flow-rate fraction through one tank goes to zero (as well as when $\hat{\tau}_2 = \hat{\tau}_1$ for any value of f_1), and at the other extreme by a value of one as the flow rate is diverted to the IM with the larger RT. A RT ratio of $100:1$ brings H within a few percent of an upper bound for each value of f_1 (Fig. 3.2-8).

The holdback for a MSCTR is a function of the power-law index (21). There is very little change in H when n is increased beyond 1.0, and the limiting maximum value of 0.296 is closely approximated when $n = 10$. As n is decreased below one, H drops rapidly towards a value of zero, characteristic of plug flow. The maximum

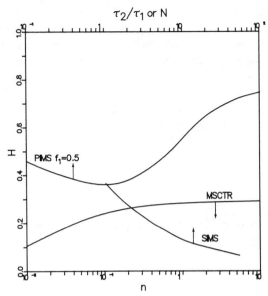

Figure 3.2-8. Holdback H versus appropriate reactor parameter for some of the configurations listed in Table 3.2-1 (PIMS = parallel IM's).

holdback is less than that of an IM indicating that the broadest RTD for a tube is still narrower than that for a completely macromixed vessel.

Cintron (21, 26) investigated flow profiles that might give rise to a more disperse RTD than the power-law fluid. He chose the RTD of an IM as the standard and equated its density function to the streamline flow fraction:

$$2\hat{v}(\hat{r})\hat{r}\,d\hat{r} = \exp(-\hat{\tau})\,d\hat{\tau} \qquad (3.2\text{-}21)$$

where $\hat{v} \equiv v/\langle v\rangle$, $\hat{r} \equiv r/R_a$, and $\hat{\tau} \equiv \tau/\bar{\tau}$. This was solved for a parametric family of velocity profiles.

$$\left(\frac{1 + \hat{v}^{-1}}{1 + \hat{\tau}_0}\right)\exp(\hat{\tau}_0 - \hat{v}^{-1}) = 1 - \hat{r}^2 \qquad (3.2\text{-}22)$$

Typical velocity profiles for a range of values of $\hat{\tau}_0$, the dimensionless RT at the axis, are shown in Fig. 3.2-9. These profiles resemble the "fingering" flow or channeling characteristic of some tubular polymerizers. However, the holdback for all these reactors will still be less than 0.363 since $\hat{\tau}_0$ for the MSIM is zero. Notwithstanding the fact that the function $e(\hat{\tau})$ is the same for both reactors, the lower limiting RT is larger for the tubular reactors. As $\hat{\tau}_0 \to 0$ the flow profile would approach zero velocity everywhere but along the center line. Such extreme velocity profiles achieve breadth of RTD through bimodalization (stagnancy and channeling) an undesirable state of mixedness.

It is clear that RTD dispersion (H) alone is inadequate to characterize goodness of macromixing in either the longitudinal or the transverse direction with respect to flow. Although it is true that good longitudinal macromixing can disperse a RTD,

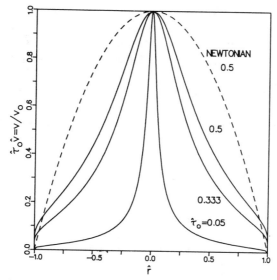

Figure 3.2-9. Synthesized velocity profiles for tubular vessels having identical RTD's to that of the IM, with centerline RT $(\hat{\tau}_0)$ as parameter.

poor macromixing can also disperse it, and to an even greater degree. Examples of the latter are found in agitated vessels with inadequate agitation and streamline vessels with channeling and abrupt variation in dimensions. Concerning transverse mixing, we note that a narrow RTD (plug-like flow) is compatible with both extremes (none and complete). Furthermore, RTD alone yields no information on the extent of micromixing that accompanies macromixing.

An important question remaining is whether the RTD is capable of revealing more about the causes of its dispersion than indicated by H. If a single number such as H cannot distinguish good macromixing from poor macromixing, then perhaps another characteristic can. Noar and Shinnar (24) proposed a function, in contrast to a single number. They named it the intensity function and defined it as follows:

$$\Omega(\alpha) \equiv \frac{e(\alpha)}{\bar{\tau}i(\alpha)}. \tag{3.2-23}$$

This equation may be derived using statistical arguments. The probability that a particle entering a vessel will remain there for a time between α and $\alpha + d\alpha$, and then exit, is $e(\alpha)\,d\alpha$. This quantity must also equal the combined probability that an entering particle will remain at least for time α [probability $1 - E(\alpha)$] and then exit immediately [probability $\Omega(\alpha)\,d\alpha$]

$$e(\alpha)\,d\alpha = [1 - E(\alpha)]\Omega(\alpha)\,d\alpha. \tag{3.2-24}$$

Thus the quantity $\Omega(\alpha)\,d\alpha$ represents the probability that any particle in the vessel with age α chosen at random will exit during the next instant. Equation 3.2-24 is obtained by combining equations 3.2-4 and 3.2-23. In dimensionless form $\Omega(\hat{\alpha}) = e(\hat{\alpha})/i(\hat{\alpha})$.

Equal exit probabilities for particles of all ages, of course, signifies complete mixing. For "good" macromixing in general young particles (small α) should not have higher exit probabilities than old ones (large α). When mixing is poor, however, and short-circuiting exists they will have higher exit probabilities. Thus, in general, the slope of a graph of $\Omega(\alpha)$ versus α (Figs.s 3.2-10–3.2-16) yields information about macroscopic mixing. Its magntidue measures extent of macromixing and its sign indicates quality.

The IM represents the ideal case in which all particles have an equal probability of exit while for the PFR all particles in the exit stream have a single age and the probability of exit is zero for all other ages. The PFR may be considered to have complete transverse mixing, whereas the IM has complete longitudinal mixing. We suspect that these two mixing processes combined in series and parallel might give rise to intensity functions of widely varying characteristics. In support of this, we recall that the RTD of a PFR can be simulated by placing many IMs in series (Fig. 3.3-1), thereby reducing longitudinal mixing. The RTD of a MSIM in turn can be simulated by placing many PFRs of varying RT's in parallel, thereby reducing transverse mixing.

Since stagnancy and channeling are manifestations of parallel pathways through the reactor system, we might expect the intensity function for such a system to have a similar appearance to other parallel flow schemes such as the MSCTR. In general, we propose that in graphs of $\Omega(\hat{\alpha})$ versus $\hat{\alpha}$ slopes with large magnitudes indicate little longitudinal mixing but possibly extensive transverse mixing while slopes with small magnitudes indicate extensive longitudinal mixing. The extremes are, of course, complete transverse mixing or plug-flow (vertical line) and complete longitudinal mixing (horizontal line). Negative slopes are manifestations of channeling and stagnancy whereas positive slopes suggest a "better quality" of mixing (Fig. 3.3-2).

Intensity functions for the various reactor configurations of Table 3.2-1 are shown in Figs. 3.2-10–3.2-16. It is clear that the IM is the only reactor in which $\Omega(\hat{\alpha})$ is

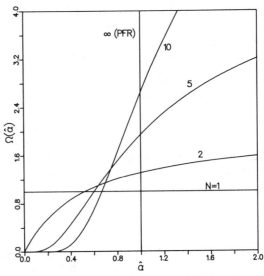

Figure 3.2-10. Intensity functions $\Omega(\hat{\alpha})$ for IM's in series (SIMS) with number of IM's (N) as parameter.

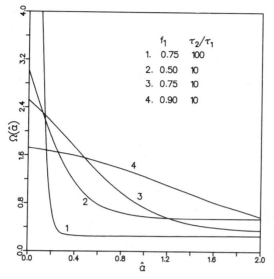

Figure 3.2-11. Intensity functions $\Omega(\hat{\alpha})$ for IM's in parallel with flow-rate fraction f_1 and residence time ratio (τ_2/τ_1) as parameters.

independent of $\hat{\alpha}$; that is, particles of all ages have an equal probability of exit. Coupling IM's in series reduces the efficiency of longitudinal mixing. It is evident in Fig. 3.2-10 that, as the number of series elements is increased, the slope of the intensity function increases in magnitude while remaining positive. There is no evidence of stagnancy here, as older particles have an increasing probability of escape. The ideal case is the PF vessel, for which $\Omega(\hat{\tau})$ is vertical.

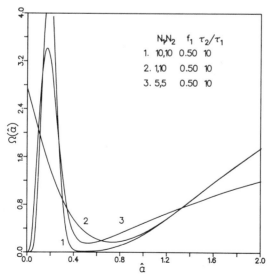

Figure 3.2-12. Intensity functions $\Omega(\hat{\alpha})$ for SIMS in parallel with flow-rate fraction (f_1), RT ratio (τ_1/τ_2), and number of IM's $(N_1$ and $N_2)$ as parameters.

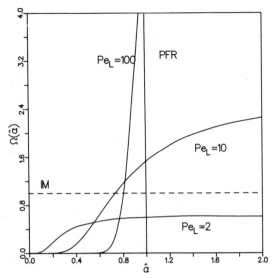

Figure 3.2-13. Intensity functions $\Omega(\hat{\alpha})$ for the dispersion model with longitudinal Peclet number (Pe_L) as parameter.

When IM's are placed in parallel, however, the intensity function changes radically. The slopes in Fig. 3.2-11 have become negative, and have increased in magnitude as a result of either an increase in the amount of flow spearation or an increase in the ratio of RT's for each element. Note that the plots of $e(\hat{\tau})$ and $i(\hat{\alpha})$ for these reactor configurations do not clearly distinguish the stagnancy from the ideal case, while the intensity function clearly identifies this condition. Stagnancy

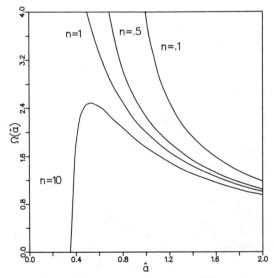

Figure 3.2-14. Intensity functions $\Omega(\hat{\alpha})$ for the microsegregated (MS) continuous tubular reactor (CTR) with power-law index n as parameter.

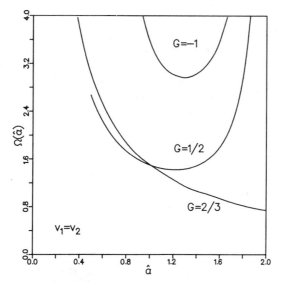

Figure 3.2-15. Intensity functions $\Omega(\hat{\alpha})$ for the microsegregated (MS) continuous drag-flow reactor (CDFR) with two moving plates ($v_1 = v_2$) with G as parameter.

can be made more extreme by placing trains of SIMS in parallel and varying the flow fraction and mean RT of each subsystem. Figure 3.2-12 illustrates the resultant effect on the intensity function. In the limit, as the number of IM's in each subsystem is increased, we describe the behavior of parallel PFR's. Parallel SIMS have decreased longitudinal and transverse mixing. We therefore conclude that intensity functions do not necessarily change monotonically with $\hat{\alpha}$ and therefore

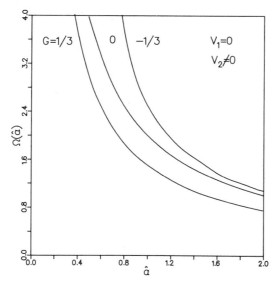

Figure 3.2-16. Intensity functions $\Omega(\hat{\alpha})$ for the microsegregated (MS) continuous drag-flow reactor (CDFR) with one moving plate ($v_1 = 0$) with G as parameter.

may not be simply described as lying between the ideal case of the IM (horizontal line) and the PFR (vertical line). The potentially bimodal nature of the RTD is also clearly apparent, as are regions of both positive and negative slope.

The results in Fig. 3.2-13 for the dispersion model also show the effect of increasing longitudinal mixing. As axial dispersion increases, the value of Pe_L decreases. The associated intensity functions range from pluglike to IM-like as Pe_L decreases. The curves are very similar to those for SIMS. One should bear in mind that the dispersion model should only be used for $Pe_L > 10$, and that results shown for lesser values are merely to illustrate qualitative trends.

Analysis of streamline-flow reactors in which the sole mixing process is laminar requires that one consider the effects of both transverse and longitudinal mixing. Since plug flow represents an extreme case of the former and the absence of the latter, it is somewhat contradictory to speak in terms of "good" versus "bad" mixing in laminar flow. Indeed, the objective of good laminar mixing might be to create sufficient longitudinal mixing (simple shear strain or backflow) so that diffusion can complete the micromixing process. If we conceptualize streamline reactors as parallel PFR's with a distribution of flow fractions and RT's, or even many parallel subsystems of SIMS, the preceding results would lead us to expect that the intensity function for these reactors would have a negative slope.

This is confirmed for MSCTR's in Fig. 3.2-14. As $n \to \infty$, longitudinal dispersion improves, notwithstanding the impression of stagnancy and channeling conveyed by the spikelike appearance of the velocity profile and the increasing values of H.

The most unique vessels in terms of their RTD's and intensity functions appear to be the CDFR's, particularly when both plates move. These vessels can have a finite value for the longest RT, in addition to a nonzero value for the shortest one, because both walls move. Between these limits, the intensity function exhibits a minimum (Fig. 3.2-15). As $\hat{\alpha}$ approaches its shortest value, $\Omega(\hat{\alpha})$ becomes very large owing to the combined effects of $e(\hat{\alpha})$ and $i(\hat{\alpha})$, the first of which approaches infinity (Fig. 3.2-5). Furthermore, as $\hat{\alpha}$ approaches its longest value, $\Omega(\hat{\alpha})$ must again approach infinity since all fluid elements of that age must exit immediately.

Table 3.2-2.   RTD for a Single-Screw Extruder (22)

$$e(\hat{\tau})\,d\hat{\tau} \;=\; e(\hat{y})\,d\hat{y}$$

where

$$e(\hat{y}) \;=\; \frac{3\hat{y}[1 - \hat{y} + (1 + 2\hat{y} - 3\hat{y}^2)^{1/2}]}{(1 + 2\hat{y} - 3\hat{y}^2)^{1/2}}$$

and

$$\hat{\tau} \equiv \tau/\bar{\tau} \;=\; \frac{3\hat{y} - 1 + 3(1 + 2\hat{y} - 3\hat{y}^2)^{1/2}}{6y[1 - y + (1 + 2y - 3y^2)^{1/2}]}$$

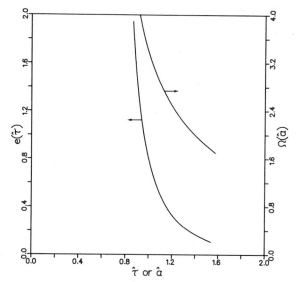

Figure 3.2-17.   External $e(\hat{\tau})$ age distribution, and intensity function $\Omega(\hat{a})$ for a single-screw extruder.

When only one plate moves (Fig. 3.2-16), the intensity functions look very much like that of the MSCTR, reflecting the stagnancy of the fluid layers near the stationary wall.

The RTD for the parallel-plate model of the single-screw extruder according to Pinto and Tadmor (22) is listed in Table 3.2-2 and illustrated in Fig. 3.2-17. Its corresponding intensity function has also been plotted in Fig. 3.2-17. It is instructive to compare the latter with the intensity function for our simple CDFR with one moving plate (Fig. 3.2-16). This comparison shows that the more pluglike flow in the extruder, which is frequently cited, is not devoid of stagnancy and channeling as one might expect owing to convective transverse mixing (cross-channel flow). By contrast, a similar comparison of Fig. 3.2-13 ($Pe_L = 100$) with Fig. 3.2-14 ($n = 1$) shows that Taylor-type molecular transverse mixing does indeed have a beneficial effect upon the quality of macromixing in pressure-flow tubes.

Example 3.2-3.   Macromixing in the MSIM and MSCTR.   Compute the holdback H and the intensity function $\Omega(\alpha)$ for the IM and the MSCTR.

Solution.   For the IM $e(\tau) = \bar{\tau}^{-1}\exp(-\tau/\bar{\tau}) = i(\alpha)$ and $E(\tau) = 1 - \exp(-\tau/\bar{\tau})$. Therefore, from equation 3.2-17, we obtain

$$ H = \frac{\int_0^{\bar{\tau}} E(\tau)\,d\tau}{\bar{\tau}} = e^{-1} $$

and from equation 3.2-20

$$ \Omega(\alpha) = \frac{e(\alpha)}{\bar{\tau}i(\alpha)} = \bar{\tau}^{-1} $$

which is constant as expected.

For the MSCTR, it has been shown in Example 3.2-1 that $e(\tau) = \bar{\tau}^2/2\tau^3$, $I(\alpha) = 1 - \bar{\tau}/4\alpha$ and

$0 < \alpha < \bar{\tau}^2$	$i(\alpha) = \bar{\tau}^{-1}$
$\alpha > \bar{\tau}/2$	$i(\alpha) = \bar{\tau}/4\alpha^2$

Therefore, from equation 3.2-16, we obtain

$$H = 1 - I(\bar{\tau}) = 1/4$$

and from equation 3.2-20

$$\Omega(\alpha) = \frac{2}{\alpha}$$

which has a negative slope. Thus we conclude that laminar mixing in the tubular reactor consists of a combination of channeling and stagnancy.

Example 3.2-4. Characteristics of a MSCTR. Compute H, $\Omega(\hat{\alpha})$, $f(\gamma)\,d\gamma$, $[\gamma]$ for a laminar, power-law MSCTR, and evaluate the first and last quantities for $n = 1/2, 1, 2$.

What can we conclude about RTD, channeling, uniformity of mixing, and extent of mixing as n increases?

Solution. Drawing on Table 3.2-1 and using relationship 3.2-4 $[i(\hat{\alpha}) = 1 - E(\hat{\tau})]$, we find

$$H \equiv \int_{\hat{\tau}_0}^{1} E(\hat{\tau})\,d\hat{\tau} = \int_{\hat{\tau}_0}^{1}\left(1 + \frac{2n}{(3n+1)\hat{\tau}}\right)\left(1 - \frac{n+1}{(3n+1)\hat{\tau}}\right)d\hat{\tau}$$

Using the substitution $u = 1 - (n+1)/[(3n+1)\hat{\tau}]$ reduces the preceding equation to

$$H = \frac{(n+1)}{3n+1}\int u^{2n/(n+1)}(1 + 2n(1-u)/(n+1))(1-u)^{-2}\,du$$

which, when solved, transformed back to the $\hat{\tau}$ domain and evaluated between limits yields

$$H = \left(1 - \frac{n+1}{3n+1}\right)^{(3n+1)/(n+1)}$$

Evaluating H for $n = 1/2, 1$, and 2, we find that $H = 0.217, 0.250$, and 0.271, respectively. Thus one concludes that pseudoplastic behavior ($n < 1$) provides the narrowest RTD and would be "least" mixed from that viewpoint.

By definition (3.2-23),

$$\Omega(\hat{\alpha}) = \frac{e(\hat{\alpha})}{\bar{\tau}i(\hat{\alpha})}$$

Thus, using the functions for $e(\hat{\alpha})$ and $i(\hat{\alpha})$ from Table 3.2-1, we find

$$\Omega(\hat{\alpha}) = \frac{\dfrac{2n}{3n+1}\left[1 - \dfrac{n+1}{(3n+1)\hat{\alpha}}\right]^{(n-1)/(n+1)}}{\bar{\tau}\hat{\alpha}^3\left\{1 - \left(1 + \dfrac{2n}{(3n+1)\hat{\alpha}}\right)\left[1 - \dfrac{n+1}{(3n+1)\hat{\alpha}}\right]^{2n/(n+1)}\right\}}$$

The strain at the tube exit for any streamline is simply $\gamma(r) = |\dot{\gamma}_{rz}(r)|\tau$ since $\dot{\gamma}_{rz}$ is independent of axial position (Example 3.1-3). As a result, the distributions of residence time and strain are the same

$$f(\gamma)\, d\gamma = e(\tau)\, d\tau$$

and when expressed in terms of radial position rather than RT yield

$$f(\gamma)\, d\gamma = 2\hat{v}_x(\hat{r})\hat{r}\, d\hat{r}$$

where

$$\hat{v}_z(\hat{r}) = \frac{3n+1}{n+1}(1 - \hat{r}^{(n+1)/n})$$

and $\hat{v}_x = v_z/\langle v\rangle$. Thus

$$\gamma(r) = |\dot{\gamma}_{rz}(r)|\tau = \left|\frac{dv_z}{dr}\right|\left(\frac{L}{v_z(r)}\right)$$

or, in dimensionless form,

$$\gamma(\hat{r}) = \left(\frac{L}{\text{Ra}}\right)\left|\frac{d\hat{v}_z}{d\hat{r}}\right|\hat{v}_z^{-1}$$

and, as in Example 3.1-3 and as defined in Appendix F, the flow-average strain $[\gamma]$ is found via

$$[\gamma] = \int_0^{\text{Ra}} \gamma(r) f(\gamma)\, d\gamma$$

$$= \int_0^1 2\frac{L}{\text{Ra}}\left|\frac{d\hat{v}_z}{d\hat{r}}\right|\hat{r}\, d\hat{r}$$

$$= \frac{2\dfrac{L}{\text{Ra}}(3n+1)}{2n+1}$$

Evaluating this for a pseudoplastic ($n = 1/2$), Newtonian ($n = 1$), and dilatant ($n = 2$) fluid, we find:

n	$[\gamma]$
1/2	2.50(L/Ra)
1.0	2.67(L/Ra)
2.0	2.80(L/Ra)

The limiting extremes are $[\gamma] \to 2L/\text{Ra}$ as $n \to 0$ and $[\gamma] \to 3L/\text{Ra}$ as $n \to \infty$. We are led to the conclusion that pseudoplasticity works against shear mixing by creating pluglike velocity profiles. This is in accord with the trends observed in the reactor holdback.

> ***Example 3.2-5.   Holdback and Relative Dispersion.***   Verify that equation 3.2-18 is equivalent to the definition of H in equation 3.2-17.

Solution.   Begin with equation 3.2-19 and note that the argument of the absolute value is positive when $\tau \geqslant \bar{\tau}$ and negative when $\tau < \bar{\tau}$; thus,

$$\sigma' \equiv \int_0^\infty |\tau - \bar{\tau}| e(\tau) \, d\tau = -\int_0^{\bar{\tau}} (\tau - \bar{\tau}) e(\tau) \, d\tau + \int_{\bar{\tau}}^\infty (\tau - \bar{\tau}) e(\tau) \, d\tau$$

Addition and subtraction of the first term yields

$$\sigma' = -2\int_0^{\bar{\tau}} (\tau - \bar{\tau}) e(\tau) \, d\tau + \int_0^\infty (\tau - \bar{\tau}) e(\tau) \, d\tau$$

Integration of the second term on the RHS shows it to be identically equal to zero. Integrating the first term by parts shows

$$\sigma' = -2\left[(\tau - \bar{\tau}) E(\tau)|_0^{\bar{\tau}} - \int_0^{\bar{\tau}} E(\tau) \, d\tau \right] = 2\int_0^{\bar{\tau}} E(\tau) \, d\tau$$

Thus

$$H \equiv \frac{\int_0^{\bar{\tau}} E(\tau) \, d\tau}{\bar{\tau}} = \frac{\sigma'}{2\bar{\tau}} = \frac{\text{RD}'}{2}$$

3.3.   MIXING AND CHEMICAL REACTION

In physical mixing processes, such as certain polymer processing operations, composition uniformity is the main objective. It is often irrelevant whether uniformity is achieved by mixing in the direction transverse to flow or in the longitudinal direction (backflow). It does matter when excessively long RT's caused by backflow result in polymer degradation. Degradation, however, is actually a chemical process.

In chemical reactors mixing, per se, is not necessarily desirable. For that reason, it is not always induced intentionally. Witness the streamline vessel versus the agitated vessel discussed in Section 1.3.

To assess the effects of mixing in polymerization reactors, we must distinguish between transverse and longitudinal mixing, and between macromixing and micromixing. An important characteristic of longitudinal macromixing, which we will use extensively in our analysis, is the RTD. A consequence of micromixing, which we will also use, is the phenomenon of backmixing (BM). The former describes the distribution among the ages of fluid elements, and is generally associated with macromixing. The latter describes the intermingling of molecules of different ages (micromixing) made possible by the action of macromixing. Such macromixing (longitudinal) is associated with the third consideration raised in Section 3.1, that is, the juxtaposition of fluid elements of different age groups. Although extensive macromixing is clearly a necessary condition for BM, it is not sufficient. The possibility of achieving good macromixing, but with complete microsegregation (MS) at the molecular level (27), is real, especially in polymeric systems.

Several illustrative examples have been cited previously. Macromixing due to shear strain in laminar-flow streamline reactors could be undesirable. It manifests itself as a RTD. Transverse mixing would then be desirable to oppose the adverse effects of the RTD upon a chemical reaction by restoring to some degree the desirable characteristics of plug-flow. A well-known example of such transverse mixing in the tubular reactor is "Taylor" diffusion. Its counterpart in the extruder would be cross-channel flow. Whereas the former is molecular (microscopic) in nature, the latter is convective (macroscopic).

Longitudinal macromixing, on the other hand, may or may not be desirable, Its presence will certainly enhance the likelihood of BM by bringing fluid elements of varying ages into one another's proximity. Such mixing is induced in agitated vessels with impellers and in drag-flow vessels with reverse pressure flow (backflow), as previously noted.

In summary, macromixing should be avoided only when pluglike flow is feasible. When it is not feasible, macromixing might be induced intentionally, either to diminish stagnancy and channeling, or to promote micromixing if its effect on the reaction in question is beneficial.

3.3.1. Classification of Reactors

Several characteristics of mixing are related qualitatively in Table 3.3-1. For all practical purposes, scale of segregation and RTD may be associated with macromixing (first row), whereas intensity of segregation and BM may be associated with micromixing (second row). On the other hand, our reactor triangle (Fig. 3.3-1), which is a convenient device for bounding and juxtaposing the relative mixing characteristics of various reactors (28), is based on extents of macroscopic and microscopic mixing (first column). Similarly, reactors are traditionally classified according to measures of goodness of mixing based in essence on scale and intensity (second column). Examples include Zwietering's scale for maximum mixedness (20) which will be reviewed subsequently, and our segregation number S (Section 3.2). However, in our attempts to estimate the effects of mixing on the performance of a

Table 3.3-1. Relationship between Mixing Processes and Their Manifestations

Macromixing	Scale of Segregation	RTD
Micromixing	Intensity of Segregation	BM

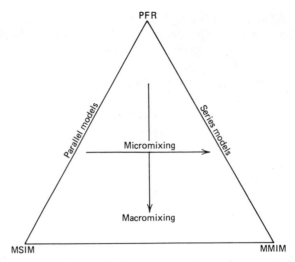

Figure 3.3-1.   The reactor triangle: a classification scheme for mixing in continuous-flow reactors.

reactor by computation, we shall utilize the RTD and the concepts of complete MS and complete BM (column 3).

Following Danckwerts (27) and Zwietering (20), we shall isolate the effects of RTD and BM conceptually for any real reactor in question having a given RTD by associating with it two idealized reactors having the identical RTD. One is the so-called microsegregated reactor (MSR), whose fluid elements satisfy the given RTD but remain completely segregated at the molecular level. In terms of our dimensionless segregation number, $S \gg 1$ for such reactors. The other is the so-called maximally micromixed reactor (MMR) whose fluid elements obey the same RTD but interact on the molecular level as completely as they can, consistent with the given RTD. For such reactors, $S < 1$ as proposed in Definition 3.1-11.

Zwietering (20) defined the state of maximum mixedness for each RTD in terms of the minimum value of Danckwerts' intensity of segregation, J_{min}, compatible with that RTD. Although all vessels are capable of complete microsegregation ($J = 1$), only the IM can achieve complete micromixing ($J_{min} = 0$). Thus, in principle, the scale $0 < J_{min} \leqslant 1$ serves to rank the degree of mixromixing possible in a given vessel and in this respect is analogous to the segregation number S (8).

According to Zwietering, maximum micromixedness obtains when all molecules, irrespective of their age (α), that are destined to exit together (same life expectancy, $\varepsilon \equiv \tau - \alpha$) are also brought together as early as possible (consistent with RTD) after entering the vessel. The question of early micromixedness is related to the second and third considerations expressed in the Section 3.1, that is, early scale reduction of fluid elements and intermingling among all ages. The only value of J that can be computed is J_{min}. Zwietering developed an expression for computing J_{min} from the RTD alone. The interested reader is referred to the literature (20).

On the basis of this discussion, logical choices for the apexes of our reactor triangle are the plug-flow reactor (PFR), the microsegregated ideal mixer (MSIM), and the maximally micromixed ideal mixer (MMIM). The first represents the absence of longitudinal macromixing and therefore micromixing as well. The second and third represent the most complete longitudinal macromixing and the most complete micromixing possible, respectively. Thus the vertical scale of the triangle is associated with degree of longitudinal macromixing [l_{seg}, $\Omega(\alpha)$, etc.] and the horizontal scale with degree of micromixing (S, J_{min}, etc.). We do not associated the vertical scale directly with dispersion (H) of the RTD for reasons that will be given shortly.

By comparing the performance of the MSIM with the PFR, we can isolate the effects of macromixing, and by comparing the MMIM with the MSIM we can isolate the effects of micromixing. This will be done for our model polymerization sequences. All MSR's lie along the left side of the triangle. Each may be viewed for convenience as a composite of many BR's or PFR's in parallel whose ages are determined by the prevailing RTD. Vessels with channeling and stagnancy, which in essence are parallel phenomena, should thus lie on or near the left side as well. The cascade of N maximally micromixed ideal mixers in series, on the other hand, must lie on the right side of the triangle. As N passes from 1 to ∞, it moves from the bottom to the top. The agitated vessel and the streamline vessel with excessive backflow are located along the base of the triangle between the MSIM and MMIM when micromixing is incomplete. Streamline vessels without backflow, such as the CTR, must lie near the left side, depending on their degree of BM, which is small in general.

An interesting example is the MSCTR, whose RTD stems only from its velocity profile, as noted. This vessel has been shown to sustain partial macromixing due to laminar shear [Section 3.1, especially Example 3.1-3]. Since such mixing is essentially longitudinal, this vessel must lie somewhere between the PFR and the MSIM on the left side of our triangle. If BM occurs (dispersion model), its location moves toward the right. If microsegregation prevails and power-law index n increases, we conclude from Example 3.2-4 and Fig. 3.2-14 that longitudinal macromixing improves. Thus as n increases from 0 to ∞, we would expect the vertical location of the MSCTR to move from the top apex (PFR) of our triangle toward the bottom left apex (MSIM). This picture presents us with a dilemma, however. Since laminar-shear mixing in the MSCTR is characterized by channeling (Example 3.3-3), we must interpret channeling as a manifestation of longitudinal macromixing, albeit of "low quality."

To avoid the dilemma of interpreting channeling as a manifestation of macromixing, it is tempting to use for the vertical scale a measure of breadth of RTD dispersion instead (28, 29), such as holdback H. Such a scale would range from 0 for the PRF to 0.368 for the MSIM (Example 3.2-3), and span all H-values for the power-law MSCTR up to 0.296 when $n \to \infty$ (Example 3.2-4). This has at least one additional advantage. It is believed that the detrimental effect of RTD dispersion on chemical reactor performance is generally monotonic, whereas no such simple relationship is known with respect to extent of macromixing. However, it also presents a dilemma of a different kind, since vessels with severe channeling and stagnancy can have RTD's that are more disperse (greater value of H) than that of the IM, which presumably would lie above them in the triangle (27). Thus the appropriate choice of a model vessel for the lower left apex having maximum channeling and stagnancy would become a problem, and the symmetry of the triangle would be destroyed.

For now, suffice it to say that if we interpret the vertical scale of Fig. 3.3-1 as a measure of longitudinal macromixing, we must include as an integral part of that scale its quality as well as its magnitude in terms of slope of the intensity function Ω. The significance of that slope is summarized in Fig. 3.3-2.

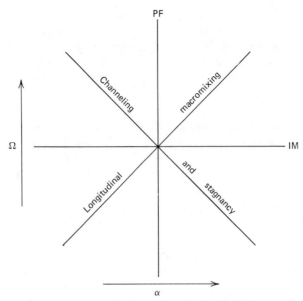

Figure 3.3-2. Significance of the slope of the intensity function $\Omega(\hat{\alpha})$.

To illustrate the utility of the model reactors we have chosen to represent continuous polymerization reactors, we can cite the following examples. Bulk polymerization might be simulated in a pressure-flow pipe reactor by the MSCTR and in an extruder by the PFR or the CDFR, provided, of course, that the precautions to be discussed in Chapter 5 are exercised. The continuous stirred tank reactor (CSTR), on the other hand, might be adequately simulated by the MMIM for solution polymerizations and by the MSIM for suspension polymerizations, if coalescence is neglected. In general, the MSR would appear to be especially appropriate for polymerization reactors in view of the high viscosities and concomitantly low molecular mobilities that can occur.

3.3.2. Reactor Calculations

When designing a continuous reactor for a particular chemical reaction, certain reactor calculations are essential. The computational method employed will depend on the reactor type and the prevailing mixing processes, as discussed in the preceding sections. It will also most certainly depend on the characteristics of the reaction in question. In order to make use of our classification scheme and our dimensionless criteria (Section 3.1) it will be necessary to introduce an additional CT, namely, that for the reaction λ_r. Thus, to achieve maximum micromixing in the presence of chemical reaction, it is reasonable to require in lieu of inequality 3.1-10 that

$$\lambda_{mic} < \lambda_r \qquad (3.3\text{-}1)$$

or, in terms of the Damköhler number of the second kind, that

$$Da_{II} < 1 \qquad (3.3\text{-}2)$$

where $Da_{II} \equiv \lambda_{mic}/\lambda_r$. An application of this criterion is given in Example 3.3-1. A useful identity relating this dimensionless group to the other Damköhler number (first kind) and the segregation number is

$$Da_{II} = S\,Da \qquad (3.3\text{-}3)$$

The more rapid the reaction is (large Da), the more effective micromixing (smaller S) must be. It is possible to deduce two useful limiting effects of mixing in terms of these dimensionless groups, which should apply to all reactor calculations of interest, that is, those pertaining to product distributions as well as to reactor performance. Intuitively, we expect that when conversion is kept low (Da $\ll$ 1) the effects of BM for any given RTD should diminish, leaving RTD as the primary factor. In terms of the dimensionless groups in equation 3.3-3 this condition implies that $Da_{II} \ll S$, which suggests that, regardless of how effective micromixing is (small S) the reaction is sufficiently slow (smaller Da_{II}) to render BM unimportant. Consequently, we would expect the behavior of each MMR to approach that of its corresponding MSR, whose characteristics are determined by its RTD only. Conversely, when conversion is very high (Da $\gg$ 1) the effects of RTD should diminish. The condition $Da_{II} \gg S$ suggests that the reaction is sufficiently fast to negate the effects of RTD imposed by the condition of microsegregation (large S). Consequently, we would expect the behavior of each MSR to approach that of the PFR, regardless of H.

Quantities that are generally used to characterize the outcome of a chemical reaction are conversion and selectivity. For polymerizations we are interested in monomer and initiator conversion, and product characteristics DP, CC, DPD, CCD, and so on, as previously discussed. A quantity that may be used to characterize reactor performance is the specific output rate, introduced in Section 1.3. In addition to their obvious dependence on space time (RT), these quantities also depend on mixing conditions, as characterized by RTD and BM. We shall examine the effects of mixing first on reactor performance and then on product properties.

Toward this end, it will be convenient to use the dimensionless quantity Φ_k/Da_k (for reactant k, say) as a measure of specific output rate (Examples 1.3-2 and 1.3-3), where $Da_k = \lambda_v/\lambda_k$. A steady-state balance equation for any continuous reactor with a single inlet and outlet, neglecting density changes, is (Appendix F)

$$\dot{V}[c_k] - \dot{V}(c_k)_0 = V\langle r_k \rangle \qquad (3.3\text{-}4)$$

and in dimensionless form it becomes

$$-\Phi_k = \frac{\lambda_v}{\lambda_k}\langle \hat{r}_k \rangle \equiv Da_k\langle \hat{r}_k \rangle \qquad (3.3\text{-}5)$$

where $\lambda_k = (c_k)_0/(r_k)_0$ and $\hat{r}_k$ is dimensionless reaction rate (Appendix E). Thus we see the origin of the quantity in question and its equivalent $\langle \hat{r}_k \rangle$.

In Section 1.3, all reactor calculations were classified into two basic methods. This classification applies whether we are interested in computing conversion, specific output rate, or selectivity. In the first method, flow-average output $[c_k]$ is computed directly (equation 1.3-9) from reactor exit conditions, which can be obtained from solutions of the transport balances for streamline reactors. In the second method, volume-average rate $\langle r_k \rangle$ is computed (equation 1.3-10). Both methods are related by a material balance, such as in equation 3.3-4. When our reactor is treated as microsegregated (MS), we can calculate $[c_k]$ for any component k as required in the first method by

$$[c_k] = \int_0^\infty c_k(\tau) e(\tau)\, d\tau \qquad (3.3\text{-}6)$$

in lieu of equation 1.3-9, where $c_k(\tau)$ is the solution of the batch rate equation for substance k ($dc_k/dt = r_k$) evaluated at the exit. No transport equations are required. Similarly, for MSR's, we can calculate $\langle r_k \rangle$ by the following equation in lieu of 1.3-10

$$\langle r_k \rangle = \int_0^\infty r_k(\alpha) i(\alpha)\, d\alpha \qquad (3.3\text{-}7)$$

where $r_k(\alpha)$ is the rate function for substance k into which has been substituted the solutions $c_c(t)$ of the required rate equations; thus $r_k[c_c(t)]$. As expected, equations 3.3-6 and 3.3-7 are related via a material balance (Example 3.3-2).

To examine the effects of RTD alone, it will be necessary to assume complete MS ($S \gg 1$). All desired calculations can be performed for any MSR [left side or lower left apex of the reactor (triangle)] knowing only its RTD. Obviously, such a calculation can also be made for the PFR (upper apex). To examine the effects of BM alone, we must fix RTD and assume maximum micromixedness. This is most conveniently done for the MMIM (lower right apex). We shall therefore restrict our analyses to three model reactors: the PFR, the MMIM, and various MSR's. They will provide us with bounds on reactor performance and product properties for comparison with actual reactors. For certain other reactors whose mixing states lie between the MS reactor and the MMIM and whose characteristics are deterministic (e.g., the dispersion model), such calculations are also possible. Zwietering has proposed a computational method for reactors in the state of maximum mixedness, which requires knowledge of the RTD only. The interested reader is referred to the literature (20).

Conversion of arbitrary reactant c_k has been computed for half-, first- and second-order reactions in some model reactors under steady-state conditions, neglecting density variations. The results, which are applicable to many polymerizations, are listed in Table 3.3-2, together with corresponding values for holdback H. The equations in the table were also plotted in Figs. 3.3-3 and 3.3-4 to emphasize the effects of RTD and BM on reactor performance. The reader will recall that addition polymerizations are frequently first-order or pseudo first-order. An example of the latter for which k represents monomer is conventional ($\alpha_k < 1$) chain-addition polymerization. Random polymerizations are sometimes second order. In this case, k might represent a functional group or by-product. Sample computations based on the results in Table 3.3-2 are shown in Example 3.3-3.

Example 3.3-1. Diffusion-Controlled Condensation Polymerization. In the catalyzed interchange formation of polyethylene terephthalate (PET), the polymerization is reversible and high molecular weights can only be achieved by removing by-prod-

Table 3.3-2. Conversion and Holdback for Various Reactor Models and Reaction Orders

n	λ_k	Φ_k PFR	Φ_k MSCTR (Newtonian)	Φ_k MSIM	Φ_k MMIM
1/2	$(c_k)_o^{1/2}/k$	$Da - (Da/2)^2$ $(Da \leq 2)$	$Da - 3Da^2/16 + (Da^2/8)\ln(Da/4)$	$Da - (Da^2/2)[1 - \exp(-2/Da)]$	$(Da/2)(4 + Da^2)^{1/2} - Da^2/2$
1	$1/k$	$1 - \exp(-Da)$	$1 - (1 - Da/2)\exp(-Da/2) - (Da/2)^2 E_i(Da/2)$	$Da/(1 + Da)$	$Da/(1 + Da)$
2	$1/k(c_k)_o$	$Da/(1 + Da)$	$Da - (Da^2/2)\ln[(2 + Da)/Da]$	$1 - Da^{-1}(\exp Da^{-1})E_1(Da^{-1})$	$1 - [(1 + 4Da)^{1/2} - 1]/2Da$
H		0	$1/4$	e^{-1}	

where $\quad r_k = -kc_k^n \quad$ and $\quad E_1(z) \equiv \int_1^\infty \zeta^{-1}\exp(-z\zeta)d\zeta \quad$ for $\quad z > 0$.

uct ethylene glycol (m_0 in our random propagation scheme). Owing to rapid reaction (small λ_r), the rate can be controlled by the transport rate of glycol out. Consequently, the reaction is frequently conducted in thin films. Using the values $\lambda_r = 5$ s (Example 2.2-6) and $D = 1.66 \times 10^{-4}$ cm^2/s (30), estimate the maximum film thickness (l_{seg}) beyond which diffusion would be expected to control the reaction rate.

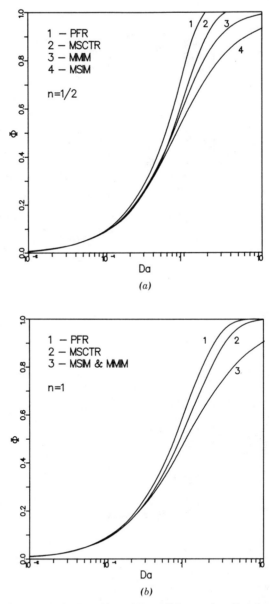

Figure 3.3-3. Conversion versus the logarithm of Damköhler number (Da) for various model reactor types (PFR, MSCTR, MSIM, MMIM) and reaction orders: 1/2 (*a*), 1 (*b*), and 2 (*c*).

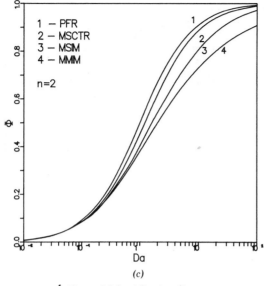

(c)

Figure 3.3-3. (*Continued*).

Solution. To avoid diffusion control, we would require that $\lambda_{mic} < \lambda_r$ where $\lambda_{mic} = l_{seg}^2/D$. Substituting λ_r and D into the inequality yields a maximum thickness $l_{seg} = 0.29$ mm which is in general agreement with experimental results (31, 32).

Example 3.3-2. Output Rate and Specific Output Rate for MS Reactors. Balance equation 3.3-4 relates output rate (LHS) and specific output rate $\langle r_k \rangle$ of

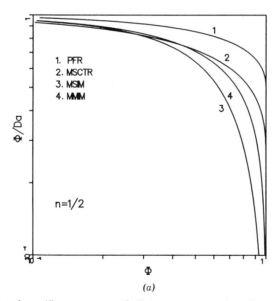

(a)

Figure 3.3-4. Reduced specific output rate ($\Phi/$Da) versus conversion (Φ) for various reactor types (PFR, MSCTR, MSIM, MMIM) and reaction orders: $1/2$ (*a*), 1 (*b*), and 2 (*c*).

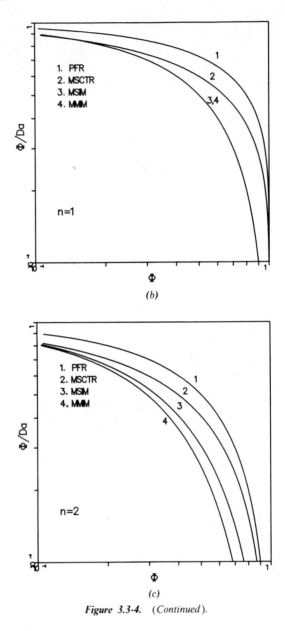

Figure 3.3-4. (*Continued*).

reactant k. Prove the validity of the following equation, which is the equivalent relationship to 3.3-4 for MS reactors:

$$\dot{V}\int_0^\infty c_k(\alpha)e(\alpha)\,d\alpha - \dot{V}(c_k)_0 = V\int_0^\infty r_k(\alpha)i(\alpha)\,d\alpha$$

Solution. After substituting equation 3.2-4 into the RHS of the preceding equation

$$V \int_0^\infty i(\alpha) \frac{dc_k}{d\alpha} d\alpha = \dot{V} \int_0^\infty [1 - E(\alpha)] \, dc_k$$

and integrating by parts, we obtain

$$\left\{ \dot{V}[1 - E(\alpha)] c_k + \dot{V} \int c_k e(\alpha) \, d\alpha \right\}_0^\infty = \dot{V} \int_0^\infty c_k(\alpha) e(\alpha) \, d\alpha - \dot{V}(c_k)_0$$

which is identical to the LHS of the equation in question.

Example 3.3-3. *Effects of RTD and BM.* Calculate and compare the values of conversion Φ_k for all reaction-reactor combinations in Table 3.3-2 under identical circumstances Da = 1. What do you conclude about the separate effects of RTD and BM on reactor performance?

Solution. By substituting Da = 1 into each equation in Table 3.3-2, the results shown in the following chart are obtained for Φ_k

n	PFR	MSCTR	MSIM	MMIM
1/2	0.750	0.639	0.508	0.618
1	0.632	0.557	0.500	0.500
2	0.500	0.451	0.404	0.382

Corresponding values of H for these reactors, which are listed in Table 3.3-2, indicate that dispersion of RTD increases from left to right as listed in the table. Therefore, by comparing the MSR's with the PFR we conclude that RTD reduces reactor performance for all reaction orders. By comparing the MMIM with the MSIM, we conclude that BM further reduces performance when $n > 1$, but it improves performance when $n < 1$.

Example 3.3-4. *Conversion in a Series of MMIM's.* Use equation 3.3-4 to find the conversion of a first-order reaction in N maximally micromixed ideal mixers in series. Show that in the limit as $N \to \infty$ the result agrees with that expected for a PFR.

Solution. Generalizing 3.3-4 and dropping parentheses (complete BM) yields

$$c_j - c_{j-1} = \frac{V_j}{\dot{V}} k c_j = \frac{\mathrm{Da}}{N} c_j$$

or

$$c_j = \frac{c_{j-1}}{1 + \dfrac{\mathrm{Da}}{N}}$$

Conversion is given by

$$\Phi = 1 - \frac{c_N}{c_0}$$

so

$$\Phi = 1 - \left(1 + \frac{Da}{N}\right)^{-N}$$

As $N \to \infty$, $Da/N \to 0$. Using the approximation (Appendix C)

$$(1 + Da/N)^{-N} \cong \exp(-Da)$$

we conclude that $\Phi = 1 - \exp(-Da)$ as $N \to \infty$.

Example 3.3-6.   Output Calculations.   Verify relation 1.3-11 using equations 3.3-6 and 3.3-7 for a second-order reaction conducted in a Newtonian MSCTR.

Solution.   For the MSCTR

$$e(\hat{\tau}) \, d\hat{\tau} = \begin{cases} 0 & \hat{\tau} < \dfrac{1}{2} \\[2ex] \dfrac{d\hat{\tau}}{2\hat{\tau}^3} & \hat{\tau} \geqslant \dfrac{1}{2} \end{cases}$$

$$i(\hat{\alpha}) \, d\hat{\alpha} = \begin{cases} d\hat{\alpha} & \hat{\alpha} < \dfrac{1}{2} \\[2ex] \dfrac{d\hat{\alpha}}{4\hat{\alpha}^2} & \hat{\alpha} \geqslant \dfrac{1}{2} \end{cases}$$

If equation 1.3-11 is written in dimensionless form:

$$\hat{O} = -\Delta[\hat{c}] = -Da\langle \hat{r} \rangle$$

From equation 3.3-6

$$-\Delta[\hat{c}] = 1 - \int_{1/2}^{\infty} \hat{c}(\hat{\tau}) e(\hat{\tau}) \, d\hat{\tau}$$

$$= 1 - \int_{1/2}^{\infty} \frac{d\hat{\tau}}{(1 + Da\hat{\tau})2\hat{\tau}^3} \tag{i}$$

while from equation 3.3-7

$$-Da\langle \hat{r} \rangle = Da \int_0^{\infty} \hat{c}^2(\hat{\alpha}) i(\hat{\alpha}) \, d\hat{\alpha} = Da\left\{ \int_0^{1/2} \frac{d\hat{\alpha}}{(1 + Da\hat{\alpha})^2} + \int_{1/2}^{\infty} \frac{d\hat{\alpha}}{(1 + Da\hat{\alpha})^2 4\hat{\alpha}^2} \right\} \tag{ii}$$

Integrating (i) by parts once and integrating the first term of (ii) gives the identical result:

$$\hat{O} = \left(1 + \frac{\text{Da}}{2}\right)^{-1} - 1 - \text{Da} \int_{1/2}^{\infty} \frac{d\hat{\tau}}{(1 + \text{Da}\hat{\tau})^2 (4\hat{\tau}^2)}$$

This may be further integrated to yield the result in Table 3.3-2.

3.4. EFFECT ON REACTOR PERFORMANCE

In the previous section, we concluded that dimensionless specific output rate (Φ_k/Da) is a suitable index for reactor performance. In Section 1.3, the effects of RT, staging, and recycle on specific output were briefly discussed. We shall now examine in some detail the effects of RTD and BM separately.

3.4.1 *Some Generalizations*

It has been stated without proof that the specific output rate of a reactor showed decline with increasing dispersion of its RTD (e.g., as measured by H). This conclusion may be deduced for nonautoacceleratory reactions by comparing areas under reaction rate curves, such as the one in Fig. 3.4-1, and it is borne out in Examples 3.3-3 and 3.4-1. All results in the former example may be interpreted in terms of specific output rate (Φ_k/Da) since all were computed for a fixed value of Da. Similarly, by comparing areas in Fig. 3.4-2 (Example 3.4-2), we conclude that complete backmixing generally lowers specific output rate compared to PF. This was also concluded in Section 1.3. It is important at this point to recall that increased RTD dispersion does not necessarily signal improved mixing. Thus, while the reactor triangle attempts to classify reactors according to their mixing characteristics, it does not rank their performance commensurately.

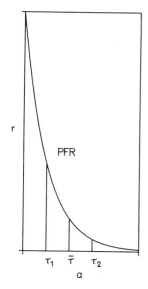

Figure 3.4-1. Reaction rate function r versus age α in a plug flow reactor (PFR).

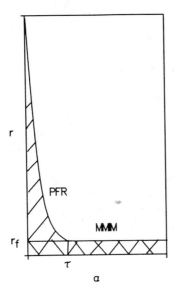

r

PFR

MMIM

r_f

τ

α

Figure 3.4-2. Reaction rate function r versus age α in a plug flow reactor (PFR) and in a maximally mixed ideal mixer (MMIM)

Concerning the effect of BM on specific output rate, we conclude from Examples 3.3-3 and 3.4-3 that it depends on the nature of the reaction (n, α_K). It can be detrimental when $n > 1$ and beneficial when $n < 1$. This conclusion is readily deduced for IM's and simple rate functions of the type

$$r = kc^n \tag{3.4-1}$$

using a general theorem for inequalities among convex and concave functions and their weighted means (33). Referring to Fig. 3.4-3, the function $r(c)$ is convex when $n > 1$ and concave when $n < 1$. Specifically, the reaction rate for a MSIM is $\langle r(c) \rangle$, whereas for a MMIM it is $r(\langle c \rangle)$. Let us now apply the theorem to convex functions. Since the function of the mean value of the argument cannot exceed the mean value of the function, that is, $f(\bar{x}) \leqslant \bar{f}(x)$, we conclude that when $n > 1$

$$r(\langle c \rangle) \leqslant \langle r(c) \rangle \tag{3.4-2}$$

which states that the MSIM is superior to the MMIM. Conversely, from the theorem applied to concave functions, we conclude that the MMIM is superior to the MSIM when $n < 1$

$$\langle r(c) \rangle \leqslant r(\langle c \rangle) \tag{3.4-3}$$

Proof of the theorem uses arguments involving arcs and chords similar to those used in Example 3.4-3.

These conclusions are readily apparent from the figures in Section 3.3-2. Conversion, Φ (reaction efficiency), versus Da has been plotted in Fig. 3.3-3 and specific output rate, Φ/Da (reactor efficiency), versus Φ in Fig. 3.3-4. Comparing the MMIM and the MSIM for half-, first- and second-order kinetics shows clearly that

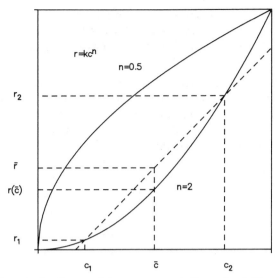

Figure 3.4-3. Reaction rate function $r = kc^n$ versus concentration c for half- and second-order reactions ($n = 0.5$ and $n = 2$, respectively).

BM leads to higher efficiencies of both kinds for half-order kinetics. There is no difference between the MMIM and the MSIM for first-order kinetics, as expected, whereas MS is superior for second-order kinetics. Concerning the effects of RTD alone, we note that the MSCTR with its narrower RTD than the MSIM gives better performance for all but the lowest conversions, where the two are essentially identical.

Examination of Fig. 3.3-3 reveals that the conversions do not differ appreciably from one another at any given value of Da, regardless of reaction order or reactor type. To achieve a high conversion, however, can require a much larger Da for reactors with RTD and BM than for PF, as vividly illustrated in Fig. 3.3-4. This signifies a larger space time λ_v, (i.e., reactor volume) and is a consequence of inferior reactor efficiency. While it is not apparent from the figures, these effects are seen to be most profound for second-order kinetics. Therefore, we expect that mixing will cause the greatest variation in random polymerizations.

Although we expect MS reactors to approach PF reactors in performance when Da is large (Table 3.4-1), it would appear from Figs. 3.3-3 and 3.3-4 that for small values of Da (low Φ), all reactors give the same performance. The trend at high conversions is confirmed by inspection of the figures for all reactions except half-order. The reason for the apparent exception is that this reaction reaches $\Phi = 1$ at finite values of Da in contrast to first- and second-order reactions, which converge assymptotically as Da $\rightarrow \infty$. The trend at low conversions can be verified analytically. One may show, for instance, that in all cases $\Phi \rightarrow$ Da as Da $\rightarrow 0$. Some convergence calculations are demonstrated in Example 3.4-4. The expressions for the MSIM and MMIM converge most rapidly at small values of Da, consistent with the observation that the effects of BM are eliminated at short RT's. In Section 3.5, we shall see, however, that differences in polymer properties persist among reactors with different RTD's at low conversions.

A summary of general conclusions concerning the effects of RTD and BM on reactor performance is shown in Table 3.4-1. Verification of the limiting cases Da → 0 and Da → 1 applied to Table 3.3-2 is left as an exercise for the reader.

Example 3.4-1.   PF versus RTD.   Figure 3.4-1 shows reaction rate as a function of age α for a nonautoaccelerating reaction. Age can signify time spent in a BR or residence time in a continuous reactor. Consider a reactor whose volume elements are all of uniform final RT's (τ) and another whose elements are distributed among two final RT's, τ_1 and τ_2, such that their respective volume fractions are e_1 and e_2 and their mean residence time $\bar{\tau}$ is identical to that of the first reactor ($\bar{\tau} = \tau$). Prove that the volume-specific output (Section 1.3) of the vessel with the RTD must be smaller than that of the one without a RTD. Since reaction times are identical for both vessels, this also proves that the specific output rate must also be smaller.

Solution.   The problem reduces to proving the following inequality between the specific outputs of the first reactor (LHS) and second reactor (RHS)

$$\int_0^{\bar{\tau}} r\,d\alpha > e_1 \int_0^{\tau_1} r\,d\alpha + e_2 \int_0^{\tau_2} r\,d\alpha$$

Table 3.4-1.   Summary of Mixing Effects on Reactor Performance for Non-autoacceleratory Reactions

RT

$$\Phi_k/Da \downarrow \quad as \quad Da \uparrow$$

RTD (H) versus PF

$$\Phi_k/Da \downarrow \quad as \quad H \uparrow$$

BM versus PF

$$\Phi_k/Da \downarrow \quad as \quad BM \uparrow$$

MS (S >> 1) versus MM (S << 1)(identical RTD)

| when n > 1 | MSR > MMR | (Φ/Da greater for MSR) |
| when n < 1 | MMR > MSR | (Φ/Da greater for MMR) |

Limiting cases

| MMR ⟶ MSR | when Da → 0 | (BM disappears) |
| MSR ⟶ PFR | when Da → ∞ | (RTD disappears) |

where $\tau_1 < \bar{\tau} < \tau_2$, $e_1 + e_2 = 1$, and $e_1\tau_1 + e_2\tau_2 = \bar{\tau}$. The proof for the special case $e_1 = e_2$ is obvious in terms of areas under the rate curve (Fig. 3.4-1). Clearly the sum of the areas for periods 0 to τ_1 and 0 to τ_2 is less than twice the area for the period 0 to $\bar{\tau}$. Concerning the general case, combining the last two equations yields

$$e_1(\bar{\tau} - \tau_1) = e_2(\tau_2 - \bar{\tau})$$

Following rearrangement of the RHS of the inequality

$$\int_0^{\bar{\tau}} r\, d\alpha > \int_0^{\bar{\tau}} r\, d\alpha - \left(e_1 \int_{\tau_1}^{\bar{\tau}} r\, d\alpha - e_2 \int_{\bar{\tau}}^{\tau_2} r\, d\alpha \right)$$

the problem is seen to be equivalent to proving that the expression in brackets is positive. After rewriting this expression in terms of the mean values $\bar{r}_1$ and $\bar{r}_2$ for the intervals $(\tau_1, \bar{\tau})$ and $(\bar{\tau}, \tau_2)$, respectively, and substituting our first equation into the result, we obtain

$$e_1 \bar{r}_1 (\bar{\tau} - \tau_1) - e_2 \bar{r}_2 (\tau_2 - \bar{\tau}) = (\bar{r}_1 - \bar{r}_2) \bar{e}_1 (\bar{\tau} - \tau_1) > 0$$

providing, of course, that $\bar{r}_1 > \bar{r}_2$, that is, that the reaction is not autoacceleratory. This completes the proof.

Example 3.4-2. BM versus PF. Figure 3.4-2 shows reaction rate versus age α for a nonautoaccelerating reaction, where age represents either time spent in a BR or residence time in a continuous reactor. Consider a batch or continuous PF vessel with uniform residence time τ, and a continuous IM with mean residence time $\bar{\tau}$. Prove that when volume-specific outputs are identical, $\bar{\tau} > \tau$. This proves simultaneously that the specific output rate of the IM is smaller than that of the PFR.

Solution. Neglecting density changes, the specific output of the PFR is

$$\int_0^\tau r\, d\alpha \equiv \Delta c$$

and for the IM it is

$$\bar{r}\bar{\tau} \equiv \Delta c$$

where Δc represents outlet minus inlet concentration of the same arbitrary substance. For identical inlet and outlet concentrations, it is clear that the effluent of concentration c_f in the PFR must be identical to the mean concentration $\bar{c}$ in the IM. Consequently $\bar{r} = r(\bar{c})$ must equal r_f as shown in Fig. 3.4-2. Since the preceding equations require that the hatched areas in the figure are equal, it follows that $\bar{\tau} > \tau$.

Example 3.4-3. MS versus BM. Consider the convex rate function ($r = kc^n$ with $n > 1$) illustrated in Fig. 3.4-3. Prove that

$$\bar{r} > r(\bar{c})$$

where

$$\bar{r} \equiv i_1 r_1 + i_2 r_2$$
$$\bar{c} \equiv i_1 c_1 + i_2 c_2$$

and i_1 and i_2 are volume fractions of the vessel contents subject to the constraint $1 = i_1 + i_2$.

Solution.   The problem is equivalent to proving that $\bar{r}$ lies on the chord shown in Fig. 3.4-3. We combine the first and third equations,

$$i_1(\bar{r} - r_1) = i_2(r_2 - \bar{r})$$

then the second and third,

$$i_1(\bar{c} - c_1) = i_2(c_2 - \bar{c})$$

and then we combine the results

$$\frac{\bar{r} - r_1}{\bar{c} - c_1} = \frac{r_2 - \bar{r}}{c_2 - \bar{c}}$$

which completes the proof.

From this we conclude that a microsegregated vessel gives a higher specific output rate for convex reactions ($n > 1$) than one that is micromixed and has the same RTD.

Example 3.4-4.   *Limiting Conversion in Reactors with BM or RTD.*   Show that in general, as Da $\to 0$, $\Phi \to$ Da.

Solution.

$$\Phi_m \equiv -\frac{\Delta c_m}{(c_m)_0}$$

$$-\Delta c_m = \int r \, dt \approx (r)_0 \Delta t$$

$$\Phi_m \approx \frac{(r)_0}{(c_m)_0} \Delta t$$

but

$$\lambda_r = \frac{(c_m)_0}{(r)_0}$$

$$\lambda_v = \Delta t$$

so

$$\Phi_m \cong \frac{\lambda_v}{\lambda_r} = \text{Da}$$

This confirms the utility of CT's as a means of uniting the behavior of reactions of various reaction order.

3.4.2. Model Polymerizations

The rate functions describing our model step polymerizations ($\lambda_p \sim \lambda_r$) are second order ($r = kc_p^2$) for simple irreversible random propagation

$$x \geq 1, y \geq 1 \qquad\qquad m_x + m_y \xrightarrow{k} m_{x+y} \qquad\qquad (2.2\text{-}12)$$

and first order [$r = k(c_0)_0 c_m$] for ideal step-addition,

$$x \geq 1 \qquad\qquad m_x + m \xrightarrow{k} m_{x+1} \qquad\qquad (2.3\text{-}1)$$

as they are ($r \cong k_{ap}(c_0)_0^{1/2} c_m$) for "very conventional" ($\alpha_K \ll 1$) free-radical chain-addition polymerizations.

$$m_0 + m \xrightarrow{k_i} m_1^* \qquad\qquad (1.10\text{-}5)$$

$$x \geq 1 \qquad\qquad m_x^* + m \xrightarrow{k_p} m_{x+1}^* \qquad\qquad (1.10\text{-}6)$$

$$y \geq 1 \qquad\qquad m_x^* + m_y^* \xrightarrow{k_t} \text{products} \qquad\qquad (1.10\text{-}8)$$

Thus, the analyses and results discussed earlier apply to these systems (Example 3.4-5). For extreme D-E reactions ($\alpha_K \gg 1$), the rate function for monomer consumption is half-order (also discussed) but with respect to initiator [$r \simeq k_{ap}(c_m)_0 c_0^{1/2}$], not monomer. For intermediate cases, consumption of both monomer and initiator must be taken into account. The result is a rate function that is still kinetically simple (Appendix E) $r = k_{ap} c_m c_0^{1/2}$, whereas the overall reaction is no longer stoichiometrically simple, which means that separate rate equations are required for monomer and initiator.

For free-radical polymerizations in a MMIM, the steady-state rate equations for monomer and initiator are, neglecting density changes and applying the LCA:

$$\frac{-\left[c_m - (c_m)_0\right]}{\lambda_v} \underset{\text{LCA}}{\cong} r_p = k_p c_m c^* \qquad\qquad (3.4\text{-}4)$$

and

$$\frac{-\left[c_0 - (c_0)_0\right]}{\lambda_v} = k_d c_0 \qquad\qquad (3.4\text{-}5)$$

If the QSSA is valid as well (Section 1.11), then the concentration of intermediates c^* is given by the usual expression, which is the solution of the following population balance:

$$\frac{c^*}{\lambda_v} \underset{\text{QSSA}}{\cong} 0 = 2k_d c_0 - k_t (c^*)^2 \qquad\qquad (3.4\text{-}6)$$

Consequently, our equation for specific output rate in dimensionless form is

$$\Phi Da_m^{-1} = (1 + Da_m \alpha_K)^{-1/2}(1 - \Phi) \equiv \hat{r} \qquad (3.4\text{-}7)$$

where Da_m is the Damköhler number of the first kind for monomer.

$$Da_m \equiv \frac{\lambda_v}{\lambda_m} \qquad (3.4\text{-}8)$$

Parameters α_K and λ_m have been defined earlier (Table 2.7-1). Graphs of the RHS of equation 3.4-7 have been plotted in Fig. 3.4-4 for $Da_m = 1$ and for various values of α_K (curves 1–3). The decline in specific output rate with increasing rate of initiator consumption (α_K) is apparent.

The conversion behavior of model polymerization sequences might be expected to follow the trends displayed by simple nth order kinetics as detailed in the first part of this section, and illustrated in Figs. 3.3-3 and 3.3-4. Since idealized step- and chain-addition polymerizations are pseudo first-order with respect to monomer conversion, one would expect BM to have little effect on efficiency. Figures 3.4-5 and 3.4-6 verify this hypothesis. For a given value of Da, the MSIM and MMIM produce the same conversion. Both are lower than that for the PFR and the separation widens as higher conversions are pursued. The MSCTR with a narrower RTD than the MSIM shows performance that is closer to that of the PFR and is thus consistent with the trends for first-order kinetics shown in Fig. 3.3-3. Also shown is the conversion behavior of a reactor with stagnancy as simulated by parallel MMIM's in which 25 percent of the flow rate is diverted into a vessel with 10 times the average RT of the other. It can be seen that this clearly has a drastic

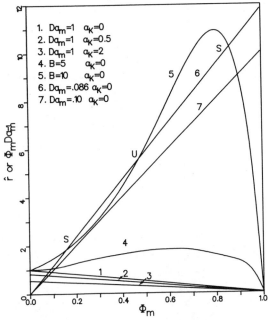

Figure 3.4-4. Dimensionless specific output rate ($\hat{r}$ or ΦDa_m^{-1}) versus conversion (Φ) for various values of Damköhler number (Da_m), kinetic parameter α_K, and gel-effect (G-E) parameter B.

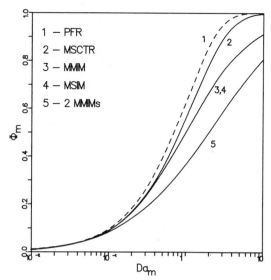

Figure 3.4-5. Conversion (Φ) versus Damköhler number (Da_m) for conventional chain-addition ($\alpha_K \ll 1$) polymerization in various reactors.

effect in reducing the efficiency. This is an instance where the RTD has been broadened ($n = 0.565$) in a nonuniform way.

The random polymerization shown in Fig. 3.4-7 follows the trends expected for a second-order reaction. The RTD of the IM's lowers the efficiency relative to the PFR, but BM has an added deleterious effect. This has already been discussed and is predicted from the analysis in Example 3.4-3.

Although the effect of mixing on the conversion of idealized polymerizations is readily interpreted in the context of simple first- and second-order kinetic schemes, some interesting deviations become apparent when more realistic conditions are

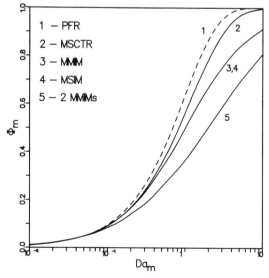

Figure 3.4-6. Conversion (Φ) versus Damköhler number (Da) for Poisson step-addition ($a_K \gg 1$) polymerization in various reactors.

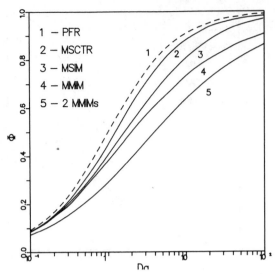

Figure 3.4-7. Conversion (Φ) versus Damköhler number (Da_m) for model irreversible random polymerization in various reactors.

imposed. Some of the variations are manifested in the polymer properties and will be discussed later. Of particular importance to this section, however, is the role of chain initiation in both step- and chain-addition polymerizations. The model reaction schemes neglect initiator variations and thus reduce to simple single-component kinetic expressions. For such systems, the PFR will always offer superior specific output rate and efficiency. When initiator variations are introduced, the notion of monomer conversion as a measure of efficiency becomes inadequate. Selectivity and yield (See Section 1.3) are more appropriate considerations when two or more reactants are present in a complex reaction scheme. Addition polymerizations are both concurrent and consecutive in nature and their response to mixing effects may vary depending on kinetic parameters. It is known, for instance, that BM can improve selectivity and yield over the PFR for certain parallel reactions (34).

Consider, first, chain-addition polymerization. When initiation is slow, the reaction may be considered to be first-order. When initiation is rapid compared to propagation in a BR or PFR, the reaction may stop short of complete monomer conversion as a result of dead-ending (Section 2.8). In these cases one might suspect that BM could be used to increase the selectivity of monomer over initiator and, as a result, the MMIM could show an advantage over the PFR and the MSIM. Figure 3.4-8 bears out this suspicion. BM eliminates the phenomenon of D-E, allowing for complete monomer conversion even when $\alpha_K > 1$. As a result, the specific output rate exceeds the value for the PFR at high conversions.

Step polymerizations also benefit from BM over certain ranges of conversion. Recall that rate of monomer conversion is proportional to the concentration of polymer. Slow rates of initiation prevent this concentration from reaching its largest value early in the course of polymerization. Therefore, although c_m decreases monotonically, c_p increases monotonically, and the potential exists for a maximum in the rate. Such autoacceleration is pseudo-catalytic. Figure 3.4-9 shows that

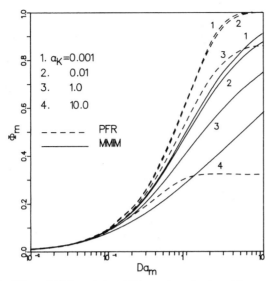

Figure 3.4-8. Effect of BM on chain-addition polymerization with α_K as parameter.

decreasing values of a_K give rise to a maximum in the specific output rate, which occurs at correspondingly higher values of conversion.

BM might be used advantageously in cases of low a_K to increase the number of active chains present and thus to boost the rate. Figure 3.4-10 shows just such an effect. In the initial stages of reaction where the PFR has small values of c_p and therefore low rates, BM offers superior efficiency of output. Although this advantage is ultimately lost at higher conversions, results presented in Section 3.5.2 show that the effect of BM on product properties remains constant.

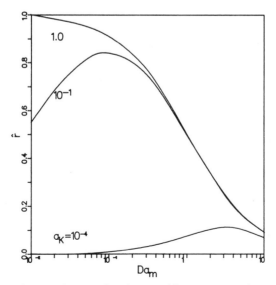

Figure 3.4-9. Dimensionless reaction rate function (specific output rate) for step-addition polymerization in a PFR with $a_K = k_i/k_p$ as parameter.

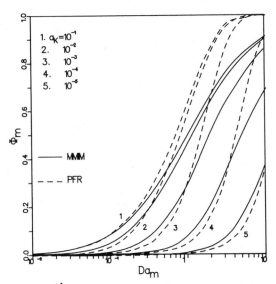

Figure 3.4-10. Effect of BM on step-addition polymerization with a_K as parameter.

The behavior of reversible random polymerization shows little deviation of note with respect to BM effects. Ultimate conversion is of course reduced as K_{eq} decreases but the MMIM always gives poorer performance than the MSIM (Fig. 3.4-11). Both are inferior to the PFR from the standpoint of reactor efficiency.

Cintron (21, 26) examined the behavior of ideal step-addition and random polymerizations in MSCTR's with laminar power-law flow profiles. A computational note should be made with respect to RTD functions for power-law fluids. When the

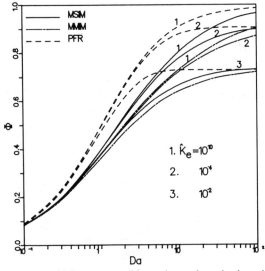

Figure 3.4-11. Effect of BM and RTD on reversible random polymerization with K_e as parameter.

power-law index is less than one, $e(\tau)$ is undefined at the minimum residence time $\hat{\tau}_0$. Quadrature techniques can give unreliable results when evaluating equation 3.3-6. One may try to find limiting behavior as the lower RT is approached (which was Cintron's method) or equations 3.3-5 and 3.3-7 may be used. Since $i(\hat{\alpha})$ is defined for all $\hat{\alpha}$, the latter technique eliminates the discontinuity at the lower limit of integration.

Cintron found that conversion-dependence on power-law index was confined to a small range of values of the index n for both kinetic mechanisms. Figure 3.4-12 shows that conversion reaches its upper and lower limiting values at $n = 10$ and 0.10, respectively. Smaller values of n give rise to larger conversions as plug-flow behavior is approximated. Note that each point along either line in the figure represents output conditions at a uniform RT, as well as a uniform $\bar{x}_N$. This is not the same as fixing the kinetic parameters and varying the RT (i.e., tracking the response to variations in Da), but is instead a variation in kinetics to maintain constant values of the polymer product properties.

Example 3.4-5. ***The Continuous Tubular Reactor.*** A continuous tubular reactor (CTR) is to be used to produce an addition polymer in the liquid phase (melt or solution). Estimate monomer conversion when Da = 2, using the following three alternative models for the laminar-flow CTR: (a) PFR; (b) MSCTR; (c) dispersion model for Taylor "axial" diffusion with $Pe_L = 10$.

Solution. Assume that rate function r is first-order or pseudo first-order. From Table 3.3-2 with $n = 1$, we obtain for the PFR

$$\Phi = 1 - \exp(-Da) = 0.865$$

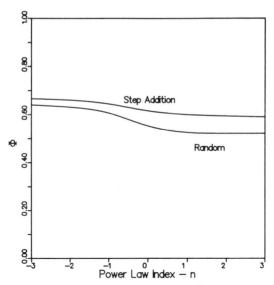

Figure 3.4-12. Effect of RTD on ideal step-addition and random polymerizations in laminar flow MSCTR's with power-law index n as parameter.

and for the MSCTR

$$\Phi = 1 - \left(1 - \frac{Da}{2}\right)\exp\left(-\frac{Da}{2}\right) - \left(\frac{Da}{2}\right)^2 E_i\frac{Da}{2} = 0.781$$

where $E_i(1) = 0.2194$. Finally, from equation 1.4-9 for the dispersion model

$$\Phi = 1 - \hat{c}$$

$$= 1 - \frac{A}{\left\{\left(\dfrac{1+A}{2}\right)^2\exp\left[\dfrac{-Pe_L(1-A)}{2}\right] - \left(\dfrac{1-A}{2}\right)^2\exp\left[\dfrac{-Pe_L(1+A)}{2}\right]\right\}}$$

$$= 0.823$$

where

$$A \equiv \left(1 + \frac{Da}{Pe_L}\right)^{1/2} = 1.342$$

From these results we confirm our earlier conclusions about the detrimental effects on reactor performance ($\Phi = 0.781$ vs 0.865) of a RTD with microsegregation, which is presumed to prevail in streamline polymerization reactors having high reaction viscosities (e.g., bulk polymerizations), and about the potentially beneficial effect ($\Phi = 0.823$ vs 0.781) of transverse diffusion if it were to occur. It is also noteworthy that the aforementioned RTD is the result of externally pressurized flow.

Example 3.4-6. Continuous Free-Radical Bulk Polymerizations in Well-Mixed Vessels.   Assuming that we can achieve complete microscopic mixing in a continuous-flow vessel (MMIM), estimate the value of Da_m required to achieve 80% monomer conversion in one MMIM and in two equal-volume MMIM's in series. Use the value $\alpha_K = 1$ to minimize variation of $\bar{x}_N$ with Da_m (Fig. 3.5-13).

Solution.   Rearranging equation 3.4-7 yields for monomer conversion

$$\Phi_m = \frac{Da_m}{Da_m + \left(1 + \alpha_K Da_m\right)^{1/2}}$$

and writing equation 3.4-5 in dimensionless form yields for initiator conversion, after rearrangement,

$$\Phi_i = \frac{\alpha_K Da_m}{1 + \alpha_K Da_m}$$

From these results we conclude that $Da_m = 16$ is required to achieve $\Phi_m = 0.8$. The corresponding initiator conversion is $\Phi_i = 0.94$.

For the jth vessel in a cascade, equation 3.4-4–3.4-6 take on the following dimensionless forms, where $Da_m \equiv N\lambda_v/\lambda_m$ (Example 1.3-2) and $N = 2$:

$$\hat{c}_{mj} - \frac{\hat{c}_{mj-1}}{1 + (Da_m/N)(\hat{c}_{0j})^{1/2}} = 0$$

and

$$\hat{c}_{0j} - \frac{\hat{c}_{0j-1}}{1 + \alpha_K Da_m/N} = 0$$

Successive solution leads to $Da_m = 8$ when $\Phi_m = 1 - c_{m2} = 0.8$. Corresponding, $\Phi_i = 1 - \hat{c}_{02} = 0.96$.

The values 16 and 8 for Da_m should be compared with 3.2 for the PFR (Example 1.11-1). Using $\lambda_m \cong 2 \times 10^4 s$ from Example 4.3-5 for DTBP-initiated styrene polymerization at 120°C suggests RT's of the order of 3.2×10^5 s and 1.6×10^5 s, respectively.

Although we have neglected thermal polymerization and gel-effect, both of which would act to increase reaction speed, this example nevertheless serves to illustrate a major problem in the design of continuous reactors, especially backmixed reactors for free-radical polymerizations, viz., the need for long RT's. In fact, specific output rates $[\Phi_m(c_m)_0 M_0 / Da_m \lambda_m]$ for industrial polystyrene reactors have been reported (49) to be less than 1 lb/h gal.

3.4.3. With Hindered Termination

As discussed in Section 2.9, autoacceleration of reaction rate can be caused by polymer product that is soluble in the reaction medium. It is well known that this gel-effect (G-E) can be used advantageously to increase specific output rate in chain-addition polymerizations. Hamielec et al. (35), for instance, accomplished this by mixing some polymer product with the monomer feed through recycling. O'Driscoll et al. (36) showed, however, that such autoacceleration can lead to multiple steady states in isothermal continuous reactors with feedback (BM), analogous to those that lead to thermal instability (Chapter 4) in continuous reactors with feedback. We shall examine this phenomena briefly using the simple model for G-E introduced in Section 2.9.

The transient monomer balance for a MMIM, neglecting density changes and applying the LCA, is

$$\frac{dc_m}{dt} \underset{LCA}{\cong} \left[(c_m)_0 - c_m \right]/\lambda_v - r_p \tag{3.4-9}$$

where $r_p = k_p c_m c^*$ as usual for free-radical propagation. It will be convenient to write this expression in a dimensionless form that is similar to those to be used in our analysis of autothermal systems (Chapter 4).

$$\frac{d\Phi}{d\hat{t}} = \underset{\substack{\text{rate of polymer} \\ \text{generation}}}{\hat{r}} - \underset{\substack{\text{rate of polymer} \\ \text{removal}}}{Da_m^{-1}\Phi} \tag{3.4-10}$$

where $\hat{t} \equiv t/\lambda_m$ and

$$\hat{r} = k_p(1 - \Phi)\lambda_m c^* \tag{3.4-11}$$

The concentration of intermediates may be obtained from a quadratic balance equation by applying the RSSA only (Section 1.11),

$$\frac{c^*}{\lambda_v} = 2k_d c_0 - (k_t)_0 G(\Phi)(c^*)^2 \tag{3.4-12}$$

where (Section 2.9),

$$G(\Phi) \equiv \exp(-B\Phi) \tag{3.4-13}$$

if it is feared that intermediates will build up as a consequence of autoacceleration (large B). On the other hand, if the QSSA is applied, we set the LHS of equation 3.4-12 to zero and obtain the following expression for rate of polymer generation (specific output rate)

$$\hat{r} = (1 + Da_m a_K)^{-1/2}(1 - \Phi)\exp\left(\frac{B\Phi}{2}\right) \tag{3.4-14}$$

in lieu of the RHS of equation 3.4-7. This function has been plotted in Fig. 3.4-4 for several values of parameter B (curves 4 and 5). The removal rate function ΦDa_m^{-1}, as previously noted, has been plotted separately on the same graph for two values of Da_m (curves 6 and 7). Their intersections represent steady-state solutions of balance equation 3.4-10. It is apparent that for sufficiently intense autoacceleration (large B), at least three steady states are possible, the upper and lower ones being stable (s) and the center one being unstable (u). It is also apparent that for such autoacceleration the upper steady state, which represents very high conversion, can be attained with substantially shorter space times ($Da_m \ll 1$) than would normally be required ($Da_m \sim 1$). This is a direct consequence of the enhancement of specific output rate by hindered termination.

3.5. EFFECT ON PRODUCT PROPERTIES

One of the major objectives of this book is to examine the effects of mixing on the selectivity of polymerization reactions. Mixing characteristics will again be separated into RTD and BM. As measures of selectivity we are especially interested in the mean value of certain polymer properties, such as DP, DB, and CC, as well as their dispersion (DPD, DBD, CCD). Intuitively, we might anticipate that mixing effects will depend strongly on the mechanism of the polymerization in question, especially its chain growth reaction (step propagation vs chain propagation). Thus RTD will probably exacerbate property dispersion for most polymerizations, as it does reactor performance for most types of reactions. However, we might expect its effect on chain polymerizations ($\lambda_p \ll \lambda_r$) to be more detrimental than its effect on step polymerizations ($\lambda_p \sim \lambda_r$) because RTD enhances drift dispersion, which dominates in the former. On the other hand, we might expect BM to exacerbate product dispersion more for step reactions than for chain reactions, because integrating local molecular populations that differ from one another in history could radically alter the statistical dispersion (Section 2.6) of step polymers during their growth. Conversely, we might anticipate that BM in concert with the steady state could actually diminish product dispersion in chain polymerizations by reducing drift dispersion (Section 2.7). The effects of backmixing were reported by Denbigh (37) as early as 1947 and the effects of microsegregation by Tadmor and Biesenberger (36, 37) in 1966. The generalizations proposed by these authors are summarized in Table 3.5-1 and their origins will be examined in detail in this section.

Table 3.5-1. Summary of Mixing Effects on Polymer Property Dispersion (DPD, DBD, CCD)

RTD (H) versus PF

        Dispersion ↑ as H ↑

MS (S >> 1) versus MM (S << 1) (identical RTD)

    when $\lambda_p \sim \lambda_r$ ($\gamma_r \cong 1$)

        MMR > MSR > PFR (dispersion greatest from MMR)
        statistical dispersion altered

    when $\lambda_p << \lambda_r$ ($\gamma_r >> 1$)

        MSR > PFR > MMR (dispersion greatest from MSR)
        drift dispersion enhanced

Limiting cases

      MMR → MSR when Da → 0 (BM disappears)
      MSR → PFR when Da → ∞ (RTD disappears)

3.5.1. *Residence Time Distribution*

We begin by studying the effects of RTD in MSR's on our three model polymerizations (Section 1.6). Using the first computational method outlined in Sections 1.3 and 3.3, the primary equation to be solved is that for flow-average concentration of product x-mer. Neglecting density changes,

$$x \geqslant 1 \qquad\qquad [c_x] = \int_0^\infty c_x(\tau) e(\tau)\, d\tau \qquad\qquad (3.5\text{-}1)$$

where $c_x(\tau)$ is the solution of the corresponding batch balance equation

$$x \geqslant 1 \qquad\qquad \frac{dc_x}{dt} = r_x \qquad\qquad (3.5\text{-}2)$$

subject to the initial condition $c_x = (c_x)_0$ [feed concentration is usually $(c_x)_0 = 0$ for all $x \geqslant 1$] and evaluated at residence time τ. The appropriate RTD $e(\tau)$ and rate function r_x must be chosen, depending on the reactor type and reaction sequence being examined. Dependence by r_x on reactants such as monomer and initiator would, of course, necessitate the simultaneous solution of equation 3.5-2 with all appropriate, coupled balance equations prior to substitution into equation 3.5-1. It would also introduce corresponding quadratures of the latter kind for $[c_m]$, $[c_0]$, and so on.

Since the complete solution of equation 3.5-1 for all $x \geqslant 1$ is frequently overwhelming with respect to information contained as well as complexity and since property dispersion can be suitably represented in terms of only a couple of average properties, it is customary to use the method of moments (Appendix C) and to seek instead the quantities

$$\left[\mu_c^k\right] = \sum_k x^k [c_x] \tag{3.5-3}$$

for several values of k $(0, 1, 2)$. By substituting equation 3.5-1 into 3.5-3 and exchanging the order of summation and integration operations, it is evident that we can compute the required quantities $[\mu_c^k]$ directly from the solutions of the corresponding batch moment equations for $\dfrac{d\mu_c^k}{dt}$ $(k = 0, 1, 2)$ by quadrature. The latter differential equations are, of course, identical to those solved in Chapter 2, and the units of μ_c^k are again molar concentration.

For simple irreversible random polymerization (sequence 2.2-12) in a MSIM, the reduced flow-average concentration of x-mer leaving the reactor may be computed with the aid of batch kinetics (equation 2.2-20) from the following equation (38)

$$\left[\hat{c}_x\right] = \int_0^\infty \left(1 - \hat{c}_p\right)^{x-1} \hat{c}_p^2 \exp(-\hat{\tau}) \, d\hat{\tau} \tag{3.5-4}$$

and the fraction of unreacted functional groups from the corresponding equation

$$\left[\hat{c}_p\right] = \int_0^\infty \hat{c}_p \exp(-\hat{\tau}) \, d\hat{\tau} \tag{3.5-5}$$

where $\hat{c}_p = \mathrm{Da}^{-1}/(\mathrm{Da}^{-1} + \hat{\tau})$ from equation 2.2-24, $\mathrm{Da} \equiv \lambda_v/\lambda_r$ and $\hat{\tau} \equiv \tau/\lambda_v$. Following integration, the last equation yields an expression for conversion

$$\Phi = 1 - \left[\hat{c}_p\right] = 1 - \mathrm{Da}^{-1} \exp(\mathrm{Da}^{-1}) E_1(\mathrm{Da}^{-1}). \tag{3.5-6}$$

where E_1 is the tabulated exponential integral E_i with $i = 1$ (Table C-1).

$$z > 0, i = 0, 1, 2 \ldots \qquad E_i(z) \equiv \int_1^\infty \zeta^{-i} \exp(-z\zeta) \, d\zeta \tag{3.5-7}$$

The DPD's were obtained by numerical integration (39) of equation 3.5-4 with the aid of reduced moments; thus $[n_x] = [\hat{c}_x]/[\hat{\mu}^0]$ and $[w_x] = x[\hat{c}_x]/[\hat{\mu}^1]$. After substituting the following finite binomial series (x is finite) into equation 3.5-4,

$$\left(1 - \hat{c}_p\right)^x = \sum_{j=0}^x \binom{x}{j}\left(-\hat{c}_p\right)^j \tag{3.5-8}$$

exchanging the order of integration and summation operations, and integrating termwise, that equation may be rewritten as a sum

$$\left[\hat{c}_x\right] = \mathrm{Da}^{-1} \exp(\mathrm{Da}^{-1}) \sum_{j=2}^{x+1} \binom{x-1}{i-2}(-1)^i E_i(\mathrm{Da}^{-1}) \tag{3.5-9}$$

from which DPD's could also be computed, but with less convenience than from the integral. The flow-average moments were computed from the batch moments as outlined earlier

$$\left[\hat{\mu}^k\right] = \int_0^\infty \hat{\mu}^k \exp(-\hat{\tau})\, d\hat{\tau} \tag{3.5-10}$$

using the sums tabulated in Appendix C. Expressions for DPD, moments, and conversion are listed in Table 3.5-2 together with the corresponding expressions for the PFR. Graphs of DPD and dispersion index have been plotted in Figs. 3.5-1 and 3.5-2 for comparison.

For simple step-addition polymerization (sequence 2.3-1) in a MSIM, monomer conversion may be computed from batch kinetics (equation 2.3-9) with the aid of equation 2.3-14

$$\Phi = 1 - \int_0^\infty \exp\left[-(1 + \mathrm{Da})\hat{\tau}\right] d\hat{\tau}$$

$$= \frac{\mathrm{Da}}{(1 + \mathrm{Da})} = \frac{z}{x_0} \tag{3.5-11}$$

where $z \equiv x_0 - [c_m]/(c_0)_0$. As in Section 2.3, all concentrations, including c_m, have been reduced with $(c_0)_0$. Thus $\lambda_r^{-1} = k(c_0)_0$ as before, and mole fraction DPD is given by the flow-average, reduced x-mer concentration. After substituting equation 2.3-12 into the required integral and expressing the result in terms of conversion in lieu of RT we obtain

$$x \geqslant 1 \quad [n_x] = [\hat{c}_{x-1}] = \frac{x_0^{x-1}}{\mathrm{Da}(x-1)!} \int_0^1 \Phi^{x-1}(1-\Phi)^{\mathrm{Da}^{-1}-1}\exp(-x_0\Phi)\, d\Phi$$

$$\tag{3.5-12}$$

where m_0 has been counted as 1-mer. A closed-form solution was not found for this integral; instead, it was evaluated numerically (39). It can also be written as a slowly converging infinite series by expanding $\exp - x_0\Phi$ in a Taylor series, integrating termwise, and expressing the result as a sum of tabulated beta functions $B(r, s)$ (Table C-1),

$$[n_x] = \frac{x_0^{x-1}\exp(-x_0)}{\mathrm{Da}(x-1)!} \sum_{j=0}^\infty \frac{x_0^j}{j!} B(\mathrm{Da}^{-1} + j, x) \tag{3.5-13}$$

which may be evaluated alternatively from associated gamma functions $\Gamma(r)$ (factorial function) with the aid of a mathematical identity (Table C-2) and Stirling's approximation. Moments were computed by the method previously outlined. DPD's moments and averages are listed in Table 3.5-3 and graphs of DPD and dispersion index have been plotted in Figs. 3.5-3 and 3.5-4.

It is evident from these results that the emerging DPD under MS conditions lies between that for the PFR (Poisson) and that for the MMIM (most probable). This is reflected in the expression for DPD in a MSIM in Table 3.5-3. Examination of this

Table 3.5-2. Mixing Effects on Random Polymerizations

$$m_x + m_y \xrightarrow{\ k\ } m_{x+y}$$
$$x \geq 1$$
$$y \geq 1$$

	PFR	MSIM	MMIM
n_x	$\Phi^{x-1}(1-\Phi)$	$[1/E_1(\bar{Da}^{-1})]\displaystyle\int_0^1 \Phi^{x-1}\exp[-\bar{Da}^{-1}(1-\Phi)^{-1}]d\Phi$	$a_x\Phi^{x-1}(1+\Phi)^{2x-1}$
w_x	$x\Phi^{x-1}(1-\Phi)$	$\bar{Da}^{-1}\exp(\bar{Da}^{-1})\displaystyle\int_0^1 x\Phi^{x-1}\exp[-\bar{Da}^{-1}(1-\Phi)^{-1}]d\Phi$	$xa_x\Phi^{x-1}(1-\Phi)/(1+\Phi)^{2x-1}$
Φ	$Da/(1+Da)$	$1 - \bar{Da}^{-1}\exp(\bar{Da}^{-1})E_1(\bar{Da}^{-1})$	$1 - [(1+4Da)^{1/2}-1]/2Da$

	PFR	MSIM	MMIM
$\hat{\mu}_0$	$1 - \Phi$	$Da^{-1} \exp(Da^{-1}) E_1(Da^{-1})$	$1 - \Phi$
$\hat{\mu}_1$	1	1	1
$\hat{\mu}_2$	$(1 + \Phi)/(1 - \Phi)$	$1 + 2Da$	$(1 + \Phi^2)/(1 - \Phi)^2$
$\hat{\mu}_3$	$(1 + 4\Phi + \Phi^2)/1 - \Phi)^2$	$1 + 6Da + 12\,Da$	$(1 + 2\Phi + 6\Phi^2 + 2\Phi^3 + \Phi^4)/(1-\Phi)^4$
$\bar{x}_N$	$1/1 - \Phi)$	$1/Da^{-1} \exp(Da^{-1}) E_1(Da^{-1}) = 1/(1-\Phi)$	$1/(1 - \Phi)$
$\bar{x}_w$	$(1 + \Phi)/(1 - \Phi)$	$1 + 2D$	$(1 + \Phi^2)/(1 - \Phi)^2$
D_N	$1 + \Phi$	$(1 + 2Da)[D^{-1} \exp(Da^{-1}) E_1(Da^{-1})]$	$(1 + \Phi^2)/(1 - \Phi)$

Note: All moments were reduced with feed monomer concentration

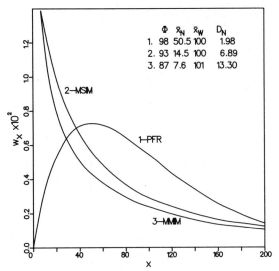

Figure 3.5-1. Weight fraction DPD (w_x) for random polymerization in a PFR (1), a MSIM (2), and a MMIM (3) when $\bar{x}_W \cong 100$.

expression reveals that, unlike the PFR (or the MMIM), the MSIM yields a DPD that is not uniquely determined by a single parameter, say z. Instead, Φ may be adjusted independently of z by manipulation of x_0 (cf. equation 3.5-11). Thus, in principle, polymers with the same mean DP ($\bar{x}_N$) can be prepared having DPD's that range from Poisson ($D_N \rightarrow 1$) to most probable ($D_N \rightarrow 2$). This behavior is reflected in Figs. 3.5-3 and 3.5-4 by the adjustable parameter Φ and is used to illustrate an interesting special case in Example 3.5-1. From Fig. 3.5-4, we conclude

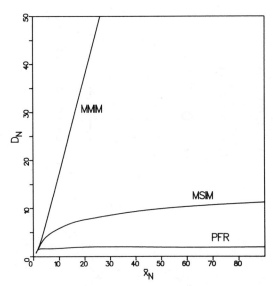

Figure 3.5-2. Dispersion index (D_N) versus DP ($\bar{x}_N$) for random polymerization in a PFR, a MSIM and a MMIM.

Table 3.5-3. Mixing Effects on Step-Addition Polymerizations

$$m_x + m \xrightarrow{k} m_{x+1} \qquad x \geq 0$$

	PFR	MSIM	MMIM
n_x	$(x_o\Phi)^{x-1}(\exp -x_o\Phi)/(x-1)!$	$[x_o^{x-1}/Da(x-1)!]\int_0^1 \Phi^{x-1}(1-\Phi)^{Da^{-1}-1}\exp(-x_o\Phi)\,d\Phi$	$[x_o\Phi/(1+x_o\Phi)]^{x-1}[1/(1+x_o\Phi)]$
w_x	$x(x_o\Phi)^{x-1}(\exp -x_o\Phi)/(1+x_o\Phi)(x-1)!$	$[xx_o^{x-1}/Da(1+x_o\Phi)(x-1)!]\int_0^1 \Phi^{x-1}(1-\Phi)^{Da^{-1}-1}\exp(-x_o\Phi)\,d\Phi$	$x[x_o\Phi/(1+x_o\Phi)]^{x-1}[1/(1+x_o\Phi)]^2$
Φ	$1 - \exp(-Da)$	$Da/(1 + Da)$	$Da/(1 + Da)$

333

Table 3.5-3. (*Continued*)

	PFR	MSIM	MMIM
$\hat{\mu}^0$	1	1	1
$\hat{\mu}^1$	$1 + z$	$1 + z$	$1 + z$
$\hat{\mu}^2$	$1 + 3z + z^2$	$1 + 3z + 2z^2/(1 + \Phi)$	$1 + 3z + 2z^2$
$\hat{\mu}^3$	$1 + 7z + 6z^2 + z^3$	$1 + 7z + 12z^2/(1+\Phi) + 6z^3/(1+\Phi)(1+2\Phi)$	$1 + 7z + 12z^2 + 6z^3$
$\bar{x}_N$	$1 + z$	$1 + z$	$1 + z$
$\bar{x}_w$	$(z+2) - 1/(1+z)$	$(2z + 1)/(1 + \Phi)$	$(2z + 1)$
D_N	$(1+3z + z^2)/(1+z)^2$	$(2z + 1)/(1 + z)(1 + \Phi)$	$(2z + 1)/(z + 1)$
z	x_o^{Φ}	x_o^{Φ}	x_o^{Φ}

NOTE: All moments were reduced with feed initiator concentration.

334

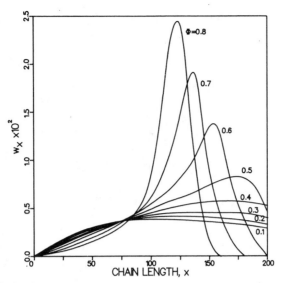

Figure 3.5-3. Weight fraction DPD (w_x) for step-addition polymerization in a MSIM at various values of monomer conversion (Φ) with $\bar{x}_N = 100$.

that dispersity is relatively insensitive to $\bar{x}_N$ when the latter exceeds 30, which suggests that high polymers of varying DP but fixed D_N could also be prepared by adjusting x_0 and λ_v to maintain Φ constant.

For conventional chain-addition polymerizations with $\alpha_K \ll 1$, analytical expressions giving batch concentrations c_x are available. As noted in Section 2.11, it is convenient to substitute monomer concentration for time in the batch kinetic

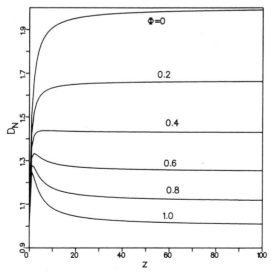

Figure 3.5-4. Dispersion index (D_N) versus average number of monomer units per chain (z) for step-addition polymerization in a MSIM at various values of monomer conversion (Φ).

expressions. The corresponding expression for RTD in a MSIM is (38)

$$\exp(-\hat{\tau}) = -\mathrm{Da}^{-1}(\hat{\phi})^{1-\mathrm{Da}^{-1}}d(\hat{\phi}^{-1}) \qquad (3.5\text{-}14)$$

where (cf. Section 2.10) $\hat{\phi} \equiv \phi/\phi_0$ and

$$\phi \equiv \frac{(\nu_N)_0 \hat{c}_m}{1 + (\nu_N)_0 \hat{c}_m} \qquad (2.10\text{-}11)$$

This reduces to the expression used in equation 3.5-12

$$\exp(-\hat{\tau}) = \mathrm{Da}^{-1}(1 - \Phi)^{\mathrm{Da}^{-1}-1}d\Phi \qquad (3.5\text{-}15)$$

when the LCA is valid, $(\nu_N)_0 \gg 1$. Moments may be calculated as usual, from batch moments. Thus the DPD for free-radical polymerization in a MSIM with termination by combination and without chain transfer may be obtained by solving the following equation numerically (39)

$$[n_x] = \frac{[c_x]}{[\mu_c^0]} = \left[\frac{\mathrm{Da}^{-1}}{1 - (1 - \phi_0)^{\mathrm{Da}^{-1}}} \right]$$

$$\times \left\{ \left[x^{-1} + (\nu_N)_0^{-1} \right](\phi_0)^x \mathrm{Da}^{-1} \int_0^1 \left[x^{-1} + (\nu_N)_0^{-1} \hat{c}_m \right] \phi^x c_m^{\mathrm{Da}^{-1}-1} d\hat{c}_m \right\}$$

$$(3.5\text{-}16)$$

Equation 3.5-10 and all moments except $[\mu_c^0]$ and $[\mu_c^1]$ were obtained with the aid of the LCA with virtually no loss of accuracy. Similarly, monomer conversion in the MSIM is given by

$$\Phi = 1 - \frac{\mathrm{Da}^{-1}}{1 + \mathrm{Da}^{-1}} \left\{ 1 - \mathrm{Da}(\nu_N)_0^{-1} \left[1 - \left(1 + (\nu_N)_0^{-\mathrm{Da}^{-1}} \right) \right] \right\} \underset{\mathrm{LCA}}{\cong} \frac{\mathrm{Da}}{1 + \mathrm{Da}}$$

$$(3.5\text{-}17)$$

where $\mathrm{Da} \equiv \lambda_v/\lambda_m$ and (Table 2.7-1) $\lambda_m = k_{ap}(c_0)_0^{1/n}$. Without the LCA the upper limit of integration must be $\lambda_m \ln[1 + (\nu_N)_0]$ because segregated elements attaining this age will have been depleted of monomer. Since conventional chain-addition polymerizations are pseudo first-order, it should not be surprising that the last result is identical to equation 3.5-11. DPD's, moments, and averages are listed in Table 3.5-4, and the results are illustrated in Figs. 3.5-5–3.5-7.

3.5.2. Backmixing

Next we shall attempt to isolate the effects of BM (MM) on our three model sequences for comparison with the effects of RTD alone (MS). To accomplish this, it is essential that the RTD's be identical. Therefore, we choose the MMIM. The primary equation to be solved is the steady-state balance for product x-mer, c_x (cf. equation 3.5-2). Using the second method outlined in Sections 1.3 and 3.3, and

Table 3.5-4. Mixing Effects on Chain-Addition Polymerizations

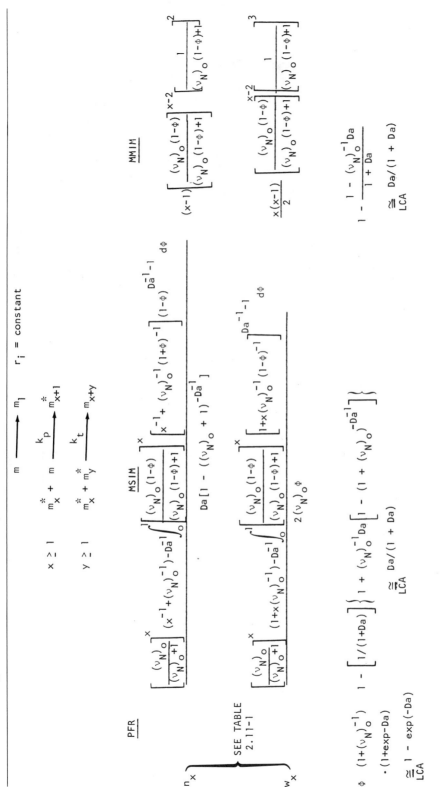

Table 3.5-4. (*Continued*)

	PFR	MSM	MMIM
$\hat{\mu}_0$	$\dfrac{x_o}{2(\nu_N)_o}\ell n\left[\dfrac{(\nu_N)_o+1}{(\nu_N)_o(1-\Phi)+1}\right]$	$\dfrac{x_o Da}{2(\nu_N)_o}\left[1-(1+\nu_N)_o^{-Da^{-1}}\right]$	$\dfrac{x_o^\Phi}{2(\nu_N)_o\left[(\nu_N)_o^{-1}+(1-\Phi)\right]}$
$\hat{\mu}_1$	x_o^Φ	x_o^Φ	x_o^Φ
$\hat{\mu}_2$	$x_o^\Phi\left[\dfrac{3(2-\Phi)(\nu_N)_o}{2}+2\right]$	$x_o^\Phi\left[\dfrac{3(\nu_N)_o}{1+\Phi}+2\right]$	$x_o^\Phi[3(\nu_N)_o(1-\Phi)+2]$
$\hat{\mu}_3$	$x_o^\Phi\left[4(3-3\Phi+\Phi^2)(\nu_N)_o^2+\dfrac{15(2-\Phi)(\nu_N)_o}{2}+4\right]$	$x_o^\Phi\left[\dfrac{12(\nu_N)_o^2}{1+2\Phi}+\dfrac{15(\nu_N)_o}{1+\Phi}+4\right]$	$x_o^\Phi\left[12(\nu_N)_o^2(1-\Phi)^2+15(\nu_N)_o(1-\Phi)+4\right]$
$\bar{x}_N$	$\dfrac{2(\nu_N)_o^\Phi}{\ell n\left[\dfrac{(\nu_N)_o+1}{(\nu_N)_o(1-\Phi)+1}\right]}$	$\dfrac{2(\nu_N)_o^\Phi}{Da\left[1-(1+(\nu_N)_o)^{-Da^{-1}}\right]}$	$2(\nu_N)_o(1-\Phi)+2$
$\bar{x}_w$	$\dfrac{3(\nu_N)_o(2-\Phi)}{2}+2$	$\dfrac{3(\nu_N)_o}{1+\Phi}+2$	$3(\nu_N)_o(1-\Phi)+2$
D_N	$\dfrac{\left[4(\nu_N)_o^{-1}+3(1-\Phi)\right]\ell n\left[\dfrac{(\nu_N)_o+1}{(\nu_N)_o(1-\Phi)+1}\right]}{4\Phi}$	$\dfrac{Da\left[4(\nu_N)_o^{-1}-2(\nu_N)_o^{-1}(1-\Phi)+3\right]\left[1-(1+(\nu_N)_o)^{-Da^{-1}}\right]}{2(1+\Phi)}$	$\dfrac{2(\nu_N)_o^{-1}+3(1-\Phi)}{2(\nu_N)_o^{-1}+2(1-\Phi)}$

NOTE: All moments were reduced with feed initiator concentration.

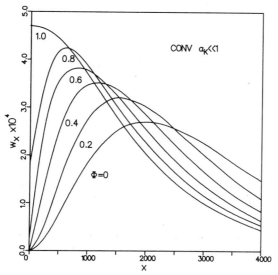

Figure 3.5-5. Weight fraction DPD (w_x) for conventional ($\alpha_K \ll 1$) chain-addition polymerization with $(\nu_N)_0 = 10^3$ in a MSIM at various values of monomer conversion (Φ).

neglecting density changes, this balance is simply

$$x \geqslant 1 \qquad\qquad \frac{\Delta c_x}{\lambda_v} = r_x \qquad\qquad (3.5\text{-}18)$$

where $\Delta c_x \equiv c_x - (c_x)_0$ and for most applications $(c_x)_0 = 0$. The same rate functions will be used for r_x as before.

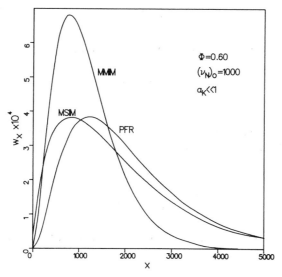

Figure 3.5-6. Weight fraction DPD (w_x) for conventional ($\alpha_K \ll 1$) chain-addition polymerization with $(\nu_N)_0 = 10^3$ in a PFR, a MSIM, and a MMIM at monomer conversion $\Phi = 0.60$.

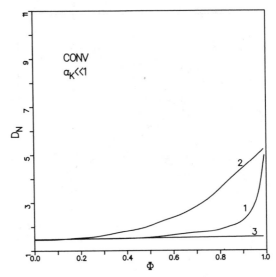

Figure 3.5-7. Dispersion index (D_N) versus monomer conversion (Φ) for model chain-addition polymerization ($\alpha_K \ll 1$) with $(\nu_N)_0 = 10^3$ in a PFR (1), a MSIM (2), and a MMIM (3).

Thus, for simple, irreversible random polymerization (equation 2.2-12) in a MMIM, the DPD may be obtained by solving the following set of dimensionless balance equations (Example 3.5-7):

$$x \geqslant 1 \qquad \frac{\hat{c}_x}{\text{Da}} = -2\hat{c}_x\hat{c}_p + \sum_{j=0}^{x} \hat{c}_j\hat{c}_{x-j} \qquad (3.5\text{-}19)$$

and

$$\frac{1 - \hat{c}_p}{\text{Da}} = \hat{c}_p^2 \qquad (3.5\text{-}20)$$

Division to eliminate Da leads to a difference equation in x

$$x > 1 \qquad \hat{c}_x = \frac{1 - \hat{c}_p}{\hat{c}_p} \sum_{j=0}^{x} \hat{c}_j\hat{c}_{x-j} \qquad (3.5\text{-}21)$$

subject to the initial condition $\hat{c}_1 = \hat{c}_p/(2 - \hat{c}_0)$. Successive solution of this equation, whose batch equivalent is 2.2-20, is

$$\hat{c}_x = \frac{a_x\hat{c}_p(1 - \hat{c}_p)^{x-1}}{(2 - \hat{c}_p)^{2x-1}} \qquad (3.5\text{-}22)$$

where constant coefficients a_x are obtained from their precursors by the convolution sum

$$x > 1 \qquad a_x = \sum_{j=1}^{x-1} a_j a_{x-j} \qquad (3.5\text{-}23)$$

starting with $a_1 = 1$. The solution of equation 3.5-23 should give a_x as a function of x only. It is (40)

$$a_x = 2^{x-1} \prod_{j=1}^{x-1} \frac{2j-1}{x!} = \frac{(2x-2)!}{x!(x-1)!} \tag{3.5-24}$$

The DPD's and average DP's corresponding to equations 3.5-22 and 3.5-24 are listed in Table 3.5-2. It is noteworthy that they are unique functions of conversion, $\Phi = 1 - \hat{c}_p$. Some moments are listed in Table 3.5-2 for comparison with those for the PFR and the MSIM. Graphs of w_x and D_N have also been plotted in Figs. 3.5-1 and 3.5-2 and contrasted with the corresponding graphs for the PFR and the MSIM.

It is evident that the DPD for the backmixed reactor (MMIM) is the broadest of the three. In fact, the relative dispersion diverges at high conversions ($D_N \to \infty$ as $\Phi \to 1$) in contrast with its convergent behavior in the PFR ($D_N \to 2$ as $\Phi \to 1$). Whereas the number DPD from a PFR has no maximum (most probable), even the weight DPD from an IM has no maximum (Fig. 3.5-1; cf. triangle $f_x = w_x$ in Fig. 2.3-6). This indicates that much more low-DP polymer is formed in the IM, especially the MMIM (Fig. 3.5-2) than in the PFR. For example, when $\bar{x}_W = 100$, D_N is 1.98 from a PFR as contrasted with 133 from a MMIM. A further comparison is made in Example 3.5-2. It must be remembered, however, that these conclusions are based on the simple sequence, 2.2-12. If rearrangement reactions occur, as they often do in practice, then we would expect all DPD's to approach the most probable distribution (Section 2.2).

Returning to the MMIM, it is instructive to examine the DPD from a statistical viewpoint (28). The probability ϕ that any polymer molecule (m_x) in a MMIM propagates rather than exits, assuming that only monomer (m_1) enters, is given by

Table 3.5-5. Coefficients for Distribution 3.5-22

x	a_x	x	a_x
1	1	10	$48{,}620{,}000 \times 10^{-4}$
2	1	20	$17{,}672{,}628 \times 10^{2}$
3	2	30	$10{,}022{,}411 \times 10^{8}$
4	5	40	$68{,}042{,}387 \times 10^{13}$
5	14	50	$50{,}955{,}055 \times 10^{19}$
6	42	60	$40{,}594{,}296 \times 10^{25}$
7	132	70	$33{,}748{,}317 \times 10^{31}$
8	429	80	$28{,}944{,}758 \times 10^{37}$
9	1,430	90	$25{,}422{,}145 \times 10^{43}$
10	4,862	100	$22{,}750{,}589 \times 10^{49}$

the ratio of the rate of propagation to that of propagation plus exit

$$\phi \equiv \frac{2kc_x \sum_{j=1}^{\infty} c_j V}{2kc_x \sum_{j=1}^{\infty} c_j V + c_x \dot{V}} = \frac{2(1 - \Phi)}{2(1 - \Phi) + \mathrm{Da}^{-1}} \qquad (3.5\text{-}25)$$

where fractional conversion is defined, as before, $\Phi \equiv 1 - \sum_{j=1}^{\infty} c_j/(c_1)_0 = 1 - \hat{c}_\mathrm{p}$. With the aid of equation 3.5-20 the probability of propagation in a MMIM may be written simply as

$$\phi = \frac{2\Phi}{1 + \Phi} \qquad (3.5\text{-}26)$$

and is therefore not equal to fractional conversion as it is in the PFR (Section 2.6). The probability of exit is $1 - \phi$. Thus, by combinatorial analysis, the probability that entering 1-mer, chosen at random, will leave as x-mer in the effluent, is $1 - \phi$ when $x = 1$, $\phi n_1 (1 - \phi)$ when $x = 2$, $\phi n_1 \phi n_1 (1 - \phi) + n_2 (1 - \phi)$ when $x = 3$, and so on, where n_x is the probability that our growing molecule reacts with x-mer. The key to this analysis is realizing that the fraction represented by the combined probability $xc_x/(c_1)_0$ is the weight fraction w_x and not the mole fraction as in the PFR (Section 2.6). Thus, for example, accounting for all possible ways to grow 4-mer, we must write: $w_4 = \phi^3 n_1^3 (1 - \phi) + (\phi^2 n_1 n_2 + \phi^2 n_2 n_1)(1 - \phi) + \phi n_3 (1 - \phi)$. Mole fraction n_x can be transformed into weight fraction via the general expression $w_x = xn_x(1 - \phi)$. After substituting this result for n_x together with equation 3.5-26 for ϕ into our expressions for combined probability, we obtain the weight fraction DPD listed in Table 3.5-2. Details are left as an exercise for the reader.

It is noteworthy that the coefficients a_x represent the number of ways of producing x-mer from 1-mer by all distinguishable combinations of molecules having DP's less than x. Thus a_x in equation 3.5-24 represents the number of ways integer x can be constructed by combining integers less than x, if we require of each combination that 1 be first in position, that the sum of all integers used be x, that they be distinguished from one another on the basis of the order in which the integers appear, and that each combination containing integers greater than 1 be weighted by the number of possible ways those integers can be constructed from smaller integers by the same process (Example 3.5-3). A partial listing of these coefficients appears in Table 3.5-5. Using Stirling's approximation for large values of x (Appendix B), we can express a_x in a more convenient form than equation 3.5-24,

$$x \gg 1 \qquad\qquad a_x \cong \frac{2^{2x-1}}{(2x - 1)(\pi x)^{1/2}} \qquad (3.5\text{-}27)$$

which gives c_{10}, for example, with less than 2-percent error.

Statistical analyses of MMIM's enhance our basic understanding of the physical factors responsible for the resulting DPD's; in the case of random polymerization, the large fraction of low ends and the long high-DP "tail" (Fig. 3.5-1). The two opposing effects responsible are evident in the expression for w_x. As x increases at any fixed conversion level, the number of allowable molecular combinations increases rapidly (Table 3.5-5), but the combined probability that the required reaction sequence will occur decreases more rapidly. Statistical arguments break down at high conversions ($\Phi \rightarrow 1$), as with batch systems (Section 2.6), because the total number of molecules ceases to be a large number.

As noted in Chapter 1, the analysis of step-addition polymerizations

$$m_0 + m \xrightarrow{k_i} m_1 \tag{2.3-19}$$

$$x \geqslant 1 \qquad\qquad m_x + m \xrightarrow{k_p} m_{x+1}$$

is of theoretical interest, especially the idealized sequence for which $k_i = k_p \equiv k$. We shall learn presently that the DPD for the latter sequence is transformed from a Poisson distribution in the PFR (also BR) to a most probable distribution in the MMIM (41). The set of algebraic balance equations to be solved (cf. 3.5-19) for sequence 2.3-19 in a MMIM is, in dimensionless form

$$x > 1 \qquad\qquad \frac{\Delta \hat{c}_x}{\text{Da}} = \hat{c}_m(\hat{c}_{x-1} - \hat{c}_x) \tag{3.5-28}$$

$$x = 1 \qquad\qquad \frac{\Delta \hat{c}_1}{\text{Da}} = \hat{c}_m(a_K \hat{c}_0 - \hat{c}_1) \tag{3.5-29}$$

$$x = 0 \qquad\qquad \frac{\Delta \hat{c}_0}{\text{Da}} = -a_K \hat{c}_m \hat{c}_0 \tag{3.5-30}$$

and

$$\frac{\Delta \hat{c}_m}{\text{Da}} = -\hat{c}_m(a_K \hat{c}_0 + \hat{c}_p) \tag{3.5-31}$$

where $a_K \equiv k_i k_p$, $\text{Da} \equiv \lambda_v/\lambda_m$ ($\lambda_m = 1/k_p(c_0)_0$; Section 2.3) and the feed conditions are $(\hat{c}_x)_0 = 0$ for all $x \geqslant 1$, $(\hat{c}_0)_0 = 1$, and $(\hat{c}_m)_0 = x_0$. Thus the DPD sought is the solution of equations 3.5-29 and 30, which may be rewritten as the familiar difference equation in x,

$$x \geqslant 1 \qquad\qquad \hat{c}_{x+1} - \phi \hat{c}_x = 0 \tag{3.5-32}$$

subject to initial condition $\hat{c}_1 = \phi a_K \hat{c}_0$, whose solution $\hat{c}_x = a_K \hat{c}_0 \phi^x$ yields the most probable distribution

$$x \geqslant 1 \qquad\qquad n_x = \frac{\hat{c}_x}{\hat{c}_p} = \phi^{x-1}(1 - \phi) \tag{3.5-33}$$

where $\hat{c}_p = \sum_{x=1}^{\infty} \hat{c}_x = a_K \hat{c}_0 \phi/(1 - \phi)$. This procedure facilitates identification of the probability of propagation (41)

$$\phi \equiv \frac{\hat{c}_m}{\hat{c}_m + \text{Da}^{-1}} \tag{3.5-34}$$

as well as the following quantity,

$$\text{Da } \hat{c}_m = k_p c_m \lambda_v \equiv z \tag{3.5-35}$$

which we recognize as the equivalent of the eigenzeit transformation (Section 2.3) for the MMIM. From the solution of the initiator and monomer balances, an expression for monomer conversion may be obtained (Example 3.5-6). For the special case $a_K = 1$ or $a_K \rightarrow \infty$ (Poisson sequence) this expression obeys first-order kinetics

$$\Phi = \frac{\text{Da}}{1 + \text{Da}} \tag{3.5-36}$$

and thus z has the same physical interpretation as for the BR

$$\bar{x}_N = \frac{1}{1 - \phi} = x_0 \Phi = z \tag{3.5-37}$$

From this result we conclude that large DP's are attainable only at high conversions where ϕ has a value close to unity [$\phi = z/(1 + z)$], similar to random polymers. Furthermore, if m_0 is counted as a repeat unit, as in Section 2.3, the expression for its concentration, $\hat{c}_0 = (1 + a_K z)^{-1}$, from equation 3.5-30 is also given by $\hat{c}_x = n_x$ with $x = 1$ for the special case $a_K = 1$ or $a_K \rightarrow \infty$. It is noteworthy that DPD 3.5-33 obtains whether or not k_i and k_p are equal and whether or not m_0 is counted as repeat unit. By comparing these results with those for the PFR and MSIM, we conclude that the MMIM produces the broadest DPD for step-addition polymers of three reactors examined, as it did for random polymers.

The transformation of the DPD from Poisson in the PFR to most probable in the MMIM can perhaps be best understood by revealing the probability of propagation in the MMIM as the ratio of the rate of propagation to that of propagation plus exit

$$\phi = \frac{V k_p c_m c_p}{\left(V k_p c_m c_p + \dot{V} c_p\right)} \tag{3.5-38}$$

and by recalling the statistical derivation of the most probable DPD given in Section 2.6 for the BR. The primary factor here is the probability of exit from the MMIM, $1 - \phi$, which is constant at steady state and is assumed to be independent of DP. Its role is thus equivalent to that of the probability of termination in the BR. The consequences of termination (exit) with respect to DPD dispersion were discussed in detail in Chapter 2. The statistical arguments leading to equation 3.5-33 are otherwise identical to those in Section 2.6 (41).

Finally, we examine chain-addition polymerization in a MMIM. At steady state, neglecting density changes, the balance equations to be solved are:

$$x \geqslant 1 \qquad\qquad \frac{\Delta c_x}{\lambda_v} = r_x \tag{3.5-39}$$

where r_x is the appropriate rate function listed in Table 1.10-4. The required intermediate concentrations are obtained from the steady-state intermediate balances by applying the RSSA (Section 1.11) and assuming that $(c_x^*)_0 = 0$ for all $x \geqslant 1$.

$$\frac{c_1^*}{\lambda_v} = r_1^* \tag{3.5-40}$$

$x > 1$
$$\frac{c_x^*}{\lambda_v} = r_x^* \tag{3.5-41}$$

$$\frac{c^*}{\lambda_v} = r^* \tag{3.5-42}$$

By neglecting transfer to polymer, branching, and secondary termination reactions, as usual, these balances lead to the familiar difference equation in x

$x \geqslant 1$
$$c_{x+1}^* - \phi c_x^* = 0 \tag{3.5-43}$$

subject to the IC

$$c_1^* = \frac{r_i}{D'} \tag{3.5-44}$$

where ϕ is the probability of propagation (cf. equation 2.2-91)

$$\phi \equiv \frac{k_p c_m}{D'} \leqslant 1 \tag{3.5-45}$$

and the denominator D' is defined (cf. equation 2.2-45)

$$D' \equiv k_p c_m + k_{fm} c_m + k_{fs} c_s + k_t c^* + k_{ts} + \lambda_v^{-1} \tag{3.5-46}$$

The solution, of course, yields the most probable DPD for intermediates, $n_x^* = c_x^*/c^*$.

Expressions for product DPD (n_x, w_x) follow immediately from our analyses of the BR (PFR) in Sections 2.2.7 and 2.6. In fact, the cumulative product DPD's from the MMIM are identical to the corresponding instantaneous product DPD's from the BR. For example, when $(c_x)_0 = 0$ for all $x \geqslant 1$,

$$n_x \equiv \frac{c_x}{\sum\limits_x c_x} = \frac{r_x}{\sum\limits_x r_x} \tag{3.5-47}$$

with the aid of equation 3.5-39. This expression is identical to equation 1.13-6 for $(n_x)_{inst}$. There are, however, several important differences between the MMIM and the PFR. First, while the numerator in expression 3.5-45 $(k_p c_m)$ must be close in value to the denominator (both D' and D) in order to obtain high-DP polymer, their quotient (ϕ) is subject to drift in the BR and the PFR, and therefore also the MSIM, but not in the MMIM. Consequently, drift dispersion is absent from the latter reactor, but not from the others. Second, D' differs from D only by the last term in equation 3.5-46, which derives from the exit rate for intermediates, $c^* \dot{V}$. Its presence represents the probability that exit will deprive polymer chains of further growth, rather than the various chemical termination alternatives available. When the QSSA is valid ($\gamma_v \gg 1$), which is frequently the case (Section 1.11), then D' and D are equal in value (Example 3.5-4). This reflects the unlikeliness that active intermediates will exit before they complete reaction.

Thus the DPD for conventional chain-addition polymerization ($\alpha_K \ll 1$) in a MMIM with termination by combination and without chain transfer, after applying the QSSA, is (38)

$$n_x = (x - 1)\phi^{x-2}(1 - \phi)^2 \tag{3.5-48}$$

where the probability of propagation is

$$\phi \equiv \frac{(\nu_N)_0 \hat{c}_m}{1 + (\nu_N)_0 \hat{c}_m} = \frac{(\nu_N)_0(1 - \Phi)}{1 + (\nu_N)_0(1 - \Phi)} \tag{3.5-49}$$

This result corresponds to the DPD in Table 2.10-1 for the BR (PFR) and equation 3.5-16 for the MSIM. Expressions for $\bar{x}_N$ and $\bar{x}_W$ are simple (Example 3.5-5). For monomer conversion, neglecting chain transfer and assuming that termination is either exclusively spontaneous ($n = 1$) or exclusively bimolecular ($n = 2$), we obtain

$$\Phi = \frac{1 + (\nu_N)_0 \text{Da}}{1 + \text{Da}} \underset{\text{LCA}}{\cong} \frac{\text{Da}}{1 + \text{Da}} \tag{3.5-50}$$

where $\text{Da} = \lambda_v/\lambda_m$ and λ_m is defined as before (equation 3.5-17). The last result is identical to equation 3.5-17 because first-order reactions are unaffected by micromixing. Graphs of w_x and D_N are compared in Figs. 3.5-6 and 7 for all three reactors. The corresponding DPD's moments and averages are listed in Table 3.5-4.

3.5.3. *Some Generalizations for Linear Polymers*

From these results, we shall draw several general conclusions. First, when $\lambda_p \ll \lambda_r$, or $\gamma_r \gg 1$ (chain-addition polymerizations), MS produces polymer with the broadest property dispersion, and complete MM yields a product with less dispersion than PF. The reasons for this are that the RTD exacerbates the drift dispersion inherent in such systems whereas BM diminishes it. On the other hand, when $\lambda_p \sim \lambda_r$, or $\gamma_r \cong 1$ (step polymerizations), BM causes the most dispersion by altering the molecular statistics of chain growth through integration of reactor elements containing molecules with vastly different histories and concomitant properties. These generalizations will be shown to apply to DBD and CCD as well as DPD. They have been summarized in Table 3.5-1.

The effects of RTD and BM on DPD dispersion must conform to the limiting behavior at low conversion ($\text{Da} \ll 1$) and high conversion ($\text{Da} \gg 1$) discussed in Section 3.3.2 and summarized in Tables 3.4-1 and 3.5-1. Thus, as $\Phi \to 0$, we would expect the effects of BM to disappear, and as $\Phi \to 1$, we would expect the effects of RTD to disappear. These limiting conditions are most easily verified for random polymerizations and step addition. By letting $\Phi \to 0$, the corresponding moment equations for the MSIM approach those for the MMIM in Tables 3.5-2 and 3.5-3, reflecting the disappearance of the effects of BM. Similarly, by letting $\Phi \to 1$, the moment equations for the MSIM approach those for the PFR, reflecting the disappearance of the effects of RTD. This limiting behavior is also clearly exhibited in Fig. 3.5-4 by D_N for step-addition polymers. Further demonstrations of its validity on the remaining polymer property characteristics are left as exercises for the interested reader.

Introduction of slightly more complex kinetics does give rise to some behavioral trends not evident in the simple model reaction schemes. The moments of the polymer DPD may be calculated by the same method as described for the model reactions. Complex kinetics (rate expressions are given in Chapter 2) generally require double numerical integration to solve for moments in MS reactors, using equation 3.3-6. That is, quadrature with unequally spaced base points must be used to approximate the integrals, and batch rate equations must be evaluated numerically to determine the value of the moments and concentrations at each base point. The following calculations were performed with the aid of a high-speed computer.

Consider, first, the behavior of reversible random polymerizations. Degree of polymerization and dispersion are shown as functions of mean residence time in Figs. 3.5-8 and 3.5-9. It is clear that MM not only gives the most disperse polymer, but also leads to exceptionally low values of $\bar{x}_N$. Microsegregated reactors, however, show cases in which there is a marked deviation from PF behavior at long RT's. For very large values of the equilibrium constant (i.e., irreversible propagation), D_N would appear to rise without bound for the MSCTR and plateau for the MSIM. This is apparently a consequence of the accelerating change in molecular weight that occurs with each increment of conversion during random polymerization. Changes in $\bar{x}_N$ do not damp out with conversion, as they do for step- and chain-addition polymerizations, so even a small spread in RT can greatly increase the breadth of the DPD.

When propagation is reversible, there is a limiting or equilibrium value of chain length. As a result, sufficiently long RT's allow all segregated elements to achieve

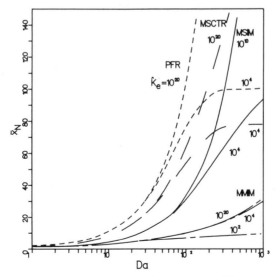

Figure 3.5-8. Effect of BM and RTD on chain length $\bar{x}_N$ for reversible random polymerization with equilibrium constant ($\hat{K}_e$) as parameter.

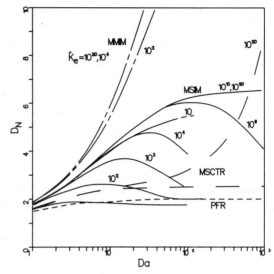

Figure 3.5-9. Effect of BM and RTD on dispersion index (D_N) for reversible random polymerization, with equilibrium constant ($\hat{K}_e$) as parameter.

equilibrium conditions. The effect is to produce a maximum in D_N versus Da. This phenomenon is evident for rather large values of the equilibrium constant. Thus, for values of K_e as large as 10^6, little difference from irreversible kinetics is apparent with respect to conversion or $\bar{x}_N$, but significant variation in product dispersion does occur. This is worth noting since the irreversibility of random polymerizations often depends on the ability to remove by-products. Assumption of irreversibility may lead to significant errors.

Retarding chain-initiation rate in step-addition polymerizations has the most significant effect on micromixed reactors. While the PFR generally gives the largest value of $\bar{x}_N$ at any value of Da, step-addition in a MMIM with small values of a_K represent an exception. This is another consequence of the enhanced selectivity of the MMIM referred to in the last section. It is interesting to note that this increased DP is achieved without a sacrifice to polydispersity. It has been shown analytically that regardless of whether or not $a_K = 1$, the DPD in a MMIM is most probable (equation 3.5-33). As shown in Fig. 3.5-10, decreasing a_K delays the onset of chain growth for the PFR but accelerates it for the MMIM at sufficiently small values of Da_m. The latter phenomenon is due to the effect of BM on the autoacceleratory behavior of these reactions described in Section 3.4-1. It causes the crossover point (below which the PFR gives a higher DP) to shift to larger values of Da_m as a_K decreases and to disappear altogether when a_K is approximately 10^{-3}, the value associated with initiator depletion (Section 2.3.2). The detrimental effects of RTD on DP are illustrated in Fig. 3.5-11. Curves for the MMIM and MSIM are identical because chain growth is first-order and MM has no effect on first-order reactions. Product dispersion, shown in Fig. 3.5-12, conforms nicely to the general trends. The MMIM shows the greatest dispersion and the PFR shows the least. The MSIM and the MMIM are the same for small values of Da while the MSIM and the PFR converge for large values.

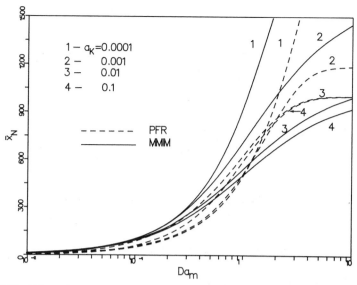

Figure 3.5-10. Effect of BM on chain length $\bar{x}_N$ for step addition with $a_K \equiv k_i/k_p$ as parameter.

Chain-addition polymerizations are also affected by the balance between initiation and propagation rates. BM eliminates dead-ending (D-E) in another manifestation of the enhanced selectivity that is possible in the MMIM. Figure 3.5-13 shows that the PFR gives the largest DP when $\alpha_K < 1$, followed by the microsegregated reactors. Under D-E conditions ($\alpha_K > 1$) the trend reverses, and the MMIM gives the largest DP. Furthermore, the MMIM has the narrowest DPD (Fig. 3.5-14) with D_N remaining independent of Da and equal to the instantaneous (statistical) dispersion.

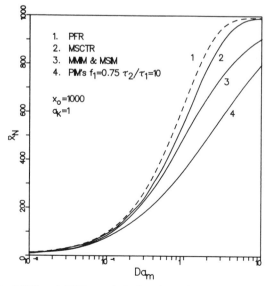

Figure 3.5-11. Effect of BM and RTD on chain length $\bar{x}_N$ for model step-addition polymerization. (PIM's = parallel IM's)

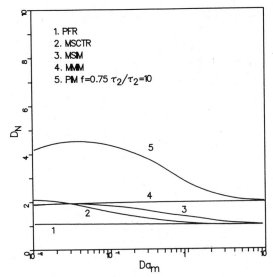

Figure 3.5-12. Effect of BM and RTD on product dispersion index (D_N) for step-addition polymerization. (PIM's = parallel IM's)

3.5.4. Series of Ideal Mixers

It is well known that the performance of the cascade of N maximally micromixed ideal mixers in series can be made to approach that of the PFR for a given total reaction volume V by increasing N and concomitantly reducing the volume of each IM, V_j ($1 \leq j \leq N$). This can be readily demonstrated in terms of the output of

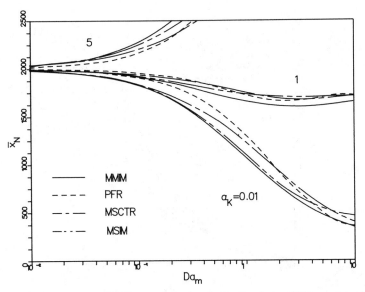

Figure 3.5-13. Effect of BM and RTD on chain length $\bar{x}_N$ for chain-addition polymerization with termination by combination. Drift parameter α_K is a parameter.

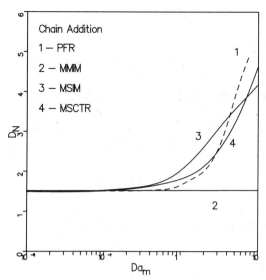

Figure 3.5-14. Effect of BM and RTD on product dispersion (D_N) for dead-end model chain-addition polymerization ($\alpha_K = 5$) with termination by combination.

arbitrary substance k from the cascade

$$\Delta c_k \equiv (c_k)_N - (c_k)_0 = \sum_{j=1}^{N} (r_k)_j \tau_j \qquad (3.5\text{-}51)$$

which must approach the output of the PFR

$$\Delta c_k \equiv \int_0^\tau r_k \, d\alpha \qquad (3.5\text{-}52)$$

in the limit as $N \to \infty$ with constant $\tau \equiv \sum_{j=1}^{N} \tau_j$ by virtue of the definition of the Riemann integral.

Thus we can remove the effects of BM to any desired degree by the appropriate selection of tank RT's τ_j ($1 \leqslant j \leqslant N$) and tank number N, while retaining the benefits of the IM, which include extensive mixing and simplicity of design (at least in the case of the CSTR).

We shall investigate in some detail the model step-addition reaction, sequence 2.3-1 ($a_K = 1$), in a series of MMIM's because the results are of theoretical interest.

A steady-state monomer balance on the jth vessel yields

$$1 \leqslant j \leqslant N \qquad \qquad \frac{-(\Delta c_m)_j}{\lambda_j} = k(c_m)_j(c_0)_0 \qquad (3.5\text{-}53)$$

The probability that any polymer chain in the jth vessel propagates by adding monomer in preference to leaving is

$$\phi_j = \frac{k(c_m)_j \lambda_j}{1 + k(c_m)_j \lambda j} = \frac{z_j}{1 + z_j} \tag{3.5-54}$$

where $Z_j \equiv k(c_m)_j \lambda_j$ is the eigenzeit transformation variable for the jth vessel. Thus, from combinatorial analysis the probability that y-mer entering vessel j from vessel $j - 1$ propagates $x - y$ times and then exits as x-mer is given by $(n_y)_{j-1} \phi_j^{x-y}(1 - \phi_j)$. Therefore, for all y we obtain the following difference equation in j

$$1 \leqslant y \leqslant x \qquad (n_x)_j = \sum_{y=1}^{x} (n_y)_{j-1}^{x-y}(1 - \phi_j) \tag{3.5-55}$$

whose solution for the Nth vessel yields the DPD (41)

$$(n_x)_N = \frac{\displaystyle\sum_{j=1}^{N} z_j^{N+x-2}}{(1 + z_j)^x \displaystyle\prod_{\substack{j=1 \\ l \neq j}}^{N} (z_j - z_l)} \tag{3.5-56}$$

Again, we have counted m_0 as a repeat unit $[n_x = c_{x-1}/(c_0)_0]$ and assumed that no polymer enters the first vessel $[(c_x)_0 = 0$ for all $x \geqslant 1]$. The corresponding average DP's are:

$$(\bar{x}_N)_N = 1 + \sum_{j=1}^{N} z_j \tag{3.5-57}$$

$$(\bar{x}_W)_N = \frac{1 + 3 \displaystyle\sum_{j=1}^{N} z_j + \displaystyle\sum_{j=1}^{N} z_j^2 + \left(\displaystyle\sum_{j=1}^{N} z_j \right)^2}{\left(1 + \displaystyle\sum_{j=1}^{N} z_j \right)} \tag{3.5-58}$$

In the limit as $N \to \infty$ and $\lambda_v = \sum_{j=1}^{N} \lambda_j$ remains finite, equation 3.5-56 must take the form of the Poisson distribution, equation 2.3-12 with $z = [(c_m)_0 - c_m]/(c_0)_0$, because BM disappears and the system behaves like a PFR (BR).

A special case of theoretical interest is that in which the probability of propagation is the same in all vessels: $\phi_1 = \phi_2 = \cdots = \phi_j = \cdots = \phi_N \equiv \phi$. This is not equivalent to equal-size vessels. Instead, it requires that vessels increase monotonically in size: $\lambda_1 < \lambda_2 < \cdots < \lambda_j < \cdots < \lambda_N$. Now, the probability of collecting x-mer at random from the effluent of the Nth vessel is given by the negative binomial distribution (41)

$$(n_x)_N = \binom{x + N - 2}{x - 1} \phi^{x-1}(1 - \phi)^N \tag{3.5-59}$$

This may be deduced from the outcome of $x - 1 + N$ Bernoulli trials (Appendix B) in which $x - 1$ are propagations and N are exits, and only the first $x - 1 + N - 1$ trials may be permuted because the last event must be an exit from tank N. It is noteworthy that if m_0 were not counted as a repeat unit ($x \to x - 1$), then equation 3.5-56 with $N = 2$ would be identical to the instantaneous DPD for chain-addition polymer with termination by combination (Section 2.2-7). In any case, when $N = 1$, equation 3.5-59 is identical to the instantaneous DPD for batchwise chain-addition polymer with termination by disproportionation. These results are consistent with our knowledge that combination termination reduces the breadth of the DPD of chain-addition polymers produced in batch reactors and that adding vessels in series reduces the breadth of the DPD of step-addition polymers produced in continuous MMIM's.

The DPD's from one, two, three, four, and five MMIM's of equal size, and therefore equal residence times λ_v, in series have been plotted in Fig. 3.5-15. Conditions were chosen to make the monomer conversion 50 percent in the case of one vessel and $\bar{x}_N = 101$ in all cases (39). The DPD is seen to sharpen as N increases and to approach the Poisson DPD with $\bar{x}_N = 101$.

The effect on step-addition polymers of reducing BM is thus a reduction in DPD dispersion. A similar result would be expected for random polymers, which are also grown by step-type propagation reactions (Table 1.6-2). The common reason is the partial restoration of equal growth opportunities for each polymer molecule throughout its lifetime commensurate with its age as opposed to equal opportunity for all molecules regardless of age, which is provided by complete BM. In other words, we are altering the statistical dispersion process.

For chain-addition polymers and copolymers, on the other hand, we would expect the opposite result: the broadening of both DPD and CCD. Furthermore, we would expect two IM's in series to produce bimodal DPD's and CCD's, three to produce trimodal distributions, and so on, until N becomes large enough to approximate the continuous smearing effect that has been associated with batch drift dispersion. We

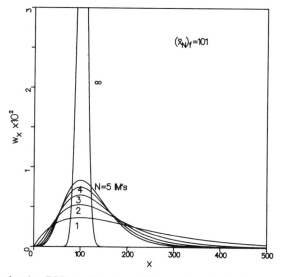

Figure 3.5-15. Mole fraction DPD (n_x) for step-addition polymerization in one (1), two (2), three (3), four (4), five (5), and infinite (6) MMIM's in series when $\bar{x}_N = 101$.

shall demonstrate this effect in terms of DPD. By virtue of the QSSA, the cumulative DPD produced in the jth vessel, $(n_x)_j$, is approximately equal to the corresponding instantaneous DPD, $(n_x)_{\text{inst}}$ (Example 2.2-12) for the prevailing probability of propagation

$$\phi_j \cong \frac{k_p(c_m)_j}{D_j} \tag{3.5-60}$$

where denominator D_j (equation 2.2-45) refers to conditions (concentrations) in reaction vessel j. Since the distribution and hence the average DP's depend directly on ϕ (Example 2.2-12) and since ϕ_j must vary with j in discrete fashion, it follows that the DPD produced in each vessel could have distinctly different values of $\bar{x}_N$ and $\bar{x}_W$ and could therefore be dispersed to significantly different degrees. The DPD and average DP of the final product, for example,

$$n_x = \sum_{j=1}^{N} (n_p)_j (n_x)_j \tag{3.5-61}$$

and

$$\bar{x}_N = \sum_{j=1}^{N} (n_p)_j (\bar{x}_N)_j \tag{3.5-62}$$

respectively, will be composites of the individual values and will be weighted according to the fraction of the total polymer produced in each vessel $(n_p)_j$.

Similar conclusions may be drawn for the CCD, except that the individual distributions, $[w_p(y)]_j$, will be very narrowly dispersed, thus rendering the final DPD, $n(y)$, a composite consisting of weighted "spikes."

Example 3.5-1. Effect of Microsegregation on Step-Addition Polymers. Show that DPD 3.5-13 reduces to the rectangular distribution (Section 2.3.2)

$$n_x = x_0^{-1} \exp(-x_0) \sum_{j=x}^{\infty} \frac{x_0^j}{j!}$$

when Φ is precisely equal to 0.5.

Solution. From equation 3.5-11, $Da = 1$ when $\Phi = 0.5$ and from Table C-2

$$B(1+j, x) = \frac{j!(x-1)!}{(j+x)!}$$

Therefore,

$$n_x = x_0^{-1} \exp(-x_0) \sum_{j=0}^{\infty} \frac{x_0^{j+x}}{(j+x)!} = x_0^{-1} \exp(-x_0) \sum_{j=x}^{\infty} \frac{x_0^j}{j!}$$

This distribution also represents the DPD for sequence 2.3-28 (Example 2.3-3) in a

BR (PFR) as well as sequence 2.3-1 in a semibatch reactor with continuous makeup initiator feed entering at a rate that is proportional at all times to monomer concentration in the reactor (39).

Example 3.5-2. Effect of Micromixing on Random Polymers. Compare $\bar{x}_N$ and D_N of a random polymer from a PFR with those from a MMIM at the following conversions: $\Phi = 0, 0.25, 0.50, 0.75, 0.9, 0.99, 0.999$

Solution. After substituting the values of Φ into the appropriate equations in Table 3.5-2, we obtain the results in the following chart. The introduction of low ends and dramatic broadening of the DPD that is evident is due to the alteration of the random molecular process by BM as described in the text.

| | $\bar{x}_N$ | D_N | |
| | Both | PFR | MMIM |
Φ			
0	1	1	1
0.25	1.33	1.25	1.89
0.50	2	1.50	5
0.75	4	1.75	25
0.90	10	1.90	181
0.99	100	1.99	19,801
0.999	1000	1.999	1,998,001

Example 3.5-3. The Coefficient a_x. Construct a_4 by the procedure described in the text.

Solution.

$$a_4 = \underset{A}{1 \cdot 1 \cdot 1 \cdot 1} + \underset{B}{1 \cdot 1 \cdot 1} + \underset{C}{1 \cdot 1 \cdot 1} + \underset{D}{1 \cdot 2} = 5$$

where

$$A : 4 \text{ ones}$$
$$B : 2 \text{ ones, 1 two}$$
$$C : 1 \text{ one, 1 two, 1 one}$$
$$D : 1 \text{ one, 1 three}$$

We note that three may be constructed 2 ways (D), whereas two may be constructed only one way (C).

Example 3.5-4. Effect of Micromixing on Chain-Addition Polymers. Verify the validity of using the QSSA (as opposed to the RSSA, Section 1.11) by comparing D' with D using the kinetic values for BP-initiated styrene polymerization at 90°C and $(c_0)_0 = 0$. Assume $Da_m = 1$ (thus $\Phi_m = 0.50$).

Solution.   From Section 1.11, we know that γ_m should characterize the validity of the QSSA for the kinetic scheme, and γ_v should characterize its applicability to a flow system. Using the kinetic constants in Example 1.11-3 and the definitions in Table 1.11-1,

$$\gamma_m \equiv \frac{\lambda_m}{\lambda_p} = \frac{k_t}{k_p} = 2.1 \times 10^5 \gg 1$$

This implies that we are justified in using the QSSA in a closed reactor system. Using equation 1.11-21,

$$\gamma_v \equiv \frac{\lambda_v}{\lambda_p} = \frac{\gamma_m}{\mathrm{Da}_m} = 2.1 \times 10^5$$

which implies that we should also be justified in using the QSSA (rather than the more restrictive RSSA) in the kinetic expressions for flow reactors. As a check, examine D' from the expression for ϕ in an IM (neglecting transfer reactions):

$$D' = k_p c_m + k_t c^* + \frac{\dot{V}}{V}$$

or

$$D' = \lambda_p^{-1} \left[\frac{(c_m)_0}{(c^*)_0} \lambda_m^{-1} \hat{c}_m + \hat{c}^* + \gamma_v^{-1} \right]$$

The only difference between the denominator for the closed system (D) and for the open one (D') is γ_v^{-1}, which is clearly small relative to either of the other two terms. Note that $(c_m)_0/(c^*)_0$ is typically on the order of 10^{10}, so its product with γ_m^{-1} is still quite large.

Example 3.5-5.   Effect of Micromixing on Chain-Addition Polymers.   Using the instantaneous properties of Section 2.2-7 for free-radical polymers with termination by combination only and constant initiator concentration, show that for the MMIM (Table 3.5-4):

$$\bar{x}_N = 2(\nu_N)_0(1 - \Phi) + 2$$

and

$$\bar{x}_W = 3(\nu_N)_0(1 - \Phi) + 2$$

Solution.   From Example 2.2-12

$$(\bar{x}_N)_{inst} = \frac{2}{1 - \phi}$$

and

$$(\bar{x}_W)_{inst} = \frac{2 + \phi}{1 - \phi}$$

where

$$\phi \cong \frac{k_p c_m}{k_p c_m + k_t c^*}$$

$$= \frac{\dfrac{c_m}{(c_m)_0}}{\dfrac{c_m}{(c_m)_0}} + \frac{[2fk_d k_t (c_0)_0]^{1/2}}{k_p (c_m)_0}$$

from equations 3.5-45 and 3.5-46 if we employ the QSSA (Section 1.11). Finally, with the aid of Table 2.7-1

$$\phi = \frac{(1 - \Phi)(\nu_N)_0}{(1 - \Phi)(\nu_N)_0 + 1}$$

which, after substitution into the expressions for instantaneous properties, yields $(\bar{x}_N)_{inst}$ = cumulative $\bar{x}_N$ and $(\bar{x}_W)_{inst}$ = cumulative $\bar{x}_W$ for the MMIM owing to the combined effects of steady-state and chain-reaction behavior. It should be pointed out that $\bar{x}_N$ and $\bar{x}_W$ both converge to the value 2 at complete conversions ($\Phi \to 1$). The low DP is a consequence of constant initiator concentration, that is, constant initiation rate, in the face of declining monomer concentrations at high conversions. The specific value of 2 is a consequence of termination by combination.

The purpose of this exercise was to demonstrate that the steady-state cumulative properties of chain-addition polymers in MMIM's are identical to the instantaneous properties discussed in Chapter 1. Statistical dispersion is unaffected by BM and drift is eliminated by the steady state.

Example 3.5-6. Step-Addition Polymerization in a MMIM. Solve balance equations 3.5-28–3.5-31 using DPD 3.5-33 to find the general expression for monomer conversion. Show that this is simplified to expression 3.5-36 when $a_K \to \infty$ and $x_0 \to \infty$.

Solution. Equation 3.5-30 yields

$$\hat{c}_0 = (a_K \mathrm{Da}\, \hat{c}_m + 1)^{-1}$$

Substitution of this, along with 3.5-33, into monomer balance 3.5-31 gives:

$$\hat{c}_m = x_0 - \frac{\hat{c}_m a_K \mathrm{Da}(1 + \mathrm{Da}\, \hat{c}_m)}{a_K \mathrm{Da}\, \hat{c}_m + 1}$$

Rearrangement results in a quadratic equation in $\hat{c}_m$:

$$0 = \mathrm{Da}(\mathrm{Da} + 1)\hat{c}_m^2 - \left[\mathrm{Da}(x_0 - 1) - a_K^{-1}\right]\hat{c}_m - a_K^{-1} x_0$$

with the following solution:

$$\hat{c}_m = \frac{\left[\mathrm{Da}(x_0 - 1) - a_K^{-1}\right] + \left[(\mathrm{Da}(x_0 - 1) - a_K^{-1})^2 + 4x_0 \mathrm{Da}(\mathrm{Da} + 1)a_K^{-1}\right]^{1/2}}{2\,\mathrm{Da}(\mathrm{Da} + 1)}$$

where the negative root has been discarded since the square root in the numerator is always larger than the other term. When $a_K \to \infty$

$$\Phi \equiv 1 - \frac{\hat{c}_m}{x_0} = 1 - \frac{(x_0 - 1)}{x_0(Da + 1)} = \frac{Da + x_0^{-1}}{Da + 1}$$

For values of x_0 characteristic of high polymer [recall $(\bar{x}_N)_f \cong x_0$] the preceding equation reduces to 3.5-36. When $a_K = 1$ is substituted into the derived expression for $\hat{c}_m$, one arrives at equation 3.5-36 without further approximation.

Example 3.5-7. *The z-Transform Method.*   Using the z-transform method (Appendix C) verify that distribution 3.5-22 is the DPD for random sequence 2.2-12 at steady state in an IM.

Solution.   Define the z-transform as (Table C-1; omitting primes for convenience)

$$F(z, Da) \equiv \sum_{x=1}^{\infty} \hat{c}_x(Da) z^{-1}$$

The dimensionless balance equations are

$$\frac{(1 - \hat{c}_1)}{Da} = -2\hat{c}_1\hat{c}_p$$

$x > 1$

$$\frac{\hat{c}_x}{Da} = -2\hat{c}_x\hat{c}_p + \sum_{j=1}^{x-1} \hat{c}_j\hat{c}_{x-j}$$

$$\frac{1 - \hat{c}_p}{Da} = \hat{c}_p^2$$

when $Da \equiv \lambda_v/\lambda_r$, $\lambda_r \equiv 1/k(c_1)_0$, and all concentrations are reduced with $(c_1)_0$. After multiplying the first equation by z^{-1} and the second by z^{-x} and summing, we obtain

$$\sum_{x=1}^{\infty} \hat{c}_x z^{-x} + 2\,Da\,\hat{c}_p \sum_{x=1}^{\infty} \hat{c}_x z^{-x} - z^{-1} - Da \sum_{x=2}^{\infty} \sum_{j=1}^{x-1} \hat{c}_j z^{-j}\hat{c}_{x-j} z^{-(z-j)} = 0$$

which becomes, with the aid of the definition of F and the special property of transformed convolution sums (Table C-2),

$$Da\,F^2 - (2\,Da\,\hat{c}_p + 1)F - z^{-1} = 0$$

The solution of this equation

$$F(z, Da) = \frac{1 + 2\,Da\,\hat{c}_p}{2\,Da} \pm \left[\left(\frac{1 + 2\,Da\,\hat{c}_p}{2\,Da} \right)^2 - \frac{z^{-1}}{Da} \right]^{1/2}$$

may be expressed as follows, after substituting the monomer balance and discarding the root with the positive sign because $F(z, \text{Da})$ cannot exceed unity:

$$F(z, \text{Da}) = \frac{\hat{c}_p(2 - \hat{c}_p)}{2(1 - \hat{c}_p)} \left\{ 1 - \left[1 - \frac{4(1 - \hat{c}_p)}{(2 - \hat{c}_p)^2} z^{-1} \right]^{-1/2} \right\}$$

Finally, by expanding this equation in an infinite series (39) and comparing the result

$$F(z, \text{Da}) = \frac{\hat{c}_p}{1!(2 - \hat{c}_p)} z^{-1} + \frac{2!}{1!2!} \frac{\hat{c}_p(1 - \hat{c}_p)}{(2 - \hat{c}_p)^3} z^{-2} + \frac{4!}{2!3!} \frac{\hat{c}_p(1 - \hat{c}_p)^2}{(2 - \hat{c}_p)^5} z^{-3} + \cdots$$

$$+ \frac{(2x - 2)!}{(x - 1)!x!} \frac{\hat{c}_p(1 - \hat{c}_p)^{x-1}}{(2 - \hat{c}_p)^{2x-1}} z^{-x} + \cdots$$

with the definition of $F(z, \text{Da})$, we conclude that equation 3.5-22 (with coefficients 3.5-24) is indeed the DPD of sequence 2.2-12.

3.5.5. Copolymers

Next, we shall examine the effects of RTD and BM on the CC and CCD of chain-addition copolymers with termination, beginning with the MS reactor. A significant dispersive effect should be expected from the RTD in view of the extensive composition drift that is possible in batch copolymerization. The computational method to be used is that of O'Driscoll et al. (42) which, in essence, is based on the monotonic relationship between instantaneous CC and time (Section 2.11) and on the assumption that instantaneous CCD is monodisperse (Section 2.6).

Consider a MS reactor under steady-state conditions. Suppose, for the moment, that the composition of interest (fraction of monomer A) increases monotonically with Φ (or τ) as before (Section 2.11.2). Then each reaction element passing through ceases to contribute instantaneous copolymer product to the cumulative CCD function precisely at the age τ_{ν_A} that its composition reaches the value ν_A corresponding to the upper limit in the function, equation 2.11-15. Thus we may separate the CCD into two parts:

$$[\hat{W}_p(\nu_A)] = [\Phi_f]^{-1} \left(\int_0^{\tau_{\nu_A}} \Phi(\tau) e(\tau) \, d\tau + \left(\Phi(\tau_{\nu_A}) \int_{\tau_{\nu_A}}^{\infty} e(\tau) \, d\tau \right) \right) \qquad (3.5\text{-}63)$$

contribution of elements younger than τ_{ν_A} contribution of elements older than τ_{ν_A}

where the final conversion is given by

$$[\Phi_f] = \int_0^{\infty} \Phi(\tau) e(\tau) \, d\tau \qquad (3.5\text{-}64)$$

and the conversion at any residence time $\Phi(\tau)$ may be obtained by combining equations 2.7-17 and 2.7-19, as before. The average CC is (Section 2.7)

$$[y] = \frac{(n_A)_0 - [n_A](1 - [\Phi_f])}{[\Phi_f]}$$ (3.5-65)

where (Example 3.5-8)

$$[n_A] = (1 - [\Phi_f])^{-1} \int_0^\infty n_A(\tau)[1 - \Phi(\tau)] e(\tau)\, d\tau$$ (3.5-66)

O'Driscoll et al. (42) performed these calculations numerically with the aid of a computer for free-radical copolymerization in a MSIM, for which the CC distribution function is

$$[\hat{W}_p(\nu_A)] = [\Phi_f]^{-1}\left[\int_0^{\tau_{\nu_A}} \Phi(\tau)\lambda_v^{-1}\exp\left(-\frac{\tau}{\lambda_v}\right) d\tau + \Phi(\tau_{\nu_A})\exp\left(-\frac{\tau_{\nu_A}}{\lambda_v}\right)\right]$$
(3.5-67)

and the corresponding density function, obtained by differentiation (Example 3.5-9) is

$$[\hat{w}_p(\nu_A)] = \frac{d[\hat{W}_p]}{d\nu_A} = [\Phi_f]^{-1}\frac{d\Phi(\nu_A)}{d\nu_A}\exp\left(-\frac{\tau_{\nu_A}}{\lambda_v}\right)$$ (3.5-68)

They used the following relationship in equation 2.7-19

$$\int_0^t c^* \, dt = 2\left[\frac{2f(c_0)_0}{k_d k_t}\right]^{1/2}\left[1 - \exp\left(-\frac{k_d t}{2}\right)\right]$$ (3.5-69)

which assumes that the composition of the radical population does not drift [i.e., $c_A^* = (c_A^*)_0$ and $c_B^* = (c_B^*)_0$]. As a result, k_t is independent of composition. The results for the system listed in Table 3.5-6 are shown in Figs. 3.5-16–3.5-18. The corresponding results for the PFR and the MMIM are included for comparison.

Table 3.5-6.   **Properties and Parameters for a Sample Copolymerization (42)**

r_A	r_B	k_{pAA} (ℓ./mol.s.)	k_{pBB} (ℓ./mol.s.)	k_t (ℓ./mol.s.)	fk_d (s^{-1})	$(c_0)_0$ (mol./ℓ.)
20	0.015	5000	300	10^7	6×10^{-6}	0.05

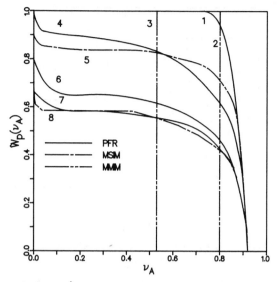

Figure 3.5-16. Cumulative CCD ($\hat{W}_p$) for various values of $(n_A)_0$ in several reactors (42): (1) $(n_A)_0 = 0.89$, $\Phi = 0.35$; (2) $(n_A)_0 = 0.74$, $\Phi = 0.35$; (3) $(n_A)_0 = 0.74$, $\Phi = 0.54$; (4) $(n_A)_0 = 0.55$, $\Phi = 0.73$; (5) $(n_A)_0 = 0.49$, $\Phi = 0.82$; (6) $(n_A)_0 = 0.49$, $\Phi = 0.73$; (7) $(n_A)_0 = 0.83$, $\Phi = 0.35$; (8) $(n_A) = 0.54$, $\Phi = 0.73$.

All are distribution functions. The corresponding density functions must be obtained by differentiation. Note that some exhibit minima (Section 2.11). Figure 3.5-17 shows the change in CCD with conversion and Fig. 3.5-18 the effect of feed composition on azeotropic copolymerization. It is apparent that CCD dispersion is greater from a reactor with a RTD than from a reactor without one (PFR) at comparable conversions. This is due to exacerbation of CC drift dispersion by the RTD, as previously discussed. The differences between the MSR and the PFR

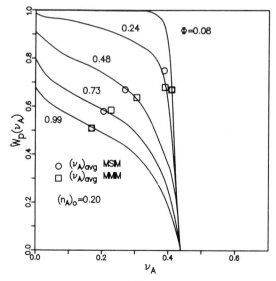

Figure 3.5-17. Cumulative CCD ($\hat{W}_p$) showing drift with conversion (42).

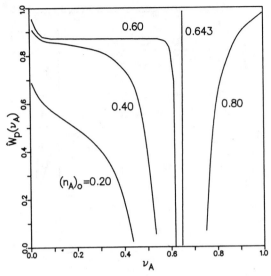

Figure 3.5-18. Cumulative CCD ($\hat{W}_p$) at complete conversion in a MSIM with $(n_A)_0 =$ as parameter (42).

should increase with the magnitude of the drift dispersion index β_K. The disappearance of BM at low conversions and the disappearance of the RTD effect at high conversions are seen in Figs. 3.5-16 and 3.5-17 (cf. limiting cases in Table 3.5-1).

On the other hand, when BM is complete, as in the MMIM, we can take advantage of the uniformity among concentration histories experienced by each reaction element at steady state, as we did earlier, and produce copolymer product without drift dispersion. Thus the cumulative CCD will be identical to the instantaneous CCD corresponding to the prevailing reactor concentrations; and that CCD, as the reader will recall (Section 2.6), is virtually monodisperse at sufficiently high values of DP. To verify this, we shall prove that cumulative average CC leaving a MMIM, y, is identical to the instantaneous CC within it, $y_{inst} = \nu_A$, at steady state. By definition (Sections 1.7 and 2.7), neglecting density changes,

$$y \cong \frac{(c_A)_0 - c_A}{(c_m)_0 - c_m} \tag{3.5-70}$$

where c_A and c_m represent concentrations of monomer A and total monomer, respectively, in the MMIM and its effluent. With the RSSA (assuming steady-state) and the LCA (Section 1.11), the component and total material balances are

$$\frac{[c_A - (c_A)_0]}{\lambda_v} \cong -r_{pA} \tag{3.5-71}$$

and

$$\frac{[c_m - (c_m)_0]}{\lambda_v} \cong -r_p \tag{3.5-72}$$

respectively, where the rate functions on the RHS are listed in Table 1.12-3. After substitution of these balances into definition 3.5-70 and application of the symmetry relation $r_{pjk} = r_{pkj}$ (Example 1.13-4), we obtain the familiar copolymer composition equation (equation 1.13-14)

$$y = \frac{R_A c_A^2 + c_A c_B}{R_A c_A^2 + 2 c_A c_B + R_B c_B^2} = \nu_A \qquad (3.5\text{-}73)$$

Details are left for the reader.

To determine the CCD for intermediates, we write a steady-state balance for each intermediate

$$\frac{c_{a,b}^{j*}}{\lambda_v} = r_{a,b}^{j*} \qquad (3.5\text{-}74)$$

where the rate functions on the RHS are listed in Table 1.12-3. The resulting expressions for $c_{a,b}^{j*}$ are identical to equations 2.5-1 and 2.5-2, only the denominator in the probabilities ϕ_{jk} now contain an extra term (λ_v^{-1}) to reflect the probability of exit (cf. Table 2.5-1).

$j = A, B;\ k = A, B;\ j \neq l = A, B$

$$\Phi_{jk} = \frac{k_{pjk} c_k}{\left(k_{pjj} + k_{fjj}\right) c_j + \left(k_{pjl} + k_{fjl}\right) c_l + k_{fsj} c_s + k_{tjj} c^{j*} + k_{tjl} c^{l*} + \lambda_v^{-1}} \qquad (3.5\text{-}75)$$

For more on the effects of mixing on the CCD, the reader is referred to the literature (29, 42–45).

Example 3.5-8. *Flow-Average Comonomer Composition.* Derive equation 3.5-66.

Solution. Flow-average mole fraction of A in the effluent of a MS reactor is given by

$$[n_A] = \frac{[c_A]}{[c_m]}$$

where

$$[c_m] = (c_m)_0 \left(1 - [\Phi_f]\right)$$

and

$$[c_A] = \int_0^\infty n_A(\tau) c_m(\tau) e(\tau)\, d\tau$$
$$= \int_0^\infty n_A(\tau)(c_m)_0 [1 - \Phi(\tau)] e(\tau)\, d\tau$$

Division of $[c_A]$ by $[c_m]$ yields the desired result.

Example 3.5-9. *Copolymer Composition Distribution (CCD).* Derive equation 3.5-68 for the MSIM from equation 3.5-67.

Solution. By differentiating equation 3.5-67 with the aid of Leibnitz's rule for differentiating an integral (Appendix C), we obtain

$$
\frac{d}{d\nu_A}\left[\hat{W}_p(\nu_A)\right] = \left[\Phi_f\right]^{-1}\left[\lambda_v^{-1}\Phi(\tau_{\nu_A})\exp\left(-\frac{\tau_{\nu_A}}{\lambda_v}\right)\right]
$$

$$
+ \frac{d\Phi(\nu_A)}{d v_A}\exp\left[-\frac{\tau_{\nu_A}}{\lambda_v} - \lambda_v^{-1}\Phi(\nu_A)\exp\left(-\frac{\tau_{\nu_A}}{\lambda_v}\right)\right]
$$

$$
= \left[\Phi_f\right]^{-1}\frac{d\Phi(\nu_A)}{d\nu_A}\exp\left(-\frac{\tau_{\nu_A}}{\lambda_v}\right)
$$

3.5.6. Branched Polymers

Kinetic analysis of long-chain branching was introduced in Section 2.10. More generally, we seek the degree of branching (DB) and its distribution (DBD), in addition to DP and DPD. Therefore, it will now be necessary to use the bivariate distributions introduced in Section 1.5: $m_{x,b}$ and $m_{x,b}^*$. Furthermore, to include propagation with polymer we must distinguish between product x-mer with and without terminal double bonds, $m'_{x,b}$ and $m''_{x,b}$, respectively. Corresponding intermediates that have been reactivated are designated by $m'^*_{x,b}$ and $m''^*_{x,b}$. Unprimed symbols (cf. Table 1.10-3) represent the total population comprising both kinds of chains, as do the corresponding concentrations

$$
c_{x,b} = c'_{x,b} + c''_{x,b} \tag{3.5-71}
$$

Table 3.5-7. Long-Chain Branching Reactions

$x > b \geq 1$	$m^*_{x,b} + m \xrightarrow{k_p} m^*_{x+1,b}$	propagation with monomer
$y > \beta \geq 1$	$m^*_{x,b} + m^1_{y,\beta} \xrightarrow{k'_p} m_{x+y,b+\beta+1}$	propagation with polymer
	$m^*_{x,b} + m \xrightarrow{k_{fm}} \begin{cases} m_{x,b} + m'^{\,*}_{1,o} \\ m'_{x,b} + m^*_{1,o} \end{cases}$	chain transfer to monomer
	$m^*_{x,b} + m_{y,\beta} \xrightarrow{k_{fp}} m_{x,y} + m^*_{y,\beta+1}$	chain transfer to polymer
	$m_{x,b} + S \xrightarrow{k_{fs}} m_{x,b} + m^*_{1,o}$	chain transfer to foreign substance

and

$$c^*_{x,b} = c'^*_{x,b} + c''^*_{x,b} \tag{3.5-72}$$

and their sums (cf. Table 1.10-4):

$$c_p = \sum_x c_x = \sum_x \sum_b c_{x,b} \tag{3.5-73}$$

and

$$c^* = \sum_x c^*_x = \sum_x \sum_b c^*_{x,b} \tag{3.5-74}$$

Graessley et al. (46–48) have analyzed branching in free-radical polymerization computationally for the three vessels of interest, the PFR, the MSIM, and the MMIM, using the methods discussed earlier in this section. The results were compared with experiments on vinyl acetate polymerization. It was assumed that the distributions among intermediates were determined by the five reactions in Table 3.5-7, as discussed in Section 2.10. Chain transfer to solvent was included because the results were compared with experiments on vinyl polymerizations that were conducted in solution. The rate functions for products and intermediates are listed in Table 3.5-8.

Table 3.5-8. Rate Functions for Long-Chain Branching

$$
r'^*_{x,b} = k_p c_m c'_{x-1,b} + x k_{fp} c'_{x,b-1} c^* + k'_p \sum_{y<x} \sum_{\beta<b} c'^*_{y,\beta} c'_{x-y,b-\beta-1}
$$

$$
- \left(k_p c_m + k_{fm} + k_{fp} \sum_x x c_x + k_{fs} + k'_p c_p \right) c'^*_{x,b}
$$

$$
r''^*_{x,b} = k_p c_m c''^*_{x-1,b} + x k_{fp} c''_{x,b-1} c^* + k'_p \sum_{y<x} \sum_{\beta<b} c'''^*_{y,\beta} c'_{x-y,b-\beta-1}
$$

$$
- \left(k_p c_m + k_{fm} c_m + k_{fp} \sum_x x c_x + k_{fs} c_s + k'_p c_p \right) c''^*_{x,b}
$$

$$
r^*_x = \sum_b \left(r'^*_{x,b} + r''^*_{x,b} \right)
$$

$$
r'_{x,b} = \left(k_{fm} c_m + k_{fp} \sum_x x c_x + k_{fs} c_s \right) c'^*_{x,b} - \left(x k_{fp} + k''_{cp} \right) c'_{x,b} c^*
$$

$$
r''_{x,b} = \left(k_{fm} c_m + k_{fp} \sum_x x c_x + k_{fs} c_s \right) c''^*_{x,b} - x k_{fp} c''_{x,b} c^*
$$

$$
r^*_x = \sum_b \left(r'_{x,b} + r''_{x,b} \right)
$$

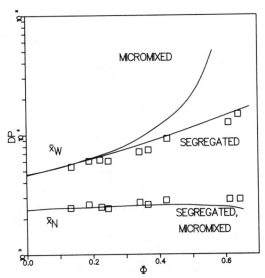

Figure 3.5-19. DP of branched polymers as a function of conversion with chain-transfer constant $R_{fm} = 3.2 \times 10^{-4}$ (47).

Intermediate concentrations $c'^{*}_{x,b}$ and $c''^{*}_{x,b}$ were computed by applying the QSSA $r'^{*}_{x,b} = 0 = r''^{*}_{x,b}$ in the usual way and solving the resulting difference equations. The method of moments was used, as discussed earlier, to calculate quantities such as $\bar{b}_N$ and $\bar{x}_N$, $\bar{x}_W$, and D_N. The associated sums involved concentrations $c'_{x,b}$ and $c''_{x,b}$ and were otherwise of the type listed in Table 1.5.1. The results are shown in Figs. 3.5-19 and 3.5-20.

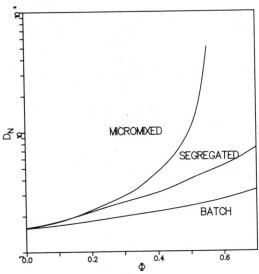

Figure 3.5-20. Dispersion index of branched polymers as a function of conversion with chain-transfer constants (46): $R_{fm} = 2.5 \times 10^{-4}$ and $R_{fp} = 2.36 \times 10^{-4}$.

From the computational results, we conclude that MM leads to the broadest dispersion. Ostensibly, this would appear to be inconsistent with our generalization (Table 3.5-1) for chain-addition polymerizations (29), for which $\lambda_p \ll \lambda_r$ ($\gamma_r \gg 1$) when they are linear. It is not, however, when one considers, as suggested in Section 2.10, that branched polymer molecules continue to grow as long as they remain in the reactor (48). Actually, therefore, $\lambda_p \sim \lambda_v$ ($\gamma_v \cong 1$) and BM among molecules of various ages broadens the product distributions by altering the molecular statistics. Agreement between experiments and MSIM analyses also suggests that the CSTR used in the experiments was possibly microsegregated, although this conclusion was not deemed to be altogether satisfactory by the authors (48).

REFERENCES

1. P. V. Danckwerts, *Appl. Sci. Res. Sect. A*, **3**, 279 (1952).

2. P. V. Danckwerts, *Chem. Eng. Sci.*, **8**, 93 (1958).

3. P. V. Danckwerts, *Chem. Eng. Sci.*, **2**, 1 (1953).

4. T. S-J. Chang, "Mixing Analysis," a Special Problem in Chemical Engineering, Stevens Institute of Technology, Hoboken, NJ (1980).

5. J. M. McKelvey, *Polymer Processing*, Wiley, New York (1962).

6. S. Middleman, *Fundamentals of Polymer Processing*, McGraw-Hill, New York (1977).

7. Z. Tadmor and C. G. Gogos, *Principles of Polymer Processing*, Wiley, New York (1979).

8. E. B. Nauman, *Chem. Eng. Sci.*, **30**, 1135 (1975).

9. H. Tennekes and J. L. Lumley, *A First Course in Turbulence*, MIT, Cambridge, Mass. (1972).

10. L. J. Lee, J. M. Ottins, W. E. Ranz, and C. W. Macosko, *Polym. Eng. Sci. Polymer Topics*, **20**, 868 (1980).

11. C. L. Tucker and N. P. Suh, *Polym. Eng. Sci. Polymer Topics*, **20**, 875 (1980).

12. W. D. Mohr, R. L. Saxton, and C. H. Jepson, *Ind. Eng. Chem.*, **49**, 1855 (1957).

13. R. S. Spencer and R. H. Wiley, *J. Colloid Sci.*, **6**, 133 (1951).

14. R. B. Bird, W. E. Stewart, and E. N. Lightfoot, *Transport Phenomena*, Wiley, New York (1960).

15. L. Erwin, SPE 36th ANTEC Preprints (1978), p. 488.

16. G. I. Taylor, *Proc. R. Soc. London Ser. A*, **219**, 186 (1953).

17. J. A. Biesenberger and A. Ouano, *J. Appl. Polym. Sci.*, **14**, 471 (1970).

18. K. G. Denbigh and J. C. R. Turner, *Chemical Reactor Theory*, 2nd ed., Cambridge University, Cambridge, England (1979).

19. O. Levenspiel, *Chemical Reaction Engineering*, Wiley, New York (1962).

20. T. N. Zwietering, *Chem. Eng. Sci.*, **11**, 1 (1959).

21. R. Cintron-Cordero, R. A. Mostello, and J. A. Biesenberger, *Can. J. Chem. Eng.*, **46**, 434 (1968).

22. G. Pinto and Z. Tadmor, *Polym. Eng. Sci.*, **10**, 279 (1970).

23. D. Bigg and S. Middleman, *I.E.C. Fund.*, **13**, 66 (1974).

24. P. Naor and R. Shinnar, *I.E.C. Fund.*, **2**, 278 (1963).

25. K. G. Denbigh, *Trans. Faraday Soc.*, **40**, 352 (1944).

26. R. Cintron-Cordero, master's thesis in chemical engineering, Stevens Institute of Technology, Hoboken, NJ (1967).

27. P. V. Danckwerts, *First European Symposium on Chemical Engineering—Chemical Reaction Engineering*, Pergamon, London (1957).

28. J. A. Biesenberger and Z. Tadmor, *Polym. Eng. Sci.*, **6**, 299 (1966).

29. E. B. Nauman, *J. Macromol. Sci. Rev. Macromol. Chem.*, Part C(10) **1**, 75 (1974).

30. T. M. Pell and T. G. Davis, *J. Polym. Sci.*, **11**, 1671 (1973).

31. H. D. Schumann, *Faserforsch. Textiltech.*, **22**(8), 389 (1971).

32. R. C. Forney, L. K. McCune, N. C. Pierce, and R. Y. Thompson, *Chem. Eng. Prog.*, **62**, 88 (1966).

33. G. H. Hardy, J. E. Littlewood, and G. Pólya, *Inequalities*, Cambridge University, Cambridge, England (1952).

34. H. Kramers and K. R. Westerterp, *Elements of Chemical Reactor Design and Operation*, Academic, New York (1963).

35. A. W. T. Hui and A. E. Hamielec, Sixty-third National A. I. ChE. Meeting, St. Louis (1968); *I.E.C. Proc. Des. Dev.*, **8**, 105 (1969).

36. K. F. O'Driscoll, W. Wertz, and A. Husar, *J. Polym. Sci.*, A-1, **5**, 2159 (1967).

37. K. G. Denbigh, *Trans. Faraday Soc.*, **43**, 648 (1947); *J. Appl. Chem.*, **1**, 227 (1951).

38. Z. Tadmor and J. A. Biesenberger, *I.E.C. Fund.*, **5**, 336 (1966).

39. Z. Tadmor, Ph.D. thesis in chemical engineering, Stevens Institute of Technology, Hoboken, NJ (1966).

40. J. A. Biesenberger, *AIChE J.*, **11**, 369 (1965).

41. J. A. Biesenberger and Z. Tadmor, *J. Appl. Polym. Sci.*, **9**, 3409 (1965).

42. K. F. O'Driscoll and R. Knorr, *Macromol.*, **2**, 507 (1969).

43. T. T. Szabo and E. B. Nauman, *AIChE J.*, **15**, 575 (1969).

44. J. C. Mecklenburgh, *Can. J. Chem. Eng.*, **48**, 279 (1970).

45. W. H. Ray, T. L. Douglas, and E. Godslave, *Macromolecules*, **4**, 166 (1971).

46. K. Nagasubramanian and W. W. Graessley, *Chem. Eng. Sci.*, **25**, 1549 (1970).

47. K. Nagasubramanian and W. W. Graessley, *Chem. Eng. Sci.*, **25**, 1559 (1970).

48. W. W. Graessley, Symposium on *Polymer Reactor Engineering*, Symposium at Fifty-fifth Chemical Conference and Exhibition, Chemical Institute of Canada, Quebec (1972).

49. L. F. Albright, *Processes for Major Addition-Type Plastics and Their Monomers*, McGraw-Hill, New York (1974).

CHAPTER FOUR

Thermal Effects

Among the many physical factors affecting reactor performance and product properties, thermal effects are probably the most significant. They frequently overshadow mixing and other effects in magnitude.

In the formation of high-molecular-weight polymer product in particular, as we have stated numerous times, chain propagation is the dominant reaction. The free energy change for this reaction is

$$\Delta G_p = \Delta H_p - T\Delta S_p \tag{4.0-1}$$

Spontaneous reaction at any temperature T requires that $\Delta G_p < 0$ (Appendix D). Since chain-building reactions impart configurational order to their participants (mers), one would expect that $\Delta S_p < 0$. It follows from the preceding equation, therefore, that polymerizations should be exothermic, $\Delta H_p < 0$. In fact, for most addition polymerizations ΔH_p falls within the range -15 to -25 kcal/ base mol and ΔS_p has a value in the neighborhood of -25 e.u. While most polymerizations are indeed highly exothermic and irreversible, there are exceptions. Reversible propagation has been discussed in Chapter 2 in connection with random polymerizations. By LeChatlier's principle, we would expect rising reaction temperatures to enhance reversibility in all polymerizations. This phenomenon will be discussed in Section 4.1.

In general, at least two significant consequences of exothermicity on reactor performance and product properties should be apparent. First, inadequate removal of reaction exotherm could obviously result in thermal autoacceleration and lead to reactor instability. Second, temperature variations could adversely affect polymer properties such as DP, DPD, CC, and CCD. Both will be examined, with particular emphasis on product properties. We will learn that two temperatures, T_{cr} and $\hat{T}_{ad}$, are key factors in determining whether or not large DP's and thermally stable reactors can be achieved, respectively. At the risk of oversimplification, we can estimate these temperatures to a first approximation using the thermodynamic properties that comprise ΔG_p.

$$T_{cr} = \frac{\Delta H_p}{\Delta S_p^* + R_g \ln(c_m)_0} \tag{4.0-2}$$

$$\hat{T}_{ad} \equiv \frac{T_{ad} - T_0}{T_0} = \frac{-\Delta H_p(c_m)_0}{\rho c_p T_0} \tag{4.0-3}$$

Thermal effects may be classified in two broad categories, based on the nature of the temperature variation. If reaction temperature were completely controllable, spatially as well as temporally, one could regard it as a parameter. A special case of such idealized behavior is a spatially uniform temperature programmed in time $T(t)$. The simplest example is a constant temperature (isothermal), $T(t) =$ constant. Another is an optimal temperature policy $T(t)$, which maximizes a target reactant conversion or product property. In Section 4.2, we shall presume that reaction temperature is a parameter.

The opposite extreme of complete control is, of course, the absence control. Temperature is then a reaction variable, $T(\mathbf{x}, t)$, subject to the conservation laws (material and energy balances); only its feed value T_0 is a true parameter. The thermal characteristics of such nonisothermal reactors depend on the reactor type. They constitute the subject matter of the remaining sections. Temperature nonuniformities manifest themselves temporally and spatially. "Hot spots" frequently occur first and subsequently propagate into the entire system. Nevertheless, it is customary for purposes of analysis to subdivide reactors into "lumped parameter" and "distributed parameter" systems. This will be done in Section 4.3. In reactors without feedback (BM or recycle), such as PFR's (BR's), thermal instability is associated with a sensitivity by the system to small changes in the operating parameters.

In continuous reactors with feedback, on the other hand, multiple steady states are possible for a single set of operating parameters. This distinction constitutes the basis for separating our analyses into Thermal Runaway (Section 4.3) and Thermal Instability (Section 4.5). Thus, Section 4.3 will deal only with systems whose potential instability arises from unique operating temperatures profiles that are parametrically sensitive, and Section 4.5 will deal with systems whose steady states could include unstable operating temperatures. In both, the system responses will be assumed to be "natural" (uncontrolled).

The effects of nonuniform temperatures on polymer and copolymer properties will be discussed in Section 4.4.

4.1. EFFECT OF ISOTHERMAL TEMPERATURE LEVEL

The maximum conversion attainable in polymerizations, as in all other chemical reactions, is ultimately dictated by chemical equilibrium. Expressions for equilibrium conversion are therefore also useful when designing polymerization reactors. However, owing to special significance of product DP in polymerizations, its maximum attainable value is probably more important than that of conversion. In general, equilibrium DP drops abruptly and dramatically with rising temperature level in a manner that has prompted Dainton et al. (1) and Tobolsky et al. (2) to describe the "critical temperature" T_{cr} at which this occurs as a transition temperature. Dainton et al. have likened the chemical aggregation process of propagation to the physical aggregation process of phase change.

4.1.1. A Critical Reaction Temperature

We begin our analysis with simple random polymerization

$$x, y \geqslant 1 \qquad\qquad \mathrm{m}_x + \mathrm{m}_y \overset{K}{\rightleftharpoons} \mathrm{m}_{x+y} + \mathrm{m}_0 \qquad\qquad (1.8\text{-}16)$$

The expressions developed in Section 1.8 apply with one additional constraint. Here we have a closed system and therefore do not permit the removal of by-product m_0. Thus, by equating c_0 to the concentration of pairs of reacted groups, a simple material balance (cf. constraint equation 1.7-2)

$$c_{\mathrm{RU}} = c_{\mathrm{p}} + c_0 \qquad\qquad (4.1\text{-}1)$$

leads to the conclusion that dimensionless m_0 concentration is identical to conversion.

$$(\hat{c}_0)_{\mathrm{e}} \equiv \frac{(c_0)_{\mathrm{e}}}{c_{\mathrm{RU}}} = \Phi_{\mathrm{e}}$$

After combining this result with the expression for the equilibrium constant (Example 2.2-1), we obtain a quadratic equation in Φ_{e}

$$K = \frac{\Phi_{\mathrm{e}}^2}{\left(1 - \Phi_{\mathrm{e}}\right)^2} \qquad\qquad (4.1\text{-}2)$$

whose solution yields the following simple expressions for conversion

$$\Phi_{\mathrm{e}} = \frac{K^{1/2}}{1 + K^{1/2}} \qquad\qquad (4.1\text{-}3)$$

and DP

$$(\bar{x}_{\mathrm{N}})_{\mathrm{e}} = 1 + K^{1/2} \qquad\qquad (4.1\text{-}4)$$

We turn next to addition polymerizations at equilibrium

$$\mathrm{m}_0 + \mathrm{m} \overset{K_i}{\rightleftharpoons} \mathrm{m}_1 \qquad\qquad (1.8\text{-}17)$$

$$x \geqslant 1 \qquad\qquad \mathrm{m}_x + \mathrm{m} \overset{K_p}{\rightleftharpoons} \mathrm{m}_{x+1} \qquad\qquad (1.8\text{-}18)$$

The distribution of equilibrium concentrations $(c_x)_{\mathrm{e}}$ has been derived in Section 2.2.

After substituting this distribution, equation 2.2-8, into the RHS of constraint equations 1.7-6 and 1.7-7 for monomer and initiator, respectively, and summing we obtain the following coupled equations:

$$(c_m)_0 - (c_m)_e = \sum_{x=1}^{\infty} x(c_x)_e = \frac{K_i(c_0)_e(c_m)_e}{\left[1 - K_p(c_m)_e\right]^2} \tag{4.1-5}$$

$$(c_0)_0 - (c_0)_e = \sum_{x=1}^{\infty} (c_x)_e = \frac{K_i(c_0)_e(c_m)_e}{1 - K_p(c_m)_e} \tag{4.1-6}$$

Solving these simultaneously for $(c_m)_e$ and $(c_0)_e$ and substituting the solutions into the definitions

$$\Phi_e \equiv 1 - \frac{(c_m)_e}{(c_m)_0} \tag{4.1-7}$$

and

$$(\bar{x}_N)_e \equiv \frac{(c_m)_0 - (c_m)_e}{(c_0)_0 - (c_0)_e} \tag{4.1-8}$$

yields the desired functional dependence of Φ_e and $(\bar{x}_N)_e$ on K_i and K_p with $(c_m)_0$ and $(c_0)_0$ as parameters.

The temperature dependence of the equilibrium DP, $(\bar{x}_N)_e$, of addition sequences 1.8-17 and 1.8-18 has been plotted in Fig. 4.1-1 using a van't Hoff relationship for

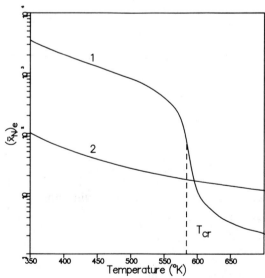

Figure 4.1-1. Equilibrium DP versus temperature for addition sequence 1.8-17 and 1.8-18 (curve 1) and random sequence 1.8-16 (curve 2) for the values: $\Delta H_i^0 = 2.25$ kcal/mol, $\Delta S_i^0 = -7.4$ e.u., $\Delta H_p^0 = -4.03$ kcal/mol, $\Delta S_p^0 = -6.9$ e.u., $(c_m)_0 = 1$, and $(c_0)_0 = 10^{-3}$.

each reaction (j).

$$K_j = \exp\left(\frac{-\Delta G_j^*}{R_g T}\right) \tag{4.1-9}$$

The values used for ΔH_j^* and ΔS_j^* correspond to ε-caprolactam with a standard state of 1 mol/kg. (3). Dilute feed concentrations were deliberately chosen to lower the transition temperature. The equilibrium DP's for random sequence 1.18-16 were also plotted for comparison. The equilibrium constant K was computed from K_i and K_p for convenience using the relationship (Example 2.2-4)

$$K = \frac{K_p}{K_i} \tag{4.1-10}$$

It is apparent from the figure that equilibrium addition polymerizations exhibit transitionlike behavior, even if they are accompanied by side reactions of the random-propagation type (Example 2.2-4). It is equally apparent that equilibrium random polymerization by itself does not.

Following Tobolsky (2), we define the critical temperature for the transition by equating the dimensionless propagation constant to unity

$$\hat{K}_p \equiv K_p(c_m)_0 = 1 \tag{4.1-11}$$

and solving the resulting equation, after substituting equation 4.1-9 for K_p, to obtain

$$T_{cr} = \frac{\Delta H_p^*}{\Delta S_p^* + R_g \ln(c_m)_0} \tag{4.1-12}$$

It is clear from the figure that T_{cr} so defined satisfactorily characterizes the transition. The physical implication of definition 4.1-11 may be brought into focus by combining it with the following equilibrium equation for DP (Section 2.2)

$$(\bar{x}_N) = \frac{1}{1 - K_p(c_m)_e} \tag{4.1-13}$$

This yields an alternative transition criterion

$$\Phi_e(\bar{x}_N)_e = 1 \tag{4.1-14}$$

which characterizes equilibrium as consisting of either a large amount of low-DP polymer or a small amount of high-DP polymer.

The effect of initiation parameters on the shape of the transition is shown in Figs. 4.1-2–4.1-4. In all cases, the critical temperature was defined by equation 4.1-12. From these graphs and numerous others like them, a generalization governing the sharpness of transitions emerges. It appears that a sharp transition requires a very large value for the dimensionless group $(K)_{cr}x_0$, where $(K)_{cr}$ is the ratio of constants ($K \equiv K_p/K_i$), defined earlier, evaluated at T_{cr}. To satisfy this condition, it

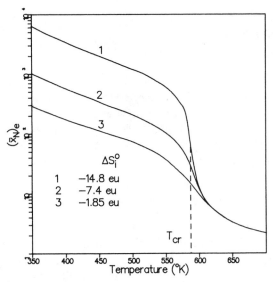

Figure 4.1-2. Parametric dependence of equilibrium DP versus temperature for addition polymerization with the following values: $\Delta H_i^0 = 2.24$ kcal/mol, $\Delta H_p^0 = -4.03$ kcal/mol, $\Delta S_p^0 = -6.9$ e.u., $(c_m)_0 = 1$, and $(c_0)_0 = 10^{-2}$. Values of K_i at T_{cr} are: (1) 8.45×10^{-5}; (2) 3.5×10^{-3}; (3) 5.72×10^{-2}.

is not necessary that $(K)_{cr} \gg 1$, as in Figs. 4.1-2 and 4.1-3. In Fig. 4.1-4 the special case $K_p = K_i$ was plotted using a different set of thermodynamic values for variety, and the evolution of the transition with rising values of x_0 is again apparent.

4.1.2. The Ceiling Temperature

For chain-addition polymerizations, Dainton defined the temperature above which polymer could not be formed as

$$T_c \equiv \frac{\Delta H_p}{\Delta S_p^* + R_g \ln c_m} \tag{4.1-15}$$

and he called it the ceiling temperature (1). The major difference between T_c and T_{cr} is that T_{cr} has a unique value for each $(c_m)_0$, whereas T_c represents a locus of critical values, one for each c_m. The reason for this distinction is that T_c is actually defined as the temperature at which chain propagation approaches equilibrium

$$m_x^* + m \overset{K_p}{\rightleftharpoons} m_{x+1}^* \tag{4.1-16}$$

irrespective of chain initiation and termination reactions. These reactions often remain virtually irreversible at T_c, especially for free-radical intermediates. To obtain equation 4.1-15, we define K_p in the usual way

$$K_p \equiv \frac{c_{x+1}^*}{c_x^* c_m} \tag{4.1-17}$$

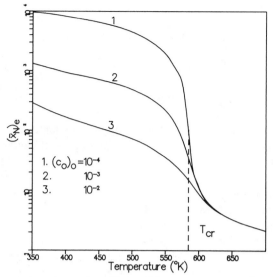

Figure 4.1-3. Parametric dependence of equilibrium DP versus temperature for addition polymerization with the following values: $\Delta H_i^0 = 2.24$ kcal/mol, $\Delta S_i^0 = -1.85$ e.u., $\Delta H_p^0 = -4.03$ kcal/mol, $\Delta S_p^0 = -6.9$ e.u., and $(c_m)_0 = 1$.

and then use the sum of the solution of this difference equation ($K_p c_m^x$) to rewrite the original definition as,

$$K_p = \frac{(c^* - c_1^*)}{c^* c_m} \cong c_m^{-1} \tag{4.1-18}$$

assuming that c_1^* is negligible compared to c^*. Applying the criterion for chemical

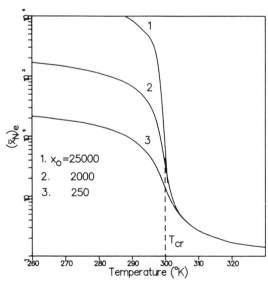

Figure 4.1-4. Parametric dependence of equilibrium DP versus temperature for addition polymerization with $K_i = K_p$ and the following values: $\Delta H_p^0 = -7.72$ kcal/mol, $\Delta S_p^0 = -27.6$ e.u., and $(c_m)_0 = 2.5$.

equilibrium (Appendix D)

$$0 = \Delta G_p = \Delta H_p - T\left(\Delta S_p^* + R_g \ln K_p\right) \tag{4.1-19}$$

and substituting the RHS of approximate equation 4.1-18 for K_p leads to equation 4.1-15.

The locus of equation 4.1-15 actually describes the reaction conditions required to make the formation of instantaneous high-polymer impossible, regardless of reaction history and polymer product formed prior to the establishment of such conditions. Replacing c_m with $(c_m)_0$ gives the critical value of T_c at which temperature it would be impossible to produce polymer from the outset. Since temperatures generally rise during the course of most chain-addition polymerizations, the locus $T_c(c_m)$ provides a valuable estimate of the upper temperature limits that must be avoided during a nonisothermal polymer reaction. In Fig. 4.1-5, graphs of this locus (equation 4.1-15) and the corresponding adiabatic temperature profile have been plotted in dimensionless form for bulk styrene polymerization.

It is instructive to examine the behavior of $(\bar{x}_N)_{inst}$ with rising temperatures from a kinetic viewpoint. To accomplish this, it is customary to define $(\bar{x}_N)_{inst}$ as follows:

$$
\begin{aligned}
(\bar{x}_N)_{inst} &\cong \frac{r_p}{\left(k_{tD} + \dfrac{k_{tc}}{2}\right)(c^*)^2} \\
&= \frac{2r_p}{(2 - R)k_t(c^*)^2}
\end{aligned}
\tag{4.1-20}
$$

where

$$r_p = \left(k_{pf}c_m - k_{pr}\right)c^* \tag{4.1-21}$$

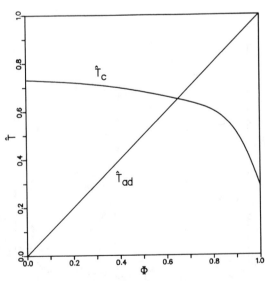

Figure 4.1-5. Dimensionless adiabatic temperature $\hat{T}_{ad}$ and ceiling temperature $\hat{T}_c$ for bulk styrene polymerization (42) where $\hat{T} \equiv (T - T_0)/(T_{ad} - T_0)$ and D-E is presumed not to occur.

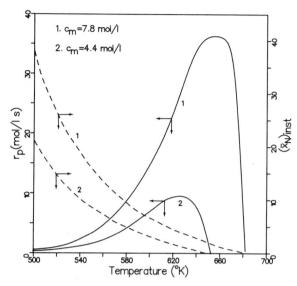

Figure 4.1-6. Rate of polymerization r_p and instantaneous DP, $(\bar{x}_N)_{inst}$ as a function of reaction temperature for two values of c_m^* and the following fixed parameters: $c_0 = 2.1 \times 10^{-4}$ mol/l, $f = 0.5$, $k_d = 1.58 \times 10^{15} \exp(-30,800/R_gT)$ s^{-1}, $k_{pf} = 1.05 \times 10^7 \exp(-23,760/R_gT)$ 1/mol s, $k_{pr} = 2.82 \times 10^{12} \exp(-7060/R_gT)$ s^{-1}, and $k_t = 1.23 \times 10^9 \exp(-1,680/R_gT)$ 1/mol s.

The reader is cautioned, however, that this definition contains the LCA (Section 1.13) and that the expression for r_p contains the assumption used earlier that c_1^* is negligible compared to c^*. By applying the QSSA, $c^* \cong (2fk_dc_0/k_t)^{1/2}$, we may obtain instantaneous DP and r_p as functions of temperature for any concentrations c_m and c_0. Such functions have been plotted in Fig. 4.1-6 for termination by disproportionation ($R = 0$).

The points of intersection of $(\bar{x}_N)_{inst}$ (or r_p) with the abscissa constitute the basis of Dainton's kinetic definition of the ceiling temperature. In Fig. 4.1-7, we have plotted the loci of such intersections as functions of c_m (upper scale) for several values of c_0. Each point on every curve corresponds to an entire $(\bar{x}_N)_{inst}$ or r_p curve in Fig. 4.1-6. The bottom curve in Fig. 4.1-7 is a graph of 4.1-15, which was obtained by assuming the following relationships between kinetics and thermodynamics

$$\frac{A_{pf}}{A_{pr}} = \exp(\Delta S_p^*/R_g) \qquad (4.1\text{-}22)$$

$$E_{pf} - E_{pr} = \Delta H_p \qquad (4.1\text{-}23)$$

It represents the locus of ceiling temperatures excluding the effect of initiator.

While the curves in Fig. 4.1-7 were actually based on arbitrary, independent values of c_m and c_0, the lower abscissa was labeled conversion Φ_m to remind us that the locus T_c can refer to a single polymerization in progress, as it did in Fig. 4.1-5. Furthermore, by comparing these curves we can estimate the sensitivity of T_c to changes in initiator concentration. The variables c_m and c_0 are, of course, actually

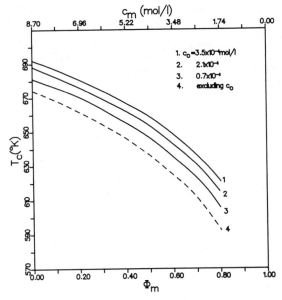

Figure 4.1-7. The effect of initiator concentration c_0 on the ceiling temperature-monomer conversion locus T_c versus Φ_m.

coupled to each other and to temperature during the course of a nonisothermal polymerization.

We shall now examine the effects of the assumption that c_1^* is negligible compared to c^* (59, 60). The precise rate functions for reversible propagation of free-radical intermediates are:

$$r_1^* = r_i - k_{pf}c_m c_1^* + k_{pr}c_2^* - k_t c_1^* c^* \qquad (4.1\text{-}24)$$

$$x > 1 \qquad r_x^* = k_{pf}c_m\left(c_{x-1}^* - c_x^*\right) - k_{pr}\left(c_x^* - c_{x+1}^*\right) - k_t c_x^* c^* \qquad (4.1\text{-}25)$$

$$r^* = r_i - k_t\left(c^*\right)^2 \qquad (4.1\text{-}26)$$

Application of the QSSA overall, $r^* = 0$, and in detail, $r_x^* = 0$ for all $x \geqslant 1$, yields

$$\frac{c_1^*}{c^*} = \frac{1 - \phi_f - \phi_r \xi}{1 - \phi_r \xi} \qquad (4.1\text{-}27)$$

for $x = 1$, and the following second-order difference equation for $x \geqslant 2$:

$$\phi_r c_{x+2}^* - c_{x+1}^* + \phi_f c_x^* = 0 \qquad (4.1\text{-}28)$$

with boundary conditions $c_\infty^* = 0$ and

$$c_2^* = \left(\frac{1 - \phi_r}{\phi_r}\right)c_1^* - \left(\frac{1 - \phi_f - \phi_r}{\phi_r}\right)c^* \qquad (4.1\text{-}29)$$

where probabilities of propagation and depropagation are respectively,

$$\phi_f \equiv \frac{k_{pf}c_m}{k_{pf}c_m + k_{pr} + k_t c^*} \tag{4.1-30}$$

and

$$\phi_r \equiv \frac{k_{pr}}{k_{pf}c_m + k_{pr} + k_t c^*} \tag{4.1-31}$$

and where

$$\xi \equiv \frac{1 - (1 - 4\phi_f\phi_r)^{1/2}}{2\phi_r} \tag{4.1-32}$$

The solution of equation 4.1-28 gives the intermediate DPD

$$x \geqslant 2 \qquad \frac{c_x^*}{c^* - c_1^*} = \xi^{x-2}(1 - \xi) \tag{4.1-33}$$

Substitution of equations 4.1-27 and 4.1-33 into the rate function $r_x = k_{tD}c_x^*c^*$, assuming irreversible termination by disproportionation only, and the result into the precise definition for instantaneous DP (Section 1.13) yields

$$(\bar{x}_N)_{inst} \equiv \frac{\Sigma x r_x}{\Sigma r_x} = \frac{2 - \xi - \dfrac{c_1^*}{c^*}}{1 - \xi} \tag{4.1-34}$$

$$(\bar{x}_W)_{inst} \equiv \frac{\Sigma x^2 r_x}{\Sigma x r_x} = \frac{(4 - 3\xi - \xi^2) - (3 - \xi)\dfrac{c_1^*}{c^*}}{(1 - \xi)\left(2 - \xi - \dfrac{c_1^*}{c^*}\right)} \tag{4.1-35}$$

It can be shown that at the ceiling temperature T_c, defined earlier by the condition $k_{pf}c_m = k_{pr}$ and now by the equivalent condition $\phi_f = \phi_r$, the instantaneous DP's are functions of a single parameter

$$\zeta \equiv \frac{\phi_r}{\phi_t} = \frac{k_{pr}}{k_t c^*} \tag{4.1-36}$$

where ϕ_t is the probability of termination, $\phi_t = 1 - \phi_f - \phi_r$. Thus,

$$\xi|_{T_c} = \frac{1 + 2\zeta - (1 + 4\zeta)^{1/2}}{2\zeta} \tag{4.1-37}$$

and

$$\left(\frac{c_1^*}{c^*}\right)\bigg|_{T_c} = \frac{1 + \zeta(1 - \xi)}{1 + \zeta(2 - \xi)} \tag{4.1-38}$$

The results have been plotted in Fig. 4.1-8, which show that the definition of T_c as the temperature at which the production of large DP is impossible is accurate only

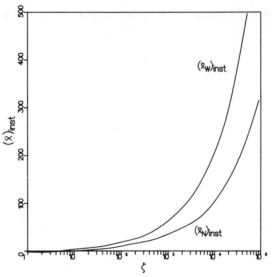

Figure 4.1-8.   The effect on the instantaneous DP's $(\bar{x}_N)_{inst}$ and $(\bar{x}_W)_{inst}$, at the ceiling temperature, T_c, of the assumption of equal probabilities of propagation and depropagation, $\phi_f = \phi_r$.

for small values of ζ. For large values of ζ, large DP's are still attainable at T_c. More specifically, $(\bar{x}_N)_{inst} < 10$ and $(\bar{x}_W)_{inst} < 20$ when $\zeta < 10^2$, but $(\bar{x}_N)_{inst} > 100$ and $(\bar{x}_W)_{inst} > 200$ when $\zeta > 10^4$. It is noteworthy that $c_1^*/c^* \to 1$ and $(D_N)_{inst} \to 1$ as $\zeta \to 0$ but $c_1^*/c^* \to 0$ and $(D_N)_{inst} \to 2$ as $\zeta \to \infty$. A small value of ζ signifies rapid termination of chains before they can grow.

Another temperature that is occasionally of concern in polymerization engineering is the glass transition temperature T_g of the polymer being produced. It is

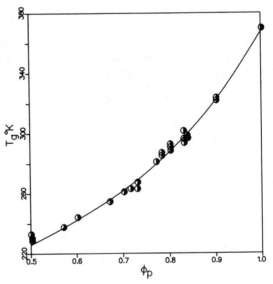

Figure 4.1-9.   Glass transition temperature T_g versus volume fraction of polymer ϕ_p for solutions of polystyrene ($\bar{M}_W = 190{,}000$) in styrene (4.)

prudent to compare T_g to the reaction temperature T chosen. If $T < T_g$, it is possible that reaction will cease at a limiting conversion due to hindered propagation. This occurs when T_g of the reacting solution (polymer in monomer or diluent; or monomer in polymer) reaches the value of the reaction temperature T as the solution becomes richer in polymer. At this point, monomer cannot diffuse rapidly enough to sustain polymerization because the reacting solution essentially freezes (vitrifies).

Such limiting conversion has been observed in the homogeneous polymerization of MMA (Section 2.9-1) as well as in the polymer-rich phase of precipitous VC polymerization (Section 2.9-2). The effects of concentration on T_g are illustrated in Fig. 4.1-9 for solutions of polystyrene in ethylbenzene. The experimental values of T_g were measured on a differential scanning calorimeter (DSC) and are presented along with a computed curve (4). The latter was computed using the principle of additivity of free volumes (total free volume is the sum of the free volumes of monomer and polymer).

4.2. OPTIMAL TEMPERATURE POLICIES

The highly exothermic nature of polymerizations coupled with the high viscosities of the reaction media makes temperature control very difficult to implement. It is therefore infrequent that temperature can be treated as a parameter to be manipulated in accordance with a predetermined policy. There are, however, special circumstances in which the polymer product properties must be maintained to such close tolerances that the design complications created by temperature programming are outweighed by the potential benefits. Of equal importance in our consideration of optimal temperature paths is the hope that we will discover that certain types of commonly occurring temperature histories are nearly optimal with respect to some desirable design criterion (e.g., minimum D_N).

When temperature is treated as a parameter, the objective is to define a temperature trajectory, $T(t)$, that minimizes some objective function subject to practical constraints. For instance, we might wish to minimize reaction time, or breadth of the DPD subject to a target value of the final average degree of polymerization $(\bar{x}_N)_f$. Frequently there are multiple objectives, and weighing the relative import of each may require a certain amount of subjective judgment. It must also be remembered that there may be practical limitations imposed on the temperature policy such as upper and lower bounds, $T_S \leq T \leq T_L$. Coolant temperatures, phase transitions, ceiling temperatures, degradation temperatures, and explosive decompositions could all constrain the range of temperatures through which control is possible. There may also be constraints on the state variables. For instance, a polymerization might experience D-E before a desired conversion level is reached. The mathematical formulation of a performance criterion subject to certain constraints is the starting point for our discussion of optimization.

The performance criterion I, which represents the quantity to be minimized or maximized, may take on a variety of forms. The objective may be simply to minimize reaction time,

$$I = t_f = \int_0^{t_f} dt \qquad (4.2\text{-}1)$$

or, we might seek to minimize the dispersion index, D_N, for a fixed value of $(\bar{x}_N)_f$. If $\bar{x}_N$ is fixed at a given level of conversion then μ^1 and μ^0 are determined (Section 1.7), and D_N is directly proportional to μ^2. The performance index would therefore be:

$$I = \mu_2(t_f) = \int_0^{t_f} \left(\frac{d\mu_2}{dt} \right) dt \qquad (4.2\text{-}2)$$

The measure of performance might be the cumulative deviation of a specified property from a predetermined set point. For instance, if it is desired to keep $\bar{x}_N$ constant, at some reference value $(\bar{x}_N)_r$, the problem could be stated as

$$I = |\varepsilon|^2 = \int_0^{t_f} \left[\bar{x}_N - (\bar{x}_N)_r \right]^2 dt \qquad (4.2\text{-}3)$$

Finally, the performance might be the sum of several weighted objectives. During plant start-up, one might wish to minimize the time to reach steady state while also minimizing the deviation from the desired final value of $\bar{x}_N$. This would require a performance criterion of the form:

$$I = \alpha_1 t_f + \alpha_2 |\varepsilon|^2 = \int_0^{t_f} \left\{ \alpha_1 + \alpha_2 \left[\bar{x}_N - (\bar{x}_N)_r \right]^2 \right\} dt \qquad (4.2\text{-}4)$$

where the α_1 and α_2 are coefficients reflecting the relative importance of each particular objective.

It is clear that for most cases the performance index to be minimized cannot be expressed as an explicit function of the state variables. More often it is a functional, that is, a function of certain parametric functions, such as a set of control policies $w_j(t)$. Once the performance index has been established, there are a variety of approaches for finding the control strategy that gives a minimum (5–7). The most frequently used technique is Pontryagin's principle (8), the details of which are outlined in Appendix C. The strength of this method is that intermediate calculations frequently yield important information about the nature of the optimal path. For instance, treatment of the minimum dispersion problem will show that the temperature history that keeps $(\bar{x}_N)_{inst}$ constant is an optimal path. This is an observation that has been advanced by several investigators (9, 10) based on intuitive arguments and is borne out mathematically by Pontryagin's principle.

This section will focus on two particular applications of this principle. First, we shall consider temperature histories that minimize reaction time, and second, histories that either maximize or minimize the breadth of the DPD.

4.2.1. Minimum Time Policies

The minimum time problem has been examined by Yoshimoto (11), Sacks et al. (6, 12), and Chen (13), among others (14, 15), for free-radical polymerizations. Each investigator used slightly different kinetic models. Yoshimoto examined the thermal polymerization of styrene. Sacks looked at chemically initiated free-radical polymerization with and without gel-effect, as did Chen, who also optimized makeup initiator feed concentration in addition to temperature path. Laurence (16) recently reexamined the work of Yoshimoto using different initiation kinetics and offered experimental results.

Let us first consider free-radical thermal polymerization in a batch reactor. This is the simplest mechanism since there is only one reactant, the monomer.

The state variables are $\hat{c}_m$, $\hat{\mu}^0$, $\hat{\mu}^1$, and $\hat{\mu}^2$. The constraint equations (Section 1.7) show that $\hat{\mu}^1 = 1 - \hat{c}_m$, so we need not consider $\hat{\mu}^1$. For minimum time, the performance index is given by the negative of equation 4.2-1, since maximizing $-t_f$ is the same as minimizing t_f. The Hamiltonian is written as (Appendix C)

$$H = -1 + z_1 f_1 + z_2 f_2 + z_3 f_3 = 0 \qquad (4.2\text{-}5)$$

where:

$$f_1\left(\hat{c}_m, \hat{T}\right) \equiv \frac{d\hat{c}_m}{dt} \qquad (4.2\text{-}6)$$

$$f_2\left(\hat{c}_m, \hat{T}\right) \equiv \frac{d\hat{\mu}^0}{dt} \qquad (4.2\text{-}7)$$

$$f_3\left(\hat{c}_m, \hat{T}\right) \equiv \frac{d\hat{\mu}^2}{dt} \qquad (4.2\text{-}8)$$

and the functional forms of these derivatives are given by the balance equations in Table 4.2-1. Let us consider, for the moment, polymerization in the absence of gel-effect, for which $g \equiv 1$. By definition (Appendix C)

$$\frac{dz_3}{dt} \equiv \frac{\partial H}{\partial \hat{\mu}^2} = 0 \qquad (4.2\text{-}9)$$

by virtue of the fact that $H(\hat{c}_m, \hat{T})$ is not an explicit function of $\hat{\mu}^2$. Furthermore, since D_N is unspecified at the end of the reaction $\hat{\mu}^2(t_f)$ is also unspecified. Thus we are free to pick the final value of the corresponding adjoint variable z_3, and we choose $z_3(t_f) = 0$. From this result, together with equation 4.2-9, we conclude that $z_3(t)$ is zero for all t. Consequently, the Hamiltonian (equation 4.2-5) simplifies to:

$$H = 1 + z_1 f_1 + z_2 f_2 \qquad (4.2\text{-}10)$$

The rate of thermal initiation can generally be expressed (Section 2.8.4) as:

$$r_i = k_i(T) c_m^p \qquad (4.2\text{-}11)$$

Yoshimoto (11) let $p = 2$ while Laurence (16) used the model of Hui and Hamielec (17) with $p = 3$. Both are special cases of the balances in Table 4.2-1. Note that a CT for chain transfer to monomer, not previously defined, has been included. Its definition follows directly from the procedure described in Appendix G.

$$\lambda_{fm} = \frac{(c_m)_0}{(r_{fm})_0} \qquad (4.2\text{-}12)$$

Table 4.2-1. Semidimensionless Balance Equations for Thermally Initiated Chain-Addition Polymerization

$$\frac{d\hat{c}_m}{dt} = -\lambda_m^{-1}\hat{r}_p - \lambda_{fm}^{-1}\hat{r}_{fm} - \lambda_i^{-1}\hat{r}_i$$

$$\frac{d\hat{\mu}^o}{dt} = \left(\frac{2-R}{2}\right)\lambda_i^{-1}\hat{r}_i + \lambda_{fm}^{-1}\hat{r}_{fm}$$

$$\frac{d\hat{\mu}^1}{dt} = -\frac{d\hat{c}_m}{dt}$$

$$\frac{d\hat{\mu}^2}{dt} = (1+R)\lambda_i^{-1}\hat{r}_i + 3\lambda_m^{-1}\hat{r}_p + \frac{\lambda_i^{-1}\hat{r}_i\left[2R\,\lambda_i^{-1}\lambda_m^{-1}\hat{r}_i\hat{r}_p + (2+R)\lambda_m^{-2}\hat{r}_p^2\right]}{(\lambda_i^{-1}\hat{r}_i + \lambda_{fm}^{-1}\hat{r}_{fm})^2}$$

$$+ \lambda_{fm}^{-1}\hat{r}_{fm}\left\{1 + \frac{2R\,\lambda_i^{-1}\lambda_m^{-1}\hat{r}_i\hat{r}_p + 2\lambda_m^{-2}\hat{r}_p^2}{(\lambda_i^{-1}\hat{r}_i + \lambda_{fm}^{-1}\hat{r}_{fm})^2}\right\}$$

$$\hat{r}_i = \hat{c}_m^P \exp \hat{E}_i\hat{T}/(1+\hat{T})$$

$$\hat{r}_p = \frac{\hat{c}_m^{1+p/2}}{g^{1/2}} \exp \hat{E}_{ap}\hat{T}/(1+\hat{T})$$

$$\hat{r}_{fm} = \frac{\hat{c}_m^{1+p/2}}{g^{1/2}} \exp \hat{E}_{af}\,\hat{T}/(1+\hat{T})$$

and

$$\hat{E}_{af} = \hat{E}_{ap} - \hat{E}_p + \hat{E}_{fm}$$

$$\hat{E} \equiv E/R_g T_o$$

$$\hat{T} \equiv (T - T_o)/T_o$$

$$g \equiv g(\hat{c}_m, \hat{\mu}^o, \hat{\mu}^2, \hat{T})$$

The reader might recognize the ratio λ_m/λ_{fm} as the chain transfer constant at the feed temperature [i.e., $(k_{fm})_0/(k_p)_0$].

Consider first the reaction sequence neglecting chain transfer. To verify that the unconstrained policy forms part of the optimal solution, we must check the sign of $\partial^2 H/\partial \hat{T}^2$ (Appendix C), when $H = 0 = \partial H/\partial \hat{T}$. Performing these operations on equation 4.2-10 verifies that the Hamiltonian is maximized for all values of temperature, and that it therefore describes the optimal policy between the bounds $\hat{T}_S \leqslant \hat{T} \leqslant \hat{T}_L$.

Useful physical information can be gained by examining the expressions for the adjoint variables z_j. Since the zeroth moment of the DPD, $\hat{\mu}^0$, does not appear explicitly in the Hamiltonian,

$$\frac{dz_2}{dt} \equiv \frac{\partial H}{\partial \hat{\mu}^0} = 0 \tag{4.2-13}$$

from which we conclude that z_2 remains constant, as we did earlier for z_3. However, since the final values of $\hat{c}_m$ and $\bar{x}_N$ are specified, and thus $\hat{\mu}^1$ and $\hat{\mu}^0$ are as well, $z_2(t_f)$ cannot be arbitrarily chosen.

Simultaneous solution of $H = 0$ and $\partial H/\partial \hat{T} = 0$ for $z_2(t)$, which must be a constant, yields

$$z_2(t) = z_2(0) = \frac{\dfrac{2}{2-R}\left(\dfrac{\hat{E}_d \lambda_i^{-1} \hat{r}_i}{\hat{E}_{ap} \lambda_m^{-1} \hat{r}_p} + 1\right)}{\lambda_i^{-1} \hat{r}_i \left(1 - \dfrac{\hat{E}_d}{\hat{E}_{ap}}\right)} \tag{4.2-14}$$

Making use of the LCA ($\hat{r}_p \gg \hat{r}_i$), the preceding equation can be further simplified and subsequently rearranged to show that

$$\hat{r}_i = \frac{2\lambda_i}{(2-R)z_2(0)\left(1 - \dfrac{\hat{E}_d}{\hat{E}_{ap}}\right)} \tag{4.2-15}$$

All parameters on the RHS of equation 4.2-15 are independent of time; indeed the RHS reduces to unity, implying that the optimal policy maintains the rate of a initiation constant and equal to its initial value.

Yoshimoto et al. (11) treated the thermal polymerization of styrene using second-order initiation kinetics while also allowing for chain transfer to monomer. We recast their result for z_2, but extend it to include pth order initiation and termination by combination or disproportionation to show that ($\hat{E}_{af}$ is defined in Table 4.2-1)

$$z_2 = \frac{\hat{E}_d \lambda_i^{-1} \hat{r}_i + \hat{E}_{ap} \lambda_m^{-1} \hat{r}_p + \hat{E}_{af} \lambda_{fm}^{-1} \hat{r}_{fm}}{\dfrac{2-R}{2} \lambda_i^{-1} \lambda_m^{-1} \hat{r}_i \hat{r}_p (\hat{E}_{ap} - \hat{E}_d) + \dfrac{R}{2} \lambda_i^{-1} \lambda_m^{-1} \hat{r}_i \hat{r}_{fm} (\hat{E}_d - \hat{E}_{af}) + \lambda_m^{-1} \lambda_{fm}^{-1} (\hat{E}_{ap} - \hat{E}_{af})} \tag{4.2-16}$$

This result reduces to equation 4.2-14 when transfer is absent. Furthermore, if the temperature dependence of the transfer coefficient is small [$E_{fm} \cong E_p$ so that $k_{fm}/k_p \cong$ constant $= (k_{fm}/k_p)_0$], use of the LCA gives the same result (viz., that temperature programming should keep the rate of initiation constant). Since monomer concentration decreases monotonically with time, it follows that temperature

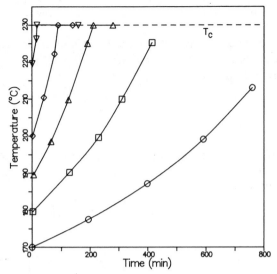

Figure 4.2-1. Optimal temperature policies for minimum reaction time of styrene thermal polymerization with initial temperature as parameter (11).

must rise so that an increasing value of $k_i(T)$ will balance the reactant decay and maintain the overall rate of initiation constant.

Figure 4.2-1 shows the temperature profiles obtained by Yoshimoto using the kinetic constants for thermally initiated styrene bulk polymerization carried to 80-percent conversion. To evaluate the effectiveness of temperature programming, the authors chose values of feed temperature T_0. Isothermal polymerizations were simulated and the values of $(\bar{x}_N)_f$ and t_f were identified as those values occurring when $\Phi_m = 0.80$. Temperature was then programmed starting from T_0, and the time required to reach $(\bar{x}_N)_f$ at $\Phi_m = 0.80$ was computed. The authors found that temperature programming consistently reduced the reaction time by roughly 30 percent compared to the corresponding isothermal reaction.

Note that not all the curves in Fig. 4.2-1 are unconstrained. The authors estimated the ceiling temperature for styrene polymerization (Section 4.1) to be 230°C. Consequently, a different policy is required when the temperature computed from the unconstrained policy exceeds $T_L = 230°C$. The condition that must be met is $\partial H / \partial \hat{T} < 0$, and this is met by maintaining $T = T_L = T_c$.

Sacks et al. (6) examined the minimum time problem for chemically initiated, free-radical polymerizations. Transfer reactions were not considered and initiator efficiency was assumed constant, but the gel-effect (G-E) was simulated by making the termination constant dependent on monomer conversion. Its functional dependence was found by curve-fitting experimental data within the range of conditions used in the simulation work.

The rate functions are listed in Table 4.2-2. The QSSA was applied and initiator efficiency was assumed to be constant. The G-E function here is different from the simple exponential form used in Chapter 2. For styrene, the authors used

$$g = \begin{cases} 1.0 & 0.7 < \hat{c}_m < 1.0 \\ \left(1.522 - 1.818(1 - \hat{c}_m)\right)^2 & 0.2 < \hat{c}_m < 0.7 \end{cases} \qquad (4.2\text{-}17)$$

while for MMA they used

$$
g = \begin{cases}
1 & 0.85 < \hat{c}_m < 1.0 \\
\left[1 - 28.72(1 - \hat{c}_m) + 228.2(1 - \hat{c}_m)^2 \right. & 0.40 < \hat{c}_m < 0.85 \\
\left. - 239.9(1 - \hat{c}_m)^3\right]^2 &
\end{cases} \quad (4.2\text{-}18)
$$

The state variables were $\hat{c}_0$, $\hat{c}_m$, $\hat{\mu}^0$, and $\hat{\mu}^2$ and the pertinent balance equations are listed in Table 4.2-2. As before, the second moment balance can be eliminated from the Hamiltonian since $\hat{\mu}^2$ does not appear explicitly in any of the balances and the final value is not constrained. Note that if $(D_N)_f$ had been specified, $\hat{\mu}^2$ would have been required as a state variable. Also, since $\hat{\mu}^0$ is proportional to $(1 - \hat{c}_0)$ (Section

Table 4.2-2. Semidimensionless Balances for Chemically Initiated Chain-Addition Polymerization

$$
\frac{d\hat{c}_o}{dt} = -\lambda_i^{-1}\hat{r}_o
$$

$$
\frac{d\hat{c}_m}{dt} = -\frac{\hat{r}_p}{\lambda_m} - \frac{\hat{r}_i}{x_o \lambda_i}
$$

$$
\frac{d\hat{\mu}_o}{dt} = \left(\frac{2 - R}{2}\right)\frac{\hat{r}_i}{x_o \lambda_i}
$$

$$
\frac{d\hat{\mu}^1}{dt} = -\frac{d\hat{c}_m}{dt}
$$

$$
\frac{d\hat{\mu}^2}{dt} = (1 + R)\frac{\hat{r}_i}{x_o \lambda_i} + (3 + 2R)\frac{\hat{r}_p}{\lambda_m} + (2 + R)\frac{x_o \lambda_i \hat{r}_p^2}{\lambda_m^2 \hat{r}_i}
$$

DIMENSIONLESS RATE FUNCTIONS

$$
\hat{r}_o = \hat{c}_o \exp \hat{E}_i \hat{T}/(1 + \hat{T})
$$

$$
\hat{r}_i = 2f\hat{r}_o
$$

$$
\hat{r}_p = \hat{c}_m \hat{c}_o^{1/2} \exp \hat{E}_{ap} \hat{T}/(1 + \hat{T})
$$

1.7) we can eliminate it as well. This reduces the expression for the Hamiltonian to:

$$H = -1 - z_1\lambda_i^{-1}\hat{r}_0 - z_2\left(\lambda_m^{-1}\hat{r}_p + x_0^{-1}\lambda_i^{-1}\hat{r}_i\right) \tag{4.2-19}$$

Specifying the target values for conversion and $(\bar{x}_N)_f$ give the boundary conditions:

$$\hat{c}_m(0) = 1 \qquad\qquad \hat{c}_m(t_f) = (\hat{c}_m)_f \tag{4.2-20}$$

$$\hat{c}_0(0) = 1 \qquad\qquad \hat{c}_0(t_f) = 1 - \frac{1 - (\hat{c}_m)_f}{(2 - R)f(\bar{x}_N)_f} \tag{4.2-21}$$

As in the case of thermal polymerization, the examination of the equations for the adjoint variables yields important information. In this case, the solution for z_2 shows that

$$\frac{z_2\hat{c}_m}{g^{1/2}} = \text{const} \tag{4.2-22}$$

where

$$g \equiv \frac{k_t}{(k_t)_0} \tag{4.2-23}$$

When combined with $H = 0$ and $\partial H/\partial\hat{T} = 0$, the constrained policy shows that:

$$\frac{\hat{r}_p}{\dfrac{\hat{c}_m}{g^{1/2}}} = k_{ap}(\hat{T})\hat{c}_0^{1/2} = \text{const} \tag{4.2-24}$$

Again, the optimal temperature policy acts to compensate for the decline in initiation rate due to reactant consumption, by increasing the lumped-rate constant. Along the optimal path, the monomer balance can be recast in the following form by virtue of equation 4.2-24:

$$\frac{d\hat{c}_m}{dt} = \frac{K[\hat{T}(t)]\hat{c}_m}{g^{1/2}(\hat{c}_m)} \tag{4.2-25}$$

where

$$K[\hat{T}(t)] = \lambda_m^{-1}\hat{c}_0^{1/2}\exp\frac{\hat{E}_{ap}\hat{T}}{1 + \hat{T}} = K(t = 0) = \lambda_m^{-1} \tag{4.2-26}$$

In the absence of G-E, the optimal policy is one that creates a pseudo-first-order, isothermal reaction. From equation 4.2-26, it should be clear that when temperature is unconstrained, $T(t)$ will rise monotonically in response to the decay of $\hat{c}_0$. Figures 4.2-2 and 4.2-3 show examples of computed optimal temperature histories. In each case, $(\bar{x}_N)_f = 1000$ for $(\hat{c}_m)_f = 0.50$.

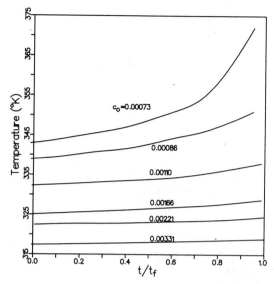

Figure 4.2-2. Optimal temperature policies for minimum reaction time for initiated styrene polymerization with $(\bar{x}_N)_f = 1000$ and $(\hat{c}_m)_f = 0.5$. Feed initiator concentration is a parameter (13).

Several generalizations can be made. Neglect G-E for the moment. If the polymerization is "conventional" (i.e., $\alpha_K \ll 1$), then the optimal policy is essentially isothermal. This follows intuitively from equation 4.2-26. If initiator concentration remains essentially constant of its own accord, then there is no need to counterbalance the effect of rising temperatures. If the reaction is a dead-end (D-E) polymerization, then an increasing temperature profile not only measurably reduces

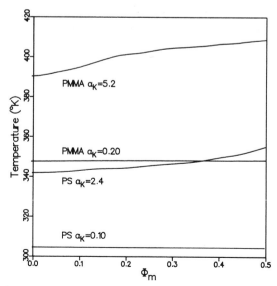

Figure 4.2-3. Optimal temperature policies for minimum reaction time with and without gel-effect, and $(\bar{x}_N)_f = 1000$, $(\hat{c}_m)_f = 0.5$ (6).

reaction time, but the resulting polymer is less disperse than that from an isothermal reaction. The temperature increase offsets the isothermal drift dispersion discussed in Chapter 2. It is interesting to note that thermally initiated systems can be much like D-E polymerizations with respect to DP drift:

$$(\bar{x}_N)_{inst} = \left(\frac{2}{2-R}\right)\frac{k_p}{\sqrt{k_i k_t}}c_m^{1-n/2} \tag{4.2-27}$$

Provided that $n \geq 2$, $(\bar{x}_N)_{inst}$ will either remain constant or increase with conversion. Thus, as long as $E_{ap} < E_d$ (which is generally true), the rising temperature along the programmed path will counteract the effect of drift dispersion.

A more recent work by Chen and Jeng (13) confirms the suspicion that D-E polymerizations offer the most advantageous way of reducing reaction time. These workers realized that the optimal path depends on the choice of $(c_0)_0$, which follows from equation 4.2-26. Figure 4.2-4 shows how the reaction time to reach target $(\bar{x}_N)_f$ and $(\hat{c}_m)_f$ decreases in proportion to $(c_0)_0$. There is a limitation in that initiator concentration must be large enough to prevent D-E prior to attainment of $(\hat{c}_m)_f$. Thus these workers concluded that the shortest time is achieved when the initiator feed concentration is fixed at a value that produces D-E just as $(\hat{c}_m)_f$ is reached. Having chosen a value of $(c_0)_0$, their analysis for $\hat{T}(t)$ is identical to that of Sacks (6), and the same qualitative observations apply.

Chen and Jeng performed experiments using styrene initiated by AIBN. They approximated the optimum temperature history by a series of small step changes in temperature. Figure 4.2-5 shows a comparison between experimental observations and theoretical predictions of conversion history for both the best isothermal policy and the optimal policy. Considering the lack of precision in the G-E model, the

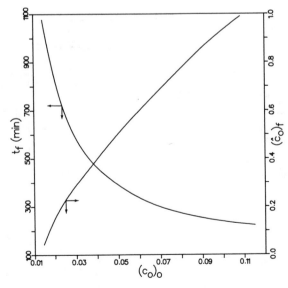

Figure 4.2-4. Optimal reaction time as a function of feed initiator concentration when $(\bar{x}_N)_f = 500$ and $(\hat{c}_m)_f = 0.5$ (13).

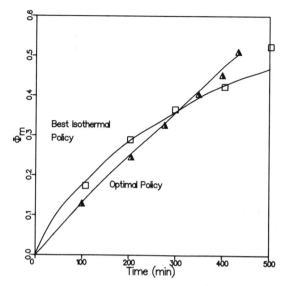

Figure 4.2-5. Comparison of isothermal polymerization conversion history with that resulting from optimal temperature programming, $(\bar{x}_N)_f = 1000$, $(\hat{c}_m)_f = 0.5$ (13).

agreement is excellent. Note further that temperature programming affords a 20-percent decrease in reaction time.

For most free-radical polymerizations, the reaction time is minimized by a gently increasing temperature history. However, the sharply increasing temperature policies required for polymerizations with strong D-E might be impractical to control. The required profiles, in fact, resemble those of a mild runaway reaction. This point will be examined again later.

4.2.2. Optimal Property Policies

In certain polymer product applications meeting and controlling property specifications is of paramount importance. Minimizing the deviation among these specifications from the desired values could thus be more critical than reducing reaction time. As an example, we shall apply Pontryagin's principle to the problem of minimizing or maximizing the dispersion index D_N.

From the discussion of dispersion in Chapter 2, we know that the best one can do with chain-addition polymerizations to obtain a small value of cumulative D_N is to eliminate the dispersion due to concentration drift and to "settle" for the value of $(D_N)_{inst}$ imposed by statistical dispersion. Several authors (9, 10) have argued that when $(\bar{x}_N)_{inst}$ is maintained constant at $(\bar{x}_N)_{inst}|_0$, D_N will be minimized (cf. Section 2.7). Pontryagin's principle will confirm this and will additionally specify the policies that maximize D_N quantitatively.

Sacks et al. (12, 18) used this principle to find temperature histories that produce extrema in D_N for free-radical polymerization with G-E. The state variables were $\hat{c}_0$, $\hat{c}_m$, $\hat{\mu}^0$, and $\hat{\mu}^2$. As in the minimum time problem, we can eliminate $\hat{\mu}^0$ since it is directly proportional to $\hat{c}_0$. If target values of $(\hat{c}_m)_f$ and $(\bar{x}_N)_f$ are specified, stoichiometry implies that $(\hat{\mu}^0)_f$ and $(\hat{\mu}^1)_f$ are essentially fixed. By expressing D_N as

$$D_N = \frac{\hat{\mu}^2/\hat{\mu}^1}{\hat{\mu}^1/\hat{\mu}^0} \tag{4.2-28}$$

it should be clear that an extremum in $(D_N)_f$ corresponds to an extremum in $(\hat{\mu}^2)_f$. Thus, if $(\hat{\mu}^1)_f$ and $(\hat{\mu}^0)_f$ are specified, the performance index to be maximized is

$$I(t_f) = \pm\hat{\mu}^2(t_f) = \int_0^{t_f} \pm \frac{d\hat{\mu}^2}{dt}\, dt \tag{4.2-29}$$

Obviously, the upper sign refers to maximum D_N. Since minimizing $\hat{\mu}^2$ is the same as maximizing $-\hat{\mu}^2$, the lower sign shall hereafter refer to minimum D_N. The Hamiltonian for both maximum and minimum can be expressed as

$$H = \frac{-z_1 \hat{r}_0}{\lambda_i} - z_2 \left(\frac{\hat{r}_p}{\lambda_m} + \frac{\hat{r}_i}{x_0 \lambda_i} \right)$$
$$+ (z_3 \pm 1) \left[\frac{(1 \mp R)\hat{r}_i}{x_0 \lambda_i} + \frac{(3+2R)\hat{r}_p}{\lambda_m} + \frac{(2+R)x_0 \lambda_i \hat{r}_p^2}{\lambda_m^2 \hat{r}_i} \right] \tag{4.2-30}$$

When $H = 0$ and $\dfrac{\partial H}{\partial \hat{T}} = 0$, we find that

$$\frac{\partial^2 H}{\partial \hat{T}^2} = \pm \hat{E}_d \hat{E}_{ap} \tag{4.2-31}$$

Thus the sign of $\partial^2 H/\partial \hat{T}^2$ is negative for minimum D_N and positive for maximum D_N. This means that the unconstrained policy can only be used to achieve minimum D_N (Appendix C). To obtain maximum D_N, the temperature policy must consist of step changes between T_S and T_L ("bang-bang" control policy).

Looking, first, at the solution for minimum dispersion, we find that along the optimal path:

$$\frac{2x_0}{(2-R)} \frac{\hat{r}_p}{\hat{r}_i} = \frac{z_2 - 2 - 2R}{2+R} \tag{4.2-32}$$

which combines with the definition of $(\bar{x}_N)_{inst}$ to give

$$(\bar{x}_N)_{inst} = \frac{z_2 - 2 - 2R}{(2+R)(2-R)} + \frac{2}{(2-R)} \tag{4.2-33}$$

Furthermore, substituting $H = 0$ and $\partial H/\partial \hat{T} = 0$ into the expression for dz_2/dt shows the latter to be zero, so z_2 is a constant with time. Thus equation 4.2-33 verifies the observation that D_N can be minimized by programming temperature to maintain a constant $(\bar{x}_N)_{inst}$. The constrained policy can be found in (12), but since the magnitude of the temperature variations required were always found to be rather small, the bounding temperatures T_S and T_L were never encountered by the authors in their simulations.

Maintaining a constant value of $(\bar{x}_N)_{inst}$ will require qualitatively different temperature policies for D-E than for CONV polymerizations. From previous discussions on drift, we know that isothermal $(\bar{x}_N)_{inst}$ can drift upward due to D-E or strong G-E, downward when polymerization is CONV, or downward followed by upward when G-E is moderate (Section 2.8 and 2.9). We must therefore require that all optimum policies reverse the drift in $(\bar{x}_N)_{inst}$ that would otherwise occur under isothermal conditions due to composition drift. Thus, rising optimal temperature policies are expected for D-E polymerizations to offset the tendency of isothermal $(\bar{x}_N)_{inst}$ to rise, and declining optimal temperature policies are expected for CONV polymerizations to offset the tendency of isothermal $(\bar{x}_N)_{inst}$ to drop.

Table 4.2-3 summarizes the qualitative behavior of the various optimal temperature policies, and Figs. 4.2-6–4.2-8 present the quantitative results of Sacks using

Table 4.2-3.   Minimum D_N

Case I – Decreasing isothermal $\left(\bar{x}_N\right)_{inst}$ as in CONV polymerization

   (a) $T_o = T_S$; $T_{opt} = T_S$ throughout

   (b) $T_S < T_o < T_L$

   T_{opt} decreases to maintain $\left(\bar{x}_N\right)_{inst}$ constant until $T_{opt} = T_S$, after which it remains at this value

   (c) $T_o = T_L$; $T_{opt} = T_L$ until (iii)* becomes an equality; if it does then T_{opt} decreases to maintain $\left(\bar{x}_N\right)_{inst}$ constant until $T_{opt} = T_S$, after which it remains at this value

Case II – Increasing isothermal $\left(\bar{x}_N\right)_{inst}$ as with strong D-E or G-E

   (a) $T_o = T_S$; $T_{opt} = T_S$ until (i) becomes an equality; if it does then T_{opt} increases to maintain $\left(\bar{x}_N\right)_{inst}$ constant until $T_{opt} = T_L$, after which it remains at this value.

   (b) $T_S < T_o < T_L$; T_{opt} increases to maintain $\left(\bar{x}_N\right)_{inst}$ constant until $T_{opt} = T_L$, after which it remains at this value

   (c) $T_o = T_L$; $T_{opt} = T_L$ throughout

Table 4.2-3   (*Continued*)

Case III -   Decreasing followed by increasing $\left(\bar{x}_N\right)_{inst}$ as with moderate D-E or G-E

(a) $T_o = T_S$; $T_{opt} = T_S$ as in *Case I* (a) until (i) becomes an equality; if it does, then T_{opt} increases as in *II* (a) to a value not in excess of T_L

(b) $T_S < T_o < T_L$; T_{opt} decreases as in *Case I* (b), to a value no less than T_S ; then it increases as in *II* (b) until $T_{opt} = T_L$, after which it remains at this value

(c) $T_o = T_L$; $T_{opt} = T_L$ as in *Case I* (c) until (iii) becomes and equality; if it does then T_{opt} decreases to a value no less than T_S and subsequently increases again to a value not in excess of T_L .

---

*where:

(i) $$\frac{2\hat{r}_p x_o}{(2-R)\hat{r}_i} > \frac{Z_2 - 2 - 2R}{(2+R)(2-R)} \qquad \hat{T} = \hat{T}_S$$

(ii) $$\frac{2\hat{r}_p x_o}{(2-R)\hat{r}_i} = const \qquad \hat{T}_S \leq \hat{T} \leq \hat{T}_L$$

(iii) $$\frac{2\hat{r}_p x_o}{(2-R)\hat{r}_i} < \frac{Z_2 - 2 - 2R}{(2+R)(2-R)} \qquad \hat{T} = \hat{T}_L$$

---

kinetic constants for styrene and methyl methacrylate initiated by AIBN. The most significant improvement in D_N was obtained with strong G-E (MMA) and with moderate G-E in combination with D-E, in which case isothermal D_N could be reduced by as much as 50 percent. The foregoing analysis was predicated on the assumption that the specified target values of $(\hat{c}_m)_f$ and $(\bar{x}_N)_f$ were physically attainable. This might actually not be the case. If $(c_0)_0$ is too large and the isothermal drift is downward, then a high value of $(\bar{x}_N)_f$ may never be reached. On the other hand, if $(c_0)_0$ is too small, D-E might occur before $(\hat{c}_m)_f$ is reached. These constraints may be formalized as limits of controllability, given by:

$$\bar{x}_N|_{(c_m)_f, \hat{T}_L} \leq (\bar{x}_N)_f < \bar{x}_N|_{(\hat{c}_m)_f, \hat{T}_S} \tag{4.2-34}$$

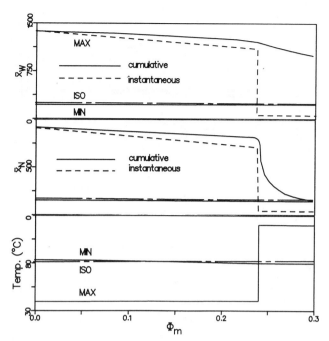

Figure 4.2-6. Optimal temperature program with resulting DP behavior for an isothermal conventional polymerization (12).

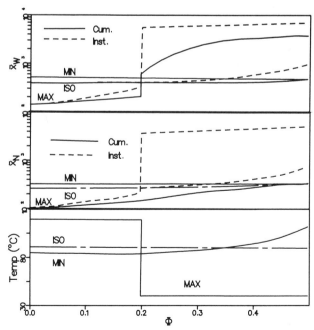

Figure 4.2-7. Optimal temperature histories with resulting DP behavior for an isothermal D-E polymerization (12).

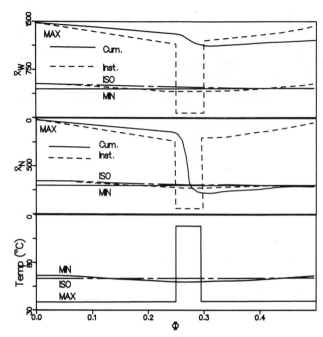

Figure 4.2-8. Optimal temperature histories with resulting DP behavior for a polymerization with G-E (12).

and

$$x_0 \leqslant \frac{f(2 - R)(\bar{x}_N)_f}{1 - (\hat{c}_m)_f} \qquad (4.2\text{-}35)$$

where

$$\bar{x}_N|_{(\hat{c}_m)_f, \hat{T}_L \text{ or } \hat{T}_S}$$

$$= x_0 f^{-1}(2 - R)^{-1}\left(1 - (\hat{c}_m)_f\right)\left\{1 - \left[1 - \left[\frac{\alpha_K \exp\dfrac{\hat{E}_d \hat{T}}{(1 + \hat{T})}}{2\left(\dfrac{f}{f_0}\right)^{1/2}\exp\dfrac{\hat{E}_{ap}\hat{T}}{(1 + \hat{T})}}\right]\int_{(\hat{c}_m)_f}^{1} g^{1/2}\hat{c}_m^{-1}\,d\hat{c}_m\right]\right\}$$

$$(4.2\text{-}36)$$

In the event that D_N is to be maximized, "bang-bang" control is necessary, since equation 4.2-31 is always negative and therefore the unconstrained policy is not optimal. In accord with the expected isothermal drift and the initial temperature, different policies are required as summarized in Table 4.2-4. In each case, the temperature starts at one its limiting values and is either held there throughout or subjected to subsequent step changes between the two limits. The characteristic profiles obtained by Sacks are shown in Figs. 4.2-6–4.2-8 for the three types of

Table 4.2-4. Maximum D_N

Case I – Decreasing isothermal $(\overline{x}_N)_{inst}$ as in CONV polymerization

    (a) $T_o = T_S$

        $T_{opt} = T_S$ until (i) becomes an equality; if it does then

        T_{opt} switches to T_L and remains at this value.

    (b) $T_o = T_L$ throughout

Case II – Increasing isothermal $(\overline{x}_N)_{inst}$ as with strong D-E or G-E

    (a) $T_o = T_S$; T_{opt} throughout

    (b) $T_o = T_L$

        $T_{opt} = T_{.L}$ until (ii) becomes an equality; if it does then

        T_{opt} switches to T_S and remains at this value.

CASE III – Decreasing followed by increasing $(\overline{x}_N)_{inst}$ as with moderate
D-E or G-E

    (a) $T_o = T_S$; $T_{opt} = T_S$ as in *Case I* (a) until (i) becomes
an equality; if it does then T_{opt} switches to T_L until
(ii) becomes an equality as in *II* (b); if it does, then
T_{opt} switches back to T_S and remains at this value.

    (b) $T_o = T_L$

        $T_{opt} = T_L$ as in *Case I* (b) until (ii) becomes an equality;
if it does, then T_{opt} switches to T_S and remains at this
value.

*where

    (i)
$$\frac{2x_o \hat{r}_p}{(2-R)\hat{r}_i} \leq \frac{(Z_2 + 2 + 2R)}{(2+R)(2-R)} \qquad\qquad \hat{T} = \hat{T}_S$$

    (ii)
$$\frac{2x_o \hat{r}_p}{(2-R)\hat{r}_i} \geq \frac{(Z_2 + 2 + 2R)}{(2+R)(2-R)} \qquad\qquad \hat{T} = \hat{T}_L$$

isothermal drift behavior. It is only in the G-E case, for which drift changes direction, that the optimal policy consists of more than one step-change in temperature.

In the CONV case (Fig. 4.2-6), the optimal policy begins at the minimum possible temperature T_S. This maximizes drift dispersion by enhancing propagation over initiation to the greatest extent possible. Prior to the sharp drop in $(\bar{x}_N)_{inst}$, characteristic of the high-conversion regime in CONV reactions, temperature is stepped to the maximum value T_L. This creates a step-change decrease in both $(\bar{x}_N)_{inst}$ and $(\bar{x}_W)_{inst}$ and promotes D-E conditions for the final half of the reaction.

When the isothermal drift is D-E (Fig. 4.2-7), temperature is set to the maximum values, enhancing natural drift dispersion. Prior to the complete consumption of initiator, temperature is quickly dropped to the lower bounding value. This makes $(\bar{x}_N)_{inst}$ jump from its minimum possible value to the maximum value achievable within the permissible temperature limits. At the same time CONV-like conditions are imposed, reversing the previous drift and preventing the complete consumption of initiator.

Finally, in the case of a mild G-E, where the isothermal drift changes direction from downward to upward as conversion increases, the optimal policy for maximum D_N requires two step-changes in temperature. In a sense, the policy is a combination of a CONV policy early in the reaction, and a D-E policy in the latter portion when the G-E upward drift in $(\bar{x}_N)_{inst}$ begins.

We have seen in Chapter 3 that a MSIM produces the broadest DPD for an isothermal free-radical polymerization. Sacks compared his result for maximum-D_N programming to the D_N from an MSIM producing polymer with the same conversion and $\bar{x}_N$. In general, the programmed BR yielded a larger value of D_N. Values were the same for conventional polymerizations. Temperature programming maximizes the dispersion by creating a bimodal DPD, and this appears to be more effective than the simple spreading caused by mixing effects in the IM.

In unpublished work (18), Sacks also examined the maximum and minimum D_N problem for thermal polymerization and for step growth addition kinetics. The approach is identical to that described for free-radical kinetics, so we will merely highlight the results.

Thermal polymerization follows much the same trends as a chemically initiated reaction. When D_N is to be minimized and temperature is unconstrained, the policy that maintains $(\bar{x}_N)_{inst}$ constant is optimal. Note that for second-order initiation kinetics, there is no isothermal drift and programming would not be necessary (Example 2.7-1). If the initiation rate is greater than second-order, the isothermal drift is upwards and the policy is one which causes $k_{ax}(T)$ to decrease with time. Since E_d is generally larger than E_{ap}, rising temperatures cause k_{ax} to decrease, Thus thermal polymerization is analogous to the D-E case in Table 4.2-3. The policies that maximize D_N are also the same as those for chemically initiated reactions (Table 4.2-4).

In step-addition polymerizations, there are no instantaneous molecular properties. Chain growth occurs throughout the course of reaction. The spread in the DPD results from chain initiation taking place continuously. In the absence of termination, chain length increases with chain lifetime. Thus the spreading out of chain births over time produces a spread in the DPD. Intuitively we would expect a temperature policy that minimizes D_N to require all initiation to take place very early in the reaction.

As with free-radical mechanisms, three state variables are used: $\hat{c}_m$, $\hat{c}_0$, and $\hat{\mu}^2$. Note that $\hat{\mu}^1 \propto 1 - \hat{c}_m$ and $\hat{\mu}^0 \propto 1 - \hat{c}_0$ (Section 1.7). Provided that the system is within the limits of controllability, i.e., as long as the target values of $(\hat{c}_m)_f$ and $(\bar{x}_N)_f$ can be reached, application of Pontryagin's principle is straightforward.

By solving the isothermal kinetic equations we find

$$\left.\frac{k_p}{k_i}\right|_{T_S \text{ or } T_L} \leqslant \frac{(\bar{x}_N)_f^{-1} - 1}{(\bar{x}_N)_f^{-1} + \left[x_0\left(1 - (\hat{c}_m)_f\right)\right]^{-1}\ln\left[1 - (\bar{x}_N)_f^{-1}x_0\left(1 - \hat{c}_m\right)_f\right)\right]}$$

$$\leqslant \left.\frac{k_p}{k_i}\right|_{T_S \text{ or } T_L} \tag{4.2-37}$$

The limits of the inequality are expressed this way to allow for either $E_p > E_i$ or $E_p < E_i$. Figure 4.2-9 graphically illustrates the range of controllability. It should be quite apparent that the largest possible value of $(\bar{x}_N)_f$ is x_0.

In solving the maximum and minimum D_N problems, Sacks found that when $H = 0$ and $\partial H/\partial T = 0$, $\partial^2 H/\partial T^2$ was also zero, which meant that the unconstrained policy might not be optimal. Indeed, further examination of the equations for the adjoint variables showed a contradiction with the state equations when the Hamiltonian and its temperature-derivative were zero. Therefore, the optimum policy for both maximum and minimum dispersion must consist of step changes between the bounding temperatures. In fact, Sacks found that the policy contained at most one switch between the temperature limits. The physical nature of the policy was to start at the bounding temperature that maximized the rate of initiation relative to propagation (T_L when $E_i > E_p$ and T_S when $E_p > E_i$). Depending on the value of x_0 and the targets $(\hat{c}_m)_f$ and $(\bar{x}_N)_f$, a switch to the other temperature may be required after most of the initiator has been exhausted. The policy creates essentially monodisperse polymer when $(\bar{x}_N)_f$ is large.

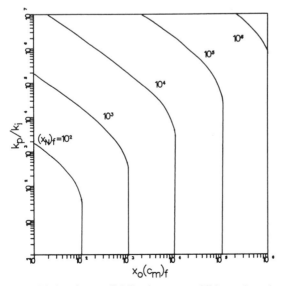

Figure 4.2-9. Limits of controllability for a step-addition polymerization.

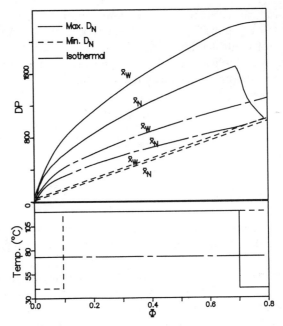

Figure 4.2-10. Optimal temperature histories with resulting DP behavior for step-addition polymerization.

The maximum D_N problem is essentially the converse of the minimum. It can be shown that there is at most one step-change between the constraining temperatures. The initial temperature is chosen to make the propagation rate dominate the initiation rate to the largest possible extent. Only a few chains will be formed at the start of the reaction, and these will grow to be quite long. When the temperature is switched, chain initiation will be favored and many shorter chains will form producing a bimodal DPD that maximizes D_N.

Figure 4.2-10 illustrates the results of numerical simulations (18). It should be pointed out that these are based on fictitious kinetics since the author was unable to find any real systems for which the parameters placed the system within the range of controllability at reasonable temperatures.

4.2.3. Miscellaneous Optimal Policies

An early study by Ray et al. (15) considered the problem of optimizing reactor start-up for a condensation polymerization as well as a chain-addition polymerization. In much of this work, the authors allowed for simultaneous control of temperature and either initiator feed concentration (BR) or flow rate (MMIM). This complicated the problem to the extent that numerical techniques would not converge to the optimal policy. When temperature alone was the variable, results were somewhat better. These authors also used a fairly demanding performance criterion in which they attempted to minimize the deviation of $\bar{x}_N$, D_N, and conversion from the steady-state operating point of the MMIM. This is not a minimum time problem, per se, but the net effect is that the system reaches steady state more rapidly than in the controlled case.

For a condensation polymerization with pure monomer in the feed and in the reactor, initially, the state equations are

$$\frac{d\hat{c}_m}{d\hat{t}} = (1 - \hat{c}_m) - k(\hat{T})\hat{c}_m \hat{\mu}^0 \tag{4.1-38}$$

$$\frac{d\hat{\mu}^0}{d\hat{t}} = (\hat{\mu}_0^0 - \hat{\mu}^0) - \frac{1}{2}k(\hat{T})(\hat{\mu}^0)^2 \tag{4.2-39}$$

$$\frac{d\hat{\mu}^2}{d\hat{t}} = -\hat{\mu}^2 + k(\hat{T}) \tag{4.2-40}$$

It is easy to show that the control policy must be "bang-bang" control. That is, the condition $\partial H/\partial \hat{T} = 0$ results in the wrong sign for $\partial^2 H/\partial \hat{T}^2$, so $\hat{T}$ must continually alternate between $\hat{T}_S$ and $\hat{T}_L$ to maintain $H = 0$. The numerical techniques by Ray were unable to converge on a single policy, but the results from two different methods were quite similar and should approximate the true solution. Figure 4.2-11 shows the resulting behavior of the state variables. Note that the time to reach steady state is roughly half that required when the controls are fixed at their steady-state value.

For the chain-addition polymerization, a similar benefit was observed by programming a linearly decreasing temperature until steady state was achieved. It should be noted that initiator concentration was taken to be constant. Also, the reduced time to reach steady state was achieved at the expense of a drop in $\bar{x}_N$ during start-up. Weighting the components of the performance index differently would change the result.

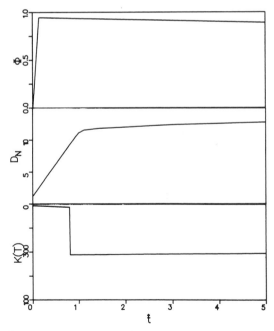

Figure 4.2-11. Temperature programming for optimal start-up of a polycondensation (15).

It has been suggested that temperature programming be used to minimize the spread of CCD in copolymerization (19). While the best available data suggests that the temperature dependence of the CC is weak (Section 4.4-2), the concept is nonetheless valid. Optimal policies would keep $(\bar{y})_{\text{inst}}$ constant, much the same as holding $(\bar{x}_N)_{\text{inst}}$ constant minimized D_N. Pontryagin's principle is not needed if the preceding assumption is made. It follows that if $(\bar{y})_{\text{inst}}$ is constant (Section 1.13) and f_1 and f_2 are defined as:

$$f_1[T(t)] \equiv n_A(t)$$

$$= -\frac{[2(\bar{y})_{\text{inst}}(1 - R_B) - 1] \pm [(\bar{y})_{\text{inst}}[(\bar{y})_{\text{inst}} - 1](1 - 4R_A R_B) + 1]^{1/2}}{2[(\bar{y})_{\text{inst}}(R_A + R_B - 2) - (R_A - 1)]}$$

$$(4.2\text{-}41)$$

and

$$f_2[n_A(t), T(t)] \equiv \frac{dn_A}{dt} = \frac{r_{pA}}{c_m} - \frac{n_A r_p}{c_m} \qquad (4.2\text{-}42)$$

then

$$\frac{dT}{dt} = \frac{\dfrac{dn_A}{dt}}{\dfrac{dn_A}{dT}} = \frac{f_2}{\dfrac{df_1}{dT}} \qquad (4.2\text{-}43)$$

Gromley and Tirell (19) applied this concept to free-radical copolymerization, using phi factor kinetics (Chapter 1) and testing the system styrene–acrylonitrile. Figure 4.2-12 shows some computed trajectories. Several comments are appropriate here. First, the phi factor approach is generally not valid, even for isothermal

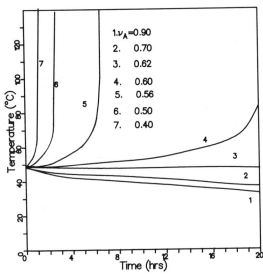

Figure 4.2-12. Temperature programming to minimize CCD dispersion, $(c_0)_0 = 0.01$, for SAN (19).

kinetics. Many systems show both a composition- and temperature-dependent phi. Second, styrene–acrylonitrile polymerization is not homogeneous when the acrylonitrile content exceeds approximately 40 percent. Consequently, even more sophisticated kinetics, such as the penultimate effect model, do not apply. If a system were to behave in accordance with the kinetics used in reference 19, the temperature histories required to keep $(\bar{y})_{\text{inst}}$ constant when drift is largest ($n_A <$ 0.70, cf. CC diagram, Fig. 1.13-5) would appear to be severe runaway profiles. Although such a control scheme would be difficult to implement, the profile does suggest that a runaway polymerization might lead to a narrower CCD than an isothermal one for certain feed compositions. In this work, no constraints were placed on $\bar{x}_N$ or D_N, and it is quite possible that the control policies suggested might preserve the composition distribution at the expense of producing high polymer or acceptably narrow DPD's.

4.3. THERMAL RUNAWAY

Thermal runaway (R-A) is a distributed-parameter phenomenon, beginning generally as a hot spot at a point and subsequently propagating throughout the reaction. It is associated with chemical explosions, whose origins actually fall into two distinct categories: one is thermal ignition (IG) and the other is kinetic chain branching. The exothermic nature of polymerizations together with their poor heat transfer characteristics render them prime candidates for IG, which may or may not result in an explosion, depending on additional factors such as amount of reaction limiting reactant available, level of pressure attainable through expansion of gases, present or produced, and the ability of the vessel to contain the pressure. Beyond its role as a potential cause of reactor instability, thermal R-A can have additional, important consequences that are specified to polymerization reactions. It can alter polymer product properties (DP, DPD, CC, and CCD) as noted in the preceding section.

Thermal R-A phenomena have been studied extensively: most notably by Semenov and Frank-Kamenetskii (20) in connection with thermal explosions; by Barkelew (21) and Amundson (34) in connection with reactor stability for reactions describable by a single concentration variable; by Biesenberger et al. (22–29) for polymerizations and copolymerizations; and by numerous others in connection with combustion and flames (30–33). Barkelew and Biesenberger et al. were concerned primarily with thermal R-A and reactor instability in industrial reactions, which generally have lower reaction exotherms and lower activation energies than explosions.

In this chapter, we seek to establish criteria for R-A and IG, as well as for polymer and copolymer property changes. A Semenov-type "theoretical" approach, as well as computer simulation methods similar to those of Barkelew, will be utilized.

Thermal R-A has been identified in a variety of ways by different investigators. When generalized beyond the specifics of the reaction mechanism considered the unifying and consistent requirements are a rapid rise in temperature with respect to time,

$$\frac{dT}{d\alpha} \gg 0 \tag{4.3-1}$$

and an acceleration of that rise characterized by an inflection in the temperature profile:

$$\frac{d^2T}{d\alpha^2} > 0 \qquad (4.3\text{-}2)$$

In chemical engineering parlance, stability is a term generally associated with steady-state behavior. A stable system is one that returns to an initial steady state even when subjected to substantial variations in system parameters. A batch system has no unique steady state, per se. However, it may exhibit parametric sensitivity. That is, the temporal behavior of a batch system may change radically in response to small changes in certain process parameters. Such systems will be regarded as potentially unstable. Thermal R-A accompanied by sensitivity is the phenomenon previously called thermal ignition (IG). In summary, IG requires rapid temperature acceleration in response to small changes in reaction parameters.

Exotherm is generated from every point throughout the volume of a reaction medium. It is removed by a transport process, which is by nature surface-area dependent. Since volume and area do not scale proportionally by simple scaling of geometric dimensions, alternative scaling criteria for reaction stability are sought. It has been demonstrated that dimensionless criteria formulated from the ratios of pertinent characteristic times (CT's) are useful in this regard. The requirement of a balance between appropriate CT's for heat generation and heat removal should therefore lead to the desired criteria.

Avoiding R-A in exothermic reaction media is tantamount to maintaining temperatures reasonably uniform in the face of rapid evolution of heat from distributed point sources. We therefore reason, as in Section 3.1 when dealing with composition nonuniformities and mixing, that temperature nonuniformities within local reacting domains must ultimately depend on molecular heat transport for their elimination. Thus, if conduction is the dominant mechanism for heat transfer, it is reasonable to expect that R-A can be avoided only when a certain CT for heat conduction (Appendix G)

$$\lambda_H \equiv \frac{l^2}{\alpha_T} \qquad (4.3\text{-}3)$$

is small relative to an appropriate CT for heat generation by reaction, where l is the characteristic length for an arbitrary reaction zone of particular interest. That zone can range in size from the entire reactor, in the absence of mixing, to the scale of mixing as defined in Chapter 3.

In practice, heat transfer in a chemical reactor usually involves convection in addition to conduction. The CT for heat removal by convection is (Appendix G)

$$\lambda_R \equiv \frac{\rho c_p}{U s_V} \qquad (4.3\text{-}4)$$

These two removal mechanisms frequently occur in series; that is, heat is conducted through the reaction medium and then removed through the walls (jacket, fins, etc.), cooling coils, or other forms of heat exchangers. In many instances, one process dominates the other and, consequently, a single CT may be used as the appropriate measure of heat removal capacity. When both λ_H and λ_R are of the same order of magnitude, a composite CT accounts for both. Because the mechanisms act in series, the CT is the sum of the component parts (Example 4.3-9).

$$\Lambda_R = \lambda_R + \lambda_H = \lambda_R(1 + Nu) \tag{4.3-5}$$

where, in the limit of one mechanisms dominating, Λ_R converges to the value of the CT for that mechanism.

Flow is another mechanism for heat removal. In an open system, the flow process can give rise to a net outflow of thermal energy that would occur in parallel with the other two mechanisms. For instance, in a PFR, longitudinal convection occurs in parallel with transverse conduction. Similarly, in an IM, the flow and (wall) losses act in parallel. The composite CT for heat removal may be expressed in general by (Example 4.3-9)

$$\Lambda_R = \left[(\lambda_R + \lambda_H)^{-1} + \lambda_v^{-1}\right]^{-1} \tag{4.3-6}$$

which may then be reduced to subcases of special interest.

Various CT's for heat generation have been suggested. They reduce to a choice amongst λ_r, λ_G and λ_{ad} (Appendix G). Both λ_G and λ_{ad} have physical significance in relation to the production of heat by chemical reaction. Thus, when conductive heat transfer dominates heat removal, one might expect a suitable criterion for avoiding R-A to be

$$\lambda_H \ll \lambda_G \tag{4.3-7}$$

or

$$\lambda_H \ll \lambda_{ad} \tag{4.3-8}$$

and when convective heat transfer dominates,

$$\lambda_R \ll \lambda_G \tag{4.3-9}$$

or

$$\lambda_R \ll \lambda_{ad} \tag{4.3-10}$$

Alternatively, we could define dimensionless groups (Appendix G) for the conduction case

$$\frac{\lambda_H}{\lambda_G} \equiv Da_{IV} \tag{4.3-11}$$

or

$$\frac{\lambda_H}{\lambda_{ad}} \equiv \delta_H \tag{4.3-12}$$

and the convection case

$$\frac{\lambda_R}{\lambda_G} \equiv \gamma_T^{-1} \tag{4.3-13}$$

or

$$\frac{\lambda_R}{\lambda_{ad}} \equiv a^{-1} \tag{4.3-14}$$

and require that they be small ($\ll 1$) to avoid R-A. The first one is a Damköhler number, the second and fourth were used by Frank-Kamenetski, and the third one was introduced by Biesenberger et al. (22, 26). It should be noted that expressions 4.3-3 and 4.3-7 are the counterparts of 3.1-12 and 3.3-1. Here, too, the crucial quantity is l, whose value can again range from the overall dimensions of the reaction vessel to the scale of segregation l_{seg} of reacting "particles" within the vessel. In either case it is sometimes instructive to estimate the critical value of l which must not be exceeded if R-A is to be avoided (Examples 4.3-3 and 4.3-5). Obviously, since heat removal rate is proportional to transfer area S and heat generation rate is proportional to volume V, a concomitant reduction of S/V as l increases is a root cause of R-A, as noted earlier.

The proportionality between CT's depends on the order of the associated process (zero-, first-, second-order, etc.). Heat removal and generation are not generally of the same order. Therefore, while a criterion such as

$$\delta_H = \frac{\lambda_H}{\lambda_{ad}} < 1 \tag{4.3-15}$$

may provide a useful estimate, further analysis is required to give the inequality a more precise numerical value.

Of course, CT's λ_G and λ_{ad} both depend on the specific kinetic mechanism (Appendix G). We shall begin our analysis by considering the simplest mechanism, nth order kinetics, with rate functions that can be expressed as separable functions of dimensionless concentration and temperature (Appendix E)

$$\hat{r}_G = \hat{c}^n \hat{g}_e(\hat{T}) \tag{4.3-16}$$

where the dimensionless heat generation function is $\hat{g}_e(\hat{T}) = \hat{k}(\hat{T})$. The simplest of these is the zero-order reaction ($n = 0$), which is equivalent to the early runaway approximation (ERA), $\hat{c}^n = 1$, invoked by Semenov and Frank-Kamenetskii to facilitate their analytical treatment of lumped- and distributed-parameter systems, respectively.

The nature of most polymerizations, however, is unlike that of the gaseous explosions considered by these investigators, as noted earlier. There is additionally a significant consumption of reactants during the preignition induction period. Therefore, the IG criteria deducted by them must be modified to account for the time dependence of composition

$$\hat{c}^n = f(\hat{t}) \tag{4.3-17}$$

This generally involves the simultaneous numerical solution of the component balance for $\hat{c}$ and the thermal energy balance for $\hat{T}$ to establish a relationship between the critical values for the R-A criterion and the dimensionless groups that will characterize reactant consumption. It is proposed that criteria for the latter be based on a measure of the time scale for reactant consumption vis-à-vis the time scale for temperature rise. This suggests two possible forms for such dimensionless groups

$$\frac{\lambda_r}{\lambda_G} = He \tag{4.3-18}$$

or

$$\frac{\lambda_r}{\lambda_{ad}} = b \tag{4.3-19}$$

where λ_r is a CT for isothermal reactant consumption.

The situation becomes more complex when multiple reactants and reactions are considered [e.g., $f(\hat{t}) = \prod_k \hat{c}_k^{n_k}$]. This is typically the case for polymerization in which one must consider both monomer and initiator effects on sensitivity. Copolymerizations have even more complex kinetic forms that are generally not separable in concentration and temperature. Nonetheless, we shall attempt to show that if the correct expressions for CT's are used, then the dimensionless criteria developed for simple kinetics can be readily extended to the more complex cases.

Distributed parameter systems will also be examined, including situations in which there is an appreciable resistance to heat transfer at the reactor wall. These systems, too, will be characterized by dimensionless groups.

4.3.1. Thermally Simple Systems

For a broad class of polymerizations, heat generation can be ascribed primarily to one kinetic step, the propagation reaction. Thus these reactions are thermally simple (Appendix E). Furthermore, many polymerizations of industrial interest are adequately described by rate functions that conform to the notion of kinetically simple reactions (Appendix E). Certainly the model polymerizations presented in Chapter 2 are kinetically simple. The thermal energy balance for simple reactions in a PFR (BR) takes on the special form (Appendix F)

$$\rho c_p \frac{dT}{d\alpha} = -\Delta H_r r - Us_V (T - T_R) \tag{4.3-20}$$

where T_R is the reservoir (coolant) temperature, and α may represent time t (BR) or space-time λ_v (PFR).

When transforming this equation into dimensionless form, several alternative definitions are available for dimensionless temperature. First, there are two obvious choices for reference temperature T_r. These are feed (or initial) temperature T_0 and reservoir temperature T_R. Second, there are two distinct forms for dimensionless temperature (Appendix E) that arise naturally:

$$\hat{T} \equiv \frac{T - T_r}{T_r} \tag{4.3-21}$$

and

$$\theta \equiv \frac{E}{R_g T_r} \hat{T} = \hat{E}\hat{T} \qquad (4.3\text{-}22)$$

where E is the activation energy of the lumped rate constant for the reaction in equation 4.3-20.

In semidimensionless form (Appendix G), equation 4.3-20 becomes

$$\frac{d\hat{T}}{d\alpha} = \underset{g_e}{\lambda_G^{-1}\hat{r}_G} - \underset{r_e}{\lambda_R^{-1}(\hat{T} - \hat{T}_R)} \qquad (4.3\text{-}23)$$

where $\lambda_G^{-1} \equiv -\Delta H_r(r)_r/\rho c_p T_r$ and $\lambda_R^{-1} \equiv U s_V/\rho c_p$. Thus λ_G^{-1} may be interpreted as an initial, semi-dimensionless heat generation rate only if $T_r = T_0$. Using T_0 for T_r, however, causes $\hat{T}_R$ (or θ_R) to appear in the resulting balance equation, as in equation (4.3-23), while using T_R will produce initial conditions $\hat{T}_0$ (or θ_0) and $(\hat{r})_0$ that are nonuniform and different from unity. From a practical viewpoint, the use of T_R has an advantage. In many reactor systems, the feed is cooled well below the reaction operating temperature, whereas the reservoir (coolant) temperature is generally much closer to the reaction temperature. As a result, CT's and dimensionless groups that are based on T_R will more accurately reflect the system behavior than those based on T_0.

Table 4.3-1 lists the semi-dimensionless and dimensionless transport balances for a PFR (BR) using T_R as the reference temperature. Table 4.3-2 lists the corresponding dimensionless rate functions for simple reactions in general and for free-radical polymerizations in particular. Note that the rate of polymerization r is kinetically simple and the rate of heat generation r_G is thermally simple by virtue of the dominance of the propagation reaction as a consumer of monomer (the LCA) and a producer of heat, respectively. Note also that λ_k^{-1} for kinetically simple reactions may be interpreted as the initial time gradient for substance k, $(d\hat{c}_k/d\alpha)|_0$, only if $T_r = T_0$. The consequences of kinetic and thermal simplicity will be important in our subsequent analyses of copolymerizations, which are neither kinetically nor thermally simple.

For simple reactions the explicit temperature dependence and implicit time dependence (via concentration) may be separated in the functional representation of dimensionless rate:

$$r_G(\hat{\alpha}, \theta) = f(\hat{\alpha})\hat{g}_e(\theta) \qquad (4.3\text{-}24)$$

where $\hat{\alpha} \equiv \alpha/\lambda_{ad}$, $\hat{g}_e(\theta) = \exp\theta/(1 + \varepsilon\theta)$, and ε is the reciprocal of the dimensionless activation energy ($\hat{E}^{-1}$). Forcing function

$$f(\hat{\alpha}) = \prod_k \hat{c}_k^{n_k} \qquad (4.3\text{-}25)$$

decreases monotonically with time from an initial value of $\hat{f}(0) = 1$, assuming that the reactions are not autocatalytic. We shall use θ in lieu of $\hat{T}$ for now in anticipation of large activation energies and concomitantly small values of ε, which validate the useful approximation $\hat{g}_e(\theta) \cong \exp\theta$.

Table 4.3-1. **Balance Equations for Lumped Parameter Systems with No Backmixing**

Balance Equations

$$\frac{dc_k}{d\alpha} = r_k$$

$$\rho c_p \frac{dT}{d\alpha} = r_G - Us_v(T - T_R)$$

Semi-Dimensionless Forms

$$\frac{d\hat{c}_k}{d\alpha} = \lambda_k^{-1} \hat{r}_k$$

$$\frac{d\hat{T}}{d\alpha} = \lambda_G^{-1} \hat{r}_G - \lambda_R^{-1}\hat{T}$$

or

$$\frac{d\theta}{d\alpha} = \lambda_{ad}^{-1} \hat{r}_G - \lambda_R^{-1} \theta$$

where $\hat{c}_k \equiv c_k/(c_k)_o$ and $\lambda_{ad} = \varepsilon\lambda_G$

Dimensionless Forms

$$\frac{d\hat{c}_k}{d\hat{\alpha}} = b_k^{-1} \hat{r}_k$$

$$\frac{d\theta}{d\hat{\alpha}} = r_G - a\theta$$

where $\hat{\alpha} \equiv \alpha/\lambda_{ad}$ and $b_k \equiv \lambda_k/\lambda_{ad}$

Free Radical Polymerizations

$$- \frac{d\hat{c}_o}{d\hat{\alpha}} = b_i^{-1} \hat{r}_i$$

$$- \frac{d\hat{c}_m}{d\hat{\alpha}} = b_m^{-1} \hat{r} + (\nu_N)_o^{-1} b_m^{-1} \hat{r}_i$$

Table 4.3-2.   Rate Functions for Kinetically and Thermally Simple Reactions

General

$$\hat{r}_j = \left(\prod_k \hat{c}_k^{n_{jk}}\right) \exp \hat{E}_j \hat{T}/(1 + \hat{T}) = \left(\prod_k \hat{c}_k^{n_{jk}}\right) \exp \varepsilon \hat{E}_j \cdot \theta/(1 + \varepsilon\theta)$$

$$\hat{r}_G = \left(\prod_k c_k^{n_{jk}}\right) \exp\theta/(1 + \varepsilon\theta)$$

where $\theta \equiv \hat{E}_1 \hat{T}$ and $\hat{E}$ corresponds to the thermally controlling re-

action

Free-Radical Polymerizations

$$\hat{r}_i = \hat{c}_o \exp \hat{E}_d \hat{T}/(1 + \hat{T}) = \hat{c}_o \exp \varepsilon \hat{E}_d \theta/(1 + \varepsilon\theta)$$

$$\hat{r} \cong \hat{c}_m \hat{c}_o^{1/2} \exp \hat{E}_{ap} \hat{T}/(1 + \hat{T}) = \hat{c}_m \hat{c}_o^{1/2} \exp\theta/(1 + \varepsilon\theta)$$

$$\hat{r}_G \cong \hat{c}_m \hat{c}_o^{1/2} \exp \hat{E}_{ap} \hat{T}/(1 + \hat{T}) = \hat{c}_m \hat{c}_o^{1/2} \exp\theta/(1 + \varepsilon\theta)$$

where $\theta \equiv \hat{E}_{ap} \hat{T}$.

The most useful dimensionless form of equation 4.3-20 is

$$\frac{d\theta}{d\hat{\alpha}} = \underset{\hat{r}_G}{\underbrace{f(\hat{\alpha})\hat{g}_e(\theta)}} - \underset{\hat{r}_e}{\underbrace{a\theta}} \tag{4.3-26}$$

where $a \equiv \lambda_{ad}/\lambda_R$. If the time scale for temperature rise is much shorter than that for reactant consumption, then we might presume that $f(\hat{\alpha})$ remains fixed at its initial value until ignition occurs. Thus $f(\hat{\alpha}) = 1$ and $\hat{r}_G = \hat{g}_e$. This is just the early runaway approximation (ERA) and was used by early researchers in the field of combustion (20, 30, 31). When applied to thermal energy balance 4.3-26, one obtains the autonomous differential equation

$$\frac{d\theta}{d\hat{\alpha}} = \hat{g}_e(\theta) - \hat{r}_e(\theta) \tag{4.2-27}$$

If the limitation $\varepsilon \ll 1$ mentioned earlier is added so that $\hat{g}_e(\theta) \cong \exp \theta$, then equation 4.3-27 reduces to that treated by Semenov in his nonstationary theory of thermal ignition.

In general, the autonomous balance equation, 4.3-27, admits a steady state

$$\frac{d\theta}{d\hat{\alpha}} = 0 \qquad (4.3\text{-}28)$$

for which temperature θ may be obtained by solving the equation:

$$\hat{g}_e(\theta) = \hat{r}_e(\theta) \qquad (4.3\text{-}29)$$

The function $\hat{g}_e$ is sigmoidal, while $\hat{r}_e$ is linear. The two are shown in Fig. 4.3-1 for several values of ε. Focus in particular on the curve for $\varepsilon = 0.04$ as this is a value characteristic of polymerization reactions. Note that there are three potential intersections between $\hat{g}_e$ and $\hat{r}_e$ which correspond to steady-state solutions of equation 4.3-27. These have been labeled θ_{s1}, θ_u, and θ_{s2}. When a is very large, $\hat{r}_e$ intersects $\hat{g}_e$ at only one point, θ_{s1}, a relatively low value of θ. As a is decreased, a tangent point θ_{t2} is added, and with subsequent decrease the three intersections θ_{s1}, θ_u, θ_{s2} occur. Continued decrease of a causes θ_{s1} and θ_u to converge until they ultimately combine at tangent point θ_{t1}. Further decrease of a gives only one intersection at the relatively large value of θ_{s2}. When $\theta(\hat{\alpha})$ lies in one of the regions on either side of θ_{s1} or θ_{s2}, the system will respond by moving toward one of these values. Hence these are called stable steady states. When $\theta(\hat{\alpha})$ lies to the immediate left or right of θ_u, it will spontaneously drift away from θ_u toward θ_{s1} or θ_{s2}, respectively. Thus the steady-state θ_u is called unstable. Remember that, depending on the value of a, there may exist from one to three steady-state solutions.

The relationship between a and the steady-state values of θ can be determined analytically when $\hat{r}_G$ has the form of 4.3-24 and $\hat{r}_e = a\theta$. From the solution

$$a = \theta^{-1}\exp\frac{\theta}{1 + \varepsilon\theta} \qquad (4.3\text{-}30)$$

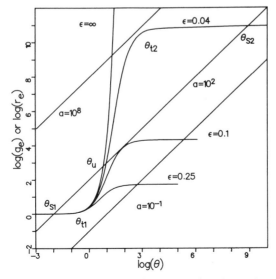

Figure 4.3-1. Generation and removal functions with various values of ε and a.

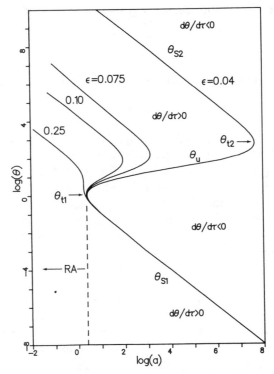

Figure 4.3-2. Locus of intersections of g_e and r_e for various values of ε.

it is possible to construct the plot of the locus of intersections of $\hat{g}_e$ and $\hat{r}_e$ that is shown in Fig. 4.3-2. Note that, for typical values of ε, when

$$\log a < 0.43 \tag{4.3-31}$$

or

$$a < 2.72 \tag{4.3-32}$$

the only steady-state solution possible is θ_{s2}. Even for fairly large values of ε, θ_{s2} has a very high value. Therefore, for small values of a, the solution $\theta(\hat{a})$ will continue to rise in pursuit of θ_{s2}. This value will not be reached however, for in practice the adiabatic temperature maximum, θ_{ad}, is far less than θ_{s2}. Consequently, we shall restrict our attention to the lower portion of $\hat{g}_e$ as shown in Fig. 4.3-3.

The effect of activation energy, as manifested through ε, is apparent in all three representations of $\hat{g}_e$ (Figs. 4.3-1–4.3-3). This is an effect not considered by researchers in the field of combustion, where ε is very small, but which may be important for reactions that are not as "hot." From Fig. 4.3-4, it is clear that ε governs the slope or steepness of the generation curve. There also seems to be a certain amount of sensitivity, as the slope remains relatively unaffected by ε until it reaches a value of approximately 0.075, after which there is a rapid decay in the slope. Note the extreme curves for $\varepsilon = 0.25$ in Figs. 4.3-1 to 4.3-3. In the first figure

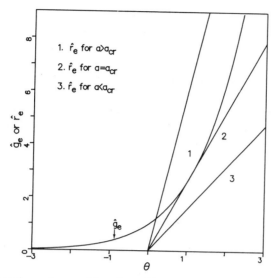

Figure 4.3-3. Generation and removal functions with a varying to eliminate intersections, $\varepsilon = 0.04$.

it appears that the maximum slope of the generation curve is close to that of the removal curve, and that θ_{s2} has dropped to a value very close to θ_{s1}. From the next figure (4.3-2), it becomes apparent that the inflection points in the ignition envelope at θ_{t1} and θ_{t2} disappear when ε exceeds 0.25. Rather than a step change of 10 orders of magnitude in θ when a drops below 2.72 (as with $\varepsilon = 0.04$), it is now possible for the system to steadily track ascending values of steady-state temperature as a is decreased below its critical value. Therefore, for systems with very low activation energies R-A and parametric sensitivity should not be a problem. It is important to

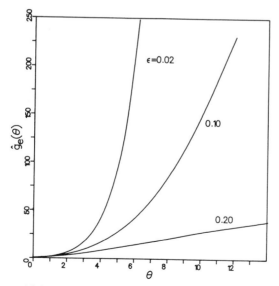

Figure 4.3-4. Generation function g_e versus θ with ε as parameter.

note, however, that free-radical initiators generally have activation energies of 25–40 kcal. Therefore, the lumped activation energy for polymerization has a value of at least half that. To achieve $\varepsilon = 0.25$ would require a reference temperature of the order of 1500°C, which is well beyond all practical limitations. The phenomenon is nonetheless worthy of mention, since it might have application to other kinetic systems.

Figure 4.3-5 shows the various juxtapositions of $\hat{g}_e$ and $\hat{r}_e$ for values of ε characteristic of polymerizations. It should be clear that a governs the slope of $\hat{r}_e$ and thus influences the number of possible intersections between $\hat{g}_e$ and $\hat{r}_e$. Note that by choosing $T_r = T_R$, the removal curve will pass through the origin, whereas the intercept of $\hat{g}_e$ will be unity. However, θ_0 will be a parameter, so the system will start at the origin only for the special case $T_0 = T_R$. Figure 4.3-5 illustrates three classes of potential juxtapositions, labeled A, B, and C. Symbols A1–A3 denote the possible locations of θ_0 for Class A, while B1 and B2 denote them for Class B. It should be clear that case C will always result in ignition since $\hat{g}_e > \hat{r}_e$ for all θ_0 and the system will attempt to attain the upper steady-state temperature θ_{s2}. This is also true of situations A3 and B2. For A1 and A2, the system will converge to the lower steady-state θ_{s1}, and will not run away. Class B clearly represents a transition

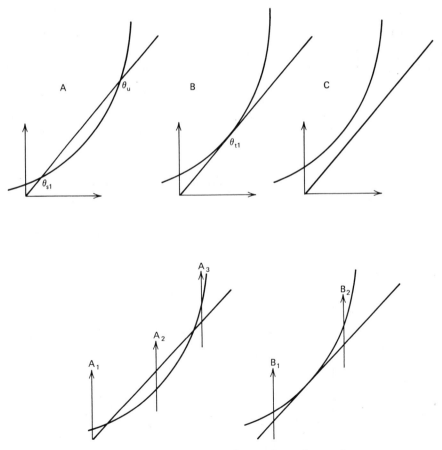

Figure 4.3-5. Possible juxtapositions of generation and removal curves.

between A and C. Just beyond this tangency (critical) condition, temperature will strive toward θ_u. Even the slightest perturbation could cause a shift that would place the system in condition B2. In conclusion, then, assuming that $\theta_0 < \theta_u$, it is evident that R-A will not occur until the slope of $\hat{r}_e$ is sufficiently reduced to cause a transition from A to B or C. When θ_0 is greater than θ_u, R-A will always occur, regardless of the slope of $\hat{r}_e$.

In other words, when $\hat{g}_e > \hat{r}_e$, it follows that

$$\frac{d\theta}{d\hat{\alpha}} > 0 \tag{4.3-33}$$

which is one of the requisite conditions for R-A. The fact that $\hat{g}_e$ diverges rapidly from $\hat{r}_e$ guarantees that

$$\frac{d^2\theta}{d\hat{\alpha}^2} > 0 \tag{4.3-34}$$

which was the second condition stipulated. As we have already demonstrated, ε must take on uncharacteristically large values for these features to be abolished.

Having established the qualitative criteria for R-A, the task of developing a mathematical statement remains. Consider first those systems with $\theta_0 \leqslant \theta_{s1}$. These include most practical cases for which $\theta_0 \leqslant 0$ additionally. The tangency θ_{t1} illustrated in Fig. 4.3-5 represents the critical point of transition from A1 (no R-A) to C (R-A). That is, the value of a, which not only makes generation and removal equal ($\hat{g}_e = \hat{r}_e$), but tangent as well,

$$\frac{d\hat{g}_e}{d\theta} = \frac{d\hat{r}_e}{d\theta} \tag{4.3-35}$$

defines the so-called ignition point. Inserting the functional form for $\hat{g}_e$ (equation 4.3-24) and $\hat{r}_e(a\theta)$ into the preceding relationships and eliminating a between them, yields the values of θ at the tangent point (26).

$$\theta_{t_2, t_1} = \frac{1 \pm \sqrt{1 - c^2}}{\varepsilon c} \tag{4.3-36}$$

where

$$c \equiv \frac{2\varepsilon}{1 - 2\varepsilon} \tag{4.3-37}$$

The two solutions correspond to the points so labeled on Figs. 4.3-1 and 4.3-2. We have already discarded the upper tangent as being well above the temperature range of practical interest, so

$$\theta_{cr} = \frac{1 - \sqrt{1 - c^2}}{\varepsilon c} \tag{4.3-38}$$

One may take advantage of the fact that c must be small, since ε is very small, and expand the square-root term in a Taylor series about $c = 0$.

$$\theta_{cr} = \frac{1}{\varepsilon c} - \frac{1}{\varepsilon c}\left(1 - \frac{c^2}{2} \cdots\right) \tag{4.3-39}$$

and thereby obtain

$$\theta_{cr} \cong \frac{1}{1 - 2\varepsilon} \tag{4.3-40}$$

To prevent R-A when $\theta_0 \leqslant \theta_{cr}$ (B1 in Fig. 4.3-5), the physical constraint is

$$\hat{g}_e(\theta_{cr}) < \hat{r}_e(\theta_{cr}) \tag{4.3-41}$$

Substitution of equation 4.3-40 into 4.3-41 with subsequent rearrangement yields the criterion

$$a > (1 - 2\varepsilon)\exp\left(\frac{1}{1 - \varepsilon}\right) \equiv a_{cr} \tag{4.3-42}$$

When $\theta_0 > \theta_{cr}$ (B2 in Fig. 4.3-5), the constraint is

$$\hat{g}_e(\theta_0) < \hat{r}_e(\theta_0) \tag{4.3-43}$$

and the resulting criterion is

$$a > \theta_0^{-1}\exp\left(\frac{\theta_0}{1 + \varepsilon\theta_0}\right) \equiv a_{cr} \tag{4.3-44}$$

Satisfying 4.3-44 is tantamount to placing the system below the central branch of the sigmoid, θ_u, in Fig. 4.3-2. These general criteria reduce to the special one of Semenov when his assumptions $\theta_0 = 0$ and $\varepsilon \cong 0$ are applied to equation 4.3-42. Thus

$$a > e = 2.72 \equiv a_{cr} \tag{4.3-45}$$

It should be noted that for given kinetic constants a single numerical value of a_{cr} applies for all $\theta_0 < (1 - 2\varepsilon)^{-1}$. Had the definition of θ been based on $T_r = T_0$, θ_R would have appeared as a parameter in equation 4.3-42 as well as equation 4.3-44, and a different value of a_{cr} would have resulted for each θ_R (26). The second criterion [equation (4.3-44)] takes effect when the initial temperature exceeds the coolant temperature by roughly 5 percent for the typical value $\varepsilon = 0.04$.

By examining a rather simple system, we have arrived at a numerical criterion for R-A in terms of parameter a, rather than γ_T. Recalling that these dimensionless parameters are not merely a random grouping of kinetic properties, but a ratio of CT's for the competing processes of heat generation and removal, should assist us in extending our results to more complex systems. In pursuit of this objective, let us look more closely at the physical significance of the CT's involved.

It is important to note that λ_G^{-1} may be interpreted as a relative measure of the magnitude of the heat generation rate at the origin of the $g_e(\hat{T})$ curve, $g_e(\hat{T})|_r$, where the term origin signifies a reference state that is initial in time [$T = T_r = T_0$ and $c_k = (c_k)_0$] or is mixed [$T = T_r = T_R$ and $c_k = (c_k)_0$].

Comparison with λ_R^{-1} as in the ratio $\gamma_T = \lambda_G/\lambda_R$, therefore, merely reflects the relative magnitudes of the heat generation and removal rates at the origin. We have seen, however, that the steepness of g_e relative to r_e is also an important factor in R-A behavior. Consequently, a parameter that compares the temperature gradients of these functions, such as $(\partial r_e/\partial \hat{T})/(\partial g_e/\partial \hat{T})$, might be more appropriate. Indeed, by interpreting λ_{ad}^{-1} and λ_R^{-1} as temperature gradients (Example 4.3-2)

$$\lambda_{ad}^{-1} = \left.\frac{\partial g_e}{\partial \hat{T}}\right|_r \tag{4.3-46}$$

$$\lambda_R^{-1} = \left.\frac{\partial r_e}{\partial \hat{T}}\right|_r \tag{4.3-47}$$

we see at once that parameter a is precisely that ratio, evaluated under reference conditions

$$a = \frac{\left.\dfrac{\partial r_e}{\partial \hat{T}}\right|_r}{\left.\dfrac{\partial g_e}{\partial \hat{T}}\right|_r} \tag{4.3-48}$$

Thus we can expect R-A to be avoided if the reference gradient of r_e is amplified relative to that of g_e by a factor at least as large as a_{cr}

$$\left.\frac{\partial r_e}{\partial \hat{T}}\right|_r > a_{cr} \left.\frac{\partial g_e}{\partial \hat{T}}\right|_r \tag{4.3-49}$$

For the systems studied by Semenov this amplification factor was found to be e (2.72). Later we will demonstrate that the same R-A criterion, with a slight modification of the numerical value of a_{cr} is directly applicable to far more complex systems if the CT's are properly defined. In particular, this will be the key to analyzing the R-A behavior of copolymerization reactions with kinetic sequences that are neither kinetically nor thermally simple.

Characteristic times for homopolymerizations introduced up to this point are summarized in Table 4.3-3. Formal definitions stemming from the general procedure outlined in Appendix G are listed together with equivalent definitions linking them to corresponding semidimensionless balance equations. These CT's in the context of a Semenov-type analysis have helped us to formulate criteria that quantify our initial R-A requirements that $d\theta/d\hat{\alpha} \gg 0$ and $d^2\theta/d\hat{\alpha}^2 > 0$. The first criterion requires a large, positive initial value of $g_e - r_e$, which is related to λ_G/λ_R because λ_G and λ_R may be interpreted as initial values of g_e and r_e. The second requires that $d\hat{g}_e/d\theta > d\hat{r}_e/d\theta$ and is characterized by λ_{ad}/λ_R. By extending Semenov's analysis, it was shown that insufficiently small values of ε (or large $\hat{E} = \varepsilon^{-1}$) can flatten $\hat{g}_e$ and thereby remove the sensitivity of R-A.

One serious limitation of the foregoing analysis, however, is the ERA, that is, the assumption of zero reactant consumption prior to R-A. Although this may be valid

Table 4.3-3.   Characteristic Times for Homopolymerization

CT	Definition	Equivalent	Process
λ_i	$\left(c_o\right)_o / \left(r_o\right)_r$	$\left(-\dfrac{d\hat{c}_o}{dt}\right)_r^{-1}$	initiator consumption
λ_m	$\left(c_m\right)_o / \left(r_p\right)_r$	$\left(-\dfrac{d\hat{c}_m}{dt}\right)_r^{-1}$	monomer consumption
λ_G	$\rho c_p T_r / \left(r_G\right)_o$	$\left(g_e\right)_r^{-1}$	heat generation
λ_{ad}	$\epsilon \rho c_p T_r / \left(r_G\right)_r$	$\left(\dfrac{\partial g_e}{\partial \hat{T}}\right)_r^{-1}$	adiabatic induction
λ_R	$\rho c_p / U s_v$	$\left(\dfrac{\partial r_e}{\partial \hat{T}}\right)_r^{-1}$	heat removal

where subscript r is either o (initial) or R (reservoir), consistent with the definition of $\hat{T}$.

for systems with extremely small values of ε, such as those studied by Semenov and others, it might not be for the milder conditions that prevail in most chemical reactions of industrial interest. Reactant consumption of 10–30 percent prior to R-A can occur. Remembering that the heat generation term is the product of the time-decaying concentration term $f(\hat{\alpha})$ and the explicit temperature term $\hat{g}_e(\theta)$, it should be clear that waning concentrations have the effect of diminishing the rate of heat output. A system that should run away, based on initial conditions, might actually remain stable owing to declining heat output with rising conversion. In terms of our graphical analysis, we can no longer limit ourselves to a single generation curve, $\hat{g}_e$, but we must instead consider a series of "instantaneous" generation curves, one for each age $\hat{\alpha}$ given by the product $\hat{r}_G = f(\hat{\alpha})\hat{g}_e(\theta)$, as illustrated in Fig. 4.3-6. The rate of temperature change is proportional to the distance between $\hat{r}_G(\alpha, \theta)$ and $r_e(\theta)$. Any reaction for which $\theta_0 = 0$ will start with $\hat{r}_G > \hat{r}_e$ (Fig. 4.3-5). As a result, $\theta(\hat{\alpha})$ will rise. The presence of decaying $f(\hat{\alpha})$ will cause the separation between curves to decrease with time and rising values of θ should accentuate the decline of $f(\hat{\alpha})$ with time. The trajectory $\theta(\hat{\alpha})$ is described by the nonautonomous D-E.

$$\frac{d\theta}{d\hat{\alpha}} = \hat{r}_G(\alpha, \theta) - \hat{r}_e(\theta) \tag{4.3-50}$$

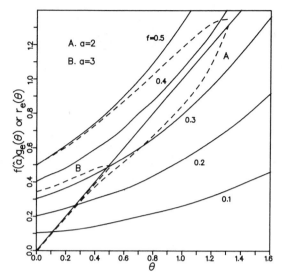

Figure 4.3-6. Effect of concentration decay on the thermal generation rate, with resulting trajectories (26).

Temperature θ will rise until $\hat{r}_G$ drops to $\hat{r}_e$, at which point the equation $d\theta/d\hat{\alpha} = 0$ is satisfied and θ will have attained its maximum value θ_{max}. Thereafter, as $f(\hat{\alpha})$ continues to fall, θ decreases, since $\hat{r}_e > \hat{r}_G$. The dashed loops in Fig. 4.3-6 represent actual temperature trajectories obtained from numerical solutions of the following simple kinetic model for free-radical polymerizations:

$$\hat{r}_G = \hat{c}_m \hat{c}_0^{1/2} \exp\left(\frac{\theta}{1 + \varepsilon\theta}\right) = \hat{r} \tag{4.3-51}$$

with values of parameter a that lie on either side of the critical value, as determined from the criterion developed earlier in this section. They may be viewed as the loci of intersections of autonomous-equivalent curves, $f\hat{g}_e$ and $\hat{r}_e$, for various values of f.

For the subcritical case ($a = 3$), we see that $\theta(\hat{\alpha})$ rises only slightly until the trajectory intersects $\hat{r}_e$. After that point, θ falls back toward the origin. Such behavior is called quasi-isothermal (Q-I) and is only an approximation to an isothermal state, which is actually not possible for systems described by nonautonomous equation 4.3-50.

The supercritical case ($a = 2$) demonstrates the strong damping effect that $f(\hat{\alpha})$ can have on $\hat{g}_e$ and how concentration decay can actually suppress potential R-A. From the initial conditions, one would expect this system to experience R-A since $\hat{r}_e$ does not intersect $\hat{g}_e$. However, by the time $\theta = 1.3$, the concentration will have dropped sufficiently to cause r_G to cross the removal curve, and temperature will subsequently drop toward $\theta = 0$. The maximum temperature, $\theta_{max} = 1.3$, is not high and a plot of $\theta(\hat{\alpha})$ for this case would show no inflection. Clearly then, the criteria developed earlier in this section provide a conservative estimate of the R-A potential of a given system. Concentration decay tends to have a stabilizing effect not taken into account in that analysis.

Up to this point, we have seen that parameter ε alters the slope of $\hat{g}_e$ and that a small value of ε is a necessary condition for R-A. Furthermore, a small value of parameter a ensures that there is a sufficient imbalance between heat generation and removal to permit R-A to occur. Next, we seek a dimensionless parameter, or set of parameters, that will characterize the extent of concentration decay for each of the reactants, and its corresponding effect on the nature of the R-A behavior of the system. One might postulate that the ratio of CT's for reactant consumption and temperature rise should be a suitable parameter. The closer the time scales for reactant decay and temperature increase are, the less accurate is the ERA and the less parametrically sensitive the R-A is likely to be. Thus, a priori, we can expect our parameter for R-A sensitivity to evolve from a comparison between CT's of the type λ_k and λ_{ad}, and to take the following form:

$$\lambda_k \gg \lambda_{ad} \tag{4.3-52}$$

or

$$\lambda_k/\lambda_{ad} \gg 1 \tag{4.3-53}$$

Characteristic time λ_k is a measure of the rate of consumption of component k, and arises in the balance equations as shown in Table 4.3-1. Characteristic time λ_{ad}, on the other hand, has been shown to be a measure of the time that elapses before the sudden rise in temperature associated with R-A (30). Temperature during this induction period is relatively isothermal and thus λ_k is a valid measure of reactant consumption during this interval. Therefore, when inequality 4.3-53 is satisfied, the ERA should be a valid approximation.

For stoichiometrically simple reactions (e.g., nth order kinetics), a single parameter has been shown to be a measure of sensitivity (33). This parameter is

$$b = \lambda_r/\lambda_{ad} \tag{4.3-54}$$

where λ_r is the CT for the single reaction, and the corresponding sensitivity criterion, $b \gg 1$, takes the form of inequality 4.3-53. We have seen that model random and addition polymerizations can in some instances be approximated by simple second- and first-order kinetics, respectively, so analysis of stoichiometrically simple reactions could prove to be useful.

For first-order kinetics, both $\hat{r}$ and $\hat{r}_G$ are given by $\hat{c} \exp \theta/(1 + \varepsilon\theta)$. Using this expression in the dimensionless balances of Table 4.3-1, the thermal energy and component balances may be combined to give

$$\frac{d\theta}{d\hat{c}} = -b \frac{\hat{c} \exp\left(\dfrac{\theta}{1 + \varepsilon\theta}\right) - a\theta}{\hat{c} \exp\left(\dfrac{\theta}{1 + \varepsilon\theta}\right)} \tag{4.3-55}$$

Since concentration decreases monotonically with time, a maximum in θ with respect to $\hat{a}$ will correspond to a maximum with respect to $\hat{c}$. That is, we can state that θ_{max} is achieved when equation 4.3-55 is set equal to zero, with the result that at the

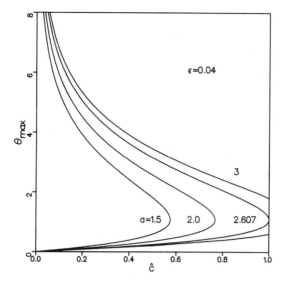

Figure 4.3-7. Locus of maxima for various values of a.

maximum temperature:

$$\hat{c} = a\theta \, \exp\!\left(\frac{-\theta}{1 + \varepsilon\theta}\right) \tag{4.3-56}$$

For a fixed value of a, this defines a curve known as the "locus of maxima." For several values of a, this curve is presented in Fig. 4.3-7. Note that the solution $\theta(\hat{a})$ starts at $\hat{c} = 1$, $\theta = \theta_0$. Provided $\theta_0 \leqslant (1 - 2\varepsilon) \simeq 1$ and $a > a_{cr}$ it is impossible to achieve R-A. This is consistent with our findings earlier in this section. The trajectory will intersect the locus at $\theta_{max} \leqslant (1 - 2\varepsilon)$ and rise no higher. However, it may not be said for $a < a_{cr}(\varepsilon)$ that R-A is ensured.

Figure 4.3-8 shows the locus of maxima for $a = 2$ along with several computed temperature profiles (obtained via numerical solution of equation 4.3-55). It is clear that decreasing the value of parameter b, defined in equation 4.3-54, can eliminate R-A behavior from a subcritical case ($a < a_{cr}$). When $b = 16$, the reactant decay is sufficiently rapid to cause $\hat{r}_G$ to cross $\hat{r}_e$ prior to the onset of R-A.

The analytical treatment of Adler and Enig (33) helps to quantify this effect. They identified R-A as the situation for which $d^2\theta/d\hat{c}^2 > 0$, which corresponds to our equation 4.3-2. Therefore, the "locus of inflections," or the solution set for $d^2\theta/d\hat{c}^2 = 0$, serves to define the R-A boundary. Those thermal trajectories that intersect the locus of inflections before the locus of maxima, will experience R-A. Those that cross the locus of maxima first will subsequently drop in temperature, never crossing the locus of inflections, and R-A will be averted. The equations for the locus of inflections is

$$\hat{c} = b^{-1}\left[\frac{\theta}{\dfrac{\theta}{(1 + \varepsilon\theta)^2} - 1}\right] + a\exp\!\left(\frac{-\theta}{1 + \varepsilon\theta}\right) \tag{4.3-57}$$

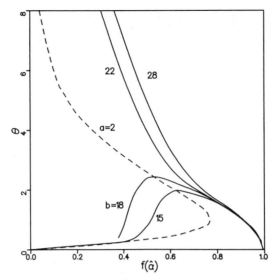

Figure 4.3-8.   Locus of maxima for $a = 2$ with computed thermal trajectories as a function of b.

Noting that the second term on the RHS is the equation for the locus of maxima, it should be clear that the inflection curves will always lie above the maximum curves provided that

$$\frac{\theta}{\dfrac{\theta}{(1 + \varepsilon\theta)^2} - 1} > 0 \tag{4.3-58}$$

which is true if $\varepsilon < 0.25$. This is consistent with our conclusions earlier in this section, where $\varepsilon = 0.25$ was shown to be a critical value for the existence of R-A. Furthermore, for any value of a, the larger the value of b is, the smaller the distance between the two locus curves.

Figure 4.3-9 shows the two locus curves for $\varepsilon = .04$ and $a = 2$. For clarity only the locus of inflections corresponding to $b = 15$ is shown since the curves for $b = 10$ and $b = 20$ are essentially the same. It is important to note the three significantly different temperature histories. For the largest value of b, $\theta(\hat{c})$ rises quickly, intersecting the locus of intersections, and R-A occurs. For the lowest b, $\theta(\hat{c})$ strikes the locus of maxima and there is no R-A. The middle case represents a transition in much the same sense that the critical tangency of $\hat{g}_e$ and $\hat{r}_e$ did in the previous section. The value of b that causes $\theta(\hat{c})$ to be just tangent to the locus of inflections is that which turns a potential R-A into a non R-A reaction.

A few qualitative observations are in order before attempting to quantify the critical condition. It should be noted that b serves a dual role. It shifts the locus of inflections to the right, away from the locus of maxima. This allows temperatures the opportunity to rise higher and still not intersect the inflection curve. Parameter b, however, is also the initial slope of $\theta(\hat{c})$. The smaller b is, the more likely $\theta(\hat{c})$ is to intersect the maxima locus before the inflection locus. Adler and Enig showed for

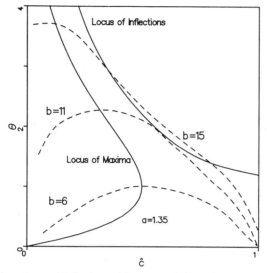

Figure 4.3-9. Loci of maxima and inflections with computed thermal trajectories for various values of *b* (33).

$\varepsilon = 0$ that even under adiabatic conditions there is a value of *b* below which $\theta(\hat{c})$ will never intersect the locus of inflections.

We shall extend their results to include the case $\varepsilon \neq 0$. By taking the derivative of equation 4.3-57 with respect to θ, equating this to the reciprocal of equation 4.3-55 (the slope of the thermal trajectory), and, finally, substituting equation 4.3-57 to eliminate $\hat{c}$, we determine θ_{cr} at the tangent point:

$$\theta_{cr} = \frac{(1 - 3\varepsilon) - \sqrt{\varepsilon^2 - 6\varepsilon + 1}}{2\varepsilon^2} \tag{4.3-59}$$

For very small values of ε, $\theta_{cr} \cong 2$, which has been used by some as a definition for the onset of R-A. Under adiabatic conditions, θ is linearly related to conversion:

$$\hat{c} = 1 - \frac{\theta}{b} \tag{4.3-60}$$

When this is equated to the locus of inflections, equation 4.3-57 with $a = 0$ (adiabatic), the critical value of *b* is determined by:

$$b_{cr} = \frac{\theta_{cr}^2}{\theta_{cr} - (1 + \varepsilon\theta_{cr})^2} \tag{4.3-61}$$

Table 4.3-4 shows the variation of b_{cr} with ε. Below these values, the second requirement for R-A, equation 4.3-2, will never be met. When $\varepsilon > 0.171$, there is no real-valued solution for θ_{cr}, so concentration effects further depress the previously determined limit of $\varepsilon = 0.25$.

Table 4.3-4.   Critical Values of b for the Elimination of IG Under Adiabatic Conditions

ε	θ_{cr}	b_{cr}
0	2	4
0.01	2.06	4.17
0.04	2.28	4.77
0.075	2.64	5.78
0.10	2.98	6.86
0.125	3.51	8.55
0.15	4.44	11.85
0.17	6.85	21.60

The case of nth order kinetics for adiabatic systems was also examined (33) with the result that, for $\varepsilon = 0$,

$$b_{cr} = (1 + n^{1/2})^2 \tag{4.3-62}$$

under more moderate thermal conditions, b_{cr} will be larger. It is quite clear that for kinetic schemes such as those discussed earlier, the key sensitivity parameter is indeed the ratio of CT's defined in equation 4.3-54.

Two problems arise when attempting to generalize this analysis to encompass complex polymerization sequences and nonadiabatic conditions. First, the criterion is too conservative. Avoiding R-A is not the same as avoiding IG. An inflection in the temperature profile might not present a problem if θ_{max} remains relatively low. Thus R-A without IG is not necessarily detrimental. Second, the analytical technique is limited to stoichiometrically simple systems. It does not apply when there are multiple reactants, each with a different activation energy in its rate expression, particularly if the kinetically and thermally controlling reactions are different. This is clearly the case for the broad class of free-radical polymerizations that are not self-initiated.

In response to these problems, we shall adopt a technique patterned after that of Barkelew (21). He studied R-A in first- and second-order reactions with product inhibition and acceleration in a PFR (BR) via numerical solutions of the mass and energy balances. In dimensionless form, these balances are identical to our batch (PFR) equations (noting that he used the approximation $\varepsilon \cong 0$) for a stoichiometrically as well as thermally and kinetically simple system. His approach was to systematically vary b and determine θ_{max} as a function of a. Realizing that under adiabatic conditions, the maximum dimensionless temperature attainable, θ_{ad}, is equal to parameter b (equation 4.3-60 with $\hat{c} = 0$ and Example 4.3-4), he reduced

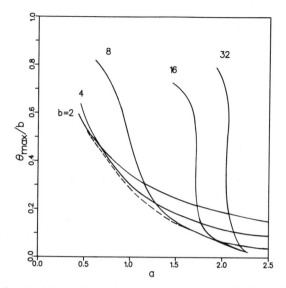

Figure 4.3-10. θ_{max}/b versus a with b as parameter for an irreversible first-order reaction (21).

each value of θ_{max} with b. Some results are shown in Fig. 4.3-10. The family θ_{max}/b rises asymptotically toward 1 as an upper value and approaches a common lower curve as well. The steepness in the change of θ_{max}/b with respect to small changes in a is taken as a measure of parametric sensitivity. Note that at the value $b = 4$ deemed critical by Adler and Enig, the curve is indistinguishable from the lower asymptote. More importantly, note that when $b < 16$ the ascent from the base curve is no longer abrupt. It takes relatively larger changes in a to produce substantial changes in θ_{max} as b drops below 16. Thus, even though R-A occurs as described by conditions 4.3-1 and 4.3-2 and θ_{max} is relatively large, the parametric sensitivity is absent and IG does not occur. It may be entirely feasible to operate a reactor in this regime, where the benefit of accelerated rates is achieved without the liability of reactor instability.

A chemically initiated addition polymerization requires two independent component balances, with rate functions like those listed in Table 4.3-2. Even if the LCA is used to make $\hat{r} = \hat{r}_G$, the reaction scheme is not stoichiometrically simple. When $\hat{E}_d$ is large, the rate of reactant consumption could be dominated by the rate of initiator decay rather than monomer decay. The initiation step cannot be ignored given its disproportionately greater temperature sensitivity.

It might be possible to emulate Barkelew's approach by fixing either dimensionless parameter b_i or b_m in the component balance equations (Table 4.3-1) and varying the value of the other to determine $\theta_{max}(a)$. This method was used (23) for values of b_i and b_m characteristic of free-radical vinyl polymerizations. Figures 4.3-11 and 4.3-12 illustrate the results for such reactions. They give a clear picture of the evolution of sensitivity with b_i and b_m and a concomitantly well-defined a_{cr}. From the figures, when $b_i > 70$ and $b_m > 20$, it is easy to identify a single value of a to characterize the onset of R-A. From the locus of a_{cr}, one can construct dimensionless R-A boundaries in the form a_{cr} versus b_m or b_i for various values of the other R-A parameters. Such boundaries are illustrated in Figs. 4.3-13–4.3-16. The first

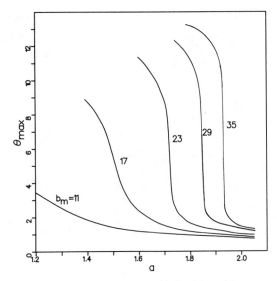

Figure 4.3-11. θ_{max} versus a for chain-addition with fixed b_i and b_m as parameters (23).

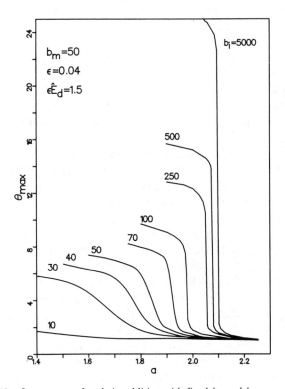

Figure 4.3-12. θ_{max} versus a for chain-addition with fixed b_m and b_i as parameters (23).

426

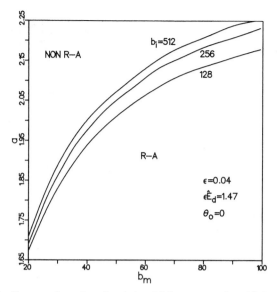

Figure 4.3-13.   Runaway boundary for chain-addition a versus b_m with b_i as parameters (23).

three are similar to plots by Adler and Enig and by Barkelew, as both authors demonstrated that a_{cr} was a function of b. It is important to note that in each of the first three figures a_{cr} is not significantly affected by changes in the respective parameter. Moreover, the range of values of b_m generally encountered in actual polymerizations is smaller than that used to construct Fig. 4.3-16. On the other hand, to represent adequately the full span of b_i's actually encountered, a logarithmic scale was needed in the latter figure. Note also that a small change in b_m can cause a substantial change in a_{cr} for a fixed b_i, but only when b_i is sufficiently small. For larger values of b_i, each curve ultimately becomes horizontal, and further increases in b_i bring about no change in a_{cr}. Such "critical" values of b_i depend on b_m, but all seem to be below $b_i = 1000$ for all cases shown.

The reader will note that $\varepsilon b_m = \hat{T}_{ad}$ where $\hat{T}_{ad}$ is defined by equation 4.0-3. With reference to Barkelew's analysis, previously described, the corresponding value of θ_{ad} is identical to b_m, provided that all the monomer can polymerize prior to consumption to initiator. From Fig. 4.3-12, however, it is apparent that none of the curves converges to a common value of θ_{ad} (actual) that is even close in value to b_m. They certainly do not converge to b_i either. Consequently, neither parameter seems appropriate to use in reducing θ_{max} via the method of Barkelew. The alternative is to compute θ_{ad} by solving the batch balance equations in Table 4.3-1 numerically for the special case $a = 0$. After performing several such computations, it is clear that θ_{ad} is functionally dependent on b_i, b_m, ε, and $\varepsilon \hat{E}_d$ in a complicated way, as contrasted with the dependence found for simple nth order kinetics. Figure 4.3-17 shows $\theta_{ad}(b_i, b_m)$ for the characteristic values of $\varepsilon = .04$ and $\varepsilon E_d = 1.5$. It is immediately apparent that there are two distinct regions of behavior; one in which the value of b_m dictates θ_{ad} (i.e., $\theta_{ad} = b_m$) and the other in which θ_{ad} is independent of b_m for given values of b_i. Unfortunately, the transition region is just broad enough to span the range of b_m values of practical interest. It is important to note, however,

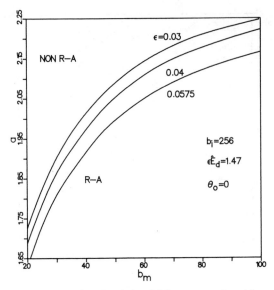

Figure 4.3-14. Runaway boundary for chain-addition a versus b_m with ε as parameter (23).

that unless b_m is very small, θ_{ad} is very close to the value expected for $b_m \cong \infty$. This shows that the large temperature coefficient for the initiator balance plays a dominant role in reactant consumption by stopping reactions at temperatures far below those expected on the basis of initial reactant ratios. Therefore, initiator effects cannot be ignored, and the assumption of first-order behavior is not valid in the face of rising temperatures.

Bearing this in mind, we shall digress for a moment and attempt to gain some insight into the general situation in which the kinetic step that is primarily responsi-

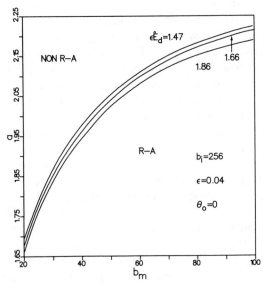

Figure 4.3-15. Runaway boundary for chain-addition a versus b_m with $\varepsilon \hat{E}_d$ as parameter (23).

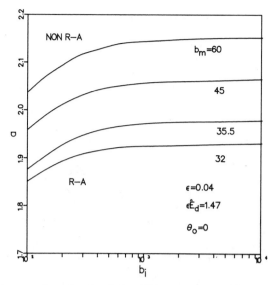

Figure 4.3-16. Runaway boundary for chain-addition a versus b_i with b_m as parameters (23).

ble for consuming the reaction-limiting reactant is different from the kinetic step that is the primary source of heat evolution. To simplify our analysis, we shall examine a fictitious reaction consisting of two such independent steps, each of which is kinetically and thermally simple (Appendix E), whose dimensionless material and energy balances are

$$\frac{d\hat{c}}{d\hat{\alpha}} = -b^{-1}\hat{c}^n\exp\frac{\varepsilon\hat{E}\theta}{1 + \varepsilon\theta} \tag{4.3-63}$$

and

$$\frac{d\theta}{d\hat{\alpha}} = \hat{c}^m\exp\frac{\theta}{1 + \varepsilon\theta} \tag{4.3-64}$$

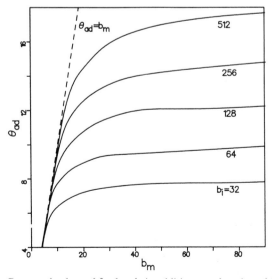

Figure 4.3-17. Computed values of θ_{ad} for chain-addition as a function of b_m and b_i.

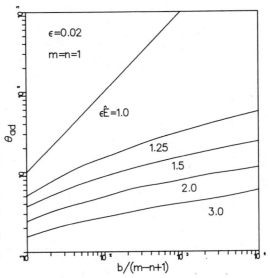

Figure 4.3-18. Computed values of θ_{ad} for reaction schemes 4.3-63 and 4.3-64 with $m = n = 1$ and $\varepsilon = 0.02$.

where $m \neq n$. A special free-radical polymerization (severe D-E) is represented by this system if we set $m = 1/2$, $n = 1$, $b = b_i$, and $\hat{E} = \hat{E}_d$. It occurs when b_m is very large ($b_m \rightarrow \infty$), and therefore $\hat{c}_m$ remains constant ($\hat{c}_m = 1$) throughout the reaction. If ε is small enough ($\varepsilon = 0$) to be approximated by zero, the preceding two equations can be solved analytically to obtain θ_{ad}:

$$\theta_{ad} = \frac{\ln\left(1 + \dfrac{b(\varepsilon\hat{E} - 1)}{m - n + 1}\right)}{\varepsilon\hat{E} - 1} \tag{4.3-65}$$

This result illustrates that θ_{ad} is not a simple ratio of two CT's unless $\varepsilon\hat{E} = 1$, in which case simultaneous solution of equations 4.3-63 and 4.3-64 gives

$$\theta_{ad} = \frac{b}{m - n + 1} \tag{4.3-66}$$

However, it is clear that simple reaction schemes of arbitrary order (values of m and n) can readily be unified in this manner. Even if $\varepsilon \neq 0$, it follows that

$$\frac{b}{m - n + 1} = \int_0^{\theta_{ad}} \exp\frac{(\varepsilon\hat{E} - 1)\theta}{(1 + \varepsilon\theta)} d\theta \tag{4.3-67}$$

Computer solutions of integral 4.3-67 for different ratios of activation energy (i.e., $\varepsilon\hat{E}$) are shown in Fig. 4.3-18. Note that only a slight increase in the reactant activation energy ($\hat{E}$) over the generation activation energy (ε^{-1}) brings about a large decrease in θ_{ad}.

The qualitative effects of differing activation energies are apparent from the figures. In order to apply this information to our analysis of reactor sensitivity, we again turn to a Barkelew-type analysis, utilizing instead the computed values of θ_{ad} to normalize θ_{max}, obtained by integrating equations 4.3-63 and 4.3-64 with $a = 0$. Figures 4.3-19 and 4.3-20 show graphs of this ratio versus a with b as parameter for

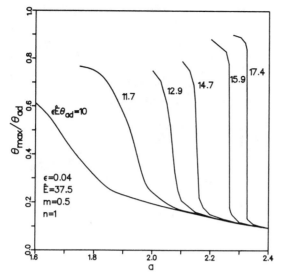

Figure 4.3-19. θ_{max}/θ_{ad} versus a for schemes 4.3-63 and 4.3-64 with $m = 0.5$, $n = 1$, $\hat{E} = 37.5$, and $\varepsilon = 0.04$.

two different schemes. Qualitatively one would accept the maximum steepness of any of these curves as a measure of a system's sensitivity. A large change in θ_{max} with respect to a small change in a certainly is an indication of a sensitive R-A. With this in mind, we note that the curves in Figs. 4.3-19 and 4.3-20 having the same value of $b/(m - n + 1)$ also have slopes that are about the same. Furthermore, there appears to be a sudden transition from curves with virtually vertical slopes to those with much lesser slopes, which occurs within the range $60 < b/(m - n + 1) < 80$. From

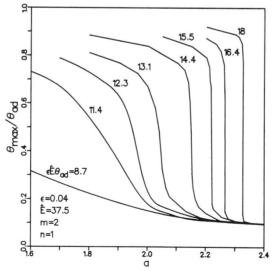

Figure 4.3-20. θ_{max}/θ_{ad} versus a for schemes 4.3-63 and 4.3-64 with $m = 2$, $n = 1$, $\hat{E} = 37.5$, and $\varepsilon = 0.04$.

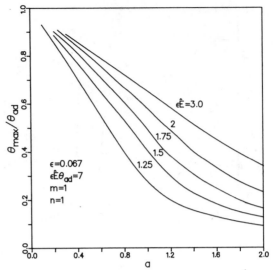

Figure 4.3-21. $\theta_{max}\theta_{ad}$ versus a for chain-addition with various values of $\varepsilon \hat{E}_d$ with a common value of $\varepsilon E_d \theta_{ad}$.

Fig. 4.3-18, it is clear that a given value of $b/(m + n - 1)$ corresponds to a unique value of θ_{ad} when ε and $\varepsilon \hat{E}$ are fixed.

We shall therefore adopt the slope $[d(\theta_{max}/\theta_{ad})/da]_{max}$ as a measure of sensitivity and seek a single parameter to characterize it. The rapid onset of decay of that slope observed in the figures suggests that our sensitivity parameter itself may be parametrically sensitive. Figures 4.3-21 and 4.3-22 exemplify the results of our search. It appears that for a wide variety of combinations of ε and $\varepsilon \hat{E}$, when $\varepsilon \hat{E} \theta_{ad}$ is constant

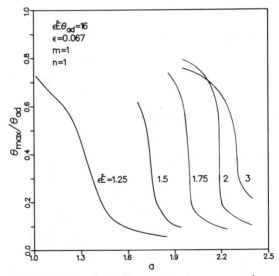

Figure 4.3-22. θ_{max}/θ_{ad} versus a for chain-addition with various values of $\varepsilon \hat{E}_d$ with a common value of $\varepsilon E_d \theta_{ad}$.

the slope is relatively constant. Furthermore, plotting the maximum slope as a function of $\varepsilon\hat{E}\theta_{ad}$, as in Fig. 4.3-23, exposes the sensitivity of the sensitivity. Beyond the value of $\varepsilon\hat{E}\theta_{ad} \cong 15$ the maximum value of the slope increases dramatically with ensuing increases in $\varepsilon\hat{E}\theta_{ad}$.

These results are not inconsistent with previous work. First, we simplify the sensitivity parameter determined earlier to show that

$$\hat{E}\hat{T}_{ad} = \varepsilon\hat{E}\theta_{ad} \tag{4.3-68}$$

For the stoichiometrically simple kinetics considered by Adler and Enig, as well as Barkelew, $\varepsilon = \hat{E}^{-1}$ or $\varepsilon\hat{E} = 1$ so that $\theta_{ad} = b$ (equation 4.3-55). Thus they were able to characterized sensitivity with the single ratio of CT's, b. Quantitatively, Barkelew saw sensitivity wane when $8 < b < 16$, which is consistent with our value of $b < 15$.

An important ramification of this analysis concerning chain-addition polymerization sequences is that there are two criteria for sensitivity. When $b_m \gg b_i$, $\hat{E} = \hat{E}_d$ and sensitivity requires that

$$\varepsilon\hat{E}_d\theta_{ad} = \hat{E}_d\hat{T}_{ad} > 15 \tag{4.3-69}$$

On the other hand, when $b_m \ll b_i$, then $\hat{E} = \varepsilon^{-1}$ and $\theta_{ad} = b_m$, and our criterion becomes

$$b_m > 15 \tag{4.3-70}$$

The first criterion covers the broad range in which initiator kinetics dominate sensitivity. The second, which corresponds to Barkelew's criterion and involves the quantity $\hat{T}_{ad}$ alluded to in the introduction to this chapter, actually ensures that monomer depletion is not so rapid that it overtakes initiator decay. Criterion 4.3-69

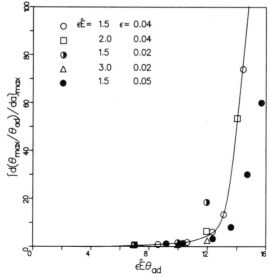

Figure 4.3-23. Maximum slope of θ_{max}/θ_{ad} versus a as a function of $\varepsilon\hat{E}_d\theta_{ad}$.

is inconvenient in that it requires knowledge of the actual value of θ_{ad} (or $\hat{T}_{ad}$). Fortunately, however, we can link θ_{ad} exclusively to b_i over a broad range of values of b_m. From Fig. 4.3-17, we conclude that $\varepsilon \hat{E}_d \theta_{ad} = 15$ corresponds to $b_i = 70$. Thus we may express our specific sensitivity criteria as

$$b_m > 15 \tag{4.3-71}$$

$$b_i > 70 \tag{4.3-72}$$

The dimensionless criteria for nonisothermal reactor performance developed up to this point are summarized in Table 4.3-5. All dimensionless groups are expressed at ratios of the CT's defined in Table 4.3-3. Thermal runaway (R-A) will occur only when both inequalities listed are satisfied. The value of a_{cr} ranges from 2.72 to 2 depending on the value of ε, which must be less than 0.25. Parametrically sensitive runaway (IG) requires that all inequalities involving b_k be satisfied, in addition to those for R-A. For polymerizations involving a single independent reaction variable (such as many random polymerizations), a single sensitivity parameter, $b_k = b$, will suffice. Its critical value depends on the value of ε, as shown in Table 4.3-4. For a large class of chain-addition polymerizations, two sensitivity parameters are required, b_m and b_i. Their critical values are given in inequalities 4.7-71 and 4.7-72. Violation of either criterion is sufficient to eliminate R-A sensitivity.

The numerical value of the critical dimensionless parameter associated with each criterion is the result of many numerical simulations of homopolymerizations, especially the chain-addition type. They are intended for use merely as a guide in the design and operation of polymerization reactors. An important question remaining, of course, concerns their ability to characterize the behavior of real reactors.

Numerous R-A experiments with styrene polymerization have been conducted (24, 26, 45) to test these criteria. A small well-mixed BR of known heat transfer characteristics was used so that all the pertinent CT's for R-A analysis could be computed. Figures 4.3-24 and 4.3-25 show excellent quantitative agreement between the thermal behavior and the values of R-A parameters a, b_i, and b_m. In Figure 4.3-24, the transition to R-A behavior is fairly abrupt relative to changes in a. The non R-A profiles are quasi-isothermal (Q-I) and tightly clustered. Reducing a causes the onset of rapidly rising temperatures, but subsequent reduction of a does not cause significant alteration of the maximum temperature encountered. Note that the

Table 4.3-5. Dimensionless Criteria for Nonisothermal Reactor Performance

PHENOMENON	PARAMETER		DEFINITION	CRITERION
R-A	lumped	ε	λ_{ad}/λ_G	$\varepsilon \ll 1$
		a	λ_{ad}/λ_R	$a < a_{cr}$
	distr	Δ	λ_{ad}/λ_H	$\Delta < \Delta_{cr}$
IG/ERA		b_k	λ_k/λ_{ad}	$b_k > (b_k)_{cr}$
	for all k			

values of both b_m and b_i are above the critical limits established by analysis of simulation results, and the value of a at the R-A transition is consistent with expectations.

In Figure 4.3-25, high temperatures and small feed values of initiator concentration combine to reduce R-A sensitivity to the point where large decreases in a are needed to bring about any significant change in the thermal behavior of the system. The values of b_i are below the threshold of sensitivity and the loss of an identifiable transition point from Q-I to R-A behavior is quite apparent. Further supportive data can be found in reference 24. In all cases the R-A transition, when identifiable, occurred at a value of approximately 2 for a, and the critical value decreased as b_i decreased. In the limit of very low values of b_i, R-A sensitivity was eliminated.

4.3-2. Thermally Complex Systems

The preceding R-A analysis has been confined to thermally simple systems (Appendix E). Most homopolymerization kinetic schemes fit this description, since the heat evolution can be attributed mainly to the propagation step, and this can generally be expressed as a kinetically simple reaction. Not all reaction schemes can be expressed in this fashion, however. When multiple reactions occur in tandem and no single reaction dominates the thermal behavior, the foregoing analysis cannot be applied in a straightforward manner. A method for characterizing the composite thermal behavior of a complex system is needed. The subsequent analysis is aimed toward that end. Chain-addition copolymerization will be used as a vehicle since it contains all the aforementioned features, that is, kinetically and thermally complex sequences. Semi-dimensionless balance equations for nonisothermal chain-addition polymerization are listed in Table 4.3-6 and pertinent CT's and dimensionless groups are

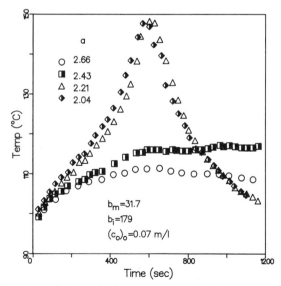

Figure 4.3-24. Experimentally determined temperature histories for styrene polymerization (24).

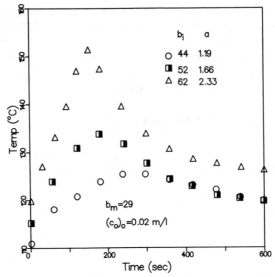

Figure 4.3-25. Experimentally determined temperature histories for styrene polymerization (24).

defined in Tables 4.3-7 and 4.3-8. These equations are based on the kinetic models posed in Sections 1.12 and 2.9, and their complex nature is readily apparent.

As in the preceding section, we seek to reduce the balance equations to a more compact form that is independent of termination kinetics. We suggest the following form.

$$-\frac{d\hat{c}_m}{d\alpha} = \sum_j \sum_k (n_k)_0 \lambda_{mjk}^{-1} \hat{r}_{jk} = \sum_j \sum_k (n_k)_0 \lambda_{mjk}^{-1} \hat{c}_k \hat{c}_0^{1/2} \hat{h}_j \exp\frac{\hat{E}_{jk}\hat{T}}{1+\hat{T}} \quad (4.3\text{-}73)$$

The dimensionless functions $\hat{h}_j$, listed in Table 4.3-7, contain specific details on the particular termination mechanism involved. In fact $\hat{h}_j$ can be interpreted as the dimensionless concentration of chain radicals in which the active end is repeat unit j (A or B). It comes from making dimensionless the expressions for intermediate concentrations (specifically, $c_j H$) as given in Table 1.12-4. Note that at "reference" conditions $[c_j = (c_j)_0$ and $T = T_r]$, $\hat{h}_j = 1$.

Monomer is consumed in four distinct propagation steps, as represented by each of the terms in the preceding summation. As in homopolymerization, the heat of reaction is attributed to the propagation reaction. Each propagation step in copolymerization can have a different enthalpy of reaction. Therefore, the generation term in the thermal energy balance must include the contribution of each of these four concurrent reactions. As a result, it is not thermally simple. In general, the generation term would have the form

$$j = A, B$$
$$k = A, B$$
$$r_G = -\sum_j \sum_k \Delta H_{jk} r_{jk} \quad (4.3\text{-}74)$$

Table 4.3-6.　Semidimensionless Balance Equations for Nonisothermal Copolymerization

$$-\frac{d\hat{c}_o}{d\alpha} = \lambda_i^{-1}\hat{r}_o$$

$$-\frac{d\hat{c}_k}{d\alpha} = \lambda_{m_{Ak}}^{-1}\hat{r}_{Ak} + \lambda_{m_{Bk}}^{-1}\hat{r}_{Bk}$$

$$-\frac{d\hat{c}_m}{d\alpha} = (n_A)_o\lambda_{m_{AA}}^{-1}\hat{r}_{AA} + (n_A)_o\lambda_{m_{BA}}^{-1}\hat{r}_{BA} + (n_B)_o\lambda_{m_{AB}}^{-1}\hat{r}_{AB} + (n_B)_o\lambda_{m_{BB}}^{-1}\hat{r}_{BB}$$

$$\frac{d\hat{T}}{d\alpha} = \lambda_{G_{AA}}^{-1}\hat{r}_{AA} + \lambda_{G_{AB}}^{-1}\hat{r}_{AB} + \lambda_{G_{BA}}^{-1}\hat{r}_{BA} + \lambda_{G_{BB}}^{-1}\hat{r}_{BB} - \lambda_R^{-1}(\hat{T} - \hat{T}_R)$$

where

$$\hat{r}_{jk} = \hat{c}_o^{1/2}\hat{c}_k\hat{h}_j \exp^{\cdot}\hat{E}_{jk}\hat{T}/(1 + \hat{T})$$

$$\hat{r}_o = \hat{c}_o \exp\hat{E}_d\hat{T}/(1 + \hat{T})$$

but we have seen that a consequence of the QSSA is that $r_{AB} = r_{BA}$. Thus the cross-reaction contributions are indistinguishable.

$$j \neq k \qquad -\Delta H_{AB}r_{AB} - \Delta H_{BA}r_{BA} = -(\Delta H_{AB} + \Delta H_{BA})r_{jk} \qquad (4.3\text{-}75)$$

If we were to express 4.3-77 as a thermally simple function, then the heat of copolymerization would be

$$-\Delta H_c = \frac{r_G}{r_p} \qquad (4.3\text{-}76)$$

As such, ΔH_c would be independent of any particular model for termination kinetics and would be a function of composition, reactivity ratios, and the ΔH_{jk}:

$$\Delta H_c = \frac{\Delta H_{AA}R_A n_A^2 + (\Delta H_{AB} + \Delta H_{BA})n_A n_B + \Delta H_{BB}R_B n_B^2}{R_A n_A^2 + 2n_A n_B + R_B n_B^2} \qquad (4.3\text{-}77)$$

It is clear that composition drift can cause the value of ΔH_c to change with conversion during reaction. Table 4.3-9 lists some experimentally determined values of the cross-propagation enthalpy sum $-(\Delta H_{AB} + \Delta H_{BA})$ for a variety of systems (36, 37). Using these values and equation 4.3-77 one can compute $\Delta H_c(n_A)$ for some

Table 4.3-7. Dimensionless Intermediate Concentrations

Geometric Mean

$$\hat{h}_k = \frac{(A + B)\hat{n}_k}{A\hat{n}_A \exp \hat{E}_{pt_{BA}} \hat{T}/2(1 + \hat{T}) + B\hat{n}_B \exp \hat{E}_{pt_{AB}} \hat{T}/2(1 + \hat{T})}$$

Phi Factor

$$\hat{h}_k = \frac{(A^2 + 2\ AB + B^2)^{1/2}\ \hat{n}_k}{\left\{(An_A)^2 \exp \hat{E}_{pt_{BA}} \hat{T}/2(1 + \hat{T}) + 2\ ABn_An_B \exp(\hat{E}_{pt_{AB}} + \hat{E}_{pt_{BA}})\hat{T}/(1 + T) + (Bn_B)^2 \exp \hat{E}_{pt_{AB}} \hat{T}/2(1 + \hat{T})\right\}^{1/2}}$$

Penultimate Effect

$$\hat{h}_k = \frac{\left\{A\left[\frac{R_An_A + R_An_B}{R_An_A + n_B}\right]_r + B\left[\frac{R_Bn_B + R_Bn_A}{R_Bn_B + n_A}\right]_r\right\}^{\hat{n}_k}}{A\hat{n}_A \exp \hat{E}_{pt_{BA}} \hat{T}/2(1 + \hat{T})\left[\frac{\mathbf{R}_An_A + \mathbf{R}_An_B}{\mathbf{R}_An_A + n_B}\right] + B\hat{n}_B \exp \hat{E}_{pt_{AB}} \hat{T}/2(1 + \hat{T})\left[\frac{\mathbf{R}_Bn_B + \mathbf{R}_Bn_A}{\mathbf{R}_Bn_B + n_A}\right]}$$

$A \equiv (n_A)_0 (R_A)_r \lambda_A$

$B \equiv (n_B)_0 (R_B)_r \lambda_B$

$R_j \equiv (R_j)_r \exp \hat{E}_{R_j} \hat{T}/(1 + \hat{T})$

$\mathbf{R}_j \equiv (R_j)_r \exp \hat{E}_{\mathbf{R}_j} \hat{T}/(1 + \hat{T})$

and

$(\mathbf{R}_j)_r \equiv (k_{t\ell jj\ell}/k_{tjjjj})_r \qquad \hat{n}_k \equiv n_k/(n_k)_0$

438

Table 4.3-8. CT's and Dimensionless Groups for Individual Copolymerization Propagation Steps

$$\lambda_k = k_{p_{kk}}^{-1} \left[r_i / k_{t_{kk}} \right]^{-1/2} \qquad k = A, B$$

$$\lambda_{m_{jk}} = \left[r_{p_{jk}} / c_k \right]^{-1} \qquad j = A, B \quad k = A, B$$

$$\lambda_{G_{jk}} = \lambda_{m_{jk}} / He_{jk} \qquad j = A, B \quad k = A, B$$

$$\lambda_{ad_{jk}} = \lambda_{G_{jk}} / \hat{E}_{jk} \qquad j = A, B \quad k = A, B$$

$$He_{jk} = -\Delta H_{jk}(c_k)_o / \rho c_p T_r \qquad j = A, B \quad k = A, B$$

$$\gamma_{T_{jk}} = \lambda_{G_{jk}} / \lambda_R \qquad j = A, B \quad k = A, B$$

$$\hat{E}_{jk} = \left[E_{p_{jk}} + E_{p\ell j} + \frac{1}{2}(E_d - E_{t_{jj}} - E_{t\ell\ell}) \right] / R_g T_r \qquad \begin{array}{l} j \neq \ell = A, B \\ j = A, B \end{array}$$

$$\hat{E}_{Pt_{jk}} = (2E_{P_{jk}} - E_{t_{jj}}) / R_g T_r \qquad j = A, B \quad j \neq k = A, B$$

$$\hat{E}_{R_j} = (E_{p_{jj}} - E_{p_{jk}}) / R_g T_r \qquad j = A, B \quad j \neq k = A, B$$

$$\hat{E}_{R_j} = (E_{t_{kjjk}} - E_{t_{jjjj}}) / R_g T_r \qquad j = A, B \quad j \neq k = A, B$$

representative cases. Figures 4.3-26–4.3-28 show the results. Note that it is possible for the heat of copolymerization to exceed the heat of homopolymerization of either comonomer.

The reactivity ratios R_A and R_B are temperature-dependent in the usual way (Arrhenius function). Activation energies and pre-exponential factors for some typical systems are listed in Table 4.3-10 (38–41). The activation energies are not very large indicating that the R_j's are relatively temperature insensitive.

In an attempt to duplicate the homopolymerization analysis, we could write the thermal energy balance in semidimensionless form similar to the balances in Table 4.3-1.

$$\frac{d\hat{T}}{d\alpha} = \sum_{j=A}^{B} \sum_{k=A}^{B} \frac{\hat{r}_{jk}}{\lambda_{G_{jk}}} - \frac{\hat{T}}{\lambda_R} \tag{4.3-78}$$

Note that we are forced to use $\hat{T}$ rather than θ for dimensionless temperature, owing to the absence of a single representative activation energy $\hat{E}$ with which to define $\theta = \hat{E}\hat{T}$. We can make age dimensionless with λ_R and express the resulting dimen-

Table 4.3-9. Cross-Propagation Enthalpy Sum for Chain-Addition Copolymers

System	$\Delta H_{AB} + \Delta H_{BA}$ (kcal/mole)
S/MMA	32.74
VA/MMA	36.61
S/VA	17.95
AN/MMA	28.40
S/AN	34.67
VA/AN	38.15

sionless generation term as follows:

$$\hat{r}_{G} = \sum_{j} \sum_{k} \hat{f}_{jk}(\hat{\alpha}) \hat{g}_{e_{jk}}(\hat{T}) \hat{h}_{j}(\hat{\alpha}, \hat{T}) \qquad (4.3\text{-}79)$$

where

$$\hat{f}_{jk}(\hat{\alpha}) = \hat{c}_{k} \hat{c}_{0}^{1/2} / \gamma_{T_{jk}} \qquad (4.3\text{-}80)$$

$$\hat{g}_{e_{jk}}(\hat{T}) = \exp \frac{\hat{E}_{jk} \hat{T}}{1 + \hat{T}} \qquad (4.3\text{-}81)$$

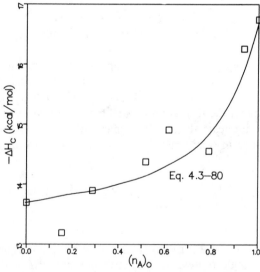

Figure 4.3-26. Variation in ΔH_c with feed composition of AN/MMA (36).

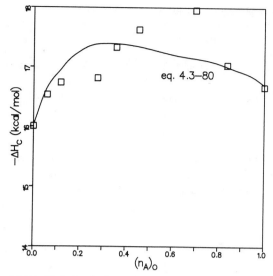

Figure 4.3-27. Variation in ΔH_c with feed composition for AN/S (36).

and $\hat{h}_j$ has been defined in Table 4.3-3. Both f_{jk} and $\hat{g}_{e_{jk}}$ have the same interpretation as in the previous sections. Functions $\hat{h}_A$ and $\hat{h}_B$ are dependent on both temperature and time (implicitly via composition). It is an important note that $\hat{h}_j$ depends on composition but not concentration. Concentration will always drift with time, but composition could remain relatively constant, notwithstanding large changes in conversion (e.g., azeotropic copolymerization or a horizontal CC curve). In these cases, $\hat{h}$ will be reduced to a function of temperature only.

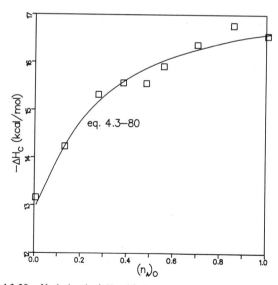

Figure 4.3-28. Variation in ΔH_c with feed composition for MMA/S (37).

Table 4.3-10. Temperature Dependence of Reactivity Ratios for Vinyl Monomers

Co-monomer	R_A	R_B
S-AN	$2.56 \, \exp(-1190/R_g T)$	$6.67 \times 10^{-5} \, \exp(4340/R_g T)$
S-MMA	$1.83 \, \exp(-895/R_g T)$	$1.274 \, \exp(-675/R_g T)$
AN-MMA	$1.935 \, \exp(-1820/R_g T)$	$0.586 \, \exp(516/R_g T)$
AN-VA	$9.75 \, \exp(-536/R_g T)$	$18 \, \exp(-3780/R_g T)$
S-αMS	$1.74 \times 10^{-2} \, \exp(2769/R_g T)$	$1.185 \times 10^{-4} \, \exp(5630/R_g T)$

The essence of a Semenov-type analysis is the ERA and the tangency condition, i.e., equating $\hat{g}_e = \hat{r}_e$ and $d\hat{g}_e/d\hat{T} = d\hat{r}_e/d\hat{T}$ to find $\hat{T}_{cr}$, the critical R-A temperature. This technique, when applied to the copolymer thermal energy balance, yields (27):

$$\sum_j \sum_k \gamma_{Tjk}^{-1} \hat{h}_j(\hat{T}) \exp \frac{\hat{E}_{jk}\hat{T}}{1 + \hat{T}} = \hat{T} \qquad (4.3\text{-}82)$$

$$\sum_j \sum_k \gamma_{Tjk}^{-1} \left[\frac{\hat{E}_{jk}\hat{h}_j(\hat{T})}{(1 + \hat{T})^2} + \frac{d\hat{h}_j}{d\hat{T}} \right] \exp \frac{\hat{E}_{jk}\hat{T}}{(1 + \hat{T})} = 1 \qquad (4.3\text{-}83)$$

It is obvious that the exponential terms cannot be eliminated between the two equations and that the presence of $\hat{h}_j(\hat{T})$ further reduces the possibility of attaining an analytical solution.

Throughout this book we have stressed the importance of interpreting dimensionless groups as ratios of CT's. Thus quantitative criteria based on such groups have significance rooted in the competition between the physical processes represented by the CT's. With this in mind, we shall examine equations 4.3-46–4.3-48 for clues leading to the formulation of R-A criteria for copolymerizations.

Recall that although homopolymerization CT's arose naturally as coefficients in semi-dimensionless balance equations associated with specific processes, it was possible to interpret them from another viewpoint. For example the CT's associated with reactions were interpreted as semi-dimensionless gradients evaluated at reference conditions. We shall now extend this approach to copolymerization balances. Thus, by adopting the "equivalent" definition of λ_m in Table 4.3-3, we may define a composite CT for copolymerization as follows:

$$\Lambda_m = \left(\frac{d\hat{c}_m}{dt} \right)_r^{-1} = \left(\sum_{j=A}^{B} \sum_{k=A}^{B} (n_k)_0 \lambda_{m_{jk}}^{-1} \right)^{-1} \qquad (4.3\text{-}84)$$

This CT reflects the time scale for total monomer consumption (as opposed to comonomer A or B) and is in every way analogous to λ_m for homopolymerization. It can be used in the formulation of dimensionless criteria as well, so that for copolymerization $\alpha_K = \Lambda_m/\lambda_i$. Composite CT's for the remaining reactions may be defined analogously. They are listed in Table 4.3-11. Using these working definitions we can operate on the dimensionless balance equations in Table 4.3-6. The resulting functional forms of the CT's are presented in Table 4.3-12. Several additional CT's arise because in addition to an overall monomer balance, there are balance equations associated with each of the comonomers. Note that all composite CT's, denoted by Λ, reflect the time scale for an overall process composed of many subprocesses acting in parallel (e.g., monomer conversion effected by four distinct propagation steps acting in tandem). Unlike homopolymerization CT's, the Λ's do not appear explicitly in the balance equations as separable coefficients. They are complex in form and most probably would not have presented themselves without the guidance provided by the "equivalent" interpretation of the original CT's. Next we shall substitute these composite CT's into our dimensionless parameters. The results are summarized in Table 4.3-13. Implicit in this procedure is the hypothesis that our dimensionless criteria previously developed for homopolymerizations will apply directly to copolymerizations.

Many computer simulations were performed using kinetic constants from several different copolymer systems and the three kinetic models for termination described in Sections 1.12 and 2.9. In all cases, the results supported this hypothesis. Figure 4.3-29 shows a homopolymer boundary, a_{cr} versus b_i. Also shown are a number of pairs, a_{cr}, b_i, corresponding to two copolymer systems under conditions such that b_m and ε are the same as the homopolymer system (note: $\varepsilon \equiv \Lambda_{ad}/\Lambda_G$). The copolymer transitions occur at virtually the same points as determined by analysis of homopolymerization kinetics. There is a small deviation and this can be attributed to composition drift, a feature not present in homopolymerization. When composition drifts so as to favor the more exothermic propagation reactions, R-A will occur at slightly higher values of a than predicted by the homopolymer curve. Conversely, drift favoring the less exothermic reactions postpones the onset of R-A to somewhat lower values of a.

Table 4.3-11. Characteristic Times for Copolymerizations

CT	DEFINITION	EQUIVALENT	PROCESS
Λ_k	$(c_k)_o/(r_{pk})_r$	$\left(\dfrac{d\hat{c}_k}{dt}\right)_r^{-1}$	co-monomer k consumption
Λ_m	$(c_m)_o/(r_o)_r$	$\left(\dfrac{d\hat{c}_m}{dt}\right)_r^{-1}$	total monomer consumption
Λ_G	$\rho c_p T_r/(r_G)_r$	$\left(g_e\right)_r^{-1}$	heat generation
Λ_{ad}		$\left(\dfrac{\partial g_e}{\partial \hat{T}}\right)_r^{-1}$	adiabatic induction

Table 4.3-12. Working Definitions of Composite CT's for Copolymerizations

$$\Lambda_A = \left(\sum_{j=A}^{B} \lambda_{m_{jA}}^{-1}\right)^{-1} = P\left[R_A n_A + n_B\right]_r^{-1}$$

$$\Lambda_B = \left(\sum_{j=A}^{B} \lambda_{m_{jB}}^{-1}\right)^{-1} = P\left[R_B n_B + n_A\right]_r^{-1}$$

$$\Lambda_m = \left(\sum_{j=A}^{B}\sum_{k=A}^{B} (n_j) \circ m_{kj} \lambda^{-1}\right)^{-1} = P\left[R_A n_A^2 + 2 n_A n_B + R_B n_B^2\right]_r^{-1}$$

$$\Lambda_G = \left(\sum_{j=A}^{B}\sum_{k=A}^{B} \lambda_{G_{jk}}^{-1}\right)^{-1} = P\left[R_A n_A He_{AA} + n_A He_{AB} + n_B He_{BA} + R_B n_B He_{BB}\right]_r^{-1}$$

$$\Lambda_{ad} = \left(\sum_{j=A}^{B}\sum_{k=A}^{B} \lambda_{ad_{jk}}^{-1} + \lambda_{G_{jk}}^{-1}\frac{\partial \hat{h}_j}{\partial \hat{T}}\right)^{-1} = P\left[R_A n_A He_{AA}\left(\hat{E}_{AA} + \frac{\partial \hat{h}_A}{\partial \hat{T}}\right) + n_B He_{AB}\left(\hat{E}_{AB} + \frac{\partial \hat{h}_A}{\partial \hat{T}}\right) + n_A He_{BA}\left(\hat{E}_{BA} + \frac{\partial \hat{h}_B}{\partial \hat{T}}\right)\right.$$

$$\left. + R_B n_B He_{BB}\left(\hat{E}_{BB} + \frac{\partial \hat{h}_B}{\partial \hat{T}}\right)\right]_r^{-1}$$

where:

$$P = \lambda_A \lambda_B R_A R_B r_i^{1/2} \Big/ (c_m H)_r$$

Table 4.3-13.   Dimensionless R-A Parameters for Copolymerizations

PARAMETER	DEFINITION
α_k	Λ_m / Λ_i
β_k	Λ_A / Λ_m
a	Λ_{ad} / λ_R
b_m	Λ_m / Λ_{ad}
b_i	λ_i / Λ_{ad}
ϵ	Λ_{ad} / Λ_G

Table 4.3-14 lists further results for several comonomer and initiator systems. In all cases, the trends in the values of R-A parameters are consistent with our expectations from consideration of homopolymer systems.

Nonisothermal property drift may also be characterized with appropriate CT's. In Chapter 2, the direction of the isothermal composition drift was shown to be characterized by β_K, where β_K was the ratio of a CT for conversion of one comonomer to a CT for total monomer conversion. This parameter also relates to

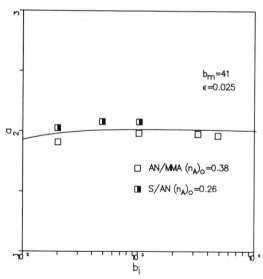

Figure 4.3-29.   Homopolymerization runaway boundary for $b_m > 41$ with computed runaway transitions for two copolymer systems (27).

Table 4.3-14. Simulation Results

Case		$(n_A)_o$	T_o, °K	$(c_o)_o$ mole/ℓ.	U_{s_v} sec °K	b_m	E	b_i	a	R-A
1	S/AN/BP	0.5	398	0.163	8.97×10^{-2}	35.2	25.7	195	1.97	no
2					8.92				1.96	yes
3		0.6	383	0.0763	2.04×10^{-2}	35.8	26.8	195	1.96	no
4					2.03				1.95	yes
5		0.7	373	0.0517	7.07×10^{-3}	35.6	27.6	195	1.935	no
6					7.05				1.93	yes
7		0.8	363	0.0358	2.32×10^{-3}	35.6	28.6	195	1.918	no
8					2.29				1.90	yes
9	AN/MMA/BP	0.2	322	1.33×10^{-4}	1.57×10^{-5}	45	28.6	300	2.025	no
10					1.55				1.998	yes
11		0.4	336	8.33×10^{-4}	1.03×10^{-4}	45	26.5	300	1.986	no
12					1.02				1.976	yes
13		0.6	353	7.75×10^{-3}	1.01×10^{-3}	45	24.8	300	1.975	no
14					0.96				1.96	yes
15		0.8	378	0.119	1.61×10^{-2}	45	23.5	300	1.998	no
16					1.60				1.985	yes
17	S/MMA/BP	0.2	318	1.21×10^{-3}	8.99×10^{-6}	45	31.3	300	2.075	no
18					8.88				2.05	yes
19		0.4	314	9.76×10^{-4}	4.81×10^{-6}	45	31.8	300	2.075	no
20					4.75				2.05	yes
21		0.6	309	7.27×10^{-4}	2.42×10^{-6}	45	32.6	300	2.075	no
22					2.39				2.05	yes
23		0.8	306	7.44×10^{-4}	1.54×10^{-6}	45	33.5	300	2.05	no
24					1.52				2.026	yes
25	S/AN/DTBP	0.7	403	0.05	1.04×10^{-2}	36	30	400	1.985	no
26					1.03				1.975	yes
27	S/AN/BP	0.7	373	0.07	8.27×10^{-3}	35.6	27.6	229	1.95	no
28		0.7	373		8.24	35.6	27.6		1.94	yes
29		0.7	373	0.025	4.86×10^{-3}	35.6	27.6	136	1.915	no
30		0.7	373		4.82	35.6	27.6		1.90	yes
31		0.7	373	0.01	3.01×10^{-3}	35.6	27.6	86	1.875	no
32		0.7	373		2.97	35.6	27.6		1.85	yes
33	S/AN/AIBN	0.7	373	0.10	3.55×10^{-2}	36.0	28.0	42	1.77	no
34		0.7	373		3.49	36.0	28.0		1.74	yes
35		0.7	373	0.05	2.37×10^{-2}	36.0	28.0	30	1.67	no
36		0.7	373		2.34	36.0	28.0		1.65	yes

the direction of nonisothermal drift owing to the fact that CC curves do not change shape significantly with respect to temperature. Figure 4.3-30 shows an example of nonisothermal drift for the system S/MMA. Parameter β_K is greater than one when drift is toward styrene-rich compositions and less than one when drift is toward increased amounts of methacrylate in the copolymer. More will be said relative to nonisothermal composition drift in Section 4.4. We now seek an additional dimensionless group that characterizes the change in the heat of reaction with comonomer composition. This will quantify the deviation in copolymerization R-A behavior from that predicted from homopolymerization R-A analysis. Qualitatively one suspects that if $-\Delta H_c$ is composition-independent, then the thermal behavior of the system will not drift from its initial value when composition changes. If, on the other hand, composition drift disproportionately favors one of the more exothermic propagation steps, then the apparent behavior of the system will be more R-A prone than predicted based on initial conditions. In a straightforward way, we can assess the drift in thermal response by examining the change in $-\Delta H_c$ with n_A. Referring to equation 4.3-76,

$$\frac{\partial(-\Delta H_c)}{\partial n_A} = \frac{r_G}{r_p}\left(r_G^{-1}\frac{\partial r_G}{\partial n_A} - r_p^{-1}\frac{\partial r_p}{\partial n_A}\right) \qquad (4.3\text{-}85)$$

If $-\Delta H_c$ is to increase with increasing n_A, then $\partial(-\Delta H_c)/\partial n_A > 0$ and equation 4.3-85 requires that

$$r_G^{-1}\frac{\partial r_G}{\partial n_A} > r_p^{-1}\frac{\partial r_p}{\partial n_A} \qquad (4.3\text{-}86)$$

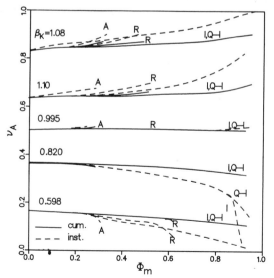

Figure 4.3-30. Composition drift in S/MMA for feed values of $(n_A)_0 = 0.10, 0.30, 0.50, 0.70, 0.90$.

Thus a criterion for exothermicity to increase with drift toward increasing compositions of A is:

$$\zeta_K \equiv \frac{r_G^{-1} \dfrac{\partial r_G}{\partial n_A}}{r_p^{-1} \dfrac{\partial r_p}{\partial n_A}} > 1 \qquad (4.3\text{-}87)$$

Evaluation of 4.3-87 proves to be independent of the kinetic model for termination
as shown below:

$$\zeta_K = \frac{2(R_A)_r(n_A)_0 \Delta H_{AA}\left[(n_A)_0 + (R_B)_r(n_B)_0\right] + (\Delta H_{AB} + \Delta H_{BA})(R_B)_r(n_B)_0^2}{2(R_B)_r(n_B)_0 \Delta H_{BB}\left[(n_B)_0 + (R_A)_r(n_A)_0\right] + (\Delta H_{AB} + \Delta H_{BA})(R_A)_r(n_A)_0^2}$$

$$(4.3\text{-}88)$$

    Characterization of nonisothermal drift involves both β_K and ζ_K. When $\beta_K > 1$
and $\zeta_K > 1$, composition drifts towards A-rich compositions, and as this drift occurs
the heat evolved from each increment of monomer conversion will increase. When
$\beta_K < 1$ and $\zeta_K < 1$, drift occurs toward B-rich compositions and, again, heat
evolution will increase relative to the feed. In both cases, the system will be easier to
provoke into R-A than a homopolymerization would be with the same values of a,
b_i, b_m, and ε. If on the other hand, $\beta_K > 1$ and $\zeta_K < 1$, then the drift occurs in the
direction of increasing composition of A, but it is accompanied by decreasing
exothermicity. When $\beta_K < 1$ and $\zeta_K > 1$, the thermal consequence is the same, but
drift occurs toward B-rich compositions. These two combinations predict a copolymerization that is more stable (less likely to R-A) than a comparable homopolymerization. Returning to Fig. 4.3-29, we see that the S/AN system experienced R-A
sooner (higher value of a for fixed b) than the comparable homopolymer, while
AN/MMA remained stable through more provocative conditions. The values of β_K
and ζ_K associated with each are consistent with the observed behavior.
    The excellent agreement among homopolymer and copolymer R-A criteria leads
one to speculate about the potential of our unified model to predict the detailed
thermal behavior of nonisothermal copolymerizations. In particular, we seek to
approximate complete copolymerization kinetics with a quasi-homopolymerization
model based on our composite CT's. Thus we shall make use of kinetic constants for
copolymerization to evaluate composite CT's required to evaluate the R-A parame-
ters a, b_i, b_m, and ε, and then use these values in the dimensionless homopolymerization mass and energy balances of Table 4.3-1. We shall refer to the alternative set of
balances as the copolymer approximate form (CPAF). The semidimensionless CPAF
balances are listed in Table 4.3-15.
    Figures 4.3-31–4.3-36 show simulation results comparing solutions of exact
kinetics to those of the CPAF. The agreement between the two sets of temperature
and conversion histories is often quite close. The magnitude of the deviation is
governed by the intensity of the composition drift and the change in heat generation
associated with the drift. For the 70/30 S/AN system, agreement is quite good
owing to the absence of drift. This composition is near the azeotrope, and the value

**Table 4.3-15.   Semidimensionless CPAF Balance
Equations**

$$-\frac{d\hat{c}_o}{d\alpha} = \lambda_i^{-1}\,\hat{c}_o\,\exp\,\hat{E}_d\hat{T}/(1 + \hat{T})$$

$$-\frac{d\hat{c}_m}{d\alpha} = \Lambda_m^{-1}\,\hat{c}_o^{1/2}\hat{c}_m\,\exp(\Lambda_G/\Lambda_{ad})\hat{T}/(1 + \hat{T})$$

$$\frac{d\hat{T}}{d\alpha} = \Lambda_G^{-1}\,\hat{c}_o^{1/2}\hat{c}_m\,\exp(\Lambda_G/\Lambda_{ad})\hat{T}/(1 + \hat{T}) - \lambda_R^{-1}\hat{T}$$

$\beta_K \cong 1.0$ predict that drift will be minimal. For other systems, like AN/MMA, where drift is large and there is a significant difference between the heats of reaction for each monomer, agreement with the CPAF is not as close. The disparities are magnified in the region of parametric sensitivity. When parameter a is either much larger or much smaller than a_{cr}, the detailed profiles tend to converge. Just as the value of a_{cr} may be above or below that for the analogous homopolymer system, detailed profiles may be "hotter" or "cooler" than the CPAF prediction owing to drift. If the system experiences R-A at higher values of a than expected (e.g., SAN in Fig. 4.3-31), then the actual temperature and conversion profiles will be somewhat higher than the CPAF profiles. The parameters measuring the drift, and hence the deviation from the CPAF, have already been discussed. Thus ζ_K may be used as a measure of applicability of the CPAF, in much the same sense as α_K is a measure of the applicability of pseudo-first-order kinetics to homopolymerization.

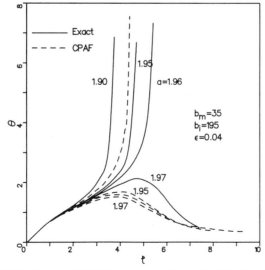

Figure 4.3-31. Exact and CPAF temperature history for "hot" system (27).

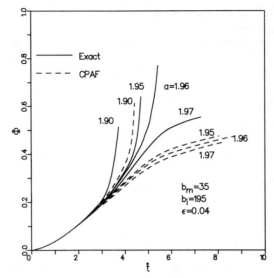

Figure 4.3-32. Exact and CPAF conversion history for "hot" system (27).

Experimental R-A studies of the comonomer pair S/AN have been conducted (25). The validity of the preceding R-A parameters could ostensibly be verified by choosing one of the kinetic termination models and then substituting the system parameters for that model into the CT definitions in Table 4.3-12. The difficulty is that none of the current models for termination kinetics adequately describe the conversion rate behavior of the S/AN system (43) and hence CT's evaluated for such models would be quantitatively incorrect. However, we may be able to take advantage of the physical interpretation of the composite CT's (Table 4.3-11) to

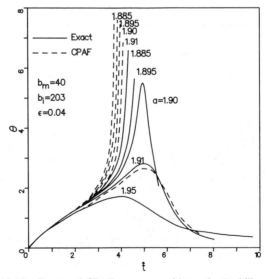

Figure 4.3-33. Exact and CPAF temperature history for "cold" system (27).

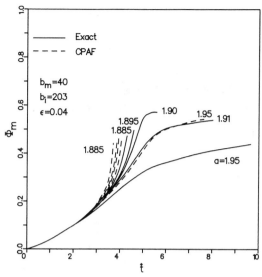

Figure 4.3-34. Exact and CPAF conversion history for "cold" system (27).

deduce their numerical values from experimental data, rather than computing them from analytical expressions. For example, isothermal kinetic studies were performed at all levels of composition (43, 44). From the initial rate data, Λ_m can be evaluated for any given feed composition and initial temperature via

$$\Lambda_m = \frac{(c_m)_0}{(r)_0}$$

(4.3-89)

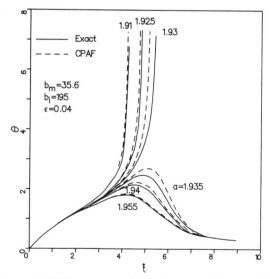

Figure 4.3-35. Exact and CPAF temperature history for composition near the azeotrope (27).

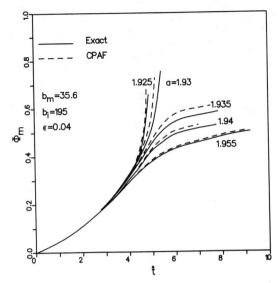

Figure 4.3-36. Exact and CPAF conversion history for composition near the azeotrope (27).

where the initial rate can be directly measured (e.g., with differential scanning calorimetry) or computed from the initial slope of the conversion history: $(r)_0 = (c_m)_0 \, d\Phi_m/dt$.

The characteristic time Λ_G can be similarly determined from the initial gradient of the temperature trajectory when $T_0 = T_R$.

$$\Lambda_G = T_0 \left(\frac{dT}{dt} \right)_0^{-1} \tag{4.3-90}$$

Computationally, this can be done more effectively by plotting experimental data in a partially transformed state. Consider a series of experiments in which feed (and coolant) temperature as well as comonomer feed composition remain the same, and initiator concentration is varied as the parameter. Realizing that Λ_G is proportional to $(r_0)_0^{-1/2}$, which in turn makes it proportional to $(c_0)_0^{-1/2}$, and plotting all the temperature curves versus $t(c_0)_0^{1/2}$ should yield a single initial trajectory. This technique is illustrated in Figs. 4.3-37 and 4.3-38 and the uniform initial slope is additionally noted on the latter.

The characteristic time Λ_{ad} is somewhat more difficult to evaluate. One could, of course perform separate adiabatic reactions and use the length of the pre R-A induction period as the value of the CT. Alternatively, one could adopt the view that for homopolymerization:

$$\lambda_{ad} \propto \frac{T_0^2}{\exp\left(\dfrac{-E}{R_g T_0} \right)} \tag{4.3-91}$$

In the range of interest for polymerizations, this ratio is a slowly varying function of

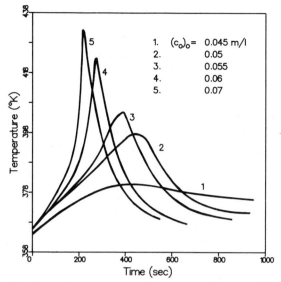

Figure 4.3-37. Experimentally determined temperature histories for 80-percent SAN (25).

T_0. Meanwhile, the basic definition of Λ_{ad} from Table 4.3-11 can be rearranged to show that when $T_0 = T_R$:

$$\Lambda_{ad} = \left(\frac{dg_e}{d\hat{T}} \right)_0^{-1} = \left[\frac{\partial}{\partial T_0} \left(\frac{dT}{dt} \right)_0 \right]^{-1} \qquad (4.3\text{-}92)$$

Thus, given values of the initial temperature slope at several different initial

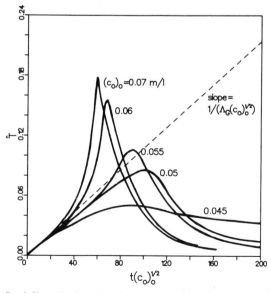

Figure 4.3-38. Semi-dimensionless plot of temperature histories shown in Fig. 4.3-37 (25).

temperatures, we can estimate the bracketed quantity in equation 4.3-92 as follows:

$$\Lambda_{ad} \approx \left[\frac{\Delta\left(\frac{dT}{dt}\right)_0}{\Delta T_0} \right]^{-1} \qquad (4.3\text{-}93)$$

Runaway parameters from expressions 4.3-89, 4.3-90, and 4.3-93 were evaluated for a variety of S/AN copolymerizations. Table 4.3-14 lists the results together with a categorization of the R-A behavior (25). Note that the values of b_m, and even more importantly b_i, are sufficiently small that sensitive R-A is not expected. The behavior actually observed was consistent with this prediction. The agreement between these (admittedly crude) estimations of the R-A parameters and the system behavior suggests that our dimensionless parameters, notwithstanding their simplistic origins, have acquired credibility by virtue of their ability to reflect correctly the underlying physical processes responsible for R-A.

4.3.3.   *Distributed Parameter Systems*

The preceding analysis of lumped parameter systems provides the underlying concepts for our consideration of distributed systems in which temperature is a spatial variable. Although the assumption of uniform mixing will identify the operating parameters for which a system cannot be stable, a more conservative criterion is needed when resistance to heat removal is not confined to the periphery of the reaction media.

   Thermal runaway is believed to occur as a result of the propagation of local "hot-spots." The remaining bulk of reaction fluid acts as a resistance to the removal of the excess reaction exotherm. The mechanism of conductive heat removal is therefore of paramount importance in distributed parameter systems. We are unable, however, to characterize the thermal response of partially mixed systems. We shall, therefore, combine a study of unmixed systems with the results of the previous section to define upper and lower bounds on the runaway behavior.

   As our model reactors, we shall consider a distributed parameter batch reactor without flow and a continuous plug-flow reactor with transverse heat conduction and diffusion. The former is equivalent to the latter in the same way that the conventional BR is the equivalent of the usual PFR with transverse uniformity. The results from our distributed parameter BR should be generalizable to a CTR, even if the latter exhibits a velocity profile, providing that flow is very slow compared to transverse transport rates, which is to say that:

$$\lambda_H \ll \lambda_v \qquad (4.3\text{-}94)$$

$$\lambda_D \ll \lambda_v \qquad (4.3\text{-}95)$$

   Table 4.3-16 lists the nonisothermal balance equations for our model reactors (Appendix F) where, again, α represents generalized age. They are directly analogous to the lumped parameter equations of Table 4.3-1. The age variable α has been used as before to represent either time in a batch system or longitudinal position in a continuous one. The equations have been given in a form that can be applied to a

Table 4.3-16.   Balance Equations for Distributed Parameter Systems

BALANCE EQUATIONS

$$\frac{\partial c_k}{\partial \alpha} = D_k \left[\frac{\partial^2 c_k}{\partial x^2} + \frac{K}{x} \frac{\partial c_k}{\partial x} \right] + r_k$$

$$\rho c_p \frac{\partial T}{\partial \alpha} = k \left[\frac{\partial^2 T}{\partial x^2} + \frac{K}{x} \frac{\partial T}{\partial x} \right] + r_G$$

SEMI-DIMENSIONLESS BALANCES

$$\frac{\partial \hat{c}_k}{\partial \alpha} = \lambda_{D_k}^{-1} \left[\frac{\partial^2 \hat{c}_k}{\partial \hat{x}^2} + \frac{K}{\hat{x}} \frac{\partial \hat{c}_k}{\partial \hat{x}} \right] + \lambda_r^{-1} \hat{r}_k$$

$$\frac{\partial \hat{T}}{\partial \alpha} = \lambda_H^{-1} \left[\frac{\partial^2 \hat{T}}{\partial \hat{x}^2} + \frac{K}{\hat{x}} \frac{\partial \hat{T}}{\partial \hat{x}} \right] + \lambda_G^{-1} \hat{r}_G$$

or

$$\frac{\partial \theta}{\partial \alpha} = \lambda_H^{-1} \left[\frac{\partial^2 \theta}{\partial \hat{x}^2} + \frac{K}{\hat{x}} \frac{\partial \theta}{\partial \hat{x}} \right] + \lambda_{ad}^{-1} \hat{r}_G$$

DIMENSIONLESS BALANCES

$$\frac{\partial \hat{c}_k}{\partial \hat{\alpha}} = b_k^{-1} \left\{ Da_{IIk}^{-1} \left[\frac{\partial^2 \hat{c}_k}{\partial \hat{x}^2} + \frac{K}{\hat{x}} \frac{\partial \hat{c}_k}{\partial \hat{x}} \right] + \hat{r}_k \right\}$$

$$\frac{\partial \theta}{\partial \hat{\alpha}} = \Delta \left[\frac{\partial^2 \theta}{\partial x^2} + \frac{K}{\hat{x}} \frac{\partial \theta}{\partial \hat{x}} \right] + \hat{r}_G$$

$$\hat{\alpha} \equiv \alpha/\lambda_{ad}$$

$\hat{r}_k$, $\hat{r}_G$ defined in Table 4.3-1, and

$$K = \begin{cases} 0 & \text{slab} \\ 1 & \text{cylinder} \\ 2 & \text{sphere} \end{cases}$$

slab, a cylinder, or a sphere by setting K to 0, 1, or 2, respectively (Section 6.2), where the transverse coordinate represents a slab half-thickness in rectangular geometry, and a radius in the other two systems.

For all practical purposes, the CT for radial (molecular) diffusion of either monomer or initiator in an unmixed reactant solution is far larger than the time scale for reaction ($\lambda_{D_k} \ll \lambda_k$). Thus we are dealing with segregated systems and the component balances have the same form as for batch kinetics:

$$\frac{d\hat{c}_k}{d\hat{a}} = \frac{1}{b_k} \hat{r}_k \tag{4.3-96}$$

except that $\hat{c}_k$ and $\hat{r}_k$ are implicitly functions of transverse position. The transverse gradients in the thermal energy balance cannot, however, be disregarded inasmuch as they provide the only mechanism for heat removal. Again, we must resolve the dilemma of choosing a reference temperature for dimensionless variables $\hat{T}$ and θ. For the moment, we shall assume a constant wall temperature T_R and use this for T_r in equations 4.3-19 and 4.3-20. Subsequently, we shall remove the constraint of constant wall temperature and allow for heat removal to a coolant at temperature T_R.

If ERA is applied to the thermal energy balance in Table 4.3-16, the resulting equation has the same form as the lumped parameter balance equation 4.3-25. The generation function is the same, but removal becomes

$$\hat{r}_e = \Delta \left[\frac{\partial^2 \theta}{\partial \hat{x}^2} + \frac{K}{\hat{x}} \frac{\partial \theta}{\partial \hat{x}} \right] \tag{4.3-97}$$

where $\Delta \equiv \lambda_{ad}/\lambda_H$. The Semenov-type analysis in Section 4.3-1 involved examination of the solution to the unsteady-state equations in the absence of concentration drift. For distributed parameter systems, Frank-Kamenetskii used a slightly different approach (20). He postulated that R-A would occur if no steady-state solutions were possible for the thermal energy balance. Thus one must find the solution set for

$$\frac{\partial^2 \theta}{\partial \hat{x}^2} + \frac{K}{\hat{x}} \frac{\partial \theta}{\partial \hat{x}} = -\Delta^{-1} \hat{g}_e \tag{4.3-98}$$

Generation function $\hat{g}_e$ is identical to the one used for lumped parameter systems in equation 4.3-24. Frank-Kamenetskii assumed a sufficiently large activation energy to justify approximating $\hat{g}_e$ by $\exp \theta$ and treated rectangular geometry so $K = 0$, reducing 4.3-98 to

$$\frac{\partial^2 \theta}{\partial \hat{x}^2} = -\Delta^{-1} \exp \theta \tag{4.3-99}$$

which has the following solution provided $\theta(\hat{x} = 1) = 0$ and $d\theta(\hat{x} = 0)/d\hat{x} = 0$:

$$\theta = \ln \left[\frac{A}{\cosh^2 \left(\hat{x} \sqrt{\frac{A}{2\Delta}} \right)} \right] \tag{4.3-100}$$

A is a constant of integration which is implicitly defined, using the boundary condition $\theta(\hat{x} = 1) = 0$, as:

$$A = \cosh^2 \sqrt{\frac{2A}{\Delta}}$$ (4.3-101)

We may say that a stable temperature distribution is possible for all real values of Δ that permit solution for A. Furthermore, it should be clear that the centerline temperature at steady state is uniquely related to A:

$$\theta_{CL} = \theta(\hat{x} = 0) = \ln A$$ (4.3-102)

We can solve equation 4.3-101 for A by assuming a value for the argument of cosh, calculating A, and finally determining Δ. Then, using A, θ_{CL} can be evaluated yielding the results shown in Fig. 4.3-39. It is evident that steady-state solutions are available only for values of

$$\Delta \geqslant 1.14$$ (4.3-103)

Note that in the region of stability, there are two values of θ_{CL} for each value of Δ. Conceptually, these correspond to the stable upper and lower intersections of removal and generation curves described for the lumped parameter system. The system will seek the lower stable state provided the initial temperature is below the lower branch of the curve, and, of course, $\Delta > 1.14$. For the normal situation of $T_0 \leqslant T_R$ or $\theta_0 \leqslant 0$, this will always be true. With superheated feeds, the upper steady-state solution might be realized.

The preceding analytical procedure can be used for cylindrical and spherical coordinates, although the latter ultimately requires numerical solution. The critical value of Δ is geometry-dependent. We find that stability for the two alternative

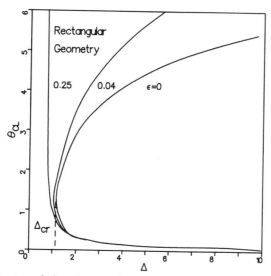

Figure 4.3-39. Steady-state solutions for centerline temperature in a rectangular slab as a function Δ.

geometries is predicted by

$$\text{cylindrical geometry} \qquad \Delta \geqslant 0.50 \qquad (4.3\text{-}104)$$

$$\text{spherical geometry} \qquad \Delta \geqslant 0.30 \qquad (4.3\text{-}105)$$

Just as for the lumped parameter system the key dimensionless parameter for characterizing stability is the ratio of CT's for the generation and removal processes. In fact, replacing λ_R with λ_H in the definition of a leads to the expression for Δ.

$$\Delta = \lambda_{ad}/\lambda_H \qquad (4.3\text{-}106)$$

The critical values of Δ determined by Frank-Kamenetskii carry the same limitations as the critical value of $a = e$ for lumped parameter systems. The assumptions of $\varepsilon \cong 0$ and $b_k \to \infty$ are both of limited applicability to polymerization reactions. To isolate the effect of ε, the exact form of $\hat{g}_e$ should be used in equation 4.3-98. The equation is subject to the split boundary conditions:

$$\frac{\partial \theta(0)}{\partial \hat{x}} = 0 \qquad (4.3\text{-}107)$$

$$\theta(1) = 0 \qquad (4.3\text{-}108)$$

and in this form it is not conveniently solved numerically or analytically. It can be rendered tractable by several transformations. If we define

$$\Theta = \theta_{CL} - \theta \qquad (4.3\text{-}109)$$

$$\chi = \frac{\hat{x}}{\sqrt{\Delta}} \qquad (4.3\text{-}110)$$

then equations 4.3-98, 4.3-107, and 4.3-108 become

$$\frac{\partial^2 \Theta}{\partial \chi^2} + \frac{K}{\chi} \frac{\partial \Theta}{\partial \chi} = \exp\left[\frac{\theta_{CL} - \Theta}{1 + \varepsilon(\theta_{CL} - \Theta)} \right] \qquad (4.3\text{-}111)$$

$$\Theta(0) = 0 \qquad (4.3\text{-}112)$$

$$\frac{\partial \Theta(0)}{\partial \chi} = 0 \qquad (4.3\text{-}113)$$

If one assumes a value for θ_{CL}, this equation can be numerically integrated from the centerline outward until Θ reaches the value θ_{CL}. This corresponds to $\hat{x} = 1$ and thus Δ can be found from equation 4.3-110:

$$\Delta = \chi^{-2}(\Theta = \theta_{CL}) \qquad (4.3\text{-}114)$$

Figure 4.3-40 shows the relationship between θ_{CL} and Δ for values of ε characteristic of polymerizations. It is apparent that Δ_{cr} is not strongly influenced by ε. Moreover, when ε is increased even further, we note that, as with lumped parameter systems, criticality is eliminated. At the same value of $\varepsilon = 0.25$ as in the lumped parameter analysis, we find that there is no transition in temperature, but rather a continuous monotonic rise in θ_{CL} as Δ is decreased.

It was apparent from the lumped parameter studies that violation of the ERA could be significant for polymerizations. Thus we turn our attention to the influence of reactant decay on distributed parameter R-A.

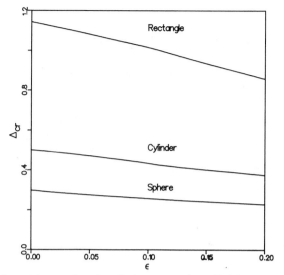

Figure 4.3-40. Values of Δ_{cr} as a function of ε for rectangular, cylindrical, and spherical geometries.

The presence of radial gradients does little to alter the reactor sensitivity with respect to reactant conversion. In point of fact, diffusion is so slow compared to the other transport processes that each laminar layer is essentially a batch reactor that exchanges only thermal energy with its neighbors. Once the centerline undergoes R-A, the phenomenon propagates outwards to the wall. Therefore, we might expect our lumped parameter sensitivity criteria to apply to the distributed parameter system as well.

Figure 4.3-41 shows a series of plots similar to Figs. 4.3-11 and 4.3-12. In these figures, θ_{max} is the maximum centerline temperature. Quite clearly, decreasing b_i

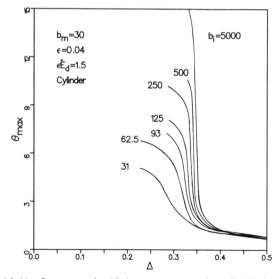

Figure 4.3-41. θ_{max} versus Δ with b_i as a parameter for cylindrical geometry.

reduces the sensitivity of the R-A transition. Ultimately, b_i is so small that sensitivity is lost. This occurs in the same range of values of b_i observed for the lumped parameter system ($b_i \cong 70$), which is not surprising in light of the fact that the component balances are essentially the same.

An additional similarity to lumped parameter R-A is the observation that Δ_{cr} is actually lower than the value predicted by ERA analysis, owing to reactant consumption. Just as a_{cr} dropped from 2.72 to roughly 2.0 for typical values of b_i and b_m, we see in Fig. 4.3-41 that Δ_{cr} is reduced from 0.5 to roughly 0.35 for cylindrical geometry. One might expect a similar decrease in rectangular and spherical geometries.

The boundary condition of constant wall temperature was useful in isolating reaction effects vis-à-vis heat removal by molecular conduction. Of more general interest, however, is the case in which the temperature of the reaction vessel is maintained by contact with a coolant fluid at temperature T_R. The vessel wall and outside heat transfer coefficient combine to form an additional resistance to the removal of reaction exotherm. Equation 4.3-11 is replaced by the flux boundary condition.

$$\frac{\partial \theta(1)}{\partial \hat{x}} = -\mathrm{Nu}\left[\theta(1) - \theta_R\right] \qquad (4.3\text{-}115)$$

If this resistance is comparable to, or greater than, that posed by the reaction fluid itself, then, clearly, runaway will occur at conditions less severe than those predicted by constant wall-temperature analysis. As discussed earlier, the overall CT for heat removal is a function of the CT's of three different removal processes, equation 4.3-6. If for the moment we neglect convection, either because the fluid is not flowing (BR) or the flow is very slow, then equation 4.3-6 reduces to equation 4.3-5. We expect this to be the appropriate CT to use in formulating a R-A criterion, and accordingly introduce a new R-A parameter analogous to a, that is,

$$\mathbf{a} \equiv \frac{\lambda_{ad}}{\Lambda_R} \qquad (4.3\text{-}116)$$

This parameter should converge to either a or Δ, depending on which heat removal path is rate-controlling. The dimensionless Nusselt number Nu should aid us in assessing the relative importance of each process. Nu may be viewed as the ratio of two CT's:

$$\mathrm{Nu} = \frac{\lambda_H}{\lambda_R} \qquad (4.3\text{-}117)$$

We expect large values of Nu to indicate that the main obstacle to heat removal is molecular conduction. Thus wall and outside resistances are relatively unimportant and a distributed parameter model with constant wall temperature might apply. Conversely, small values of Nu imply that wall resistance dominates heat removal, and a lumped parameter approach might adequately approximate thermal behavior. Definition 4.3-116 can be rewritten in two ways:

$$\mathbf{a} = \frac{a}{1 + \mathrm{Nu}} \qquad (4.3\text{-}118)$$

$$\mathbf{a} = \frac{\Delta}{1 + \mathrm{Nu}^{-1}} \qquad (4.3\text{-}119)$$

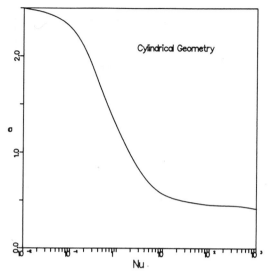

Figure 4.3-42. Critical value of **a** for runaway as a function of Nu for cylindrical geometry.

From the first it is clear that

$$\lim_{Nu \to \infty} \mathbf{a} = \Delta \qquad\qquad (4.3\text{-}120)$$

while from the second:

$$\lim_{Nu \to 0} \mathbf{a} = a \qquad\qquad (4.3\text{-}121)$$

Therefore, parameter **a** would seem to span the full range of possible behavior.

Figure 4.3-42 shows a plot of $\mathbf{a}_{cr}$ versus Nu for cylindrical geometry. The asymptotic extreme values of **a** correspond to typical values of a and Δ for

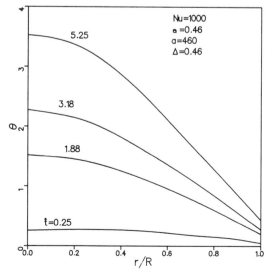

Figure 4.3-43. Radial temperature profiles at various times when molecular conduction dominates heat transfer characteristics.

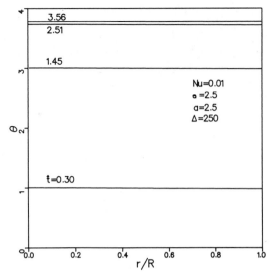

Figure 4.3-44.   Radial temperature profiles at various times when wall resistance dominates heat transfer characteristics.

polymerization. Note that the transition from a completely unmixed ($\mathbf{a} = \Delta$) to completely mixed ($\mathbf{a} = a$) state occurs in the range $0.1 < \text{Nu} < 10$. Thus, when $\text{Nu} > 10$, the wall resistance to heat transfer is sufficiently high to render the runaway behavior essentially the same as that for the lumped parameter system, even if internal temperature gradients do exist. Conversely, when $\text{Nu} < 0.1$, transmittal of thermal energy through the fluid is slow enough to cause the system to behave like the distributed parameter model with constant wall temperature. This is

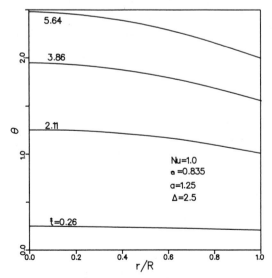

Figure 4.3-45.   Radial temperature profiles at various times when molecular conduction and wall resistance are equivalent.

shown quite clearly in Figs. 4.3-43–4.3-45, which are radial temperature profiles for runaway reactions with three different Nusselt numbers. When Nu = 0.1 and **a** = 2.5, radial gradients are essentially absent and the R-A transition occurs at $\mathbf{a}_{cr} = a_{cr}$. On the other hand, when Nu = 1000 and $\mathbf{a} \simeq 0.45$, a steep radial gradient is present. The centerline temperature rises to a value nearly 10 times as high as the wall temperature. The R-A occurs at a value of $\mathbf{a}_{cr} = \Delta_{cr} = 0.459$, the expected value for a unmixed cylinder with constant wall temperature. Figure 4.3-45 shows a case within the transition region, Nu = 1. Note that the reaction starts out with no radial gradients, but as temperature rises a small change does develop in the radial direction. This transition occurred at **a** = 1.25, which corresponds to $a = 1.25$ and $\Delta = 2.5$. The value of a is closer to a_{cr} than the value of Δ is to Δ_{cr}, and the profiles indicate that the system behaves more like a well-mixed reactor than an unmixed one.

Example 4.3-1. *Evaluation of ε For Typical Vinyl Monomers.* Use the kinetic constants for S, MMA, and AN in Example 2.7-3 to evaluate ε for each system when initiated with AIBN, (at 323, 343, 363, K), BP (at 343, 363, 383 K), and DTBP (at 393, 413, 433 K).

Solution. When the QSSA and LCA are applied to free-radical kinetics, the resulting balances are thermally and kinetically simple. The activation energy for the generation rate is the lumped activation energy for propagation.

$$E_{ap} = E_p + \frac{1}{2}(E_d - E_t)$$

and $\varepsilon = R_g T_0/E_{ap}$. Using the appropriate values, we calculate

	323	343	363
S/AIBN	0.0295	0.0313	0.0331
MMA/AIBN	0.0326	0.0346	0.0366
AN/AIBN	0.0306	0.0325	0.0344

	343	363	383
S/BP	0.0323	0.0343	0.0361
MMA/BP	0.0359	0.0380	0.0401
AN/BP	0.0337	0.0356	0.0376

	393	413	433
S/DTBP	0.0316	0.0332	0.0348
MMA/DTBP	0.0345	0.0363	0.0380
AN/DTBP	0.0327	0.0343	0.0360

CONCLUSION

For free-radical polymerization, when initiators are used in their normal temperature range, ε is not large enough to limit R-A sensitivity.

Example 4.3-2.   An Interpretation of λ_{ad}. Prove that $\lambda_{ad}^{-1} = (\partial g_e / \partial \hat{T})|_r$ for kinetically and thermally simple reactions.

Solution.   From equation 4.3-23, $g_e \equiv \lambda_G^{-1} \hat{r}_G$, and from equation 4.3-24, $\hat{r}_G = f(\hat{\alpha}) \hat{g}_e(\hat{T})$, where $\hat{g}_e(\hat{T}) = \exp \hat{E}\hat{T}/(1 + \hat{T})$. Thus

$$\frac{\partial g_e}{\partial \hat{T}} = \lambda_G^{-1} f(\hat{\alpha}) \frac{\partial \hat{g}_e}{\partial \hat{T}}$$

and

$$\frac{\partial \hat{g}_e}{\partial \hat{T}} = \frac{\hat{E}}{(1 + \hat{T})^2} \hat{g}_e$$

Since $\hat{T} \equiv (T - T_r)/T_r$, it is clear that $\partial \hat{g}_e / \partial \hat{T} \to \hat{E}$ if $T \to T_r$ where,

$$T_r = \begin{cases} T_0 \\ T_R \end{cases}$$

Furthermore, $f(\hat{\alpha}) \to 1$, which is also required, when $c_k \to (c_k)_0$ for all k. In conclusion,

$$\left. \frac{\partial g_e}{\partial \hat{T}} \right|_r = \frac{\hat{E}}{\lambda_G} = \lambda_{ad}^{-1}$$

Example 4.3-3.   Effect of Feed Temperature on Critical Tube Radius. Calculate the critical tube radius for R-A when polyethylene is reacted by radical initiation in a PFR. The jacket temperature T_R is 160°C. Consider feed temperatures T_0 of 160, 170, and 195°C. Use the following data from Ehrlich et al., Han et al., and Lee and Marano (Section 5.3.1).

$$-\Delta H = 21.4 \text{ kcal/mol} \qquad (c_0)_0 = 8.82 \times 10^1 \text{ mol/m}^3$$

$$\rho = 516 \text{ kg/m}^3 \qquad (c_m)_0 = 1.951 \times 10^4 \text{ mol/m}^3$$

$$c_p = 0.518 \text{ kcal/kg K} \qquad U = 1 \text{ kcal/m}^2 \text{ s K}$$

$$k_d = 3.23 \times 10^{11} \exp\left(\frac{-33,000}{r_g T}\right) \text{ s}^{-1}$$

$$k_p = 2.95 \times 10^4 \exp\left(\frac{-7091}{R_g T}\right) \text{ m}^3/\text{mol s}$$

$$k_t = 1.6 \times 10^3 \exp\left(\frac{-2400}{R_g T}\right) \text{ m}^3/\text{mol s}$$

Solution.   At a value of $T_r = T_R = 160°C$,

$$\lambda_i = 1.41 \times 10^5 \text{ s} \qquad b_i = 37,170$$
$$\lambda_m = 353 \text{ s} \qquad b_m = 93$$
$$\lambda_{ad} = 3.79 \text{ s} \qquad \varepsilon = 0.038$$

The b's are sufficiently large to approximate Semenov-like conditions, consequently we choose $a_{cr} = 2.72$ when $T_0 = T_R$. Therefore,

$$(s_V)_{cr} = \frac{2.72 \dfrac{\rho c_p}{U}}{\lambda_{ad}} = 1.92 \text{ cm}^{-1}$$

so

$$(Ra)_{cr} = 1.04 \text{ cm}$$

When $T_0 = 170°C$, $\theta_0 = 0.60$. Since $\theta_0 < (1 - 2\varepsilon)^{-1} = 1.08$, the value of a_{cr} remains unchanged. Therefore, $(Ra)_{cr} = 1.04$ cm. When $T_0 = 195°C$, $\theta_0 = 2.1$, and so $\theta_0 > (1 - 2\varepsilon)^{-1}$ and equation 4.3-44 must be used to evaluate a_{cr}. The result is $a_{cr} = 3.44$. Thus

$$(s_V)_{cr} = \frac{3.44 \dfrac{\rho c_p}{U}}{\lambda_{ad}} = 2.42 \text{ cm}^{-1}$$

and

$$(Ra)_{cr} = 0.82 \text{ cm}$$

As expected, superheated feeds may increase the likelihood of R-A.

Example 4.3-4.   Relationship between b and θ_{ad}.   Show that for stoichiometrically simple reactions $\theta_{ad} = b$.

Solution.   Equation 4.3-57 follows directly from the component and thermal energy balances. When $a = 0$, which is true for adiabatic conditions, equation 4.3-57 becomes:

$$\frac{d\theta}{d\hat{c}} = -b$$

Separating variables and integrating yields

$$\int_0^{\theta_{ad}} d\theta = -b \int_1^0 dc$$

or

$$\theta_{ad} = b$$

Example 4.3-5.   Critical Parameters for Continuous Styrene Polymerizations.
Estimate the largest allowable diameter (d_t) and smallest allowable heat transfer

coefficient (U), to just avert R-A of DTBP-initiated styrene polymerization in a continuous tubular reactor (PF), when both feed and jacket temperatures are 120°C and feed monomer and initiator concentrations are 8.7 and 0.01 mol/l, respectively. Use the following thermodynamic and transport data: $\rho c_p = 0.36$ cal/cm^3 K., $-\Delta H_r = 16.7$ kcal/mol and $k = 3 \times 10^{-4}$ cal/cm s K.

Solution. We begin by allowing for internal resistance to be rate-controlling, owing to inadequate thickness d_t, and assume that U is infinite. Thus we have a distributed parameter problem and λ_H is required. Next, by choosing d_t below the critical value, external resistance becomes controlling, and we have a lumped parameter problem requiring λ_R. Using the data in Chapter 2 for this system yields

$$\hat{E} = \frac{E_{ap}}{R_g T_R} = 31.6 \quad \text{or} \quad \varepsilon = 0.0316$$

and

$$(r_G)_r = (-\Delta H_r) A_{ap} (c_m)_0 (c_0)_0^{1/2} \exp(-\hat{E}) = 7.60 \text{ cal/l s}$$

Therefore,

$$\lambda_{ad} = \frac{\varepsilon \rho c_p T_r}{(r_G)_r} = 585 \text{ s}$$

Similarly,

$$\lambda_i = \frac{(c_0)_0}{(r_0)_r} = 8.79 \times 10^4 \text{ s}$$

and

$$\lambda_m = \frac{(c_m)_0}{(r_p)_r} = 1.89 \times 10^4 \text{ s}$$

Allowing for reactant consumption, $\Delta_{cr} = 0.35$ for cylindrical systems (Section 4.3.3) where $\Delta \equiv \lambda_{ad}/\lambda_H$. Hence we require that $\lambda_H < \lambda_{ad}/0.35$ to avert R-A due to internal resistance or

$$\text{Ra} < \left(\frac{\alpha_T \lambda_{ad}}{0.35} \right)^{1/2} = 1.2 \text{ cm}$$

from which we conclude that $d_t < 2.4$ cm. Then, to avert R-A due to external resistance, we further require that $a \equiv \lambda_{ad}/\lambda_R < a_{cr}$ where $a_{cr} = 1.88$ from Fig. 4.3-16 (for $b_i = 150$ and $b_m = 32$). Consequently, $\lambda_R < \lambda_{ad}/1.88$ or

$$U > \frac{1.88 \rho c_p}{s_V \lambda_{ad}} = 6.94 \times 10^{-4} \text{ cal/cm}^2 \text{ s K}$$

where $s_V = 2/\text{Ra}$. The reader who considers critical $d_t = 2.4$ cm to be large is reminded that it was computed by assuming U to be infinite and that critical U was computed using critical d_t. Therefore, the designer must exercise discretion and make both d_t smaller and U larger than the critical values by an appropriate margin of safety. Suppose that $U = 1 \times 10^{-4}$ cal/cm^2 s K. Using $\lambda_R < \lambda_{ad}/1.88$, once again, we conclude that

$$s_V > \frac{1.88 \rho c_p}{U \lambda_{ad}} = 11.6 \text{ cm}^{-1}$$

or that critical $d_t = 0.34$ cm. This illustrates the dependence of critical d_t on U and demonstrates that external as well as internal sources of R-A can be simultaneously alleviated by reducing tube diameter (increasing s_V).

Example 4.3-6. Predictions of Drift. Compute the values of β_K and ζ_K for S/MMA at 373 K with styrene compositions of 0.7 and 0.3. Use these to predict the nature of thermal drift in this reaction.

Solution. Using the data in Tables 4.3-8 and 4.3-9, we compute:

	β_K	ζ_K
$(n_A)_0 = 0.70$	1.07	1.09
$(n_A)_0 = 0.30$	0.85	1.17

For $(n_A)_0 = 0.7$, both β_K and ζ_K are greater than one. This implies that drift is toward progressively higher styrene compositions and that these will be more exothermic than the feed condition. Thus a_{cr} will be somewhat higher than expected for a similar homopolymerization. On the other hand, when $(n_A)_0 = 0.30$, drift occurs in the other direction. The copolymerization will be "cooler" than estimated by the parameters based on feed values.

Example 4.3-7. Styrene R-A in a Plate-and-Frame Press. Styrene polymerization may be conducted in a plate-and-frame press. An industrial rule-of-thumb places 5 in. as the maximum permissible plate thickness for non R-A reactions (46). Using data for thermally initiated styrene polymerization at 90°C, calculate the critical slab thickness for pure monomer and 30-percent polymer feeds.

Given:
$$k_i = 4.38 \times 10^5 \exp\left(\frac{-27,620}{R_g T}\right) \text{ l}^2/\text{mol}^2 \text{ s}$$

$$k = 3.06 \times 10^{-4} \text{ cal}/\text{cm}^2 \text{ s K}$$

Solution. We use a critical value of $\Delta = 0.80$, since reactant consumption supresses Δ_{cr}

$$\lambda_{ad} = -\left(\frac{\Delta H_r (k_{ap})_0 (c_m)_0^{5/2} E_{ap}}{\rho c_p R_g T_0^2}\right)^{-1} = 1.742 \times 10^4 \text{ s}$$

thus
$$(\lambda_H)_{cr} = \frac{\lambda_{ad}}{\Delta_{cr}} = 2.18 \times 10^4 \text{ s}$$

and, using the definition of λ_H (see Appendix G), $(s_V^{-1})_{cr} = 4.3$ cm. This represents the slab half-width so the maximum total width is $\ell_{cr} = 8.6$ cm $= 3.4$ in., which is in good agreement with experimental results.

Example 4.3-8.   Distributed Parameter Approximation of a Laminar Flow SCTR.
Using the data of Wallis (56), justify the use of the stagnant tube R-A criterion for
his SCTR polymerizer. Compute a critical value of d_t for his reactor. Use the PFR
criterion to estimate $(Us_V)_{cr}$ by way of comparison

Given: $\lambda_v = 8106s$ $(c_0)_0 = 0.00447$ mol/l

$T_R = 353$ K $(c_m)_0 = 8$ mol/l

Solution.   Using the parameters for S/AIBN,

$$\lambda_m = 26{,}940s$$
$$\lambda_{ad} = 825s$$

An estimate of Δ_{cr} for cylindrical geometry, allowing for violation of the ERA, is
$\Delta_{cr} = 0.35$. Thus

$$(\lambda_H)_{cr} = \frac{\lambda_{ad}}{\Delta_{cr}}$$

yielding

$$\frac{\rho c_p (d_t)^2_{cr}}{4k} = \frac{825}{0.35}$$

or

$$(d_t)_{cr} > 2.8 \text{ cm}$$

Wallis's reactor was 2.36 cm in diameter. It should be noted that $\lambda_v > \lambda_m$ so the
reactions could not go to completion. Based on Wallis's value of $d_t = 2.36$, $\lambda_H = 1682$
s which yields

$$\Lambda_R = \left(\lambda_v^{-1} + \lambda_H^{-1}\right)^{-1} = 1392s$$

Thus $\Lambda_R \cong \lambda_H$ and the use of stagnant tube criterion is justified.
    A plug-flow criterion of $a_{cr} = 2$ would yield $(Us_V)_{cr} = 2\rho c_p/\lambda_{ad} = 8.73 \times 10^{-4}$
cal/cm^3 s K. Using Wallis's diameter of 2.36 cm and noting that $s_V = 4/d_t$ indicates
that an overall heat transfer coefficient to prevent R-A must be greater than

$$(U)_{cr} = 5.14 \times 10^{-4} \text{ cal/cm}^2 \text{ s K}$$

Example 4.3-9.   Composite CT's for Series and Parallel Processes.   Show that
for transport processes in series and in parallel the composite CT's are $\Lambda = \Sigma_j \lambda_j$
and $\Lambda = (\Sigma_j \lambda_j^{-1})^{-1}$, respectively, and that the largest λ_j therefore dominates the
former and the smallest dominates the latter

Solution.   For N processes, conservation of extensive property Q (Appendixes F
and G) takes the general form

$$\frac{dQ}{dt} = \dot{Q}$$

where

$$\dot{Q} = \sum_{j=1}^{N} \dot{Q}_j$$

for parallel configurations and

$$\dot{Q}_1 = \cdots = \dot{Q}_j = \cdots = \dot{Q}_N \equiv \dot{Q}$$

for series configurations. We make use of the following relations (Appendix D)

$$Q_j = C_j P_j$$

and

$$Q_j = K_j \Delta P_j$$

where P is the conjugate intensive property corresponding to Q and C and K are generalized capacitance and conductance properties. Following substitution of these relations, we obtain, with the aid of

$$\Delta P_1 = \cdots = \Delta P_j = \cdots = \Delta P_N \equiv \Delta P$$

for parallel configurations and

$$\Delta P = \sum_{j=1}^{N} \Delta P_j$$

for series configurations, the following results from the balance equations:

$$\frac{dP}{dt} = \Delta P \sum_{j=1}^{N} \lambda_j^{-1} \quad \text{(parallel)}$$

and

$$\frac{dP}{dt} \sum_{j=1}^{N} \lambda_j = \Delta P \quad \text{(series)}$$

where λ_j is a generalized CT defined as (Appendix G)

$$\lambda_j \equiv \frac{C}{K_j}$$

Clearly, from the first result, the process with the smallest value of λ_j (fastest) dominates, and from the second result the process with the largest value (slowest) dominates.

Thus equation 4.3-6 consists of two series heat transport processes in parallel with a third:

$$\Lambda_R = \left[\underbrace{(\lambda_R + \lambda_H)^{-1}}_{\text{series}} + \lambda_v^{-1}\right]^{-1}$$

$$\underbrace{}_{\text{parallel}}$$

Example 4.3-10. Critical Radius for R-A When Fluid and Wall Resistance are Comparable. Consider the catalyzed reaction of polyurethane in a cylindrical mold. The mold wall is held at 50°C and the reactants are fed at the same temperature. There is a heat transfer coefficient between the mold and the jacket coolant, $U = 0.01$ cal/cm^2 s K. Using the kinetic constants and rate expressions of Macosko et al. (35), find the critical mold radius to avert R-A. Assume that: (1) the wall resistance dominates, (2) the fluid resistance dominates, and (3) neither resistance dominates the radial heat transfer. Use the following:

$$r = 1.38 \times 10^7 \exp\left(\frac{-7775}{T}\right) c^{3/2} \text{ eq/m}^3 \text{ s}$$

$$\rho c_p = 530 \text{ kg/m}^3$$

$$-\Delta H = 14.34 \text{ kcal/eq}$$

$$(c)_0 = 2600 \text{ eq/m}^3$$

$$k = 5 \times 10^{-4} \text{ cal/cm s K}$$

Solution. From the given data, the reaction related CT's and dimensionless groups can be computed:

$$\lambda_r = 40.3 \text{ s} \qquad\qquad b = 5.2$$

$$\lambda_{ad} = 7.7 \text{ s} \qquad\qquad \varepsilon = 0.042$$

(1). When wall resistance dominates, a lumped-parameter model is applicable. The small value of b suggests that R-A sensitivity may be greatly diminished. Therefore, we use a value of $a_{cr} = 2$, realizing that it will represent a conservative estimate of $(Ra)_{cr}$.
    Rearranging the definition of a, one can solve for $(s_V)_{cr}$:

$$(s_V)_{cr} = \frac{\rho c_p a_{cr}}{U\lambda_{ad}} = 13.7 \text{ cm}^{-1}$$

For a cylinder, $s_V = 2/Ra$ and so

$$(Ra)_{cr} = 0.146 \text{ cm}$$

(2). When the fluid resistance dominates, one can use a distributed parameter model, with an isothermal wall. The value of Δ comparable to $a_{cr} = 2$ is $\Delta_{cr} = 0.368$.

Therefore,

$$(Ra)_{cr} = \left(\frac{k\lambda_{ad}}{\rho c_p \Delta_{cr}} \right)^{1/2} = 0.140 \text{ cm}$$

These two values of $(Ra)_{cr}$ are so close that it should be clear that neither resistance is dominant and as a result the true critical radius may be less than that computed in (1) or (2).

(3). When both resistances are comparable, as is evident in Fig. 4.3-42, the value of $\mathbf{a}_{cr}$ is a function of Nu. Since Nu is also a function of Ra and U, an iterative solution is required. When $0.10 < \text{Nu} < 10$, we may approximate

$$\mathbf{a}_{cr} = 1.36 - 1.04 \log \text{Nu}$$

where $\mathbf{a} = \lambda_{ad}/\lambda_R(1 + \text{Nu})$ and $\text{Nu} \equiv Us_v^{-1}/k$. Therefore, one can write

$$(1.36 - 1.04 \log \text{Nu}) \frac{\rho c_p \text{Ra}}{2U} (1 + \text{Nu}) = \lambda_{ad}$$

Varying Ra until the LHS is equal to λ_{ad} (7.7s), yields

$$(Ra)_{cr} = 0.106 \text{ cm}$$

Example 4.3-11. *Using Various T_r's When Composing CT's.* Often, CT's and dimensionless groupings are evaluated using $T_r = T_0$. However, $T_r = T_R$ has been shown to be more meaningful especially when $T_R > T_0$. Show how the two forms can be interrelated comparing $(\lambda_m)_0$ and $(\lambda_m)_R$.

Solution.

$$(\lambda_m)_r = \frac{1}{A_{ap}(c_0)_0^{1/2}} \exp \frac{E_{ap}}{R_g T_r}$$

Thus

$$\frac{(\lambda_m)_0}{(\lambda_m)_R} = \exp \frac{E_{ap}}{R_g} \left(\frac{1}{T_0} - \frac{1}{T_R} \right)$$

Thus

$$(\lambda_m)_0 = (\lambda_m)_R \exp \frac{-\hat{E}_{ap_R} \hat{T}_0}{1 + \hat{T}_0}$$

or

$$(\lambda_m)_R = (\lambda_m)_0 \exp \frac{-\hat{E}_{ap_0} \hat{T}_R}{1 + \hat{T}_R}$$

where

$$\hat{E}_{ap_R} = \frac{E_{ap}}{R_g T_R} \quad \text{and} \quad \hat{E}_{ap_0} = \frac{E_{ap}}{R_g T_0}$$

while

$$\hat{T}_R = \frac{T_R - T_0}{T_0} \quad \text{and} \quad \hat{T}_0 = \frac{(T_0 - T_R)}{T_R}$$

The reader should verify that

$$(\lambda_{ad})_R = (\lambda_{ad})_0 (1 + \hat{T}_R)^2 \exp \frac{-\hat{E}_{ap_0} \hat{T}_R}{1 + \hat{T}_R}$$

and

$$(\lambda_i)_R = (\lambda_i)_0 \exp \frac{-\hat{E}_{d_0} \hat{T}_R}{1 + \hat{T}_R}$$

4.4. EFFECT ON PRODUCT PROPERTIES

Average DP and DPD are key molecular properties that determine the ultimate physical properties of a polymer product. If there were no interaction between the thermal history of the reactor and the polymer produced, then studying nonisothermal polymerizations would be of value primarily in the interest of preventing R-A and instability. Earlier discussions of instantaneous properties and temperature programming have shown, however, that there may indeed be a strong interdependence.

4.4.1. Homopolymers

In random and step-addition polymerizations, temperature changes could activate side reactions or enhance reversibility of the kinetic steps. However, if initiation or functional group formation is presumed to be complete early during reaction, then propagation remains as the only step to be considered in both these step reactions. The situation for chain-addition, however, is quite different.

At any given moment, the polymer produced in a chain-addition polymerization has an average DP that is determined by the balance between the rate of propagation and the rate of initiation. Instantaneous dispersion (statistical dispersion) is a reflection of molecular statistics.

With this in mind, one might expect runaway temperatures to result in poor product quality. Our previous analysis of reactor sensitivity has taught us that chain-initiation rates frequently accelerate far more quickly with increasing temperatures than do chain propagation rates ($\hat{E}_i > \hat{E}_{ap}$). This might lead one to expect that, as temperatures rise without bound, instantaneous DP will continually decrease and the ultimate product will be both low in DP and broadly dispersed. Although this may be true for reactions that are intended to run isothermally but run away instead, it is not a general rule. It is possible to take advantage of the high rates afforded by nonisothermal temperature policies, and still maintain product quality.

Under isothermal conditions, the so-called conventional polymerization ($\alpha_K \ll 1$) exhibits a downward drift in DP as shown in Fig. 4.4-1 (22). This arises because monomer is continually consumed while initiator concentration remains quite close to its feed value. As a result, the rate of initiation increases relative to the rate of propagation and DP drifts downward. When a polymerization of this type experiences R-A, the drift is exacerbated by the temperature history. The rate of initiation increases relative to the rate of propagation not only as a consequence of the composition change, but also because the activation energy for initiation is larger than that for propagation. This dual effect does indeed lead to final DP's that are much lower than initial DP's and to broader DPD's. Figures 4.4-1 and 4.4-2 show the DP drift and DPD drift associated with the R-A transition of a conventional (CONV) polymerization. It is clear that the onset of R-A has a dramatic effect on properties, and the additional cases cited in Table 4.4-1 document a similar trend.

Not all polymerizations are the CONV type, however. When the initiator is consumed more rapidly than the monomer, the isothermal DP drift is upward with conversion. Since rising temperatures tend to promote a downward drift, the net result is a partial cancellation of drift, as pictured in Fig. 4.4-3 (22). We note from Table 4.4-1 that when $\alpha_K \simeq 1$, R-A does not produce dramatic increases in D_N and could even lead to a more narrowly dispersed polymer than isothermal conditions would. In fact, D_N could pass through a maximum value and subsequently decline with severity of R-A. This occurs when the reaction does not dead-end during the R-A. Temperature peaks and drops quickly back toward the reservoir temperature (T_R), and the effect is much like the step-change optimal policy for maximizing D_N (Section 4.2-7). As may be seen in Figs. 4.2-7 and 4.2-8, the sudden rise in D_N occurs

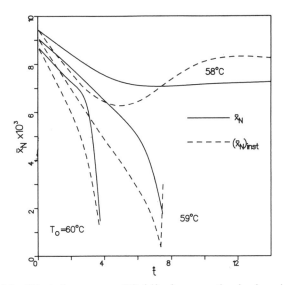

Figure 4.4-1. Effect of runaway on DP drift of a conventional polymerization (23).

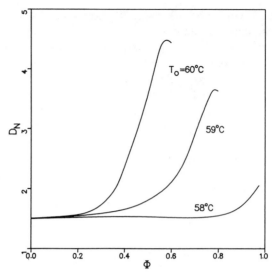

Figure 4.4-2. Effect of runaway on DPD dispersion of a conventional polymerization (23).

Table 4.4-1. Effect of R-A on Polymer Properties Simulation Results

b_m	b_i	a	$(\bar{x}_N)_o$	$(\bar{x}_N)_f$	$(D_N)_f$	R-A
35.4	3159	2.07	9732	2700	2.03	N
35.4	3350	1.95	9175	86	31.3	Y
31.8	1550	1.97	4250	1123	2.01	N
31.7	1520	1.90	4173	76	17.1	Y
35.9	456	1.99	1424	382	1.92	N
35.9	459	1.98	1415	87	6.15	Y
27.5	119	1.81	7544	2353	1.79	N
27.5	118	1.77	7472	1412	2.66	N
27.5	117	1.75	7425	1228	2.68	Y
26.8	97.6	1.80	6347	2172	1.76	N
26.7	95.2	1.72	6212	1230	2.33	Y
26.6	94.1	1.69	6154	1148	2.48	Y

just at the point where the quenching takes place. Hence it is the borderline non R-A reaction that has the largest $(D_N)_f$, since the polymer formed after the temperature maximum ($\Phi > 0.40$) sees a reaction environment similar to that of a severely dead-ended isothermal reaction. That is, temperature drops quickly to T_R and remains there, but by this time most of the initiator has been exhausted (which is why the reaction "peaks"). Progressively higher-molecular polymer is formed with each increment of conversion, resulting in a large spread in the DPD and a large D_N. The fact that R-A reactions dead-end prevents the formation of a high-molecular-weight tail. Products thus have a narrower DPD than for comparable non R-A reactions carried to higher conversion levels.

Figure 4.4-4 confirms this hypothesis. It can be seen that for conventional polymerizations, reducing a to values just below a_{cr} causes a slight narrowing of the DPD. When R-A occurs ($a < 2$), however, the value of D_N experiences an upward step-change. For D-E polymerizations, however, we see that D_N goes through a maximum at the R-A point and then drops back to values equal to, or less than, those produced by non R-A reactions.

A large spread in DPD is not always undesirable. Tubular reactors used to produce Low density polyethylene (LDPE) are aimed at optimizing conversion and molecular weight, while values for D_N ranging from 7 to 30 are not uncommon. This is not surprising since reactions are conducted in regions of high sensitivity. R-A is used to achieve high conversion rates and reaction kinetics are such that acceptable DP's are still realized. There are finite limits to the use of R-A since continued increase in temperature and/or feed initiator concentration to stimulate reaction rate in the sensitive region ultimately brings about a drop in molecular weight.

Lee and Marano (47) computed conditions that would result in maintaining the system in a R-A state. Initiator concentrations and jacket temperatures were chosen to maximize conversion, and they correspond to a point at which a is just less than a_{cr}, as discussed in Section 5.3-1.

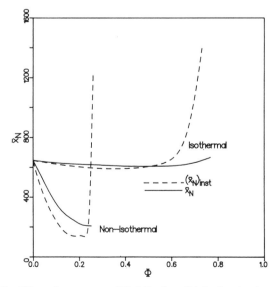

Figure 4.4-3. Effect of runaway on DP drift of a mild dead-end polymerization (22).

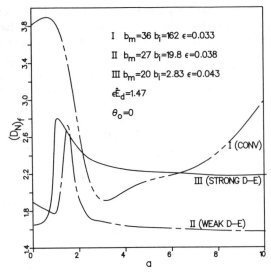

Figure 4.4-4. Effect of runaway on DPD dispersion for differing isothermal drift characteristics (23).

Figures 4.4-5 and 4.4-6 show the response of average molecular weight, $\overline{M}_N$, and dispersion D_N as optimal values of jacket temperature and initiator feed concentration are chosen to span the range of possible conversions. It should be evident that, regardless of the kinetic or transport parameter manipulated, at low levels of conversion the polymer properties are relatively unaffected by accelerated reaction rates. This region corresponds to nonsensitive R-A with early dead-ending and concomitant low conversion. When higher levels of conversion are desired, a larger θ_{max} is required and hence a more sensitive R-A. Note the rapid decay of product

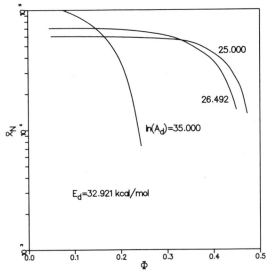

Figure 4.4-5. Variation of DP with conversion along the R-A boundary, with A_d as parameter (47).

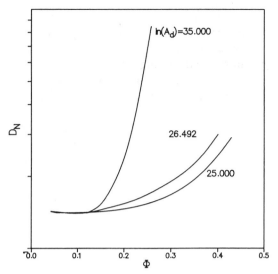

In(A_d)=35.000

26.492

25.000

D_N

Φ

0.0   0.1   0.2   0.3   0.4   0.5

Figure 4.4-6. Variation of D_N with conversion along the R-A boundary with A_d as parameter (47).

quality as soon as one attempts to raise the level of conversion past a critical point. The existence of a sensitivity of reaction sensitivity has been demonstrated in Section 4.3-1 and its effect on polymer properties is manifested here.

Dead-ending is not the only factor that can alter the molecular weight drift pattern. It has been previously shown that the presence of gel-effect can produce the same qualitative results of upward drift in the isothermal molecular weight trajectory. Analyzing the effect of G-E is not quite so straightforward, however, since its presence alters the rate of reaction and thus the rate of heat evolution. Given the same initial conditions, if a reaction with G-E is compared to one without, the temperature histories will not be the same. G-E will cause a more rapid rise in temperature as kinetic autoacceleration feeds thermal autoacceleration.

One cannot generalize on the ultimate response of the system to G-E. A CONV polymerization, when subjected to R-A, shows a downward drift in cumulative $\bar{x}_N$ and $\bar{x}_W$ because the instantaneous response of DP to rising temperatures dominates the drift response of ensuing D-E conditions. Under isothermal conditions, the instantaneous response to G-E is an upward drift in DP, but the drift response is downward since rate of monomer consumption is increased relative to initiator by kinetic autoacceleration. When G-E is strong, the drift response is dominated by the instantaneous response. When nonisothermal conditions prevail, however, rising temperatures weaken the G-E as does the reduction in DP resulting from the system's instantaneous response to rising temperatures. The drift response of G-E opposes that of rising temperature, whereas the latter, as we have seen, may actually be beneficial. The net result is that although the isothermal change in DP due to G-E looks like that of D-E, the nonisothermal response may be quite different.

There is experimental evidence to support the conclusion that the R-A need not have a dire effect on polymer properties. Table 4.4-2 lists experimental PFR data from Valsamis (48) and BR data from Sebastian (24) for bulk styrene polymerization. Both consider conditions that reflect sensitive and nonsensitive R-A, and the trends observed by each are the same. The presence of R-A in some cases actually leads to a higher-molecular-weight, although slightly more disperse, polymer. This occurs when b_i is small and $\alpha_K > 1$. The increasing severity of D-E brought about by higher initial temperatures and/or higher temperature rises enhances the drift response, causing it to dominate the instantaneous response ($\bar{x}_N \downarrow$ as $T \uparrow$). The rise in D_N as $\bar{x}_W$ climbs is a reflection of the favored production of high-molecular-weight species (see Section 2.7). This coincides with the pattern of D-E drift response. Note that for S/AIBN and S/BP, $\bar{x}_N$ is relatively unaffected by thermal history.

When R-A occurs with sensitive systems (e.g., S/DTBP), the instantaneous response dominates the drift response since D-E conditions have been weakened. As a result, $\bar{x}_W$ decreases as R-A is provoked. Note that for S/BP in a BR, R-A under sensitive conditions leads to a less disperse polymer. This may be due to the effects

Table 4.4-2. Experimental Polymer Property Data for Nonisothermal Styrene Polymerization

PFR	T_r	$\bar{M}_N$	$\bar{M}_W$	D_N	α_k	b_i
S/AIBN	125°C	10,411	25,300	2.43	7.6	
	130	9,618	27,700	2.88	10.4	
	135	9,188	28,300	3.08	13.0	
	140	10,780	35,900	3.33	14.5	
	145	10,168	36,200	3.56	16.3	
S/BP	125	16,544	31,600	1.91	0.68	
	130	15,330	37,100	2.42	0.92	
	135	17,272	43,700	2.53	1.15	
	140	18,366	46,100	2.57	1.27	
S/DTBP	140	55,149	92,650	1.68	0.36	
	150	31,876	61,840	1.94	0.38	
	160	31,696	58,320	1.84	0.41	
	165	28,350	55,000	1.94	0.54	
BR						
S/BP	97.5	23,516	65,845	2.8	0.178	179
	97.5	13,967	33,520	2.4	0.178	
	97.5	11,840	26,049	2.2	0.178	
	105	18,860	37,156	1.97	0.315	97
	105	17,199	37,667	2.19	0.315	
	105	15,036	37,892	2.52	0.315	

of D-E drift response and thermal drift response balancing each other. The analysis of the PFR is complicated by the fact that the feed temperature was increased to trigger R-A. As a result, the initial chain length decreased from one run to the next. It is thus difficult to isolate the nonisothermal drift response from the simple instantaneous response to an increase in reaction temperature.

4.4.2. Copolymers

The effect of nonisothermal temperature histories on copolymer composition is of less significance than their effect on DP and DPD. Thermal drift in CC does occur, caused by the change in reactivity ratios with temperature. The temperature dependence of R_A and R_B is shown in Table 4.3-10 for several comonomer systems. The relationship between feed and product composition is dictated by equation 1.13-14 from which CC diagrams for various initial temperatures can be computed. These are shown in Figs. 1.13-2–1.13-5 and are repeated here. Note that fairly large changes in temperature do not significantly alter the shapes of the CC diagrams. Recall from Chapter 2 that the distance between the CC curve and the 45° line governs the magnitude of the composition drift. Since this separation is not a strong function of temperature, one would not expect a nonisothermal temperature history to have a very profound affect on the product CC. Figure 4.4-7 and 4.4-8 as well as 4.3-30 show the composition drift for these systems in detail over the full range of initial compositions, including isothermal, quasi-isothermal, runaway, and adiabatic temperature histories. Note that a nonisothermal policy does not change the direction of drift (characterized by β_K). Furthermore, the cumulative CC's of product formed via the four temperature paths are not significantly different. If anything, R-A may provide a benefit in that D-E prevents the reaction from proceeding to sufficiently high conversions to cause one of the comonomers to be depleted before

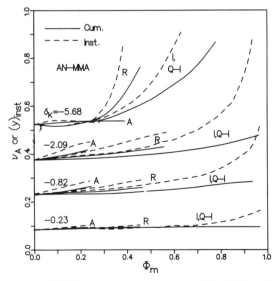

Figure 4.4-7.   Nonisothermal CC drift for nonazeotropic system AN/MMA, with $(n_A)_0 = 0.8, 0.6, 0.4, 0.2$.

the other. As a result, the bimodal CCD's that might occur under isothermal conditions (e.g., S/AN, $\beta_K = 0.2$) are avoided. An optimization study by Gromley and Tirell (19) suggested that rapidly rising temperatures were optimal with respect to minimizing CCD dispersion for S/AN copolymerization with high AN content (Section 4.2). This would seem to confirm the benefits just described to naturally occurring temperature profiles that resemble those computed by optimization.

To examine the direction of composition drift with respect to temperature, we adopt a technique similar to that used in Section 4.3-2 to determine the drift in thermal behavior with respect to composition change. Recall from Section 1.13 that instantaneous composition is given by

$$(\bar{y}_A)_{inst} = \frac{\left(R_A n_A^2 + n_A n_B\right)}{R_A n_A^2 + 2n_A n_B + R_B n_B^2} \tag{1.13-14}$$

If we take the partial derivative of 1.13-14 with respect to temperature, we can establish conditions under which $(\bar{y}_A)_{inst}$ either increases or decreases as temperature rises. The sign of the derivative can be shown to be positive if:

$$E_{RA} R_A n_A (n_A + R_B n_B) - E_{RB} R_B n_B (n_B + R_A n_A) > 0 \tag{4.4-1}$$

which yields the dimensionless criterion for drift toward increasing values of $\bar{y}_A$ as temperature rises:

$$\delta_K \equiv \left[\frac{E_{RA} R_A n_A (n_A + R_B n_B)}{E_{RB} R_B n_B (n_B + R_A n_A)} \right]_r \begin{cases} > 1, & E_{RB} > 0 \\ < 1, & E_{RB} < 0 \end{cases} \tag{4.4-2}$$

Note that δ_K is related to the composite CT's previously derived from comonomer

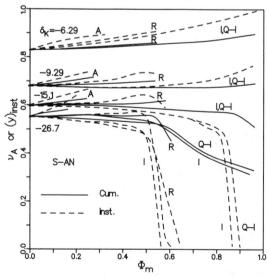

Figure 4.4-8. Nonisothermal CC drift for azeotropic system S/AN, with $(n_A)_0 = 0.8, 0.7, 0.6, 0.5$.

consumption via:

$$\delta_K = \frac{(E_{RA}R_A n_A)_r \Lambda_A}{(E_{RB}R_B n_B)_r \Lambda_B} \tag{4.4-3}$$

Several generalizations follow immediately. If $E_{RB} > 0$ and $E_{RA} < 0$, the inequality is always violated, and rising temperatures will lead to lower fractions of component A in the copolymer. None of the comonomer pairs in Figs. 1.13-2–1.13-5 fit this category. If $E_{RB} < 0$ and $E_{RA} > 0$, the inequality is always satisfied, and the content of A should increase with rising temperature. Both S/AN and AN/MMA are systems of this type and the figures verify that higher values of temperature would translate the CC curve upward over the entire composition range. Other cases require computation of δ_K. For instance, evaluating δ_K for S/MMA at 50°C for compositions of $(n_A)_0 = 0.7$ and 0.3 yields $\delta_K = 4.3$ and 0.42, respectively. Thus the CC diagram should move upward with temperature for large values of n_A, and downward for low values. Figure 1.13-4 exhibits such a pattern. In effect, rising temperatures make the CC diagram approach the 45° line and thus act to minimize composition drift. A similar computation for AN/VA would show that this system drifts towards the 45° line for small values of $(n_A)_0$ while it drifts away for larger values. Figure 4.4-9 illustrates the variation of δ_K with composition for the four of the systems given in Table 4.3-10. Note that for AN/MMA and S/AN the first inequality applies (since $E_{RB} < 0$), and for S/MMA and AN/VA the second holds. Consistent with the CC diagrams, only the latter two systems will exhibit a change in direction of thermal drift as feed composition changes.

To properly assess the nonisothermal drift response of a system, we must know both the direction of isothermal drift as characterized by β_K and the effect of temperature on the CC diagram as measured by δ_K. If, under isothermal conditions,

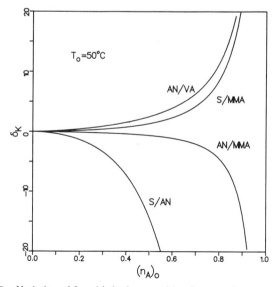

Figure 4.4-9. Variation of δ_K with feed composition for several comonomer systems.

Table 4.4-3. Dimensionless Criteria for Polymer and Copolymer Properties

TARGET	INSTANTANEOUS RESPONSE	ISOTHERMAL DRIFT RESPONSE	NONISOTHERMAL DRIFT RESPONSE
Large DP/LCA	$\left(\nu_N\right)_o \gg 1$	$\alpha_k > 1$	$\alpha_k > 1$
Narrow DPD	$\phi \ll 1$	$\alpha_k \sim 1$	$\alpha_k > 1$
Large CC (in A)	large $\left(\nu_A\right)_o$	$\beta_k > 1$	$\delta_k > 1 \ ; \ E_{RB} > 0$ $\delta_k < 1 \ ; \ E_{RB} < 0$
Narrow CCD	$R_A R_B \ll 1$	$\beta_k \sim 1$	$\delta_k = 1$

Table 4.4-4. Summary of Chain-Addition Polymer and Copolymer Properties

INSTANTANEOUS

DP $\qquad \left(\overline{x}_N\right)_{inst} = 2\, r_p/(2 - R)r_i = 2\nu_N/(2 - R)$

CC $\qquad \left(\overline{y}\right)_{inst} = r_{PA}/r_p = \nu_A$

CUMULATIVE

DP $\qquad \overline{x}_N = \mu_c^1/\mu_c^o$

$\qquad\qquad \overline{x}_W = \mu_c^2/\mu_c^1$

CC $\qquad y = \Delta c_A/\Delta c_m$

where density changes and chain transfer reactions have been neglected and

$$R = \begin{cases} 0 \quad \text{disproportionation} \\ 1 \quad \text{combination} \end{cases}$$

the CC curve lies above the 45° line, A is preferentially consumed, drift is in the direction of lesser compositions of A, and $\beta_K < 1$. If rising temperature causes the CC curve to move farther from the 45° line ($\delta_K > 1$ when $E_{RB} > 0$ and $\delta_K < 1$ when $E_{RB} < 0$), then R-A will aggravate the isothermal drift in CC and broaden the CCD. If increasing temperatures bring the CC curve closer to the 45° line, then R-A will result in a narrower CCD. Care must be taken when feed composition lies near an azeotrope. Rising temperatures could shift the position of the azeotrope sufficiently to cause a reversal in the direction of drift. This phenomenon cannot occur in isothermal copolymerization (Section 2.7). Also, the slope of the CC curve is

Table 4.4-5.   Dependence of Dimensionless Groups on Reaction Properties and Reactor Parameters (Chain-Addition Polymerizations and Copolymerizations)

PROPERTIES/PARAMETERS		α_K	β_K	γ_K	$\left(\nu_N\right)_0$	$\left(\nu_A\right)_0$	R_A	R_B	a	b_i	b_m^+ $\bullet$
INITIATOR/MONOMER											
A_d	↑	↑		-	↓	-	-	-	-	↓	-
E_d	↓	↑		-	↓	-	-	-	-	↓	↓
$\varepsilon = \hat{E}_{ap}^{-1}$	↓	↑				-	-	-	↓	↑	↑
FEED CONDITIONS											
$\left(c_o\right)_0$	↑	↓			↓	-	-	-	↓	↑	-
$\left(c_m\right)_0$	↑	-				-	-	-	↓	↑	↑
$\left(c_A\right)_0$	↑	-					-	-			
T_o	↑	↑*			↓*		-	-	↓	↓	↓
TRANSPORT PARAMETERS											
k							-	-			
Us_v	↑						-	-	↑		

*when $E_{ax} < 1$, which is typical for free-radical polymerizations

+$b_m > 20$ and approximately constant for a large class of free-radical polymerizations.

important. Discussions of isothermal drift (also in Section 2.7) pointed out that if the CC curve is horizontal, then drift in n_A has no influence on the value of $(\bar{y}_A)_{inst}$. This is characteristic of S/AN in the composition range $0.2 < n_A < 0.7$. If temperature causes the CC curve to move up or down, then a composition drift previously absent will be imposed, and the CCD will be broadened. The preceding conclusions are summarized in Table 4.4-3. The reader is cautioned that these are generalizations and that exceptional behavior, as just described, can exist.

4.5. THERMAL INSTABILITY AND MULTIPLE STEADY STATES

The reactor schemes considered in the previous sections had one common feature: the absence of any form of feedback or backmixing (BM). When there are no mechanisms for feedback, the system will have a single (unique) steady state (e.g., PFR). Hence the term stability in the previous sections actually referred to the sensitivity of that single state to changes in the system's parameters. These parameters reflect the feed conditions and material properties.

The classical notion of stability is somewhat different. It refers to a system's tendency to return to a given steady state after a perturbation in the state variables (e.g., temperature or composition) occurs while parameters remain fixed. Even the term steady state may be misleading. There may be nothing inherently steady about a solution that causes the time derivatives of the reactor balances to equal zero. As we shall show for a given set of system parameters, there may be several such solutions, some of which are never realized in an uncontrolled reactor system.

Whereas instability may be readily recognizable, stability can be more elusive to define. If one interprets the tendency to return to an initial stationary state as a measure of stability, then one must inquire about the magnitude of perturbation that permits such behavior. Will the system always return to the same state regardless of how large a disturbance is introduced? If so, this would be called global stability. If not, then what does happen? Does it return for small deviations and relocate to another state for larger ones? Does it remain in the neighborhood of the initial state without seeking a new stationary set of values of the state variables (e.g., limit cycle), or does it diverge regardless of the size of the disturbance? We shall see that, except for the latter situation, one might consider all the preceding as working definitions of stability. One word of caution is in order. Conditions exist under which a system can make large excursions in temperature and conversion before returning to the stationary state. Although such a system would satisfy even the most restrictive mathematical definitions of stability, the extremes of temperature or conversion might be well beyond the physical limitations placed on the reactor or reaction medium. A reactor that freezes solid with polymer enroute to a steady state would certainly fall outside the bounds of "practical stability."

We shall examine lumped parameter systems. In particular, the MMIM serves as a model of an open system with complete backmixing. As we observed in our earlier analyses of closed reactors, there are several apparent choices for dimensionless reactor temperature. Feed temperature is frequently chosen to reduce reactor temperature, but one might argue that the coolant temperature is just as appropriate. We prefer a choice that combines the two. It is the temperature that the reactor would reach in the absence of chemical reaction, and it represents an "equilibrium" point in much the same sense that T_R did for the batch reactor (49). Our characteristic reference temperature is defined as:

$$T_r \equiv \frac{T_0 \lambda_v^{-1} + T_R \lambda_R^{-1}}{\lambda_v^{-1} + \lambda_R^{-1}} \tag{4.5-1}$$

By using this T_r, the partially dimensionless transient equations for the MMIM can be generalized as shown in Table 4.5-1. Note that Λ_R is a composite CT combining the parallel effects of heat removal by flow and by external coolant (cf. equation

Table 4.5-1. Balance Equations for Lumped Parameter Systems with Backmixing (MMIM)

$$\frac{dc_k}{dt} = (\dot{V}/V)((c_k)_o - c_k) + r_k$$

$$\rho C_p \frac{dT}{dt} = r_G - [U s_V (T - T_R) + \rho C_p (\dot{V}/V)(T - T_o)]$$

$$\frac{d\hat{c}_k}{dt} = \lambda_v^{-1}(1 - \hat{c}_k) + \lambda_k^{-1} r_k$$

$$\frac{d\hat{T}}{dt} = \lambda_G^{-1} \hat{r}_G - \Lambda_R^{-1} \hat{T}$$

or

Dimensionless Form

$$\frac{d\theta}{dt} = \lambda_{ad}^{-1} \hat{r}_G - \Lambda_R^{-1} \theta$$

$$\frac{d\hat{c}_k}{d\hat{t}} = b_k^{-1}[Da_k^{-1}(1 - \hat{c}_k) + \hat{r}_k]$$

$$\frac{d\theta}{d\hat{t}} = \hat{r}_G - a\theta$$

$$\hat{t} \equiv t/\lambda_{ad}$$

$$\hat{r}_k, \hat{r}_G \quad \text{defined in Table 4.3-1}$$

4.3-6 and Example 4.3-9)

$$\Lambda_R^{-1} \equiv \lambda_v^{-1} + \lambda_R^{-1} \tag{4.5-2}$$

which follows from equation 4.3-6 when $\lambda_R \gg \lambda_H$. The thermal energy balance for the MMIM is similar to that used for the batch reactor (Table 4.3-1), with one important difference, that is, the addition of flow terms (Appendix F). Flow terms in the component balances provide the equations (Table 4.5-1) with true steady-state solutions when all time derivatives are set equal to zero. The algebraic equations that result may thus be reduced to a nonlinear equation in temperature (θ) only. This equation may or may not be tractable analytically for the steady-state value of θ. The nonlinearity suggests, however, that there can be more than one solution even when all system parameters are held constant, which is to say that the system could exhibit a multiplicity of steady states.

If we use free-radical polymerization as an example, the dimensionless balances may be written as:

$$\frac{d\hat{c}_0}{d\hat{t}} = b_i^{-1} \left[\frac{1 - \hat{c}_0}{\mathrm{Da}_i} - \hat{c}_0 \exp\left(\frac{\varepsilon \hat{E}_d \theta}{1 + \varepsilon\theta} \right) \right] \tag{4.5-3}$$

$$\frac{d\hat{c}_m}{d\hat{t}} = b_m^{-1} \left[\frac{1 - \hat{c}_m}{\mathrm{Da}_m} - \hat{c}_0^{1/2} \hat{c}_m \exp\left(\frac{\theta}{1 + \varepsilon\theta} \right) \right] \tag{4.5-4}$$

$$\frac{d\theta}{d\hat{t}} = \hat{c}_0^{1/2} \hat{c}_m \exp\left(\frac{\theta}{1 + \varepsilon\theta} \right) - \mathbf{a}\theta \tag{4.5-5}$$

with

$$\hat{t} \equiv \frac{t}{\lambda_{ad}}$$

Note that $\mathrm{Da}_i = \alpha_K \mathrm{Da}_m$ and that we have made use of the parameter **a**, which is analogous to dimensionless group a for lumped parameter systems without BM and which was introduced in Section 4.3 (equation 4.3-119):

$$\mathbf{a} \equiv \frac{\lambda_{ad}}{\Lambda_R} \tag{4.5-6}$$

Setting the RHS of equations 4.5-3–4.5-5 equal to zero, we find the steady-state values of $\hat{c}_0$ and $\hat{c}_m$ in terms of θ.

$$\hat{c}_0 = \left[1 + \mathrm{Da}_i \exp\left(\frac{\varepsilon \hat{E}_d \theta}{1 + \varepsilon\theta} \right) \right]^{-1} \tag{4.5-7}$$

$$\hat{c}_m = \frac{1 - \dfrac{\mathrm{Da}_m \exp \dfrac{\varepsilon \hat{E}_d \theta}{1 + \varepsilon\theta}}{(\nu_N)_0 \left[1 + \mathrm{Da}_i \exp \dfrac{\varepsilon \hat{E}_d \theta}{1 + \varepsilon\theta} \right]}}{1 + \dfrac{\mathrm{Da}_m \exp \dfrac{\theta}{1 + \varepsilon\theta}}{\left[1 + \mathrm{Da}_i \exp \dfrac{\varepsilon \hat{E}_d \theta}{1 + \varepsilon\theta} \right]^{1/2}}} \tag{4.5-8}$$

These results, when inserted into equation 4.5-5, allow the thermal energy generation rate to be expressed in terms of θ as the only state variable.

$$\hat{r}_G = \frac{\left[1 + \mathrm{Da}_i \exp\dfrac{\varepsilon E_d \theta}{1 + \varepsilon\theta} - \dfrac{\mathrm{Da}_m}{(\nu_N)_0}\exp\dfrac{\varepsilon \hat{E}_d \theta}{1 + \varepsilon\theta}\right]\exp\dfrac{\theta}{1 + \varepsilon\theta}}{\left[\left(1 + \mathrm{Da}_i\exp\dfrac{\varepsilon \hat{E}_d \theta}{1 + \varepsilon\theta}\right)^{1/2} + \mathrm{Da}_m\exp\dfrac{\theta}{1 + \varepsilon\theta}\right]\left(1 + \mathrm{Da}_i\exp\dfrac{\varepsilon \hat{E}_d \theta}{1 + \varepsilon\theta}\right)}$$

$$(4.5\text{-}9)$$

This equation is the MMIM equivalent of equation 4.3-51 for PFR's. The corresponding expression for step-addition polymerizations is derived in Example 4.5-1.

The functional dependence of $\hat{r}_G$ on θ is illustrated for a step-addition polymerization in Fig. 4.5-1 and for a variety of chain-addition polymerizations with parameter values characteristic of vinyl monomers in Figs. 4.5-2–4.5-6. The step-addition curve is typical for first-order reactions in a MMIM (50). The precise $\hat{r}_G$ function for first-order kinetics is obtained in Poisson-like step-addition polymerizations [i.e., when $a_K = 1$ or $a_K \to \infty$ (Example 4.5-1)] and in conventional chain-addition polymerizations for which the long-chain approximation applies [i.e., when $\alpha_K \to 0$ and $(\nu_N)_0 \to \infty$ (equation 4.5-9)]. The first-order function is sigmoidal, like its batch (PFR) counterpart in Fig. 4.3-1, but the concentration dependence here, which derives from a steady-state material balance, that is, $f(\mathrm{Da}_m)$ in $\hat{r}_G = f[\mathrm{Da}_m, \hat{k}_{ap}(\theta)]$, is much "stronger" than its batch counterpart $f(\hat{a})$ (Section 4.3). Consequently, the resulting sigmoid lies entirely within a temperature regime that could be experienced in practice. We should point out, however, that the upper steady state might occur at sufficiently high temperatures to negate the long-chain approximation (LCA) and thus produce polymer with very low DP in addition to large errors in the heat generation term, which assumes that propagation dominates the exotherm.

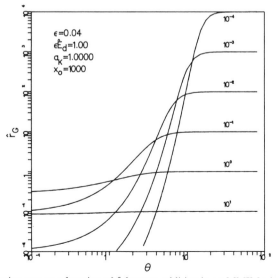

Figure 4.5-1: Generation rate as a function of θ for step-addition in an MMIM with Da_m as parameter.

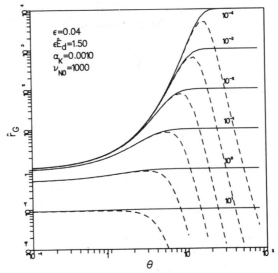

Figure 4.5-2. Generation rate as a function of θ for chain-addition in an MMIM with Da_m as parameter.

Concerning the chain-addition curves, several important features are immediately apparent. Consider Fig. 4.5-2 first. Notwithstanding its adherence to the conditions for pseudo-first-order behavior $[\alpha_K \ll 1, (\nu_N)_0 \gg 1]$, the appearance of many of the curves is most unlike the sigmoidal shape expected for $\hat{r}_G$. Two sets of curves are actually given. The dashed lines represent the solution to equation 4.5-9, in which the assumption of the LCA is implicit. That is, when the heat of polymerization is attributed solely to the propagation step, it is tacitly assumed that contribution of the initiation reaction to monomer consumption is negligible, and consequently its

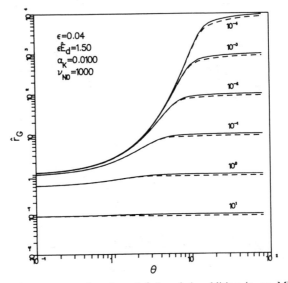

Figure 4.5-3. Generation rate as a function of θ for chain-addition in an MMIM with Da_m as parameter.

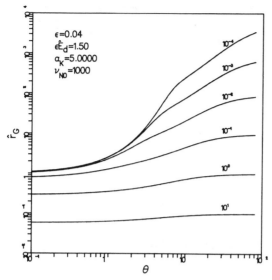

Figure 4.5-4. Generation rate as a function of θ for chain-addition in an MMIM with Da_m as parameter.

contribution to $\hat{r}_G$ is also negligible. Hence the LCA is implicit in the thermal energy balance, whether or not it is applied to the monomer rate equation. This can lead to anomalous behavior when conditions are sufficiently extreme to violate the LCA. Under such circumstances, the monomer will be primarily consumed by the initiation step, but this will contribute nothing to the production of thermal energy. At high temperatures, $\hat{r}_G$ passes through a maximum value and then returns to zero. This is an example of the anomalous high-temperature behavior referred to earlier. A

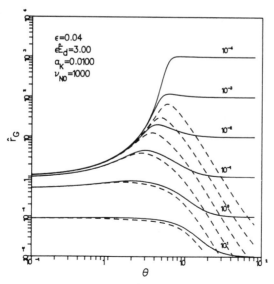

Figure 4.5-5. Generation rate as a function of θ for chain-addition in an MMIM with Da_m as parameter.

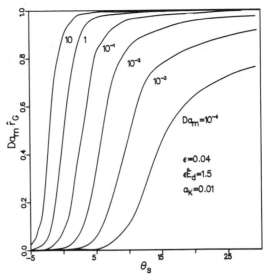

Figure 4.5-6. Generation rate normalized by its maximum value Da_m versus θ.

very small value of α_K ensures that, even at high temperatures, monomer reaches complete conversion with very little initiator conversion. As a result, the LCA is violated under these conditions. Monomer is consumed as rapidly by initiation as it is by propagation, but only the latter is reflected in $\hat{r}_G$. The consequence is that c_m drops more rapidly with temperature than $\exp \theta/(1 + \varepsilon\theta)$ rises, and $\hat{r}_G$ declines with subsequent increases in θ. If we include the initiation term in $\hat{r}_G$, then

$$\hat{r}_G = c_0^{1/2} c_m \exp\left(\frac{\theta}{1 + \varepsilon\theta}\right) + \frac{\hat{c}_0}{(\nu_N)_0} \exp \frac{\varepsilon\hat{E}_d\theta}{1 + \varepsilon\theta} \tag{4.5-10}$$

Then the solid curves in Fig. 4.5-2 result. These curves are closer in appearance to the expected sigmoid, but still exhibit anomalous high-temperature behavior. This is because the expression for the rate of initiation breaks down at high conversions. We saw in Section 2.8-3 that writing the rate of initiation as an efficiency factor times the initiator decay rate is an approximation. An exact treatment would show that the rate of initiation is a function of both monomer and initiator concentration, and that is approaches zero when either $\hat{c}_0$ or $\hat{c}_m$ falls to zero. This is not true of the expression in 4.5-10. Hence $\hat{r}_G$ is overestimated at high temperatures (which correspond to low monomer concentrations).

As α_K is increased, the deviation from first-order behavior grows, as depicted in Figs. 4.5-3 and 4.5-4. Also as α_K increases, the difference between $\hat{r}_G$ expressed with and without the LCA decreases. This is because larger values of α_K indicate that $\hat{c}_0$ is also very small when $\hat{c}_m \to 0$ and thus the rate of monomer consumption will not become unduly large by inclusion of the rate of initiation in the rate expression. When $\alpha_K = 1$, there is no difference between the two forms of $\hat{r}_G$ (equations 4.5-9 and 4.5-10).

Monomer parameters may be held constant and the temperature sensitivity of rate of initiation manipulated via $\hat{E}_d$, as shown in Fig. 4.5-5. When the initiation step is strongly temperature-dependent, however, as in Fig. 4.5-5 with $\varepsilon E_d = 3.0$, there is a substantial difference between the generation curves with and without LCA. Their behavior is also very different from the curves with $\varepsilon\hat{E}_d = 1.5$, other conditions being similar (Fig. 4.5-3). Increasing the importance of initiation, either through larger values of α_K or larger values of $\hat{E}_d$, has the effect of reducing the maximum slope of $\hat{r}_G$ with respect to θ. This is of increasing importance as Da_m is reduced. In a sense, the higher initiator conversion that is due to manipulation of α_K or $\hat{E}_d$ serves to mitigate the tendency of Da_m to decrease $(\hat{r}_G)_{max}$.

In all the figures, increasing Da_m has the effect of translating $\hat{r}_G$ to the left (towards lower values of θ) and downward, consistent with its interpretation as a specific output rate (Section 1.3). The former effect is not readily apparent from the log-log plots, so we offer a linear version of Fig. 4.5-3 in Fig. 4.5-6. In order to accommodate the large range of values of $\hat{r}_G$, the abscissa has been scaled by Da_m so that the maximum value for all curves is the same and is one (see Example 4.5-3). Now $Da_m\hat{r}_G$, which is proportional to conversion, rises with increasing Da_m. Note that there is a certain symmetry among the curves. There are two asymptotic values of $Da_m\hat{r}_G$: zero and one. At some value of θ, each curve experiences a sharp upward rise. This value of θ moves from negative values to progressively higher (and positive) values of Da_m is increased. In viewing the figures, their similarity to the BR and PFR generation curves of Fig. 4.3-6 is apparent. Increasing the value of Da_m permits a higher degree of conversion and thus is conceptually similar to decreasing the value of $f(\hat{\alpha})$ in the batch time-dependent generation rate, $f(\hat{\alpha})\hat{g}_e(\theta)$. The major difference in appearance between the generation curves for the IM and the BR is the absence of the high-temperature asymptote in the latter.

Several important features are immediately apparent. Treatment of free-radical systems often starts with the assumption that the effect of initiator concentration can be neglected. This would reduce the monomer rate equation to a simple first-order expression, making it amenable to the wealth of existing engineering knowledge concerning the model reaction $A \rightarrow B$; $r = kc_A$. Jai Singhani and Ray (51) devoted most of their work on the stability of free-radical polymerization reactors to such a model. However, it is only in the extreme case of very small values of α_K that behavior is qualitatively like that of a first-order system for all values of Da_m.

Our choice of characteristic temperature T_r makes representation of the removal curve quite simple. It is always a straight line that passes through the origin, with a positive slope of **a**. Bearing in mind that intersection of the removal line with the generation curve locates a steady-state point, it is clear from the graphs that two distinct juxtapositions may be identified by the number of intersections possible. When

$$Da_m < 0.10 \tag{4.5-11}$$

three potential steady-states are available. The reasoning is similar to that used in our theoretical analysis of the batch reactor (Section 4.3.1). Increasing the value of Da_m has the effect of shifting the inflection point of the generation curve to the left. When Da_m is greater than 0.10, the abrupt rise in the sigmoidal $\hat{r}_G$ curve occurs before the origin ($\theta = 0$). This precludes the possibility of a lower steady state. In fact, for such values of Da_m, there is a unique steady-state solution to the component and energy balances. Multiplicity of steady states demands that $Da_m <$

0.10. We wish to point out that the nature of the $\hat{r}_G$ curves for this mechanism are such that there will be at most three steady-state intersections. This is in accord with the findings of Ray (51) who considered free-radical polymerization with gel-effect. Others have claimed the existence of as many as five steady states. We feel that this is a consequence of rather unrealistic kinetic mechanisms. Goldstein and Amundson (52) considered spontaneous initiation and monomer termination, for instance, while Zweitering assumed that the monomer concentration remained constant (53). These assumptions produced $\hat{r}_G$ curves with several plateaus in their rise toward a maximum value at large values of θ. By manipulating kinetic constants through a wide range of realistic values for chain-addition polymerizations, we have encountered only the classical "sigmoidal" $\hat{r}_G(\theta)$ curve. Consequently, the maximum number of steady-state solutions possible seems to be three.

The earliest considerations of steady-state stability were based on arguments similar to those invoked in the Semenov analysis of closed systems. Van Heerden (54) claimed that stability required the slope of the removal line to exceed the slope of the generation curve at the steady-state point. In the situation in which three steady states exist, the middle one clearly violates this condition and may be deemed unstable, as discussed in Section 4.3-1. Failure to meet this condition guarantees instability of the steady state, but fulfilling it does not guarantee stability.

In classical stability analysis, the system equations are linearized and expressed in terms of small displacements from the steady state in question. Thus, if there are $N - 1$ component balances and one thermal energy balance, the results of linearization may be written compactly by defining

$$1 \leqslant j \leqslant N - 1 \qquad\qquad x_j \equiv \hat{c}_k - \hat{c}_{ks} \qquad\qquad (4.5\text{-}12)$$

$$x_N \equiv \theta - \theta_s \qquad\qquad (4.5\text{-}13)$$

and by letting function f_j represent the RHS of the jth dimensionless balance equation. The linearized system of differential equations in matrix form is

$$\frac{d\mathbf{x}}{d\hat{t}} = \mathbf{Fx} \qquad\qquad (4.5\text{-}14)$$

where $\mathbf{F}$ is the Jacobian matrix of $\mathbf{f}$ evaluated at the steady state:

$$F_{ij} = \left.\frac{\partial f_j}{\partial x_i}\right|_{\mathbf{x}_s} \qquad\qquad (4.5\text{-}15)$$

The nature of the Jacobian matrix $\mathbf{F}$ determines the dynamic behavior of the system in the vicinity of the steady-state point. In particular, the eigenvalues of $\mathbf{F}$ must all be negative or have negative real parts or else the solution of equation 4.5-13 will grow with time.

Most of the chemical engineering stability literature is devoted to the analysis of single-reaction systems. The reactant and thermal energy balances form a second-order system for which there are two eigenvalues (z_1, z_2) and consequently two stability criteria (C_1 and C_2). Using the Routh–Hurwitz criteria (Appendix C), it can be shown that necessary and sufficient conditions for stability of second-order systems are:

$$C_1 = -\operatorname{Tr} F > 0 \tag{4.5-16}$$

$$C_2 = \operatorname{Det} F > 0 \tag{4.5-17}$$

It was recognized quite early (55) that the second constraint (4.5-17) reduced to a statement of Van Heerden's observation that the slope of the removal line at the steady state must be greater than that of the generation curve if that steady state is to be stable. The first criterion is often called the dynamic criterion. When inequality 4.5-17 is satisfied but 4.4-16 is violated, the system could exhibit sustained oscillations or "limit cycles" (55). Uppal, Ray, and Poore (56) have applied the bifurcation theory of Friedrichs to first-order kinetics to characterize quantitatively the many types of dynamic behavior possible for this system. Behavior can range from a single, globally stable steady state to three steady states surrounded by a limit cycle, with many possible intermediate combinations.

Little has been done to generalize results to systems of higher dimensions. The free-radical polymerization model is clearly third-order, and we will show that step-additions give rise to a four-dimensional system. The approach of linearization is still a valid means of examining stability of higher order systems. As shown in Appendix C, the number of stability criteria is equal to the order of the system, and that two of the criteria will always be

$$C_1 = -\operatorname{Tr} F > 0 \tag{4.5-18}$$

$$C_N = (-1)^N \operatorname{Det} F > 0 \tag{4.5-19}$$

where N is the order of the system. We shall now show that the Nth criterion can always be interpreted as the "slope" condition. As we have written the balances, $\hat{r}_G = \hat{r}_p$, and the total derivative of $\hat{r}_G$ with respect to θ is

$$\frac{d\hat{r}_G}{d\theta} = \frac{\partial \hat{r}_G}{\partial \theta} + \sum_{k=1}^{N} \frac{\partial \hat{r}_G}{\partial \hat{c}_k} \frac{d\hat{c}_k}{d\theta} \tag{4.5-20}$$

At the steady state, there is a relationship among θ and the $\hat{c}_k$. For free-radical kinetics, it is

$$a\operatorname{Da}_m \theta_s = \left(1 - \hat{c}_{m_s}\right) - \frac{1 - \hat{c}_{0_s}}{x_0} \tag{4.5-21}$$

Using equations 4.5-20 and 4.5-21, it can be shown that the determinant expression of Table 4.5-2 reduces to the following for model chain-addition kinetics

$$\operatorname{Det} F = b_i^{-1} b_m^{-1} \left(\operatorname{Da}_i^{-1} + \frac{\partial \hat{r}_0}{\partial \hat{c}_0}\right)\left(\operatorname{Da}_m^{-1} + \frac{\partial \hat{r}_p}{\partial \hat{c}_m}\right)\left(\frac{d\hat{r}_G}{d\theta} - \frac{d\hat{r}_e}{d\theta}\right) \tag{4.5-22}$$

As a result, stability criterion 4.5-19 becomes

$$\frac{d\hat{r}_e}{d\theta} - \frac{d\hat{r}_G}{d\theta} > 0 \tag{4.5-23}$$

which is, indeed, the slope criterion.

Table 4.5-2. The Jacobian Matrix and Determinant for Addition Polymerization with Termination

$$\underline{\underline{F}} = \begin{bmatrix} -b_i^{-1}(Da_i^{-1} + \hat{r}_{oo}) & -b_i^{-1}\hat{r}_{om} & -b_i^{-1}\hat{r}_{o\theta} \\[2ex] -b_m^{-1}(\hat{r}_{po} + \nu_{No}^{-1}\hat{r}_{oo}) & -b_m^{-1}(Da_m^{-1} + \hat{r}_{pm} + \nu_{No}^{-1}\hat{r}_{om}) & -b_m^{-1}(\hat{r}_{p\theta} + \nu_{No}^{-1}\hat{r}_{o\theta}) \\[2ex] \hat{r}_{po} & \hat{r}_{pm} & \hat{r}_{p\theta} - a \end{bmatrix}$$

$$\text{Det } \underline{\underline{F}} = b_i^{-1}b_m^{-1}\left\{(Da_i^{-1}\hat{r}_{oo})(Da_m^{-1} + \hat{r}_{pm} + \nu_{No}^{-1}\hat{r}_{om}) - \hat{r}_{om}(\hat{r}_{po} + \nu_{No}^{-1}\hat{r}_{oo})\right\} \cdot \left\{\hat{r}_{p\theta} + \hat{r}_{po}\frac{d\hat{c}_o}{d\theta} + \hat{r}_{pm}\frac{d\hat{c}_m}{d\theta} - a\right\}$$

where:

$$\frac{d\hat{c}_o}{d\theta} = \frac{(\hat{r}_{p\theta} + \nu_{No}^{-1}\hat{r}_{o\theta})}{(\hat{r}_{po} + \nu_{No}^{-1}\hat{r}_{oo})} - \frac{(Da_m^{-1} + \hat{r}_{pm} + \nu_{No}^{-1}\hat{r}_{om})}{(\hat{r}_{po} + \nu_{No}^{-1}\hat{r}_{oo})}\frac{d\hat{c}_m}{d\theta}$$

$$\frac{d\hat{c}_m}{d\theta} = \frac{r_{o\theta}(\hat{r}_{po} + \nu_{No}^{-1}\hat{r}_{oo}) - (Da_i^{-1} + \hat{r}_{oo})(\hat{r}_{p\theta} + \nu_{No}^{-1}\hat{r}_{o\theta})}{(Da_i^{-1} + \hat{r}_{oo})(Da_m^{-1} + \hat{r}_{pm} + \nu_{No}^{-1}\hat{r}_{om}) - \hat{r}_{om}(\hat{r}_{po} + \nu_{No}^{-1}\hat{r}_{oo})}$$

and

$$\hat{r}_{ko} = \frac{\partial\hat{r}_k}{\partial\hat{c}_o} \qquad \hat{r}_{km} = \frac{\partial\hat{r}_k}{\partial\hat{c}_m} \qquad \hat{r}_{k\theta} = \frac{\partial\hat{r}_k}{\partial\theta}$$

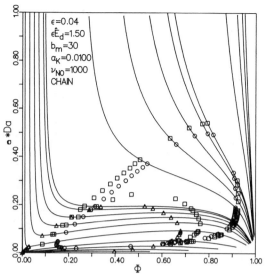

Figure 4.5-7. Regions of stability for conventional chain-addition polymerization: C_1 (circle), C_2 (square), C_3 (triangle); solid lines with Da_m values left to right of 10^{-n}, 3×10^{-n}, 5×10^{-n}, 7×10^{-n}, 9×10^{-n} for $n = -3, -2, -1, 0$ successively.

Clearly, then, violation of the determinant criterion C_N is confined to the middle steady state when there is multiplicity (three steady states). When there is only one steady state, the determinant condition will be satisfied automatically.

To examine the range of applicability of the other criteria, we extend the approach used by Hugo and Wirges (49) for single reactant, first-order kinetics. We can solve the steady-state balances to find the steady-state conversion of monomer

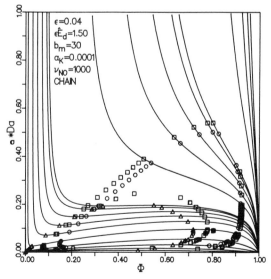

Figure 4.5-8. Regions of stability for conventional chain-addition polymerization, annotation as in Figure 4.5-7.

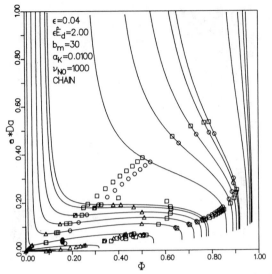

Figure 4.5-9. Regions of stability for conventional chain-addition polymerization, annotation as in Figure 4.5-7.

as a function of the system parameters. Figures 4.5-7–4.5-11 show the steady-state conversion as a function of the product $\mathbf{a}Da_m$ with Da_m as a parameter. We could equally have chosen $\mathbf{a}$, rather than $\mathbf{a}Da_m$, but curves plotted in this manner tend to converge and become indistinguishable for $Da_m > 0.10$. Furthermore, the parameter group $\mathbf{a}Da_m$ is not without physical significance. Recall that the maximum value of the generation rate under pseudo-first-order conditions ($\alpha_K \ll 1$) is Da_m^{-1} (Example 4.5-3). Thus, from equation 4.5-5 it is apparent that:

$$\theta_{max} = (\mathbf{a}Da_m)^{-1} \qquad (4.5\text{-}24)$$

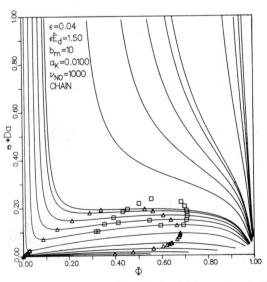

Figure 4.5-10. Regions of stability for conventional chain-addition polymerization, annotation as in Figure 4.5-7.

Rearranging the CT's composing **a** and Da_m shows:

$$\theta_{max} = \frac{\Lambda_R}{\lambda_v} b_m \qquad (4.5\text{-}25)$$

We saw in Section 4.3-1 that b_m could be identified as the adiabatic temperature maximum for a first-order reaction. It was also shown that the dimensionless adiabatic temperature maximum was an important parameter with respect to the sensitivity of the system. Equation 4.5-25 indicates that for open (flow) systems, the maximum attainable temperature is less than that for an equivalent reaction in a closed system, since b_m is scaled by a fraction. It may not be immediately evident that $\Lambda_R/\lambda_v < 1$. Using the definition of the composite CT, Λ_R (equation 4.5-2), we find

$$\frac{\Lambda_R}{\lambda_v} = \frac{1}{1 + \dfrac{\lambda_v}{\lambda_R}} \qquad (4.5\text{-}26)$$

which must be a fraction. This may be argued from a physical viewpoint as well. Since Λ_R is characteristic of the time scale for parallel heat removal processes (cf. Example 4.3-9), it will be no larger than the CT for the most rapid removal process and must therefore also be smaller than the CT for any other component removal processes. The maximum value of Λ_R/λ_v is reached when $\lambda_v \to 0$, in which case $\Lambda_R \to \lambda_v$ and is obviously unity. In the bifurcation work of Ray et al. (56), a key stability parameter is $(1 + \beta)$. This parameter is the inverse of the ratio Λ_R/λ_v. From equations 4.5-25 and 4.5-26, we see that the value of θ_{max} is bounded by zero when $\lambda_v \gg 1$ and by b_m when $\lambda_v \ll 1$. In Chapter 3, we saw that when $Da \ll 1$, the MMIM approached both the MSIM and PFR in behavior with respect to the effects of mixing on conversion. Similarly, when residence time is extremely short ($\lambda_v \ll 1$)

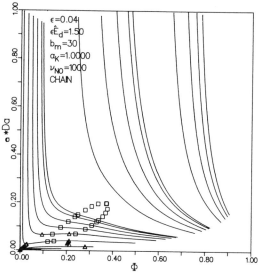

Figure 4.5-11. Regions of stability for conventional chain-addition polymerization, annotation as in Figure 4.5-7.

relative to the CT for reaction, the thermal behavior of the MMIM is like that of the PFR, and $\theta_{max} = b_m$. When residence time is very long compared to the CT's of the rate processes, reaction should be slow ($\hat{r}_G \to 0$ as $\hat{c} \to 0$) and the expected maximum temperature rise will be quite low, provided of course that λ_R is not unreasonably large.

Let us return to the discussion of Figs. 4.5-7–4.5-11. As one moves along a line of constant Da_m in the direction of increasing Φ_m, one observes a unique value of aDa_m associated with each Φ_m. Although not shown, there are, of course, values of θ_s and Φ_i corresponding to each pair, Φ_m, aDa_m. As one travels along a given line, the system parameters may violate one or more of the Routh–Hurwitz criteria. Noting the points on a line where a criterion changes sign, and connecting these points from line (of constant Da_m) to line, encloses regions in which the various stability criteria are violated. Thus each figure shows three distinct regions within which the stability criterion indicated is violated.

Several general features of the figures are apparent. As the steady-state value of conversion increases, the associated value of $\mathbf{a}$ decreases (actually aDa_m decreases while Da_m is held constant). When $Da_m > 0.1$ the lines of constant Da_m decrease monotonically with increasing Φ_m, which implies that, for a given value of the parameter groups $\mathbf{a}$ and Da_m, there is a unique steady-state solution $(\Phi_m)_s$. When $Da_m < 0.1$, the curves do not decrease monotonically. Instead, as larger values of Φ_m are sought, aDa first decreases (when $\Phi_m \leq 0.20$), then increases slightly ($0.20 \leq \Phi_m \leq 0.80$), before finally falling to zero. The curves do not all reach $\Phi_m = 1.0$, since excessively high temperatures are required to achieve complete conversion, and the kinetic assumptions (thermal simplicity, LCA, constant initiator efficiency) are no longer valid. Owing to the nature of these curves, three steady-state solutions are possible for any specified combination of aDa_m and Da_m. Also shown on the figures is the region in which the third (determinant) stability criterion, 4.5-18, is violated. This region is roughly bounded by the curve for $Da_m = 0.10$. Both observations confirm the conclusion drawn from the $\hat{r}_G$ versus θ curves that multiplicity of steady states (and thus instability in the classic sense) requires $Da_m < 0.10$. The figures impose an additional constraint on multiplicity. Note that when $Da_m = 0.10$ but $\Phi_m < 0.20$, there is only one steady state. Multiplicity is only possible when $aDa_m < 0.20$. Physically interpreted, $Da_m < 0.10$ ensures that the generation curve is moved sufficiently far to the right of the origin to permit multiplicity, and $aDa_m < 0.20$ ensures that the slope of the removal line ($\hat{r}_e = \mathbf{a}\theta$) is sufficiently small to allow multiple intersections. Figure 4.5-12 depicts the juxtapositions. It should be made clear that the condition $aDa_m < 0.20$, for a given value of Da_m such that $Da_m < 0.10$, does not mandate multiplicity. When the slope of $\hat{r}_e$ is made very small, it is possible to reestablish the condition of a single steady state, where the upper stable steady state is now the unique solution. However, one can say unequivocally that when $aDa_m \geq 0.20$ multiplicity is avoided.

Also shown on the figures are the regions in which the first (trace) and second stability criteria (Appendix C) are violated. These generally extend beyond the region circumscribed by the third (slope) stability criterion. In general, the region of violation of the trace criterion is enveloped by the region of violation of the second criterion. This is in concert with the observation that the Routh–Hurwitz criteria may exhibit some redundancy (57). Figure 4.5-13 shows the boundaries of violation of the second criterion as a function of b_m. The region of violation for the slope

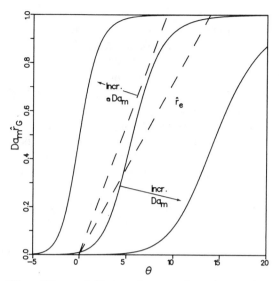

Figure 4.5-12. Effect of Da_m and aDa_m on the generation and removal curves for MMIM's.

criterion is also shown, and it is independent of the value of b_m. Although not all are shown for the sake of clarity, the lines of constant Da_m are independent of b_m as well. Based on this figure, one concludes that increasing b_m not only increases the value of aDa_m at which instability may occur, but it also extends this region outward to higher levels of conversion. In general, it appears that the upper steady state (when multiplicity occurs) will exhibit dynamic instability. Reducing the value of b_m diminishes the range of conversions over which this instability is evident, but only

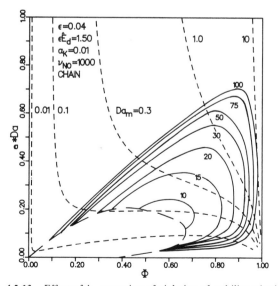

Figure 4.5-13. Effect of b_m on region of violation of stability criterion C_2.

lowering b_m to 10 admits stable upper steady states at a reasonable temperature level. Ray (51) was disappointed by the lack of "richness of dynamic behavior" exhibited in his study of vinyl polymerization with gel-effect. The conditions he chose, however, maintained $b_m = 12$. Note that when $b_m = 10$, not only is the region of violation of the second criterion small, but the trace criterion (C_1) is never violated. We shall show presently that some of the interesting limit cycle behavior is restored when a wider range of parameters is examined.

Figures 4.5-14 and 4.5-15 summarize the effect of initiation on the stability boundaries. Both α_K and $\hat{E}_d$, unlike b_m, alter the size of the slope criterion region. As initiator decay rate increases, the upper bound in conversion on the instability region decreases. It also causes the region of violation of the first and second criteria to shift to higher values of aDa_m, such that at the more extreme conditions the region of multiplicity is exceeded and the upper steady states become stable.

Step-addition polymerization requires four independent state variables ($\hat{c}_0$, $\hat{\mu}^0$, $\hat{c}_m$, and θ), and thus four stability criteria must be satisfied. Figure 4.5-16 shows that in general the regions of stability look similar to those for chain-addition. The fourth criterion is the slope criterion. It defines a region similar in shape to C_3 for chain-addition. It would appear that criteria C_2 and C_3 for step-addition substitute for C_1 and C_2, respectively, for chain-addition, while the trace criterion, C_1, for step-polymerization defines a new region. For the most part, however, the region of violation of C_1 is contained wholly within the regions bounded by the other criteria and thus seems to offer little extra information. Additionally, many parameter combinations were encountered in which criterion C_2 was always satisfied.

We turn our attention now to the types of dynamic behavior characteristic of the different instability regions. It was postulated that for most chain-addition polymerizations, when multiplicity was evident, not only would the middle steady state be unstable in the classical sense, but the upper steady state should also exhibit

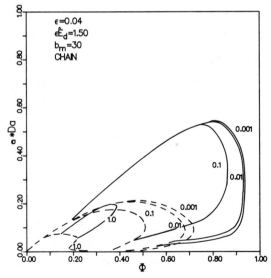

Figure 4.5-14.   Effect of α_K on region of violation of stability criterion C_2 (solid) and C_3 (dashed).

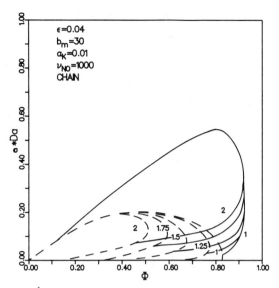

Figure 4.5-15. Effect of $\varepsilon \hat{E}_d$ on region of violation of stability criterion C_2 (solid) and C_3 (dashed).

instability owing to violation of the dynamic stability criteria (C_1 and/or C_2). In Fig. 4.5-17, we show phase space trajectories for just such a system. The three steady-state solutions are identified by circles and the dashed lines give their height in the θ-direction above the $\Phi_i - \Phi_m$ plane. Note that, regardless of its initial state, the system will always move to the lower steady state. When the system is perturbed from the middle steady state, it may attempt to relocate at the upper steady state, but on arriving there it will find that conditions of instability still prevail and it will spiral outward, ultimately finding the lower steady state. A system placed initially near the upper steady-state would also spiral out toward the lower steady state.

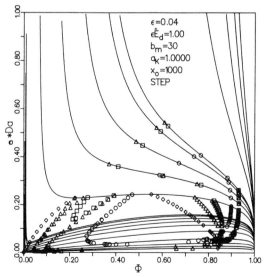

Figure 4.5-16. Stability regions for step-addition polymerization: C_1 (circle), C_2 (square), C_3 (triangle), C_4 (diamond); solid lines as in Figure 4.5-7.

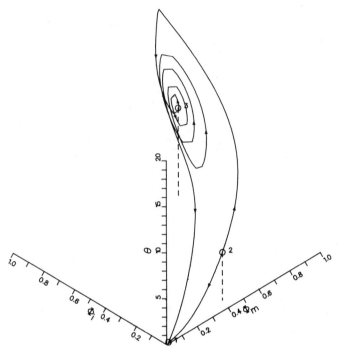

$\epsilon=0.04$ $b_m=30$ $a_K=0.0100$ $_aDa=0.0800$

$\acute{E}_d=1.50$ $\nu_{NO}=1000$ $Da=0.0100$

1    $C_1{>}0$ $C_2{>}0$ $C_3{>}0$
2    $C_1{<}0$ $C_2{>}0$ $C_3{<}0$
3    $C_1{>}0$ $C_2{<}0$ $C_3{>}0$

Figure 4.5-17. Dynamic behavior of a system with three steady-state solutions, where steady-state 3 violates C_2.

As previously mentioned, the instability of the upper steady state may be eliminated by reducing aDa_m. There is a very narrow window of values for which this is possible, since reducing aDa_m too much removes the system from the region of multiplicity, whereas insufficient reduction leaves the upper steady state within the bounds of the region in which C_1 and C_2 are violated. Figure 4.5-18 shows that reducing aDa_m from 0.08 (Fig. 4.5-17) to a value of 0.075 is sufficient to stabilize the upper steady state and present the classical picture of MMIM stability. The upper and lower steady states are stable, whereas the middle one will spontaneously yield to either of the two in response to the smallest perturbation.

An additional type of instability coupled with multiplicity is shown in Fig. 4.5-19. In this case, the upper steady state is surrounded by a limit cycle. When the upper steady state is perturbed, the system will respond by spiraling outward, ultimately locking into the closed loop shown. Although the middle state is not pictured for the sake of clarity, its response to certain perturbations will cause the system to spontaneously relocate to the same closed-loop path.

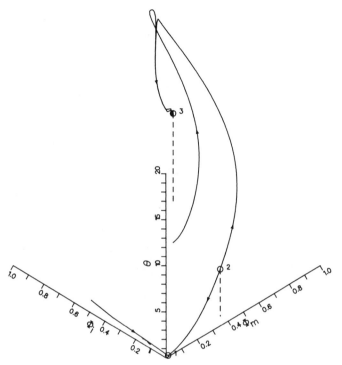

ϵ=0.04 b_m=30 a_K=0.0100 $_a$Da=0.0780

$\epsilon\hat{E}_d$=1.50 ν_{NO}=1000 Da=0.0100

1	$C_1{>}0$ $C_2{>}0$ $C_3{>}0$
2	$C_1{<}0$ $C_2{>}0$ $C_3{<}0$
3	$C_1{>}0$ $C_2{>}0$ $C_3{>}0$

Figure 4.5-18. Dynamic behavior of a system with three steady-state solutions, where steady states 1 and 3 violate no criteria.

It should be clear from the mapping of stability regions in Figs. 4.5-7–4.5-11 that there are no cases in which the lower steady state is unstable. There is, however, a large region in which unique but dynamically unstable steady states should exist. These represent type of conditions that were not apparent in the study of Jai Singhani and Ray (51). Figure 4.5-20 shows the simulation results for a typical case chosen from Fig. 4.5-7. Pictured as a phase-plane portrait, θ versus Φ_m, the behavior is very unlike that observed for one-component systems. Clearly a "limit-cycle" is established—that is, regardless of the initial state of the system, all trajectories decay toward a single closed loop that appears to enclose the steady-state solution. Uncharacteristic is the apparent crossing of phase-plane trajectories. This is an artifact, due not to inaccuracies in numerical simulation, but to the projection of a trajectory in three-dimensional phase-space onto a two-dimensional phase-plane. In phase-space (Fig. 4.5-21), the limit cycle is apparent. The system spirals upward from the steady-state solution until it reaches the closed-loop trajectory of the limit

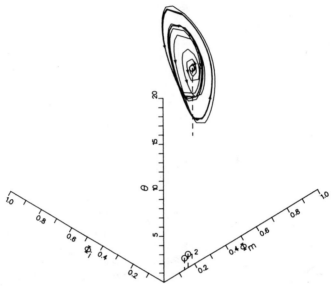

$\epsilon=0.04 \quad b_m=30 \quad a_K=0.0100 \quad {}_a Da=0.1130$

$\acute{\epsilon}\acute{E}_d=1.50 \quad \nu_{NO}=1000 \quad Da=0.0500$

1	$C_1>0$	$C_2>0$	$C_3>0$
2	$C_1>0$	$C_2>0$	$C_3<0$
3	$C_1>0$	$C_2<0$	$C_3>0$

Figure 4.5-19. Dynamic behavior of a system with three steady-state solutions, where steady state 3 is surrounded by a limit cycle.

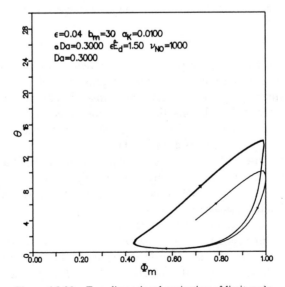

$\epsilon=0.04 \quad b_m=30 \quad a_K=0.0100$

$_a Da=0.3000 \quad \acute{\epsilon}\acute{E}_d=1.50 \quad \nu_{NO}=1000$

$Da=0.3000$

Figure 4.5-20. Two-dimensional projection of limit cycle.

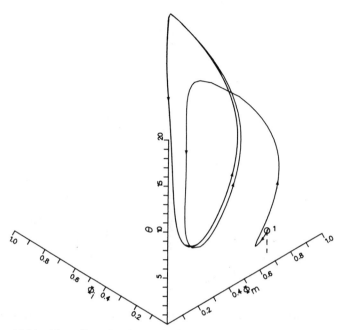

ϵ=0.04 b_m=30 a_K=0.0100 $\circ$ Da=0.3000

$\hat{E}_d$=1.50 v_{NO}=1000 Da=0.3000

1   C_1<0 C_2<0 C_3>0

Figure 4.5-21.   Three-dimensional portrait of Fig. 4.5-20, where the system violates C_2.

cycle. What is remarkable is that the limit cycle is totally removed in phase-space from the steady-state solution. Unlike single-component systems, for which limit-cycles represent orbits around the steady-state node, higher-order systems may seek an ultimate trajectory that has no proximity whatever to the mathematical solution of the steady-state equations.

Figures 4.5-22–4.5-25 document the progression from a single, globally stable steady state to one that spontaneously decays to a limit cycle. As the value of aDa_m is increased, the first transition moves the system just within the region of violation of C_2. Dynamic behavior is transformed from inwardly spiraling trajectories to outwardly directed spirals that ultimately converge on a limit cycle. As aDa_m is increased further, the number of oscillations required to go from the steady state to the limit cycle decreases. In the final figure of the series, both C_1 and C_2 are violated, although no substantive change in dynamic behavior is apparent.

The stability diagrams we have presented are a convenient means for quickly assessing the dynamic behavior of a system. They do not provide the type of detailed information on the character of the instability that the bifurcation analysis used by Ray (51, 56) is capable of yielding, To date, however, successful application of bifurcation theory has been confined to second-order systems. We have seen that initiator effects are sufficiently important, particularly in the face of large temperature variations, that they cannot be ignored. Hence chain-addition polymerization

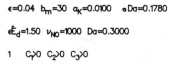

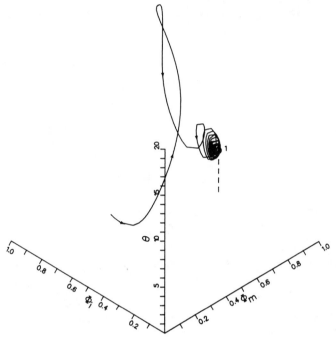

Figure 4.5-22. Dynamic behavior of system with single stable steady state.

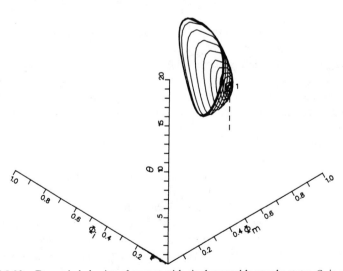

Figure 4.5-23. Dynamic behavior of system with single unstable steady state; C_2 is violated.

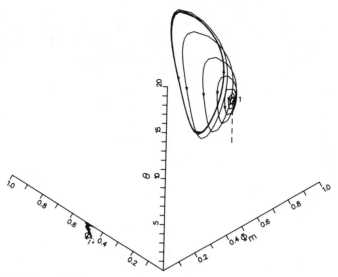

ϵ=0.04 b_m=30 a_K=0.0100 $_a$Da=0.1900

$\epsilon\hat{E}_d$=1.50 v_{NO}=1000 Da=0.3000

1   C_1>0 C_2<0 C_3>0

Figure 4.5-24. Dynamic behavior of system with single unstable steady-state; C_2 is violated.

ϵ=0.04 b_m=30 a_K=0.0100 $_a$Da=0.2000

$\epsilon\hat{E}_d$=1.50 v_{NO}=1000 Da=0.3000

1   C_1<0 C_2<0 C_3>0

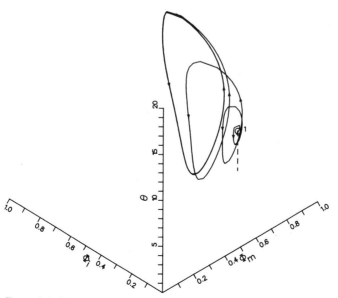

Figure 4.5-25. Dynamic behavior of system with single unstable steady state; C_2 and C_3 are violated.

should not be treated as a pseudo-one-component system. The establishment of a link between our stability analysis and the formalism of bifurcation theory is left as a challenge to future investigators.

Example 4.5-1.   Heat Generation Rate for Step-Addition Polymerizations.   Derive the steady-state dimensionless generation function $\hat{r}_G = f(\text{Da}, \theta_s)$ analogous to equation 4.5-9 for step-addition polymerization sequence 2.3-19 and show that it reduces to the following form, which is characteristic of first-order reactions, when $k_i = k_p$ (Poisson sequence 2.3-1)

$$\hat{r}_G = \frac{\exp \dfrac{\theta_s}{1 + \varepsilon\theta_s}}{1 + \text{Da} \exp \dfrac{\theta_s}{1 + \varepsilon\theta_s}}$$

Solution.   Referring to Section 2.3, the steady-state initiator and monomer balances are

$$-\frac{\Delta c_0}{\lambda_v} = k_i c_0 c_m$$

and

$$-\frac{\Delta c_m}{\lambda_v} = k_i c_0 c_m + k_p c_m c_p$$

where $c_p = (c_0)_0 - c_0$ from constraint equation 1.7-7. In dimensionless form, these equations become

$$\frac{1 - \hat{c}_0}{\text{Da}} = a_K x_0 \hat{c}_0 \hat{c}_m \hat{k}_i$$

and

$$\frac{1 - \hat{c}_m}{\text{Da}} = a_K \hat{c}_0 \hat{c}_m \hat{k}_i + \hat{c}_m (1 - \hat{c}_0) \hat{k}_p$$

where $\hat{k}_i \equiv \exp \varepsilon \hat{E}_i \theta / (1 + \varepsilon\theta)$, $\hat{k}_p \equiv \exp \theta / (1 + \varepsilon\theta)$, $\varepsilon \equiv \hat{E}_p^{-1}$, $\theta \equiv \hat{E}_p \hat{T}$, $\hat{E}_p \equiv E_p / R_g T_r$, $\hat{E}_i \equiv E_i / R_g T_r$, $\text{Da} \equiv \lambda_v / \lambda_m$, $\lambda_m \equiv 1/(k_p)_r (c_0)_0$, and $a_K \equiv (k_i)_r / (k_p)_r$. These equations are the nonisothermal versions of equations 3.5-30 and 3.5-31, respectively, with the one important difference, that both concentrations here have both been reduced with their feed values. Thus $\hat{c}_m \equiv c_m / (c_m)_0$, unlike $\hat{c}_m$ in equations 3.5-30 and 3.5-31, with the consequence that x_0 appears in the preceding equations. Their solution leads to

$$\hat{c}_m = \frac{\left[\hat{k}_i \text{Da}(x_0 - 1) - a_K^{-1}\right] + \left[\left(\text{Da}\hat{k}_i(x_0 - 1) - a_K^{-1} + 4x_0 \text{Da}\hat{k}_i(\text{Da}\hat{k}_p + 1)a_K^{-1}\right)\right]^{1/2}}{2x_0 \text{Da}\hat{k}_i(\text{Da}\hat{k}_p + 1)}$$

$$= 1 - \Phi$$

It is noteworthy that the last result reduces to the corresponding expression for $\hat{c}_m$ (i.e., $x_0 \hat{c}_m$) in Example 3.5-6 under isothermal conditions, that is, $\hat{k}_i = 1 = \hat{k}_p$. For the Poisson sequence under nonisothermal conditions, $\hat{k}_i = \hat{k}_p \equiv \hat{k}$ and $a_K = 1$, the heat generation rate is thermally simple (Appendix E) without assuming that the propagation step dominates heat generation

$$\hat{r}_G = \hat{k}\hat{c}_0\hat{c}_m + \hat{k}\hat{c}_m(1 - \hat{c}_0) = \hat{k}\hat{c}_m$$

and the preceding expression for c_m reduces to

$$\hat{c}_m = \frac{1}{Da\hat{k} + 1}$$

where $\hat{k} \equiv \exp \theta/(1 + \varepsilon\theta)$. Substitution of $\hat{c}_m$ into $\hat{r}_G$ yields the first-order generation function.

Example 4.5-2. Use of Multiplicity Criteria. Consider solution polymerization of methyl methacrylate which is 30 mole percent in benzene. The feed and coolant temperature at 50°C. The initiator is AIBN at a feed concentration of 0.373 mol/l. Specify design constraints to avoid multiplicity if the average residence time is 150 min, 15 min.

Solution. Using the kinetic constants for MMA/AIBN evaluated at 50°C, we find

$$\lambda_i = 1.85 \times 10^6 \text{ s}$$
$$\lambda_m = 1.85 \times 10^4 \text{ s}$$
$$\lambda_{ad} = 1493 \text{ s}$$

so

$$\varepsilon = 0.034 \qquad b_m = 12.4 \qquad b_i = 1240$$

When $\lambda_v = 150$ min, Da $= 0.50$; thus multiplicity is avoided regardless of the value of **a**. However, when $\lambda_v = 15$ min, Da $= 0.05$, which is less than the critical value of 0.10, so **a** must be adjusted via Λ_R to avoid multiplicity. To satisfy **a** > 2, we require

$$\Lambda_R < \frac{\lambda_{ad}}{2} = 747 \text{ s}$$

but

$$\Lambda_R^{-1} = \lambda_R^{-1} + \lambda_v^{-1}$$

so

$$\lambda_R < 4394 \text{ s}$$

Using the definition of λ_R, we compute

$$Us_v > 6.5 \times 10^{-5} \text{ cal / cm}^3 \text{ s K}$$

must be maintained to avoid multiplicity.

Example 4.5-3. Maximum Value of the Heat Generation Rate.   Determine the maximum value of the dimensionless heat generation rate, $\hat{r}_G$ for a single independent reaction in a MMIM.

Solution.   For a single independent reaction, $\hat{r}_G = \hat{r}$. At steady state, solution of the thermal energy balance shows:

$$\hat{r}_s = \mathbf{a}\theta_s$$

Solution of the component balance yields

$$\hat{r}_s = \frac{\Phi_s}{\mathrm{Da}}$$

Therefore,

$$\theta_s = \frac{\Phi_s}{\mathbf{a}\mathrm{Da}}$$

and the maximum value of θ_s, θ_{max}, will occur when $\Phi_s = 1$:

$$\theta_{max} = (\mathbf{a}\mathrm{Da})^{-1}$$

Finally,

$$(\hat{r}_G)_{max} = (\hat{r})_{max} = \mathbf{a}\theta_{max}$$

so

$$(\hat{r}_G)_{max} = \mathrm{Da}^{-1}$$

Example 4.5-4. Stability of a Random Polymerization.   A CSTR (MMIM) is being considered for a second-order random polymerization at 156°C starting with 5 mols per liter of functional group A in the feed and $T_0 = 127°C = T_R$. A conversion of 90 percent is to be achieved with a Da = 11.5. Is the steady state stable at this operating temperature? Use the following properties and parameters: $\rho c_p = 0.4$ cal/cm³ K, $-\Delta H_r = 25.6$ cal/mol, $\mathrm{Us}_V = 2.79 \times 10^{-4}$ cal/cm³ s K, and $k = 1.44 \times 10^7 \exp(-20,000/R_g I)$ l/mol s.

Solution.   The dimensionless balance equations are:

$$\frac{d\hat{c}}{d\hat{t}} = \frac{1 - \hat{c}}{b\mathrm{Da}} - \frac{\hat{r}}{b}$$

$$\frac{d\theta}{d\hat{t}} = \hat{r} - \mathbf{a}\theta$$

where $\hat{r} = \hat{c}^2 \exp \theta/(1 + \varepsilon\theta)$, and at steady state, $1 - \hat{c}_s = \mathbf{a}\mathrm{Da}\theta_s$. We define displacement variables $x_1 \equiv \hat{c} - \hat{c}_s$, $x_2 \equiv \theta - \theta_s$, and expand $\hat{r}$ as follows:

$$\hat{r} - \hat{r}_s \cong 2\hat{c}_s(\hat{c} - \hat{c}_s)\exp\frac{\theta_s}{1 + \varepsilon\theta_s} + \frac{\hat{c}_s^2(\theta - \theta_s)}{(1 + \varepsilon\theta_s)^2}\exp\frac{\theta_s}{1 + \varepsilon\theta_s}$$

Thus

$$\frac{dx_1}{d\hat{t}} = F_{11}x_1 + F_{12}x_2$$

$$\frac{dx_2}{d\hat{t}} = F_{21}x_1 + F_{22}x_2$$

where

$$F_{11} = \frac{1}{b\mathrm{Da}} - \frac{2\hat{c}_s}{b}\exp\frac{\theta_s}{1 + \varepsilon\theta_s}$$

$$F_{12} = -\left[\frac{\hat{c}_s^2}{b(1 + \varepsilon\theta_s)^2}\right]\exp\frac{\theta_s}{1 + \varepsilon\theta_s}$$

$$F_{21} = 2\hat{c}_s\exp\frac{\theta_s}{(1 + \varepsilon\theta_s)}$$

$$F_{22} = \left[\frac{\hat{c}_s^2}{(1 + \varepsilon\theta_s)^2}\right]\exp\frac{\theta_s}{1 + \varepsilon\theta_s} - a$$

Furthermore, since $T_r = T_0 = T_R$, we obtain for heat removal

$$\Lambda_V = \left(\lambda_V^{-1} + \lambda_R^{-1}\right)^{-1}$$

Also, $\lambda_r = c_0/r_r = 1/k_{ap}c_0 = 1134$ s, so $\lambda_V = \mathrm{Da}\lambda_r = 13041$ s, and $\lambda_R = \rho c_p/Us_V = 1434$ s.

Thus, $\Lambda_V = 1293$ s. For heat generation,

$$\lambda_{ad} = \rho c_p T_r/kc_0^2(-H_r)\hat{E} = 55.3 \text{ s}$$

Therefore,

$$a = \frac{\lambda_{ad}}{\Lambda_R} = 0.0428$$

and

$$b = \frac{\lambda_r}{\lambda_{ad}} = 20.5$$

The steady-state variables are $\hat{c}_s = 1 - 0.9 = 0.1$ and $\theta_s = \hat{E}(T_s - T_0)/T_0 = 1.83$. To check our steady state, we compute $a\mathrm{Da}\theta_s = 0.9$, which is obviously equal to $1 - \hat{c}_s$ and is therefore consistent with our steady-state balance equation. To check our multiplicity criterion, we observe that $a\mathrm{Da} = 0.492 > 0.2$, which is consistent with the existence of a single steady state. Finally,

$$F_{11} = -0.0495 \qquad\qquad F_{12} = -7.67 \times 10^{-4}$$

$$F_{21} = 1.10 \qquad\qquad F_{22} = 5.02 \times 10^{-3}$$

so that

$$-\mathrm{Tr}\,\mathsf{F} = 0.0445 > 0$$

and

$$\mathrm{Det}\,\mathsf{F} = 4.96 \times 10^{-4} > 0$$

from which we conclude that our steady state is stable.

REFERENCES

1. F. S. Dainton and K. J. Ivin, *Quart. Rev. (London)*, **12**, 61 (1955).
2. A. V. Tobolsky and A. Eisenberg, *J. Colloid Sci.*, **17**, 49 (1962).
3. A. V. Tobolsky and A. Eisenberg, *J. Am. Chem. Soc.*, **81**, 2302 (1957).
4. J. C. Yang, Ph.D. thesis in chemical engineering, Stevens Institute of Technology, Hoboken, NJ (1978).
5. J. S. Shastry, L. T. Fan, and L. E. Erickson, *J. Appl. Polym. Sci.*, **17**, 3101 (1973).
6. M. E. Sacks, S-I. Lee, and J. A. Biesenberger, *Chem. Eng. Sci.*, **27**, 2281 (1972).
7. W. H. Ray, *Can. J. Chem. Eng.*, **45**, 356 (1967).
8. M. Denn, *Optimization by Variational Methods*, McGraw-Hill, New York (1969).
9. K. G. Denbigh, *Trans. Faraday Soc.*, **43**, 648 (1947); *J. Appl. Chem.*, **1**, 227 (1951).
10. Z. Tadmor and J. A. Biesenberger, *Ind. Eng. Chem. Fund.*, **5**, 336 (1966).
11. Y. Yoshimoto, H. Yanagawa, T. Suzuki, Y. Inaba, and T. Araki, *Kagaku Kogaku*, **32**, 595 (1968).
12. M. E. Sacks, S-I. Lee, and J. A. Biesenberger, *Chem. Eng. Sci.*, **28**, 241 (1973).
13. S-A. Chen and W-F. Jeng, *Chem. Eng. Sci.*, **33**, 735 (1978).
14. Y. D. Kwon, L. B. Evans, and J. J. Noble, First International Symposium on Chem. React. Eng., Washington, D.C. (1970).
15. J. Hicks, A. Mohan, and W. H. Ray, *Can. J. Chem. Eng.*, **47**, 598 (1969).
16. G. Wu, L. A. Denton, and R. L. Laurence, *Polym. Eng. Sci.*, **22**, 1 (1982).
17. A. W. Hue and A. E. Hamielec, *J. Appl. Polym. Sci.*, **16**, 749 (1972).
18. M. E. Sacks, Ph.D. thesis in chemical engineering, Stevens Institute of Technology, Hoboken, NJ (1972).
19. M. Tirrell and K. Gromley, *Chem. Eng. Sci.*, **36**, 367 (1981).
20. D. A. Frank-Kamentskii, *Diffusion and Heat Exchange in Chemical Kinetics*
21. C. R. Barkelew, *Chem. Eng. Prog. Symp. Ser.*, No. 25, **55**, 37 (1959).
22. J. A. Biesenberger and R. Capinpin, *Polym. Eng. Sci.*, **14**, 737 (1974).
23. J. A. Biesenberger, R. Capinpin, and J. C. Yang, *Polym. Eng. Sci.*, **16**, 101 (1976).
24. D. H. Sebastian and J. A. Biesenberger, *Polym. Eng. Sci.*, **16**, 117 (1976).
25. D. H. Sebastian and J. A. Biesenberger, *Polym. Eng. Sci.*, **19**, 190 (1979).
26. J. A. Biesenberger, R. Capinpin, and D. Sebastian, *Appl. Polym. Symp.*, **26**, 211 (1975).
27. D. H. Sebastian and J. A. Biesenberger, *J. Appl. Polym. Sci.*, **23**, 661 (1979).
28. D. H. Sebastian and J. A. Biesenberger, Chapter 15 in *Chemical Reaction Engineering—Houston*, ACS Symposium Series 65, American Chemical Society, Washington, D.C. (1978).
29. J. A. Biesenberger, *Polymerization Reactors and Processes*, ACS Symposium Series 104, American Chemical Society, Akron (1979).
30. O. K. Rice, A. O. Allen, and H. C. Campbell, *J. Am. Chem. Soc.*, **57**, 2212 (1935).
31. P. Gray and M. J. Harper, *Trans. Faraday Soc.*, **55**, 581 (1959).
32. W. Squire, *Combust. Flame*, **7**, 1 (1963).

33. J. Adler and J. W. Enig, *Combust. Flame*, **8**, 97 (1967).

34. O. Bilous and N. Amundson, *AIChE J.*, **2**, 117 (1956).

35. E. B. Richter and C. W. Macasko, *Polym. Eng. Sci.*, **18**, 1012 (1978).

36. H. Miyama and S. Fujimoto, *J. Polym. Sci.* **54**, 532 (1961).

37. M. Suzuki, H. Miyama, and S. Fujimoto, *J. Polym. Sci.* **31**, 212 (1958).

38. G. Goldfinger and M. Steidlitz, *J. Polym. Sci.*, **3**, 786 (1948).

39. R. M. Joshi and S. L. Kapur, *J. Sci. Ind. Res.* (*Hardwar, India*), **16B**, 379 (1957).

40. R. H. Wiley and E. E. Sale, *J. Polym. Sci.*, **42**, 479 (1960).

41. H. K. Johnston and A. Ruding, *Polym. Prepr. Am. Chem. Soc. Div. Polym. Chem.*, **11** (1970).

42. J. A. Biesenberger, *Polymer Reaction Engineering*, Symposium at Fifty-fifth Chemical Conference and Exhibition, Chemical Institute of Canada, Quebec (1972).

43. D. H. Sebastian and J. A. Biesenberger, *J. Macromol. Sci. Chem.*, **A15**, 553 (1981).

44. D. H. Sebastian, Ph.D. thesis in chemical engineering, Stevens Institute of Technology, Hoboken, NJ (1977).

45. D. H. Sebastian, Master's thesis in chemical engineering, Stevens Institute of Technology, Hoboken, NJ (1975).

46. S. Kovenklioglu, *Polym. Eng. Sci.*, **20(12)**, 816 (1980).

47. K. H. Lee and J. P. Marano, *Polymerization Reactors and Processes*, ACS Symposium Series 104, American Chemical Society, Washington, D.C. (1979), p. 221.

48. L. N. Valsamis, Ph.D. thesis in chemical engineering, Stevens Institute of Technology, Hoboken, NJ (1977).

49. P. Hugo and H. P. Wirges, Chapter 41, *Chemical Reaction Engineering-Houston*, ACS Symposium Series 65, American Chemical Society, Washington, D.C. (1978).

50. O. Bilous and N. Amundson, *AIChE J.*, **1**, 513 (1955).

51. R. Jai Singhani and H. Ray, *Chem. Eng. Sci.*, **32**, 811 (1977).

52. R. P. Goldstein and N. Amundson, *Chem. Eng. Sci.*, **20**, 477 (1965).

53. P. J. Hoftyzer and Th.N. Zwietering, *Chem. Eng. Sci.*, **14**, 241 (1961).

54. C. Van Hearden, *Ind. Eng. Chem.*, **45**, 1242 (1953).

55. S. Liu and N. Amundson, *Ber. Bunsenges. Phys. Chem.*, **65**, 276 (1967).

56. A. Uppal, H. Ray, and A. B. Poore, *Chem. Eng. Sci.*, **26**, 967 (1974).

57. N. R. Amundson, *Mathematical Methods in Chemical Engineering*, Prentice-Hall, Englewood Cliffs, NJ (1966).

58. J. P. A. Wallis, R. A. Ritter, and H. Andre, *A.I.Ch.E. J.*, **21**, 686 (1975).

59. G. Lowry, ACS Meeting, Houston (1970).

60. R. Capinpin, Ph.D. thesis in chemical engineering, Stevens Institute of Technology, Hoboken, NJ (1976).

CHAPTER FIVE

Flow Phenomena

In Chapter 3, we examined the effects of mixing on polymerizations and polymer properties. Various states of mixing were represented with model reactors. No attempt was made to simulate the concomitant flow patterns or to describe the pumping (conveying) characteristics of the fluids involved. Such flow characteristics constitute the objectives of this chapter.

The root cause of the difficulties experienced when attempting to mix or pump polymerizing fluids is, of course, the rapidly rising viscosities that accompany reaction, especially in bulk polymerizations. To simulate the associated flow phenomena, it is necessary to deal with the coupling between reaction extent and viscosity. Reaction viscosity depends on the concentration of polymer product in solution and its molecular weight and the temperature of the reaction medium. It rises with increasing concentration and molecular weight and decreases with rising temperature. The magnitude of its sensitivity to these variables, however, seems to decrease in the order listed; that is, viscosity appears to be much more sensitive to concentration and molecular weight than to temperature, in general.

Polymer concentration must increase with extent of reaction when the product is soluble, as should temperature, owing to rapid reaction exotherm and poor heat transfer (Chapter 4). Thus we can expect rising reaction viscosities that are mitigated somewhat by rising temperatures. Molecular weight, on the other hand, could increase or decrease with reaction, as we learned in Chapter 2, depending on the polymerization type (Section 1.6). Consequently, the shapes of viscosity growth profiles can be varied and diverse.

Reaction viscosity profiles can profoundly affect design strategies for polymerization processes. Decisions concerning the pumping mode (external versus internal pressurization) and the mixing mode (agitation versus shear) to be used are critically dependent on reaction viscosity. Perhaps the most vivid examples of the importance of this dependence are found in reactive processing.

5.1. REACTIVE PROCESSING

In the manufacture of products made from synthetic polymers (plastics, rubbers, fibers, films, etc.) at least two major operations are involved: reaction and processing. Traditionally, they have been separate and distinct. Polymerization reactors structure monomer molecules into polymer molecules, and polymer processing equipment structures polymer molecules into shaped products.

Reactive processing attempts to combine these operations. One may define it more specifically as processing in the presence of chain-building reactions. The ultimate goal is to introduce monomer and catalyst at one end of the process and withdraw molded products at the other end.

Historically, equipment design for polymer processing lay predominantly in the domain of the mechanical engineer, whereas material considerations were shared by the material scientist and the chemical engineer. Polymerization engineering, on the other hand, appears to be the exclusive property of the latter. Reactive processing is an activity that spans these disciplines, and one in which their respective practitioners can learn much from one another (1).

Two examples, which may be viewed as representing extreme ends of the reactive processing spectrum are reactive injection molding (RIM) and reactive extrusion (REX). The first takes advantage of the low viscosities of monomers and pre-polymers and uses conventional mixing (e.g., impingement) and pumping (external pressurization) modes prior to polymerizing. The second takes advantage of the high viscosity of advanced polymerizing media and utilizes drag flow (internal pressurization) to convey it. The drag-flow principle is still somewhat novel to the chemical engineer.

Earlier we proposed a new category of reactor, the continuous drag-flow reactor (CDFR), which would parallel the traditional continuous stirred tank reactor (CSTR). Instead of mixing by agitation, it would be accomplished by the combined action of shear (laminar mixing) and backflow (pressure flow). These ideas were discussed briefly and illustrated in Section 1.2. Flow aspects of specific streamline reactors will be examined in greater depth in Sections 5.3 (pressure flow) and 5.4 (drag flow).

5.2. REACTION VISCOSITY

The design of conventional (nonreactive) polymer processing equipment is complicated by the non-Newtonian nature (shear-rate dependence) of polymer melt viscosity. When attempting to design equipment to process reacting fluids, one is faced with an even more formidable task: accounting for viscosities that change with conversion, temperature, and DP nonuniformities within the equipment. As previously mentioned, the viscosity of a reaction that takes a monomeric feed to a polymer melt product may change five to six decades in magnitude. Although a great deal of effort has been spent in describing flow behavior of dilute polymer solutions (intrinsic viscosity) as well as that of polymer melts, relatively little attention has been paid to the broad range of polymer concentrations between these extremes. Nonetheless, it is this region that is of importance in describing reacting mixtures. Coupled with the temperature gradients created by highly exothermic reactions, drastic spatial variations in viscosity may be expected for reacting mixtures. An additional complication arises when one considers the contrasting DP growth characteristics of random, step-addition, and chain-addition polymerizations and copolymerizations (Chapter 2).

5.2.1. *Without Reaction*

In an attempt to introduce order to the complex dependence of viscosity on the key system variables, composition, DP, temperature, and shear rate it is reasonable to use weight fraction of polymer, w_p, and weight-average DP, $\bar{x}_W$, to represent the first two variables, and to separate shear rate from the others in the generally accepted way, that is,

$$\eta\left(w_p, \bar{x}_W, T, \dot{\gamma}\right) = f\left(w_p, \bar{x}_W, T\right)g(\dot{\gamma}) \tag{5.2-1}$$

where function f is commonly referred to as the zero-shear viscosity η_0. The separate function $g(\dot{\gamma})$ represents any one of a number of relations that can be used to describe generalized Newtonian fluids or GNF's (2) such as the power-law model or the Bird–Carreau model (3). We shall see (Example 5.2-1) that for shear rates encountered by reacting solutions, the shear-rate dependence of η for solutions is not very significant compared to the composition effects. Therefore, we concentrate on the variation of η_0 in response to changes in conversion, DP, and temperature.

The variables in the function of $\eta_0 = f(w_p, \bar{x}_W, T)$ may be related to our kinetic rate equations via the constraint equations in Section 1.7. Thus, for addition polymerizations, we may write

$$\eta_0 = f'\left(\Phi, \mu_c^2, T\right) \tag{5.2-2}$$

It is noteworthy that for condensation polymerizations in bulk w_p is always unity, formally, because monomer is counted as polymer (1-mer), and $\bar{x}_W$ is a function only of the conversion of functional groups, Φ. Consequently, $\eta_0 = f'(\Phi, T)$ for such systems.

For the time being, however, we shall leave our viscosity function in terms of the variables w_p and $\bar{x}_W$, since they are indigenous to the rheological investigations we must draw on to formulate a suitable model for simulating reaction viscosities. Ideally, we would like to simplify function f even further by seeking totally separable functionalities of composition, DP, and temperature

$$\eta_0 = K(T)A(w_p)B(\bar{x}_W) \tag{5.2-3}$$

We consider temperature dependence first. The temperature dependence of η_0 for polymer melts depends on the system's proximity to the glass transition temperature of the polymer. In the vicinity of T_g (to as high as $T_g + 100°C$) a William–Landel–Ferry (WLF) type relation is often used (4)

$$\log \frac{\eta(T)}{\eta(T_g)} = \frac{C_1(T - T_g)}{C_2 + C_1(T - T_g)} \tag{5.2-4}$$

However, for temperatures well above T_g, an Arrhenius relation with temperature-independent activation energy, as expected from Eyring's rate theory (5), seems to be adequate.

$$\frac{\ln \eta(T)}{\eta(T_r)} = \frac{E_v}{R_g}\left(\frac{1}{T} - \frac{1}{T_r}\right) \tag{5.2-5}$$

The two expressions overlap in range of applicability, as shown in Fig. 5.2-1 for polystyrene melt. One must bear in mind, however, that the reaction mixture is actually a solution of polymer in monomer. The quantity of importance is T_g of the solution, which is considerably less than T_g of the polymer melt. This value can be measured experimentally (6) or calculated via free-volume theory (7). Experimental results for styrene in ethylbenzene were shown in Fig. 4.1-6. It is evident that only at high-polymer-weight fractions does the T_g of the solution approach that of the polymer melt. Realistic operating conditions would generally place the reaction temperature well above the range of the WLF relation. Consequently, the functional relationship of temperature will probably most often conform to an Arrhenius equation.

$$K(T) = K_0 \exp\left(\frac{E_v}{R_g T}\right) \tag{5.2-6}$$

The interaction between the effects of molecular-weight and polymer-weight fraction on viscosity is more difficult to analyze. For polymer melts, the DP dependence has been fairly well established (8)

$$\eta_0 \propto \bar{x}^\beta \tag{5.2-7}$$

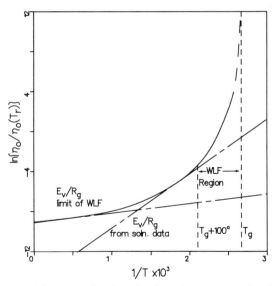

Figure 5.2-1. Viscosity shift factor as a function of reciprocal temperature, showing regions of WLF and simple fluid behavior.

The exponent β may be 3.4 or 1 depending on whether the polymer is above or below the critical molecular weight for entanglement. When the polymer is not monodisperse, best results are obtained by letting $x = \bar{x}_W$. Onogi et al. (9, 10) proposed modifying function 5.2-7 to account for composition and to facilitate conformity to relation 5.2-3 in the following way:

$$\eta_0 = K(T)w_p^\alpha \bar{x}_W^\beta \tag{5.2-8}$$

The value of α was found to be 4.7 and β either 1 or 3.4, as described earlier. The authors chose a WLF type expression for $K(T)$, and used the expression to characterize mixtures of polystyrene in toluene. The correlation was only applied to solutions of viscosities of 3 poise or greater. Values of $\alpha = 4.7$ and $\beta = 3.4$ were reported, indicating that the critical DP for entanglement had been exceeded under these conditions.

Note that equation 5.2-8 may be regrouped as

$$\eta_0 = K(T)\left(w_p \bar{x}_W^{0.72}\right)^{4.7} \tag{5.2-9}$$

This is very close to the functional dependence predicted by the hydrodynamic model of Williams (11) which suggests that η_0 is a function of $w_p \bar{x}_W^{0.625}$. Solution data (12) for polyisobutylenes, spanning a decade in molecular weight and dissolved in toluene, conform to a similar dependence, $w_p \bar{x}_W^{0.68}$, when the product is greater than 100. One might suspect that the deviation for lower values of $\bar{x}_W$ can be attributed to the lack of entanglements.

Lynn and Huff (13) presented an empirical correlation for anionic butadiene polymerization. Although it represented something of a curve-fit to existing data, it is interesting to note that polymer-weight fraction and molecular weight appear in the numerator coupled in a manner similar to that suggested by Spencer and Williams.

$$\eta = \frac{\eta_s \exp\left(8.56 \times 10^{-3} w_p \overline{M}_W^{0.76}\right)}{\exp\left[\dfrac{w_p}{\rho^{0.46}}\left(1.219 + 3.42 \times 10^{-5} \overline{M}_W^{0.76} - .0013T\right)\right]} \tag{5.2-10}$$

where η_s is the solvent viscosity and T is in degrees Celsius. Polymers formed by step-addition often differ in structure from those obtained from the same monomer by a free-radical mechanism. Nevertheless, the technique to be described below for formulating a viscosity model should apply equally to either mechanism.

A relationship like equation 5.2-9 indicates that a plot of $\log \eta$ versus $\log w_p$ for fixed $\bar{x}_W$ should be linear. Although Onogi's data follow this trend, collected viscosity data for PS in various solvents, shown in Fig. 5.2-2 cannot be so generalized. We focus on some data for PS in ethyl benzene (14) replotted to emphasize the temperature dependence. We see in Fig. 5.2-3 that at each level of polymer-weight fraction, the activation energy appears to be temperature-independent, yet the value is clearly different for each value of composition. The dependence can be fitted to an exponential relation, $E_v = 2300 \exp(2.4 w_p)$. As $w_p \to 0$, the flow activation energy is

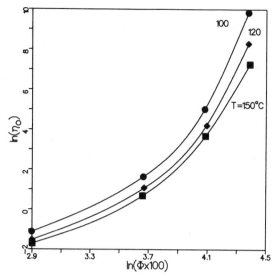

Figure 5.2-2. Zero-shear viscosity of polystyrene in solution as a function of polymer-weight fraction with temperature as parameter.

that of monomer, while as $w_p \to 1$ a value of 25 kcal is approached. The latter is close to values reported for polystyrene melt. The data of Spencer and Williams for PS in isopropyl benzene (15) also fit the preceding relationship, as shown in the figure. If at any temperature and composition, the viscosity is normalized by the exponential of $E_v(w_p)/R_gT$, then the remainder should be a function of w_p and $\bar{x}_W$ only. Since only one value of $\bar{x}_W$ is used in a given set of experiments, one can extract the composition dependence from this remainder.

$$\frac{\eta}{\exp\left[\dfrac{E_v(w_p)}{R_gT}\right]} = K_0 A(w_p)\bar{x}_W^{\beta} \qquad (5.2\text{-}11)$$

The value of β (1 or 3.4) is determined by comparing the weight-average DP to the critical value for entanglement. For polystyrene, it has been found that (16):

$$\left(\bar{M}_W\right)_{cr} = \frac{40,000}{w_p} \qquad (5.2\text{-}12)$$

Clearly as polymer-weight fraction increases, the critical value for entanglement decreases, causing β to switch from 1 to 3.4 at a lower value of $\bar{M}_W$.

For a polymer of $\bar{M}_W = 215,000$, $\bar{M}_W > (\bar{M}_W)_{cr}$ provided that $w_p > 0.20$. Since all data lie in that region, we may factor $\bar{x}_W^{3.4}$ from the viscosity expressions. Thus, if the molecular-weight and weight fraction functionalities are separable:

$$K' \equiv \frac{\eta}{\bar{x}_W^{\beta}\exp\left[E_v\left(\dfrac{w_p}{R_gT}\right)\right]} = K_0 A(w_p) \qquad (5.2\text{-}13)$$

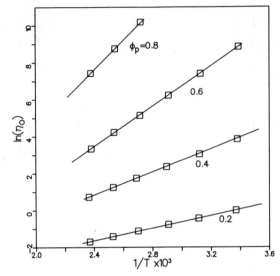

Figure 5.2-3. Temperature dependence of zero-shear viscosity of polystyrene in solution with polymer-weight fraction as parameter.

If the molecular weight of the polymer is held constant, the RHS of equation 5.2-13 is a function of w_p only and should be obtainable from experimental data once the activation energy relationship has been established. If the simple Onogi relationship (5.2-8) holds, a plot of the LHS versus the RHS on log-log coordinates should be linear with a slope of α. Figure 5.2-4 shows such a plot for polystyrene in ethyl benzene, and clearly the relationship is nonlinear. By way of reference, lines of slope 1 and 4.7 are also shown on the figure since these are the values of α suggested for

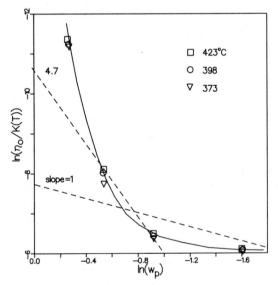

Figure 5.2-4. Polymer-weight fraction dependence of zero-shear viscosity normalized with respect to temperature dependence.

low and high concentration regimes. The dependency in Fig. 5.2-4 would seem to be weaker at low concentrations ($\alpha < 1$) and stronger at high concentrations ($\alpha > 4.7$). Without a fundamental model to provide further guidance, one must resort to a curve fit of this "remnant" portion of the viscosity function. Although any functional form could conceivably be chosen, we have found the following to provide a good fit to data over a wide range of polymer-weight fraction:

$$K' = K_0 w_p^{4.7} \exp\left(\sum_{i=1}^{N} a_i \ln^i w_p \right) \tag{5.2-14}$$

$$K' = K_0 w_p^{4.7} \exp\left(\sum_{i=1}^{N} a_i w_p^i \right) \overline{M} \tag{5.2-15}$$

$$K' = K_0 \exp\left(\sum_{i=1}^{N} a_i \ln^i w_p \right) \tag{5.2-16}$$

$$K' = K_0 \exp\left(\sum_{i=1}^{N} a_i w_p^i \right) \tag{5.2-17}$$

A relationship of form 5.2-14 was used by Valsamis (17), although he used a different functional form than presented here to describe the dependence of E_v on w_p. The only practical objection to 5.2-14 and 5.2-15 is that as $w_p \to 0$ K' and thus η must also go to zero. To be used in a simulation, these functions must be set to some constant value (i.e., "switched off") at some predetermined value of w_p to avoid viscosities of zero. For instance, one might not choose to use the correlation unless the computed viscosity exceeds that of monomer, or alternatively, exceeds a threshold value of several centipose. The motivation for a form like 5.2-14 and 5.2-15 is that it preserves the relationship suggested by Onogi, modified by a correcting factor (the terms in the exponential). However, given the fact that even Onogi suggests that the 4.7 exponent doesn't apply at low polymer concentrations, one might as readily opt for a form like that shown in equations 5.2-16 or 5.2-17. The various alternatives are all presented in Table 5.2-1 as calculated for the Dow data.

Valsamis (17) used a slightly different approach. He, too, found a weight-fraction-dependent form of the activation energy for flow. He then chose a single temperature and molecular weight (373 K, 238,000) and curve-fit the data of Spencer and Williams (15) for polystyrene in isopropyl-benzene to find the composition-dependence. This reference viscosity is determined from:

$$\eta_r = -5.345 + 0.778 \ln w_p - 0.814 \ln^2 w_p + 0.026 \ln^3 w_p \tag{5.2-18}$$

Viscosity at other conditions is them computed relative to the reference case via:

$$\eta = \eta_r \exp\left[-\frac{E_v(w_p)}{R_g T_r} \right] \exp\left[\frac{E_v(w_p)}{R_g T} \right] \overline{M}_W^\beta f(w_p) \tag{5.2-19}$$

The functional form of $f(w_p)$ is like that in equation 5.2-14, and the parameters are given in Table 5.2-1.

Table 5.2-1. Viscosity Correlations

$$\text{GENERAL FORM:} \quad \eta = A \exp(E_v/R_g T) f(w_p) \overline{M}_W^{\beta}$$

I. *POLYSTYRENE*

    a. $E_v = 2300 \exp(2.4 \, w_p)$

$$\beta = \begin{cases} 1 & \overline{M}_W < 40{,}000/w_p \\ 3.4 & \overline{M}_W > 40{,}000/w_p \end{cases}$$

$$Af(w_p) = \begin{cases} 2.51 \times 10^{-21} \exp(-6.76 \, w_p + 52.7 \, w_p^2 - 61.6 \, w_p^3 + 24.8 \, w_p^4) \\ 4.352 \times 10^{-26} \exp(-28.9 \, \ell n w_p - 30.2 \, \ell n^2 w_p - 13.9 \, \ell n^3 w_p - 2.3 \ell n^4 w_p \\ 1.7 \times 10^{-14} \, w_p^{4.7} \exp(.67 \, w_p + 158 \, w_p^2 - 199 \, w_p^3 + 83 \, w_p^4 \\ 7.41 \times 10^{-2} \, w_p^{4.7} \exp(-1.52 \, \ell n w_p - 5.06 \, \ell n^2 w_p - 2.08 \, \ell n^3 w_p - 0.027 \, \ell n^4 w_p) \end{cases}$$

    b.   (From ref. 17)

$$E_v = \begin{cases} 2250 & 0 < w_p < 0.06 \\ 0.15 + 42.9 \, w_p + 140 \, w_p^2 + 170 \, w_p^3 & 0.06 < w_p < 0.75 \\ 23000 & 0.75 < w_p < 1.0 \end{cases}$$

$$\beta = \begin{cases} 1 & \overline{M}_W < 40{,}000/w_p \\ 3.4 & \overline{M}_W > 40{,}000/w_p \end{cases}$$

$$Af(w_p) = 1.769 \times 10^{-11} \, w_p^{4.526} \exp(5.886 \, \ell n w_p + 2.455 \, \ell n^2 w_p + 0.269 \, \ell n^3 w_p)$$

II. *POLYURETHANE*

    a.   (From ref. 18)

$$E_v = -23{,}490 + 5006 \, \ell n \overline{M}_W$$

$$\beta = -5.3176$$

$$Af(w_p) = 1.266 \times 10^8$$

    b.   (From ref. 19)

$$E_v = \begin{cases} -61{,}353 + 11{,}039 \, \ell n \overline{M}_W & \overline{M}_W < 1500 \\ 17{,}388 & \overline{M}_W \geq 1500 \end{cases}$$

$$\beta = \begin{cases} -10.51 & \overline{M}_W < 1500 \\ 3.44 & \overline{M}_W \leq 1500 \end{cases}$$

$$Af(w_p) = \begin{cases} 2.93 \times 10^{22} & \overline{M}_W > 1500 \\ 3.427 \times 10^{-12} & \overline{M}_W \leq 1500 \end{cases}$$

Viscosity correlations for random polymers may not include the weight fraction of polymer as a variable. Unlike addition polymers, where one may consider the polymer to be in a solution of monomer, random addition monomers are considered to be polymer of DP = 1 and thus such systems are not treated as solvent/solute systems. The values of $\overline{M}_N$ and $\overline{M}_W$ reflect the fraction of monomer present. If the reaction is conducted in bulk and no side products are formed, the viscosity should correlate solely to the temperature and average molecular weight. This is the approach adopted by Macosko (18, 19) in describing the viscosity of urethane network polymerizations used in reactive injection molding (RIM) processes. He proposed relations of the form

$$\eta = K_0 \exp\left(\frac{E_{v_1}}{RT}\right)\left[\frac{\overline{x}_W}{(\overline{x}_W)_0}\right]^{\frac{E_{v_2}}{R_gT}-s} \tag{5.2-20}$$

Relationship 5.2-20 can be recast in a form wholly compatible with 5.2-2. We can rewrite it as

$$\eta = K_0'\overline{x}_W^\beta \exp\left[\frac{E_v(\overline{x}_W)}{R_gT}\right] \tag{5.2-21}$$

where

$$K_0' = \frac{K_0}{(\overline{x}_W)_0^s} \tag{5.2-22}$$

$$\beta = s \tag{5.2-23}$$

and

$$E_v(\overline{x}_W) = E_{v_1} + E_{v_2}\ln\left[\frac{\overline{x}_W}{(\overline{x}_W)_0}\right] \tag{5.2-24}$$

Some values of the constants reported by the authors are presented in Table 5.2-1. Note that in one case the flow activation energy becomes molecular weight-independent for large values of $\overline{x}_W$. In that case, the molecular weight exponent, $\beta = 3.44$, is close to the expected value for entangled polymers. If we return to the form of the equation in 5.2-20, we might interpret the exponent on $\overline{x}_W$ as a temperature-dependent β. For the systems tested by Macosko, thus interpreted, β ranged from 2.1 to 2.6 for two systems and from 3.4 to 6.7 for a third. The exponent decreased within these bounds as temperature increased.

Macosko arrived at these relationships by fitting experimental data for viscosity and conversion taken in independent measurements. The value of molecular weight was computed from functional group conversion and was not directly measured. Figure 5.2-5 shows an example of the viscosity molecular weight relation thus developed. To verify that the viscosity measurements were taken in the Newtonian plateau region, tests were conducted over a range of shear rates. Figure 5.2-6 shows the result and leads one to conclude that shear-rate dependence in a polymer solution is not highly significant, even at relatively advanced stages of reaction.

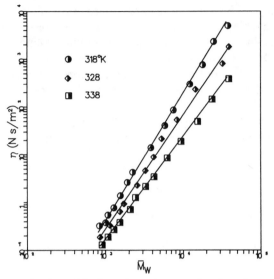

Figure 5.2-5. Viscosity-molecular-weight dependence for network polyurethanes with temperature as a parameter (18).

Because a broad spectrum of shear rates are present in reactive processing and because the solutions are on the borderline of shear dependence at best, a simple power-law model is inappropriate. The power-law, or Oste-de-Wale, model does not plateau at low shear rates and thus grossly overestimates the viscosity in this region. One model that attempts to fit this transitional region is the Bird-Carreau model (2, 3). At low shear rates, η is Newtonian, while at higher rates its behavior conforms to the power-law model. Domine and Gogos used this model (20) in simulating RIM

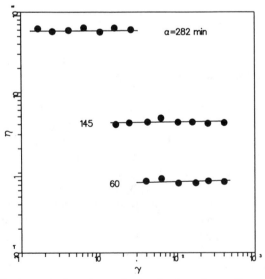

Figure 5.2-6. Viscosity-shear-rate dependence of network polyurethane with reaction time (thus conversion) as parameter (18).

of urethanes. The functional form of dependence on shear rate γ is given by:

$$\eta = \eta_0\left(1 + (\lambda\dot{\gamma})^2\right)^{(n-1)/2} \tag{5.2-25}$$

where λ is a relaxation time constant that determines the shear rate at which non-Newtonian behavior begins and n is the power-law constant. The authors used a relationship like equation 5.2-21 for η_0. The activation energy was greater than the critical entanglement weight (and of course $\beta = 3.4$). When molecular weight was less than the critical value, $\beta = 1$ and $E_v(w_p)$ in equation 5.2-21 is given by:

$$E_v = E_{v_f}\left(\frac{E_{v_p}}{E_{v_f}}\right)^{[(x_{cr}+\bar{x}_{N_f})/(x_{cr}-\bar{x}_{N_f})](\bar{x}_N)_f(1/\bar{x}_{N_f}-1/\bar{x}_N)} \tag{5.2-26}$$

where E_{v_f} and E_{v_p} are flow activation energies for the feed and pure polymer.

5.2.2. With Reaction

Although one could synthesize polymer/monomer solutions with various values of w_p and $\bar{x}_W$ and measure viscosity over a broad range of temperatures, this would not give a picture of how viscosity changes with reaction. Recall from Chapter 2 that each model mechanism (random, step-addition, and chain-addition) had unique chain-growth and conversion characteristics. Therefore, we must expect the viscosity history of a step reaction to be quite different from that of a chain reaction. The DP of the former will either wait until complete conversion to grow to a large value or grow steadily throughout reaction, whereas the latter will attain its largest value very early and may subsequently even decline (Section 2).

To examine the viscosity growth pattern of our model mechanisms, we shall adopt the viscosity functions of Macosko (18) for random (urethane) polymers, while our own correlation given in Section 5.2-1 will be used for addition polymers. Note that Macosko's work was with network polymers, but we use his results for conversions below the gel point to describe linear condensations as well. Both step- and chain-addition reactions may be regarded as producing polymer in a solution of monomer, so there is no fundamental reason why the same viscosity relationship cannot be applied to each. For example, ionically initiated polystyrene of $\bar{x}_N = 1000$ in a 40-percent solution of its own monomer should have the same viscosity as that formed by free-radical initiation when $\bar{x}_N$ and w_p are the same. The potential stereospecificity of the step-growth polymer, and the influence of breadth of DPD might lead to some deviation from this assumption.

To simulate model polymerization kinetics, the more exact forms presented in Chapter 2 were set to conditions that would produce "ideal" behavior. For example, the equilibrium constant for reversible random polymerization was set to an extremely large value ($K = 10^{20}$). For step-addition, initiation and propagation rate constants were set equal ($a_K = 1$) while for chain-addition, conventional conditions were created by letting $\alpha_K = 0.01$. Note that under these assumptions the conversion rates are second-order for the random reaction and first-order for the two addition schemes.

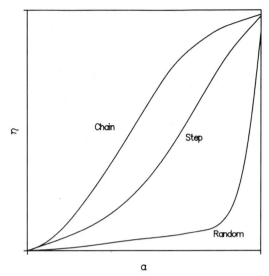

a

Figure 5.2-7.   Viscosity as a function of reaction time for model polymerizations.

The viscosity behavior shown in Fig. 5.2-7 is directly attributable to the characteristics of the individual kinetic mechanisms. Free-radical polymerizations form high-molecular-weight product immediately, and the absolute magnitude of drift in DP with conversion is relatively small. Therefore, viscosity growth is primarily due to increasing polymer-weight fraction during reaction. In random polymerizations, only oligmers are present until extremely advanced degrees of conversion. At high conversions, however, molecular weight grows suddenly and rapidly. As a result, viscosity remains quite low until very late in the reaction and then increases in almost steplike fashion. Step-addition falls between the extremes of behavior demonstrated by chain-addition and random reactions. Low molecular weights are present initially, but DP grows in proportion to conversion, so the combined effects of polymer-weight fraction and product DP lead to a rapid increase in viscosity after the initial stages of reaction.

One must be constantly mindful of reaction viscosity in mind when discussing flow reactors. Not only do bulk-polymerizing fluids increase in viscosity by up to eight decades in magnitude, but the path from monomeric viscosity to that of polymer melt is dictated by reaction kinetics. As we shall see, pressure flow requires low viscosities while drag flow needs high viscosities. Although a pressure-flow device might be suitable for a random polymerization at all but the highest conversion, the viscosity growth of a chain-addition reaction indicates a substantial viscosity even at moderate conversions, suggesting that drag flow might be more efficient. These are the kinds of considerations to bear in mind when examining the specific flow reactor schemes discussed later.

5.3.   CONTINUOUS EXTERNALLY PRESSURIZED REACTORS (CEPR'S)

When designing a continuous reactor, the traditional choices able to the chemical engineer are the stirred tank (CSTR) and the pressure-flow tube (CTR). From our discussions in Chapter 3 on mixing and in Chapter 4 on thermal effects, one might quickly suspect that the stirred tank would be fraught with difficulties when used as a reactor for bulk polymerization. The extremely high viscosity of polymer solutions effectively restrict us to laminar mixing, whereas stirred tanks are designed for turbulent mixing. This is exacerbated by the small surface-to-volume ratio of tanks and results in inherently poor heat transfer characteristics for this class of reactor. The tubular reactor seems to offer hope of improved heat transfer. The characteristic length for heat transfer is reduced from the order of feet (or tens of feet) for the tank to the order of inches or fractions of an inch for the tube. This virtue combined with simplicity of design explain why the tubular reactor remains a focus of continued interest in chemical reaction engineering. Despite its potential as a continuous polymerizer, however, its actual use industrially has been limited. The most notable application is the high-pressure process for low-density polyethylene (LDPE). A major drawback of the tubular polymerizer is, of course, plugging, or flow occlusion, which occurs in the slow-moving region near the walls. A number of research publications have examined the feasibility of extending existing technology on tubular polymerizers to systems other than LDPE (13, 17, 21–23).

In the design of flow reactors in general, and tubular reactors in particular, we must consider the intimate linkage of the fluid dynamics and the reaction kinetics. As discussed in the preceding section, viscosity is a function of temperature, polymer molecular weight, and polymer-weight fraction. All three parameters could be subject to large changes during the course of reaction and thus viscosity could sustain large spatial variations, both longitudinal and transverse to flow. As a result, transverse velocity profiles and longitudinal pressure gradients will deviate significantly from their forms at the reactor inlet.

Although frequent reference is made to the tubular reactor, there is no reason to restrict our analysis of pressure-flow reactors to cylindrical geometry. Flow in the gap between narrowly spaced parallel plates is another application of pressure flow, one that is sometimes used to simulate mold-filling in reaction injection molding (RIM).

5.3.1. Lumped Parameter Models

The simplest approach to modeling a flow reactor is the so-called lumped parameter technique. Velocity in the axial direction is presumed to be independent of transverse position. Such a model might be appropriate for the high-pressure LDPE process where flow of the supercritical monomer is turbulent. One could also use it to describe the flow in a single-screw extruder, if leakage over the flights is small, and thus there is minimal backmixing. Although the cross- and down-channel components of extruder velocity are frequently represented as simple parallel plate drag flow, it should be stressed that the actual composite flow pattern experienced by the fluid is more like a "helix within a helix" (Chapter 3). Recall from Section 3.2, the disparate RTD's for drag flow and for the extruder and how the latter more closely resembles that of a PFR instead.

The plug-flow model is also frequently used to approximate the behavior of streamline reactors in which transport of mass and heat occurs primarily in the direction transverse to the direction of flow. To represent transverse conduction with a lumped parameter model, we approximate the true transverse flux as follows:

$$-k\nabla^2 T = h_{\text{loc}} s_V ([T] - T_R) \tag{5.3-1}$$

where $[T]$ is the longitudinally local flow-average transverse temperature (Appendix F) and h_{loc} is the local heat transfer coefficient. Both may be functions of axial position. It can be shown (Example 5.3-1) that in the absence of chemical reaction, flow-average axial temperature profiles exactly match plug-flow profiles provided that h_{loc} used in the lumped parameter model is defined by

$$h_{\text{loc}}(x_1) = \frac{-k \dfrac{\partial T(x_1, h)}{\partial x_2}}{[T(x_1)] - T_R} \tag{5.3-2}$$

where x_1 is the flow direction, x_2 is the transverse direction, and h is the maximum (wall) value of x_2. Furthermore, h_{loc} ultimately attains a limiting or asymptotic value and, consequently, its axial dependence may be justifiably neglected. For constant flux at the wall and constant temperature at the wall the limiting values of Nusselt number (based on tube radius) are, respectively (24):

$$Nu = 2.18 \tag{5.3-3}$$

and

$$Nu = 1.83 \tag{5.3-4}$$

The reader should bear in mind, however, that these calculations are based upon nonreactive, Newtonian flow. When reaction distorts the temperature profile, the flux at the wall (numerator of equation 5.3-2) increases and h_{loc} will deviate from predictions. Nonetheless, many investigators have found the PF model to adequately describe the conversion, temperature, and average molecular properties of experimental tubular reactors (17, 21, 23, 25). The equations used to describe the PFR have already been presented (cf. Table 4.3-1). They are the same as for the BR, only α represents space time, x_1/v_1. To be more general, however, we might include transient behavior and thereby make the balances applicable to flow in both the runners and the mold during reaction injection molding, for instance. These equations are listed in Table 5.3-1.

Plug flow models have been used to characterize the tubular polymerizers used in the commercial high-pressure process for LDPE (25–27). Flow in these reactors is turbulent and thus a plug-flow model is appropriate. Han (26) suggested the use of a dispersion model (Chapter 3) to account for the axial mixing induced by the sometimes practiced technique of pulsating the feed stream to prevent deposition of product on the reactor wall. The value of Pe_L he chose to characterize axial mixing ($Pe_L = 100$) is sufficiently large, however, to diminish the difference between the dispersion model and PFR (Sections 1.5 and 3.2). Thus we can discuss the PF results together with those from the dispersion model for the same reaction.

Figures 5.3-1 and 5.3-2 present axial temperature profiles computed by Ehrlich (27) and Han (26), respectively, for LDPE in tubular polymerizers. Each of their kinetic models included a variety of branching reactions, but realizing that the propagation step is primarily responsible for the evolution of heat, we can treat the reaction kinetics as a model chain-addition polymerization (Section 3.4.2). Using the authors' kinetic data, CT's and dimensionless R-A parameters for the trajectories

Table 5.3-1. Transient Plug-Flow Balance Equations

$$\frac{\partial c_k}{\partial t} = -v_1 \frac{\partial c_k}{\partial x_1} + r_k$$

$$\rho c_p \frac{\partial T}{\partial t} = -\rho c_p v_1 \frac{\partial T}{\partial x_1} + r_G - US_v(T - T_R)$$

SEMIDIMENSIONLESS FORM

$$\frac{\partial \hat{c}_k}{\partial t} = -\lambda_v^{-1} \frac{\partial \hat{c}_k}{\partial \hat{x}_1} + \lambda_k^{-1} \hat{r}_k$$

$$\frac{\partial \hat{T}}{\partial t} = -\lambda_v^{-1} \frac{\partial \hat{T}}{\partial \hat{x}_1} + \lambda_G^{-1} \hat{r}_G - \lambda_R^{-1}(\hat{T} - \hat{T}_R)$$

$$\frac{\partial \theta}{\partial t} = -\lambda_v^{-1} \frac{\partial \theta}{\partial \hat{x}_1} + \lambda_{ad}^{-1} \hat{r}_G - \lambda_R^{-1}(\theta - \theta_R)$$

DIMENSIONLESS FORM

$$\frac{\partial c_k}{\partial \hat{t}} = -A \frac{\partial \hat{c}_k}{\partial \hat{x}_1} + b_k \hat{r}_k$$

$$\frac{\partial \theta}{\partial \hat{t}} = -A \frac{\partial \theta}{\partial \hat{x}_1} + \hat{r}_G - a(\theta - \theta_R)$$

$$\hat{t} = t/\lambda_{ad}$$

$$A = \lambda_{ad}/\lambda_v$$

in Figs. 5.3-1 and 5.3-2 were computed and are listed in Table 5.3-2. We see that the BR criteria for R-A developed in Chapter 4 correctly identify the thermal behavior of Ehrlich's model. There appears to be an inconsistency, however, with the results of Han. One can estimate values for his CT's based on quasi-isothermal profiles. If we consider initiator decay to be first-order, then the initiator conversion at the exit would suggest a value of $\lambda_i \cong 370$ s. This compares favorably with the value

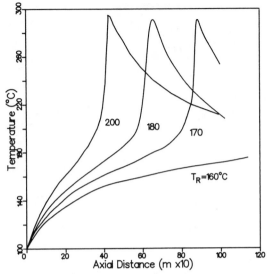

Figure 5.3-1. Simulated axial temperature profiles for high pressure LDPE tubular reactor (27).

computed from the kinetic constants, $\lambda_i = 280$ s. Approximating the monomer consumption rate by pseudo-first-order kinetics suggests that:

$$\frac{\alpha}{\lambda_m} \cong -\ln(1 - \Phi_m) \qquad (5.3\text{-}5)$$

Using the exit conversion of 0.12 when $\alpha = 165$ s gives $\lambda_m = 1300$ s, which is quite

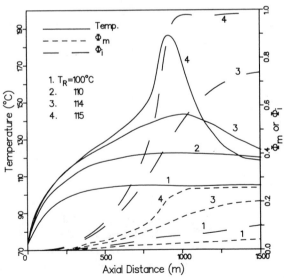

Figure 5.3-2. Simulated axial temperature and reactant conversion profiles for high pressure LDPE tubular reactor (26).

Table 5.3-2. CT's and Dimensionless R-A Groupings for High-Pressure LDPE Simulations

I. Ehrlich *et al.* (27)

	T_R = 433 $^\circ$K	443	453	473
λ_i	1511s	552	211	34.7
λ_m	2481s	1284	685	211
λ_{ad}	32.6s	17.7	9.9	3.3
λ_R	18.7s	18.7	18.7	18.7
a	1.75[*]	0.95[≠]	0.53[≠]	0.18[≠]
b_i	46	31	21	11
b_m	76	73	69	64

II. Han *et al.* (26)

	Calculated from kinetic constants	Estimated from axial profiles
λ_i	280s	370
λ_m	19370s	1290
λ_{ad}	323s	25
b_i	0.87	14.8
b_m	60	52
$a(Us_V = 8.6 \times 10^{-2}$ cal/cm^{3}s $^\circ$K)	46[*]	3.57[*]
$a(Us_V = 4.3 \times 10^{-2})$	23[≠]	1.77[≠]
$a(Us_V = 4.1 \times 10^{-2})$	21.6[≠]	1.67[≠]

[*] No R-A observed in simulation

[≠] R-A observed in simulation

different from the computed value of 19370 s. If we use the author's values of physical constants as well as the lumped activation energy, the resulting λ_{ad} and dimensionless R-A parameters based on λ_m computed from 5.3-5 and listed in the table also fall within the range expected from the BR analysis in Chapter 4. Note that the small value of b_i suggests that sensitivity should be limited. The depressed value of a_{cr} observed is in line with this supposition.

Lee and Marano (25) presented extensive characterization of the tubular LDPE process. A Barklew-type sensitivity analysis much like that used in Section 4.3-1 was conducted. In the LDPE reactor, R-A was the desired objective rather than a condition to be avoided. The concomitantly rapidly rising temperatures raised conversion levels. Subsequent D-E was not detrimental from a design viewpoint, owing to the phase separation of ethylene monomer from the reactor effluent, which facilitated recycle of the unconverted reactant. As a result, the authors defined optimum operating conditions as those that just provoke R-A. For proprietary reasons, the authors omit some key data, which precludes calculation of CT's and dimensionless groups. Qualitative agreement with our observations on R-A, however, is apparent.

Figure 5.3-3 shows parametric sensitivity curves used to define the optimum operating line, and Fig. 5.3-4 shows the molecular-weight-conversion behavior used to justify denoting this region as optimum. When initiator concentration is fixed, raising jacket temperature in a non R-A regime causes an increase in conversion from accelerated rates, but a decrease in molecular weight. Just beyond the R-A point, conversion continually decreases owing to progressively stronger D-E as jacket temperatures are raised. Thus the authors conclude that the onset of R-A gives the maximum in conversion with only a small penalty in $\overline{M}_N$. Furthermore, the value of $\overline{M}_N$ starts to drop rapidly as feed initiator concentration is increased. Although Figs. 5.3-3 and 5.3-4 represent different parametric values, comparison of the two leads one to suspect that the sharp fall-off in $\overline{M}_N$ observed when conversion approached 40 percent corresponds to the development of sensitive R-A (IG). The maximum $\overline{M}_N$ along the optimum operating line appears to occur as insensitive R-A is encountered. Again, this is consistent with our observation (Section 4.4) that R-A has a less drastic effect on polymerizations with isothermal D-E drift behavior. That is, the conditions that produce a loss of sensitivity (rapid initiator decay) also give rise to an upward drift in $\overline{M}_N$, which mitigates the downward drift produced by rising R-A temperatures. Thus we might add to the authors' conclusion about the

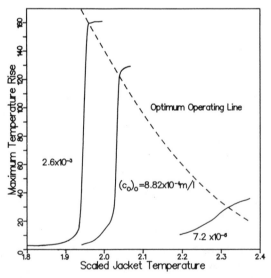

Figure 5.3-3. Maximum temperature rise as a function of normalized coolant temperature, with initiator concentration as parameter (25).

existence of an optimum operating line that there is also an optimum operating point on that line where R-A sensitivity disappears. The conditions at that point would seem to give the best combination of conversion and molecular weight and would correspond to an initiator concentration of about 20 ppm in Fig. 5.3-4.

In modeling their experimental results on continuous styrene polymerizations in tubes, both Valsamis (17) and Wallis (21) found that plug flow fit their data virtually as well as a more sophisticated distributed parameter model. In terms of predicting gross thermal behavior, conversion, and average molecular properties at the reactor exit, the simple model matched the simulation predictions of the more complicated model. Use of the more complicated distributed parameter model would only seem to be justified if details of the flow pattern and axial pressure profiles were needed.

Both Valsamis and Wallis were simulating reactors operating in the laminar-flow regime. Each used a constant value for heat transfer coefficient. Wallis used the limiting value of Nu [equation (5.3-3)] to estimate h. Note that equation 5.3-4, and not equation 5.3-3, is actually the appropriate one to use when a constant wall-temperature boundary condition is imposed, but either relationship gives the same order-of-magnitude estimate for h. Valsamis determined the limiting value of h experimentally from the cooling rates of nonreacting fluids. His experiments showed that an asymptotic value of h was reached within the first 10 percent of the reactor length. For a 1/4-in.-diameter tube the value of h was found to be 0.002 cal/cm^2 s K, which compares favorably with the value computed via equation 5.3-4 of 0.0017 cal/cm^2 s K.

Table 5.3-3 compares the experimental values and plug-flow simulations of both Wallis and Valsamis. Also included are Wallis' results from a distributed parameter model. He did not consider the variation of viscosity with conversion, temperature, or molecular weight, and only slow flow was treated. These conditions present the least challenge to the applicability of the plug-flow approximation of a distributed

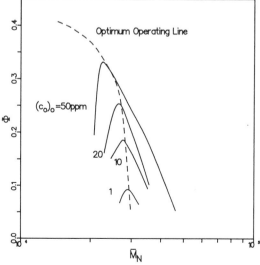

Figure 5.3-4. Change in molecular weight with conversion along lines of constant $(c_0)_0$ in Figure 5.3-3 (25).

Table 5.3-3.   Experiments and Simulations of Continuous Styrene Polymerization in Tubes

Wallis - PS/AIBN tubular reactor

Experimental				PFR		Diffusion Model	
Φ_m	$\overline{M}_N$		Φ_m	$\overline{M}_N$		Φ_m	$\overline{M}_N$
20.1	77.7×10^3		21.2	77.1×10^3		18.9	76.3×10^3
27.8	55.7		29.1	51.2		26.2	51.4
32.1	53.6		28.5	44.9		29.7	47.6
10.3	127.4		11.0	146.0		9.7	145.6
19.2	75.2		21.2	77.1		18.9	76.3

Valsamis - PS/DTBP tubular reactor

Experimental			PFR	
Φ_m	$\overline{M}_N$		Φ_m	$\overline{M}_N$
40.1	77.4×10^3		41.1	106.9×10^3
35.6	96.1		35.5	113.5
42.	92.6		40.2	96.1
58.8	53.8		69.6	51.6

parameter model. It is readily apparent from the table that plug-flow models can be used to estimate the behavior of laminar-flow polymerizers, even when the fluid resistance to heat transfer is important.

5.3.2.   Distributed Parameter Models

The lumped parameter approximation fails to provide certain information that is important to the design of flow polymerizers. In particular, estimation of the longitudinal pressure profile requires knowledge of the transverse velocity profile. Although the distributed parameter approach may also yield much extraneous information on distributed variables of various sorts, it is nonetheless necessary when a combined solution of the fluid-dynamic and reaction-kinetic equations is sought. These two sets of equations are linked explicitly through the viscosity. As discussed in Section 5.2, the viscosity is a strong function of temperature, polymer-weight fraction, and molecular weight. These factors are all subject to large changes in both the axial and transverse directions. One would suspect that such large changes could substantially alter the velocity profile, the relative importance of conduction and convection in the transport balances and the linearity of the pressure gradient.

While we emphasize the importance of viscosity variation, in the absence of a priori evidence to the contrary, it is possible that variations in other physical

Table 5.3-4.   Full-Balance Equations

$$\frac{\partial \rho}{\partial t} + \overline{\nabla} \cdot \rho \overline{v} = 0$$

$$\rho \frac{\partial \overline{v}}{\partial t} + \rho \overline{v} \cdot \nabla \overline{v} + \overline{\nabla} p - \overline{\nabla} \cdot \eta \dot{\overline{\overline{\gamma}}} + \rho \overline{g} = 0$$

$$\frac{\partial c_k}{\partial t} + \overline{\nabla} \cdot \overline{v} c_k - \overline{\nabla} \cdot D_k \overline{\nabla} c_k + r_k = 0$$

$$\rho c_p \frac{\partial T}{\partial t} + \rho c_p \overline{v} \cdot \overline{\nabla} T - \overline{\nabla} \cdot k \overline{\nabla} T + \frac{1}{T}(\frac{\partial P}{\partial T}) \overline{\nabla} \cdot \overline{v} + r_G + \eta \dot{\overline{\overline{\gamma}}} : \overline{\nabla} \overline{v}$$

properties such as diffusivity, heat capacity, thermal conductivity, and even density, could also have profound effects on the reactor behavior. Furthermore, the shear-rate dependence of viscosity, so important in flow characterization of polymer melts, could certainly be a factor in reacting solutions. In the face of so many concurrent physical and chemical processes, we seek a framework in which to formulate an engineering model.

The general balance equations for a flow reactor are listed in Table 5.3-4. They apply to Generalized Newtonian Fluids (GNF's), that is, fluids whose non-Newtonian behavior is completely attributable to a shear-rate dependence of viscosity. Note that diffusivity and thermal conductivity as written are not necessarily molecular properties. They could represent effective dispersion coefficients, which are not isotropic in general (e.g., longitudinal and transverse dispersion, D_L and D_T, respectively).

In most instances there is one primary flow direction and one primary flow gradient. We consider these to be longitudinal and transverse, respectively, and denote them with subscripts 1 and 2. Thus, in tube flow, $x_1 = z$ and $x_2 = r$, since the primary flow is in the z-direction and there is a transverse (radial) gradient with respect to $v_1 \equiv v_z$. Since mass must be conserved, integration of the longitudinal mass flow rate across the transverse dimension is constant at steady-state

$$\dot{m} = \int_s \rho v_1 \, ds \tag{5.3-6}$$

However, this does not constrain the longitudinal velocities along individual streamlines from varying in the longitudinal direction, provided there is no net change in flow across the cross section. As a consequence of the continuity equation, one must consider the possibility of secondary flows in the transverse direction. One would expect convective transport to dominate molecular transport in the longitudinal direction, while in the transverse direction transport will occur primarily by molecular mechanisms. Validation of these intuitive conclusions can be achieved through comparison of simulation results for models with and without these second-order

Table 5.3-5. Balances for Unidirectional Constant Density Flow

$$\frac{\partial v_1}{\partial x_1} = 0$$

$$\rho \frac{\partial v_1}{\partial t} + \rho v_1 \frac{\partial v_1}{\partial x_1} + \frac{\partial p}{\partial x_1} - \frac{\partial}{\partial x_2} \eta \frac{\partial v_1}{\partial x_2} + \rho g_1 = 0$$

$$\frac{\partial c_k}{\partial t} + v_1 \frac{\partial c_k}{\partial x_1} - \frac{\partial}{\partial x_2} D_k \frac{\partial c_k}{\partial x_2} + r_k = 0$$

$$\rho c_p \frac{\partial T}{\partial t} + \rho c_p v_1 \frac{\partial T}{\partial x_1} - \frac{\partial}{\partial x_2} k \frac{\partial T}{\partial x_2} + r_G + \eta \left(\frac{\partial v_1}{\partial x_2}\right)^2 = 0$$

effects. Cintron (28, 29) performed such a comparison for a tubular polymerizer. Correlating experimental data for viscosity and density variations with respect to temperature, composition, and molecular weight, and allowing for radial flows, he integrated the balance equations in Table 5.3-4. He compared the results to a much less sophisticated model that assumed constant density and neglected radial flows, but preserved the full dependence of viscosity on temperature, composition, molecular weight, and shear rate (Table 5.3-5). Some typical radial profiles are given in Fig. 5.3-5 as evidence of his conclusion that the simpler transport model posed in Table 5.3-5 preserves the most important kinetic and flow details. The observed errors were less than 3 percent and temperature profiles seemed to be insensitive to the small

Table 5.3-6. Semidimensionless Balances for Distributed Parameter Model

$$\frac{\partial \hat{v}_1}{\partial \hat{x}_1} = 0$$

$$\frac{\partial \hat{v}_1}{\partial t} + \lambda_v^{-1} \hat{v}_1 \frac{\partial \hat{v}_1}{\partial \hat{x}_1} + \lambda_F^{-1} \frac{\partial \hat{p}}{\partial \hat{x}_1} - \lambda_G^{-1} \frac{\partial}{\partial \hat{x}_2} \hat{\eta} \frac{\partial \hat{v}_1}{\partial \hat{x}_2} + \lambda_g^{-1} \hat{g}_1 = 0$$

$$\frac{\partial \hat{c}_k}{\partial t} + \lambda_v^{-1} \hat{v}_1 \frac{\partial \hat{c}_k}{\partial \hat{x}_1} - \lambda_D^{-1} \frac{\partial^2 \hat{c}_k}{\partial \hat{x}_2^2} + \lambda_k^{-1} \hat{r}_k = 0$$

$$\frac{\partial \hat{T}}{\partial \hat{t}} = \lambda_v^{-1} \hat{v}_1 \frac{\partial \hat{T}}{\partial \hat{x}_1} - \lambda_H^{-1} \frac{\partial^2 \hat{T}}{\partial \hat{x}_2^2} + \lambda_G^{-1} \hat{r}_G + \lambda_G^{-1} \hat{\eta} \left(\frac{\partial \hat{v}_1}{\partial \hat{x}_2}\right)^2 = 0$$

Table 5.3-7. Distributed Parameter Dimensionless Balance Equations

$$\frac{\partial \hat{v}_1}{\partial \hat{x}_1} = 0$$

$$\frac{\partial \hat{v}_1}{\partial \hat{t}} + A \left\{ \hat{v}_1 \frac{\partial \hat{v}_1}{\partial \hat{x}_1} + Eu \frac{\partial \hat{p}}{\partial \hat{x}_1} - Re^{-1} \frac{\partial}{\partial \hat{x}_2} \hat{\eta} \frac{\partial \hat{v}}{\partial \hat{x}_2} - Fr^{-1} \hat{g}_1 \right\} = 0$$

$$\frac{\partial \hat{c}_k}{\partial \hat{t}} + A \left\{ \hat{v}_1 \frac{\partial \hat{c}_k}{\partial \hat{x}_1} - Pe^{-1} \frac{\partial^2 \hat{c}_k}{\partial \hat{x}_2^2} \right\} + b_k^{-1} \hat{r}_k = 0$$

$$\frac{\partial \hat{T}}{\partial \hat{t}} + A \left\{ \hat{v} \frac{\partial \hat{T}}{\partial \hat{x}_1} - Pe_H^{-1} \frac{\partial^2 \hat{T}}{\partial \hat{x}_2^2} - Pe_H^{-1} Br \hat{\eta} \left(\frac{\partial \hat{v}_1}{\partial \hat{x}_2} \right)^2 \right\} - \varepsilon \hat{r}_G = 0$$

where $\quad \hat{v} \equiv v/v_r$

$\hat{t} \equiv t/\lambda_{ad}$

$A \equiv \lambda_{ad}/\lambda_v$

fluctuations in velocity profile present in the results of the full transport model. Wallis (21), too, included the effect of density variation in his simulation of tubular styrene polymerizers and concluded that only minor differences in reactor output are observed with its inclusion. Wallis did not, however, allow for viscosity variation, but assumed instead a parabolic velocity profile at the inlet that changed only in response to local density variations.

If flow is treated as incompressible and cross-sectional dimensions are uniform, the existence of axial gradients in the axial velocity, which arise as a consequence of the flow profile rearranging in response to viscosity changes, would seem to imply a violation of the continuity equation. There are, in fact, downstream velocity changes, but the corresponding gradients are assumed to be relatively small, $\partial v_1/\partial x_1 \cong 0$. This also allows elimination of the convective transport term from the momentum balance. An order-of-magnitude estimate presented in Example 5.3-2 shows that, indeed, pressure gradients dominate convection in the axial direction as the primary driving force for axial momentum transport.

For the sake of completeness, the viscous energy dissipation term has been included in the thermal energy balance. Although this mechanism of heat production is important in the flow of polymer melts, it is not likely to dominate heat generation from highly exothermic reactions. In support of this supposition, we examine the CT's for these two mechanisms of heat production. The Brinkman number (Br) is frequently cited as the important dimensionless group for quantifying viscous heat

Table 5.3-8. Boundary Conditions for Distributed Parameter Model

I. *Velocity*

$$\hat{v}_1(\hat{x}_2 = 0) = f(\hat{x}_2)$$

where f is generally the Newtonian profile

and

$$\frac{\partial \hat{v}_1}{\partial \hat{x}_2}(\hat{x}_2 = 0) = 0$$

symmetry at the centerline

or

$$\hat{v}_1(\hat{x}_2 = 1) = 0$$

no slip at the wall

II. *Composition*

$$\hat{c}_k(\hat{x}_1 = 0) = 1$$

$$\frac{\partial \hat{c}_k}{\partial \hat{x}_2}(\hat{x}_2 = 0) = 0$$

symmetry at the centerline

and

$$\hat{c}_k(\hat{x}_2 = 1) = 0$$

complete conversion at the wall due to stagnancy

or

$$\frac{\partial \hat{c}_k}{\partial \hat{x}_2}(\hat{x}_2 = 1) = 0$$

no flux through the wall

III. *Temperature*

$$\hat{T}(\hat{x}_1 = 0) = T_o$$

(where $T_r = T_R$)

$$\frac{\partial \hat{T}}{\partial \hat{x}_2}(\hat{x}_2 = 0) = 0$$

symmetry at the centerline

and

$$\hat{T}(\hat{x}_2 = 1) = 0$$

isothermal wall

or

$$\frac{\partial \hat{T}}{\partial \hat{x}_2}(\hat{x}_2 = 1) + Nu[\hat{T}(\hat{x}_2 = 1)] = 0$$

Newton's law of cooling at the wall

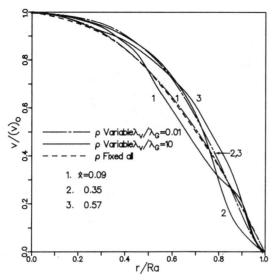

Figure 5.3-5. Radial profiles with and without assumption of constant density.

production

$$Br = \frac{\lambda_H}{\lambda'_G} \tag{5.3-7}$$

where λ'_G is the CT for viscous heat generation. Written this way, we see that Br compares the CT for shear heat generation to that for heat removal by molecular conduction. Of more import to flow reactors is the ratio

$$\frac{\lambda_G}{\lambda'_G} \tag{5.3-8}$$

which compares the chemical heat generation to that produced by shear. For a fluid with a viscosity of 1 poise at 373 K subjected to a shear rate of 100 s^{-1}, $\lambda'_G \cong 6 \times 10^5$ s. By comparing this to typical values of λ_G, which are two or more orders of magnitude smaller, one concludes that viscous heat generation may be neglected in all but high-conversion, high-shear-rate regions. This is an assumption that may need to be revalidated for some extreme cases in reactive processing where viscous prepolymer feeds and high shear rates may combine effects to make shear generation significant.

With the preceding assumptions, the pertinent CT's in the thermal energy and component material balances are: λ_v, λ_H, λ_{ad}, λ_k, and λ_D. If we focus on the component balances, we find that the number of CT's can be further reduced. In Chapter 4, it was shown that transverse diffusion played little or no role in shaping the thermal behavior of reactions in nonflow distributed parameter systems. With diffusivities in polymer systems on the order of 10^{-5} cm^2/s and lower, the transverse reactor dimension must typically be 0.1 cm or less to make λ_D comparable in magnitude to the CT's for reaction.

In the present chapter, unlike Chapter 4, we must take flow into account; more specifically, the role of the velocity profile as an amplifier for longitudinal convective transport, which competes with both transverse and longitudinal molecular transport. In a streamline flow reactor, longitudinal concentration gradients always exist owing to the presence of reaction, which renders the upstream portion richer in reactants and leaner in products than the downstream portion. Longitudinal temperature gradients could result. Consequently, longitudinal convective $[v_1(\partial/\partial x_1)]$ as well as molecular $(\partial^2/\partial x_1^2)$ transport will occur, even when the material is conveyed as a plug. Certainly transverse temperature gradients are likely because of the reaction, which could in turn produce transverse concentration gradients, even in plug flow. It is generally acknowledged, however, that convective longitudinal transport is far more rapid that its parallel molecular process, and on that basis longitudinal diffusion and thermal conduction are generally neglected to simplify analysis. Thus we are left with longitudinal convection and transverse conduction. Furthermore, if λ_H, λ_k, and λ_{ad} are all much smaller than λ_v, we may liken our streamline flow reactor to the nonflow, distributed parameter batch reactor analyzed in Chapter 4, and apply the results of the latter to the former.

Actual streamline reactors are generally not pluglike, however, especially when the walls are stationary. A transverse velocity gradient then exists and each streamline may experience a very different mix of longitudinal and transverse transport. The degree of variation will depend not only on the magnitude of the maximum velocity, but also on the nature of the transverse distribution of velocities. A power-law fluid with a very small power-law index ($n < 1$) will have a pluglike profile and should therefore be influenced by convection less than a Newtonian fluid ($n = 1$). "Finger-flow" profiles, on the other hand, such as those shown in Fig. 3.4-9, should show a greater effect on convection than either of the preceding cases at the same flow rate.

Cintron (28, 29) performed a comprehensive numerical study of free-radical polymerization in tubular reactors. His kinetic model had some deficiencies, however, especially the assumption of constant rate of initiation (i.e., $\alpha_K = 0$ or $b_i = \infty$). He included gel-effect and attempted to correlate viscosity with conversion, molecular weight, and temperature. For an insulated (adiabatic) reactor, he found that IG could be avoided in either of two ways.

As shown in Chapter 4, small values of b_m indicate sufficiently reduced sensitivity to suppress IG. In addition, convective cooling can prevent the onset of IG. Clearly, if $\lambda_v \ll \lambda_{ad}$, then IG should not occur, simply because the residence time is less than the induction period for thermal runaway. Cintron's results, presented in Fig. 5.3-6, take this postulate a step further. As apparent from the figure, IG occurs at a single value of b_m under most conditions. This value is the same as that observed by Barkelew for first-order reactions and is therefore consistent with Cintron's assumptions, which render the polymerization scheme first order. When Da reaches a value of one, however, the critical value of b_m for IG seems to rise without bound. Thus, regardless of the inherent "hotness" of the reaction as measured by b_m, when the residence time and reaction time are the same, IG is averted.

One might generally expect that a distributed parameter analysis would show the maximum temperature to occur along the center streamlines. Some published results show radial temperature profiles that run contrary to this expectation. Instead, the temperature profile has a maximum value somewhere between the centerline and the

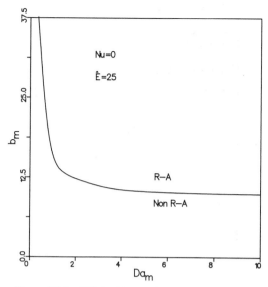

Figure 5.3-6. Effect of convection on R-A boundary.

wall, as shown in Figs. 5.3-7–5.3-9. These figures were computed by Lynn and Huff for ionic polymerization (13), and by Cintron (28) and Mostello (30) for free-radical polymerization. Cintron also observed such profiles for ficticious systems in which the heat-producing source was independent of temperature or composition [e.g., $r_G = (r_G)_0$]. Although Lynn and Huff did not provide sufficient data to calculate the pertinent CT's, the other cases seem to show a clear trend. Figures 5.3-7–5.3-9 show that the temperature maximum moves outward from the centerline when λ_G is decreased relative to λ_v. Mostello (30) also observed that while decreasing λ_{ad} brought on R-A, further decreases caused a separation of the maximum from the centerline. Since λ_G and λ_{ad} are proportional when ε is fixed, the result is probably attributable to the same cause. When flow rates are relatively slow, a substantial fraction of streamlines have residence times that are greater than the R-A induction period. If radial heat transfer is sufficiently poor, these streamlines will experience R-A before the center streamlines do, and the "hot spot" will propagate toward the center (as in Figs. 5.3-7–5.3-9).

Valsamis (17), Wallis (21), and Husain (22), on the other hand, modeled systems with substantial heat transfer to the walls. Their results show the maximum temperature at the center. Cintron's results also showed that the maximum temperature may be shifted away from the wall toward the centerline by either decreasing the wall temperature or by increasing radial heat removal through an improved heat transfer coefficient. It would appear then that rippled temperature profiles require a combination of slow flows and poor radial heat transfer coupled with heat generation rates sufficient to produce runaway.

Up to this point, nothing has been said about the role of heat transfer at the wall. Much of Cintron's analysis involved adiabatic tubes. Many other investigators have assumed a constant wall temperature that is equal in value to the coolant temperature. We saw in Chapter 4 that for stagnant systems the condition of constant wall

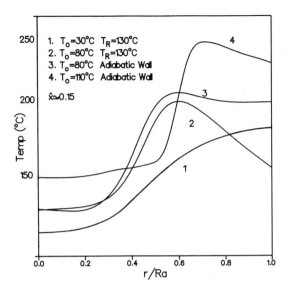

Figure 5.3-7. Radial temperature profile for distributed parameter model of butadiene polymerization.

temperature can be simulated with large values of $Nu \equiv \lambda_H/\lambda_R$. Lumped parameter behavior (no radial variations) can be simulated with $Nu \to 0$, which also corresponds to an adiabatic wall. Cintron examined the effect of Nu on IG boundaries. As shown in Fig. 5.3-10, the R-A transition reaches an asymptotic value for large values of Nu. When flow is rapid relative to reaction (small Da), attainment of the asymptotic value requires larger values of Nu. These observations are in accord with the results of Chapter 4 for distributed parameter R-A. For nonflow systems, constant wall temperature was approximated with $Nu > 10$.

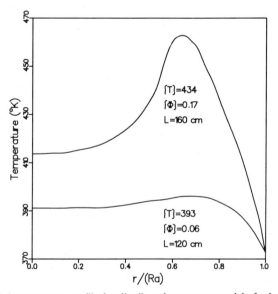

Figure 5.3-8. Radial temperature profile for distributed parameter model of tubular polymerizer.

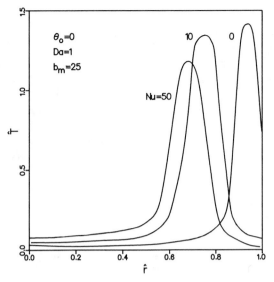

Figure 5.3-9. Radial temperature profile for distributed parameter model of tubular polymerizer.

Concerning the potential for convection to be a major source of cooling, alluded to earlier, we speculate that this phenomenon is not likely to occur frequently in practice. In support of this hypothesis, we shall relate the CT for convection, λ_v, to those for transverse heat removal, λ_R and λ_H, by utilizing several previous definitions of key dimensionless groups.

$$\lambda_v = ab_m Da_m \lambda_R \qquad (5.3-9)$$

$$\lambda_v = \Delta b_m Da_m \lambda_H \qquad (5.3-10)$$

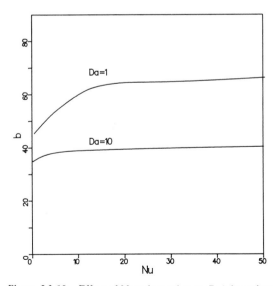

Figure 5.3-10. Effect of Nusselt number on R-A boundary.

From these expressions, we conclude that λ_v will be roughly two orders of magnitude larger than either λ_R or λ_H unless Da_m is extremely small. Should Da_m be small, however, then conversion will also be small and the reactor will not be very efficient. Since efficient systems have been the main objects of our simulations, the role of high convection regions has not been investigated explicitly in detail.

Thus we presume that the thermal behavior may be evaluated from the quantitative criteria developed for nonflow systems of similar geometry. In support of this, we offer the evidence in Table 5.3-9. It summarizes the thermal behavior of tubular reactors from a number of different published sources, and includes chemically and thermally initiated free-radical polymerizations. Note from the table that $\lambda_H < \lambda_v$ and $\lambda_R < \lambda_v$, as required.

Perhaps the major motivation for modeling the MSCTR as a distributed parameter system is the desire to examine the effect of the flow profile on output (via RTD) and on the axial pressure profile. Both effects may be deduced from a solution of the microscopic momentum balance (Appendix F), which is implicitly linked to the thermal energy and component material balances via the functional form of viscosity. Because viscosity becomes dependent on both the axial and transverse position, the momentum balance cannot be solved analytically by separation of variables. In cylindrical coordinates, the microscopic momentum balance is

$$\frac{\partial p}{\partial z} = \frac{1}{r}\frac{\partial}{\partial r}\left(\eta r \frac{\partial v_z}{\partial r}\right) \tag{5.3-11}$$

If the pressure gradient is independent of radial position, we can integrate twice and apply the boundary conditions $v_z(Ra) = 0$ and $dv_z(0)/dr = 0$ to find (42):

$$v_z(r, z) = \frac{\partial p}{\partial z}\bigg|_z \left(\int_0^r \frac{r}{\eta}dr - \int_0^{Ra}\frac{r}{\eta}dr\right) \tag{5.3-12}$$

Table 5.3-9.   CT's and Dimensionless R-A Parameters for Distributed Parameter Tubular Reactor Models

	λ_m	λ_{ad}	λ_H	ε	b_m	b_i	Δ	R-A
Valsamis	1157	41	257	.035	28	126	.16	N
	818	30	257	.035	28	103	.12	Y
	235	9	70	.036	26	74	.13	N
	171	7	70	.036	25	62	.10	Y
Mostello	359	10	294	.037	35	∞	.04	Y
	359	10	29	.037	35	∞	.35	N
	3585	10	294	.037	35	∞	.04	N
Wallis	28140	967	1867	.032	29	8	.52	N
Husain	94682	4171	5432	.037	23	none	.77	N
	94682	4171	21728	.037	23	(thermal)	.19	N
	94682	4171	48889	.037	23		.09	Y
	94682	4171	86914	.037	23		.05	Y

For incompressible fluids (constant density), the volumetric flow rate must be constant, and so, after integrating by parts, we can write:

$$\dot{V} = \int_0^{Ra} v_z 2\pi r \, dr = \pi r^2 v_z \Big|_0^{Ra} - \pi \int_0^{Ra} r^2 \frac{dv_z}{dr} \, dr \qquad (5.3\text{-}13)$$

The first term on the RHS is always zero, and dv_z/dr can be replaced by the derivative of the velocity profile to enable us to solve for the local pressure gradient:

$$\frac{dp}{dz}\bigg|_z = -\dot{V} \Big/ \pi \int_0^{Ra} \frac{r^3}{\eta} \, dr \qquad (5.3\text{-}14)$$

This, in turn, permits the velocity profile to be written more compactly.

$$v_z(r, z) = \frac{\dot{V}\left[\int_0^{Ra} \frac{r}{\eta} \, dr - \int_0^r \frac{r}{\eta} \, dr\right]}{\pi \int_0^{Ra} \frac{r^3}{\eta} \, dr} \qquad (5.3\text{-}15)$$

The reader can verify that for a constant viscosity the above reduces to the familiar parabolic velocity profile. Solution for the radial velocity profile and axial pressure difference requires an iterative technique. An inlet velocity profile is assumed, and temperatures and compositions are computed at the next axial station by a numerical solution of the balance equations. Viscosity is then computed at each radial position and used in equation 5.3-15 to determine the velocity profile. The algorithm is repeated until the profile calculated converges at each radial point. The pressure gradient is then computed from equation 5.3-14, and the newly-computed velocity profile serves as input to the next axial station. An extension of this technique for noncylindrical systems is described in the Section on drag flow reactors.

This procedure allows one to compute velocity profiles without the restrictive assumption that the momentum balance can be de-coupled from the other balance equations. Earlier, brief reference was made to the somewhat unusual velocity profiles that nonisothermal polymerizations can produce. We concluded that temperature maxima displaced from the flow axis were peculiar to adiabatic reactors or those reactors with very poor radial heat transfer characteristics. Of more universal concern is the possible development of channeling or finger-flow patterns, in which stagnant layers build up on the reactor walls, while the balance of the material speeds through the reactor core. In Chapter 3, we presented such a velocity profile, which was synthesized by Cintron (31) rather than being the result of a reactor simulation. In conjunction with attempts to match experimental and simulated behavior of polystyrene tubular polymerizers, Valsamis (17) examined the development of radial flow profiles in both runaway and non R-A reactors. Figures 5.3-11 and 5.3-12 show extensive simulation results for both radial and axial variations occurring in a non R-A polymerization. The development of stagnancy at the wall is quite evident. At little more than 30 percent of the tube length, occlusion has reduced the effective tube radius to 80 percent of its initial value. Midway through

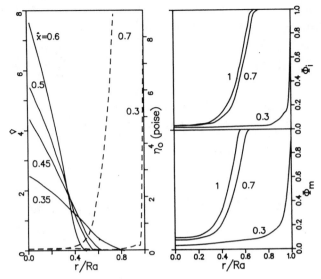

Figure 5.3-11. Radial temperature, composition, and viscosity profiles simulated for styrene tubular reactor.

the reactor, the effective radius is merely half of the tube radius. One can show from the simple Hagen–Poiseuille relation that such a restriction in flow cross section to cause a significant pressure rise. Thus, for constant volumetric flow rate of a Newtonian fluid:

$$-\frac{dp}{dz} = \frac{8\eta\dot{V}}{\pi Ra^4} \qquad (5.3\text{-}16)$$

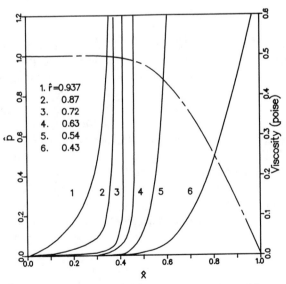

Figure 5.3-12. Axial temperature, viscosity, and pressure profiles simulated for styrene tubular reactor. Dimensionless pressure calculated as $\hat{p} = (p_0 - p)/(p_0 - p_L)$.

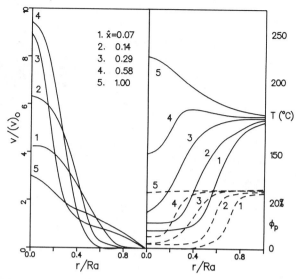

Figure 5.3-13. Radial temperature and conversion profiles simulated for butadiene step-addition in a tubular reactor.

which leads one to the conclusion that even if the axial variation in viscosity were ignored, the local pressure gradient would experience a sixteenfold increase in response to a 50 percent reduction in tube radius. The radial viscosity profiles indicate that the material near the centerline has essentially a uniform viscosity, and it is a value very close to the feed value. Near the walls, however, viscosity takes an abrupt upward change, and this corresponds to the point where flow essentially ceases.

Viewing the axial pressure profile, one can see that virtually all the pressure drop occurs in the final half of the reactor. Indeed, one might approximate the pressure profile with two linear segments. The reduced flow area in the downstream portion gives rise to such a large velocity increase that very little additional conversion occurs in the region surrounding the centerline. Radial profiles in Fig. 5.3-11 verify that the completion of the centerline changes very little between 30 and 70 percent of the reactor length.

Qualitatively similar results were observed by Lynn and Huff in their simulation of the step-addition polymerization of butadiene in solution. Note in Fig. 5.3-13 the development of flow segregation. This condition is aggravated by a supercooled feed, which leads to higher temperatures and, consequently, greater conversions in the already slower streamlines near the wall. It is not until the midway point of the reactor that radial heat transfer penetrates to the core, and this stimulates reaction exotherm and increases centerline conversions. These conditions should be contrasted with those of Valsamis where radial heat transfer was better and no late occurring temperature rise was apparent.

The axial pressure profile (Fig. 5.3-14) is similar to that observed by Valsamis for chain-addition kinetics. Flow segregation and occlusion shape the profile, and the actual growth of viscosity with conversion doesn't seem to have much effect since the radial conversion profile is essentially bimodal. That is, conversion is either complete

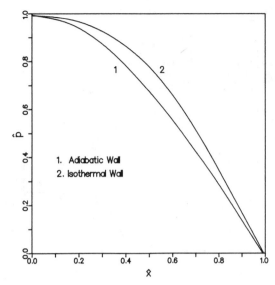

Figure 5.3-14. Axial pressure profile simulated for butadiene step-addition in a tubular reactor. Dimensionless pressure calculated as $\hat{p} = (p_0 - p)/(p_0 - p_L)$ (13).

(near the wall) or very low (near the center), and the different growth pattern of step-versus chain-addition kinetics does not play an important role in the development of the axial pressure drop. Based on these results, one is drawn to the conclusion that laminar-flow tubular reactors have a limited region of practical applicability. Stagnancy at the reactor walls leads to occlusion, which in turn causes an abrupt increase in the axial pressure gradient required to maintain constant flow rates. When this point is reached, very little further conversion seems to take place.

5.3.3. Experimental Studies

As previously noted, the production of LDPE in a tubular reactor is an established commercial process (32). Several experimental studies have been reported in the literature on attempts to use tubular reactors for the continuous polymerization of other vinyl monomers, in particular styrene (17, 21, 23).

One of the most widely quoted studies is that of Wallis (21). The authors' objective was to use a tubular reactor as a styrene prepolymerizer. Such reactors are used, e.g., to feed plate- and frame-press reactors. Conversions of about 30 percent with molecular weights greater than 10^5 were sought. These objectives were indeed met for styrene initiated by AIBN using reaction temperatures in the vicinity of 80°C. Residence times were quite long however (1–3 h), and throughput fairly low ($\dot{V} < 3000$ cm^3/h). More importantly, the conditions chosen were strong dead-end polymerizations (Section 2.8-1). The limiting conversions computed by the authors (Fig. 5.3-15) are a consequence of the initiator being consumed (D-E), which is in accord with the prediction based on the drift parameter, $\alpha_K = 5$.

The D-E drift in molecular weight tends to be balanced by a slightly increasing axial temperature rise and the result is a product of remarkably narrow DPD, with D_N quite close to the limit imposed by statistical dispersion. To the debit, D-E

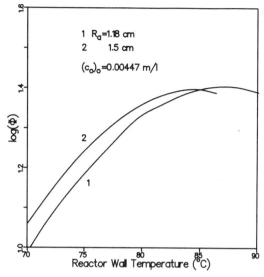

Figure 5.3-15. Exit conversion as a function of coolant temperature for polystyrene tubular reactor operating in the D-E regime (21).

makes the effectiveness of this reactor as a prepolymerizer somewhat limited. The reactor effluent would need to be replenished with fresh initiator if higher conversions were desired. Blending the viscous reactor product with low-viscosity initiator solution poses one of the most difficult mixing problems. Use of this reactor as a first stage, therefore, seems less desirable than other schemes that produce comparable monomer conversion but substantially lower initiator conversion.

The study by Valsamis and Biesenberger (17, 23) had a different objective from that of Wallis. It was to examine the feasibility of producing vinyl polymers with high conversion and acceptable DP's (e.g., $\overline{M}_W > 10^5$) in time scales that are characteristic of polymer processing equipment (minutes instead of hours). As such, it qualifies as a study in reactive polymer processing in the broad interpretation of that activity (1). Styrene was used as the test monomer. The stringent time requirement dictated the use of high jacket temperatures and the potential for thermal R-A was therefore ever present.

Valsamis found that neither AIBN nor BP were suitable initiators for achieving the stated objectives. When temperatures were raised in an effort to promote increased rates of conversion, D-E became a serious problem. Typical axial temperature profiles are shown in Figs. 5.3-16–5.3-18, which show the characteristics of insensitive R-A discussed in Chapter 4. In addition, the stimulated rate of initiation at high temperatures led to low DPD's. It was observed, however that ignition was not a problem. In light of the fact that $b_i < 5$ for AIBN and $b_i < 30$ for BP, it is not surprising that sensitivity should be lacking in these reactions.

In an effort to obtain the high molecular weights known at low temperatures, a slower initiator (DTBP) was chosen for the high-temperature reaction on the basis of similarity analysis (Section 2.12). Temperatures were increased further with hopes of reducing R-A sensitivity (b_i ranged from 60 and 300). Conversions from 50 to 70 percent were achieved in five minutes residence time. Molecular weights were still a

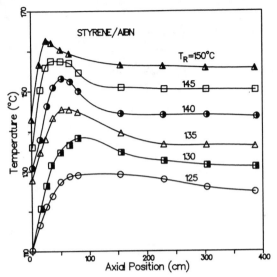

Figure 5.3-16.   Experimental axial temperature profiles for AIBN initiated styrene in a tubular reactor (23).

bit low with $\overline{M}_N \cong 40,000$. Although initiators suitable for even higher temperatures could have been selected, no improvement in product DP was anticipated. This is due to styrene's particularly strong tendency to polymerize thermally (Section 2.8-4). Although thermal initiation is slow enough below 100°C to be ignored relative to catalytic initiation, above 150°C it can become rate-controlling. Thus reducing $(c_0)_0$ or choosing an initiator with a smaller value of $(k_d)_0$ will not significantly alter the final $\overline{M}_N$.

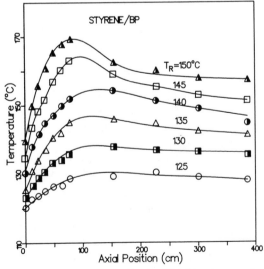

Figure 5.3-17.   Experimental axial temperature profiles for BP initiated styrene in a tubular reactor (23).

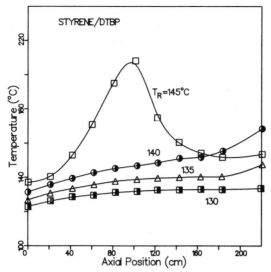

Figure 5.3-18. Experimental axial temperature profiles for DTBP initiated styrene in a tubular reactor (23).

Valsamis observed that a small-diameter reactor (0.25 in OD) was not prone to IG, while a larger reactor (0.375 in OD) did experience IG for high-temperature DTBP-initiated polymerization. This is, again, a manifestation of sensitivity. The small-diameter reactor required provocation to higher temperatures to cause heat generation rate to exceed removal rate. As temperature is increased, not only does Δ_{cr} decrease, making R-A more difficult to achieve, but b_i approaches the limit of sensitivity, eliminating IG.

5.3.4. Reaction Injection Molding

Although pressure-flow tubes (CEPR's) may not have wide applicability as polymerizers, pressure flow is nonetheless important in certain polymerizations, specifically the reaction injection molding (RIM) process. Runners and sprues leading to a mold may be treated as pressure-flow tubular reactors, and the mold itself may be modeled as a parallel-plate reactor (Chapter 3). An important distinction between the molding process and the flow reactors previously discussed is the transient nature of the flow. As the melt front advances through the runners and ultimately through the mold cavity, the length of the flow path is continually increasing. At the same time, if the fluid is reacting, its material properties will change significantly with both time and position. If the time to fill the mold is appreciably less than the time for reaction, we would expect the reaction kinetics to play only a small role in influencing the fluid mechanics. As a result, analysis of the RIM process can be subdivided into two separate problems. The first is flow without reaction (mold-filling) and the second is reaction in a stagnant medium (curing). With fill-times on the order of seconds, or even less, one might expect such an assumption to be valid for polymerizations with CT's typically an order of magnitude greater or more. However, the success of RIM processes demands fast reactions to minimize cycle time, and urethane polymerizations frequently used for RIM are sufficiently rapid (Example 5.3-3) that fill time and λ_r are about equal.

One important motivating aspect of the RIM process as opposed to conventional injection molding is the substantial reduction in pumping costs made possible by the use of low-viscosity prepolymer feeds. If the filling phase can be completed prior to advanced reaction, a reduction in conventional injection pressures typically greater than 10^3 psi to as little as 10 psi can be realized. In the interest of diminished cycle times, the closed mold curing time must not be excessively long, however. Thus effective design requires treading a fine line between delaying reaction sufficiently long to preclude excessive pressure rise during fill, while still attaining a high enough conversion rate to satisfy the demands of a short cycle time.

We have already seen (Section 5.2) that random polymerizations have viscosity-growth characteristics that appear to be suited to this end. Viscosity remains quite low and essentially independent of polymer-weight fraction until very high conversions are reached. As a result, high conversions can be achieved during the flow phase without substantially affecting the flow behavior. The final conversion achieved during curing then creates an increase of as much as five orders of magnitude in viscosity. The presence of two virtually distinct viscosity growth regimes seems to fit the process requirements better than the continued rise and early rise characteristics of step- and chain-addition polymerizations, respectively.

The CT's for flow and reaction developed in previous chapters can be used as a guide in assessing the interplay of these processes during fill and curing. The space time, $\lambda_v = V/\dot{V}$, may now be interpreted as the mold-filling time. Evaluation of the CT's associated with reaction requires a choice for the reference temperature, T_r. As indicated in Section 4.3-1, either feed temperature (T_0) or the coolant (wall) temperature (T_R) may be used. The choice here is not as clear-cut as with more conventional reactors, where reaction times are generally much longer than heat-up times, thus making T_R the more appropriate choice. If feed and wall temperatures are unequal, while transverse heat transfer is slow, then in the absence of reaction the mold contents are likely to remain close to their feed temperature for the duration of fill. If, on the other hand, transverse conduction is rapid, then the contents should quickly approach the mold wall temperature. In terms of CT's this can be stated as follows:

$$\text{if } \lambda_v \ll \lambda_H \qquad \text{then } T_r \equiv T_0 \tag{5.3-17}$$

$$\text{if } \lambda_v \gg \lambda_H \qquad \text{then } T_r \equiv T_R \tag{5.3.18}$$

A dimensionless group, the Fourier number, combines these two CT's:

$$\text{Fo} \equiv \frac{\lambda_v}{\lambda_H} \tag{5.3-19}$$

Thus criteria 5.3-16 and 5.3-17 can be expressed as Fo $\ll$ 1 and Fo $\gg$ 1, respectively. Macosko (33) recently reported a similar approach to choosing the reference temperature.

With an appropriate measure of reaction time (λ_r evaluated at T_r), one can evaluate the importance of reaction during the filling process. Clearly, if the fill time,

λ_v, is much shorter than the time for reaction, λ_r, then one could justifiably approximate the flow as that of a fluid with constant properties. A criterion for this condition may be written in terms of $Da \equiv \lambda_v/\lambda_r$:

$$Da \ll 1 \qquad (5.3\text{-}20)$$

Based on the relationship governing Poisieulle flow of a Newtonian fluid in a rectangular duct, pressure drop can be expressed as

$$\Delta p = \frac{3\eta \dot{V} L}{2 W h^3} \qquad (5.3\text{-}21)$$

where h is the slit half-thickness and W is the width. At constant volumetric flow rate with constant viscosity, the pressure should rise linearly with the length of the flow path. When the cross-sectional area of the mold is constant, length is proportional to time and so, it follows, is pressure:

$$\Delta p = \frac{3\eta \dot{V}^2}{4 W^2 h^4} t \qquad (5.3\text{-}22)$$

Experimental data of Castro and Macosko (33) for urethane network polymerization is shown in Figs. 5.3-19–5.3-22. A word is in order concerning the author's value of Da. The CT for reaction is the characteristic time associated with reaching the gel point for the network reaction and is the product of λ_r for an irreversible second-order reaction and the ratio $C_{gel}/(1 - C_{gel})$.

In Fig. 5.3-19, criteria 5.3-17 and 5.3-20 are both met and the expected linear pressure rise is observed. The linear pressure rise calculated from 5.3-22 is included

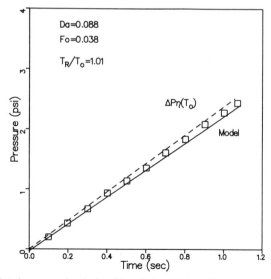

Figure 5.3-19. Simulated pressure rise during fill cycle for RIM urethane reaction. Fill time is much less than reaction or heat transfer time (33).

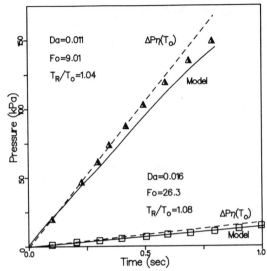

Figure 5.3-20. Simulated pressure rise during fill cycle for RIM urethane reaction. Heat transfer time is much less than fill time and $T_R > T_0$ (33).

on each figure for reference. In Fig. 5.3-20, criterion 5.3-18 is in effect. Since $T_R > T_0$ and conduction is rapid, the viscosity of the fluid decreases during the fill cycle and thus pressures lower than that expected from equation 5.3-22 are observed. In contrast, Fig. 5.3-21 represents a case in which heat conduction is important, but $T_R < T_0$. As result, the fluid viscosity increases with time and pressure rises above expectation. In Fig. 5.3-22, criterion 5.3-20 is violated and so reaction is important

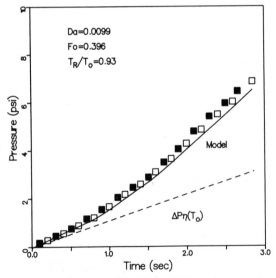

Figure 5.3-21. Simulated pressure rise during fill cycle for RIM urethane reaction. Heat transfer time is much less than fill time and $T_R < T_0$ (33).

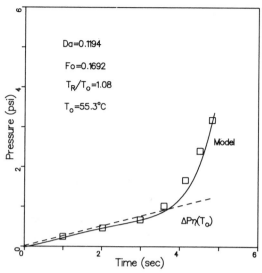

Figure 5.3-22. Simulated pressure rise during fill cycle for RIM urethane reaction. Reaction time is less than fill time (33).

during the filling process. The authors concluded that reaction is important if Da > 0.10, when Da is based on the gelation time.

Understanding the role of reaction in shaping flow characteristics requires some discussion of transient flow behavior during the mold-filling process. One might expect that the fluid advances through the mold progressively, with the fluid-front entering at one end of the mold and being conveyed the full axial distance. This would lead to the situation in which the oldest fluid elements are at the far end of the mold and the youngest are at the gate or entrance, after filling is complete. Tracer studies of injection molding show that this is not the case (38). Material that first enters the mold along the mold's centerline is immediately deposited at the wall, close to the gate. The following material is subsequently deposited on the wall at locations progressively farther downstream. Thus the oldest particles may be found near the mold entrance while the youngest ones may be located at the extreme axial position. The mechanism is a flow rearrangement at the flow front called the "fountain effect." Pictured in Fig. 5.3-23, it results in the relocation of central streamlines to a new position at the wall. The axial dimension of the flow front is small compared to the overall length of the flow path and thus its influence on the axial pressure profile may be minimal. When reaction is significant during the filling process, however, the fountain effect can produce a profound effect on the flow field and pressure buildup. Consider the first fluid element to enter the mold along the center streamline. It soon finds itself at the mold wall close to the gate region. It remains there, reacting, growing in conversion and generating heat for the duration of the fill cycle. If the mold wall is hotter than the feed, the fluid element will have been positioned even closer to the source of heat than it was at its entry point. This serves to further accelerate reaction. Given this scenario, one could envision a layer of high-conversion, high-viscosity material building up at the wall and causing occlusion much like that described for tubular reactors in Section 5.3-2. The effect of

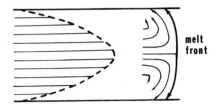

Figure 5.3-23. Fountain flow streamlines characteristic of flow in the melt front region during injection molding.

occlusion is clear from equations 5.3-20 and 5.3-21 with regard to its dependence on *b*.

Both Domine and Gogos (20) and Manzione (34) have included algorithms approximating the fountain effect in their simulations of RIM. Both simulated mold-filling processes in which λ_H was several times larger than λ_v, and thus reaction characteristics for the fill process could be estimated by using $T_r = T_0$. Based on this assumption, CT's for both authors were calculated and are listed in Table 5.3-10. For Manzione's system, it is estimated that Da is quite small and inequality 5.3-20 is ensured. One would expect that at fill time most of the fluid would still be at the feed temperature and very little conversion would be evident. The complete temperature profile shown in Fig. 5.3-24 confirms the observation.

If reaction is important during the fill cycle, then based on our knowledge of the exothermic nature of polymerizations (Chapter 4), we would expect heat generation to be an important consideration. The key CT that characterizes runaway (R-A) behavior was shown to be λ_{ad}. If the mold wall temperature is held constant, then R-A can be averted if

$$\Delta \equiv \frac{\lambda_{ad}}{\lambda_H} > \Delta_{cr} \qquad (5.3\text{-}23)$$

and, as shown in Section 4.3-3, Δ_{cr} for rectangular geometry has a value of 1.14. If

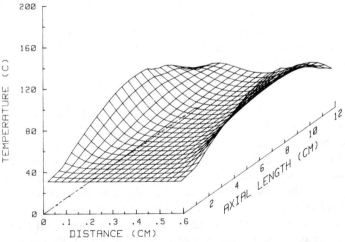

Figure 5.3-24. Spatial temperature distribution at fill time for RIM process with small values of Da and Fo [from L. T. Manzione, *Polym. Eng. Sci.*, **21**, 1234 (1981)].

the mold wall temperature is determined by a coolant fluid, it is conceivable that heat transfer from the mold to the fluid might dominate the transverse heat transfer characteristics. This phenomenon was discussed in Section 4.3-3 and is signaled by a small value of $Nu \equiv \lambda_H / \lambda_R$. In such cases, λ_{ad} should be compared to λ_R via

$$a \equiv \frac{\lambda_{ad}}{\lambda_R} > a_{cr} \qquad (5.3\text{-}24)$$

if R-A is to be averted. The value of a_{cr} is approximately 2 for polymerizations if $b_m \cong 30$.

The extreme temperatures of R-A may pose no problem provided they occur after the mold is filled. That is, if the onset of R-A is postponed long enough to prevent occlusion levels of conversion, and if the associated temperature rise has no deleterious effects on polymer product properties, then R-A might be advantageous in promoting shorter cycle times. This condition requires that criterion 5.3-23 be violated and, to ensure that R-A is delayed until after fill, that the following criterion be satisfied:

$$\lambda_v \ll \lambda_{ad} \qquad (5.3\text{-}25)$$

In addition, if R-A, but not ignition (IG) is encountered, it is possible that extreme temperature rises will be diminished by rapid reactant consumption. From Chapter 4, we estimate that $b \equiv \lambda_r / \lambda_{ad}$ should be less than 15 to remove parametric sensitivity from R-A. The values of b given in Table 5.3-10, as well as the constants measured by Lipshitz and Macosko [19], indicate that parametric sensitivity is not a significant problem in methane polymerizations. At temperatures near 50°C, $b < 5$. This may be attributed to the low heat effect number, $He \equiv \lambda_m / \lambda_G$ (PES I or II), which is small due to the low reactant concentration. The feed is generally a prepolymer, so many of the functional groups have already been converted prior to the start of the RIM process.

We have yet to discuss the importance of viscous energy dissipation in the thermal behavior of the molding process. Although this mechanism is dominant in the flow of molten polymers, one suspects that the low viscosity of polymer solutions would render dissipation insignificant relative to reaction exotherm. Following our usual procedure (Appendix G), a CT for heat generation by dissipation may be defined as follows:

$$\lambda'_G = \frac{\rho c_p T_r}{\eta \dot{\gamma}^2}$$

Note that if η has a value of 1 poise, and $\dot{\gamma}$ is 1 s^{-1}, then the volume-specific energy production rate is 2.4×10^{-8} cal/cm^3 s. Thus λ'_G would have a value of the order of 10^6 s which is far greater than the time scale of any other transport mechanism in the RIM process. However, if an appreciable amount of reaction takes place during the filling, these numbers many change drastically. Domine found that in occluded flows $\dot{\gamma}$ could rise to a value as high as 10^4 s^{-1}, and values of 10^3 s^{-1} were not uncommon [20]. It is not difficult to see that, even if η were to remain near its feed

Table 5.3-10. CT's and Dimensionless Groupings for Urethane RIM Reactions

T_o	$30°C^*$	$40°C^*$	$60°C^{\#}$	$60°C^{\#}$	$60°C^{\#}$	$60°C^{\#}$	$60°C^{\#}$
λ_v	12s	12	4.8	2.4	9.6	2.4	9.6
λ_H	61s	61	25	25	25	25	25
λ_r	197s	109	3.6	1.8	7.3	7.3	1.0
λ_{ad}	35s	21	2.8	1.4	5.6	5.6	1.4
b	5.6	5.3	1.3	1.3	1.3	1.3	1.3
Da	0.06	0.11	1.33	1.33	1.33	0.33	5.33
Fo	0.20	0.20	0.19	0.10	0.38	0.10	0.38

*Manzione (34)

$^{\#}$Gogos and Domine (20)

value of 1 poise, λ'_G would then only be reduced to the same order of magnitude as the other CT's for flow and reaction. However, heating induced by the high shear rates encountered as fluid passes through occluded regions on its way to the flow front may become significant.

Example 5.3-1. Comparison of Distributed and Lumped Parameter Thermal Energy Balances. Show that the axial flow-average behavior of a laminar-flow tube can be described by a plug-flow model if the heat transfer coefficient is given an axial dependence dictated from the distributed parameter profiles by:

$$h(z) \equiv \frac{-k \left.\dfrac{\partial T}{\partial r}\right|_{r=R}}{[T] - T_R}$$

Solution. By considering an energy balance on an axial section of the tube, we find that at steady state, in the absence of reaction and assuming constant ρ and c_p,

$$0 = \rho c_p \iint v_z(r) T(r) \Big|_z dA - \rho c_p \iint v_z(r) T(r) \Big|_{z+\Delta z} dA + k \left.\frac{\partial T}{\partial r}\right|_{r=R} 2\pi R \, \Delta z$$

However:

$$\iint v_z(r)T(r)\,dA = \frac{\iint v_z T\,dA}{\iint v_z\,dA}\frac{\iint v_z\,dA}{\iint dA}\iint dA = [T]\{v_z\}\pi R^2$$

so the original equation can be more compactly expressed as:

$$0 = \rho c_p\left[\frac{\{v_z\}[T]|_z - \{v_z\}[T]|_{z+\Delta z}}{\Delta z}\right] + s_V\left(k\left.\frac{\partial T}{\partial r}\right|_{r=R}\right)$$

Taking the limit as $\Delta z \to 0$, and realizing that for a constant density fluid in tube flow, $\{v_z\} \neq f(z)$

$$0 = \rho c_p\{v_z\}\frac{d[T]}{dz} - s_V\left(k\left.\frac{\partial T}{\partial r}\right|_{r=R}\right)$$

If the heat transfer coefficient is defined as postulated in the statement of the problem, the flux at the tube wall can be expressed by a Newton's law of cooling, yielding:

$$0 = \rho c_p\{v_z\}\frac{d[T]}{dz} + s_V h(z)([T] - T_R)$$

Identifying v_z in the plug-flow model with $\{v_z\}$ and T with $[T]$ completes the solution.

Example 5.3-2. Relative Importance of Convective Momentum. Estimate the relative importance of pressure and convection in the axial transfer of momentum, using the following data (17) for a polystyrene tubular reactor.

1. $x_1 = 498$ cm
   $p = 5.70 \times 10^4$ dynes/cm^2
   $\{v_1\} = 3.65$ cm/s

2. $x_1 = 538$ cm
   $p = 1.04 \times 10^5$ dynes/cm^2
   $\{v_1\} = 4.36$ cm/s

Solution.

$$\frac{dp}{dx_1} \cong \frac{\Delta p}{\Delta x_1} = 116 \text{ g/cm}^2 \text{ s}^2$$

$$\rho v_1\frac{\partial v_1}{\partial x_1} \cong \rho v_1\frac{\Delta v_1}{\Delta x_1} = 0.20 \text{ g/cm}^2 \text{ s}^2$$

CONCLUSION

Even though viscosity changes give rise to a $\partial v_1/\partial x_1$, and thus a convective flux, the effect is small enough to justify its neglect relative to pressure momentum flux.

Example 5.3-3.   Reaction Characteristics of RIM Urethanes.   Using the data of Macosko for catalyzed urethane polymerization, evaluate λ_r and λ_{ad}. If the fill time is 5 s and the mold is 1 cm thick, comment on the expected pressure rise behavior.

Data:

$$-\Delta H = 14.34 \text{ kcal/eq} \qquad\qquad (c)_0 = 2600 \text{ eq/m}^3$$

$$\rho c_p = 0.53 \text{ g/cm}^3 \qquad\qquad T_0 = 50°C$$

$$k = 3 \times 10^{-4} \text{ cal/cm s K} \qquad T_R = 50°C$$

$$r = 1.033 \times 10^{12} \exp\left(\frac{-7775}{T}\right) c^{1.5} \text{ eq/m}^3 \text{ s}$$

Solution.

$$\lambda_r = (c)_0/(r)_0 = 40.3 \text{ s}$$

$$\lambda_{ad} = \frac{\rho c_p T_0^2 \lambda_r}{\dfrac{-\Delta HE}{R_g}} = 7.7 \text{ s}$$

$$\lambda_H = \rho c_p l^2/k = 442 \text{ s}$$

$$\lambda_v = 5 \text{ s}$$

Since $\lambda_v \ll \lambda_r$ and $\lambda_v < \lambda_{ad}$, reaction will not have an appreciable effect on the flow behavior during filling, and thus pressure rise should be linear. Since $\lambda_H \gg \lambda_{ad}$ (i.e., $\Delta > \Delta_{cr}$), R-A may be expected, but it will occur subsequent to complete filling.

5.4. CONTINUOUS DRAG-FLOW REACTORS

Our discussion of the continuous externally pressurized reactor (CEPR) has brought to light several deficiencies associated with its use as a polymerizer. The high viscosity of polymerizing solutions requires excessively high inlet pressures to maintain flow. The buildup of polymer near the wall can lead to channeling and occlusion. To maintain adequate heat transfer, a small transverse dimension is required. Therefore, to achieve sufficient residence times for high conversions, long reactor lengths are necessary, further aggravating the pressure-drop problem.

In processing polymer melts, it has long been recognized that pumping efficiency actually improves with high viscosity if the drag-flow principle is utilized. Thus, instead of opposing viscous drag with large pressure drops, we can take advantage of it to generate a pressure rise via internal pressurization (35). It has been estimated that viscosities of roughly 500 poise (36) signal the point at which drag flow becomes a feasible alternative to pressure flow in single-screw extruders.

In a pressure-flow vessel, the flow is induced by application of an external force in the longitudinal direction to the fluid at the inlet. Longitudinal momentum enters and some is transported in the transverse direction toward the stationary wall(s) by shear stresses between moving streamlines. Since viscous forces are responsible for this transfer, the higher the viscosity of the fluid, the greater the fraction of momentum in the feed that is transmitted in the nonflow direction. Consider simple Newtonian flow in a tube. The average velocity is given by:

$$\langle v \rangle = \left(\frac{\Delta p}{L} \right) \frac{d_t^2}{16\eta} \tag{5.4-1}$$

It should be readily apparent that to maintain a specific flow rate, the applied pressure must increase with the viscosity of the fluid.

In contrast, drag-flow devices impart an external force in the direction of flow via motion of one or more of the vessel walls oriented in the longitudinal direction. This longitudinal motion is then transmitted in the transverse direction by shear stresses. The effectiveness of such a device therefore depends on the fluid's viscosity. The use of the term "drag flow" should not leave the impression that pressure gradients are absent from such configurations. Indeed, a significant feature of drag-flow devices is their ability to induce flow in the face of an opposing pressure gradient and a concomitant opposing pressure flow. The single-screw extruder is an obvious example of a drag-flow device in which a single moving wall induces flow. Examples with two moving walls include the twin-screw extruder, the two-roll mill, and the Diskpack (37).

We shall analyze drag-flow devices via a conceptual model introduced in previous chapters (Section 1.2 and Chapter 3). The continuous drag-flow reactor (CDFR) pictured in Fig. 5.4-1 simply consists of two infinite parallel plates. The upper plate moves with velocity v_2, and the lower plate moves with velocity v_1. A pressure gradient may be imposed in either direction. When one plate is stationary, we have simple parallel-plate drag flow. Such a model is sometimes used to describe flow in a single-screw extruder (40, 41) and thereby reactive extrusion (REX). The reader should be aware, however, that, while the downstream component of flow may indeed be visualized as simple drag flow, there is also a cross-stream component owing to the pitch of the screw flight. Consequently, the actual flow path experienced by the material in the channel is a composite of the two and resembles a helix. In fact, the residence time distribution in the absence of reaction (no viscosity change), discussed in Chapter 3, appears pluglike owing to the cross-channel (transverse) flow component. It is also noteworthy that, while "pressure flow" opposes drag flow in the down-channel direction as a result of internal pressuri-

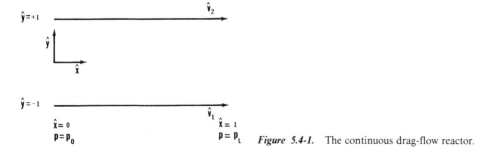

Figure 5.4-1. The continuous drag-flow reactor.

zation, it is believed that no reverse flow exists in the true sense; that is, no material actually travels upstream (35).

Notwithstanding the absence of cross-channel flow in our simple drag-flow model, we shall find it useful in providing information on the effects of changing reaction viscosity on external pressurization (pressure profiles) and flow patterns in reactive processing in general.

The longitudinal component of the momentum balance for a GNF ($\tau_{ij} = -\eta \, dv_j/dx_i$) may be written in dimensionless form as

$$\text{Po} \frac{\partial \hat{p}}{\partial \hat{x}} = \frac{\partial}{\partial \hat{y}} \hat{\eta} \frac{\partial \hat{v}_x}{\partial \hat{y}} \tag{5.4-2}$$

where $\text{Po} \equiv \lambda_M/\lambda_f$ is the Poiseuille number. The characteristic velocity v_r used to make v_x dimensionless is the mean plate velocity, $v_m = 1/2(v_1 + v_2)$. There are many options available for defining $\hat{p}$. Someone could reduce p with a "kinetic pressure," $(1/2)\rho[v]^2$ (39). Alternatively, one could reduce p with feed pressure p_0, or with the magnitude of the overall pressure change $|p_L - p_0|$. Since the magnitude of Δp is generally not known at the outset, a workable alternative is to use the theoretical value for Newtonian Flow. This can be computed via equation 5.3-16 or equation 5.3-21, for example. We denote this with the subscript N to signify "Newtonian" as opposed to the actual pressure change. We shall use

$$\hat{p} \equiv \frac{p - p_0}{|p_L - p_0|_N} \tag{5.4-3}$$

which has a value of 0 at the inlet. The value at the exit is the ratio of the actual pressure drop and the Newtonian value, and may be positive or negative depending on whether the pressure rises or drops, respectively. When the fluid is incompressible, the volumetric flow rate remains constant. This provides an additional constraint on the velocity profile:

$$2\langle \hat{v} \rangle = \int_{-1}^{1} \hat{v}_x \, d\hat{y} = \text{constant} \tag{5.4-4}$$

By analogy to the method for tubular reactors (5.3-2), this integral can be expanded via integration by parts

$$2\langle \hat{v} \rangle = \hat{v}_x \hat{y}|_{-1}^{1} - \int_{-1}^{1} \hat{y} \frac{d\hat{v}_x}{d\hat{y}} \, dy \tag{5.4-5}$$

The momentum balance can be solved to find an expression for $d\hat{v}_x/d\hat{y}$ and this result, together with equation 5.4-5 and no-slip boundary conditions $\hat{v}_x(-1) = \hat{v}_1$ and $\hat{v}_x(1) = \hat{v}_2$, yields the integral expression for the velocity profile listed in Table 5.4-1. Note that when $v_m \equiv (v_1 + v_2)/2$ is chosen as the characteristic velocity for reducing all velocities to dimensionless form, the parameter $\hat{v}_m = (\hat{v}_1 + \hat{v}_2)/2$ in the balances reduces to a value of one.

To gain insight into the flow characteristic of CDFR's, we consider first a Newtonian fluid so that viscosity remains spatially constant, and $\hat{\eta} = 1$. The flow

Table 5.4-1.   Flow Equations for CDFR

CONTINUITY EQUATION

$$\frac{\partial \hat{v}_x}{\partial \hat{x}} = 0$$

MOMENTUM EQUATION

$$Po\,\frac{\partial \hat{p}}{\partial \hat{x}} = \frac{\partial}{\partial \hat{y}}\,\hat{\eta}\,\frac{\partial \hat{v}_x}{\partial \hat{y}}$$

VELOCITY PROFILE

$$\hat{v}_x - \hat{v}_1 = \frac{(2(\hat{v}_m - \{\hat{v}\})(I_o - \hat{v}_d I_1)(C_1 I_o - C_o I_1))}{I_2 I_o^2 - I_1^2 I_o} + \hat{v}_d\,\frac{C_o}{I_o}$$

where

$$I_k = \int_{-1}^{1} (\hat{y}^k/\hat{\eta})\,d\hat{y} \qquad C_k = \int_{-1}^{\hat{y}} (\hat{y}^k/\hat{\eta})\,d\hat{y}$$

and $\hat{\eta} = \hat{\eta}(\hat{x}, \hat{y})$

PRESSURE PROFILE

$$P_o\,\frac{d\hat{p}}{\partial \hat{x}} = \frac{2(\hat{v}_m - \{\hat{v}\})(I_o - \hat{v}_d I_1)}{I_2 I_o - I_1^2}$$

---

equations in Table 5.4-1 can be solve analytically to find the velocity profile:

$$\hat{v}_x = \frac{s\,Po}{2}(\hat{y}^2 - 1) + \frac{\hat{v}_D}{2} + \hat{v}_m \tag{5.4-6}$$

If the mean plate velocity is used for reference velocity v_r, then $\hat{v}_m = 1$, while $\hat{v}_D$ is the dimensionless difference in plate velocities, $\hat{v}_2 - \hat{v}_1$. When viscosity is constant, the pressure gradient is linear. If pressure increases ($p_L > p_0$), then $s \equiv (p_L - p_0)/|\Delta P| = 1$. If pressure drops ($p_L < p_0$), then $s = -1$.

The restriction of constant volumetric flow rate links the parameters Po, $\hat{v}_m$, and $\{\hat{v}\}$:

$$\tfrac{1}{3}s\,Po = \hat{v}_m - \{\hat{v}\} \tag{5.4-7}$$

Furthermore, if $v_r = v_m$, then $\{\hat{v}\}$ and Po are uniquely related.

$$\{\hat{v}\} = 1 - \frac{s\,\text{Po}}{3} \tag{5.4-8}$$

For the Newtonian fluid, the velocity profile (equation 5.4-4) is actually the superposition of two profiles. One reflects pure drag flow:

$$(\hat{v}_x)_d = \frac{\hat{v}_D}{2}\hat{y} + 1 \tag{5.4-9}$$

and the second, pure pressure flow:

$$(\hat{v}_x)_p = \frac{s\,\text{Po}}{2}(\hat{y}^2 - 1) \tag{5.4-10}$$

Integrating the velocity profile in the transverse direction yields the volumetric flow rate and shows it, too, to be the sum of a drag-flow rate $(\dot{V}_d)$ and a pressure flow rate $(\dot{V}_p)$. More specifically, the ratio of the two components is given by:

$$\frac{\dot{V}_p}{\dot{V}_d} = -\frac{s\,\text{Po}}{3} \tag{5.4-11}$$

Note that in Chapter 3 we identified this flow ratio as G, a symbol used by Tadmor and Gogos (35). The relationship between G and Po is straightforward: $G = s\text{Po}/3$. We shall use the Poiseuille number together with s to denote the sign of the pressure change. As the CT's composing Po would suggest, this dimensionless group is a measure of the magnitude of pressure-induced flow relative to drag flow. Note however, that the two components need not be complementary, as when $s = 1$ ($p_L > p_0$), in which case pressure flow opposes drag flow.

Some representative velocity profiles for the Newtonian CDFR are shown in Fig. 5.4-2. Cases ii and vi pose a special problem in that they sustain backflow (reverse flow). Without a detailed knowledge of the flow pattern at the ends of the vessel ($x = 0$ and $x = L$), we cannot describe the properties of the reverse streamlines and, therefore, we cannot compute the RTD or the reactor output. A similar problem exists when attempting to model cross-channel flow in reactive extrusion (REX) since drag flow in one direction is exactly counterbalanced by pressure flow in the opposite direction. As in a practical note, cases ii and vi sacrifice throughput for the sake of raising the exit pressure and will dictate in the limit the maximum pressure gradient that can be overcome. Cases ii and iv also present a computational problem because they lack symmetry. Unlike cases v–vii, the maximum/minimum in the velocity profile does not lie along an axis of symmetry. This is primarily a problem when attempting to transform velocity profiles into residence time distributions, as in Chapter 3.

The values of Po identified with transitions amongst the various cases in Fig. 5.4-2 stem from analysis of the velocity profile described by equation 5.4-6. For reverse flow to occur, there must be a minimum in the velocity profile between the

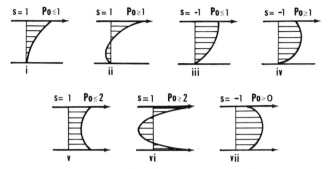

Figure 5.4-2. Possible velocity profiles in the CDFR.

two plates, which is identified by

$$\frac{d\hat{v}_x(\hat{y}_{min})}{d\hat{y}} \tag{5.4-12}$$

and

$$-1 \leqslant \hat{y}_{min} \leqslant 1 \tag{5.4-13}$$

and leads to the condition

$$\hat{v}_D = 2s\,\text{Po} \tag{5.4-14}$$

The latter is the requirement for an extremum to occur between the walls. If the pressure drops ($s < 0$), the preceding condition represents a velocity maximum. For flow reversal, $s > 0$ and we conclude that

$$\hat{v}_D \leqslant 2\text{Po} \tag{5.4-15}$$

However, this is not a sufficient condition for reverse flow, only for a minimum in the flow profile. Note that the individual plate velocities relate to $\hat{v}_D$ through the following equations

$$\hat{v}_1 = 1 - \frac{\hat{v}_D}{2} \tag{5.4-16}$$

$$\hat{v}_2 = 1 + \frac{\hat{v}_D}{2} \tag{5.4-17}$$

Although we have considered the cases of one fixed plate and one moving plate ($\hat{v}_1 = 0$, $\hat{v}_2 = 2 = \hat{v}_D$) or both plates moving at the same velocity ($\hat{v}_D = 0$), we need not be restricted to these conditions. To ensure that the plate velocities are greater tan zero, equations 5.4-16 and 5.4-17 restrict us to $0 \leqslant \hat{v}_D \leqslant 2$, if $\hat{v}_2 > \hat{v}_1$. Even within these constraints, it is possible to create conditions for a minimum in the velocity profile that is non-negative (e.g., case v, Fig. 5.4-2). If we require that $\hat{v}_x(\hat{y}_m) \geqslant 0$, solution of the velocity profile when $s = 1$ gives:

$$\hat{v}_D^2 \leqslant 4\text{Po}(2 - \text{Po}) \tag{5.4-18}$$

Clearly, if Po > 2, the preceding condition can never be satisfied and reverse flow will occur regardless of $\hat{v}_D$. To summarize:

Po ⩽ 1	no reverse flow
1 ⩽ Po ⩽ 2	no reverse flow if $v_D^2 \leqslant 4\text{Po}(2 - \text{Po})$
Po > 2	reverse flow for all $\hat{v}_D$

The special case of closed discharge is tantamount to having a reverse pressure flow precisely equal in magnitude but opposite in direction to the forward drag flow. Setting the quantity defined by equation 5.4-11 equal to -1 for $s = 1$ yields the condition

$$\text{Po} = 3 \qquad (5.4\text{-}19)$$

for zero net flow.

As discussed in Chapter 3, the largest and smallest residence times in the CDFR need not be infinity and zero, respectively. When there is a pressure drop ($s = -1$) and both walls move with $v_D = 0$, the shortest RT occurs at the centerline and the largest at the moving wall. The RT at any streamline, when made dimensionless by the mean RT, is

$$\hat{\tau}(\hat{y}) = \frac{\langle \hat{v} \rangle}{\hat{v}_x(\hat{y})} \qquad (5.4\text{-}20)$$

which, by virtue of equation 5.4-8, can be expressed as:

$$\hat{\tau}(\hat{y}) = \frac{1 - \dfrac{s\,\text{Po}}{3}}{\hat{v}_x(\hat{y})} \qquad (5.4\text{-}21)$$

For the pressure-flow reactor ($s = -1$) with $\hat{v}_D = 0$, the preceding equation yields

$$\hat{\tau}_0 = \frac{1 + \dfrac{\text{Po}}{3}}{1 + \dfrac{\text{Po}}{2}} \qquad (5.4\text{-}22)$$

and

$$\hat{\tau}_{\max} = 1 + \frac{\text{Po}}{3} \qquad (5.4\text{-}23)$$

When $s = -1$ and $\hat{v}_D = 0$, Po is unconstrained in magnitude (no flow-reversal is possible). In the limit as Po gets large, $\hat{\tau}_0 \to 2/3$ and $\hat{\tau}_{\max} \to \infty$. This situation is equivalent to that which prevails in a CEPR (stationary walls) and is consistent with the observation that a large value of Po indicates that pressure flow dominates drag flow.

·When $s = 1$, Po is bounded by a maximum value of 2 before the onset of flow-reversal. The minimum RT is at the moving wall while the maximum is at the centerline. From equation 5.4-20, we find that

$$\hat{\tau}_0 = 1 - \frac{Po}{3} \tag{5.4-24}$$

and

$$\hat{\tau}_{max} = \frac{1 - \dfrac{Po}{3}}{1 - \dfrac{Po}{2}} \tag{5.4-25}$$

The expression for $\hat{\tau}_{max}$ indicates that the onset of stagnancy occurs when Po = 2 (since $\hat{\tau}_{max} = \infty$). Clearly, if Po $\rightarrow$ 0, the maximum and minimum RT converge to one since plug flow is achieved.

When only one plate is set in motion and Po $\leqslant$ 1, regardless of the value of s, there will be no maximum or minimum in the velocity profile. The longest RT will be infinity (at the stationary plate), and the shortest RT will occur at the moving wall and will be given by

$$\hat{\tau}_0 = \frac{1}{2}\left(1 - \frac{s\,Po}{3}\right) \tag{5.4-26}$$

This expression has limiting values of $1/3$ and $2/3$ for a pressure rise and pressure drop, respectively, when Po = 1. As expected, for pure drag flow, Po = 0 and $\hat{\tau}_0 = 1/2$. Two obstacles to the use of pressure-flow reactors for polymerizations were the buildup of stagnant layer of polymer near the walls and the channeling of unreacted material along the centerline of the vessel. Case v of Fig. 5.4-2 is of special interest in this regard. The motion of the walls prevents stagnancy. The slowest streamlines and thus those with potentially the highest polymer conversions are at the center. Temperature, however, might attain a maximum value at the centerline, helping to mitigate transverse gradients in viscosity caused by the RTD. As a result, we expect that longitudinal distortion of the velocity profile due to variations in the viscosity of the reaction medium will be less intensity than for the pressure flow reactor.

The balance equations presented for the CDFR (Tables 5.3-5 and 5.4-1) have been solved for several test cases. Free-radical kinetics and the corresponding viscosity model discussed in Section 5.2 were used. The two-moving-wall reactor was simulated presuming an initial profile such as Case v in Fig. 5.4-2 (less than the maximum back-pressure before the onset of backflow). The single-wall reactor was simulated for pure drag flow only. Figures 5.4-3–5.4-5 illustrate the resulting velocity profile development. In each case the tendency is for the initial velocity profile to become pluglike as conversions increase. The trend is diminished by severe R-A conditions.

It is clear that the growth of viscosity with conversion serves to unify the flow profile in the CDFR rather than segregate it as was the case with the CEPR. The

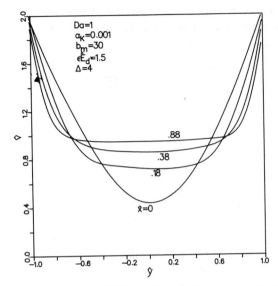

Figure 5.4-3. Radial velocity profile in a CDFR with both walls moving and maximum pressure rise without backflow at the inlet. Chain-addition kinetics with no R-A.

extent to which the CDFR is successful in conveying the reacting solution and building a pressure in the longitudinal direction is dependent on the nature of the viscosity growth. A random polymerization builds viscosity only at high conversions. Thus one expects small pressure changes over most of the reactor length, as well as potentially large radial gradients in flow behavior owing to low conversions near the moving walls. A free-radical polymerization, on the other hand, with a continuous and substantial rise in viscosity should show a greater axial pressure rise and less

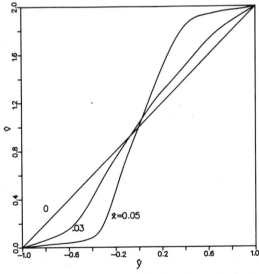

Figure 5.4-4. Radial velocity profile in a CDFR with both walls moving and maximum pressure rise.

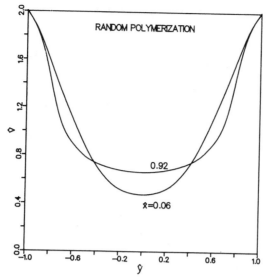

Figure 5.4-5. Radial velocity profile in a CDFR with both walls moving and maximum pressure rise without backflow at the inlet. Random kinetics with no R-A.

radical variations. Figure 5.4-6 offers, at best, preliminary verification of this conclusion. The free-radical polymerization shows twice the pressure that the random polymerization does. Also, the core velocity (profiles in Figs. 5.4-3 and 5.4-4) in the chain-addition case as one approaches the reactor exit ($\hat{x} = 1$) is closer to the wall velocity than in the random case. The random polymerization shows substantial slip near the wall, where lower conversions (due to both lower temperatures and shorter RT's) create a large viscosity gradient between the core and the wall.

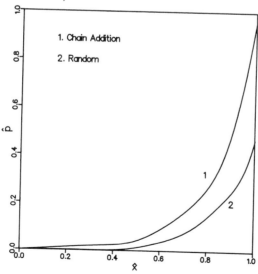

Figure 5.4-6. Axial pressure profiles for CDFR with random- and chain-addition model polymerizations. Dimensionless pressure calculated as $\hat{p} = (p - p_0)/(p_L - p_0)$.

Example 5.4-1.   CDFR Velocity Profile for Constant Viscosity.   Show that when $\eta = \text{const}$ (i.e., $\hat{\eta} = 1$) the formulas in Table 5.4-1 reduce to the velocity profile given by equation 5.4-6.

Solution.   When η is constant,

$$I_k = \hat{y}^k \, d\hat{y} = \left.\frac{y^{k+1}}{k+1}\right|_{-1}^{1} = (k+1)^{-1}\left[1^{k+1} - (-1)^{k+1}\right]$$

and

$$C_k = \hat{y}^k \, d\hat{y} = (k+1)^{-1}\left[\hat{y}^{k+1} - (-1)^{k+1}\right]$$

Thus the specific values for $k = 0, 1,$ and 2 are

$$I_0 = (1 + 1) = 2$$
$$I_1 = \frac{1}{2}(1 - 1) = 0$$
$$I_2 = \frac{1}{3}(1 + 1) = \frac{2}{3}$$
$$C_0 = (\hat{y} + 1)$$
$$C_1 = \frac{\hat{y}^2 - 1}{2}$$

so

$$\hat{v}_x - \hat{v}_1 = \frac{(\hat{v}_m - \{\hat{v}\})(\hat{y}^2 - 1)\hat{\eta}^{-2}}{2/3} + \hat{v}_d\left(\frac{\hat{y} + 1}{2}\right)$$

which gives

$$\hat{v}_x = \frac{3}{2}(\hat{v}_m - \{\hat{v}\})(\hat{y}^2 - 1) + \frac{\hat{v}_d}{2}\hat{y} + \left(\hat{v}_1 + \frac{\hat{v}_d}{2}\right)$$

Noting that

$$\hat{v}_D \equiv \hat{v}_2 - \hat{v}_1$$

and

$$\hat{v}_m \frac{\hat{v}_1 + \hat{v}_2}{2}$$

We find

$$\hat{v}_1 + \frac{\hat{v}_d}{2} = \hat{v}_m$$

Also from Table 5.4-1, realizing that $\partial \hat{p} / \partial \hat{x} = s$

$$s\text{Po} = \frac{2(\hat{v}_m - \{\hat{v}\})}{I_2}$$

$$= 3(\hat{v}_m - \{\hat{v}\})$$

Thus

$$\hat{v}_x = \frac{s\text{Po}}{2}(\hat{y}^2 - 1) + \frac{\hat{v}_d}{2}\hat{y} + \hat{v}_m$$

REFERENCES

1. J. A. Biesenberger and C. G. Gogos, *Polym. Eng. Sci.*, *Polymer Topics*, **20**, 837 (1980).
2. R. B. Bird, R. C. Armstrong, and O. Hassager, *Dynamics of Polymeric Liquids*, *Vol. 1*, *Fluid Mechanics*, Wiley, New York (1977).
3. P. J. Carreau, Ph.D. thesis, University of Wisconsin, Madison, WI (1968).
4. M. L. Williams, R. F. Landel, and J. D. Ferry, *J. Am. Chem. Soc.*, 77, 3701 (1955).
5. S. Glasstone, K. J. Laidler, and H. Eyring, *Theory of Rate Processes*, McGraw-Hill, New York (1941).
6. J. C. Yang, Ph.D. thesis in Chemical Engineering, Stevens Institute of Technology, Hoboken, NJ (1978).
7. F. Bueche, *Physical Properties of Polymers*, Interscience, New York (1961).
8. V. R. Allen and T. G. Fox, *J. Chem. Phys.*, **41**, 337 (1964).
9. S. Onogi, T. Kabayashi, Y. Kojima, and Y. Taniguchi, *J. Appl. Polym. Sci.*, **7**, 817 (1963).
10. S. Onogi, S. Kinuta, T. Kato, T. Masuda, and N. Miyanagia, *J. Polym. Sci.*, **C15**, 381 (1966).
11. M. Williams, *AIChE J.*, **13**, 534 (1967).
12. J. Dunleavy and S. Middleman, *Trans. Soc. Rheol.*, **10**, 157 (1966).
13. S. Lynn and J. E. Huff, *AIChE J.*, **17**, 475 (1971).
14. Dow Chemical Corp., private communication.
15. R. S. Spencer and J. L. Williams, *J. Colloid. Sci.*, **2**, 117 (1947).
16. T. G. Fox and V. R. Allen, *J. Chem. Phys.*, **41**, 344 (1964).
17. L. Valsamis, Ph.D. thesis in Chemical Engineering, Stevens Institute of Technology, Hoboken, NJ (1978).
18. S. D. Lipshitz and C. W. Macosko, *Polym. Eng. Sci.*, **16**, 803 (1976).
19. E. B. Richter and C. W. Macoslo, *Polym. Eng. Sci.*, **20**, 921 (1980).
20. J. D. Domine and C. G. Gogos, *Polym. Eng. Sci.*, **20**, 13 (1980).
21. J. P. Wallis, Ph.D. thesis in Chemical Engineering, University of Calgary (1973).
22. A. Hussain and A. E. Hamielec, *AIChE. Symp. Ser. 160*, **72**, 112 (1976).
23. L. Valsamis and J. A. Biesenberger, *AIChE Symp. Ser. 160*, **72**, 18 (1976).
24. W. M. Rohsenow and H. Y. Choi, *Heat Mass and Momentum Transfer*, Prenctice-Hall, New Jersey (1961).
25. K. H. Lee and J. P. Marano, *Polymerization Reactors and Processes* ACS Symposium Series 104, American Chemical Society, Washington, D.C. (1979), p. 221.
26. S. Agrawal and C. D. Han, *AIChE J.*, **21**, 449 (1975).
27. C. H. Chen, J. G. Vermeychuk, J. A. Howell, and P. Ehrlich, *AIChE J.*, **22**, 463 (1976).
28. R. Cintron-Cordero, Ph.D. thesis in Chemical Engineering, Stevens Institute of Technology, Hoboken, NJ (1971).

29. R. Cintron-Cordero, R. A. Mostello, and J. A. Biesenberger, *Can. J. Chem. Eng.*, **46**, 434 (1968).

30. R. A. Mostello, Master's thesis in Chemical Engineering, Stevens Institute of Technology, Hoboken, NJ (1967).

31. R. Cintron-Cordero, Master's thesis in Chemical Engineering, Stevens Institute of Technology, Hoboken, NJ (1967).

32. L. F. Albright, *Processes for Major Addition Type Plastics and Their Monomers*, McGraw-Hill, New York (1974).

33. J. M. Castro and C. W. Macosko, "Proceedings of the First International Conference of Reactive Processing of Polymers," University of Pittsburgh, Pittsburgh (1980).

34. L. T. Manzione, "Proceedings of the First International Conference of Reactive Processing of Polymers," University of Pittsburgh, Pittsburgh (1980); L. T. Manzione, *Polym. Eng. Sci.*, **21**, 1234 (1981).

35. Z. Tadmor and C. G. Gogos, *Principles of Polymer Processing*, Wiley, New York (1979).

36. W. A. Mack, "Proceedings of the First International Conference of Reactive Processing of Polymers," University of Pittsburgh, Pittsburgh (1980).

37. Z. Tadmor, P. Hold, and L. Valsamis, "The DISKPACK Polymer Processor; A Novel Polymer Processing Machine," 39th SPE ANTEC, New Orleans (1979); P. Hold, Z. Tadmor, and L. Valsamis, "Applications and Design of the DISKPACK Polymer Processor," 39th SPE ANTEC, New Orleans (1979).

38. L. R. Schmidt, *Polym. Eng. Sci.*, **14**, 797 (1974).

39. R. B. Bird, W. E. Stewart, and E. N. Lightfoot, *Transport Phenomena*, Wiley, New York (1960).

40. J. T. Lindt, *Polym. Eng. Sci.*, **21**, 424 (1981).

41. J. T. Lindt, "Flow of Polymerizing Fluid Between Rotating Concentric Cylinders," Eigth International Congress on Rheology, Naples (1980).

42. A. Ya. Malkin, *Polym. Eng. Sci.*, **20**, 1035 (1980).

CHAPTER SIX

Polymer Devolatilization

Virtually all polymerization products require a postreactor separation process to remove residual small-molecule substances, such as unreacted monomer or diluent (e.g., solvent or water). Examples are food packaging materials and polymers whose end-use properties are adversely affected by contaminants. Reactor discharges generally contain such substances in the percent range and often require purification to the order of parts per million. Since the contaminants in most cases are volatile relative to the polymer product, they are removed from the condensed (molten or solid) phase by evaporation into a contiguous gas phase. While such separation processes are often collectively termed devolatilization (DV), at least two regimes are distinguishable. One involves gross evaporation of volatile component from the reactor effluent and the other ultimate purification to exacting specifications. They would perhaps be more aptly described as vacuum stripping when the volatile component is present in large proportion and devolatilization when it is a trace component. In the latter case we encounter solutions of small-molecule substances in polymer, which lie at the opposite end of the concentration spectrum from the more conventional "dilute" polymer solutions. Thus the solvent (or monomer) becomes the solute when polymer is the major component.

DV may be effected from films or particles consisting of melts or solids. In many industrial processes, molten polymer containing dissolved volatile component is supersaturated through a sudden reduction in external pressure P to a value less than the equilibrium partial pressure of the dissolved solute P_s under feed conditions. This procedure, which in principle invites boiling, is then repeated N times in successive "stages." Such stages are often inherent, as we shall learn, in the flow patterns that prevail in rotating processing machinery.

In the high-concentration regime, bubble transport (foam DV) is known to occur in melts. How effectively and for how long bubbles nucleate, grow, and burst depends on many factors, among them being degree of supersaturation, $P_s - P$, presence of impurities (solids, gases), diffusivity, and melt properties (viscosity and elasticity), which can cause bubbles to "freeze," leaving a quasi-stable foam. At low concentrations, diffusion should be the only remaining transport mechanism, as it is in solids. Therefore, we shall consider both diffusion-limited evaporation and foam DV. However, while diffusion theory is fairly well developed, and is thus readily applied to DV processes, this is not the case for bubble mass transfer.

In this chapter, our strategy will be to subdivide DV processes, irrespective of the mechanisms or specific equipment involved, into elementary steps that are common to all. The elementary steps will be analyzed individually and in various combinations that simulate simple DV processes or stages within complex processes. The role of these steps in the performance of a stage and, in turn, the role of stages in the performance of the process will be examined in detail. Thus to model an actual process requires the synthesis of these elements and stages into a configuration that simulates the behavior of the process in question.

We define the elementary steps in all DV processes as: (1) interfacial surface generation; (2) interfacial mass transfer; and (3) surface renewal (when possible, i.e., when $T > T_g$). The first and third steps are essentially fluid mechanical in nature and the second one is a mass transfer process. Interfacial areas can be generated by spreading melt films, growing bubbles or grinding solid particles. Mass transfer, as noted, could occur by diffusion or bubble rupture. Surface renewal primarily describes convective mixing (random or ordered rearrangement of streamlines), but it could also be applied to bubble processes. Foam DV will be further subdivided into its component steps: bubble nucleation, growth, motion, and rupture.

Three basic configurations, or evaporation elements, are distinguishable in most industrial processes. One configuration is, or course, the solid particle and the other two are melt with shear and melt without shear. Two melt configurations that appear simultaneously in all rotating DV machinery are the translating film and the rotating pool. Emphasis will be given to the last two configurations.

In our treatment of the separation process, we shall use the two approaches that are traditional to chemical engineering, that is, equilibrium and rate. The equilibrium approach gives the maximum degree of separation allowed by thermodynamics. It manifests itself in the form of partition laws for the distribution of matter between phases. The goal of the rate approach, on the other hand, is the formulation of models that describe the rate of separation and its dependence on process design and operating parameters and material properties.

6.1. EQUILIBRIUM THEORY

Before discussing equilibrium and rate theories, it is useful to review briefly some concentration scales common to each. Molar concentration (c_k for any substance k) and mass concentration (ρ_k) are commonly used in transport theory to formulate rate models (see Appendix F). For thermodynamic models, on the other hand, it is customary to use mole fraction (n_k) for small molecule solutions (see Appendix D). However, as Flory (1) has pointed out, volume fraction ϕ_k is more appropriate for polymer solutions in general. Furthermore, when dealing with devolatilization partition equilibrium in particular, weight fraction (w_k) has come into common usage in the literature (2). An attempt will be made to reconcile these various concentration scales.

Mole fraction of solvent (or monomer) is expressed by

$$n_s = \frac{N_s}{N_s + N_p} \tag{6.1-1}$$

weight fraction by

$$w_s = \frac{N_s M_s}{N_s M_s + N_p M_0 x} \tag{6.1-2}$$

and volume fraction by

$$\phi_s = \frac{N_s v_s}{N_s v_s + N_p v_p} \tag{6.1-3}$$

where v_s and v_p are molar volumes occupied by solvent and polymer molecules in solutions, respectively, It is customary (1) to write for monodisperse polymers

$$v_p = x' v_s \tag{6.1-4}$$

where x' is the number of segments of volume v_s occupied by the polymer chain, as contrasted with x which represents the number of structural repeat units. Substitution of equation 6.1-4 into 6.1-3 leads to a relationship between volume fraction and mole fraction

$$\phi_s = \frac{n_s}{n_s + x' n_p} \tag{6.1-5}$$

Alternatively, we could write

$$v_p = x v_1 \tag{6.1-6}$$

where v_1 is the molar volume occupied by a structural repeat unit. Substitution of equation 6.1-6 into 6.1-3 leads to

$$\phi_s = \frac{N_s M_s}{N_s M_s + \dfrac{\rho_s}{\rho_p} x M_0 N_p} \tag{6.1-7}$$

where the quantity $v_1 M_s / M_0 v_s$ has been replaced by the ratio of solvent density to polymer density, ρ_s / ρ_p. Comparison of equations 6.1-7 and 6.1-2 reveals that when this ratio is unity

$$\phi_s = w_s \tag{6.1-8}$$

More generally, for dilute solutions with respect to solvent (monomer), we obtain the following useful approximation to equation 6.1-7

$$\phi_s \cong \frac{\rho_p}{\rho_s} w_s \tag{6.1-9}$$

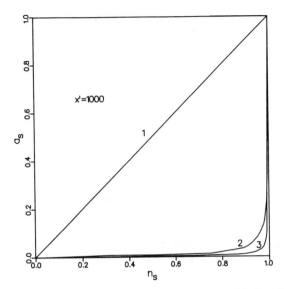

Figure 6.1-1. Variation of solute activity a_s with mole fraction n_s for: an ideal small-molecule solution, $a_s = n_s$ (curve 1); an ideal polymer solution, $a_s = \phi_s$ with $x = 1000$ (curve 2); a real polymer solution, data of Gee and Treloar in reference 1 (curve 3).

Thus we see that volume fraction is similar to weight fraction. Both concentrations in effect smooth out the disparity among large and small molecules reflected by their molar concentrations by amplifying the importance of the repeat unit in the polymer chains. This disparity is vividly illustrated in Fig. 6.1-1, which contains a graph of equation 6.1-5 for $x' = 1000$ ($a_s = \phi_s$).

6.1.1. Vapor-Liquid Equilibria

For ideal solutions composed of small molecules, the activity of any substance would be given by (3)

$$a_s = n_s \tag{6.1-10}$$

For solutions of polymers and small molecule species Flory (1) has proposed

$$a_s = \phi_s \tag{6.1-11}$$

as a condition for ideality. Both functions have been plotted in Fig. 6.1-1 together with a simulation of a_s for a "real" polymer solution. It is evident that ϕ_s is more appropriate than n_s. Nevertheless, there is an additional component of nonideality in real polymer solutions, especially in the region of interest in polymer devolatilization (dilute solutions of s), which has been amplified in Fig. 6.1-2 by replotting the curves of Fig. 6.1-1. We note in this figure that real polymer solutions become ideal in the sense of equation 6.1-10 ($a_s \rightarrow n_s$) when dilute with respect to polymer and not with respect to volatile component s.

Thermodynamic theories of polymer solutions have been developed by Flory and Huggins (4–8). They are based on a pseudo-lattice model of liquid mixtures

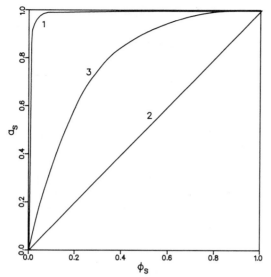

Figure 6.1-2. Variation of activity a_s with volume fraction ϕ_s for: an ideal small-molecule solution, $a_s = n_s$ (curve 1); an ideal polymer solution, $a_s = \phi_s$ with $x = 1000$ (curve 2); a real polymer solution, data of Gee and Treloar in reference 1 (curve 3).

composed of solvent (e.g., monomer) and long-chain polymer molecules, where each lattice cell may be occupied either by a molecule of solvent or by a segment of polymer chain equal in volume to the solvent molecule. For monodisperse polymer in solution Flory–Huggins theory (1) leads to the following expression for configurational entropy of mixing,

$$\Delta S_M = -R_g \left(N_s \ln \phi_s + N_p \ln \phi_p \right) \qquad (6.1\text{-}12)$$

This has been extended to polydisperse polymers

$$\Delta S_M = -R_g \left(N_s \ln \phi_s + \sum_{x'} N_{x'} \ln \phi_{x'} \right) \qquad (6.1\text{-}13)$$

where

$$\phi_{x'} \equiv \frac{x' N_{x'}}{N_s + \sum x' N_{x'}} \qquad (6.1\text{-}14)$$

According to Flory (9), this equation suffices for polydisperse systems as long as the DPD does not change, as it can with chemical reaction or interphase transport of polymer.

Equations 6.1-12 and 6.1-13 are to be compared with the expression for ideal small-molecule solutions (see Appendix D, equation D-57),

$$\Delta S_M = \sum_k N_k S_k^M = -R_g \sum_k N_k \ln n_k \qquad (6.1\text{-}15)$$

which they approach in the limit $x' \to 1$, as they must since $\phi_s \to n_s$ and $\phi_1 \to n_1$. When a Van Laar type heat of mixing term is included, the corresponding equations for Gibbs free energy of mixing are, respectively,

$$\Delta G_M = R_g T \left(N_s \phi_p \chi + N_s \ln \phi_s + N_p \ln \phi_p \right) \tag{6.1-16}$$

and

$$\Delta G_M = R_g T \left(N_s \phi_p \chi + N_s \ln \phi_s + \sum_{x'} N_{x'} \ln \phi_{x'} \right) \tag{6.1-17}$$

where χ is a molecular interaction parameter, which is supposed to be proportional to T^{-1} and independent of ϕ_p. Values of χ for various systems have appeared in the literature (2). They generally range from 0.2 to 0.5.

Differentiation of equations 6.1-16 and 6.1-17 with respect to n_s yields the chemical potentials (equations D-38 and D-52) for solvent in monodisperse polymer.

$$\mu_s = \mu_s^0 + \underbrace{R_g T \chi \phi_p^2}_{H_s^M} + \underbrace{T\left[R_g \ln \phi_s + \left(1 - \frac{1}{x'} \right) \phi_p \right)}_{-S_s^M} \tag{6.1-18}$$

and in polydisperse polymer

$$\mu_s = \mu_s^0 + \underbrace{R_g T \chi \phi_p^2}_{H_s^M} + \underbrace{T\left[R_g \ln \phi_s + \left(1 - \frac{1}{\bar{x}_N} \right) \phi_p \right]}_{-S_s^M} \tag{6.1-19}$$

respectively.

From these results, we deduce (see equation D-56) that the activity of solvent in solution

$$a_s = \exp\left(\frac{\mu_s - \mu_s^0}{R_g T} \right) \tag{6.1-20}$$

is given by the expression

$$a_s = \gamma_\phi \phi_s \tag{6.1-21}$$

when the standard state is pure solvent. The activity coefficient has been defined as

$$\gamma_\phi \equiv \exp\left[\left(1 - \frac{1}{\bar{x}_N} \right) \phi_p + \chi \phi_p^2 \right] \tag{6.1-22}$$

and we have approximated $\bar{x}'_N$ with $\bar{x}_N$ in equation 6.1-22 to expedite its application. It is customary to define the general partition equation for vapor-liquid equilibrium (VLE) as

$$P_s = P_s^0 \gamma_\phi \phi_s \tag{6.1-23}$$

For high-molecular-weight polymer ($\bar{x}_N \gg 1$), we may use the approximation

$$\gamma_\phi \approx \exp\left(\phi_p + \chi\phi_p^2\right) \tag{6.1-24}$$

Owing to the choice of standard state for s, we note that while γ_ϕ approaches unity (and hence $a_s \to \phi_s$) for dilute polymer solutions ($\phi_p \to 0$), it approaches a limiting value other than unity at low solvent concentrations (see Appendix D). When $\phi_p \to 1$, the activity coefficient is approximately given by

$$\gamma_\phi \approx \exp(1 + \chi) \tag{6.1-25}$$

It may then be incorporated into a Henry's law constant (equation D-66),

$$K_\phi \equiv P_s^0 \exp(1 + \chi) \tag{6.1-26}$$

which is independent of composition and applies in the region of ideality ($\phi_s \to 0$) of interest in DV. Thus we obtain for equation 6.1-23

$$P_s = K_\phi \phi_s \tag{6.1-27}$$

As we have stressed earlier, w_s represents a convenient approximation of ϕ_s (equation 6.1-8), which may be evaluated from the former via equation 6.1-9. It is therefore not surprising to find the partition equation

$$P_s = P_s^0 \gamma_w w_s \tag{6.1-28}$$

more widely used than 6.1-23, where an alternative activity coefficient, γ_w, has been introduced. The two activity coefficients are related via

$$\frac{P_s}{P_s^0} \equiv a_s = \gamma_\phi \phi_s = \gamma_w w_s \tag{6.1-29}$$

After assembling the appropriate relationships, we conclude that γ_w may be computed from the expression

$$\gamma_w = \underbrace{\frac{\rho_p}{\rho_s} \exp\left(\phi_p + \chi\phi_p^2\right)}_{\gamma_\phi} \tag{6.1-30}$$

and the corresponding alternative Henry's law constant from

$$K_W = P_s^0 \gamma_w \tag{6.1-31}$$

in the limit as $\phi_p \to 1$.

The effect of temperature on the partition equation enters via the temperature dependence of the interaction parameter $\chi(T)$, which is a heat-of-mixing term, and vapor pressure P^0, whose temperature dependence is given by the well-known

Table 6.1-1. **VLE Data**

System	T(°C)	χ	K_w(atm.)	Reference
PS/S	171	0.284	7.92	10
	260	Example 6.1-1	50	14
PS/EB	171	0.319	11.0	10
PE/Et	250		2050	12
PE/Hex	250	0.44	115	12
PVC/VC	90	0.97	212	13

Clausius–Clapeyron equation (3)

$$\frac{d \ln P^0}{dT^{-1}} = \frac{-\Delta H_v(T)}{R_g} \tag{6.1-32}$$

where $\Delta H_v(T)$ is the heat of vaporization, which is also temperature-dependent. Frequently χ is sufficiently constant over limited ranges of temperature for practical purposes, and the temperature dependence of K_w is dominated by equation 6.1-32.

Vapor-liquid equilibrium data for some common polymer/volatile systems are listed in Table 6.1-1. Corresponding Flory–Huggins equilibrium graphs have been plotted in Figs. 6.1-3–6.1-5. Comparisons with Henry's law, Raoult's law, and LVE data are included where appropriate. References are listed in the table. Ethyl benzene (EB) has been included because it is sometimes monitored in lieu of styrene (S). Besides being chemically similar to styrene, it is also frequently present in polystyrene (PS) as a by-product of styrene manufacture. EB has several advantages over S. It is neither depleted by thermally polymerizing, nor formed by depolymerization of PS at very high temperatures ($> 500°F$).

At any temperature and pressure, the fraction of impurity in the feed that is extractable theoretically, assuming trace quantities, is approximately $(w_0 - w_e)/w_0$, where w_e is the equilibrium concentration corresponding to that temperature and pressure, and w_0 is the initial (feed) concentration. Although equilibrium theory provides the means for estimating w_e, it does not concern itself with the time required to attain this value, which can be very long. When rate processes control DV, the final (effluent) concentration w_f is greater than w_e, and therefore the extractable fraction $(w_0 - w_f)/w_0$ is less than the theoretical value. In fact, we shall use their quotient as a basis for defining separation efficiency.

Example 6.1-1. Flory–Huggins Interaction Parameter. Using the following properties of styrene (S) and polystyrene (PS) at 260°C, estimate the value of the Flory–Huggins interaction parameter χ at this temperature; $P_s^0 = 10$ atm, $\rho_s = 0.67$ g/cm³, and $\rho_p = 0.98$ g/cm³.

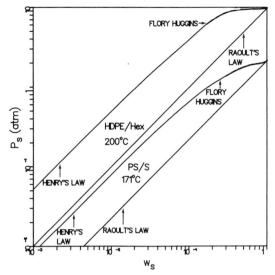

Figure 6.1-3. Logarithmic relationship of equilibrium partial pressure P_s to weight fraction w_s for the systems high-density polyethylene–hexane, HDPE/Hex, at 200°C and polystyrene–styrene, PS/S, at 171°C contrasting the regimes of Henry's law, Raoult's law, and the Flory–Huggins equation.

Solution. From Table 6.1-1, $K_W = 50$ atm. Under conditions for which Henry's law applies, $K_W = P_s^0 \gamma_w$, where $\gamma_w \cong (\rho_p/\rho_s)\exp(1 + \chi)$. Substituting the preceding properties for PS and S yields $\chi = 0.23$.

Example 6.1-2. Henry's Law Constant. Using the following properties of vinyl chloride (VC) and polyvinyl chloride (PVC) and $\chi = 0.97$, check the value of K_W for

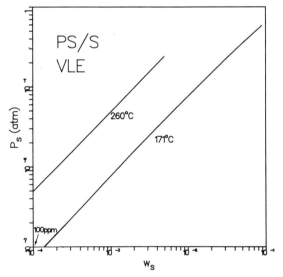

Figure 6.1-4. Equilibrium partial pressure P_s as a function of weight fraction w_s for PS/S at two temperatures, according to Flory–Huggins (equations 6.1-28 and 6.1-30).

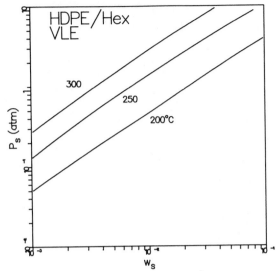

Figure 6.1-5. Equilibrium partial pressure P_s as a function of weight fraction w_s for HDPE/Hex at three temperatures, according to Flory–Huggins (equations 6.1-28 and 6.1-30).

PVC/VC at 90°C listed in Table 6.1-1: $P_s^0 = 17.8$ atm, $\rho_s(l) = 0.85$ g/cm³, and $\rho_p = 1.4$ g/cm³.

Solution. Substituting these properties into the equation

$$K_W = P_s^0 \frac{\rho_p}{\rho_s} \exp(1 + \chi)$$

yields

$$K_W = 210 \text{ atm}$$

which is close to the value listed.

Example 6.1-3. Equilibrium Composition by Henry's Law. Molten polystyrene produced in a bulk polymerization process and flashed to a styrene content of 1.0 percent requires further devolatilization at 260°C to a final content of 100 ppm. Vacuum at 10 Torr is available. Is it sufficient and, if not, what final composition is attainable theoretically at this pressure?

Solution. The required pressure from Table 6.1-1 is

$$P = K_W w_s = (50)(760)(1 \times 10^{-4}) = 3.8 \text{ Torr}$$

which is obviously lower than that available. At 10 Torr, we can expect a final composition of

$$w_s = P/K_W = \frac{10}{(760)(50)} = 2.63 \times 10^{-4} = 263 \text{ ppm}$$

at best.

Example 6.1-4. Equilibrium Pressure by Henry's Law. Owing to its toxicity, VC in PVC must be reduced in composition to 1 ppm or less. From equilibrium and rate considerations, the higher the temperature used, the better. However, since suspension polymerization is the most common mode for VC, it is most desirable to evaporate from "solid" beads, that is, at temperatures not far above T_g, say 90°C. What level of vacuum will be required?

Solution. From Table 6.1-1, we obtain

$$P \leqslant (212)(760)(0.000001) = 0.167 \text{ mm Hg}$$

which is quite low for large industrial processes.

6.1.2. Effect of Inert Substances

The beneficial effects on vapor-liquid separations of introducing fluid substances that are immiscible with the condensed phase are well known. We shall separate these effects into two categories: equilibrium effects and rate effects.

In the first category, an insoluble fluid is added which permeates the gas phase. The basic principle may be easily demonstrated in terms of Henry's law

$$w_s = \frac{n_s P}{K_W} \tag{6.1-33}$$

where n_s now represents the mole fraction of volatile component s in the gas phase. Obviously, as the fraction of inert substance in the gas phase increases at fixed total pressure P, n_s decreases, as does the corresponding equilibrium fraction of volatile component w_s in the condensed phase. In principle, n_s, and therefore w_s, may be reduced to zero by sweeping the interface with inert gas. Nitrogen would be a suitable inert gas, for example. Steam, which could initially be injected as water, is another.

In the second category, the principal mechanism involved is a rate effect. If the inert fluid that is added to the liquid solution is also a liquid, the combined vapor pressures cause the solution to boil (when the total pressure P is exceeded) at lower concentrations and temperatures than it otherwise would and thereby enhances its evaporation rate. If the fluid is dispersed gas (bubbles), the effect is also kinetic, that is, the large gas-liquid interface created by the bubbles enhances interphase mass transport by diffusion. The former is called steam distillation when the inert liquid is water, and the latter is called sparging.

These techniques have been employed in polymer melt DV (14) using both water and steam injection. One might expect water to be less difficult to disperse in molten polymer than steam, and steam bubbles to be more efficient than water bubbles for absorbing dissolved organic volatiles. The first speculation is based on the observation that, in general, the more similar the viscosities of two fluids are, the more readily they are mixed mechanically. The second speculation is based on the fact that gases are miscible in all proportions.

To estimate the equilibrium effect of inert substances, we refer to the equilibrium stage illustrated in Fig. 6.1-6. An equilibrium stage is an idealization in which all

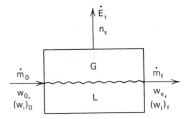

Figure 6.1-6. A vapor-liquid equilibrium (VLE) stage.

effluent streams are in equilibrium with each other. Symbols $\dot{m}$ and w represent mass flow rate and weight fraction respectively, and subscript i signifies inert substance. No subscript is used for volatile component in the melt. Total molar evaporation rate including inert substance is designated by $\dot{E}_t$. We require the quantity n_s for substitution into equation 6.1-33 to determine the final composition of volatile w_e, which must be an equilibrium composition because we have an equilibrium stage.

Two distinct situations suggest themselves. In the first, not all inert substance entering is vaporized ($P_I^0 < P$), and therefore some leaves as vapor and some as liquid. In the second, all is vaporized, so that $(w_I)_f = 0$.

In the first case n_s may be computed simply as follows:

$$n_s = 1 - \frac{P_I^0}{P} \tag{6.1-34}$$

where P_I^0 represents the vapor pressure of pure substance I. In the second case, n_s may be computed from the following material balances: For component s

$$n_s \dot{E}_t = \frac{\dot{m}_0 w_0}{M_s} - \frac{\dot{m}_f w_e}{M_s} \tag{6.1-35}$$

and for both volatile components

$$\dot{E}_t = \frac{\dot{m}_0 (w_I)_0}{M_I} + \frac{\dot{m}_0 w_0 - \dot{m}_f w_e}{M_s} \tag{6.1-36}$$

where M_s and M_I are molecular weights. Elimination of $\dot{E}_t$ between these equations yields an expression for n_s. A sample calculation when water is the inert substance is shown in Example 6.1-4.

Example 6.1-4.   VLE with Water Injection.   Will injection of 1-percent water into the feed accomplish the desired separation in Example 6.1-3?

Solution.   Assuming that all the water is vaporized, we have an example of the second case. By combining equations 6.1-33, 6.1-35, and 6.1-36 and neglecting the loss in total mass due to evaporation, that is, assuming $\dot{m}_f \cong \dot{m}_0$, we obtain

$$\frac{K_w w_e}{P} = \frac{w_0 - w_e}{w_0 - w_e + \frac{(w_I)_0 M_s}{M_I}}$$

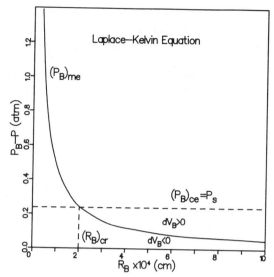

Figure 6.1-7. Relationship of excess bubble pressure, $P_B - P$, to bubble radius R_B according to the Laplace–Kelvin equation, $P_B - P = 2\sigma/R_B$, for $\sigma = 25$ ergs/cm^2.

After substituting the following values: $w_0 = 10^{-2} = (w_I)_0$, $M_s = 104$, $M_w = 18$, and $K_W = 50$ atm, we obtain the result $w_e \cong 3.9 \times 10^{-5}$ or 39 ppm, which represents a better separation than required.

6.1.3. Bubble Equilibria

In gas-liquid systems with curved interfaces, the condition for mechanical equilibrium does not require uniform pressure as it does for flat interfaces. In fact, the gas pressure (P_B) in a submerged bubble at equilibrium ($P_B = P_{me}$) must exceed the hydrostatic pressure P of the surrounding liquid. Specifically, the two pressures are related by the well-known Laplace–Kelvin equation (Example 6.1.5) for spherical bubbles

$$P_{me} = P + 2\sigma/R_B \tag{6.1-37}$$

where σ represents surface tension and R_B the radius of the bubble. The locus of mechanical equilibrium states is shown in Fig. 6.1-7 for $\sigma = 25$ ergs/cm^2.

Most nucleation theories (15) apply only to pure (single-component) fluids and ignore the effects of viscous forces on the formation of voids. In DV, we must deal with multicomponent systems in which elastic forces are generated in addition to large viscous forces. It is evident from the Laplace–Kelvin equation that even without such forces extremely large pressures due to surface tension alone are required to sustain bubbles when R_B is small. Classical theory states that when actual bubble pressure P_B exceeds P_{me}, a bubble should grow; when it is less, the bubble should collapse. In other words, the equilibrium is unstable (Example 6.1-6). These conditions are represented by the regions above and below the curve in Fig. 6.1-7, respectively.

Partly because DV is frequently conducted in thin films and partly because molten polymers are rubbery materials, we expect bubbles of two fundamentally distinct types to exist at different stages during DV. Initially, bubbles are small and may be considered to be suspended in an "infinite" medium. This is the situation associated with equation 6.1-37. Following their growth, we would ultimately expect the bubbles to become thin-walled, or balloonlike. The hoop stress (s) for such bubbles with elastic walls is related to pressure by

$$P_{\text{me}} = \frac{2st}{R_B} \tag{6.1-38}$$

where t is the wall thickness. Both types are illustrated in Fig. 6.2-1.

When the medium surrounding a gas bubble is elastic, Gent et al. (16) have proposed that

$$P_{\text{me}} = \frac{G}{2}\left(5 - 4\hat{R}_B^{-1} - \hat{R}_B^{-4}\right) \tag{6.1-39}$$

for spheres of the first type and

$$P_{\text{me}} = 2G\frac{t_0}{(R_B)_0}\left(\hat{R}_B^{-1} - \hat{R}_B^{-7}\right) \tag{6.1-40}$$

for spheres of the second type, where G is shear modulus, t_0 is wall thickness before inflation, $(R_B)_0$ is the radius in the undeformed state, and $\hat{R}_B \equiv R_B/(R_B)_0$ is the reduced radius.

Assuming that the gases inside the bubble are perfect, the bubble pressure is simply the sum of the partial pressure of the vapor that evaporates from the melt and the partial pressure of any other gas that is present. Thus, for example,

$$P_B = P_s + P_I \tag{6.1-41}$$

where P_I is the partial pressure of the other (inert) gas and P_s is the partial pressure of the volatile vapor, which is related at chemical equilibrium to the composition of the condensed phase by a partition law such as Henry's law,

$$P_s = K_W w \equiv K_c c \tag{6.1-42}$$

where

$$K_c \equiv \frac{M_s}{\rho_p} K_W \tag{6.1-43}$$

and subscript s has been dropped for convenience from weight fraction w and concentration c.

For the special case in which the bubble contains only vapor from the dissolved volatile component, $P_I = 0$ and at chemical equilibrium $P_B \equiv P_{\text{ce}} = P_s$. Thus, by neglecting viscoelastic stresses during bubble birth and by assuming that surface tension dominates (very small R_B), we can define a critical radius (Fig. 6.1-7)

$$(R_B)_{cr} \equiv \frac{2\sigma}{P_s - P} \qquad (6.1\text{-}44)$$

based on equation 6.1-37. Since the initial bubble pressure cannot exceed P_s, we expect a viable vapor bubble to have an initial radius $(R_B)_0$ at least as large as the critical value $(R_B)_0 \geqslant (R_B)_{cr}$. A bubble whose radius at birth is less than the critical value should collapse, and one whose radius exceeds it should grow spontaneously by virtue of the inequality $P_{me} < P_s$. The critical radii for a specific polymer-monomer system have been estimated in Example 6.1-7. Bubble nucleation will be discussed in more detail in Section 6.2.

If a swarm of bubbles nucleate simultaneously or successively, we must account for the loss of volatile component in the condensed phase with a material balance. If volume shrinkage of the melt is neglected and all bubbles are assumed to be uniform in size (V_B), this balance is simply

$$c_0 - c \cong \frac{N_B}{V} c_B V_B \qquad (6.1\text{-}45)$$

where N_B/V represents the bubble population density (number of bubbles per unit melt volume). Bubble and melt concentrations can be related by Henry's law in the form

$$c_B = K_c' c \qquad (6.1\text{-}46)$$

where $K_c' = K_c/R_g T = (M_s/\rho_p R_g T)K_W$ (Example 6.1-8), and bubble concentration can be related to bubble pressure with an equation of state, such as the perfect gas law

$$P_B = c_b R_g T \qquad (6.1\text{-}47)$$

After combining the last three equations with the Laplace–Kelvin equation (6.1-37), we obtain a quartic equation for the critical radii of a swarm of N_B/V bubbles of uniform size. This computation is demonstrated in Example 6.1-9 and the results are illustrated in Fig. 6.1-8 for a specific set of physical conditions with applied vacuum (P) as the parameter (37).

The loci of critical radii in Fig. 6.1-8 are seen to exhibit maximum values for number density. The reason is a competitive effect between the large pressures required by the bubbles when they are small and the large quantities of volatile component when they are large, both of which are limited by the volatile concentration of the feed. Several conclusions from Fig. 6.1-8 are noteworthy. The first stems from the double-valued critical radius for all populations that are less dense than the maximum value. The implication of this locus with respect to the simple dynamic model to be formulated in Section 6.2, in which a constant, submaximum number of uniform-size bubbles grow following a perturbation from equilibrium $(R_B)_0 > (R_B)_{cr}$, is that the initial bubble radii that lie near the left legs of the curves in Fig. 6.1-8 should approach corresponding final radii that lie on the right legs, assuming, of course, that the latter represent stable equilibrium states.

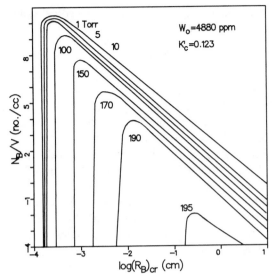

Figure 6.1-8. Number density N_B/V versus radius R_B with external pressure P as parameter for a swarm of uniform-size bubbles in chemical and mechanical equilibrium (equation in Example 6.1-9).

Another conclusion from the figure is that the critical radius for a very small population density ($\lim N_B/V \to 0$) is not much larger than that for a swarm of many bubbles. The former case is equivalent to that of a single bubble growing in an infinite medium without a final equilibrium state, which is the case most often treated in the literature. Note also that the population densities and critical radii are increasingly sensitive to applied vacuum as pressure increases. If we associate the radius of a bubble with a probability of nucleation, the smaller bubbles being more likely to nucleate than the larger ones, we would conclude that many bubbles are virtually as likely to nucleate as a single bubble, but that likelihood diminishes sharply as the degree of supersaturation declines.

Example 6.1-5.   The Laplace-Kelvin Equation.   Derive the Laplace–Kelvin equation in two different ways: from the laws of mechanics and from the laws of thermodynamics.

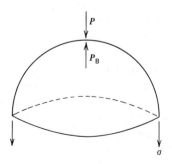

Solution. A static force balance on the bubble shown in the sketch, neglecting buoyancy, yields $\pi R_B^2 P_B = \pi R_B^2 P + 2\pi R_B \sigma$, from which we obtain the mechanical equilibrium pressure

$$P_B = P + \frac{2\sigma}{R_B} \equiv P_{me}$$

The Helmholtz free-energy function for pure phases in terms of its natural variables (Appendix D) is $A(T, V, S)$ and its differential form for an isothermal variation is $dA = -P\,dV + \sigma\,dS$, where S is surface area. Specifically, for the system consisting of a bubble in a liquid the total free energy is $A = A_B + A_L$ and the total change is $dA = dA_B + dA_L$. Substitution of differential forms into this expression yields

$$dA = -\left(P_B - P - \frac{2\sigma}{R_B}\right) dV_B$$

with the aid of the following relations: $dV_B = -dV_L$, $dV_B = 4\pi R_B^2 dR_B$, and $dS_B = 8\pi R_B dR_B = 2/R_B\,dV_B$. The criterion for thermodynamic equilibrium requires that $dA = 0$ for all variations dV_B. From this, we conclude that the expression in brackets must be zero, or equivalently that

$$P_B = P + \frac{2\sigma}{R_B}$$

Example 6.1-6. Instability of Bubble Equilibrium. Using the results of Example 6.1-5, show that when $P_{me} < P_B$ the bubble will grow spontaneously ($dV_B > 0$).

Solution. From the differential form rewritten as

$$dA = -(P_B - P_{me})\, dV_B$$

and from the thermodynamic criterion for spontaneous processes, $dA < 0$, we conclude that $dV_B > 0$ when $P_{me} < P_B$. Conversely, when $P_{me} > P_B$, the bubble will collapse ($dV_B < 0$) spontaneously.

Example 6.1-7. Critical Radius of a Single Bubble. Estimate the critical radius $(R_B)_{cr}$ for a styrene bubble in molten polystyrene at 260°C for various external pressures from 0 to 10 mm Hg when $w_s = 0.001$ (1000 ppm). Use $\sigma = 25$ ergs/cm² (dynes/cm) and the conversion factor 10^{-6} atm = 1 dyne/cm².

Solution. From equation 6.1-44,

$$(R_B)_{cr} = \frac{2\sigma}{K_W w_s - P}$$

where $K_W = 50$ atm (Table 6.1-1). The results are tabulated below. They suggest the smallest bubble possible must exceed 10^{-3} cm and that no bubbles are possible when $P \geqslant 38$ mm Hg.

P (mm Hg)	$(R_B)_{cr}$ (cm)
0	1.00×10^{-3}
10^{-1}	1.00×10^{-3}
1	1.03×10^{-3}
10	1.36×10^{-3}
38	∞

Example 6.1-8.   Vapor Versus Liquid Molar Concentrations.   Compare the molar concentration of styrene dissolved in molten polystyrene at 171°C with that of styrene vapor in a bubble at equilibrium with the melt.

Solution.   From equation 6.1-46,

$$\frac{c_B}{c} = K'_c = \frac{K_W M_s}{\rho_p R_g T}$$

Using data for styrene and polystyrene ($M_s = 104$, $\rho_p = 0.98$ g/cm³, $K_W = 7.92$ atm) and the gas constant ($R_g = 82.1$ atm cm³/g mol K), we obtain $K'_c = 2.3 \times 10^{-2}$. Thus we conclude that the volatile component is much less concentrated in the bubble than it is in the melt phase where it nucleates.

Example 6.1-9.   Critical Radius of a Swarm of Bubbles.   Derive an expression for the locus of equilibrium (critical) radii and population densities of spherical bubbles of uniform size formed from a supersaturated solution of concentration c_0 after applying a vacuum P.

Solution.   By combining equations 6.1-45–6.1-47, we obtain

$$P_0 = P_B\left(1 + \frac{K'_c V_B N_B}{V}\right)$$

where P_0 represents the equilibrium partial pressure (P_s) corresponding to feed concentration c_0. Equation 6.1-37 is then substituted for P_B and the expression $V_B = (4/3)\pi R_B^3$ is introduced. The result is a quartic equation for N_B/V as a function of R_B, which represents the critical radii of a swarm of bubbles of uniform size.

$$P_0 = \left(P + \frac{2\sigma}{R_B}\right)\left(1 + \frac{4}{3}\pi K'_c \frac{N_B}{V} R_B^3\right)$$

6.1.4. Staging

The output rate (Section 1.3) of an equilibrium stage in moles of volatile component evaporated per unit time is

$$\dot{E} = (c_0 - c_e)\dot{V} \qquad (6.1\text{-}48)$$

assuming constant density and neglecting the decline in melt volume due to loss of volatile component. Obviously, the smaller c_e is, the better the separation and the greater the output rate for a fixed throughput rate $\dot{V}$. From the thermodynamic partition law, a small c_e requires a low partial pressure P_s, which in turn implies a low total pressure P when vacuum provides the driving force for DV.

On the other hand, an alternative expression for $\dot{E}$ from the perfect gas law is $P_s \dot{V}_G / R_g T$, where $\dot{V}_G$ is the volumetric displacement rate of the gas phase provided by the vacuum pump. From this, we conclude that the higher the partial pressure P_s is, for a fixed displacement rate $\dot{V}_G$, the greater will be the evaporation rate $\dot{E}$. Therefore, the need for a low value of P_s to achieve good separation competes with the desire for a high value of P_s to alleviate the practical limitation of vacuum pump capacity. This competition suggests that perhaps the total vacuum capacity required to achieve a given separation could be reduced by devolatilizing in stages, that is, by removing large quantities of the volatile component at high pressures and only the remaining traces at low pressures.

The jth stage of an N-stage cascade is illustrated in Fig. 6.1-9. Its output rate is

$$\dot{E}_j = (c_{j-1} - c_j)\dot{V} \qquad (6.1\text{-}49)$$

where c_j is related to pressure P_j by a partition law,

$$P_j = K_{cj}c_j \qquad (6.1\text{-}50)$$

if we assume that it is an equilibrium stage. The variable coefficient K_{cj} has been included mainly to allow for the possibility of a stagewise temperature distribution $T_j = T(j)$, for which $K_{cj} = K_c(T_j)$, and subscript e has been dropped from exit concentration c_j for convenience. The volumetric displacement rate required by the jth stage is

$$\dot{V}_{Gj} = \frac{R_g T_j \dot{E}_j}{P_j} \qquad (6.1\text{-}51)$$

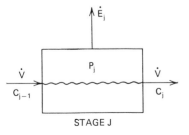

STAGE J

Figure 6.1-9. The jth stage of an N-stage cascade of separators in series.

After combining equations 6.1-49–6.1-51, we obtain

$$P_j - a_j P_{j-1} = 0 \qquad (6.1\text{-}52)$$

where

$$a_j \equiv \left[K_{cj-1} \left(K_{cj}^{-1} + \frac{\dot{V}_{Gj}}{R_g T_j \dot{V}} \right) \right]^{-1} \qquad (6.1\text{-}53)$$

For the special case $K_{cj} = K_c$ for all j ($T_j = T$ for all j), we obtain $a_j = R_g T\dot{V}/(R_g T\dot{V} + K_c \dot{V}_{Gj}) < 1$. The total vapor displacement rate is

$$\dot{V}_G = \sum_{j=1}^{N} \dot{V}_{Gj} \qquad (6.1\text{-}54)$$

where

$$\dot{V}_{Gj} = R_g T_j \dot{V} \left(\frac{1}{a_j K_{cj-1}} - K_{cj}^{-1} \right) \qquad (6.1\text{-}55)$$

from equation 6.1-53. The following quantities are often specified as constraints: c_0, c_N, P_0, and P_N. It is clear from Henry's law that these constraints fix T_0 and T_N as well. Thus, in general, we seek the pressure and temperature distributions, $P_j = P(j)$ and $T_j = T(j)$, for any given value of N that will minimize $\dot{V}_G$ subject to the constraints on the feed and effluent cited earlier.

When the temperature profile is uniform ($T_j = T$ for all j), the problem reduces to a search for the distribution $a_j = a(j)$ that minimizes the sum $\sum_{j=1}^{N}(1 - a_j)/a_j$ subject to the constraint $\pi_{j=1}^{N} a_j = P_N/P_0$. It can be shown that the optimal pressure profile under these conditions is (Example 6.1-10)

$$\frac{P_j}{P_{j-1}} = \text{constant} = \left(\frac{P_N}{P_0} \right)^{1/N} \qquad (6.1\text{-}56)$$

This result is consistent with the solution of equation 6.1-52 for the special case of constant $a_j = a$ for all j, which then reduces to the familiar first-order difference equation with constant coefficient

$$P_j - a P_{j-1} = 0 \qquad (6.1\text{-}57)$$

whose solution for the Nth stage, subject to the initial condition $P_j = P_0$ when $j = 0$, is

$$P_N = P_0 a^N \qquad (6.1\text{-}58)$$

where P_0 is the equilibrium pressure corresponding to c_0.

It is evident from equations 6.1-53 and 6.1-55 that a constant value of a_j does not necessarily imply that T_j and $\dot{V}_{Gj}$ are uniform. Many different combinations are admissible. Todd (17) stated without proof that pressure distribution 6.1-58 was the optimum one for staging vacuum. Then, using this distribution, he computed the volumetric displacement distribution and finally the total volumetric displacement required for an arbitrary rising temperature profile $T_j = T(j)$ to achieve a given separation subject to the constraints on feed and effluent previously discussed. Although this pressure distribution is not the optimum one for nonuniform temperature profiles as suggested, the practice of increasing T_j with j to partly offset the demand for declining values of P_j is reasonable. Furthermore, Todd did not neglect the mass lost by the melt during evaporation as we do (Examples 6.1-11 and 6.1-12).

Should both distributions $\dot{V}_{Gj}$ and T_j be uniform, which is likely in practice, we can deduce a simple expression that contrasts the total pump capacity required by a single stage, $(\dot{V}_G)_1$, with that of an N-stage cascade, $(\dot{V}_G)_N$, to achieve the identical final pressure $P_f = P_N$ (i.e., separation) and the identical total output rate $\dot{E}$. After equating output rates, $\dot{E}_1 = \sum_{j=1}^{N} \dot{E}_j$, substituting equations 6.1-51 and 6.1-57, and summing the result (Appendix C)

$$(\dot{V}_G)_1 P_0 a^N = \frac{(\dot{V}_G)_N P_0}{N} \sum_{j=1}^{N} a^j \qquad (6.1\text{-}59)$$

we obtain

$$\frac{(\dot{V}_G)_N}{(\dot{V}_G)_1} = \frac{Na^{N-1}(1-a)}{1-a^N} < 1 \qquad (6.1\text{-}60)$$

From this we conclude that, indeed, a reduction in vacuum capacity can be achieved under appropriate circumstances by staging (Example 6.1-11).

Example 6.1-10. Optimum Vacuum Staging. Find the distribution $a_j = a(j)$ that minimizes the sum $\sum_{j=1}^{N}(a_j^{-1} - 1)$ subject to the product $\prod_{j=1}^{N} a_j = $ constant $\equiv A$.

Solution. This problem is equivalent to finding the extremum of the following function

$$F \equiv \sum_{j=1}^{N} \left(a_j^{-1} - 1\right) + z \prod_{j=1}^{N} a_j$$

where z is a Lagrange multiplier. The requirements are that $\dfrac{\partial F}{\partial a_j} = 0$ for all a_j.

Differentiating and equating to zero yields

$$\frac{\partial F}{\partial a_l} = -a_l^{-2} + z \prod_{\substack{j=1 \\ j \neq l}}^{N} a_j = 0$$

Therefore,

$$z^{-1} = a_l^2 \prod_{\substack{j=1 \\ j \neq l}}^{N} a_j = Aa_l$$

and

$$a_l = (Az)^{-1} \equiv a \qquad \text{for all} \quad l$$

Inserting this result into the constraint equation yields

$$A = a^N$$

or

$$a = A^{1/N} = \text{constant}$$

Example 6.1-11.   Effect of Staging Vacuum.   Contrast the pump displacements required by a single stage and a three-stage cascade for reductions in volatile content of $c_f/c_0 = 0.5$ and 0.1, assuming that feed (0) and effluent (f) temperatures are identical.

Solution.   Substituting $a = (0.5)^{1/3}$ and $(0.1)^{1/3}$ into equation 6.1-60 yields the values shown in the following chart,

a^N	0.5	0.1
$(\dot{V}_G)_3/(\dot{V}_G)_1$	0.780	0.385

which indicate a significant potential reduction in vacuum capacity as a result of staging vacuum, especially for large concentration reductions.

Example 6.1-12.   Computation of Vacuum Requirements.   Polyethylene feed containing 6.7 percent by weight hexane at atmospheric pressure is to be devolatilized to a final hexane content of 0.5 percent. The highest vacuum available is 13 Torr. In order to achieve the desired final composition, it is necessary for the effluent temperature T_f to be higher than the feed temperature T_0. Values of K_W for feed and effluent, as well as for the arbitrary intermediate conditions (17), are listed in the following chart.

Stage	w_s	P (Torr)	K_W (Torr)	T(K)
0	0.0607	760	11,343	373
1		195	15,984	383
2		50	20,000	393
f	0.0005	13	25,600	403

Compute the distribution and total vacuum rate in cubic feet per hour required for 100 lb/h of polymer feed when $N = 3$. Compare the total with that when $N = 1$.

Solution.   After rewriting equation 6.1-55 as $\dot{V}_{Gj} = (R_g \dot{m} T_j/M_s)(1/aK_{W_{j-1}} - K_{W_j}^{-1})$ using the values $M_s = 86$ and $R_g = 998.6$ Torr ft^3/lb mol K, we compute the results listed in the following chart.

N	a	$\dot{V}_{G1}$ (ft^3/h)	$\dot{V}_{G2}$ (ft^3/h)	$\dot{V}_{G3}$ (ft^3/h)	$\dot{V}_{G}$ (ft^3/h)
1	0.0168	1070	—	—	1070
3	0.256	125	88.7	73.1	287

6.2. RATE THEORY

The rational design and scale-up of DV equipment requires a model for evaporation rate that contains all pertinent design and operating parameters for the system in question. To facilitate the development of such models, we shall analyze separately the evaporation elements common to all DV processes.

From equilibrium considerations alone, it is not possible to estimate the size (volume) of continuous process equipment required to achieve a given output, because time (space time, λ_v) is excluded. It is presumed that sufficient time is provided for the attainment of equilibrium regardless of how long it takes. Actually, most industrial processes are rate-limited and the volume required to achieve a given output must be determined by rate considerations. Therefore, we shall express separation efficiency as fractional output in which maximum output possible is based on thermodynamic equilibrium (Section 1.3).

Evaporation by diffusion and by flashing are both rate processes. Obviously the latter should be faster than the former. The formation of bubbles followed by flashing is the transport mechanism by which a system responds to a thermodynamic driving force (supersaturation) in much the same way as diffusion is the transport mechanism by which the system responds to another driving force (chemical potential gradient). In all our rate models, we shall employ the "penetration theory" approach to mass transport, which is based on average mass transfer rates. Whether or not bubbles play a role, molecular diffusion in the condensed phase toward a gas-liquid interface is likely to be an important, if not rate-determining, step.

The sequence of concentration profiles envisioned for both diffusive and foam DV have been sketched in Fig. 6.2-1, assuming that the external pressure is suddenly reduced from P_0 to P at $\alpha = 0$, where α is either the exposure time t in a batch process or the age x/v in a steady-state continuous process. Variables x and y represent longitudinal and transverse positions, respectively.

The concentration profile commonly used to represent interfacial mass transfer at the diffusion level is shown in Fig. 6.2-2. It is customary to assume that diffusion in the condensed phase is rate-controlling. Thus the pressure (concentration) distribution in the gas phase is uniform and partition equilibrium at the interface prevails and adjusts instantaneously to disturbances. These assumptions are generally applied to bubble growth as well.

6.2.1. Evaporation Elements

Industrially, flash evaporation is used when volatile concentrations are very high. A variety of equipment has been designed for the purpose of DV further downstream.

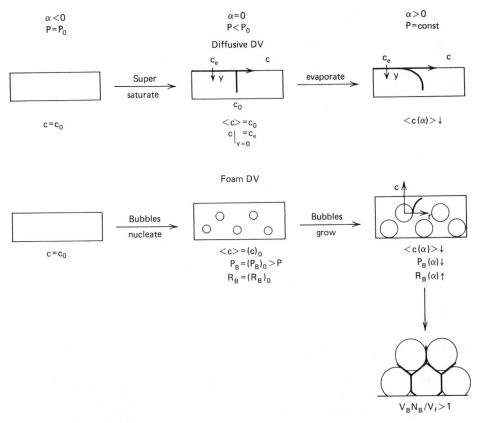

Figure 6.2-1. Schematic diagram of diffusive and foam devolatilization (DV) showing the evolution with age α of melt concentration c and its distribution, bubble pressure P_B, and bubble radius R_B, following reduction of external pressure P from its initial value P_0.

Most rotating machinery currently used for DV of polymer melts consists of two basic types: the vented extruder (single or twin screw) and the thin-film evaporator (with scraper blades to assist flow). In such equipment both melt configurations previously mentioned, that is, the translating film and rotating pool, appear to contribute to DV.

In the vented extruder, for example, DV could occur in the polymer film that is deposited on the barrel wall due to flight clearance, as well as in the bulk polymer

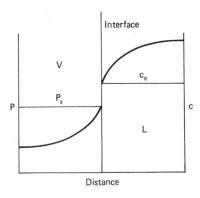

Figure 6.2-2. Vapor-phase (V) and liquid-phase (L) transport resistances in series with VLE at the interface, $P_s = P_s(c_e)$.

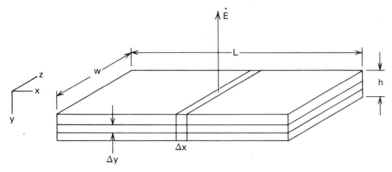

Figure 6.2-3. An evaporating melt film (f).

that flows down the partly filled screw channel while simultaneously rotating. The latter, of course, is represented by the rotating pool. A similar pool may be observed in other types of processing equipment as well, such as the bow wave (18) in thin-film evaporators with scraper blades. Which configuration, if any, dominates in its contribution to DV remains an unresolved question.

The idealized evaporating film and evaporating melt pool are illustrated in Figs. 6.2-3 and 6.2-4. Each one may be batch or continuous as shown in Figs. 6.2-5 and 6.2-6. The continuous film is assumed to move like a rigid plug (translating film). The rotating pool is formed at the junction of a moving wall–stationary wall pair. The continuous pool may either accept feed and discharge effluent from the ends or from the sides, as shown, or both. Perfect mixing is assumed within the pool as a first approximation. Gas phase G represents either overhead vapor space or gas bubbles within the melt, depending on whether diffusion or bubble transport prevails.

Thus the batch film has a concentration gradient only in the direction orthogonal to the film surface S_f. The translating film (control volume) has a uniform velocity profile and steady-state concentration gradients in both the transverse and longitudinal directions. Longitudinal mixing (backmixing) associated with the latter gradient is neglected in the idealized element. On the other hand, while the bulk concentration in the idealized stationary pool is assumed to be spatially uniform, with the exception of a gradient in the surface layer only, the flowing pool (control volume) additionally has a steady-state longitudinal concentration gradient.

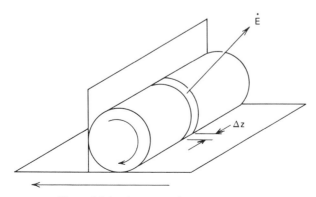

Figure 6.2-4. An evaporating melt pool (p).

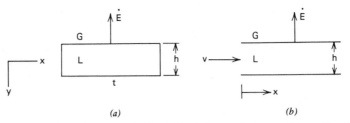

(a)   *(b)*

Figure 6.2-5.   The stationary batch film (*a*) and the translating continuous film (*b*).

To demonstrate how these elements may be used to simulate an actual process, the reader is referred to the parallel combination of evaporating film and evaporating pool illustrated in Fig. 6.2-7, which has been proposed as a basic DV model for the vented single-screw extruder (19, 20) sketched in Fig. 6.5-1.

6.2.2.   Separation Efficiency

Following our discussion in Section 1.3 on the efficiency of processes in general, we now seek a definition for separation efficiency. As mentioned, two alternative methods are available (Table 1.3-1). Neglecting density changes and volume lost due to evaporation, we define DV process efficiency as

$$E_F \equiv \frac{c_0 - c_{avg}}{c_0 - c_e} \qquad (6.2\text{-}1)$$

or equivalently, as

$$E_F \equiv \frac{e(\Lambda_v)}{c_0 - c_e} \qquad (6.2\text{-}2)$$

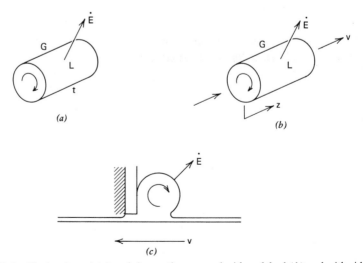

(a)   *(b)*

(c)

Figure 6.2-6.   The batch pool (*a*) and the continuous pool with end feed (*b*) and with side feed (*c*).

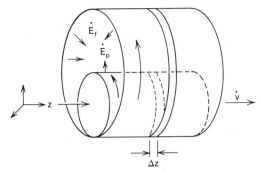

Figure 6.2-7. The evaporating film and evaporating pool in parallel.

where c_{avg} is an appropriate final average concentration, c_e is an equilibrium concentration, and $e(\Lambda_v)$ represents the amount of volatile component removed per unit volume of polymer after time Λ_v. For batch processes c_{avg} is a volume-average concentration $\langle c \rangle$, and for continuous processes it is a flow-average concentration $[c]$ (Appendix F). Thus E_F is the fraction of extractable volatile component removed. Conversely

$$\hat{c}_{avg} \equiv 1 - E_F = \frac{c_{avg} - c_e}{c_0 - c_e} \tag{6.2-3}$$

is the fraction of extractable volatile component remaining unevaporated.

Since most DV processes are complex (Section 1.3) and consist of stages, we define a stage efficiency analogously:

$$E_{fj} \equiv \frac{c_{j-1} - c_j}{c_{j-1} - c_{ej}} \tag{6.2-4}$$

or

$$E_{fj} \equiv \frac{e(\lambda_j)}{c_{j-1} - c_{ej}} \tag{6.2-5}$$

where c_{j-1} and c_j represent feed and final concentrations, respectively, c_{ej} is the equilibrium concentration corresponding to the jth stage, and $e(\lambda_j)$ represents the amount of volatile component evaporated per unit volume of polymer (V_j) in stage j after exposure time λ_j. While the subscript avg has been dropped from the nonequilibrium concentrations for convenience, the appropriate average quantity must be used. Examples of exposure time λ_j are λ_f and λ_p for melt films and melt pools, respectively.

Sometimes, under the appropriate conditions, it is possible to express $e(\lambda_j)$ as the simple integral

$$e(\lambda_j) = \int_0^{\lambda_j} r_e(\alpha)\, d\alpha \tag{6.2-6}$$

of an "instantaneous" rate function $r_e(\alpha)$, which represents the volume-specific rate of evaporation. In such cases, diffusive DV being one example (Section 6.2.3), $r_e(\alpha)$ is usually expressed in terms of a local driving force $c(\alpha) - c_e$ and a local mass transfer coefficient k, which may vary with α (time or position). Thus

$$r_e(\alpha) = k(\alpha)s_V[c(\alpha) - c_e] \qquad (6.2\text{-}7)$$

where s_V is the surface-to-volume ratio in the stage.

Because a constant driving force is more convenient to deal with in mass transfer operations than a variable one, it is customary practice in penetration theory to average r_e over λ_j and to fit the result to the following linear format

$$(s_V)_j(k_f)_j = \frac{e(\lambda_j)}{\lambda_j(c_{j-1} - c_j)} \qquad (6.2\text{-}8)$$

which defines a stage mass transfer coefficient k_f in terms of a global driving force $c_{j-1} - c_j$, whose concentrations are understood to be appropriate average values (e.g., flow-average). The stage efficiency and mass transfer coefficient are related in general by

$$\frac{E_f}{\lambda_j} = k_f s_V \qquad (6.2\text{-}9)$$

where the transcript j has been dropped for convenience. It is evident that k_f is a function of λ_j, $k_f(\lambda_j)$. The aim of penetration theory is to relate k_f (or equivalently E_f) to fundamental transport properties, such as D, which control evaporation rate.

It is important to recognize the fundamental difference between E_f, which reflects amount (fractional) of DV, and k_f, which reflects rate of DV, especially with respect to their dependence on parameters such as λ_j. From our discussion in Section 1.3 and from Fig. 1.3-2 (n vs. t), we would expect E_f to increase with λ_j and k_f to diminish.

A useful alternative interpretation of stage efficiency E_f is fractional penetration depth $\hat{\delta}_p \equiv \delta_p/h$ (21). Referring to Fig. 6.2-8, we define penetration depth δ_p as the thickness of an equivalent, imaginary surface layer whose volatile content has been depleted to the equilibrium value. To evaluate δ_p, we replace the concentration profile with a step profile and define $\delta_p(\alpha)$ as the value of coordinate y at age α where the concentration changes in step fashion from c_e to c_0.

$$\delta_p(\alpha) = \frac{c_0 h - \int_0^h c(\alpha, y)\, dy}{c_0 - c_e} \qquad (6.2\text{-}10)$$

This definition follows directly from a material balance, which requires that the amount of volatile component depleted at age α in the imaginary surface layer be identical to the amount actually depleted. To obtain the equivalence

$$\hat{\delta}_p(\lambda_j) \equiv E_f(\lambda_j) \qquad (6.2\text{-}11)$$

we must recognize the RHS of equation 6.2-10, after evaluation at $\alpha = \lambda_j$ and division by h, as the stage efficiency.

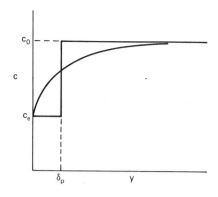

Figure 6.2-8. Sketch of the liquid-phase concentration profile $c(y)$ showing the penetration depth δ_p.

As we develop rate models for our evaporation elements in succeeding sections of this chapter, we shall perform sample calculations that are aimed at placing these models in perspective and assessing whether they are capable of simulating actual rates observed in DV processes. A yardstick against which to compare their performance would therefore be helpful. In this context, it is known, for instance, that 90 percent of the residual styrene in molten polystyrene in the thousand parts per million range can be removed in a vented extruder in less than 20 s total residence time at an operating temperature of approximately 260°C using a vacuum level not in excess of 10 mm Hg. (41).

6.2.3. Diffusing Films

The simplest evaporating film is one in which the polymer loses volatile component solely by molecular transport in the y direction (Fig. 6.2-3). Assuming that liquid-phase diffusion is rate-controlling, as is customary, and that D is constant, we may write

$$r_e(\alpha) = s_V n_n(\alpha) \tag{6.2-12}$$

where $n_n(\alpha)$ is the normal component of the diffusion flux

$$n_n(\alpha) = D \left. \frac{\partial}{\partial y} c(\alpha, y) \right|_{y=0} \tag{6.2-13}$$

and $c(\alpha, y)$ is the solution of the differential material balance for volatile component in rectangular coordinates (Appendix F)

$$\frac{\partial c}{\partial \alpha} = D \frac{\partial^2 c}{\partial y^2} \tag{6.2-14}$$

subject to initial condition $c(0, y) = c_0$ and boundary condition $c(\alpha, 0) = c_e$, together with either BC

$$c(\alpha, \infty) = c_0 \tag{6.2-15}$$

or BC

$$\frac{\partial c}{\partial y}\bigg|_{y=h} = 0 \qquad (6.2\text{-}16)$$

depending on whether we view the film as infinite or finite, respectively. The former BC is less general than the latter, but it yields a simpler expression for E_f. The choice between the infinite film model and the finite film model depends on the depth of penetration δ_p. The larger $\hat{\delta}_p$ (or E_f) is, the greater the error associated with the infinite film model.

The solution of equation 6.2-14 with BC 6.2-15 is well known

$$\frac{c(\alpha, y) - c_e}{c_0 - c_e} \equiv \hat{c}(\alpha, y) = \mathrm{erf}\left[\frac{y}{2(\alpha D)^{1/2}}\right] \qquad (6.2\text{-}17)$$

where erf is the error function (Appendix C). After substituting this result into equations 6.2-12 and 6.2-6, we obtain (cf. equation 6.2-7)

$$r_e(\alpha) = s_V\left(\frac{D}{\pi\alpha}\right)^{1/2}(c_0 - c_e) \qquad (6.2\text{-}18)$$

and

$$e(\lambda_f) = 2s_V\left(\frac{D\lambda_f}{\pi}\right)^{1/2}(c_0 - c_e) \qquad (6.2\text{-}19)$$

and finally

$$E_f = 2s_V\left(\frac{D\lambda_f}{\pi}\right)^{1/2} \qquad (6.2\text{-}20)$$

The corresponding mass transfer coefficient is also well known from penetration theory

$$k_f = 2\left(\frac{D}{\pi\lambda_f}\right)^{1/2} \qquad (6.2\text{-}21)$$

Similarly, from the solution of equation 6.2-14 with BC 6.2-16 (22) the efficiency is

$$E_f = 1 - \sum_{n=0}^{\infty}\frac{8}{[(2n+1)\pi]^2}\exp\left\{-\frac{[(2n+1)\pi]^2 s_V^2 D\lambda_f}{4}\right\} \qquad (6.2\text{-}22)$$

It can be shown that equation 6.2-20 is an excellent approximation of equation 6.2-22 for small values ($D_e \to 0$) of the dimensionless group D_e called the devolatilization number (20), that is, when

$$s_V^2 D\lambda_f \equiv D_e \leqslant 0.1 \qquad (6.2\text{-}23)$$

On the other hand, as $D_e \to 1$, the higher terms in equation 6.2-22 become negligible.

In fact, when

$$s_V^2 D\lambda_f \equiv D_e > 0.1 \tag{6.2-24}$$

the truncated form

$$E_f = 1 - \frac{8}{\pi^2}\exp\left(-\frac{\pi^2}{4}s_V^2 D\lambda_f\right) \tag{6.2-25}$$

is an excellent approximation of the finite film model. The corresponding mass transfer coefficient is

$$k_f = (s_V\lambda_f)^{-1}\left[1 - \frac{8}{\pi^2}\exp\left(-\frac{\pi^2}{4}s_V^2 D\lambda_f\right)\right] \tag{6.2-26}$$

These results may be summarized as follows (20):

$$E_f = \begin{cases} \left(\frac{4}{\pi}D_e\right)^{1/2} & \text{when } D_e \leqslant 0.1 \text{ (infinite film)} \\[2mm] 1 - \frac{8}{\pi^2}\exp\left(-\frac{\pi^2 D_e}{4}\right) & \text{when } D_e > 0.1 \text{ (finite film)} \end{cases} \tag{6.2-27}$$

Since $s_V = h^{-1}$ for rectangular geometry, we can express the devolatilization number in another way:

$$D_e = \frac{\lambda_f}{\lambda_D} \tag{6.2-28}$$

where λ_D is the familiar CT for diffusion, $\lambda_D = h^2/D$. Graphs of E_f versus D_e over the entire range of useful values of D_e spanned by equations 6.2-27 have been plotted in Fig. 6.2-9. The match of these two approximate solutions at $D_e = 1$ (or $E_f = 0.36$) is seen to be excellent.

An important conclusion from our analysis is that E_f and k_f depend only on λ_f for both finite and infinite film diffusion models when the geometry (s_V) is fixed, and therefore, they are independent of feed and equilibrium concentrations c_0 and c_e. Furthermore, their dependence shows, as we speculated earlier, that E_f rises with λ_f, whereas k_f declines, provided that s_V remains constant. The reason k_f rises as λ_f declines is apparent in the diffusion model, being that k_f is proportional to the concentration gradient at the interface, which is larger for short times than for long times.

Sometimes, in industrial DV processes, process efficiency actually increases as residence time in the equipment is decreased (by increasing the rpm of a rotating machine, for instance). It is noteworthy that such "anomalous" behavior, which was previously alluded to in Section 1.3, may be rationalized in terms of our simple diffusion model. From the expression for throughput rate $\dot{V}_f = hs_f/\lambda_f$ together with the definition De $= \lambda_f/Dh^2$ it is clear that as λ_f declines while $\dot{V}_f$ remains constant. De, and therefore E_f, will increase provided that s_f remains constant. The cause of

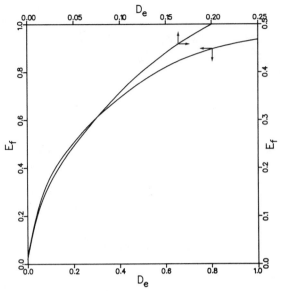

Figure 6.2-9. Variation of stage efficiency with devolatilization number, $E_f = f(D_e)$, for the diffusing film.

this increase, of course, is the decrease in film thickness h required to maintain $\dot{V}_f$ constant. In essence, we will remove more volatile component in unit time (higher k_f) from the same total quantity of polymer (constant $\dot{V}_f$).

To summarize, the efficiency expression $E_f = f(De)$ for diffusing films may be determined by first estimating the value of De relative to 0.1, and then substituting it into the appropriate function f from equations 6.2-27, or utilizing Fig. 6.2-9.

Finally, referring to Section 1.3, we shall make use of the product $E_f x$, which is a measure of the specific output rate of the film. The quantity $x \equiv \dot{V}_f / s_f$ may be viewed as a "thickness velocity" ($\dot{h}$) and, with the aid of equations 6.2-9 and 6.2-11, we can assign to the product $E_f x$ two additional useful interpretations: penetration rate $\dot{\delta}_p$ and mass transfer coefficient k_f. Thus,

$$xE_f = \begin{cases} \dot{\delta}_p \\ k_f \end{cases} \tag{6.2-29}$$

A sample computation is given in Example 6.2-1. The target of 90-percent efficiency is reasonable from experience as previously mentioned. The long time required to achieve it underscores the slowness of diffusion, notwithstanding the use of a value for diffusivity of 10^{-4} in.^{2}/min, which is of an order of magnitude that is typical for solutions containing only small molecules (10^{-5} cm^2/s). It is noteworthy that recent studies (23) show D to be strongly non-Fickian for small molecules dissolved in polymers and to plunge to values as low as 10^{-9} cm^2/s at low concentrations. However, although the concentrations used in these studies were certainly within the range of interest in DV, the temperatures were not, being considerably lower than typical processing temperatures. It is conceivable that the temperature effect could offset the concentration effect and produce diffusivities that are albeit not constant, but of the order used in the example. We shall therefore employ this value throughout this chapter for the sake of convenience and assume it to be constant to facilitate mathematical tractability.

Example 6.2-1. **The Diffusing Film.** Estimate the exposure time λ_f required to achieve $E_f = 0.9$ when $h = 50$ mils and $D = 10^{-4}$ in.^{2}/min, and compute the corresponding value of xE_f.

Solution. We conclude from Fig. 6.2-9 that De = 0.85 is required, which clearly indicates a finite film, and from equation 6.2-28 for the CT of this process

$$\lambda_D = h^2 D = 25 \text{ min}$$

that the following exposure time is needed.

$$\lambda_f = \lambda_D \text{De} = 21 \text{ min or } 0.35 \text{ h}$$

The corresponding specific output rate is

$$xE_f = \frac{E_f h}{\lambda_f} = 2.14 \text{ mils/min}$$

6.2.4. Diffusing Particles

Devolatilization is sometimes conducted more conveniently from condensed phases below T_g than from molten films. In such cases (e.g., PVC beads from suspension polymerization), we can have particles of various geometries. When transport within the particles is one-dimensional and solely molecular, the following material balance for volatile component applies in lieu of equation 6.2-14

$$\frac{\partial c}{\partial t} = D \left(\frac{\partial^2 c}{\partial \zeta^2} + \frac{K}{\zeta} \frac{\partial c}{\partial \zeta} \right) \tag{6.2-30}$$

subject to the initial condition $c(0, \zeta) = c_0$ and boundary conditions $(\partial c/\partial \zeta)|_{\zeta=0} = 0$ and $c(\pm h \text{ or Ra}, t) = c_e$ where K is a geometry parameter

$$K = \begin{cases} 0 \text{ for a rectangular slab} \\ 1 \text{ for a cylinder} \\ 2 \text{ for a sphere} \end{cases} \tag{6.2-31}$$

and ζ is the single space variable

$$\zeta = \begin{cases} y & -\dfrac{h}{2} \leqslant y \leqslant \dfrac{h}{2} \\ r & 0 \leqslant r \leqslant \text{Ra} \end{cases} \tag{6.2-32}$$

The slab is assumed to be of thickness h and "infinite" in the x and z direction. The cylinder and sphere have radius Ra and the cylinder is assumed to be "infinitely" long in the z direction.

The diffusing sphere represents a convenient approximation of liquid droplets and solid particles of polymer in which volatile solute diffuses radially outward for evaporation. Its efficiency, which can be obtained from the solution of equation 6.2-30 for $K = 2$ when D is constant (22), is

$$E_f = 1 - \sum_{n=1}^{\infty} a_n^{-1} \exp(-6a_n \text{De})$$ (6.2-33)

where $a_n \equiv (n\pi)^2/6$. To retain the physical significance of De as a ratio of CT's (De $= \lambda_f/\lambda_D$) and to obtain a surface-to-volume ratio of $s_V = 3/\text{Ra}$ and a CT for diffusion of $\lambda_D = \text{Ra}^2/D$, which we would expect intuitively, we shall define the devolatilization number as

$$\text{De} = (s_V/3)^2 D\lambda_f$$ (6.2-34)

The appearance here, as before, of the diffusion distance squared in the denominator of De underscores the importance of particle size, which, if sufficiently small, can be used to overcome the lower values for D anticipated in solid polymer particles. This is, in fact, the case for demonomerization of PVC suspension resins at temperatures not far from T_g (Example 6.2-2).

Example 6.2-2.   The Diffusing Particle.   A value of 5.6×10^{-10} cm^2/s has been reported (13) for D of VC in PVC at 90°C. Estimate the CT for diffusion from spheres of 10-μ diameter and the time required for one order of magnitude reduction in extractable VC content.

Solution.   From equation 6.2-36

$$\lambda_D = \frac{\text{Ra}^2}{D} = 7.4 \text{ min}$$

and from equation 6.2-33 after dropping higher-order terms (shown by trial and error to be virtually negligible)

$$0.1 \cong \frac{6}{\pi^2} \exp\left(-\frac{\pi^2 \lambda_f}{\lambda_D}\right)$$

we obtain De $\cong 0.18$ and consequently $\lambda_f = 1.3$ min.

Comparing this exposure time to that of Example 6.2-1, which is more than an order of magnitude longer, emphasizes the importance of a large value of s_V (15.2 mils^{-1} vs. 0.02 mils^{-1}). It is sufficient to offset the much smaller value of D (5.6×10^{-10} cm^2/s vs. 1.08×10^{-5} cm^2/s).

6.2.5.   Surface Renewal and Staging

It is evident from our analysis of the diffusing film that the most rapid diffusion flux occurs initially when the concentration gradient is greatest. This observation suggests a way to improve the efficiency of the DV process, that is, by exposing the film surface to a sequence of short periods, rather than a single long period, and by intermittently replacing the spent surface with fresh material from below which has a uniform volatile concentration. We shall term this staged pair of operations the surface renewal (SR) process. Each pair of evaporation-renewal operations constitutes a cycle (batch process) or stage (continuous process) and is repeated N times. Excluded, of course, are polymers below T_g for which surface renewal is impossible without permanent size reduction.

At least two distinct mechanisms for achieving surface renewal suggest themselves. The most obvious one is random renewal, illustrated in Fig. 6.2-10 for a simple batch film, in which mixing is induced in the direction orthogonal to the surface (y-direction) after exposure time λ_f. If mixing is complete, the film is entirely homogenized to a uniform concentration. It is then subjected to a second evaporation cycle followed by mixing and so on. Another mechanism is selective renewal (24), illustrated in Fig. 6.2-10. Here only the depleted surface layer is replaced after time λ_f with a fresh layer from beneath having a concentration c_0, which is then exposed for evaporation and subsequently replaced and so on. We contrast these mechanisms by observing that selective renewal results in a higher surface concentration at the outset of each succeeding cycle, since $c_0 > \langle c \rangle_1 > \langle c \rangle_2 > \cdots$, but only for a limited number of cycles n after which all available layers beneath the surface will have been partially depleted.

We consider random renewal first. A material balance for volatile component on the jth stage (cycle)

$$1 \leqslant j \leqslant N \qquad c_j - c_{j-1} = -E_{fj}(c_{j-1} - c_{ej}) \qquad (6.2\text{-}35)$$

yields a first-order, nonhomogeneous difference equation with variable coefficient

$$c_j - (1 - E_{fj})c_{j-1} = E_{fj}c_{ej} \qquad (6.2\text{-}36)$$

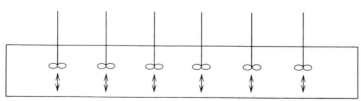

Random SR

Selective SR

Figure 6.2-10. Schematic diagram of random surface renewal (SR) and selective surface renewal.

where c_j is the appropriate average concentration ($\langle \ \rangle$ for batch and $[\ ]$ for steady-state continuous processes). The function $c_{ej} = c_e(j)$ allows for the possibility of cascading vacuums is discussed in Section 6.1.4. For example, we might use $c_{ej} = c_0 a^j$ based on optimum pump displacement, where a is given by equation 6.1-58.

For the important special case of identical stages, equation 6.2-36 in dimensionless form becomes a homogeneous equation with constant coefficients

$$\hat{c}_j - (1 - E_f)\hat{c}_{j-1} = 0 \tag{6.2-37}$$

where

$$\hat{c}_j \equiv \frac{c_j - c_e}{c_0 - c_e} \tag{6.2-38}$$

From its solution, subject to initial condition $\hat{c}_0 = 1$, we obtain the following expression for process efficiency of the N-stage SR model (20)

$$E_F = 1 - (1 - E_f)^N \tag{6.3-39}$$

The beneficial effects of SR are illustrated in Example 6.2-3 in terms of efficiency as well as specific output rate.

It is noteworthy that equation 6.2-39 may be approximated by an exponential function (Appendix B)

$$E_F \cong 1 - \exp(-NE_f) \tag{6.2-40}$$

for small values of E_f ($E_f < 0.1$) in which the two parameters appear as a single lumped parameter. Graphs of $1 - E_F \equiv \hat{c}_N$ versus E_f for equations 6.2-39 and 6.2-40 have been plotted in Fig. 6.2-11 with N as parameter. Their divergence for values of $E_f > 0.1$ is apparent.

It is clear that the SR model does not apply to non-Fickian diffusion (variable D) since E_f would then vary from stage to stage, being a function of feed concentration in addition to exposure time $E_{fj} = f(\lambda_j, c_{j-1})$.

In selective renewal the material balance for the surface layer after the ith replacement is

$$c_i' - c_0 = -E_{fi}'(c_0 - c_e) \tag{6.2-41}$$

where the primes signify layer quantities. The mean concentration of the entire film after replacement of j layers is

$$j \leqslant n \qquad c_j = \frac{\displaystyle\sum_{i=1}^{j} c_i' + (n - j)c_0}{n} \tag{6.2-42}$$

For identical layers ($c_1' = c_2' = \cdots \equiv c'$)

$$c_j = \frac{jc' + (n - j)c_0}{n} \tag{6.2-43}$$

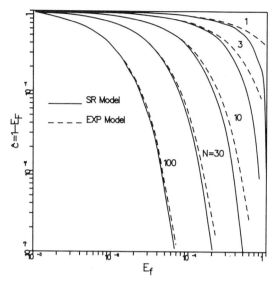

Figure 6.2-11. Logarithmic relationship between fraction of extractable volatile component remaining after N cycles (stages), $\hat{c}_N = 1 - E_F$, and stage efficiency E_f for the surface renewal (SR) model, $\hat{c}_N = (1 - E_f)^N$, and the exponential (EXP) model, $\hat{c}_N = \exp(-NE_f)$.

and the corresponding efficiency is

$$E_F' \equiv \frac{c_0 - c_j}{c_0 - c_e} = \frac{j}{n} E_f' \tag{6.2-44}$$

from which we obtain the obvious result that after precisely n replacements $E_F' = E_f'$.

It should be noted that the value of E_f' is determined by the choice of n and that n in turn is bounded by the value of E_f, which is the efficiency that the entire film would exhibit if it were exposed for the same period λ_f as each of the layers. This limit may be calculated with the aid of the penetration depth δ_p. For the special case of rectangular geometry and identical layer efficiencies,

$$n \leqslant \frac{h}{\delta_p} = \hat{\delta}_p^{-1} = E_f^{-1} \tag{6.2-45}$$

If we assume that each layer behaves like an infinite film, it follows from equation 6.2-20 that $E_f = E_f'/n$, which confirms the obvious fact that successive exposure of n films with s_V each is identical to a single exposure of one film with ns_V. Substituting this result into equation 6.2-44 yields

$$E_F' = nE_f \tag{6.2-46}$$

for $j = n$. This expression requires for its validity that $E_f \leqslant 0.36/n$ since each layer must obey the infinite-film criterion (Section 6.2.3). It is noteworthy that when $n = E_f^{-1}$ our result predicts that $E_F' = 1$, which is intuitively satisfying.

Let us now combine random renewal with selective replacement by homogenizing the entire film via mixing after n selective replacements and repeating the entire process N' times. Thus we have N' mixing cycles, each of which contains n selective replacements nested within it. The concentrations remaining after the first mixing cycle is $c_I = c_n$, after the second $c_{II} = c_I - E_f'(c_0 - c_e)$ and so on. The overall process efficiency after a total of $N = nN'$ stages is thus obtained from the solution of the following difference equation

$$1 \leqslant k \leqslant N' \qquad\qquad c_k - c_{k-1}(1 - E_f') = E_f' c_e \qquad\qquad (6.2\text{-}47)$$

It is

$$1 \leqslant n \leqslant N \qquad\qquad E_F = 1 - (1 - E_f')^{N/n} \qquad\qquad (6.2\text{-}48)$$

If we again compare E_f' and E_f for each of the N' stages with a common value of λ_f and assume infinite-film layers as before, this result yields

$$E_F = 1 - (1 - nE_f)^{N/n} \qquad\qquad (6.2\text{-}49)$$

It is noteworthy that equation 6.2-49 reduces to the following special cases: the SR model (equation 6.2-39) when $n = 1$ and the selective renewal model (equation 6.2-46) when $n = N$.

In conclusion, selective SR appears to be only slightly superior to random SR for the identical number of stages $n = N$ (Example 6.2-4). The reason is that while the fraction of volatile removed in each stage may be the same for both (E_f), the amount removed is greater for selective SR because the initial concentration is higher (c_0).

Example 6.2-3. The Effect of Surface Renewal.   Estimate the total film exposure time $\Lambda_f = N\lambda_f$ required to achieve a process efficiency of $E_F = 0.9$ in N identical stages, each consisting of a pair of diffusion-mixing operations, when $h = 50$ mils and $D = 10^{-4}$ in.^{2}/min and $N = 10, 100$. Also, compute the corresponding values for specific output rates (Section 1.3) xE_f and XE_F where $x \equiv \dot{V}/s_f$, $X \equiv \dot{V}/S_f$, and $S_f = Ns_f$.

Solution.   From equation 6.2-39

$$E_f = 1 - \log^{-1}\left[N^{-1}\log(1 - E_F)\right] = \begin{cases} 0.20 & \text{when } N = 10 \\ 0.023 & \text{when } N = 100 \end{cases}$$

and from Fig. 6.2-9 (equation 6.2-27) $D_e = 0.034$ and 0.00041, respectively. Substituting into equation 6.2-24 yields

$$\lambda_f = \frac{D_e h^2}{D} = \begin{cases} 0.85 \text{ min} & \text{when } N = 10 \\ 0.01 \text{ min} & \text{when } N = 100 \end{cases}$$

so that $\Lambda_f = 8.5$ and 1.0 min, respectively. These results should be contrasted with 25 min when $N = 1$ (Example 6.2-1).

Furthermore,

$$xE_f = \frac{hE_f}{\lambda_f} = \begin{cases} 11.7 \text{ mils/min} \\ 115 \text{ mils/min} \end{cases}$$

and

$$XE_F = \frac{hE_F}{N\lambda_f} = \begin{cases} 5.3 \text{ mils/min} \\ 45 \text{ mils/min} \end{cases}$$

The first specific output rate (xE_f) may be interpreted as a mass transfer coefficient (equation 6.2-29). The second one (XE_F) should be contrasted with the value 2.14 mils/min in Example 6.2-1 to verify that the SR process is more efficient than the simple film.

Example 6.2.4. Selective Versus Random SR. Contrast process efficiencies for random SR (equation 6.2-39) and selective SR (equation 6.2-46) for various values of $n = N$ and $E_f \leqslant 0.30$.

Solution. Substitution of E_f and N into the appropriate equations yields the values of E_F (or E_F') listed in the following chart, which confirm that selective SR is slightly more efficient than random SR. The following test for a case in which $n < N$ using equation 6.2-48 also checks: When $E_f = 0.1$, $n = 10$, and $N = 20$, we obtain $E_F = 0.190$, which lies between the corresponding values of 0.182 and 0.20 for random SR and selective SR, respectively.

	E_f	1	2	5	10	20
			N			
Random SR	0.36	0.36				
	0.1	0.1	0.19			
	0.05	0.05	0.0975	0.226		
	0.01	0.01	0.0199	0.0490	0.0956	0.182
	0.005	0.005	0.00998	0.0248	0.0489	0.0954
	0.001	0.001	0.002	0.005	0.010	0.0198
Selective SR	0.36	0.36				
	0.1	0.1	0.2			
	0.05	0.05	0.10	0.25		
	0.01	0.01	0.02	0.05	0.10	0.20
	0.005	0.005	0.010	0.025	0.05	0.10
	0.001	0.001	0.002	0.005	0.010	0.020

6.2.6. Rotating Melt Pools

The simplest model for the evaporating melt pool illustrated in Fig. 6.2-4 is

$$\frac{dc}{d\alpha} = r_e(\alpha) = -s_V k_p (c - c_e) \tag{6.2-50}$$

which assumes perfect mixing within the pool and infinite-film diffusive evaporation from the surface layer described by constant k_p, whose value depends only on the period of rotation λ_p as suggested by equation 6.2-21

$$k_p = 2\left(\frac{D}{\pi\lambda_p}\right)^{1/2} \tag{6.2-51}$$

This model applies to the batch pool when $\alpha = t$ and the steady-state continuous pool with end feed and discharge (Fig. 6.2-6) when $\alpha = z/\langle v_z \rangle$. The solution of equation 6.2-50 leads to the exponential (EXP) model for process efficiency (20)

$$\frac{c_0 - c}{c_0 - c_e} \equiv E_F = 1 - \exp(-Ex) \tag{6.2-52}$$

where we now have a single dimensionless parameter

$$Ex \equiv \alpha k_p s_V = \frac{\alpha}{\lambda_p} E_f(\lambda_p) \tag{6.2-53}$$

termed the extraction number, and where

$$E_f = 2s_V \left(\frac{D\lambda_p}{\pi}\right)^{1/2} \tag{6.2-54}$$

A comparison of equation 6.2-52 with 6.2-40 suggests that the extraction number is composed of the product

$$Ex = N_p E_f \tag{6.2-55}$$

and that the number of equivalent stages in the melt pool is given by the quotient $\alpha/\lambda_p \equiv N_p$. Thus we can interpret each revolution as a stage. In rotating processing machinery, we would expect λ_p^{-1} to be proportional to the speed of rotation N_R (rpm).

A sample computation for such an ideal melt pool is carried out in Example 6.2-5 for comparative purposes. Included also is an evaluation of Ex to give us a feeling for its magnitude. From this example and the previous ones we must conclude that diffusion models do not appear adequate to explain the DV rates observed in actual processes, even when they include surface renewal and staging. We shall therefore consider next the mechanism of bubble transport in which the interfacial surface areas for mass transfer are much larger than those considered heretofore.

One feature common to all diffusion models is their prediction that process efficiency E_F should be independent of feed composition c_0. Intuitively we would certainly not expect this to obtain with bubble transport, where the bubbles are created by application of vacuum at levels below the equilibrium partial pressure of the vapor contained in them.

Example 6.2-5. Rotating Melt Pools. Estimate the extraction number Ex and the surface-to-volume ratio s_V required by a melt pool rotating at a speed of $N_R = 60$ rpm to give the same process efficiency (0.9) in the same total (8.5 min) as the SR model in Example 6.2-3. Assume that the film is exposed during only half of its rotational period and use the value $D = 10^{-4}$ in.^{2}/min.

Solution. From equation 6.2-51

$$k_p = 2\left(\frac{D}{\pi\lambda_p}\right)^{1/2} = 124 \text{ mils/min}$$

where $\lambda_p = 0.5/60$ min and from equation 6.2-52

$$Ex = -\ln(1 - E_F) = 2.3$$

as contrasted with 2 for the SR model (Example 6.2-3). Substituting these results together with $\alpha = 8.5$ min into equation 6.2-53 yields $s_V^{-1} = 458$ mils. The last number indicates that a much thicker melt can be used in the pool than in the film (50 mils) owing to the much higher mass transfer coefficient (124 mils/min vs. 11.7 mils/min). Note that $N_p = 1020$.

The reader is reminded that these conclusions are all based on the assumption of perfect mixing, the validity of which is certainly open to serious doubt.

6.2.7. Foam Devolatilization

The process referred to here is the DV of volatile solute promoted by bubbles. Such bubbles can form when the polymer melt is suddenly exposed to a reduced pressure P that is less than the equilibrium partial pressure P_s of the volatile solute being removed or that of a third volatile component such as dissolved or entrained air (Fig. 6.2-1). This renders the solution supersaturated and thus invites boiling or cavitation. Bubbles produced in polymer melts are often actually foams owing to the high viscosity and elastic nature of these melts.

We shall discuss briefly the individual steps involved in foam DV: bubble nucleation, growth, motion and rupture.

Nucleation. The formation of gas bubbles in supersaturated liquids in general can occur by homogeneous or heterogeneous nucleation or both. It is customary to describe nucleation in terms of a critical bubble radius $(R_B)_{cr}$. A bubble whose radius at birth $(R_B)_0$ is less than $(R_B)_{cr}$ will collapse, and one whose radius exceeds it will tend to grow spontaneously. When the gas that gives rise to the bubble is the vapor of the volatile component being removed, as in Section 6.1.3, for instance, then bubble formation requires that P be less than P_s by the amount required by the surface tension to sustain curved interfacial surfaces, and the critical radius may be defined by the Laplace–Kelvin equation together with a thermodynamic partition law.

The formation of voids in homogeneous media that exceed the required minimum size is generally believed to be due to random density fluctuations that overcome an activation barrier. Thus $(R_B)_0$ may be a random variable distributed in Boltzmann-like fashion. The likelihood of nucleation can be enhanced by flow or agitation. Homogeneous nucleation is generally considered to be relatively unimportant compared to heterogeneous nucleation in the presence of solids, such as container walls and finely suspended particles (impurities), whose irregular surfaces can harbor trapped gases that act as nucleating agents. Classical nucleation theory ignores the effects of viscous forces on the formation of voids.

A complete model for DV requires knowledge of the nucleation rate function r_B, expressed as the number of bubbles per unit time per unit melt volume. The magnitude of r_B should increase markedly with degree of supersaturation, $P_s - P$, that is, with temperature, volatile concentration, and vacuum level. This expectation is based on expressions for r_B from nucleation theory (15) that take the following form for both homogeneous and heterogeneous nucleation,

$$r_B = A \exp\left\{-\frac{B\sigma^3}{T\left[(P_s - P)\right]^2}\right\} \tag{6.2-56}$$

where A and B are system constants that include physical properties and surface area per unit volume in the case of heterogeneous nucleation. With the aid of Henry's law, we can express this result in a more convenient way

$$r_B = A \exp\left[-\frac{B'(T)}{(c - c_e)^2}\right] \tag{6.2-57}$$

Growth. Since bubble growth is a rate process driven by stress differences and fed by mass transport, it is likely that neither mechanical nor chemical equilibrium conditions will prevail during growth. At the expense of oversimplification, we represent mechanical disequilibrium, that is, the amount by which actual bubble pressure P_B differs from P_{me} (Section 6.1.3) at any instant, by a time-dependent viscoelastic normal stress $\tau_{rr}(t)$

$$P_B = \underbrace{P + \frac{2\sigma}{R_B}}_{(P_B)_{me}} + \tau_{rr} \tag{6.2-58}$$

which relaxes to zero with time for any fixed value of R_B. It is customary to ascribe chemical disequilibrium entirely to a mass transfer resistance in the fluid, as we did in Sections 6.2.1–6.2.6, which is manifested by a difference between liquid concentration at the bubble interface (c_i) and that in the bulk melt (c). The surface concentration is then related to bubble pressure by a thermodynamic partition law, e.g., $P_s = K_c c_i$ (equation 6.1-42) together with Dalton's law of partial pressures (equation 6.1-41). If the bubble contains only vapor of the volatile being removed then, of course, $P_B = P_s$. At chemical equilibrium, $c_i = c$, and therefore

$$P_{ce} = K_c c \tag{6.2-59}$$

Another phenomenon, which may be viewed as a mechanism for bubble growth and is difficult to treat analytically, is coalescence. Certainly some bubbles will coalesce with one another and thus form larger product bubbles.

Motion. The motion of bubbles beyond that which accompanies growth will be disregarded, mainly because our bubbles frequently occur in viscous, foaming films. Bubble motion in general requires the action of force fields such as shear stress and hydrostatic pressure gradients, which are either due to pressurization (internal or external) or to gravity (rising bubbles in soda bottles). The influence of gravity is unimportant in viscous melts in drag flow conveyed by moving walls that may not even be horizontal.

Rupture. Owing to the elasticity and relatively shallow depth of polymer films, it is likely that bubbles will swell in balloonlike fashion (Fig. 6.2-1) prior to bursting, with tough surface skins formed by biaxial orientation and possibly aided by evaporative cooling. It is also possible that some bubbles will cease to grow because they attain equilibrium, or that they simply "freeze" kinetically as a consequence of high melt viscosities. Quasi-stable foams and frozen bubbles will obviously require rupturing by induced mechanical deformation. Thus surface renewal in foam DV could take the form of bubble-breaking to release the volatile contents into the adjacent gas phase. Failure to do this could doom subsequent evaporation to the (slower) diffusion mechanism.

Returning to the task at hand, which is to formulate a rate model for foam DV, the simplest approach is to postulate an "instantaneous" rate function r_e of the type defined by equation 6.2-6. This is not unreasonable since bubble nucleation is a volume-intensive process, as opposed to area-intensive, and in that sense is more like chemical reaction than transport (diffusion). Such a model implies that nucleation is rate-controlling and that the rate decays with time like a chemical reaction.

Thus let us postulate a rate function with constant mass transfer coefficient k

$$r_e = ks_v\big[c(\alpha) - c_e\big]^n \tag{6.2-60}$$

This equation is nonlinear ($n > 1$) in contrast to equation 6.2-7, which is linear but has a variable ("local") mass transfer coefficient $k(\alpha)$. The material balance for volatile component on a foaming film, which must now be used in lieu of equation 6.2-14,

$$\frac{dc}{d\alpha} = -ks_V(c - c_e)^n \tag{6.2-61}$$

is analogous to the rate equation for an nth order reaction. Its solution is

$$\frac{c - c_e}{c_0 - c_e} = \Big[1 - (1 - n)(c_0 - c_e)^{n-1}ks_V\alpha\Big]^{1/(n-1)} \tag{6.2-62}$$

As with chemical reaction, the linear case ($n = 1$) must be treated separately:

$$\frac{c - c_e}{c_0 - c_e} = \text{ex}(-ks_V\alpha) \tag{6.2-63}$$

Next we consider staging as in Section 6.2.5. The material balances for the jth stage (cycle) follow directly from the last two results:

$$\frac{c_j - c_{ej}}{c_{j-1} - c_{ej}} = \begin{cases} \left[1 - \dfrac{\lambda_j}{\lambda_{bj}}\right]^{1/(1-n)} & \text{when } n > 1 \\[2em] \exp\left(-\dfrac{\lambda_j}{\lambda_b}\right) & \text{when } n = 1 \end{cases} \tag{6.2-64}$$

which lead to the following definitions for stage efficiency

$$
E_{fj} \equiv \frac{c_j - c_{j-1}}{c_{j-1} - c_{ej}} = \begin{cases} 1 - \left[1 - \dfrac{\lambda_j}{\lambda_{bj}} \right]^{1/(1-n)} \\ 1 - \exp\left(-\dfrac{\lambda_j}{\lambda_b} \right) \end{cases}
\tag{6.2-65}
$$

where we have defined a characteristic time (CT) for bubble transport

$$
\lambda_{bj} \equiv \begin{cases} \left(c_{j-1} - c_{ej} \right)^{1-n} (1-n)^{-1} (ks_V)^{-1} & \text{when } n > 1 \\ (ks_V)^{-1} \equiv \lambda_b & \text{when } n = 1 \end{cases}
\tag{6.2-66}
$$

From these results, we observe that E_{fj} depends on feed concentration (c_{j-1}) in addition to exposure time (λ_j) and vacuum level (c_{ej}) when $n > 1$.

For N identical stages (cycles) with constant $\lambda_j \equiv \lambda_f$ and $c_{ej} \equiv c_e$, it is clear that the linear model satisfies balance equation 6.2-37, that is, $\hat{c}_j / \hat{c}_{j-1} = 1 - E_{fj}$. Process efficiency thus follows directly from equation 6.2-39

$$
E_F \equiv \frac{c_0 - c_N}{c_0 - c_e} = 1 - \exp\left(-\frac{N\lambda_f}{\lambda_b} \right)
\tag{6.2-67}
$$

and, as expected, is independent of N for a fixed exposure time $\Lambda_f = N\lambda_f$. By analogy, we may deduce from equation 6.2-65 that process efficiency for the nonlinear model ($n > 1$) is also independent of N

$$
E_F = 1 - \left(1 - \frac{\Lambda_f}{\lambda_b} \right)^{1/(1-n)}
\tag{6.2-68}
$$

where $\lambda_b \equiv (c_0 - c_e)^{1-n}(1-n)^{-1}(k_{sv})^{-1}$. This conclusion is intuitively satisfactory, owing to the similarity of the process at hand to an nth order reaction in a cascade of N plug-flow (PF) reactors, for which conversion depends solely on total residence time. The implication of equations 6.2-67 and 6.2-68 is, of course, that staging has no beneficial effect on a foaming film that is described by our simple rate function 6.2-60. Equation 6.2-68 has been derived in a more formal way in Example 6.2-6, following the procedure used in Section 6.2.5.

The use of an instantaneous rate function to describe DV promoted by bubbles is not realistic. It seems more reasonable to employ the methods of Volterra (25) and to express the amount of volatile evaporated after age α per unit melt volume, $e(\alpha)$, as required in our definition of efficiency (equation 6.2-5), directly as a history functional

$$
e(\alpha) = \int_0^\alpha e_B(\beta, t) r_B(t) \, dt
\tag{6.2-69}
$$

where $e_B(\beta, t)$ represents the amount of volatile component absorbed during growth period β by each of the $r_B(t)\,dt$ bubbles per unit volume born at t. A similar approach was used by Avrami (26) to formulate a general theory for phase-change kinetics and has been extensively applied to crystallization kinetics. It should be noted that the concept of stage efficiency lends itself well to the description of foam DV by functionals, especially the second definition, 6.2-5, by virtue of the fact that $e(\lambda_f)$ is expressed as an integral.

In equation 6.2-69, β represents the age of an individual bubble and must therefore satisfy the constraint condition $\beta \leqslant \lambda_f - t$. It is, in general, a function of history itself $\beta(t)$, depending on the specific circumstances surrounding bubble growth and rupture. For example, in foams where bubbles grow until they are ruptured at time λ_f, we would write $\beta = \lambda_f - t$. Function $r_B(t)$ is the nucleation rate discussed earlier.

The quantity $e_B(\beta, t)$ is the sum of the volatile contents of a bubble at birth $\beta = 0$, $(c_B)_0(V_B)_0$, and the amount absorbed during growth. It must satisfy a component transport balance on a single bubble

$$\frac{d}{dt}[c_B(t)V_B(t)] = -s_B(t)\{n_n(t)\} \tag{6.2-70}$$

where s_B and V_B are bubble surface and volume, respectively, c_B is its volatile concentration, and n_n is the diffusion flux (normal component) at the bubble surface. The latter quantity is obtained

$$n_n = -D\,\frac{\partial}{\partial r}c(r, t)\Big|_{r=R_B} \tag{6.2-71}$$

from the solution $c(r, t)$ of a component balance on the surrounding melt phase

$$r \geqslant R_B \qquad \frac{\partial c}{\partial t} = D\left[\frac{\partial^2 C}{\partial r^2} + \frac{2}{r}\frac{\partial c}{\partial r}\right] - v\frac{\partial c}{\partial r} \tag{6.2-72}$$

which is coupled to balance equation 6.2-72 by a flux BC (6.2-71) and an equilibrium BC

$$c(r, t)\big|_{r=R_B} \equiv c_i = c_B/K_c' \tag{6.2-73}$$

where K_c' is a Henry's law constant (equation 6.1-46). The remaining initial and boundary conditions commonly used are $c(r, 0) = c$ and $c(\infty, t) = c$, respectively, where c is the (uniform) bulk concentration of the melt.

The bubble concentration $c_B(t)$ in equation 6.2-70 is related by an equation of state, such as the perfect gas law $c_B + c_I = P_B/R_g T$, to bubble pressure $P_B(t)$. Pressure, in turn, must satisfy a force balance, such as equation 6.2-58 with $\tau_{rr} = (4\eta/R_B)(dR_B/dt)$, which is the so-called Rayleigh equation without the inertia term.

It should be noted that the general analysis of bubble growth requires a convection term in the liquid-phase balance. The boundary at $r = R_B(t)$ moves with velocity dR_B/dt and causes the adjacent fluid to move with velocity $v = v(r, t)$. An additional complication is due to the depletion of solute in the bulk liquid $c = c(t)$, which simultaneously reduces the equilibrium pressure P_s. The extent to which c varies depends, of course, on the number of bubbles and the melt volume V. To account for this variation, we require an interphase material balance equating the amount of volatile component absorbed by all the bubbles to the amount lost by the melt.

Many studies have been reported on the kinetics of growth (and collapse) of a single bubble in an infinite medium. We cite only a few (27–33). Most of these studies were aimed at describing the growth history of the bubble radius $R_B(t)$. Some ignored the convection term in equation 6.2-72, or the viscous resistance term τ_{rr} in equation 6.2-58, or the curvature of the interface (spherical). A summary of bubble growth dynamics may be found in a review article on foam molding (34).

It is convenient to replace equations 6.2-70–6.2-72 with the following simple transport balance (27, 33, 35)

$$\frac{d}{dt}(c_B V_B) = s_B k(c - c_i) \tag{6.2-74}$$

where k is a variable mass transfer coefficient

$$k = AD\left[R_B^{-1} + (\pi Dt)^{-1/2}\right] \tag{6.2-75}$$

and A is a constant whose value depends on the relative rates of motion of the bubble surface $R_B(t)$ (i.e., convection) and the diffusion penetration depth $\delta_p(t)$. If convection due to the expanding interface is slow compared to diffusion ($\lambda_{bc} \gg \lambda_D$) then $A = 1$ (27). In this case, diffusion penetrates into the liquid much more rapidly than the moving interface. On the other hand, if convection is swift ($\lambda_{bc} \ll \lambda_D$), then $A = (7/3)^{1/2}$ (33, 35, 36). In this case, the diffusion layer (δ_p) remains thin and the enhancement of mass transfer reflected by the larger value of A is caused by a sharpening of the liquid-phase concentration gradient (36). If the diffusion layer is sufficiently thin, the curvature of the interface can even be neglected, in which case equation 6.2-75 reduces to $k = (7D/3\pi t)^{1/2}$.

Thus our model for a growing bubble now includes transport balance 6.2-74 and the following force balance

$$\frac{dR_B}{dt} = \frac{R_B}{4\eta}(P_B - P_{me}) \tag{6.2-76}$$

where $P_{me} \equiv P + 2\sigma/R_B$. In equation 6.2-74, we can relate surface concentration c_i to c_B, and subsequently replace c_B and c with corresponding pressures P_B and $P_s \equiv P_{ce}$, respectively, by utilizing a partition law (e.g., Henry's law) and an equation of state (e.g., perfect gas law). If inert gas is again neglected ($c_I = 0$), this balance becomes

$$\frac{dP_B}{dt} = \frac{3AD}{K_c' R_B}\left[R_B^{-1} + (\pi Dt)^{-1/2}\right](P_{ce} - P_B) - \frac{3P_B}{4\eta}(P_B - P_{me}) \tag{6.2-77}$$

with the aid of equation 6.2-76 with which it is coupled. Table 6.2-1 summarizes the equations comprising our model.

Table 6.2-1. Dynamic Model for Bubble Growth

SINGLE BUBBLE

Transport balance
$$\frac{d}{dt}(c_B V_B) = s_B k(c - c_i)$$

where:
$$V_B = (4/3)\pi R_B^3 \;;\quad s_B = 4\pi R_B^2$$

$$k(t) = AD[R_B(t) + (\pi Dt)^{-1/2}]$$

$$A = 1 \;(27)\quad \text{or}\quad (7/3)^{1/2}\;(33,37)$$

$$\lambda_D = (R_B)_o^2/D$$

Partition law
$$c_i = c_B/K_c'$$

where :
$$K_c' = (M_s/\rho_p R_g T)K_w = K_c/R_g T$$

Equation of state
$$P_B = (c_B + c_I)R_g T$$

Force balance
$$\frac{dR_B}{dt} = (R_B/4\eta)(P_B - P - 2\sigma/R_B)$$

SWARM OF UNIFORM-SIZE BUBBLES

Interphase material balance

$$c_o - c = (N_B/V)[c_B V_B - (c_B)_o (V_B)_o]$$

By reducing all pressures and the bubble radius with respective initial values for the bubble, $\hat{P} \equiv P/(P_B)_0$ and $\hat{R}_B \equiv R_B/(R_B)_0$, we can write the balance equations in semidimensionless form (Appendix G)

$$\frac{d\hat{P}_B}{dt} = \left(3A/\lambda_{bc}\hat{R}_B\right)\left[R_B^{-1} + \left(\frac{\lambda_D}{\pi t}\right)^{1/2}\right]\left(\hat{P}_{ce} - \hat{P}_B\right) - \frac{3\hat{P}_B}{4\lambda_{bm}}\left(\hat{P}_B - \hat{P}_{me}\right)$$

$$(6.2\text{-}78)$$

$$\frac{d\hat{R}_B}{dt} = \frac{\hat{R}_B}{4\lambda_{bm}}\left(\hat{P}_B - \hat{P}_{me}\right) \tag{6.2-79}$$

and thereby define two CT's

$$\lambda_{bc} \equiv \frac{K_c'(R_B)_0^2}{D} \tag{6.2-80}$$

and

$$\lambda_{bm} \equiv \frac{\eta}{(P_B)_0} \tag{6.2-81}$$

The second one is clearly associated with mechanically (rheologically) controlled bubble growth and the first one, alluded to earlier, is associated with chemically controlled growth. To start the growth process, a small initial disturbance is required, e.g., $(\hat{P}_{me})_0 < (\hat{P}_B)_0 = 1$ and/or $(\hat{P}_{ce})_0 > (\hat{P}_B)_0 = 1$. The mechanical equilibrium pressure must decline as radius rises in accordance with

$$\hat{P}_{me} = \hat{P} + \frac{2\sigma(R_B)_0}{(P_B)_0} \hat{R}_B^{-1} \tag{6.2-82}$$

The chemical equilibrium pressure $\hat{P}_{ce}$ must also decline as the liquid is depleted of solute. Only for a single bubble growing in an infinite medium can it be treated as a constant, $\hat{P}_{ce} \cong (\hat{P}_{ce})_0$.

Following Langlois (32), we define the dimensionless parameter

$$B \equiv \frac{\lambda_D}{\lambda_{bm}} \tag{6.2-83}$$

When B is very small, it is reasonable to expect bubble growth to be rheologically limited, and thus $P_{me} < P_B \cong P_{ce}$. The bubble grows slowly, compared to the penetration depth due to diffusion, and the concentration in the liquid is therefore approximately uniform ($c_i \cong c$). On the other hand, if B is very large, bubble growth should be diffusion-controlled and we expect that $P_{me} \cong P_B < P_{ce}$. In this case, the concentration gradient could be confined to a thin shell (δ_p) surrounding the bubble, provided that bubble growth is sufficiently rapid compared to diffusion. However, it is still possible for the diffusion depth to expand more rapidly than the bubble interface. This is the situation in which diffusion is more rapid than convection cited earlier and investigated by Epstein and Plesset (27).

More specifically, if we replace concentrations in equation 6.2-74 with pressures, as before, only this time we set $A = 1$ and $P_B = P_{me}$, and combine the result with the Kelvin–Laplace equation $P_{me} = P + 2/R_B$ in lieu of equation 6.2-76, we obtain Epstein and Plesset's expression for bubble growth. The resulting equation is, in dimensionless form,

$$\frac{d\hat{R}_B}{dt} = \frac{\frac{3}{\lambda_{bc}} \left[\hat{R}_B^{-1} + \left(\frac{\lambda_D}{\pi t} \right)^{1/2} \right] (\hat{P}_{ce} - \hat{P}_B)}{(2\hat{P}_B + \hat{P})} \tag{6.2-84}$$

and is coupled with equation 6.2-82 via $\hat{P}_B = \hat{P}_{me}$. From definition 6.2-80, we can identify another important dimensionless parameter for chemically controlled bubble growth, that is,

$$K_c' = \frac{\lambda_{bc}}{\lambda_D} \tag{6.2-85}$$

When $K_c' \gg 1$, we can expect the situation assumed by Epstein and Plesset, in which the motion of the bubble is slow compared to diffusion. It should be noted that the

CT for chemically controlled bubble growth (convection) λ_{bc} decreases with K_c', which might appear to be contradictory at first glance. Actually, this dependence is similar to the dependence of the CT for heat conduction λ_H on specific heat c_p, that is, $\lambda_H = l^2 \rho c_p / k$ (Appendix G). Both K_c' and c_p are capacitance properties (Appendix D). The smaller c_p is, the greater the temperature change at a point per unit heat absorbed or released by conduction during a given time interval. Similarly, the smaller K_c' is, the greater the volume change of the bubble per unit mass absorbed or released by diffusion during a given time interval.

From the relation $\lambda_{bc}/\lambda_{bm} = K_c' B$ we expect K_c' together with B to determine whether growth is rheologically or chemically limited. Thus a small value of K_c' should reinforce rheological control, whereas a sufficiently large value might hinder the establishment of chemical quasi equilibrium, $P_B = P_{ce}$, during growth, notwithstanding the fact that B has a small value as well. The former appears to be the case in Example 6.2-7, where values of parameters B and K_c' and associated CT's were compared using physical properties that are typical for the system polystyrene–styrene. Similarly, for diffusion control, we would expect large values for both B and K_c'.

As previously noted, most of the simulations of bubble growth kinetics quoted here, with the exception of that by Newman and Simon (35), were limited to a single bubble in an infinite liquid. Thus the liquid concentration (pressure) could be treated as a constant. In DV, on the other hand, the decline in melt concentration $c(t)$, which determines stage efficiency, is significant and requires that we deal with a swarm of bubbles of all sizes and ages. A complete analysis of foam DV therefore requires a model for bubble nucleation and an interphase material balance. The simplest nucleation model is impulse nucleation (Fig. 6.2-1) for which

$$r_B(t) = \frac{\delta(t) N_B}{V} \tag{6.2-86}$$

where $\delta(t)$ is the Dirac impulse function

$$\delta(t) = \varepsilon^{-1} \begin{cases} \text{for} & 0 \leqslant t \leqslant \varepsilon \\ 0 & \text{for} & t < 0 \text{ and } t > \varepsilon \end{cases} \tag{6.2-87}$$

such that

$$\lim_{\varepsilon \to 0} \int_0^t \delta(t)\, dt = 1 \tag{6.2-88}$$

Substituting equation 6.2-86 into 6.2-69 and assuming that all viable bubbles nucleated at $t = 0$ are uniform in size $(V_B)_0$ yields $e(0) = (c_B)_0 (V_B)_0 N_B / V$. Thus, neglecting volume changes, concentration c may be coupled to c_B with an interphase material balance for volatile component similar to equation 6.1-45 (Table 6.2-1).

$$c_0 - c(t) = \frac{N_B}{V} \left[c_B(t) V_B(t) - (c_B)_0 (V_B)_0 \right] \tag{6.2-89}$$

which may be written alternatively in terms of dimensionless pressures as

$$\left(\hat{P}_{ce} \right)_0 - \hat{P}_{ce} = \frac{K_c' N_B}{V} \left[\hat{P}_B V_B - (V_B)_0 \right] \tag{6.2-90}$$

It should be noted that equations 6.2-89 and 6.2-90 treat c as a spatially uniform variable, which of course it is not. This points to the underlying weakness of using equation 6.2-74 (with mass transfer coefficient) in lieu of equations 6.2-70–6.2-72. A more rigorous material balance to be used in conjunction with the latter equations would be

$$e(t) = e(0) - \frac{N_B}{V} \int_0^t n_n(t) s_B(t) \, dt \tag{6.2-91}$$

where $e(t)$ contains c_B $[e(t) = c_B(t)V_B(t)N_B/V]$ and $n_n(t)$ contains c (equation 6.2-71), thereby coupling the concentrations inside and outside the bubble.

The system of equations that comprise our approximate model are 6.2-78, 6.2-79, 6.2-82, and 6.2-90. A bubble population density N_B/V is chosen, which identifies an associated critical bubble radius lying on the left leg of an appropriate curve in Fig. 6.1-8. The system of equations is then given a small disturbance from its initial equilibrium (unstable) state by setting the initial bubble radius at a value slightly in excess of the critical value $(R_B)_0 > (R_B)_{cr}$. The corresponding values of P_{me} and P_{ce} are computed from the Kelvin–Laplace (6.2-82) and interphase balance (6.2-90) equations, respectively.

Thus the initial bubble pressure in equations 6.2-78 and 6.2-79 satisfies the inequalities $P_{me} < P_B < P_{ce}$, which renders the first term in equation 6.2-78 positive and the second one negative. Therefore, while bubble radius must grow (equation 6.2-79 is positive), bubble pressure could rise slightly before it declines, which it must do ultimately. The greater the bubble population density (N_B/V) is, the more rapid the rate of decline of P_{ce}.

Results from the approximate model have been plotted in Figs. 6.2-12–6.2-15 (37). Figures 6.2-12 and 6.2-13 show the retardation of bubble growth and separa-

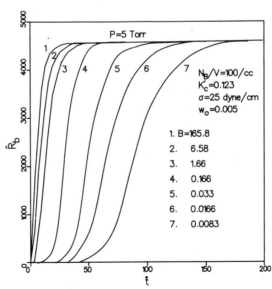

Figure 6.2-12. Variation of dimensionless radius, $R_B/(R_B)_0$, with dimensionless time, $\hat{t} = (\pi t/\lambda_D)^{1/2}$, during bubble growth with B as parameter.

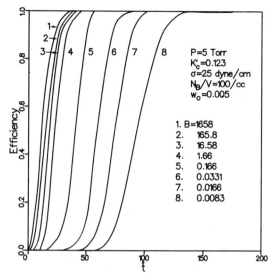

Figure 6.2-13. Variation of separation efficiency, $E_f = e(t)/(c_0 - c_e)$, with dimensionless time, $\hat{t} = (\pi t/\lambda_D)^{1/2}$, during bubble growth with B as parameter.

tion efficiency by increasing rheological resistance (decreasing B). Figures 6.2-14 and 6.2-15 have been plotted in real time to illustrate that the large separations and rapid rates observed in foam-DV can indeed be explained by bubble dynamics, notwithstanding its ultimate dependence on diffusion and viscous resistances. Certainly the enhanced mass transfer area contributed by the presence of numerous bubbles is a major factor. The physical constants and parameters used are believed to be reasonable. A bubble density of 100 cm^{-3} was chosen, which is not far from

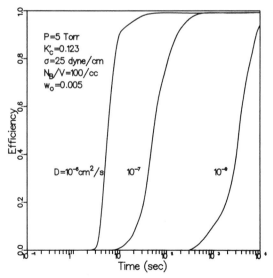

Figure 6.2-14. Variation of separation efficiency, $E_f = e(t)/(c_0 - c_e)$, with time t during bubble growth with diffusivity D as parameter.

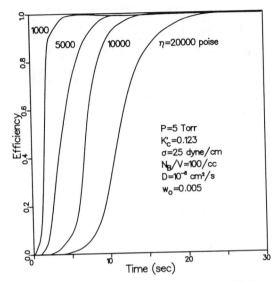

Figure 6.2-15. Variation of separation efficiency, $E_f = e(t)/(c_0 - c_e)$, with time t during bubble growth with viscosity η as parameter.

values observed experimentally and reported elsewhere (35). Diffusivities as low as 10^{-9} cm²/s were included because such values have been observed reported (23), albeit at far lower temperatures than ours, for small molecules diffusing in polymers.

In summary, while our foam-DV model predicts efficiencies within time periods that are consistent with those experienced in industrial processing equipment (Section 6.5), the reader is reminded of its incompleteness and the many assumptions on which it is based. Notwithstanding these, it was deemed appropriate for inclusion owing to the prevalence of the bubble transport mechanism in industrial DV processes.

Example 6.2.6.   Instantaneous, Nonlinear Bubble Evaporation Model.   The material balance for volatile component on the jth stage corresponding to equation 6.2-37 for our simple nonlinear foam model ($n > 1$) is

$$\hat{c}_j^{1-n} - \hat{c}_{j-1}^{1-n} = -\frac{\lambda_f}{\lambda_b}$$

Derive equation 6.2-68 from the solution of this difference equation.

Solution.   After transforming variables, $f_j \equiv \hat{c}_j^{1-n}$, we obtain the nonhomogeneous difference equation

$$f_j - f_{j-1} = -\frac{\lambda_f}{\lambda_b}$$

whose solution (Appendix C)

$$f_j = -\frac{j\lambda_f}{\lambda_b} + c$$

is just the indefinite sum $(\Delta^{-1}\lambda_f/\lambda_b)$ where c is an arbitrary constant. From the initial condition $f_0 = \hat{c}_0^{1-n} = 1$, this constant may be evaluated, $c = 1$, and the solution of the original difference equation found for $j = N$

$$\hat{c}_N^{1-n} = 1 - \frac{N\lambda_f}{\lambda_b}$$

from which equation 6.2-68 follows directly by virtue of the relation

$$E_F = 1 - \hat{c}_N$$

Example 6.2-7. Dynamic Characteristics of Bubble Growth. Contrast λ_{bc} for $K_c' = 0.02$ (Example 6.1-8) and $K_c' = 50$ (27) when $(R_B)_0 = 10^{-3}$ cm, using the value $D = 2 \times 10^{-5}$ cm^2/s. Also, estimate B when $\eta = 10,000$ poise and $P = 10$ mm Hg, using $\sigma = 25$ ergs/cm. What do you conclude?

Solution. From the definition,

$$\lambda_D = \frac{(R_B)_0^2}{D} = 0.056 \text{ s}$$

Therefore,

$$\lambda_{bc} = K_c'\lambda_D = \begin{cases} 0.001 \text{ s} & \text{when } K_c' = 0.02 \\ 2.5 \text{ s} & \text{when } K_c' = 50 \end{cases}$$

From the Laplace–Kelvin equation,

$$(P_B)_0 = P + \frac{2\sigma}{(R_B)_0} = 0.063 \text{ atm}$$

Finally,

$$\lambda_{bm} = \frac{\eta}{(P_B)_0} = 0.16 \text{ s}$$

so

$$B = \frac{\lambda_D}{\lambda_{bm}} = 0.31$$

For the combination of small B and small K_c' (PS), $\lambda_{bc}/\lambda_{bm} = 0.0062$. Thus we would expect bubble growth to be rheologically limited.

6.3. CONTINUOUS PROCESS MODELS

The equipment used for continuous DV ranges from film evaporators to vented extruders and includes specially designed vessels with large interfacial areas and with various configurations for scraping and mixing to effect surface renewal. In this section, we shall combine the elements of evaporation and surface renewal in various complex configurations to simulate the following characteristics of continuous processes: series and parallel operation, bypassing (BP), backmixing (BM), and recycling (RC). Both discrete (staged) and continuous (differential) models will be developed.

6.3.1. Staged Models

A study of staged models is helpful in acquiring an understanding of the behavior of complex DV processes, whether or not they are precise simulations of actual processes. The simplest staged model is the continuous SR model, which consists of N pairs of evaporation-mixing elements in series (20). The jth SR stage is illustrated in Fig. 6.3-1. The evaporation element could be a film (f), as shown, and the mixer a rotating melt pool for which DV is negligible compared to the film, or it could be an evaporating melt pool (p).

A steady-state material balance for volatile component on the jth stage of any N-stage continuous devolatilizer yields

$$1 \leqslant j \leqslant N \qquad \dot{V}_{j-1}[c]_{j-1} - \dot{V}_j[c]_j = \{n\}_j s_j \qquad (6.3\text{-}1)$$

where brackets [ ] and { } signify flow-average and volume-average, respectively (Appendix F). Following customary chemical engineering practice, we express the average transport flux in terms of a mass transfer coefficient and an overall concentration-difference driving force.

$$\{n\}_j = k_j\left([c]_{j-1} - c_{ej}\right)$$

By neglecting density changes ($\dot{V}_j \equiv \dot{V}$ for all j) and assuming complete transverse

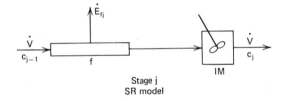

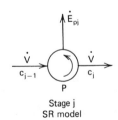

Figure 6.3-1. Schematic diagram of the jth stage of an N-stage surface renewal (SR) model, each stage consisting of a film-mixer pair or a rolling pool.

mixing ($[c]_j \equiv c_j$ for all j), balance 6.3-1 becomes, after substituting the transport flux,

$$c_{j-1} - c_j = \lambda_j k_j s_{V_j}(c_{j-1} - c_e). \tag{6.3-2}$$

For the special case of N identical stages ($\lambda_j \equiv \lambda$ and $c_{ej} \equiv c_e$ for all j), the solution of difference equation 6.3-2 gives the expected result for process efficiency.

$$E_F \equiv \frac{c_N - c_e}{c_0 - c_e} = 1 - (1 - \text{ex})^N \tag{6.3-3}$$

where stage efficiency has been replaced by stage extraction number ex, whose value depends on the nature of the stages involved.

If the stages consist of film-mixer (IM) pairs, then $\text{ex} = \lambda_f k_f s_{Vf} \equiv E_{ff}$ from equation 6.2-9, where E_{ff} is the film stage efficiency. If the stages consist of melt pools having either side-feed and discharge, or end-feed and discharge (Fig. 6.3-1), then $\text{ex} = \lambda_v k_p s_{Vp} \equiv N_p E_{fp}$, where E_{fp} is the pool stage efficiency and $N_p \equiv \lambda_v/\lambda_p$ (equation 6.2-55).

When diffusion is the prevailing mechanism, E_{ff} is given by equation 6.2-27 and E_{fp} is given by equation 6.2-54. As before, λ_v is the residence time in the pool and λ_p is the period of rotation. A comparison of these two SR models is made in Example 6.3-1. By writing $N_p E_{fp} = \lambda_v k_p s_{Vp} = k_p s_p/\dot{V}$ and noting that $k_p \propto \lambda_p^{-1/2}$, we can rationalize how E_F can increase in response to an increase in machine rpm (decrease in λ_p) when throughput $\dot{V}$ and surface area s_p remain fixed, notwithstanding a concomitant decline in residence time. This "anomalous" behavior was discussed earlier (Section 6.2.3) in the context of film diffusion.

The approximate expression for equation 6.3-3 when ex is small is, of course, the EXP model

$$E_F = 1 - \exp(-Ex) \tag{6.3-4}$$

where

$$Ex = \begin{cases} NE_{ff} & \text{film} \\ NN_p E_{fp} & \text{pool} \end{cases} \tag{6.3-5}$$

In polymer processing equipment, owing to the impracticability of creating a thin film from the entire bulk material, surface renewal is frequently accomplished by splitting from the main polymer melt stream a separate stream for film evaporation. The phenomenon of stream-splitting (SS) within the context of staged modeling allows several alternative intrastage configurations among the evaporation and IM elements, distinguishable from one another by the manner in which the side-stream is separated and subsequently reunited with the mainstream in the IM. Figure 6.3-2 shows three stage configurations with a single evaporation element and various types of stream-splitting; each stage, may, of course, be interconnected in series with others of like or different kinds to give an N-stage cascade, such as the SR model. The first configuration is termed bypass (BP), the second recycle (RC), and the third

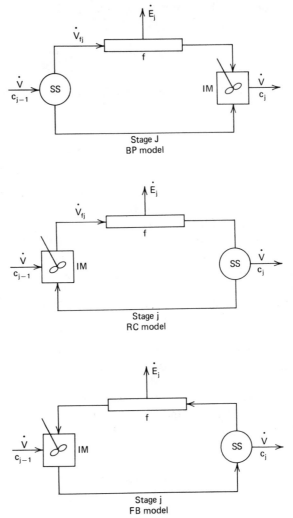

Figure 6.3-2. Schematic diagram of the jth stage of an N-stage cascade, each stage consisting of a single evaporation element (film) with stream splitting (SS): the bypass (BP) model; the recycle (RC) model; and the feedback (FB) model.

feedback (FB), for obvious reasons. We note that all three actually constitute forms of BM because "old" (low c) elements are mixed with "young" (high c) ones, thereby reducing the concentration-difference driving force for evaporation. The main distinction between the RC and FB models is the feed concentration entering the evaporation element. It is lower in the FB model than in the RC model and, consequently, so is the driving force for diffusion.

Figure 6.3-3 illustrates two-element models, one series and one parallel, each consisting of an evaporating film (f) and an evaporating melt pool (p) of either type [(b) or (c)] shown in Fig. 6.2-6. The series arrangement may be viewed as a variation of the SR model in which the mixer (melt pool) contributes to DV. The parallel arrangement may be considered to be a stagewise version of the continuous film-pool combination illustrated in Fig. 6.2-7 (20).

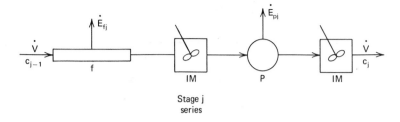

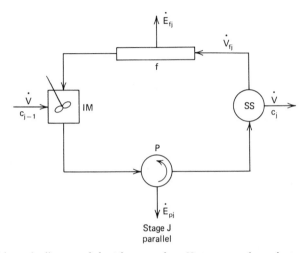

Figure 6.3-3. Schematic diagram of the jth stage of an N-stage cascade, each stage consisting of two evaporation elements (film and pool) in series or in parallel with feedback.

It can be shown (reference 20 and Example 6.3-2) that the steady-state component balance on the jth stage of all six models in Figs. 6.3-1–6.3-3 takes on the following general form with the usual assumptions

$$1 \leqslant j \leqslant N \qquad\qquad c_{j-1} - c_j = s_{Rj} E_{fj}(c^* - c_{ej}) \qquad\qquad (6.3\text{-}6)$$

where $s_{Rj} \equiv \dot{V}_{fj}/\dot{V}$ is the stream flow ratio and

$$c^* = \begin{cases} c_{j-1} & \text{for SR and BP models} \\ \bar{c}_j \equiv \left[c_{j-1} + (s_{Rj} - 1)c_j \right]/s_{Rj} & \text{for RC model} \\ c_j & \text{for FB model} \end{cases} \qquad (6.3\text{-}7)$$

For N identical stages, all models satisfy the same difference equation with a constant coefficient.

$$c_j - a c_{j-1} = (1 - a)c_e \qquad\qquad (6.3\text{-}8)$$

Stream ratio s_R and coefficient a for each model are listed in Table 6.3-1. It should be noted that s_R is related to the recycle ratio R_R used in Section 1.3 for a RC

Table 6.3-1. Parameters for Staged Models

Model	s_R	a	ex
SR	–	$1 - ex$	E_f
BP	$0 \leq s_R \leq 1$	$1 - ex$	$s_R E_f$
RC	$1 \leq s_R < \infty$	$(1 + ex)^{-1}$	$s_R E_f (1 - E_f)^{-1}$
FB	$0 \leq s_R < \infty$	$(1 + ex)^{-1}$	$s_R E_f$
SERIES	–	$1 - ex$	$1 - (1 - E_{ff})(1 - N_p E_{fp})$
PARALLEL	$0 \leq s_R < \infty$	$(1 + ex)^{-1}$	$s_R E_{ff} + (1 + s_R) N_p E_{fp} (1 - N_p E_{fp})^{-1}$

$E_f = E_{ff}$ or $N_p E_{fp}$

$E_{ff} = $ stage efficiency of film

$E_{fp} = $ stage efficiency of melt pool

reactor by $s_R = (1 - R_R)^{-1}$. In dimensionless form, equation 6.3-8 is homogeneous

$$\hat{c}_j - a\hat{c}_{j-1} = 0 \tag{6.3-9}$$

if we define $\hat{c}_j \equiv (c_j - c_e)/(c_0 - c_e)$. Its solution for $j = N$ yields

$$E_F = 1 - a^N \tag{6.3-10}$$

for process efficiency. Substitution of the appropriate definition of a listed in Table 6.3-1 gives the process efficiency for any of our staged models.

It is convenient to express our results in terms of a dimensionless stage extraction number ex. Thus process efficiency takes on two general functional forms

$$E_F = 1 - \begin{cases} (1 - ex)^N \\ (1 + ex)^{-N} \end{cases} \tag{6.3-11}$$

which have been plotted in Fig. 6.3-4. It is clear that in all our models E_F increases with N when ex is fixed, and it increases with ex when N is fixed. Inspection of Table 6.3-1 reveals that ex and therefore E_F also increase with s_R. As s_R is raised in the BP model, starting with a value less than 1, we approach the SR model by reducing bypassing. In the RC model, recycle ratio R_R rises with s_R, which could improve process efficiency (cf. Section 1.3). In the FB model, nonzero values of s_R are required for evaporation because the mass transfer component lies in the FB stream.

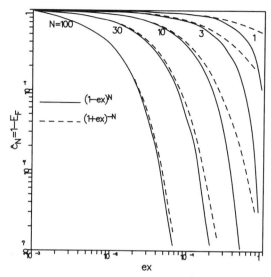

Figure 6.3-4. Logarithmic relationship between fraction of extractable volatile component remaining after N stages, $\hat{c}_N = 1 - E_F$, versus stage extraction number ex with number of stages, N, as parameter for two classes of stage functions: ($\hat{c}_N = (1 - \mathrm{ex})^N$ and $\hat{c}_N = (1 + \mathrm{ex})^{-N}$.

Both RC and FB models involve backmixing (BM), which is known to have a detrimental effect on certain chemical processes (Section 1.3 and Chapter 3). Thus, although RC is known to be beneficial to certain chemical processes (Section 1.3) and FB is required in the FB model to achieve separation, the important question is whether the benefits can offset the detrimental effects of BM. This question is examined in Examples 6.3-3 and 6.3-4 for diffusion-controlled DV under two of the conditions proposed in Section 1.3, that is, fixing V and allowing $\dot{V}$ to decline and fixing both. These examples should be compared with Example 1.3-3.

We conclude from the first set of conditions that higher process efficiencies E_F can be achieved through RC and FB at the expense of throughput $\dot{V}$ and specific output rate XE_F, as with chemical reaction. However, from the second set of conditions, it appears that in diffusion-controlled DV, unlike chemical reaction, it is possible to obtain both higher efficiencies and higher specific output rates with RC than without it (simple SR). The reason is probably that diffusion rate depends only on the concentration gradient and not on the concentration level (as reaction rate does), which is lowered in the feed stream due to BM. We have shortened λ_f per pass in the RC case and simultaneously increased the number of passes, thereby increasing evaporation rate for constant $\dot{V}$.

To summarize, process efficiency for staged models takes on the following general functional form

$$E_F = F(N, \mathrm{ex}) \tag{6.3-12}$$

where the function F depends on the particular process model. Parameter ex is a function of s_R and E_f, where

$$E_f = f(\mathrm{De}) \tag{6.3-13}$$

Function F is given by equation 6.3-11 and f is given by equation 6.2-27 for diffusion-controlled DV. For small values of ex (< 0.1), both forms of F reduce to a single one in which parameters N and ex appear as a single product $N\,\text{ex} = \text{Ex}$:

$$E_F \cong 1 - \exp(-\text{Ex}) \qquad (6.3\text{-}4)$$

Example 6.3-1.   Film SR Versus Pool SR.   Estimate the total residence time Λ_v required by an SR model with 10 melt pools to achieve $E_F = 0.9$ if each melt pool is $1/2$-in. in diameter and is rotating at 60 rpm. Compare this result with the 8.5 min required by the 10-film SR model in Example 6.2-3. Assume that evaporation occurs only during half of each revolution, and use $D = 10^{-4}$ in.^{2} min.

Solution.   From equations 6.3-3 and 6.2-9, $N_p E_{fp} = 0.2 = \lambda_v k_p s_v$ where λ_v is the RT per stage. Assuming cylindrical pools for which $s_v = 2/Ra$, we obtain $s_v = 8$ in.$^{-1}$. From equation 6.2-51, $k_p = 124$ mils/min when $\lambda_p = 0.5/60$ min. (Example 6.2-5). Therefore, $\lambda_v = 0.2/k_p s_v = 0.2$ min and $\Lambda_v = 2$ min, which is much less than the film devolatilizer, especially since we haven't taken into account the RT of the 10 mixers in the latter. It must be remembered, however, that the assumption of perfect mixing in the evaporating pools is open to serious question and, consequently, so are these results.

Example 6.3-2.   Parallel and Series Models.   Write the material balances for the parallel and series models of Fig. 6.3-3 and show that they fit the general format.

Solution.   For the parallel model, the material balance is

$$c_{j-1} + s_R c_{fj} = (1 + s_R) c_j$$

and the film and pool efficiencies are defined by

$$1 - N_p E_{fp} = \frac{c_j - c_e}{\bar{c}_j - c_e}$$

and

$$1 - E_{ff} = \frac{c_{fj} - c_e}{c_j - c_e}$$

where c_{fj} and $\bar{c}_j$ represent the stream concentrations entering and leaving the IM, respectively. Elimination of the latter two quantities among the three equations yields

$$c_j \left[(1 + s_R)(1 - N_p E_{fp})^{-1} - s_R(1 - E_{ff}) \right] - c_{j-1}$$
$$= c_e \left[(1 + s_R)(1 - N_p E_{fp})^{-1} - s_R(1 - E_{ff}) - 1 \right]$$

which clearly satisfies equation 6.3-8 and can be easily shown to define a and consequently ex correctly.

For the series model, the two efficiencies are defined by

$$1 - N_p E_{fp} = \frac{c_j - c_e}{c_{fj} - c_e}$$

and

$$1 - E_{ff} = \frac{c_{fj} - c_e}{c_{j-1} - c_e}$$

where c_j represents the concentration leaving the film and entering the pool. Elimination of this quantity yields the material balance

$$c_j - c_{j-1}\left[\left(1 - E_{ff}\right)\left(1 - N_p E_{fp}\right)\right] = c_e\left[1 - \left(1 - E_{ff}\right)\left(1 - N_p E_{fp}\right)\right]$$

which clearly conforms to equation 6.3-8.

Example 6.3-3.   Effect of RC and FB with Variable $\dot{V}$.   Examine the effects of recycle and feedback with $s_R = 1.5$ on the efficiency and specific output rate of the 10-stage SR film devolatilizer in Example 6.2-3 when $\dot{V}_f$ is held constant and $\dot{V}$ is allowed to change with s_R. Assume that h and s_f remain the same.

Solution.   Since the film volume hs_f and $\dot{V}_f$ are fixed, so is λ_f. From Example 6.2-3, $\lambda_f = 0.85$ min. Therefore, $\lambda_v = s_R\lambda_f = 1.28$ min, if we ignore the volume of the IM's, and $\Lambda_v = N\lambda_v = 12.8$ min. Since λ_f and λ_D are identical, so is $E_f = 0.20$. Thus we compute ex from Table 6.3-1 and E_F from equation 6.3-3. Finally, $XE_F = E_F/s_v\Lambda_v$. The results are summarized in the following chart.

Assuming that the SR, RC, and FB models have identical total volumes V, we conclude that recycle and feedback improve process efficiency at the expense of throughput rate ($\dot{V} = V/\Lambda_v$) and therefore specific output rate (vapor removal rate).

Model	s_R	R_R	ex	E_F	Λ_v (min)	XE_F (mils/min)
SR	1	0	0.20	0.90	8.5	5.3
RC	1.5	0.33	0.375	0.96	12.8	3.75
FB	1.5	—	0.30	0.93	12.8	3.62

Example 6.3-4.   Effect of RC and FB with Constant $\dot{V}$.   Examine the effects of recycle and feedback with $s_R = 1.5$ on the efficiency and specific output rate of the 10-stage SR film devolatilizer in Example 6.2-3 when $\dot{V}$ is held constant and $\dot{V}_f$ is allowed to change with S_R. Assume that h and s_f remain the same.

Solution.   Since the system volume and $\dot{V}$ are fixed, so is λ_v. Consequently, λ_f must change. If we ignore the volume of the IM's, $\lambda_v = 0.85$ min from Example 6.2-3. Therefore, $\lambda_f = \lambda_v/s_R = 0.567$ min, De $= 0.0227$ (infinite film), and $E_f = 0.17$. Computation of ex and E_F follow from Table 6.3-1 and equation 6.3-3, respectively. Finally, $XE_F = E_F/s_v\Lambda_v$. The results are summarized in the following chart.

We conclude that recycle can still improve process efficiency slightly, even for fixed throughput, by increasing the specific output rate (vapor removal rate). The rising value of s_R is sufficient to offset the competing effect of a decline in E_f and to produce a greater composite value of ex in both models.

Model	s_R	R_R	ex	E_F	Λ_v (min)	XE_F (mils/min)
SR	0	0	0.20	0.90	8.5	5.3
RC	1.5	0.33	0.307	0.93	8.5	5.5
FB	1.5	—	0.255	0.90	8.5	5.3

6.3.2. *Differential Models*

The continuous film in Section 6.2.3 and the continuous pool in Section 6.2.6 are examples of differential models when α is a measure of position (x/v or z/v). They are both described by differential equations in which concentration is a continuous function of position in the direction of flow (longitudinal). In this section, we shall extend these models.

We begin by rearranging our general difference equation for staged models, equation 6.3-9, as follows:

$$\hat{c}_j - \hat{c}_{j-1} = -(1-a)\hat{c}_{j-1} \qquad (6.3\text{-}14)$$

As we pass from finite stages to infinitesimal ones in the direction of flow, it is reasonable to expect stage efficiency to decrease simultaneously as the number of stages increases without limit; that is, ex $\to$ 0 as $N \to \infty$. It follows that $a \to 1$ and $1 - a \cong$ ex (Table 6.3-1). Thus we may write equation 6.3-14 as

$$\hat{c}_j - \hat{c}_{j-1} = -\text{ex}\,\hat{c}_{j-1} \qquad (6.3\text{-}15)$$

Recalling that in general Ex = Nex, we define a dimensionless (fractional) longitudinal position variable $\hat{z} \equiv z/L$ as the equivalent of N^{-1} in continuous mathematics and rewrite our stage extraction number as ex = $\Delta\hat{z}$ Ex. Following substitution into equation 6.3-15 and passage to the limit $\Delta z \to 0$ ($N \to \infty$), we obtain

$$\frac{d\hat{c}}{d\hat{z}} = -\text{Ex}\,\hat{c} \qquad (6.3\text{-}16)$$

The solution evaluated at $\hat{z} = 1$, of course, leads to the familiar EXP model

$$E_F = 1 - \text{ex}(-\text{Ex}) \qquad (6.3\text{-}4)$$

In particular, the extraction number for parallel film and pool DV as illustrated in Fig. 6.2-7 is

$$\text{Ex} = N_f E_{ff} + N_p E_{fp}$$
$$= \frac{k_f s_f + k_p s_p}{\dot{V}} \qquad (6.3\text{-}17)$$

where $N_f = \dot{V}_f/\dot{V}$, $N_p = \lambda_v/\lambda_p$, $\dot{V}$ is the volumetric flow rate through the pool longitudinally (z-direction), and $\dot{V}_f$ is the volumetric flow rate of the film along its entire length (z-direction).

Implicit in our differential model thus far is the absence of longitudinal dispersion. It is customary in chemical engineering to account for it by using the so-called dispersion model (Section 1.4) in lieu of equation 6.3-16

$$\mathrm{Pe}_L^{-1} \frac{d^2 \hat{c}}{d\hat{z}^2} - \frac{d\hat{c}}{d\hat{z}} - \mathrm{Ex}\,\hat{c} = 0 \qquad (6.3\text{-}18)$$

where $\mathrm{Pe}_L \equiv Lv/D_L$ is the longitudinal Peclet number and D_L is the longitudinal dispersion coefficient. The solution of equation 6.3-18 and graphs of $\hat{c}(1) \equiv 1 - E_F$ were presented in Section 1.4.

To summarize, process efficiency for differential models take the general form

$$E_F = F(\mathrm{Ex}, \mathrm{Pe}_L) \qquad (6.3\text{-}19)$$

where the function F depends on the particular process model. Specific examples of F that pertain to actual DV processes (vented extruders) will be developed in Section 6.5.

6.4. BATCH PROCESS MODELS

Batch DV can be useful on a laboratory scale for collecting data on limited quantities of polymer/volatile systems for the purpose of design and scale-up of continuous processes. Batch apparatus should therefore contain the elementary steps of evaporation and surface renewal, and design algorithms should be formulated in terms of continuous model parameters such as stage efficiency E_f.

Among the disadvantages of batch operation are the obvious ones: the difficulty of reproducing initial conditions, necessitated by the absence of the steady state, and the possibility of long residence times due to the addition of fill and discharge times, which could lead to polymer degradation.

The ideal batch process is illustrated in Fig. 6.4-1. It consists of a series of evaporation and surface renewal operations that occur sequentially in time. All the material passes from the evaporator to the mixer to the evaporator and so on. Although such a process appears simple in principle, it is difficult to achieve in practice. One scheme for implementing it is the batch loop devolatilizer, illustrated in Fig. 6.4-2. A problem that arises immediately with this process, and with any other batch process in which only a portion of the material is devolatilized at any given time, is the introduction of backmixing (BM). This phenomenon will be examined first in general. Then we will attempt to model the loop devolatilizer in particular.

6.4.1. Backmixing

Let us consider a general batch process consisting of an ideal mixer (IM) of volume V_M and an evaporation element (film or pool) of volume V_f, in which only a fraction (V_f/V_M) of the material is devolatilized at any time for a period λ_j. The subscript j again refers to the jth evaporation cycle. We define the volume-average volatile

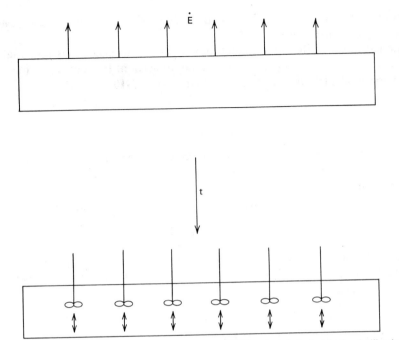

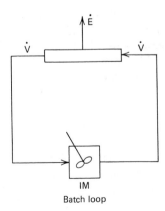

Figure 6.4-1. Schematic diagram of the surface-renewal (SR) principle in batch devolatilization.

concentration in the entire system at the end of cycle j, $\langle c_j \rangle$, in terms of the volume-average concentrations in the film, $\langle c_{fj} \rangle$, and the IM, $c_M = \langle c_{j-1} \rangle$ as follows:

$$c_j = \frac{c_{fj} + Mx c_{j-1}}{1 + Mx} \tag{6.4-1}$$

where $Mx \equiv V_M/V_f$ is a dimensionless mixing number and the brackets $\langle \ \rangle$ have been dropped for convenience. Next we introduce our stage efficiency

$$1 - E_{fj} = \frac{c_{fj} - c_{ej}}{c_{fj-1} - c_{ej}} \tag{6.4-2}$$

Batch loop

Figure 6.4-2. Schematic diagram of the batch loop devolatilizer.

and assume that E_{fj} depends only on λ_j and that all cycles are identical. Thus, after combining equations 6.4-1 and 6.4-2 and writing the result in dimensionless form, we obtain our familiar difference equation

$$\hat{c}_j - \left(1 - \frac{E_f}{1 + Mx}\right)\hat{c}_{j-1} = 0 \tag{6.4-3}$$

where $\hat{c}_j \equiv (c_j - c_e)/(c_0 - c_e)$. The solution for N cycles, of course, leads to

$$1 - \hat{c}_N \equiv E_F = 1 - \left(1 - \frac{E_f}{1 + Mx}\right)^N \tag{6.4-4}$$

with the SR model as a special case ($Mx = 0$). Clearly, however, when Mx is large, indicating that BM is important, the composite stage efficiency $E_f/(1 + Mx)$ is reduced and consequently so is the process efficiency. For sufficiently small values of $E_f/(1 + Mx)$, we can approximate our result with the EXP model

$$E_F \cong 1 - \exp\left(\frac{-NE_f}{1 + Mx}\right) \tag{6.4-5}$$

These models are applied in Examples 6.4-1–6.4-3.

6.4.2. Batch Loop Devolatilizers

The process referred to represents an abstraction of a rotating processing machine operated in zero throughput mode. It is a closed flow loop (Fig. 6.4-2) consisting of a translating film evaporator and an IM in series spatially, instead of a batch film and batch mixer in sequence temporally. Although the main purpose of the IM is to effect surface renewal (transverse mixing), a by-product is longitudinal mixing (BM). This is a consequence of the finite volume V_M, and therefore finite residence time λ_M, of the IM together with the variation with time of the film effluent concentration that feeds it.

Our differential balance equation for volatile component in the film (Appendix F)

$$\frac{\partial c_f}{\partial t} = -v_f \frac{\partial c_f}{\partial x} + D \frac{\partial^2 c_f}{\partial y^2} \tag{6.4-6}$$

now contains a convective transport term and a second space variable (x) in addition to time (t). It is thus more complex than that (equation 6.2-14) for the batch film and the steady-state continuous film, which may be viewed as a special cases. The longitudinal concentration profile here is not uniform in space nor is the inlet concentration to the film constant in time. To account for the time-dependent feed concentration, it is necessary to modify the BC's. In addition to the initial condition $c_f(x, y, 0) = c_0$ and boundary conditions $c_f(x, 0, t) = c_e$ and $(\partial c_f/\partial y)|_{y=h} = 0$ for the finite film, we must modify the BC for the infinite film

$$c_f(x, \infty, t) = c_M\left(t - \frac{x}{v_f}\right) \tag{6.4-7}$$

and add one for the feed

$$c_f(0, y, t) = c_M(t) \tag{6.4-8}$$

The concentration of volatile component in the IM, $c_M(t)$, is the dependent variable in a material balance for volatile component in the IM

$$V_M \frac{dc_M}{dt} = -v_f W h \{ c_M - [c_f(t)] \} \tag{6.4-9}$$

where $[c_f(t)]$ is the flow-average concentration leaving the film

$$[c_f(t)] = \int_{y=0}^{h} \frac{c_f(L, y, t)\, dy}{h} \tag{6.4-10}$$

The IC for equation 6.4-9 is $c_M(0) = c_0$.

It appears necessary to solve coupled equations 6.4-6 and 6.4-9 simultaneously. Before attempting this, however, we shall deduce two limiting solutions using mainly physical arguments. First, we write the two differential equations in partly dimensionless form (Appendix G):

$$\lambda_f \frac{\partial \hat{c}_f}{\partial t} = -\frac{\partial \hat{c}_f}{\partial \hat{x}} + \mathrm{De} \frac{\partial^2 c_f}{\partial \hat{y}^2} \tag{6.4-11}$$

and

$$\lambda_M \frac{d\hat{c}_M}{dt} = \int_0^1 \hat{c}_f(1, \hat{y}, t)\, d\hat{y} - \hat{c}_M \tag{6.4-12}$$

where $\lambda_M \equiv V_M / \dot{V}$ is the CT for surface renewal (mixing). Distances x and y have been rendered dimensionless through division by L and h, respectively.

The first limiting case occurs when Mx $\ll 1$, where

$$\mathrm{Mx} \equiv \frac{V_M}{V_f} = \frac{\lambda_M}{\lambda_f} \tag{6.4-13}$$

Application of this condition gives

$$\hat{c}_f(0, \hat{y}, t) = \hat{c}_M(t) = \int_{\hat{y}=0}^{\hat{y}} \hat{c}_f(1, \hat{y}, t)\, d\hat{y} \tag{6.4-14}$$

for BC 6.4-8, which signifies that the feed concentration to the film is equal to the flow-average concentration to its effluent. Thus an observer traveling with the material around the batch loop would experience sequences of film evaporation for times λ_f followed by instantaneous transverse mixing. Since the detrimental BM effect of diminished average inlet concentration has been abolished, this process is now in essence equivalent to the SR model. Therefore, we should expect its efficiency to be given by

$$1 - [\hat{c}_f]_N \cong 1 - (\hat{c}_M)_N \equiv E_F = (1 - E_f)^N \tag{6.4-15}$$

where $N = t/\lambda_f$.

The second limiting case occurs when $Mx \gg 1$. This condition suggests that the film is in a quasi steady state. The steady-state solution of equation 6.4-6 for any feed concentration c_M is, of course, well known (Section 6.2). From the associated film efficiency $E_f(\lambda_f)$, we deduce the following expression

$$c_M - [c_f] = E_f(c_M - c_e) \qquad (6.4\text{-}16)$$

which is then substituted into the RHS of equation 6.4-9. The resulting differential equation

$$\frac{dc_M}{dt} = -\lambda_M^{-1} E_f(c_M - c_e) \qquad (6.4\text{-}17)$$

yields for process efficiency after time t

$$1 - \hat{c}_M(t) \equiv E_F = 1 - \exp\left(\frac{-E_f t}{\lambda_M}\right) \qquad (6.4\text{-}18)$$

For large values of Mx ($Mx \gg 1$), it can be shown that $t/\lambda_M = N/(1 + Mx)$, which renders equation 6.4-18 identical to equation 6.4-5.

It is important to recognize the distinction between what we defined earlier as the number of evaporation cycles (stages) $N \equiv t/\lambda_f$ and the actual number of cycles (stages) $t/(\lambda_f + \lambda_M) = N/(1 + Mx)$ experienced on the average by all the material in the batch devolatilizer after time t (cf. equation 6.4-5). Obviously, they are identical when Mx is zero. However, as Mx becomes large, the material actually experiences fewer evaporation cycles per actual cycle and process efficiency must therefore decline concomitantly (Example 6.4-1).

The role of BM can perhaps be best demonstrated by comparing the limiting models, SR and EXP. By equating expressions 6.4-15 and 6.4-18, we obtain a ratio of the number of stages required to achieve the same process efficiency E_F with the same stage efficiency E_f

$$\frac{N_{EXP}}{N_{SR}} = -\frac{Mx}{E_f} \ln(1 - E_f) \qquad (6.4\text{-}19)$$

For small values of E_f (< 0.1)

$$\frac{N_{EXP}}{N_{SR}} \cong Mx \qquad (6.4\text{-}20)$$

It is apparent that $N_{EXP}/N_{SR} > 1$ owing to the detrimental effects of BM and that this ratio becomes larger as E_f increases. For a given E_f, we conclude that BM will result in longer total process time, $t_{EXP} > t_{SR}$. If we assume that constant E_f implies constant λ_f, then as λ_M increases with Mx, t_{EXP} must grow concomitantly. An important implication of this additional (mixing) time is that equipment size (V) must increase (Example 6.4-2).

One might speculate about why λ_M would be made large intentionally. Just as large λ_f enhances evaporation efficiency, large λ_M can enhance mixedness in very viscous fluids when laminar mixing is the dominant mechanism. This is so because extent of mixing increases with total shear strain (Chapter 3), which increases with time for a given mean shear rate.

We shall now attempt to develop a more general solution for the loop devolatilizer model using the concept of penetration depth (Section 6.2.2) to simplify our analysis (45). Imagine the concentration of volatile solute at the time evaporation starts, $t = 0$, to be uniform and equal in the film and the IM, $c_f = c_0 = c_M$. An instant later, $t > 0$, the value of the effluent concentration $[c_f(t)]$ is already less than c_0 as a result of evaporation. Using penetration depth $\delta_p(t)$, we may write

$$c_f(L, y, t) = \begin{cases} c_e & \text{when } 0 \leqslant y < \delta_p(t) & (6.4\text{-}21) \\ c_0 & \text{when } \delta_p(t) \leqslant y \leqslant h & (6.4\text{-}22) \end{cases}$$

At the same instant the concentration at the film entrance also drops to a lower value than c_0 because the film is fed from the IM which propagates the changing effluent concentration instantaneously. More precisely, c_f from equation 6.4-22 is valid only for times between 0 and λ_f because the material leaving during this period had entered it at a concentration c_0. For times exceeding λ_f, however, we must write

$$c_f(L, y, t) = c_M(t - \lambda_f) \qquad \text{when } \delta_p(\lambda_f) \leqslant y \leqslant h \qquad (6.4\text{-}23)$$

because the material leaving after λ_f entered at the concentration prevailing in the mixer $c_M(t - \lambda_f)$, which had changed in response to evaporation in the film. It is important to remember that this analysis presumes plug flow, no axial diffusion in the evaporator, and perfect mixing in the mixer.

Thus for the flow-average effluent concentration, we obtain

$$[c_f(t)] = \frac{\int_0^h c_f(L, y, t) \, dy}{h}$$

$$= \frac{\int_0^{\delta_p(t)} c_e \, dy + \int_{\delta_p(t)}^h c_0 \, dy}{h} = c_0 - \frac{(c_0 - c_e)\delta_p(t)}{h} \qquad (6.4\text{-}24)$$

for times $0 \leqslant t \leqslant \lambda_f$, and

$$[c_f(t)] = \frac{\int_0^{\delta_p(\lambda_f)} c_e \, dy + \int_{\delta_p(\lambda_f)}^h c_M(t - \lambda_f) \, dy}{h}$$

$$= c_M(t - \lambda_f) - \frac{[c_M(t - \lambda_f) - c_e]\delta_p(\lambda_f)}{h} \qquad (6.4\text{-}25)$$

for times $t > \lambda_f$. In dimensionless form, these results may be expressed more conveniently as

$$[\hat{c}_f(t)] = \begin{cases} 1 - \hat{\delta}_p(t) & 0 \leqslant t \leqslant \lambda \\ [1 - \hat{\delta}_p(\lambda_f)][\hat{c}_M(t - \lambda_f)] & t \geqslant \lambda_f \end{cases} \qquad (6.4\text{-}26)$$

where $\hat{\delta}_p(\lambda_f) = E_f$. It should be noted that the limits in equation 6.4-25 are independent of time, the reasons being that the depth of penetration is fixed by the film exposure time λ_f.

By the procedure just described, we have actually deduced the quantity required on the RHS of equation 6.4-9, which reduces our model of the staged batch devolatilizer to the following simple differential equation

$$\text{Mx} \frac{d\hat{c}_M}{d\hat{t}} + \hat{c}_M = \begin{cases} 1 - \hat{\delta}_p(t) & \text{when } 0 \leqslant \hat{t} < 1 \\ (1 - E_f)\hat{c}_M(\hat{t} - 1) & \text{when } \hat{t} > 1 \end{cases} \qquad (6.4\text{-}27)$$

where $\hat{t} \equiv t/\lambda_f$. For N cycles, we seek the solution $\hat{c}_M(\hat{t})$ for integer values $\hat{t} = N$.

The limiting cases (SR and EXP models) are immediately deducible from differential equation 6.4-27. For relatively small mixer volumes ($\text{Mx} \ll 1$), the first term in the differential equation may be neglected, and we obtain

$$\hat{c}_M(\hat{t}) = \begin{cases} 1 - \hat{\delta}_p(t) \\ (1 - E_f)\hat{c}_M(\hat{t} - 1) \end{cases} \qquad (6.4\text{-}28)$$

The latter result for integer values of $\hat{t}$ is equivalent to the dimensionless difference equation

$$(\hat{c}_M)_j - (1 - E_f)(\hat{c}_M)_{j-1} = 0 \qquad (6.4\text{-}29)$$

whose solution for $j = N$ yields the SR model for E_F. For relatively large mixer volumes ($\text{Mx} \gg 1$), we can set $\hat{c}_M(\hat{t} - 1) = \hat{c}_M(\hat{t})$ in the differential equation, which then reduces to the dimensionless equivalent of equation 6.4-17.

$$\text{Mx} \frac{d\hat{c}_M}{d\hat{t}} + E_f\hat{c}_M = 0 \qquad (6.4\text{-}30)$$

whose solution for $\tau = N$ yields the EXP model for E_F.

Equations 6.4-27 are first-order differential-difference equations with feedback in which the time lag corresponds to one evaporator residence time. Expanding $\hat{c}_M(\hat{t} - 1)$ in a Taylor series and neglecting terms of higher order than two leads to the dispersion model, which has been discussed in detail in Section 6.3. Repeated integration of equations 6.4-27 using the integrating factor $\exp(-\hat{t}/\text{Mx})$ yields the results listed in Table 6.4-1 (21, 45).

Example 6.4-1. The Effect of Backmixing.   Estimate $\hat{c}_N$ for two batch processes, one with $\text{Mx} = 0.1$ (SR) and the other with $\text{Mx} = 10$ (EXP), when both have identical values of $\lambda_D = 25$ min, $\lambda_f = 0.1$ min, and $t = 5$ min.

Solution.   First we compute

$$\text{De} = \frac{0.1}{25} = 0.004$$

Table 6.4-1.   Solution of Equations 6.4-27

$$\hat{C}_M(\theta_{N+1}) \;=\; \sum_{j=o}^{N} A_j(\theta_{N+1}) y_{N-j} \;+\; B(\theta_{N+1})$$

where:

$$\theta_{N+1} \;=\; \tau - N$$

$$y_j \;=\; \hat{C}_M(\theta_j = 1)$$

$$A_j(\theta_{N+1}) \;=\; \left(\frac{1 - E_f}{M_x}\right)^j \frac{\theta_{N+1}^j}{j!} \exp\left(-\frac{\theta_{N+1}}{M_x}\right)$$

$$B(\theta_{N+1}) \;=\; \int_o^{\theta_{N+1}} F(\tau)\, G(\theta_{N+1} - \tau)\, d\tau$$

$$F(\tau) \;=\; \frac{(1 - E_f)^N}{M_x^{N+1}\, N!}\, (1 - \delta_p(\tau))$$

$$G(\tau) \;=\; \tau^N \exp(-\tau/M_x)$$

from which we obtain

$$E_f = \left(\frac{4}{\pi}\text{De}\right)^{1/2} = 0.0714$$

Next, when Mx = 0.1 and $N \cong 50$, equation 6.4-4 yields

$$\hat{c}_N \cong \left(1 - E_f\right)^N = 0.0246$$

When Mx = 10 and $N/(1 + \text{Mx}) \cong 5$, equation 6.4-5 yields

$$\hat{c}_N = \exp\left(\frac{-E_f N}{1 + \text{Mx}}\right) \cong 0.700$$

These results demonstrate the effect of BM, which may be interpreted as reducing the number of actual cycles (stages) from 50 to 5 and concomitantly reducing efficiency.

Example 6.4-2. *The Effects of Surface Renewal and Backmixing.* For identical values of $\lambda_D = 25$ min, $V_f = 1250$ in.^{3} and actual number of cycles, $N\lambda_f/(1 + Mx)$ $= 5$ min, contrast $\hat{c}_N$, total time t and total volume V in three batch processes: a simple evaporating film ($N/(1 + Mx) = 0$, $Mx = 0$); a SR process ($N/(1 + Mx) = 50$, $Mx = 0.1$); and an EXP process ($N/(1 + Mx) = 50$, $Mx = 10$).

Solution. The results, obtained from equations 6.4-4 and 6.4-5, are listed in the following chart. We conclude that SR reduces volatile content significantly within a given process time (5 vs. 5.5 min); and excessive BM, while not reducing process efficiency a great deal in this case, does require a much longer time and therefore larger volume for a given throughput rate ($V/t = 250$ in.^{3}/min). The slight dissimilarity between values for $\hat{c}_N$ in the second and third processes is due to the functional differences between equations 6.4-4 and 6.4-5.

Model	$N/(1 + Mx)$	N	λ_f (min)	λ_M (min)	E_f	$\hat{c}_N$	t (min)	V (in.^{3})
DF	1	1	5	0	0.504	0.496	5	1,250
SR	50	55	0.1	0.01	0.0714	0.0249	5.5	1,370
EXP	50	550	0.1	1.0	0.0714	0.0282	55	13,750

Example 6.4-3. *Volatile Concentration As a Function of Time.* Compute $\hat{c}$ versus t for a batch DV process when $Mx = 1.25$ (intermediate value) and $D = 10^{-4}$ in.^{2}/min, $h = 50$ mils, $\lambda_f = 0.0533$ min.

Solution.

$$\text{De} = \frac{D\lambda_f}{h^2} = 0.00213$$

$$E_f = 0.0521 \text{ (caution! borderline)}$$

Using $\hat{c} = \exp[-E_f t/\lambda_f(1 + Mx)]$, we obtain the following values:

t(min)	$\hat{c}$
1.2	0.594
5	0.114
10	0.013
12	0.00546

6.5. EQUIPMENT THEORY

Heretofore we have been concerned primarily with the analysis of elementary steps in DV via models, emphasizing their effect on separation efficiency. We shall now turn to several existing DV processes and examine factors that influence their design and operation. We begin with a review of two performance criteria that were introduced in Section 1.3: process efficiency and equipment efficiency.

6.5.1. *Specific Output Rate*

The concept of equipment efficiency was discussed in general in Section 1.3 and the specific output rate was proposed as a measure of that efficiency. In particular, the quantities xE_f and XE_F were defined for simple (single-element) and complex DV processing respectively, where $x \equiv \dot{V}/s$ and $X \equiv \dot{V}/S$. Areas s and S are intended to reflect equipment cost and therefore pertain to "working" surfaces that are generated for the purpose of interphase mass transport (DV); $\dot{V}$ represents overall throughput rate. These quantities were evaluated in Section 6.2 for certain model DV processes. We shall continue to compute XE_F for actual DV processes whenever possible to acquire a "feeling" for its magnitude.

The specific output rate may be interpreted in several different ways. Comparing XE_F with its microscopic counterpart xE_f, which is equivalent to a mass transfer coefficient k_f (equation 6.2-29), suggests that XE_F may be interpreted as a sort of overall process transfer coefficient. Alternatively, the interpretation of xE_f as a penetration rate δ_p (equation 6.2-29) suggests a similar interpretation for XE_F as well. Finally, XE_F is also proportional to the vapor removal rate at steady state.

The quantity XE_F is readily evaluated from the information available in most research articles on DV processes and technical literature on DV equipment. The least accessible data are the values for equilibrium concentration c_e. Compositions are generally expressed as weight fraction (w), throughput rates as pounds per hour ($\dot{m}$) and areas (S) as film (thin-film evaporators) or heat transfer (extruders) surfaces. Consequently, we may use the quantity $\dot{m}E_F'/S$ in lieu of XE_F when w_e is comparatively small, where

$$E_F \equiv \frac{1 - \dfrac{w_f}{w_0}}{1 - \dfrac{w_e}{w_0}} \cong 1 - \frac{w_f}{w_0} \equiv E_F' \tag{6.5-1}$$

It is noteworthy that thin-film evaporators for DV are in fact rated on the basis of amount of volatile component removed per unit time per unit film area in $\dot{m}E_F/S$ (pounds per hour per square foot), which is equivalent to XE_F multiplied by density ρ_p and the value of 30 lb/h ft^2 apparently represents an efficient process for polymer melt DV. Using a density of $\rho_p = 1$ g/cm^3 leads to an equivalent value for XE_F of approximately 100 mils/min, which we shall take as our yardstick for DV performance. This number should be compared with the values for xE_f and XE_F computed in Section 6.2 for model processes.

6.5.2. *Single-Screw Extruders*

Very few detailed analytical studies have been reported on vacuum DV in extruders (19, 20, 38, 40–42). Only three (38, 40, 41–42) included experimental data. In all cases the models proposed were based on diffusion theory. The customary method for checking the validity of these models is to use the experimentally determined separation data (E_F) and the key parameters (N_R, L_B, d_B, g, H, ϕ, and h), and compute a value for the diffusion coefficient D that is consistent with the data. The

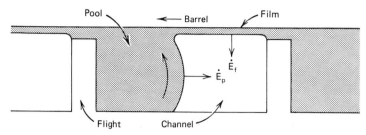

Figure 6.5-1. Vented section (cross section) of a single-screw extruder showing barrel-film and channel-pool evaporation.

general conclusion is that observed DV rates are higher by orders of magnitude than those predicted by diffusion theory, using for comparison the most favorable diffusivities ($D \sim 10^{-5}$ cm^2/s). In other words, estimated values for D are found to be higher by several orders of magnitude (19, 20, 38, 41–42) than those measured directly in independent experiments (23). This suggests, of course, that a mechanism far more rapid than diffusion (e.g., foaming) is operative.

It has been postulated that DV in vented, single-screw extruders occurs via two parallel mechanisms (19, 38): One is evaporation from the bulk polymer melt flowing in the partially filled screw channel, and the other is evaporation from the film that is continuously wiped on the barrel surface owing to the clearance ($2h$ for Newtonian fluids) between the barrel and the screw flight. The channel melt may be viewed as an evaporating melt pool that rotates as it flows owing to the angle of drag ϕ, and the barrel film may be viewed as a translating film (Section 6.2) if we imagine the screw to be stationary and the barrel to be rotating. Together, they may be represented by the parallel processes sketched in Fig. 6.2-7.

The DV extruder has been sketched in greater detail in Figs. 6.5-1 and 6.5-2. We denote screw speed in revolutions per unit time by N_R, flight angle by ϕ, channel

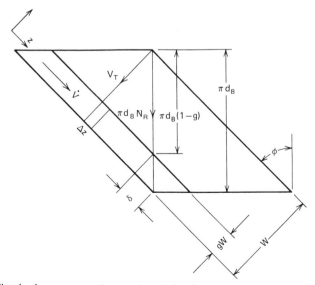

Figure 6.5-2. Sketch of an unwrapped screw channel showing pertinent geometric and flow-parameters.

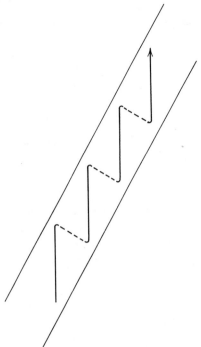

Figure 6.5-3. Particle path in the channel of a single-screw extruder (39).

width and depth by W and H, respectively, and the fraction of the channel filled with flowing polymer melt by g. $L_B (= L \sin \phi)$ and d_B are barrel length and diameter, respectively, and L is channel length measured along the z-direction.

The flow pattern in the channel melt consists of a down-channel component and a cross-channel component. The flow path of fluid elements, which is sometimes described as a helix within a helix (39), has been sketched in Fig. 6.5-3. Transverse mixing via cross-channel flow is believed to be very effective. "Backflow" due to internal pressurization is not very extensive. Thus we may view backmixing (BM) as a sort of Taylor axial dispersion, resulting from the combined action of a down-channel velocity distribution and cross-channel flow, in which radial motion occurs by convection rather than by diffusion. We shall describe such BM with coefficient D_L.

In addition to BM, another feedback (FB) mechanism is operative (19). It is caused by the polymer film that is continuously wiped from the channel melt by the flight clearance of the screw and continuously returned to it upstream. Owing to the angle ϕ and the nonzero thickness gW of the channel melt, the returning film enters the channel at a distance $\delta = \pi d_B g \cos \phi$ upstream from its departure point measured in the z-direction (Fig. 6.5-2). The stagewise analog of this mechanism is the FB model (20) described in Section 6.3.

It is evident that the extruder requires a continuous rather than a staged model. A steady-state material balance for volatile component on a cross-sectional channel element Δz (Figs. 6.2-7 and 6.5-2), neglecting changes in volumetric flow rate due to loss of material and density variations, yields

$$0 = gWH\big[n_z(z) - n_z(z + \Delta z)\big] + v_T h\,\Delta z\big[c_f(z) - c(z)\big]$$

longitudinal
transport   film evaporation

$$- k_p H\Delta z\big[c(z) - c_e\big]$$

bulk evaporation

$$(6.5\text{-}2)$$

where n_z is the longitudinal component of the total transport flux, v_T is the transverse component of the film velocity perpendicular to the channel direction z

$$v_T = \pi d_B N_R \sin\phi \qquad (6.5\text{-}3)$$

and $c_f(z)$ is the concentration in the film reentering the channel melt at point z after having departed from it at a distance δ downstream and having experienced an exposure time

$$\lambda_f = \frac{1 - g}{N_R} \qquad (6.5\text{-}4)$$

In the limit as $\Delta z \to 0$, we obtain

$$gWH\frac{dn_z}{dz} = v_T h(c_f - c_e) + k_p H(c - c_e) \qquad (6.5\text{-}5)$$

Following Latinen (40), we account for BM with a (constant) longitudinal dispersion coefficient D_L. This is equivalent to expressing n_z as the sum of a convection term and a BM term

$$n_z = cv_z - D_L\frac{dc}{dz} \qquad (6.5\text{-}6)$$

Next we must eliminate concentration c_f, which is related to concentration c at a finite distance δ away from point z, $c(z + \delta)$, by the film efficiency

$$E_{ff} = \frac{c(z + \delta) - c_f(z)}{c(z + \delta) - c_e} \qquad (6.5\text{-}7)$$

Following Roberts (19), we connect $c(z + \delta)$ to $c(z)$ via a Taylor series expansion and neglect all terms higher than second order

$$c(z + \delta) \cong c(z) + \delta\frac{dc}{dz} + \frac{\delta^2}{2}\frac{d^2c}{dz^2} \qquad (6.5\text{-}8)$$

After eliminating $c(z + \delta)$ from the last two equations and substituting the result for c_f into our material balance, we obtain

$$\left[gWHD_L + \frac{v_T h(1 - E_{ff})\delta^2}{2}\right]\frac{d^2c}{dz^2} - \big[gWHv_z + v_T h(1 - E_{ff})\delta\big]\frac{dc}{dz}$$

$$- \big[v_T h E_{ff} + k_p H\big](c - c_e) = 0 \qquad (6.5\text{-}9)$$

which is identical in form to the familiar dispersion model (Section 1.4). Certainly the presence of FB together with BM in the first term is consistent with our knowledge that longitudinal mixing generally manifests itself as a second derivative in differential models.

Writing this equation in dimensionless form yields the dispersion model

$$\text{Pe}_L^{-1} \frac{d^2 \hat{c}}{d\hat{z}^2} - \frac{d\hat{c}}{d\hat{z}} - \text{Ex}\hat{c} = 0 \tag{6.3-18}$$

and provides us with definitions for the dimensionless Peclet and extraction numbers. They are

$$\text{Pe}_L = \frac{N_f (1 - E_{ff})\hat{\delta} + 1}{\dfrac{N_f (1 - E_{ff})\hat{\delta}^2}{2} + \dfrac{D_L}{Lv_z}} \tag{6.5-10}$$

and

$$\text{Ex} = \left(N_f E_{ff} + N_p E_{fp} \right) / \left[N_f (1 - E_{ff})\hat{\delta} + 1 \right] \tag{6.5-11}$$

where $N_f \equiv \dot{V}_f / \dot{V}$ and $N_p \equiv \lambda_v / \lambda_p$ are the equivalent number of film stages and pool stages (Section 6.2), respectively, and E_{fp} is the pool efficiency, which is related to k_p by virtue of equation 6.2-9, $k_p / gW = E_{fp} / \lambda_p$. The quantity $\hat{\delta} \equiv \delta / L$ is reduced FB length, where

$$\delta = \pi d_B g \cos \phi \tag{6.5-12}$$

from Fig. 6.5-2.

It should be pointed out that film flow rate in this context refers to the film on the entire barrel wall

$$\dot{V}_f = \pi d_B h L_B N_R \tag{6.5-13}$$

rather than each "stage" (length $\Delta z \sin \phi$) as did the film flow rate in our staged models. Thus N_f in the present model is equivalent to the product $N s_R$ in our staged FB model.

To summarize, equation 6.3-18 represents a model for diffusion-controlled DV in a vented, single-screw extruder when Pe_L and Ex are defined by equations 6.5-10 and 6.5-11. Large values of Pe_L signify little BM and large values of Ex signify effective separation. Increasing N_f enhances separation (Ex), but it contributes to BM at the same time via Pe_L. Similarly, increasing δ contributes to BM (Pe_L) and simultaneously diminishes separation by reducing Ex. The solution of equation 6.3-18 was stated in Section 1.4 and plotted in Fig. 1.4-1. It is noteworthy that FB disappears when $\hat{\delta} \to 0$ or $E_{ff} \to 1$. In that case $\text{Pe}_L = Lv_z / D_L$ (Section 1.4) and equation 6.5-11 reduces to the simpler corresponding expression

$$\text{Ex} = N_f E_{ff} + N_p E_{fp} = \frac{k_f S_f + k_p S_p}{\dot{V}} \tag{6.3-17}$$

Furthermore, when FB (δ) and BM (D_L) are both small (large Pe_L), which is frequently the case, our model reduces to equation 6.3-4 (the EXP model) and Ex is again given by equation 6.3-17. In conclusion, it is not unreasonable to expect the vented DV extruder to obey the simple EXP model.

If we assume additionally that infinite-film conditions prevail [i.e., that $\lambda_f/\lambda_D < 0.1$ where $\lambda_f = (1 - g)/N_R$] and that the period for pool revolution is simply $\lambda_p = H/v_T$ (19), then equation 6.3-17 becomes

$$\text{Ex} = \frac{2L_B d_B [\pi N_R D(1 - g)]^{1/2}}{\dot{V}} + \frac{2L_B [HN_R d_B D/\sin\phi]^{1/2}}{\dot{V}} \quad (6.5\text{-}14)$$

This equation provides a link between process variable E_F and transport property D in terms of process parameters $\dot{V}$, g, N_R, and equipment parameters L_B, d_B, ϕ. Its applicability is limited, of course, to infinite-film diffusion and constant D.

An expression for throughput $\dot{V}$ may be derived from a momentum balance on the pool. The simplest one for single-plate drag flow, neglecting "shape factors," is

$$\dot{V} \cong \frac{\pi d_B N_R g W H}{2} \cos\phi \quad (6.5\text{-}15)$$

Thus two classes of experiments suggest themselves: variable-$\dot{V}$ and constant-$\dot{V}$ (Section 1.3). As N_R is increased and $\dot{V}$ grows proportionately, being unrestricted elsewhere, the degree of fill (g) remains constant. On the other hand, if $\dot{V}$ is fixed by metering the feed, then g drops in accordance with equation 6.5-15 as N_R is increased. The first case (variable-$\dot{V}$, constant-V) leads to a simple relationship between E_F and N_R via Ex, which exposes D

$$\text{Ex} = (c_1 + c_2)\left(\frac{D}{N_R}\right)^{1/2} \quad (6.5\text{-}16)$$

where c_1 and c_2 are constants:

$$c_1 = \frac{4L_B(1 - g)^{1/2}}{g\pi^{1/2}WH\cos\phi} \quad (6.5\text{-}17)$$

for the film and

$$c_2 = \frac{4L_B}{\pi g W(Hd_B\sin\phi)^{1/2}\cos\phi} \quad (6.5\text{-}18)$$

for the pool. Equations 6.5-16, 6.5-19, and 6.5-20 predict that Ex, and therefore E_F, will decline with mean residence time (λ_v) as N_R is increased, which is not unexpected. More specifically, a graph of $-\ln(1 - E_F)$ versus $N_R^{-1/2}$ should be linear.

For constant $\dot{V}$ (variable V), on the other hand, we obtain

$$\text{Ex} = c_1'(N_R - c_3)^{1/2} + c_2'N_R^{1/2} \quad (6.5\text{-}19)$$

in lieu of equation 6.5-16), which predicts that E_F will actually increase with N_R, notwithstanding a decline in mean residence time (Example 6.5-2). Such an increase would reflect the beneficial effect on E_F of surface renewal (mixing) and its ability to prevail over the competing effect of shorter residence times (Section 1.3). More specifically, from infinite-film diffusion (penetration) theory, this enhancement is due to the inverse dependence of the mass transfer coefficients of both the film and the pool (k_f, k_p) on the square root of their exposure periods (λ_f, λ_p), which are shortened in response to the increase in N_R. Such behavior is characteristic of surface renewal (mixing) alluded to earlier (Section 1.3). In particular, when $\dot{V}$ is fixed, it results in the "anomalous" increase in E_F with N_R predicted by equation 6.5-19.

A useful quantity for assessing the relative efficiency of DV equipment in general, whether diffusion or foaming is the dominant mechanism, is the area-specific output rate $(E_F \dot{V}/S)$ introduced in Sections 1.3 and 6.5.1. It has the units of velocity (e.g., mils/min) and may be interpreted as an overall mass transfer coefficient or a film penetration rate. To give this quantity numerical significance, we have estimated its value to be less than 100 mils/min for diffusion-controlled DV (Sections 6.2-5 and 6.2-6). For foam-DV of polystyrene–styrene in single-screw and twin-screw extruders, typical values computed from process data lie between 300 and 700 mils/min (Examples 6.5-1 and 6.5-3).

The variable-$\dot{V}$ and constant-$\dot{V}$ methods described in Section 1.3, especially Table 1.3-2, may be applied to DV processes relatively easily, but without the concomitant increase in L required for constant RT, owing to the experimental difficulties that that aspect would entail. Thus in both types of experiments we expect a decrease in λ_v to accompany an imposed increase in N_R.

Such experiments have been reported (41, 42) on the DV of styrene from polystyrene in a vented, 3-in. single-screw extruder. Monomer content in the feed was at the level of 5400 ppm. Process efficiency E_F was determined experimentally using gas chromatography. Screw speed (N_R) was varied and four sets of experiments were conducted: one pair (diffusion-controlled DV) under a nitrogen sweep at atmospheric pressure to ensure that the equilibrium concentration in the melt (c_e) was unambiguously zero; and the other pair under a vacuum of 1 mm Hg. In the first experiment of each pair, flow rate $(\dot{V})$ was allowed to increase with N_R in accord with equation 6.5-15 and degree of channel fill (g) remained constant. In the second experiment of each pair, flow rate was maintained constant by metering the feed and thus g was compelled to vary, presumably in inverse proportion to N_R as required by equation 6.5-15.

The results are shown in Figs. 6.5-4–6.5-6. Screw speeds were varied between 10 and 120 rpm and throughput rates ranged from 20 to 190 lb/h. Residence times were estimated from $\lambda_v = V/\dot{V}$, where $V = LHgW$, and found to lie between 5 and 100 seconds. Degree of fill (g) was computed from equation 6.5-15. Process efficiencies were computed from equation 6.5-1, using $w_e = 0$ for the nitrogen experiments and $w_e = 96.7$ ppm for the vacuum experiments (Example 6.5-1).

From the graphs in Fig. 6.5-4, it is evident that vacuum DV is much more efficient than DV into an inert gas at atmospheric pressure and that both modes of DV exhibit the "anomalous" effect of increasing process efficiency with decreasing residence time when $\dot{V}$ is fixed. From the first observation, we conclude that bubble transport is operative and that it constitutes an important mechanism in effecting

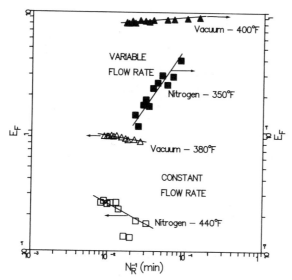

Figure 6.5-4. Relationship of process efficiency E_F to residence time (proportional to N_R^{-1}) for variable flow-rate and constant flow-rate DV experiments on PS/S in a single-screw extruder at atmospheric pressure with a nitrogen sweep and under vacuum (41, 42).

DV in the 1000–100-ppm range, and perhaps even lower. From the second observation, we conclude, on the basis of the discussions in Section 1.3 and earlier in this section, that there exists in both modes some sort of mixing (surface-renewal) effect, which has a beneficial effect on DV. Furthermore, the different kinds of rpm-dependence predicted in Example 6.5-2 from diffusion theory for variable-$\dot{V}$ and constant-$\dot{V}$ experiments are borne out in Fig. 6.5-4, not only for the diffusion modes but for the vacuum modes as well. The graph in Fig. 6.5-5 shows the specific output rate (overall transfer coefficient) for the vacuum mode to be far more sensitive to flow velocity than that for the diffusion mode. In fact, it is the sharp rise in XE_F with N_R (λ_v^{-1}) that flattens the decline in E_F with λ_v (Fig. 6.5-4), by virtue of the proportionality $E_F \propto \lambda_v XE_F$. Further speculation on the mechanistic interpretation of these results in the absence of a model for foam DV in extruders would be premature.

It should be noted that Fig. 6.5-6 was plotted in the form suggested by equation 6.5-16 for diffusion-controlled DV. The slope of this graph should give $(c_1 + c_2)D^{1/2}$, according to the simple EXP model. After computing c_1 and c_2 from equations 6.5-17 and 6.5-18, we estimate that $D = 1.4 \times 10^{-4}$ in.^{2}/min (41, 42) which is not unreasonable. However, the barrel film in the experiments did not, in fact, obey the infinite-film assumption (cf. E_{ff} in Table 6.5-2) implicit in equation 6.5-16, owing to the high efficiency of the thin film deposited by the tight flight clearance, as predicted by Roberts (19). Values of $\lambda_f/\lambda_D \equiv De$ actually ranged from 1.2 to 0.2, which exceed the limit of 0.1 for infinite films. These values were computed using $h = 2.5$ mils (from a flight clearance of 5 mils by virtue of the continuity equation for steady incompressible flow).

The inapplicability of the infinite-film model should, at most, cause only a small error in our value for D. Furthermore, the finite-film model consists of a more complex expression than the infinite-film model for film efficiency; thus it would not

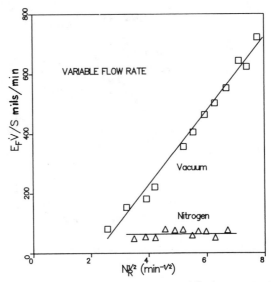

Figure 6.5-5. Variation of specific output rate XE_F with $N_R^{1/2}$ for variable flow-rate and constant flow-rate DV experiments on PS/S in a single-screw extruder at atmospheric pressure with a nitrogen sweep and under vacuum (41, 42).

yield a simple expression for Ex in terms of D that is amenable to graphical analysis, as equation 6.5-16 is.

From all available evidence, it appears that the exponential model is satisfactory to characterize process efficiency

$$E_F = 1 - \exp(-\text{Ex}) \tag{6.3-4}$$

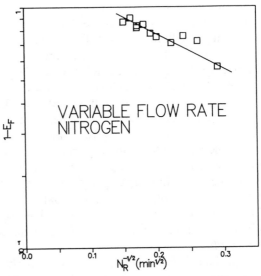

Figure 6.5-6. Relationship between the logarithm of $1 - E_F$ and $N_R^{-1/2}$, suggested by equation 6.5-16, for variable flow-rate DV experiments on PS/S in a single-screw extruder with nitrogen sweep (41, 42).

Table 6.5-1.   Results from Vented Single-Screw DV Without Vacuum

	N_R (RPM)	$\dot{m}$ (lb./hr.)	λ_v (s.)	E_F (exp'tal)	E_x^* (exp'tal)	$\hat{\delta}$	E_x^+ (theoret.)	$Pe_L^{\neq} \times 10^{-4}$ (theoret.)
	15	41	53	0.29	0.34	0.028	0.52	19.5
Variable $\dot{V}$	30	85	26	0.16	0.17	0.028	0.42	4.3
	45	126	18	0.14	0.15	0.028	0.35	2.7
	43	81	18	0.19	0.21	0.018	0.53	4.8
Constant $\dot{V}$	80	78	10	0.25	0.29	0.0098	0.79	5.4
	100	81	8	0.27	0.31	0.0081	0.86	5.8

$*$Experimental : $E_x = -\ln(1 - E_F)$

$+$Theoretical   : $E_x = [N_f E_{ff} + N_p E_{fp}]/[1 + N_f(1 - E_{ff})\hat{\delta}]$

$\neq$Theoretical   : $Pe_L \cong 2[1 + N_f(1 - E_{ff})\hat{\delta}]/\hat{\delta}^2 N_f(1 - E_{ff})$

for diffusive DV in single-screw extruders when the extraction number is defined by equation 6.3-17. In an attempt to gain some insight into the DV mechanism, we shall evaluate the key internal parameters of our model contained in equation 6.3-17, using data from DV experiments on a 3-in. single-screw extruder with PS/S under a nitrogen sweep at atmospheric pressure (41, 42). The extruder parameters appear in Example 6.5-1. A value of $D = 1.4 \times 10^{-4}$ in.^{2}/min was assumed. The results are summarized in Tables 6.5-1 and 6.5-2. It should be noted that the finite-film model was used in these computations, as required.

Table 6.5-1 lists some experimental results and some computed parameters of interest, such as feedback fraction ($\hat{\delta}$) and mean residence time (λ_v), and contrasts experimental and theoretical values for the extraction number (Ex). Also tabulated are computed Peclet numbers (Pe_L), whose values confirm that longitudinal dispersion due to FB is indeed small. Table 6.5-2 contains computed parameters that provide us with a comparison between the contributions to DV made by the barrel film and the channel pool.

From Table 6.5-1, it is clear that our idealized model consistently predicts values for Ex, and therefore E_F, that are too high. Table 6.5-2 suggests that the film generally contributes more to DV than the pool does. It is noteworthy that while film efficiencies (E_{ff}) are large compared to pool efficiencies (E_{fp}), the number of film stages (N_f) is much smaller than the number of pool stages (N_p). We must remember, however, that this analysis pertains only to DV by the diffusion mechanism. It does not imply that the channel pool is also less important than the film in foam DV. It is quite conceivable that the opposite is true (viz., that rotating melt pools are the locus for foam-induced DV).

Example 6.5-1.   The Single-Screw Extruder—Equipment Efficiency.   A vented, 3-in., single-screw extruder operating at 50 rpm, 400°F (204°C) and 1 mm Hg pressure with a throughput of 161 lb/h reduced the styrene content of a commercial

Table 6.5-2.   Model Parameters for Vented Single-Screw DV Without Vacuum

	N_R (RPM)	$\dot{V}$ (in.^{3}/min.)	g	N_F	N_p	E_{ff}^*	E_{fp}^+	$N_F E_{ff}$	$N_p E_{fp}$
Variable $\dot{V}$	15	19.1	0.18	0.33	61	0.96	0.0035	0.31	0.21
	30	39.4	0.18	0.33	61	0.82	0.0025	0.27	0.15
	45	58.7	0.18	0.33	61	0.71	0.0020	0.23	0.12
Constant $\dot{V}$	43	37.8	0.12	0.48	61	0.73	0.0030	0.35	0.18
	80	36.4	0.064	0.92	61	0.58	0.0042	0.54	0.26
	100	37.8	0.053	1.1	61	0.52	0.0046	0.58	0.28

$^*E_{ff} = 1 - (8/\pi^2) \exp[-\pi^2 D\lambda_f/4h^2]$

$^+E_{fp} = (2/gW)(D\lambda_p/\pi)^{1/2}$

polystyrene from 5400 ppm to 971 ppm (42). A single vented section of length $L_B = 17.9$ in. was used. The remaining equipment parameters were: $d_B = 3$ in., $H = 0.62$ in., $W = 2.57$ in., $\phi = 17.7°$, and the flight clearance = 5 mils. Estimate the extraction number Ex and the specific output rate $E_F \dot{V}/S$. Assume $K_W = 13.6$ atm and $\rho = 0.981$ g/cm^3.

Solution.   From Henry's law, $w_e = P/K_W = 96.7$ ppm. Thus,

$$E_F = \frac{1 - \dfrac{w}{w_0}}{1 - \dfrac{w_e}{w_0}} = 0.835$$

and

$$\mathrm{Ex} = -\ln(1 - E_F) = 1.8$$

Volumetric throughout is simply $\dot{V} = \dot{m}/\rho = 75$ in.^{3}/min. Assuming that evaporation takes place from both the barrel film and the channel pool, and neglecting the flight thickness,

$$S = S_f + S_p = \left(\pi d_B L_B - \frac{gWL_B}{\sin\phi}\right) + \frac{HL_B}{\sin\phi}$$

After substituting $g = 0.21$, computed from equation 6.5-15 for unfoamed melt, we obtain $S = 169$ in.^{2}, and therefore

$$\frac{E_F \dot{V}}{S} = 374 \text{ mils/min}$$

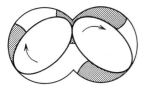

Figure 6.5-7. Cross section of corotating twin-screw extruder.

Comparing the last result with the corresponding values computed from diffusion theory (Examples 6.2-1 and 6.2-3) confirms that foaming was the dominant mechanism of DV (42).

Example 6.5-2.   The Single-Screw Extruder—Mixing Effects.   Using the simple diffusion model for the extruder developed in this section, what effect on process efficiency E_F would you expect from an increase in screw speed (N_R) when $V(g)$ remains constant? When $\dot{V}$ is fixed (Section 1.3)?

Solution.   Neglecting BM and FB effects in the channel, E_F rises with extraction number Ex in accordance with the relation $E_F = 1 - \exp(-\text{Ex})$, where

$$\text{Ex} = \frac{\dot{V}_f}{\dot{V}} E_{ff} + \frac{\lambda_v}{\lambda_p} E_{fp}$$

By virtue of equations 6.5-13 and 15, $\dot{V}_f/\dot{V}$ is independent of N_R when V is constant, whereas it rises in proportion to N_R when $\dot{V}$ is constant. On the other hand, E_{ff} varies with De $= S_f^2 D/\lambda_f \dot{V}_f^2$ (equation 6.2-27), which declines as N_R increases when V is constant, but grows when $\dot{V}$ is constant, by virtue of equations 6.5-4 and 6.5-13. Consequently, the effect of the film on Ex, and therefore E_F, is to cause it to increase with N_R when $\dot{V}$ is constant, and to decrease when V is constant. Concerning the pool, we conclude from $\lambda_v \cong gWHL/\dot{V}$ with the aid of equation 6.5-15 that $\lambda_v \propto N_R^{-1}$, whether V or $\dot{V}$ remains constant, and similarly that $\lambda_p \propto N_R^{-1}$ from the approximation (19) $\lambda_p = H/v_T \cong H/\pi d_B hL_B N_R$. Thus λ_v/λ_p is always independent of N_R. Furthermore, from (infinite film expression) $E_{fp} = (2S_p/\dot{V}\lambda_v)$ $(D\lambda_p/\pi)^{1/2}$, we conclude that E_{fp} declines with rising N_R ($E_{fp} \propto N_R^{-1/2}$) when V is constant, but rises ($E_{fp} \propto N_R^{1/2}$) when $\dot{V}$ is constant. Consequently, the effect of the pool on E_F, like the film, is to cause it to increase with N_R when $\dot{V}$ is constant and to decrease when V is constant. The reader will note for the constant-$\dot{V}$ case, that the rise in N_R is exactly offset by a decline in g (V) via equation 6.5-15. Furthermore, the concomitant increase in Ex, mutually reinforced by the film and the pool, could actually produce a significant increase in E_F, since it occurs in the exponent.

6.5.3.   Twin-Screw Extruders

Vented, twin-screw extruders of the corotating type are also commonly used for industrial DV. Figure 6.5-7 shows a cross section of such an extruder and illustrates the location of the polymer. Like the single-screw extruder, the twin-screw machine wipes films and remixes them with bulk streams. The bulk streams also exchange material with one another where the channels meet. Thus BM, in addition to dual evaporation from wiped films and melt pools, is likely as in the single-screw extruder.

No detailed mathematical model for DV in twin-screw extruders has been found in the literature to date. Data from an actual process, however, have been fitted with an EXP model (43). In Examples 6.5-3 and 6.5-4, we attempt to characterize these processes using such data.

Example 6.5-3.   The Twin-Screw Extruder—Equipment Efficiency. The following data for DV of PS/S at 280°C and 13 mm Hg were taken from the literature (43): w_0 = 1200 ppm, w_f = 490 ppm, w_e = 300 ppm (at 13 mm Hg), $\dot{m}$ = 1500 lb/h, and heat transfer area = 0.523 m² (811 in.²). Compute and contrast efficiency and specific output rate in the following two alternative forms: E_F', E_F and $\dot{m}E_F/S$, $E_F'\dot{V}/S$, respectively.

Solution.   Starting with efficiency

$$E_F' = 1 - \frac{w_f}{w_0} = 0.59$$

$$E_F = \frac{E_F'}{1 - \dfrac{w_e}{w_0}} = 0.79$$

The significant difference between E_F' and E_F underscores the importance of w_e when it is not sufficiently small. Assuming that the heat and mass transfer areas are identical, for lack of an alternative, we obtain

$$\frac{\dot{m}E_F'}{S} = 157 \text{ lb/h ft}^2$$

In terms of volumetric throughput $\dot{V} = \dot{m}/\rho = 699$ in.³/min (Example 6.5-1)

$$XE_F = \frac{E_F\dot{V}}{S} = 681 \text{ mils/min}$$

It is interesting to compare these results with the values mentioned in the text (30 lb/h ft² and 100 mils/min) and the results of Example 6.5-1 for the single-screw extruder (374 mils/min) and Example 6.5-4 for the thin-film evaporator. When making such comparisons, however, one must remember two additional, important parameters, namely, temperature and vacuum level, which were all different in the examples cited.

6.5.4.   Thin-Film Evaporators

These units are descendents of the falling film evaporator, which has been used in the chemical industry for years. They have been modified with rotating, pitched blades to assist the conveying of viscous polymer solutions, which do not flow readily under the action of gravity alone. Blade configurations and flow patterns (relative to stationary blades) are illustrated in Fig. 6.5-8 (44).

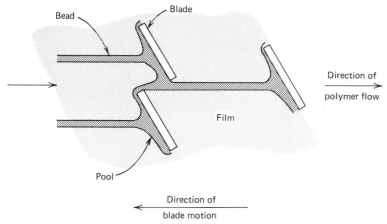

Direction of polymer flow

Direction of blade motion

Figure 6.5-8. Schematic drawing of rotating blade profiles in a continuous, thin-film evaporator showing melt flow.

It is apparent that melt pools (stationary beads and rolling banks) as well as films are present in series and parallel, and that surface renewal, stream-splitting, and backmixing all occur in the complex flow pattern that prevails (cf. staged models, Section 6.3.1). Some studies on the fluid mechanics (18) and mass transfer aspects (44) of these evaporators have been described. However, detailed models for mixing and evaporation were not found. In Example 6.5-4, some overall process characteristics are examined.

Example 6.5-4.   The Thin-Film Evaporator—Equipment Efficiency.   Estimate E_F', E_F, $\dot{m}E_F'/S$, and XE_F for continuous film DV of the system PS/S at 260°C and 15 mm Hg when (44)

$$\frac{w_f}{w_0} = 0.001/0.05 \quad \text{and} \quad \dot{m}/S = 30 \text{ lb/h ft}^2$$

Solution.

$$E_F' = 1 - \frac{w_f}{w_0} = 0.980$$

From Table 6.1-1, $K_W = 50$ atm, therefore, $w_e = 395$ ppm and

$$E_F = \frac{1 - \dfrac{w_f}{w_0}}{1 - \dfrac{w_e}{w_0}} = 0.989$$

Also

$$\frac{\dot{m}E_F'}{S} = 29.4 \text{ lb/h ft}^2$$

and

$$XE_F = 95 \text{ mils/min}$$

using

$$\rho = 0.981 \text{ g/cm}^2 \text{ (Example 6.5-1)}$$

REFERENCES

1. P. J. Flory, *Principles of Polymer Chemistry*, Cornell University, Ithaca (1953).

2. D. C. Bonner, *J. Macromol. Sci. Rev. Macromol. Chem.*, **C13(2)**, 263 (1975).

3. K. G. Denbigh, *The Principles of Chemical Equilibrium*, Cambridge University, Cambridge, England (1971).

4. M. L. Huggins, *J. Chem. Phys.*, **9**, 440 (1941).

5. M. L. Huggins, *J. Phys. Chem.*, **46**, 151 (1942).

6. M. L. Huggins, *Ann. N.Y. Acad. Sci.*, **41**, 1 (1942).

7. P. J. Flory, *J. Chem. Phys.*, **9**, 660 (1941).

8. P. J. Flory, *J. Chem. Phys.*, **10**, 51 (1942).

9. P. J. Flory, *J. Chem. Phys.*, **12**, 425 (1944).

10. F. H. Covitz and J. W. King, *J. Polym. Sci.*, **10**, 689 (1972).

11. J. L. Duda, G. K. Kimmerly, W. L. Sigelko, and J. S. Vrentas, *Ind. Eng. Chem. Fund.*, **12**, 133 (1973).

12. D. P. Maloney and J. M. Prausnitz, *AIChE J.*, **22**, 74 (1976).

13. A. R. Berens, *Angew. Chem.*, **47**, 97 (1975).

14. K. M. Hess, *Kunststoffe*, **69**, 199 (1979).

15. M. Blander and J. L. Katz, *AIChE J.*, **21**, 833 (1975).

16. R. L. Denecour and A. N. Gent, *J. Polym. Sci.*, A-2, **6**, 1853 (1968).

17. D. B. Todd, "Polymer Devolatilization," Thirty-fourth S.P.E. ANTEC, San Francisco, May (1974).

18. J. M. McKelvey and G. V. Sharps, *Polym. Eng. Sci.*, **19**, 651 (1979).

19. G. W. Roberts, *AIChE J.*, **16**, 878 (1970).

20. J. A. Biesenberger, "Polymer Devolatilization: Theory of Equipment," Thirty-seventh S.P.E. ANTEC, New Orleans (1979); also *Polym. Eng. Sci.*, **20**, 1015 (1980).

21. J. A. Biesenberger and P. S. Mehta, "Fundamental Aspects of Batch Devolatilization," Thirty-eighth S.P.E. ANTEC, New York (1980).

22. J. Crank, *The Mathematics of Diffusion*, 2nd ed., Oxford University, London (1975).

23. J. S. Vrentas and J. L. Duda, *AIChE J.*, **25**, 1 (1979).

24. Dr. Peter Hold, personal communication.

25. V. Volterra, *Theory of Functionals*, Dover, New York (1959).

26. M. Avrami, *J. Chem. Phys.*, **7**, 1103 (1939).

27. P. S. Epstein and M. S. Plesset, *J. Chem. Phys.*, **18**, 1505 (1950).

28. L. E. Scriven, *Chem. Eng. Sci.*, **10**, 1 (1959).

29. A. N. Gent and D. A. Tompkins, *J. Appl. Phys.*, **40**, 2520 (1969).

30. J. L. Duda and J. S. Vrentas, *Int. J. Heat Mass Transfer*, **4**, 395 (1971).

31. E. Zana and L. G. Leal, *I.E.C. Fund.*, **14**, 175 (1975).

32. W. E. Langlois, *J. Fluid Mech.*, **15**, 111 (1963).

33. D. S. Walia and D. Vir, *Chem. Eng. Sci.*, **31**, 525 (1976).

34. J. Throne, Proceedings of the International Conference on Polymer Processing, N. Suh and N. Sung, eds., MIT, Cambridge, MA (1977).

35. R. E. Newman and R. H. M. Simon, "A Mathematical Model of Devolatilization Promoted by Bubble Formation," Seventy-third Annual A.I.Ch.E. Meeting, Chicago (1980).

36. T. K. Sherwood, R. L. Pigford, and C. R. Wilke, *Mass Transfer*, McGraw-Hill, New York (1975).

37. S. T. Lee, master's thesis in Chemical Engineering, Stevens Institute of Technology, Hoboken, NJ (1982).

38. R. W. Coughlin and G. P. Canevari, *AIChE J.*, **15**, 560 (1969).

39. J. M. McKelvey, *Polymer Processing*, Wiley, New York (1962).

40. G. Latinen, *A.C.S. Adv. Chem. Ser.*, **34**, 235 (1962).

41. J. A. Biesenberger and G. Kessidis, "Devolatilization of Polymer Melts in Single-Screw Extruders," Fortieth S.P.E. ANTEC, San Francisco (1982).

42. J. A. Biesenberger and G. Kessidis, *Polym. Eng. Sci.*, **22**, 832 (1982).

43. M. Hess and K. Eise, "Devolatilization of Residual Styrene Monomers in Twin-Screw Extruders," Thirty-fifth S.P.E. ANTEC, Montreal (1977).

44. J. B. Lane, "Mass Transfer in the Luwa Filtruder," P.I.A. Intensive Short Course on Polymer Devolatilization, Stevens Institute of Technology, Hoboken, NJ (1980).

45. P. S. Mehta and J. A. Biesenberger, submitted for publication to *Polym Process Eng.*

APPENDIX A

Polymerization Chemistry

Polymerization reactions are usually classified according to the chemical nature of the monomer and its growth mechanism, for example, condensation versus addition. Three distinct groups of organic monomers that form the basis for most current industrial polymerizations are:

1. Linear monomers that contain two or more reactive functional groups, such as— $\overset{\overset{\displaystyle O}{\displaystyle \|}}{C}$ OH, —OH, —NH$_2$, —Cl, and that polymerize by condensation of these groups.

2. Linear monomers that contain one or more pairs of double bonds, such as C=C, C=O, and which polymerize by addition to one another.

3. Ring-type monomers, such as cyclic ethers, lactams, and lactones, which polymerize via ring scission.

Some industrially important monomers and polymers in each group are listed in Tables A-1–A-3. Specific sample polymerization sequences appear in Tables A-4–A-10.

For convenience of manipulation, all these reactions may be written in abbreviated form, using general symbols in lieu of chemical formulas. For example, polyesters and polycarbonates are formed by reacting diacids (HO $\overset{\overset{\displaystyle O}{\displaystyle \|}}{C}$ R $\overset{\overset{\displaystyle O}{\displaystyle \|}}{C}$ OH) and phosgene (Cl $\overset{\overset{\displaystyle O}{\displaystyle \|}}{C}$ Cl), respectively, with diols (HOR'OH). Thus both polymerizations involve so-called AA and BB type comonomers where

$$A = \begin{cases} -\,\overset{\overset{\displaystyle O}{\displaystyle \|}}{C}OH & \text{group for polyesters} \\ -Cl & \text{group for polycarbonates} \end{cases}$$
$$B = -OH \qquad \text{group for both}$$

Table A-1. Condensation Polymers

Monomer(s) Name	Formula	Repeat Unit	Interunit Linkage	Polymer
ethylene glycol	HOCH$_2$CH$_2$OH			
and				
terephthalic acid	HOC—⬡—COH (O=C, C=O)	⟨C(=O)—⬡—C(=O)OCH$_2$CH$_2$O⟩	—OC—⬡— (O=C)	polyester (polyethylene terephthalate)
or				
dimethyl terephthalate	CH$_3$OC—⬡—COCH$_3$ (O=C, C=O)			
bisphenol A	HO—⬡—C(CH$_3$)(CH$_3$)—⬡—OH	⟨C(=O)O—⬡—C(CH$_3$)(CH$_3$)—⬡—O⟩	—OCO— (O=C)	polycarbonate
and				
phosgene	ClCCl (O=C)			
or				
diphenyl carbonate	⬡—OCO—⬡ (O=C)			
epichlorohydrin	◁—CH$_2$Cl (O)	⟨OCH$_2$CHCH$_2$O—⬡—C(CH$_3$)(CH$_3$)—⬡—OCH$_2$CHCH$_2$O⟩ (OH)	—O—	epoxy prepolymer
or				
4,4'-dichloro-diphenyl-sulfone	Cl—⬡—S(O)(O)—⬡—Cl	⟨⬡—S(=O)(=O)—⬡—O—⬡—C(CH$_3$)(CH$_3$)—⬡—O⟩	—O—	polysulfone

CONDENSATION POLYMERS

Monomer(s) Name	Formula	Repeat Unit	Interunit Linkage	Polymer
adipic acid (or acid chloride) and hexamethylene diamine	$HOC(CH_2)_4COH$ (diacid, two C=O) $H_2N(CH_2)_6NH_2$	$\left[C(CH_2)_4CNH(CH_2)_6NH \right]$	$- NHC -$ (C=O)	polyamide (nylon 66)
α,ω amino-undecanoic acid	$H_2N(CH_2)_{10}COH$	$\left[C(CH_2)_{10}NH \right]$	$- NHC -$ (C=O)	polyamide (nylon 11)
1,4 butanediol and hexamethylene diisocyanate or tolylene 2,4 diisocyanate	$HO(CH_2)_4OH$ $O=C=N(CH_2)_6N=C=O$ $O=C=N$— (ring, CH_3) —$N=C=O$	$\left[CNH(CH_2)_6NCO(CH_2)_4O \right]$	$- NHCO$ (C=O)	polyurethane
phenol and formaldehyde	OH (benzene ring) CH_2O	OH (benzene ring) $-CH_2-$		phenol formaldehyde prepolymer

Since AA and BB appear in alternating sequence in all polymer chains, the repeat unit is AABB. For each DP, three polymer molecules are distinguishable on the basis of their end-groups: $[AABB]_x$, $[AABB]_x AA$, and $BB[AABB]_x$. The stoichiometric equations for their formation are:

$$xAA + xBB = [AABB]_x + (2x - 1)S$$

$$x \geqslant 1 \quad (x + 1)AA + xBB = [AABB]_x AA + 2xS \qquad (A-1)$$

$$xAA + (x + 1)BB = BB[AABB]_x + 2xS$$

Table A-2.   Addition Polymers

Monomer		Repeat Unit	Polymer
Name	**Formula**		
Vinyls	$\begin{array}{cc} H & H \\ C{=}C \\ H & X \end{array}$	$-\!\!\left[\begin{array}{cc} H & H \\ C - C \\ H & X \end{array}\right]\!\!-$	
	$\underline{X}$		
styrene	$-\bigcirc$		polystyrene (PS)
vinyl chloride	$- Cl$		polyvinyl chloride (PVC)
vinyl acetate	$-\overset{\overset{\textstyle O}{\|}}{O}CCH_3$		polyvinyl acetate (PVA)
Acrylics	$\begin{array}{cc} H & Y \\ C{=}C \\ H & X \end{array}$	$-\!\!\left[\begin{array}{cc} H & Y \\ C - C \\ H & X \end{array}\right]\!\!-$	
	$\underline{X} \qquad \underline{Y}$		
acrylonitrile	$- C{\equiv}N \qquad H$		polyacrylonitrile (PAN)
methyl acrylate	$-\overset{\overset{\textstyle O}{\|}}{C}OCH_3 \qquad H$		polymethyl acrylate
ethyl acrylate	$-\overset{\overset{\textstyle O}{\|}}{C}OC_2H_5 \qquad H$		polyethyl acrylate
methyl methacrylate	$-\overset{\overset{\textstyle O}{\|}}{C}OCH_3 \qquad CH_3$		polymethyl methacrylate (PMMA)

Table A-2. (*Continued*)

Monomer		Repeat Unit	Polymer
Name	Formula		

Olefins

$$\begin{array}{cc} H & R' \\ C = C \\ H & R \end{array}$$

Repeat Unit:

$$\left[\begin{array}{cc} H & R' \\ C - C \\ H & R \end{array}\right]$$

Name	R	R'	Polymer
ethylene	H	H	polyethylene (PE)
propylene	$-CH_3$	H	polypropylene
butene-1	$-C_2H_5$	H	polybutene-1
4-methyl pentene-1	$-CH \begin{array}{c} CH_3 \\ CH_3 \end{array}$	H	poly 4-methyl pentene-1
isobutylene	$-CH_3$	$-CH_3$	polyisobutylene

Miscellaneous

$$\begin{array}{cc} Z & Y \\ C = C \\ W & X \end{array}$$

Repeat Unit:

$$\left[\begin{array}{cc} Z & Y \\ C - C \\ W & X \end{array}\right]$$

Name	X	Y	W	Z	Polymer
vinylidene chloride	$-Cl$	$-Cl$	H	H	polyvinylidene chloride
tetrafluoro ethylene	$-F$	$-F$	$-F$	$-F$	polytetrafluoro ethylene (PTFE)
chlorotrifluoro ethylene	$-Cl$	$-F$	$-F$	$-F$	polychlorotrifluoro ethylene

formaldehyde	$\begin{array}{c} H \\ C = O \\ H \end{array}$	$\left[\begin{array}{c} H \\ CO \\ H \end{array}\right]$	polyoxymethylene
acetaldehyde	$\begin{array}{c} H \\ C = O \\ \| \\ CH_3 \end{array}$	$\left[\begin{array}{c} H \\ CO \\ \| \\ CH_3 \end{array}\right]$	polyacetaldehyde

Table A-2.   (*Continued*)

Monomer		Repeat Unit	Polymer
Name	Formula		

Dienes	$\begin{array}{ccc} H & X & H \\ C{=}C & - & C{=}C \\ H & Y & H \end{array}$	$\left[C - C(X){=}C(Y) - C \right]$	
	$\begin{array}{cc} X & Y \end{array}$		
butadiene	H H		polybutadiene
isoprene	$-CH_3$ H		polyisoprene

where S is a small-molecule by-product such as H_2O for polyesters and HCl for polycarbonates.

Two distinct mechanisms for the formation of x-mer are possible. One involves direct, random propagation of AA and BB monomers and the other ester interchange. In the direct route, whose stoichiometry is represented by equations A-1, four distinguishable elementary propagation steps may be identified. They are:

$$[AABB]_y AA + BB[AABB]_{x-y-1} \rightleftarrows [AABB]_x \; + \; S$$

$$[AABB]_y \; + \; [AABB]_{x-y} \quad \rightleftarrows [AABB]_x \; + \; S$$

$$x > y \geqslant 0 \tag{A-2}$$

$$[AABB]_y AA + [AABB]_{x-y} \quad \rightleftarrows [AABB]_x AA + S$$
$$BB[AABB]_y + [AABB]_{x-y} \quad \rightleftarrows BB[AABB]_x + S$$

The ester interchange route, shown in Table A-4, is actually a two-step reaction between the —OH end group of a polymer chain and the ester linkage $-\overset{\overset{O}{\|}}{C}-$ in another. The propagation step ideally starts with only BBAABB molecules, which have been formed in a previous step by reacting a di-ester RBAABR, such as di-methyl terephthalate, with an excess of monomer BB. The stoichiometric equations are:

$$\begin{array}{lll} \text{1st step} & RBAABR + 2BB = BBAABB \; + \; 2RB \\ \text{2nd step} \quad x \geqslant 1 & xBBAABB \quad = \quad BB[AABB]_x + (x-1)BB \end{array} \tag{A-3}$$

and the sequence of elementary propagation steps corresponding to the second stoichiometric equation may be represented compactly by

$$x > y \geqslant 0 \qquad BB[AABB]_{y-1}AABB + BB[AABB]_{x-y} \rightleftarrows BB[AABB]_x + BB \tag{A-4}$$

↑   ↑
linkage and end-group involved
in interchange

Table A-3. Ring Scission Polymers

Monomer Name	Formula	Repeat Unit	Interunit Linkage	Polymer
cyclic ethers				
epoxides { ethylene oxide	(epoxide ring)	$\left[\begin{array}{c} H & H \\ C - C - O \\ H & H \end{array}\right]$	$- O -$	polyethylene oxide
propylene oxide	(epoxide ring)$-CH_3$	$\left[\begin{array}{c} H & H \\ C - C - O \\ H & CH_3 \end{array}\right]$	$- O -$	polypropylene oxide (PPO)
trioxane	(trioxane ring)	$\left[\begin{array}{c} H \\ C - O \\ H \end{array}\right]$	$- O -$	polyoxymethylene
	$C\ell H_2C$ (ring) $CH_2C\ell$	$\left[\begin{array}{c} CH_2C\ell \\ CH_2 - C - CH_2 - O \\ CH_2C\ell \end{array}\right]$	$- O -$	polychloroether
cyclic amides				
ε-caprolactam	(lactam ring)	$\left[\begin{array}{c} O & H \\ \parallel & \\ C(CH_2)_5N \end{array}\right]$	$\begin{array}{c} O \\ \parallel \\ - NHC - \end{array}$	polyamide (nylon 6)
cyclic esters				
δ-valerolactone	(lactone ring)	$\left[\begin{array}{c} O \\ \parallel \\ C(CH_2)_4O \end{array}\right]$	$\begin{array}{c} O \\ \parallel \\ - OC - \end{array}$	polyester

Table A-4. Polymerization of Ethylene Glycol and Dimethyl Terephthalate

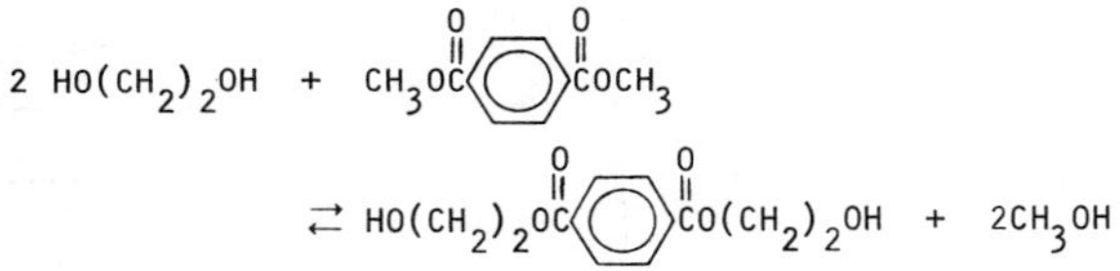

monomer formation

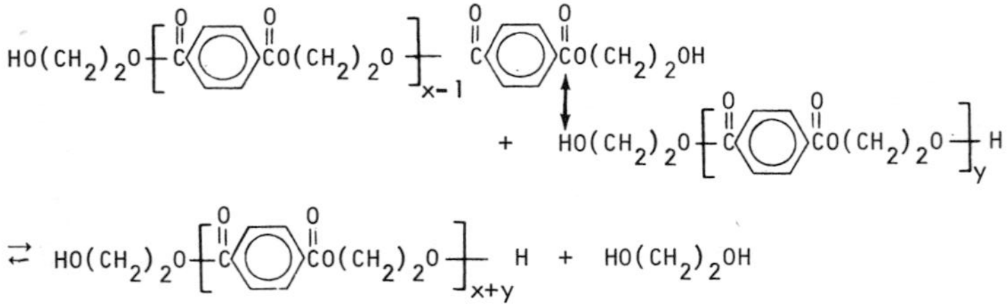

$$2\ HO(CH_2)_2OH\ +\ CH_3OC(=O)\!-\!\bigcirc\!-\!COCH_3 \;\rightleftarrows\; HO(CH_2)_2OC(=O)\!-\!\bigcirc\!-\!CO(CH_2)_2OH\ +\ 2CH_3OH$$

propagation by interchange

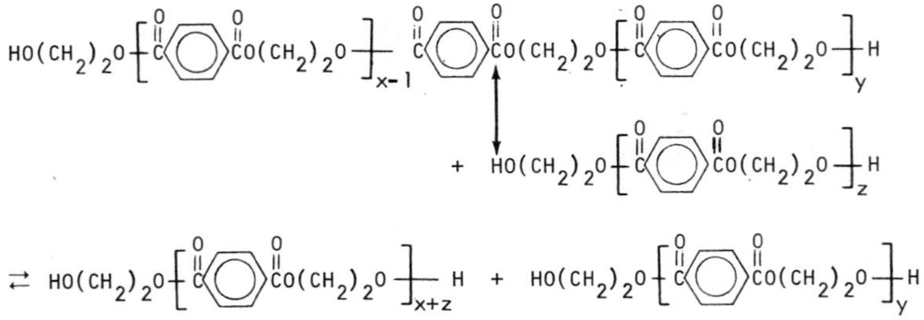

$$HO(CH_2)_2O\!\left[C(=O)\!-\!\bigcirc\!-\!CO(CH_2)_2O\right]_{x-1}\!C(=O)\!-\!\bigcirc\!-\!CO(CH_2)_2OH\ +\ HO(CH_2)_2O\!\left[C(=O)\!-\!\bigcirc\!-\!CO(CH_2)_2O\right]_y\!H$$

$$\rightleftarrows\; HO(CH_2)_2O\!\left[C(=O)\!-\!\bigcirc\!-\!CO(CH_2)_2O\right]_{x+y}\!H\ +\ HO(CH_2)_2OH$$

rearrangement by interchange

$$HO(CH_2)_2O\!\left[C(=O)\!-\!\bigcirc\!-\!CO(CH_2)_2O\right]_{x-1}\!C(=O)\!-\!\bigcirc\!-\!CO(CH_2)_2O\!\left[C(=O)\!-\!\bigcirc\!-\!CO(CH_2)_2O\right]_y\!H\ +\ HO(CH_2)_2O\!\left[C(=O)\!-\!\bigcirc\!-\!CO(CH_2)_2O\right]_z\!H$$

$$\rightleftarrows\; HO(CH_2)_2O\!\left[C(=O)\!-\!\bigcirc\!-\!CO(CH_2)_2O\right]_{x+z}\!H\ +\ HO(CH_2)_2O\!\left[C(=O)\!-\!\bigcirc\!-\!CO(CH_2)_2O\right]_y\!H$$

Table A-5.   Polymerization of Adipic Acid and Hexamethylene Diamine to Nylon 66

$$\underset{\text{HOC(CH}_2)_4\text{COH}}{\overset{\overset{O}{\|}\qquad\overset{O}{\|}}{}} \; + \; H_2N(CH_2)_6NH_2 \;\; \rightleftarrows \;\; \left\{ \underset{{}^+H_3N(CH_2)_6NH_3^+}{\overset{{}^-OC(CH_2)_4CO^-}{\overset{\overset{O}{\|}\quad\overset{O}{\|}}{}}} \right\} \quad\text{salt formation}$$

$$x \; \left\{ \underset{{}^+H_3N(CH_2)_6NH_3^+}{\overset{{}^-OC(CH_2)_4CO^-}{\overset{\overset{O}{\|}\quad\overset{O}{\|}}{}}} \right\} \;\; \rightleftarrows \;\; HO\!\!\left[\overset{\overset{O}{\|}\qquad\overset{O}{\|}}{C(CH_2)_4CNH(CH_2)_6NH}\right]_x\!\!H \;\; + \;\; (2x\!-\!1)H_2O \quad\text{polymerization}$$

SOLUTION

$$HO\!\!\left[\;\right]_x\!\!H \; + \; HO\!\!\left[\;\right]_y\overset{\overset{O}{\|}}{C(CH_2)_4}\overset{\overset{O}{\|}}{CNH(CH_2)_6NH}\!\!\left[\;\right]_z\!\!H$$

$$HO\!\!\left[\;\right]_y H \; + \; HO\!\!\left[\;\right]_{x+z} H \quad\text{interchange}$$

$$HO\!\!\left[\;\right]_{w-1}\overset{\overset{O}{\|}}{C(CH_2)_4}\overset{\overset{O}{\|}}{CNH(CH_2)_6NH}\!\!\left[\;\right]_x H \; + \; HO\!\!\left[\;\right]_y\overset{\overset{O}{\|}}{C(CH_2)_4}\overset{\overset{O}{\|}}{CNH(CH_2)_6NH}\!\!\left[\;\right]_{z-1} H$$

$$\rightleftarrows \;\; HO\!\!\left[\;\right]_{w+z} H \; + \; HO\!\!\left[\;\right]_{x+y} H \quad\text{interchange}$$

$$\left[\quad\right]\;\text{signifies repeat unit}\;\left[\overset{\overset{O}{\|}}{C(CH_2)_4}\overset{\overset{O}{\|}}{CNH(CH_2)_6NH}\right]$$

Table A-6. Water-Catalyzed Caprolactam Polymerization to Nylon 6

$$H_2O + (CH_2)_5 - NH \rightarrow HOC(CH_2)_5NH_2 \qquad \text{initiation}$$

$$HO\left[C(CH_2)_5NH\right]_x H + (CH_2)_5 - NH \rightarrow HO\left[C(CH_2)_5NH\right]_x C(CH_2)_5NH_2 \qquad \text{addition propagation}$$

$$HO\left[C(CH_2)_5NH\right]_x H + HO\left[C(CH_2)_5NH\right]_y H \rightleftharpoons HO\left[C(CH_2)_5NH\right]_{x+y} H + H_2O$$

$$\text{random propagation}$$

$$HO\left[C(CH_2)_5NH\right]_w H + HO\left[C(CH_2)_5NH\right]_x H \rightleftharpoons HO\left[C(CH_2)_5NH\right]_{w+x} H + HO\left[C(CH_2)_5NH\right]_y H$$

$$\text{interchange}$$

$$HO\left[C(CH_2)_5NH\right]_w H + HO\left[C(CH_2)_5NH\right]_x H \rightleftharpoons HO\left[C(CH_2)_5NH\right]_z H + HO\left[C(CH_2)_5NH\right]_{y} \left[C(CH_2)_5NH\right]_x H$$

$$\text{interchange}$$

Table A-7. Free-Radical Initiated Styrene Polymerization

$$\overset{O\quad O}{\underset{}{\parallel\quad\parallel}}\ \ RCOOCR \longrightarrow 2R\cdot \ +\ 2CO_2\uparrow \qquad\qquad \text{formation of initiating species}$$

$$R\cdot\ +\ CH_2{=}\underset{Ph}{CH} \longrightarrow RCH_2\overset{H}{\underset{Ph}{C}}\cdot \qquad\qquad \text{initiation}$$

$$R\,(CH_2CH)_{x-1}CH_2\overset{H}{\underset{Ph}{C}}\cdot\ +\ CH_2{=}\underset{Ph}{CH} \longrightarrow R\,(CH_2CH)_{x-1}CH_2CH_2\overset{H}{\underset{Ph}{C}}\cdot \qquad\qquad \text{propagation}$$
$$\underset{Ph}{}$$

$$R\,(CH_2CH)_{x-1}CH_2\overset{H}{\underset{Ph}{C}}\cdot\ +\ CH_2{=}\underset{Ph}{C}\cdot$$
$$R\,(CH_2CH)_{x-1}CH{=}\underset{Ph}{CH}\ +\ CH_3\overset{H}{\underset{Ph}{C}}\cdot \qquad\qquad \begin{array}{l}\text{chain transfer to}\\ \text{monomer}\end{array}$$

$$R\,(CH_2CH)_{x-1}CH{=}CH\ +\ H_2CCH_2\,(CHCH_2)_{y-1}R$$
$$\underset{Ph}{}\qquad\qquad \underset{Ph}{}\quad \underset{Ph}{} \qquad\qquad \text{termination by disproportionation}$$

$$R\,(CH_2CH)_{x-1}CH_2CH\ CHCH_2\,(CHCH_2)_{y-1}R$$
$$\underset{Ph}{}\qquad \underset{Ph}{}\ \underset{Ph}{}\qquad \underset{Ph}{} \qquad\qquad \text{termination by combination}$$

671

Table A-7. (*Continued*)

$$R \xleftarrow{}{} CH_2CH \xrightarrow{}{}_{x-1} CH_2 \overset{H}{\underset{Ph}{\overset{|}{C}}} \cdot \; + \; R \xleftarrow{}{} CH_2CH \xrightarrow{}{}_{y} CH_2CH \xleftarrow{}{} CH_2CH \xrightarrow{}{}_{z}$$

$$\underset{Ph}{} \qquad \underset{Ph}{} \qquad \underset{Ph}{}$$

$$\rightarrow \; R \xleftarrow{}{} CH_2CH \xrightarrow{}{}_{x-1} CH_2CH_2 \; + \; R \xleftarrow{}{} CH_2CH \xrightarrow{}{}_{y} CH_2\overset{\cdot}{C} \xleftarrow{}{} CH_2CH \xrightarrow{}{}_{z}$$

$$\underset{Ph}{} \qquad \underset{Ph}{} \qquad \underset{Ph}{} \qquad \underset{Ph}{}$$

$$\left.\right\} \text{chain transfer to polymer}$$

$$R \xleftarrow{}{} CH_2CH \xrightarrow{}{}_{x-1} CH_2 \overset{H}{\underset{Ph}{\overset{|}{C}}} \cdot \; + \; CC\ell_4 \; \rightarrow \; R \xleftarrow{}{} CH_2CH \xrightarrow{}{} CH_2CHC\ell \; + \; \cdot CC\ell_3$$

$$\underset{Ph}{} \qquad \qquad \underset{Ph}{} \qquad \underset{Ph}{}$$

chain transfer to foreign agent

Ph stands for a phenyl group

Table A-8. Cationically Catalyzed Isobutylene Polymerization

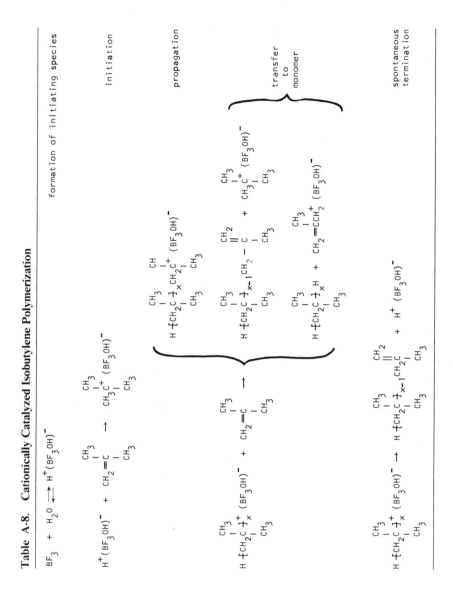

Table A-9. Anionically Initiated Ethylene Oxide Polymerization

$RONa \rightarrow RO^- + N\overset{+}{a}$	formation of initiating species
$RO^- + H_2C\overset{O}{\overset{\diagup\diagdown}{-}}CH_2 \rightarrow ROCH_2CH_2O^-$	initiation
$R(OCH_2CH_2)_xO^- + H_2C\overset{O}{\overset{\diagup\diagdown}{-}}CH_2 \rightarrow R(OCH_2CH_2)_xOCH_2CH_2O^-$	propagation
$R(OCH_2CH_2)_xO^- + ROH \rightarrow R(OCH_2CH_2)_xOH + RO^-$	transfer to foreign agent
$R(OCH_2CH_2)_wO^- + R(OCH_2CH_2)_y(OCH_2CH_2)_{x-y}OH \rightarrow$ $R(OCH_2CH_2)_{w+x-y}OH + R(OCH_2CH_2)_yO^-$	interchange

Although reaction A-4 represents growth-type interchange, it should be pointed out that interchange in general is not limited to terminal linkages, but can involve other linkages within the polymer chains as well. Such reactions may produce polymer molecules less extreme in size than x-mer and monomer; in fact, product molecules having any combination of chain lengths are possible. The general sequence for interchange, which includes reaction A-4 as a special case (when $z = 0$),

$$x \geqslant y > z \geqslant 0 \qquad BB[AABB]_{y-z-1}\underset{\uparrow}{[AABB]}[AABB]_z + \underset{\uparrow}{BB[AABB]}_{x-y+z}$$

linkage and end-group involved

$$\rightleftarrows \quad BB[AABB]_x + BB[AABB]_z \qquad (A\text{-}5)$$

is not exclusively a chain-building reaction, but includes chain-size rearrangements as well.

Other examples that are important industrially are the formation of urethane and amide polymers. Polyurethanes are formed from diisocyanates $\left(\overset{O}{\overset{\|}{C}}NRN\overset{O}{\overset{\|}{C}}\right)$ and diols (HOR′OH), and polyamides from diacids $(HO\overset{O}{\overset{\|}{C}}R\overset{O}{\overset{\|}{C}}OH)$ and diamines $(H_2NR′NH_2)$. Stoichiometric equations and reaction sequences are of the AABB-type and are therefore represented by equations A-1 and A-2, respectively, where:

$$A = \begin{cases} -N\overset{O}{\overset{\|}{C}} \quad \text{group for polyurethanes} \\[2em] -\overset{O}{\overset{\|}{C}}OH \quad \text{group for polyamides} \end{cases}$$

Table A-10. Ziegler–Natta Catalyzed Propylene Polymerization

$$\left[TiCl_3\right] + AlR_3 \rightleftarrows \left[TiCl_3 \cdot AlR_2\right]R$$

formation of initiating species by chemisorption

$$\left[TiCl_3\right] + CH_2{=}CH{\overset{\underset{\displaystyle CH_3}{|}}{}} \rightleftarrows \left[\quad\right]\!\!\cdots\!\!{\overset{\underset{\displaystyle CH_2}{\|}}{CHCH_3}}$$

competitive monomer adsorption

$$\left[\quad\right]\!\!{\overset{R}{\underset{\displaystyle CH_2}{\underset{\|}{CHCH_3}}}}\!\!\cdots \rightarrow \left[\quad\right]{-}CH_2{\overset{\underset{\displaystyle CH_3}{|}}{C}}HR$$

initiation by Langmuir-Hinshelwood mechanism

$$\left[\quad\right]{-}R + CH_2{=}CH{\overset{\underset{\displaystyle CH_3}{|}}{}} \rightarrow \left[\quad\right]{-}CH_2{\overset{\underset{\displaystyle CH_3}{|}}{C}}HR$$

initiation by Rideal mechanism

$$\left[\quad\right]\!\!{\overset{\underset{\displaystyle CH_2}{\|}}{\underset{\displaystyle CHCH_3}{{-}(CH_2CH)_x R \atop |{\scriptstyle CH_3}}}}\!\! \rightarrow \begin{cases} \left[\quad\right]{-}CH_2{\overset{CH_3}{|}}CH(CH_2{\overset{CH_3}{|}}CH)_x R & \text{propagation} \\ & \text{Langmuir-Hinshelwood} \\ \left[\quad\right]{-}CH_2CH_2{\overset{CH_3}{|}} \\ \quad + CH_2{=}C(CH_2{\overset{CH_3}{|}}CH)_{x-1}{\overset{CH_3}{|}}R & \text{chain transfer to monomer} \end{cases}$$

$$\left[\quad\right]{-}(CH_2{\overset{CH_3}{|}}CH)_x R + CH_2{=}CH{\overset{CH_3}{|}} \rightarrow \begin{cases} \left[\quad\right]{-}CH_2{\overset{CH_3}{|}}CH(CH_2{\overset{CH_3}{|}}CH)_x R & \text{propagation} \\ & \text{Rideal} \\ \left[\quad\right]{-}CH_2CH_2{\overset{CH_3}{|}} \\ \quad + CH_2{=}C(CH_2{\overset{CH_3}{|}}CH)_{x-1}{\overset{CH_3}{|}}R & \text{chain transfer to monomer} \end{cases}$$

$$\left[\quad\right]{-}CH_2{\overset{CH_3}{|}}CH(CH_2{\overset{CH_3}{|}}CH)_x R \rightarrow \left[\quad\right]{-}H + CH_2{=}C(CH_2{\overset{CH_3}{|}}CH)_x{\overset{CH_3}{|}}R$$

spontaneous termination by desorption

Chain transfer reactions involving AlR_3 and $TiCl_3$ have also been postulated but are not included here.

Square brackets $\left[\quad\right]$ here denote solid catalyst surface

---

675

$$B = \begin{cases} -OH & \text{group for polyurethanes} \\ -NH_2 & \text{group for polyamides} \end{cases}$$

$$S = \begin{cases} \text{none for polyurethanes} \\ H_2O \text{ for polyamides} \end{cases}$$

Such reactions are also often two-step processes. Polyamides generally proceed via a separate first step involving the formation of a salt intermediate (see Table A-5).

Polyamides can also be formed from amino acids, which are so-called AB-type monomers. The repeat unit is then AB and the generalized stoichiometric equations and propagation sequence are, respectively,

$$x > 1 \qquad\qquad x AB = [AB]_x + (x - 1)S \qquad\qquad \text{(A-6)}$$

and

$$x > y \geqslant 1 \qquad\qquad [AB]_y + [AB]_{x-y} \rightleftarrows [AB]_x + S \qquad\qquad \text{(A-7)}$$

where $S = H_2O$ for polyamides.

A third route to polyamides is via ring-type monomers (Lactams)

$$\left(\begin{array}{c} O \\ \| \\ C \\ \diagup \quad \diagdown \\ NH-R \end{array} \right)$$

in which rings are opened through the action of an initiator (I) such as water (Table A-6). This method involves two simultaneous but distinct reactions, namely an initiation step followed by propagation. In symbolic notation, the stoichiometric equations are

$$x \geqslant 1 \qquad\qquad 1 + x\widehat{AB} = [AB]_x \qquad\qquad \text{(A-8)}$$

and the reaction sequence is

$$x > 1 \qquad\qquad \left. \begin{array}{l} 1 + \widehat{AB} \rightarrow AB \qquad \text{initiation} \\ [AB]_{x-1} + \widehat{AB} \rightarrow [AB]_x \quad \text{propagation} \end{array} \right\} \qquad \text{(A-9)}$$

where $\widehat{AB}$ represents cyclic monomer. The nature of the propagation step is seen to be fundamentally different in kind from those of AA-BB and A-B polymerizations, although it is sometimes possible for propagation as in reaction A-7 to occur simultaneously, once linear molecules of the type $[AB]_x$, $x \geqslant 1$, are present in the reaction mixture. This is not possible in certain industrial processes where AB represents catalytic intermediates that cannot mutually condense.

Amide interchange is an alternative route to polyamides. The interchange reactions appearing in Tables A-5 and A-6 are of a more general nature than in Table A-4 since they are not restricted to terminal amine groups and amide linkages. The stoichiometric equation for such interchange among polymer molecules of the type

$[AB]_x$ is

$$x \geqslant y > z \geqslant 0 \qquad [AB]_{y-z}[AB]_z + [AB]_{x-y+z}[AB]_w = [AB]_x + [AB]_{z+w} \qquad \text{(A-10)}$$

(with "exchange" annotations connecting the terms)

It should be noted that the special case $w = 0$ is the counterpart of scheme A-5 for AB-type polymerizations. Equation A-10 includes A-7 as a further special case when $z = 0$ and $[AB]_0$ is represented by S.

If monofunctional RA or trifunctional $A\!-\!\!<^A_A$ monomers are introduced into AA-BB polymerization, the result will be a reduction in DP and a network polymer, respectively. Monofunctional monomers act as chain stoppers when R is a nonreactive group and polyfunctional monomers as crosslinking agents. On the other hand, if $A\!-\!\!<^A_B$ is introduced into A-B polymerizations, network polymers will not be formed. Obviously, $A\!-\!\!<^A_B$ and B—B or $B\!-\!\!<^B_B$ monomers will crosslink.

All remaining condensation and ring scission polymerizations cited in Tables A-1 and A-3 may be represented either by the preceding general stoichiometric equations or modifications thereof. We shall not elaborate further on specific schemes involving functional groups of rings since it would be beyond the scope of this book. Instead, we turn now to monomers containing double bonds.

Examples of monomers containing double bonds are listed in Table A-2. Their polymerization, copolymerization, and terpolymerization involve reactive intermediates instead of stable molecules with reactive functional groups. Such intermediates may be free radicals, ions, or catalyst complexes, and are formed from the decomposition of an initiating species. Mechanisms that have been proposed for some common polymerizations are shown in Tables A-7–A-10.

In symbolic form, the stoichiometric formula for catalyzed vinyl and olefin polymerizations is very simple.

$$c + xA = A_x + c \qquad \text{(A-11)}$$

However, it is evident from the tables that not all initiating species are true catalysts. One counterexample is the free-radical initiator I. In this case molecular fragments from I can remain permanently attached to the polymeric product molecules. Stoichiometric equations, which must reflect this, are further complicated by whether or not termination is bimolecular, as it is in free-radical polymerizations, and if so, whether it occurs by disproportionation or combination. For free-radical polymerizations, which currently account for a significant, if not dominant, fraction of annual polymer production, we have

$$\left(\frac{1 + R'}{2}\right)I + xA = A_x \qquad \text{(A-12)}$$

where $R' = 0$ for disproportionation and $R' = 1$ for combination. The reaction sequences of Tables A-7–A-10 have been rewritten in Table A-11 in abbreviated

Table A-11.

Initiator and Catalyst Decomposition

$$I \rightarrow 2m_o^* + 2G \uparrow$$

(Kinetic chain initiation)

$$c \rightleftarrows m_o^*$$

Polymer Chain Initiation

primary

$$m_o^* + A \rightarrow m_oA*$$

(Kinetic chain propagation)

secondary

$$\left. \begin{array}{c} S^* \\ m_oS^* \end{array} \right\} + A \longrightarrow \left\{ \begin{array}{c} SA* \\ m_oSA* \end{array} \right.$$

(Kinetic chain propagation)

Polymer Chain Propagation and Chain Transfer to Monomer

$$m_oA_x^* + A \longrightarrow m_oA_{x+1}^*$$

(Kinetic chain propagation)

Chain Transfer to Monomer

$$m_oA_x^* + A \longrightarrow \left\{ \begin{array}{c} m_oA_x + A_1^* \\ A_x + m_oA_1^* \end{array} \right.$$

(Polymer chain termination, kinetic chain propagation)

Chain Transfer to Foreign Agent

$$m_oA_x^* + S \longrightarrow \left\{ \begin{array}{c} m_oA_x + S \\ A_x + m_oS^* \end{array} \right.$$

(Polymer chain termination, kinetic chain propagation)

Polymer Chain Termination

$$m_oA_x* + m_oA_y^* \longrightarrow \left\{ \begin{array}{c} m_oA_{x+y}m_o \\ m_oA_x + m_oA_y \end{array} \right.$$

(Kinetic chain termination)

$$m_oA^* \quad \rightleftarrows \quad \left\{ \begin{array}{c} m_oA_x \\ A_x + m_o^* \end{array} \right.$$

(Kinetic chain termination)

(Kinetic chain propagation)

form, using general symbols in lieu of chemical formulas. The symbol $*$ represents active intermediate of any kind: free-radical, ion, or catalyst complex.

The stoichiometric equation for copolymerization of vinyl and olefinic monomers A and B, disregarding initiating species, may be written in abbreviated form as

$$aA + bB = A_aB_b \tag{A-13}$$

where the degree of polymerization is

$$x = a + b \tag{A-14}$$

and the copolymer composition is

$$y = \frac{a}{x} \tag{A-15}$$

The sequence of comonomers in copolymer A_aB_b can be random, alternating, or block. Rarely is the composition of A and B periodic with x-mer being simply $[AB]_x$, $[AABB]_x$, and so on. The presence of two different monomers gives rise to several distinguishable initiation, propagation, and termination steps. They are, respectively,

$$
\begin{aligned}
m_0^* + A &\to m_0 A^* \\
m_0^* + B &\to m_0 B^*
\end{aligned} \tag{A-16}
$$

$$
\begin{aligned}
[A_{a-2}B_b]A^* + A &\to [A_{a-1}B_b]A^* \\
[A_{a-1}B_{b-1}]A^* + B &\to [A_aB_{b-1}]B^* \\
[A_{a-1}B_{b-1}]B^* + A &\to [A_{a-1}B_b]A^* \\
[A_aB_{b-2}]B^* + B &\to [A_a, B_{b-1}]B^*
\end{aligned} \tag{A-17}
$$

and

$$
\begin{aligned}
[A_{a-1}B_b]A^* + {}^*A[A_{\alpha-1}B_\beta] &\to [A_{a+\alpha}B_{b+\beta}] \\
[A_{a-1}B_b]A^* + {}^*B[A_\alpha B_{\beta-1}] &\to [A_{a+\alpha}B_{b+\beta}] \\
[A_aB_{b-1}]B^* + {}^*B[A_\alpha B_{\beta-1}] &\to [A_{a+\alpha}B_{b+\beta}]
\end{aligned} \tag{A-18}
$$

Termination equations A-18 show combination only for illustration. Also they distinguish between only two different categories of active intermediates based on their ultimate repeat units. Theories have been proposed which claim that these categories are too broad and that classification by chemical behavior requires distinction on the basis of both ultimate and penultimate groups. This results in four distinguishable categories of active intermediates:

$$
\begin{aligned}
&[A_{a-2}B_b]AA^* \\
&[A_{a-1}B_{b-1}]BA^* \\
&[A_{a-1}B_{b-1}]AB^* \\
&[A_aB_{b-2}]BB^*
\end{aligned} \tag{A-19}
$$

where clearly the first two belong to one of the broader categories above and the last two belong to the other. It is evident that we now have eight and ten distinguishable propagation and termination steps, respectively.

To simplify the mathematical modeling of polymerizations and copolymerizations in the text, the A's and B's have been dropped and a more compact system of symbols introduced, which uses the letter m_x to represent x-mer together with additional subscripts and superscripts for copolymer composition, degree of branching, and so on.

APPENDIX B

Distributions

Consider a property that is distributed among a population, and let variable s be a measure of the magnitude that property. The distribution of s is described by frequency function f_s when s is a discrete variable and by density function $f(s)$ when s is continuous. Thus f_s, or $f(s)\,ds$, represents the portion (number, weight, volume, fraction, etc.) of the population whose property is precisely s (discrete) or whose property lies between s and $s + ds$ (continuous). The distribution functions corresponding to frequency and density functions are, respectively,

$$F_S = \sum_{s_0}^{S} f_s \tag{B-1}$$

and

$$F(S) = \int_{s_0}^{S} f(s)\,ds \tag{B-2}$$

where s_0 is the initial value of s, and S is an arbitrary higher value. They represent the portion of the population whose property is at most S (i.e., S or less). It follows that

$$\Delta F_s \equiv F_s - F_{s-h} = f_s \tag{B-3}$$

and

$$\frac{dF(s)}{ds} = f(s) \tag{B-4}$$

where ΔF_s is a finite difference and h is the constant spacing (period) between discrete values of s.

$$h \equiv \Delta s \tag{B-5}$$

681

For several common discrete distributions s is a positive integer and $h = 1$ as, for example, when s represents the degree of polymerization (DP), $s = x$, or stage number, $s = j$. An important continuous distribution is the residence time distribution (RTD), for which residence time $s = \tau$ is a positive real number.

CHARACTERIZATION OF DISTRIBUTIONS

Distributions are generally characterized by their shape and breadth and by some measure of the "central" value of s. These characteristics may be computed systematically through the use of moments. Two types of moments are commonly employed. One type is defined about the origin ($s = 0$) of the distribution and the other about its mean value ($s = \bar{s}$). The kth moment about the origin is defined as

$$
\mu^k \equiv \begin{cases} \displaystyle\sum_{\text{all } s} s^k f_s \\[2em] \displaystyle\int_{\text{all } s} s^k f(s)\, ds \end{cases} \tag{B-6}
$$

and the kth moment about the mean is defined as

$$
\sigma^k \equiv \begin{cases} \displaystyle\sum_{\text{all } s} (s - \bar{s})^k \hat{f}_s \\[2em] \displaystyle\int_{\text{all } s} (s - \bar{s})^k \hat{f}(s)\, ds \end{cases} \tag{B-7}
$$

where k is an integer and $\hat{f}$ signifies that the distribution has been normalized through division by the population size

$$
\hat{f} \equiv \begin{cases} \dfrac{f_s}{\displaystyle\sum_{\text{all } s} f_s} \\[2em] \dfrac{f(s)}{\displaystyle\int_{\text{all } s} f(s)\, ds} \end{cases} \tag{B-8}
$$

Thus $\hat{\mu}^k$ represents the kth moment of the normalized distribution about the origin (using $\hat{f}$ in definition B-6).

It should be noted that since σ^k takes on the value of zero when $k = 1$, σ^1 may be defined alternatively as follows

$$
\sigma^1 \equiv \int_{\text{all } s} |s - \bar{s}| \hat{f}(s)\, ds \tag{B-9}
$$

where symbol $|\ \ |$ denotes absolute value.

Moments σ^k and $\hat{\mu}^k$ are related in general by

$$\sigma^k = \sum_{j=0}^{k} \binom{k}{j} (-1)^{k-j} \hat{\mu}^j (\hat{\mu}^1)^{k-j} \tag{B-10}$$

where $\binom{k}{j}$ is the binomial coefficient

$$\binom{k}{j} \equiv \frac{k!}{j!(k-j)!} = \binom{k}{k-j} \tag{B-11}$$

Two examples of these equations are:

$$\sigma^2 = \hat{\mu}^2 - (\hat{\mu}^1)^2 \tag{B-12}$$

and

$$\sigma^3 = \hat{\mu}^3 - 3\hat{\mu}^2 \hat{\mu}^1 + 2(\hat{\mu}^1)^3 \tag{B-13}$$

Several commonly used characteristics follow. One is μ^0, which measures population size and was used in definition B-8. Another is the mean

$$\hat{\mu}^1 = \bar{s} \equiv s_{\text{avg}} \tag{B-14}$$

which represents a central value of s. Another central value is the median s_M, defined by the equation

$$\int_{s_0}^{s_f} \hat{f}(s)\, ds = \frac{1}{2} = \int_{s_M}^{s_f} \hat{f}(s)\, ds \tag{B-15}$$

where s_f is the final value of s. The median is sometimes preferred over the mean because the latter weights each member of the population in proportion to its value of s and thus favors those with large s, whereas the former does not. A characteristic of distributions that is of considerable importance to us is their breadth or dispersion. Such a characteristic is the variance σ^2 or, equivalently, the standard deviation SD. Both measure the mean deviation of a distribution about its mean value. Thus,

$$\sigma^2 \equiv \overline{(s - \bar{s})^2} = \overline{s^2} - \bar{s}^2 \equiv \text{var } s \tag{B-16}$$

and

$$\text{SD} \equiv (\text{var } s)^{1/2} \equiv \sigma \tag{B-17}$$

Standard deviation is sometimes preferred to variance because its magnitude and units are comparable to $\bar{s}$. In physical applications, however, relative dispersion RD is frequently more meaningful than absolute dispersion (Section 1.4). We define

$$\text{RD} \equiv \frac{\text{SD}}{\bar{s}} \tag{B-18}$$

and note that statisticians have termed this quantity the coefficient of variation. An alternative measure of relative dispersion involving σ^1 rather than σ is

$$\text{RD}' \equiv \frac{\sigma^1}{\bar{s}} \tag{B-19}$$

The shape of a distribution, specifically its asymmetry, is measured by skewness

$$\text{SK} \equiv \frac{\sigma^3}{(\text{SD})^3} \tag{B-20}$$

Positive, negative, and no skewness are illustrated in Fig. B-1.

THE BINOMIAL COEFFICIENT

The number of different ways r distinguishable objects can be arranged into groups of s, if order within groups is distinguished, is given by the factorial

$$(r)_s \equiv r(r-1)\cdots(r-s+1) \tag{B-21}$$

If order is not distinguished we obtain the result

$$\frac{(r)_s}{s!} = \frac{r!}{s!(r-s)!} \equiv \binom{r}{s} \tag{B-22}$$

which is obviously smaller than $(r)_s$, owing to division by the number of permutations among the s objects within each group, $s!$. The quantity $\binom{r}{s}$ is the binomial coefficient.

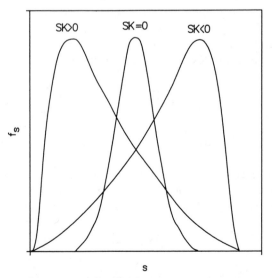

Figure B-1.   Sketch showing the relationship of skewness (SK) to symmetry.

A simple example will illustrate the distinction. Consider random drawings of cards without replacement from a well-mixed poker deck. The probability of drawing a royal flush of any suit in the precise order AKQJ10 is $(1/52)(1/51)(1/50)(1/49)(1/48)$ by combinational analysis. This result may be interpreted alternatively as the ratio of the number of favorable draws to the total number possible $1/(52)_5$. On the other hand, if a royal flush in any order will suffice, then the probability is $(5/52)(4/51)(3/50)(2/49)(1/48)$, which may also be interpreted as $5!/(52)_5$. Now 5! draws are favorable among a total of $(52)_5$. The same probability may also be written as $1/\binom{52}{5}$, where only one draw is again favorable because we have reduced the total to $\binom{52}{5}$ by not distinguishing order.

BERNOULLI TRIALS

Distributions often arise from a sequence of random experiments. The Bernoulli trials are such experiments, which lead to the binomial distribution. They involve a succession of r random experiments, each having probability of success P. The experiments are independent and therefore have identical probabilities of success. Thus, the probability of s successes in succession followed by $r - s$ failures is

$$P^s(1 - P)^{r-s} \equiv \hat{f}_s \tag{B-23}$$

whereas the probability of s successes in any order during r trials is

$$\binom{r}{s}P^s(1 - P)^{r-s} = \hat{f}_{r,s} \tag{B-24}$$

The last result is the well-known binomial distribution. Its mean is $\bar{s} = rP$ and its variance is $\text{var } s = rP(1 - P)$. The binomial coefficient represents the number of ways that r numbers 1 to r can be combined in groups of s, not distinguishing order.

To illustrate the binomial distribution, we again consider random drawings of cards from a well-mixed poker deck, this time with replacement. The probability of drawing an ace in any suit 3 times in 4 attempts is, from combinatorial analysis, $(4/52)(4/52)(4/52)(48/52) + (4/52)(4/52)(48/52)(4/52) + (4/52)(48/52)(4/52)(4/52) + (48/52)(4/52)(4/52)(4/52)$ which is equal to $(4!/3!1!)(4/52)^3(48/52)^1$ from equation B-24.

SOME SPECIFIC DISTRIBUTIONS

Some distributions encountered in the text have been listed in Table B-1. The Poisson, normal (Gaussian) and log normal distributions are compared in Fig. B-2 for equal values of $\bar{s} = 1000$.

Certain useful interrelationships exist among the distributions tabulated. The negative binomial distribution is a special case of the binomial distribution when the last trial is required to be either a success (1) or a failure; in our case, a failure. Thus only $r - 1$ trials are permutable. The most probable distribution is a further special

Table B-1.   Various Distributions

NAME	VARIABLES AND PARAMETERS	DISTRIBUTION
BINOMIAL	$0 \leq s \leq r,\ P < 1$	$\hat{f}_{r,s} = \binom{r}{s} P^s (1-P)^{r-s}$
NEGATIVE BINOMIAL	$0 \leq s \leq r,\ P < 1$	$\hat{f}_{r,s} = \binom{r-1}{s} P^s (1-P)^{r-s}$
MOST PROBABLE	$s \geq 0,\ P < 1$	$\hat{f}_s = P^s (1-P)$
OR	$s \geq 1,\ P < 1$	$\hat{f}_s = P^{s-1} (1-P)$
POISSON	$s \geq 0$	$\hat{f}_s = (t^s/s!) \exp(-t)$
OR	$s \geq 1$	$\hat{f}_s = [t^{s-1}/(s-1)!] \exp(-t)$
TANKS-IN-SERIES (TIS)	$t \geq 0$	$\hat{f}(t) = [\beta^{-1} (t/\beta)^{s-1}/(s-1)!] \exp(-t/\beta)$
EXPONENTIAL	$t \geq 0$	$\hat{f}(t) = \beta^{-1} \exp(-t/\beta)$
GAUSSIAN (NORMAL)	$-\infty \leq s \leq \infty$	$\hat{f}(s) = (\beta/\pi)^{1/2} \exp[-\beta(s-t)^2]$
LOG NORMAL	$0 \leq s \leq \infty$	$\hat{f}(s) = (\beta/\pi)^{1/2}/s] \exp[-\beta(\ln s - \ln t)^2]$

$$\text{WHERE} \quad \beta = \pi \left(\frac{dF(\ln s)}{d\ln s}\right)^2_{s=t} \quad \text{AND} \quad t \equiv s_M$$

NAME	VARIABLES AND PARAMETERS	DISTRIBUTION
RAMP	$1 \leq s \leq a$	$f_s = cs$
RECTANGULAR	$1 \leq s \leq a$	$f_s = c$
TRIANGULAR	$1 \leq s \leq a$	$f_s = b - cs$
PARABOLIC	$1 \leq s \leq \infty$	$f_s = cs^2$
QUADRATIC	$1 \leq s \leq a$	$f_s = s - s^2/a$

case, that is, when the last trial is required to be a failure and, additionally, is the only failure allowed ($s = r - 1$). In this case, no permutations are required. The reader will note that we have included alternative forms of the most probable and Poisson distributions. It is sometimes convenient to define the distributed variable s in the interval starting at 0 instead of 1. This is frequently done when $s = 1$ represents the first propagation (success) but the second repeat unit. In this way, $s = 0$ represents no successes but a DP of 1.

The reader will also note the similarity between the alternative form of the Poisson distribution, the so-called tanks-in-series (TIS) distribution (for lack of a better name) and the exponential distribution. In fact, the TIS distribution reduces to the latter when $s = 1$. These three may be written in unified form by defining the

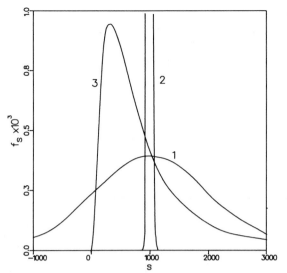

Figure B-2. Comparison of three distributions f_s with identical mean values $\bar{s}$ and the following characteristics, including dispersion index, $D = \mu^2 \mu^0 / (\mu^1)^2$ (Section 1.5): (1) normal with $\beta = 5 \times 10^{-7}$, $t = 1000$, and $D = 2$; (2) Poisson with $t = 1000$ and $D = 1.001$; (3) log normal with $\beta = 0.7215$, $t = 707.1$, and $D = 2$.

following bivariate distribution:

$$\hat{f}_s(\hat{t}) = \frac{\hat{t}^{s-1}}{(s-1)!} \exp(-\hat{t}) \tag{B-25}$$

Thus we have the alternative form of the Poisson distribution when s is the distributed variable and the TIS distribution when $\hat{t} = t/\beta$ is the distributed variable.

The Poisson and Gaussian distributions may be obtained from the binomial distribution as approximations under different conditions. These conditions can be deduced starting with the normalized binomial distribution

$$\hat{f}_{r,s} = \frac{r!}{s!(r-s)!} P^s (1-P)^{r-s} \tag{B-26}$$

and making use of the following approximations for small ε:

$$(1-\varepsilon)^N \approx \exp(-\varepsilon N) \tag{B-27}$$

when $\varepsilon < 0.1$,

$$\frac{N!}{(N-\varepsilon)!} \approx N^\varepsilon \tag{B-28}$$

when $\varepsilon \ll N$, and Stirling's approximation

$$N! \approx (2\pi N)^{1/2} N^N \exp(-N) \tag{B-29}$$

when $N \geqslant 10$. Thus it can be shown that by allowing $r \to \infty$ and $P \to 0$ simultaneously while the product rP remains fixed,

$$\hat{f}_{r,s} \to \frac{(rP)^s}{s!} \exp(-rP) \tag{B-30}$$

which is the Poisson distribution (Table B-1). Its mean and variance are equal, $\bar{s} = t = \text{var } s$ where $t = rP$. This approximation is valid when $P \leqslant 0.05$ and $r \geqslant 100$. Similarly, by allowing r to become large without requiring that P be very small, we can show that

$$\hat{f}_{r,s} \to \left[\frac{1}{2\pi rp(1-P)}\right]^{1/2} \exp\left[\frac{-(s-rP)^2}{2rP(1-P)}\right] \tag{B-31}$$

which is the Gaussian distribution (Table B-1). Its mean is $\bar{s} = t$ and its variance is $\text{var } s = 1/2\beta$ where $\beta = 1/rP(1-P)$ and $t = rP$. The Gaussian approximation is valid for $P \leqslant 0.5$ when $rP > 5$, and for $P > 0.5$ when $r(1-P) > 5$.

Finally, we can approximate the most probable distribution very accurately with the exponential distribution for large values of β, that is, when P is close to 1. This can easily be proved by setting $s = t + 1$ and $P = 1 - \beta^{-1}$ in the most probable distribution and applying approximate equation B-27.

REFERENCES

1. W. Feller, *An Introduction to Probability Theory and Its Applications*, 2nd ed., Wiley, New York (1960), Volume 1.

APPENDIX C

Mathematical Methods

Some definitions, integrals, miscellaneous relations, and summations used in the text are listed in Tables C-1, C-2, and C-3. The material was organized to emphasize the analogy between certain operations with continuous functions, such as differentiating and integrating, and the equivalent operations with discrete functions, that is, finite-differencing and summing.

The solution of the inhomogeneous, linear first-order difference equation with variable coefficient

$$f_{s+1} - a_s f_s = b_s \tag{C-1}$$

is (1)

$$f_s = \left(\prod_{j=0}^{s-1} a_j \right) \Delta^{-1} \frac{b_s}{\prod_{j=0}^{s} a_j} \tag{C-2}$$

where Δ^{-1} represents indefinite summation (Table C-1). The homogeneous form with constant coefficient

$$f_{s+1} - af_s = 0 \tag{C-3}$$

frequently arises throughout the text. Its solution

$$f_s = f_0 a^s \tag{C-4}$$

where $f_0 = c$ follows directly from Equation C-2 and Table C-1.

Table C-1.   Some Definitions

Functions

Continuous	$f(s)$	s = real variable
Discrete	f_s	s = discrete variable

Operators

Differential	$Df = \dfrac{d}{ds} f(s)$	
Indefinite integral	$D^{-1}f = \int f(s)ds + c$	c = arbitrary constant
Difference	$\Delta f = f_{s+h} - f_s$	
Indefinite sum	$\Delta^{-1}f = \Sigma f_s + c$	c = arbitrary constant

Special Functions

Beta $r,s > 0$ $\displaystyle B(r,s) \equiv \int_0^1 \zeta^{r-1}(1-\zeta)^{s-1}d\zeta$

Gamma $s > 0$ $\displaystyle \Gamma(s) \equiv \int_0^\infty \zeta^{s-1} \exp(-\zeta)d\zeta$

Delta (Dirac impulse) $\delta(s) \equiv \begin{cases} 0 & \text{when } s < 0 \text{ and } s > \varepsilon \\ 1/\varepsilon & \text{when } 0 \leq s \leq \varepsilon \end{cases}$

such that $\displaystyle \lim_{\varepsilon \to 0} \int_0^\varepsilon \delta(s)ds = 1$

Unit (step) $u(s) \equiv \begin{cases} 0 & \text{when } s < 0 \\ 1 & \text{when } s > 0 \end{cases}$

Exponential integral (tabulated) $\displaystyle E_s(r) \equiv \int_1^\infty \zeta^{-s} e^{-\zeta r}d\zeta$ s = positive integer
r = positive real number

Error function (tabulated) $\displaystyle \text{erf } s \equiv (2/\pi^{1/2}) \int_0^s \exp(-\zeta^2)d\zeta$

Convolution integral $\displaystyle I(s) \equiv \int_0^s f(\zeta)g(s-\zeta)d\zeta$

Convolution sum $\displaystyle S_s \equiv \sum_{j=0}^s f_j g_{s-j}$

Equation C-1 is analogous to the inhomogeneous, linear first-order differential equation with variable coefficient

$$\frac{df}{ds} - a(s)f = b(s) \tag{C-5}$$

whose solution by the method of integrating factor is

$$f(s) = \left(\exp D^{-1}a\right) D^{-1}\left(\frac{b}{\exp D^{-1}a}\right) \tag{C-6}$$

where D^{-1} represents indefinite integration (Table C-1).

Table C-1. (*Continued*)

Bilateral (two-sided) Laplace transform pair ; z = complex variable

transform $\qquad F(z) = \int_{-\infty}^{\infty} f(s) \exp(-sz)ds$

inverse $\qquad f(s) = -(1/2\pi i)\int_{-i\infty}^{i\infty} F(z) \exp(sz)dz \equiv \mathcal{L}^{-1} F(z)$

one-sided $\mathcal{L}f(s) = \int_{0}^{\infty} f(s) \exp(-sz)ds \equiv F'(z)$
transform

where $\qquad i = (-1)^{1/2}$

Bilateral discrete transform (Laurent series about origin); z = complex

variable

transform $\qquad F(z) = \sum_{s=-\infty}^{\infty} f_s z^s = \sum_{s=-\infty}^{\infty} f'_s z^s + \sum_{s=1}^{\infty} f''_s z^{-s} \equiv F'(z) + F''(z)$

$\qquad\qquad\qquad\qquad\qquad\qquad$ generating $\qquad$ z-transform
$\qquad\qquad\qquad\qquad\qquad\qquad$ function $\qquad$ (zero if no singularity
$\qquad\qquad\qquad\qquad\qquad\qquad\qquad\qquad\qquad\qquad$ at origin)

inverse $\qquad f_s = (1/2\pi i) \oint F(z)z^{-(s+1)}dz ; \quad \oint = $ contour
$\qquad\qquad\qquad\qquad$ annular $\qquad\qquad\qquad\qquad\qquad\qquad$ integration
$\qquad\qquad\qquad\qquad$ region

$\qquad\qquad f'_s = (1/2\pi i) \oint F(z)z^{-(s+1)}dz = (1/s!)\left[\dfrac{d^s}{dz^s} F(z)\right]_{z=0}$
$\qquad\qquad\qquad\qquad\qquad$ outer
$\qquad\qquad\qquad\qquad\qquad$ circle

$\qquad\qquad f''_s = (1/2\pi i) \int F(z)z^{-(s+1)}dz$
$\qquad\qquad\qquad\qquad\qquad$ inner
$\qquad\qquad\qquad\qquad\qquad$ circle

METHOD OF MOMENTS

Rate equations for polymer products generally take the form of a differential equation, for example,

$$x \geqslant 1 \qquad\qquad\qquad \frac{dc_x}{dt} = r_x \qquad\qquad\qquad\qquad \text{(C-7)}$$

or an integral such as (Section 1.3 and Chapter 3)

$$x \geqslant 1 \qquad\qquad\qquad c_x = \int r_x(t)\, dt \qquad\qquad\qquad\qquad \text{(C-8)}$$

Table C-2. **Some Useful Relations**

Continuous

Beta function

$$B(r,s) = \Gamma(r)\Gamma(s)/\Gamma(r + s)$$

Gamma function

$$\Gamma(s + 1) = s! \quad \text{if } s \text{ is a positive integer}$$

Error function

$$\lim_{s \to \infty} \text{erf } s = \pi^{1/2}/2$$

$$\int_{-\infty}^{\infty} \exp(-\zeta^2)d\zeta = \pi^{1/2}$$

Leibnitz's rule for differentiating integrals

$$\frac{d}{ds}\int_{a(s)}^{b(s)} f(s,\zeta)d\zeta = \int_{a(s)}^{b(s)} \frac{\partial f}{\partial s} d\zeta + f(s,b)\frac{db}{ds} - f(s,a)\frac{da}{ds}$$

Transformed (Laplace) convolution integral

$$\mathcal{L}I(s) \equiv I\check{}(z) = \mathcal{L}f(s)\int g(s) \equiv F\check{}(z)G\check{}(z)$$

Miscellaneous

$$\int_{0}^{1} \zeta^N \exp(-a\zeta)d\zeta = (N!/a^{N+1})\left[\sum_{N+1}^{\infty} a^j/j!\right]$$

$$\int_{a}^{s_N}\cdots\int_{a}^{s_1} f(s)ds \cdots ds_{N-1} = (1/(N-1)!)\int_{a}^{s} f(\zeta)(s-\zeta)^{N-1}d\zeta$$

Discrete

Transformed convolution sum

$$S\check{}(z) = F\check{}(z)G\check{}(z)$$

These equations are often intractable owing to the complexity of rate function r_x and the large number ($x = \text{DP}$) of equations involved.

Since polymer properties are generally characterized in practice in terms of only a couple of average properties anyway (Section 1.5), it is convenient to transform the preceding equations into the two or three appropriate moment equations from which these average properties may be deduced. The averages referred to are $\overline{M}_N$ and $\overline{M}_W$ and the moments are μ^0, μ^1, and μ^2. The transformation is called the method of moments.

Table C-3. Some Summations

General form

$$F(P) = \sum_{s=s_o}^{s_f} f_s(P)$$

where P = parameter and s_o, s_f are lower and upper limits

Geometric series $f_s = s^a P^s$

$a = o$ $$F(P) = \left(P^{s_o} - P^{s_f+1} \right) \bigg/ (1 - P) \equiv F_o(P)$$

$a = 1$ $$F(P) = P \frac{d}{dP} F_o(P) = \left(P^{s_o} - P^{s_f+1} \right) \bigg/ (1 - P)^2$$

Infinite series

Form $f_s = s^a P^s$ where $P < 1$, $s_o = 1$ and $s_f = \infty$

a	-1	0	1	2	3
F(P)	$\ln(1-P)^{-1}$	$P/(1-P)$	$P/(1-P)^2$	$P(1+P)/(1-P)^3$	$P(P^2+4P+1)/(1-P)^4$

Form $f_s = s^a P^s/s!$ where $P > 0$, $s_o = 0$ and $s_f = \infty$

a	0	1	2	3
F(P)	$\exp P$	$P \exp P$	$P(P+1)\exp P$	$P(P^2+3P+1)\exp P$

Form $f_s = (-1)^s P^s/s!$ where $P > 0$, $s_o = 0$ and $s_f = \infty$

$$F(P) = \exp(-P)$$

Form $f_s = s^a g_s$ where $g_s \sum_{j=s}^{\infty} P^j/j!$, $P > 0$, $s_o = 1$ and $s_f = \infty$

a	0	1	2
F(P)	$P \exp P$	$[P(P+2)/2]\exp P$	$[P(2P^2+9P+6)/6]\exp P$

Table C-3. (*Continued*)

Form $f_s = x^a g_s q^s$ where $g_s = \sum_{j=s}^{\infty} P^j/j!$, $P,q > 0$, $s_o = 1$ and $s_f = \infty$

a	0	1	2
F(P)	$[q/(1-q)][\exp P - \exp Pq]$	$[q/(1-q)^2][\exp P - (1+Pq-Pq^2)\exp Pq]$	↗

$$[q/(1-q)^3]\left\{ (1+q)\exp P - [(1+q) + Pq(1-q)^2(2+Pq) + Pq(1-q^2)]\exp Pq\right\}$$

Moments of the Convolution Sum

$$\sum_{x=2}^{\infty}\sum_{j=1}^{x-1} f_{x-j}f_j = (\mu^0)^2$$

$$\sum_{x=2}^{\infty}x\sum_{j=1}^{x-1} f_{x-j}f_j = 2\mu^1\mu^0$$

$$\sum_{x=2}^{\infty}x^2\sum_{j=1}^{x-1} f_{x-j}f_j = 2(\mu^0\mu^2 + (\mu^1)^2)$$

$$\sum_{x=2}^{\infty}x^3\sum_{j=1}^{x-1} f_{x-j}f_j = 2(\mu^0\mu^3 + 3\mu^1\mu^2)$$

Moments of the Sum $x + 1 \to \infty$

$$\sum_{x=1}^{\infty}\sum_{j=x+1}^{\infty} f_j = \sum_{x=1}^{\infty}(x-1)f_x = \mu^1 - \mu^0$$

$$\sum_{x=1}^{\infty}x\sum_{j=x+1}^{\infty} f_j = \sum_{x=1}^{\infty}(\frac{x^2-x}{2})f_x = \frac{1}{2}(\mu^2 - \mu^1)$$

$$\sum_{x=1}^{\infty}x^2\sum_{j=x+1}^{\infty} f_j = \sum_{x=1}^{\infty}\left(\frac{2x^3 - 3x^2 + x}{6}\right)f_x = \frac{1}{3}\mu^3 - \frac{1}{2}\mu^2 + \frac{1}{6}\mu^1$$

$$\sum_{x=1}^{\infty}x^3\sum_{j=x+1}^{\infty} f_j = \sum_{x=1}^{\infty}\left(\frac{x^4 - 2x + x^2}{4}\right)f_x = \frac{1}{4}\mu^4 - \frac{1}{2}\mu^3 + \frac{1}{4}\mu^2$$

In general, the method is applied as follows. First we define moments with concentration units (Appendix B)

$$\mu_c^k \equiv \sum_x x^k c_x \tag{C-9}$$

Then we differentiate both sides of this equation, exchange the order of summation and differentiation operations on the RHS, and substitute rate equation C-7. The results are usually two or three coupled differential equations for moments which are manageable, such as

$$k = 0, 1, 2 \qquad \frac{d\mu_c^k}{dt} = f_k\left(\mu_c^0, \mu_c^1, \mu_c^2\right) \qquad \text{(C-10)}$$

Alternatively, we can substitute rate equation C-8 into equation C-9 and exchange the order of summation and integration operations. For details, the reader is referred to Section 3.5.

TRANSFORM METHODS

Two one-sided transforms, which are discrete counterparts of the Laplace transform, are the generating function and z-transform. They are seen in Table C-1 to be the Taylor-series component and residue, respectively, of the Laurent-series expansion of a complex function. The generating function was described in 1956 by Scanlan (2), and later by Liu and Amundson (3), and the z-transform by Abraham (4) and Kilkson (5).

As a simple illustration of the generating function method, we shall apply it to difference equation C-3. After substituting equation C-3 into the definition (Table C-1)

$$F(z) \equiv \sum_{s=0}^{\infty} f_s z^s \qquad \text{(C-11)}$$

where the primes have been dropped for convenience, we obtain the following equation in terms of initial condition f_0

$$F - azF = f_0 \qquad \text{(C-12)}$$

Next, we expand the solution of this equation

$$F(z) = \frac{f_0}{1 - az} \qquad \text{(C-13)}$$

in an infinite series and compare the result

$$\frac{f_0}{1 - az} = f_0 \sum_{s=0}^{\infty} (az)^s \qquad \text{(C-14)}$$

with definition C-11, thereby deducing that

$$f_s = f_0 a^s \qquad \text{(C-15)}$$

which is in agreement with equation C-4.

THE LEGENDRE TRANSFORMATION

Consider the function of independent variables $x_1, x_2 \ldots, y_1, y_2 \ldots$

$$L = L(x_c, y_c) \tag{C-16}$$

with the corresponding differential form,

$$dL = u_c^T \, dx_c + v_c^T \, dy_c \tag{C-17}$$

where x_c and y_c are column matrices, u_c^T and v_c^T are the transpose (row) matrices of column matrices u_c and v_c, respectively, and where

$$u_j \equiv \frac{\partial L}{\partial x_j} \tag{C-18}$$

$$v_j \equiv \frac{\partial L}{\partial y_j} \tag{C-19}$$

If we transform the independent variables y_c to v_c, then the resulting function

$$H = H(x_c, v_c) \tag{C-20}$$

is related to the original one (L) by a Legendre transformation $(6, 7)$

$$H = L - v_c^T y_c \tag{C-21}$$

and its corresponding differential form is

$$dH = u_c^T \, dx_c - y_c^T \, dv_c \tag{C-22}$$

The quantities y_c and u_c are now related to variables x_c and v_c by the relations

$$y_j = -\frac{\partial H}{\partial v_j} \tag{C-23}$$

and

$$u_j = \frac{\partial H}{\partial x_j} = \frac{\partial L}{\partial x_j} \tag{C-24}$$

HAMILTON'S PRINCIPLE

The Euler–Lagrange equations of motion in classical mechanics

$$\frac{d}{dt}\left(\frac{\partial L}{\partial y_j}\right) - \frac{\partial L}{\partial x_j} = 0 \tag{C-25}$$

can be derived via the calculus of variations from the extremum $\delta I = 0$ of the line integral in time

$$I = \int_{t_1}^{t_2} L \, dt \tag{C-26}$$

where L is the so-called Lagrangian function

$$L = L(x_c, y_c) \tag{C-27}$$

Quantities x_j represent generalized position coordinates (and time) and y_j represent generalized velocities. Frequently, the equations of motion can be solved more readily if they are transformed into the so-called Hamiltonian form via the Legendre transformation (7). Thus the generalized velocities y_c are replaced by generalized momenta v_c

$$v_j = \frac{\partial L}{\partial y_j} \tag{C-28}$$

The resulting Hamiltonian function is (cf. equation C-21)

$$H = L - \sum_j v_j y_j \tag{C-29}$$

and the Euler–Lagrange equations (C-25) are thereby replaced by Hamilton's canonical equations (C-23–C-24).

PONTRYAGIN'S PRINCIPLE

This is a convenient variational principle for finding the matrix of parametric control functions $w_c(t)$ that will maximize or minimize a functional of the form

$$I(t_f) = \int_0^{t_f} F(x_c, w_c, t) \, dt \tag{C-30}$$

where x_j are dynamic state variables of a system described by a set of differential equations

$$\frac{dx_j}{dt} = f_j(x_c, w_c, t) \tag{C-31}$$

Pontryagin's principle states (8) that expression C-30 will be maximized (minimized) if $w_c(t)$ is adjusted at all times to maintain the following Hamiltonian function stationary ($\delta H - 0$)

$$H \equiv F + \sum_j z_j f_j \tag{C-32}$$

where z_j are adjoint variables (Lagrange multipliers). The associated canonical equations are:

$$f_j = \frac{\partial H}{\partial z_j} = \frac{dx_j}{dt} \tag{C-33}$$

$$\frac{dz_j}{dt} = -\frac{\partial H}{\partial x_j} \tag{C-34}$$

If F in equation C-30 is the Lagrangian L and the control functions are taken to be velocity variables $w_j = dx_j/dt \equiv y_j$, it can be shown (8) that the adjoint variables are generalized momenta $z_j = -v_j$, and equation C-32 is identical to the Hamiltonian of classical mechanics, equation C-29.

More generally, the optimal policy $w_c(t)$ is determined by solving the set of differential equations C-33 and C-34, subject to boundary conditions on variables x_c and z_c, together with certain criteria in terms of w_c to maintain the Hamiltonian constant and equal to zero. Usually the inital state of the system is known, $x_c(0) = (x_c)_0$, in which case the initial values of the adjoint variables are unspecified, $z_j(0) = $ free. If the final values of the state variables are specified, $x_c(t_f) = (x_c)_f$, then the adjoint variables are not, $z_j(t_f) = $ free. However, if they are unspecified, then the final values of the adjoint variables may be set equal to zero, $z_j(t_f) = 0$.

An unconstrained control policy is one in which the control variables are allowed to vary between lower and upper limits:

$$(w_j)_S \leqslant w_j \leqslant (w_j)_L \tag{C-35}$$

The aforementioned optimality criteria are then

$$\frac{\partial H}{\partial w_j} = 0 \tag{C-36}$$

together with

$$\frac{\partial^2 H}{\partial w_i \partial w_j} < 0 \tag{C-37}$$

for $I(t_f)$ to be maximized and

$$\frac{\partial^2 H}{\partial w_i \partial w_j} > 0 \tag{C-38}$$

for $I(t_f)$ to be minimized. In the event that $\dfrac{\partial H}{\partial w_j} \neq 0$, we take the optimum policy to be either one of the two limiting values of w_j as follows:

$$w_j = (w_j)_S \qquad \text{if } \frac{\partial H}{\partial w_j} > 0 \tag{C-39}$$

and

$$w_j = (w_j)_L \qquad \text{if } \frac{\partial H}{\partial w_j} < 0 \tag{C-40}$$

THE ROUTH–HURWITZ STABILITY CRITERIA

Given a system of N independent variables whose dynamic behavior is described by N ordinary differential equations, we can express the systems response in matrix form as:

$$\frac{d(x_c)}{dt} = f(x_c) \tag{C-41}$$

which is equivalent to

$$j = 1, \text{N} \qquad \frac{dx_j}{dt} = f_j(x_1, x_2 \cdots x_N) \tag{C-42}$$

In the vicinity of any solution to the steady-state problem $(x_c)_s$ defined by

$$0 = f_c\big[(x_c)_s\big] \tag{C-43}$$

the system response can be approximated by linearizing f_c via expansion in a Taylor series about the steady-state solution, neglecting the nonlinear terms. This yields

$$\frac{d}{dt}\big[x_c - (x_c)_s\big] = F_s\big[x_c - (x_c)_s\big]$$

where $F_s = F_s[(x_c)_s]$ represents the Jacobian matrix of the following system equations evaluated at the steady state:

$$F_{jk} = \left. \frac{\partial f_k}{\partial x_j} \right|_{x_c = (x_c)_s} \tag{C-45}$$

The system response can then be described in terms of arbitrary constants c_c as

$$x_c - (x_c)_s = c_c \exp(F_s t) \tag{C-46}$$

Thus x_c will only return to the solution $(x_c)_s$ as t increases if the eigenvalues of F_s have negative real parts. The eigenvalues z_j are found from the characteristic equation:

$$\text{Det}(F_s - zI_s) = 0 \tag{C-47}$$

where I_s, is the identity matrix:

$$I_{jk} = 0 \text{ when } j \neq k; \qquad I_{jk} = 1 \text{ when } j = k \tag{C-48}$$

The solution of the preceding equation yields a polynomial in z, with N roots:

$$0 = z^N + a_1 z^{N-1} + a_2 z^{N-2} \cdots + a_{N-1} z + a_N \tag{C-49}$$

The Routh–Hurwitz criteria are a means for determining the sign of the real parts of the N solutions, z_j, to equation C-49. In general, they require that each of N determinants composed of the coefficients a_j be positive. The N criteria, C_N, are given by:

$$C_1 = a_1 > 0 \tag{C-50}$$

$$C_2 = \begin{vmatrix} a_1 & a_3 \\ 1 & a_2 \end{vmatrix} > 0 \tag{C-51}$$

$$C_3 = \begin{vmatrix} a_1 & a_3 & a_5 \\ 1 & a_2 & a_4 \\ 0 & a_1 & a_5 \end{vmatrix} > 0 \tag{C-52}$$

$$C_4 = \begin{vmatrix} a_1 & a_3 & a_5 & \cdots & 0 \\ 1 & a_2 & a_4 & \cdots & 0 \\ 0 & a_1 & a_3 & \cdots & 0 \\ 0 & 1 & a_2 & \cdots & \\ \vdots & \vdots & \vdots & & \vdots \\ & & & & a_N \end{vmatrix} > 0 \tag{C-53}$$

The coefficients a_j have a direct relationship to the invariants of the matrix F_s. In fact, they are given by:

$$a_j = (-1)^j \mathcal{I}_j \tag{C-54}$$

The form of the invariants defined by C-54 is not unique (9). An alternative set $\mathcal{I}'$ is more simply defined by

$$\mathcal{I}'_N = \text{trace}\left(F_s^N\right) \tag{C-55}$$

These can then be used to compute the other form of the invariants

$$\mathcal{I}_1 = \mathcal{I}'_1 \tag{C-56}$$

$$\mathcal{I}_2 = \frac{1}{2}\left[(\mathcal{I}'_1)^2 - \mathcal{I}'_2\right] \tag{C-57}$$

$$\mathcal{I}_3 = \frac{1}{6}\left[(\mathcal{I}'_1)^3 - 3\mathcal{I}'_1 \mathcal{I}'_2 + 2\mathcal{I}'_3\right] \tag{C-58}$$

Furthermore, it should be clear that $\mathcal{I}'_1 = \mathcal{I}_1$ is always the trace of F_s which, via equations C-50 and C-54, dictates that stability criterion C_1 will always be

$$(-1)\,\text{trace}\left(F_s\right) > 0 \tag{C-50}$$

or

$$\text{trace}\,(F_s) < 0 \qquad\qquad \text{(C-60)}$$

Although not readily apparent by inspection, the Nth criterion will always reduce to:

$$(-1)^N \mathcal{I}_N > 0 \qquad\qquad \text{(C-61)}$$

This is illustrated for a third-order system in Example C-1. In addition, it can be shown that:

$$\mathcal{I}_N = \text{Det}\,(F_s), \qquad\qquad \text{(C-62)}$$

Thus, for even-order systems, $N = 2, 4, 6\ldots$, stability requires:

$$\text{Det}(F_s) > 0 \qquad\qquad \text{(C-63)}$$

while for odd-order systems $N = 3, 5, 7\ldots$,

$$\text{Det}(F_s) < 0 \qquad\qquad \text{(C-64)}$$

Example C-1. Show that for a third-order system, stability criterion C_3 reduces to:

$$\mathcal{I}_3 < 0$$

Solution. For a third-order system, the eigenvalues are the roots of the third degree polynomial:

$$0 = z^3 - \mathcal{I}_1 z^2 + \mathcal{I}_2 z = -\mathcal{I}_3$$

Thus the three stability criteria are:

$$C_1 = -\mathcal{I}_1 > 0$$

$$C_2 = \begin{vmatrix} \mathcal{I}_1 & -\mathcal{I}_3 \\ 1 & \mathcal{I}_2 \end{vmatrix} = \mathcal{I}_1 \mathcal{I}_2 + \mathcal{I}_3 > 0$$

$$C_3 = \begin{vmatrix} \mathcal{I}_1 & -\mathcal{I}_3 & 0 \\ 1 & \mathcal{I}_2 & 0 \\ 0 & \mathcal{I}_1 & -\mathcal{I}_3 \end{vmatrix} = -\mathcal{I}_3(\mathcal{I}_1 \mathcal{I}_2 + \mathcal{I}_3) > 0$$

But satisfying C_2 reduces C_3 to the simple restriction

$$-\mathcal{I}_3 > 0$$

or

$$\mathcal{I}_3 < 0$$

REFERENCES

1. G. Boole, *A Treatise on the Calculus of Finite Differences*, 2nd ed., Dover, New York (1960).
2. J. Scanlan, *Trans. Far. Soc.*, **52**, 1286 (1956).
3. S-L Liu and N. R. Amundson, *Chem. Eng. Sci.*, **17**, 797 (1962).
4. W. H. Abraham, *Indian Eng. Chem. Fund.*, **2**, 221 (1963).
5. H. Kilkson, *Indian Eng. Chem. Fund.*, **3**, 281 (1964).
6. H. B. Callen, *Thermodynamics*, Wiley, New Yrok (1963).
7. H. Goldstein, *Classical Mechanics*, Addison-Wesley, Reading, MA (1965).
8. M. M. Denn, *Optimization by Variational Methods*, McGraw-Hill, New York (1969).
9. R. B. Bird, R. C. Armstrong, and O. Hassager, *Dynamics of Polymer Liquids*, Wiley & Sons, New York (1977) Volume 1, A-14.

APPENDIX D

Thermodynamics

GENERAL

The fundamental equation (1) of thermodynamics contains all information about a system pertaining to its equilibrium properties. In its Lagrangian form (Appendix C),

$$L = L(Q_c) \tag{D-1}$$

or equivalently

$$l = l(q_c) \tag{D-2}$$

The natural variables are all extensive properties Q_c or specific extensive properties q_c of the system. Examples of fundamental equation D-1 for fluid mixtures are the energy form

$$U = U(S, V, N_c) \tag{D-3}$$

and the entropy form

$$S = S(U, V, N_c) \tag{D-4}$$

Because the L's are homogeneous functions, they may be written in intensive form. Corresponding to equation D-2, we have

$$u = u(s, v, w_c) \tag{D-5}$$

and

$$s = s(u, v, w_c) \tag{D-6}$$

where in lieu of mole fractions we have used mass fractions w_c, which must satisfy the constraint $\Sigma w_k = 1$. Thus the q's are all mass-specific intensive quantities. For solids we have

$$u = u(s, \varepsilon) \tag{D-7}$$

where ε is the strain tensor.

Energy minimum and entropy maximum principles in thermodynamics make use of variations.

$$\delta L = dL + \frac{1}{2}d^2L + \cdots \tag{D-8}$$

where L represents a thermodynamic potential function, such as U, or the entropy function S,

$$dL = \sum_i \frac{\partial L}{\partial Qi} dQ_i = P_c^T dQ_c \tag{D-9}$$

and

$$d^2L = \sum_i \sum_j \frac{\partial^2 L}{\partial Q_i \partial Q_j} dQ_i \, dQ_j = dQ_c^T M_s \, dQ_c \tag{D-10}$$

Equation D-9 defines intensive properties P_c, which are first derivatives of L

$$\frac{\partial L}{\partial Q_i} \equiv P_i = \frac{\partial l}{\partial q_i} \tag{D-11}$$

and functions of Q_c.

$$P_i = P_i(Q_c) \tag{D-12}$$

Such derivatives in general relate a thermodynamic potential to corresponding property pairs P_i, Q_i, or P_i, q_i, called conjugate properties. Equations D-12 are called equations of state, which together are equivalent to the fundamental equation and contain all its thermodynamic information. Equation D-10 defines the second derivative properties of L

$$\frac{\partial^2 L}{\partial Q_i \partial Q_j} \equiv M_{ij} = \frac{\partial P_i}{\partial Q_j} \tag{D-13}$$

They are sometimes called moduli and comprise the components of a symmetric matrix M_s, known as the stiffness matrix, which relates variations in properties Q to variations in P_c

$$dP_c = M_s \, dQ_c \tag{D-14}$$

The well-known Maxwell relations

$$\frac{\partial P_i}{\partial Q_j} = \frac{\partial P_j}{\partial Q_i} \qquad \text{(D-15)}$$

follow immediately from the symmetry of M_s.

The inverse of equation D-14

$$dQ_c = C_s \, dP_c \qquad \text{(D-16)}$$

defines the symmetric matrix C_s, whose components are known as capacity properties

$$C_{ij} = \frac{\partial Q_i}{\partial P_j} \qquad \text{(D-17)}$$

or

$$c_{ij} = \frac{\partial q_i}{\partial P_j} \qquad \text{(D-18)}$$

where C_{ij} are extensive and c_{ij} (specific capacities) are intensive. These properties are in turn the second derivatives of a function $H(P)$

$$C_{ij} = \frac{\partial^2 H}{\partial P_i \, \partial P_j} \qquad \text{(D-19)}$$

which represents the Hamiltonian form (Appendix C) of the fundamental equation with intensive properties P_c as the natural variables and is thermodynamically equivalent to equation D-1.

It is generally convenient to partition the property matrix Q_c

$$L(Q_c) = L(Q_{c1}, Q_{c2}) \qquad \text{(D-20)}$$

and transform L only partially ($Q_{c2} \rightarrow P_{c2}$), thus defining thermodynamic potential functions as Hamiltonian forms of the fundamental equation with mixed variables $H(Q_{c1}, P_{c2})$ where

$$P_{2j} = \frac{\partial L}{\partial Q_{2j}} \qquad \text{(D-21)}$$

These potential functions are related to those of form D-1 by Legendre transformations (Appendix C)

$$H = L - P_{c2}^T Q_{c2} \qquad \text{(D-22)}$$

and have corresponding differential forms

$$dH = \sum_j P_{1j} \, dQ_{1j} - \sum_j Q_{2j} \, dP_{2j} = P_{c1}^T \, dQ_{c1} - Q_{c2}^T \, dP_{c2} \qquad \text{(D-23)}$$

The first derivatives (as in D-11) which relate potential functions to their conjugate property pairs are given by Hamilton's canonical equations

$$\frac{\partial H}{\partial P_{2j}} = -Q_{2j} \tag{D-24}$$

$$\frac{\partial H}{\partial Q_{1j}} = P_{1j} = \frac{\partial L}{\partial Q_{1j}} \tag{D-25}$$

Examples of such potential functions are the Helmholtz

$$A = A(T, V, N_c) \tag{D-26}$$

and Gibbs

$$G = G(T, P, N_c) \tag{D-27}$$

free-energy functions, which are related to U via Legendre transformations,

$$A = U - TS \tag{D-28}$$

and

$$G = U + PV - TS \tag{D-29}$$

respectively. Examples of equations D-24 and D-25 are

$$\left(\frac{\partial G}{\partial T}\right)_{P, N_c} = -S \tag{D-30}$$

and

$$\left(\frac{\partial G}{\partial N_k}\right)_{T, P, N_l} \equiv \mu_k = \left(\frac{\partial U}{\partial N_k}\right)_{S, V, N_l} \tag{D-31}$$

respectively, where the latter is a so-called Maxwell relation. Both equations D-26 and D-27 have associated equations of state. A well-known example for pure fluids in general is

$$-\left(\frac{\partial A}{\partial V}\right)_T = P = P(T, V) = -\left(\frac{\partial U}{\partial V}\right)_S \tag{D-32}$$

of which the perfect gas law is a special case.

$$P = \frac{NR_gT}{V} \tag{D-33}$$

An example for elastic solids is

$$-\left(\frac{\partial A}{\partial \varepsilon}\right)_T = \tau = \tau(T, \varepsilon) = -\left(\frac{\partial U}{\partial \varepsilon}\right)_S \tag{D-34}$$

of which Hooke's law is a special case

$$\tau = G\varepsilon \qquad (D-35)$$

where G is the modulus of elasticity. Equations D-32 and D-34 show that A and U are potential functions for deformation under isothermal and adiabatic conditions, respectively.

An example of differential equation D-9 is the Gibbs equation,

$$dU = T\,dS - P\,dV + \sum_k \mu_k\,dN_k \qquad (D-36)$$

whose alternative form (equation D-23)

$$dG = -S\,dT + V\,dP + \sum_k \mu_k\,dN_k \qquad (D-37)$$

is seen to be the differential of a Hamiltonian form of the fundamental equation. Property μ_k is the chemical potential. Examples of capacity properties are listed in Table D-1.

CHEMICAL THERMODYNAMICS

The stoichiometric equation for a single chemical reaction among N substances

$$A_1 + A_2 + \cdots = \cdots + A_{N-1} + A_N \qquad (D-38)$$

can be written in abbreviated form as

$$\sum_{k=1}^{N} \nu_k A_k = 0 \qquad (D-39)$$

where the stoichiometric coefficients ν_k for reactants are negative and those for products are positive. For multiple reactions among the N substances, the jth reaction is written as

$$\sum_{k-1}^{N} \nu_{jk} A_k = 0 \qquad (D-40)$$

where ν_{jk} is the stoichiometric coefficient for substance k in reaction j.

Following Prigogine and Defay (2), we define an extent of reaction in units of moles for each independent stoichiometric equation. Thus ξ_j is the molar extent of reaction D-40 and is defined as

$$\xi_j \equiv \frac{(\Delta N_1)_j}{\nu_{j1}} = \frac{(\Delta N_2)_j}{\nu_{j2}} = \cdots = \frac{(\Delta N_k)_j}{\nu_{jk}} = \cdots \qquad (D-41)$$

where $(\Delta N_k)_j$ represents the change in moles of species k due to reaction j. By applying the law of definite proportions (material balance) to a closed system (fixed total mass), we obtain a constraint equation for each substance

$$N_k = (N_k)_0 + \sum_j \nu_{jk}\xi_j \tag{D-42}$$

From the theory of homogeneous functions, it can be shown that for any extensive property Q, there exists an associated intensive partial property

$$Q_k \equiv \left(\frac{\partial Q}{\partial N_k} \right)_{T, P, N_l} \tag{D-43}$$

or alternatively

$$Q_k' \equiv \left(\frac{\partial Q}{\partial m_k} \right)_{T, P, m_l} \tag{D-44}$$

which satisfies an equation of the type

$$Q = \sum_k N_k Q_k \tag{D-45}$$

where N_l signifies that all molar variables except N_k are held constant during differentiation.

The last result can be used together with the Gibbs equation to obtain the Gibbs–Duhem equation

$$0 = -S\,dT + V\,dP - \sum_k N_k\,d\mu_k \tag{D-46}$$

which is the differential of the completely inverted Hamiltonian form $H(P_c)$ of the fundamental equation, D-1. An alternative form of equation D-45 is

$$Q = \sum_k m_k Q_k' \tag{D-47}$$

from which we obtain the following useful result

$$q = \sum_k w_k Q_k' \tag{D-48}$$

Following Prigogine and Defay (2), we can define a change in any property Q due to reaction j at constant T and P in general as

$$(\Delta Q_r)_j \equiv \left(\frac{\partial Q}{\partial \xi_j} \right)_{T, P} \tag{D-49}$$

After inserting equation D-45 and differentiating, we obtain with the aid of D-42 and the identity $\sum_k N_k \dfrac{\partial Q_k}{\partial \xi} = 0$

$$(\Delta Q_r)_j = \sum_k \nu_{jk} Q_k \tag{D-50}$$

Two ubiquitous examples are heat of reaction ΔH_r and chemical affinity, ΔG_r, the driving force for reaction. For a single reaction

$$\Delta H_r = \sum_{k=1}^{N} \nu_k H_k \tag{D-51}$$

$$\Delta G_r = \sum_{k=1}^{N} \nu_k \mu_k \tag{D-52}$$

and for reaction j among multiple reactions

$$\Delta H_j = \sum_{k=1}^{N} \nu_{jk} H_k \tag{D-53}$$

$$\Delta G_j = \sum_{k=1}^{N} \nu_{jk} \mu_k \tag{D-54}$$

It is evident that the Gibbs free energy G is the potential function for chemical reaction and diffusion, and that μ_k is equivalent to the partial Gibbs free energy G_k.

It is customary (2) to express the functional dependence of partial properties Q_k as the sum of two component parts, a standard property Q_k^* and a mixing property Q_k^M

$$Q_k = Q_k^* + Q_k^M \tag{D-55}$$

such that the standard property Q_k^* is always independent of composition. For ideal solutions made up of small (nonpolymeric) molecules,

$$H_k^M = 0 \tag{D-56}$$

and

$$S_k^M = -R_g \ln n_k \tag{D-57}$$

for all k, where S_k^M is the partial entropy of mixing. Thus

$$\mu_k = H_k - TS_k$$
$$= \mu_k^0(T, P) + R_g T \ln n_k \tag{D-58}$$

where $\mu_k^0(T, P)$ is the chemical potential for pure substances k at T and P. For perfect gas mixtures in particular, we obtain

$$\mu_k = \mu_k^+(T) + R_g T \ln P_k \tag{D-59}$$

which is a special case of equation D-58 by virtue of the relationship

$$\mu_k^0(T, P) = \mu_k^+(T) + R_g T \ln P \tag{D-60}$$

For imperfect gas mixtures

$$\mu_k = \mu_k^+(T) + R_g T \ln f_k \tag{D-61}$$

and nonideal liquid solutions

$$\mu_k = \mu_k^*(T, P) + R_g T \ln a_k \tag{D-62}$$

fugacities f_k, activities a_k, and standard states are defined in such a way as to preserve the form of equation D-55 for chemical potential

$$\mu_k = \mu_k^* + \mu_k^M \tag{D-63}$$

Since T and P are natural variables for G, it is evident that equations D-58 and D-62 represent equations of state for multicomponent systems.

The standard state for $\mu^+(T)$ is pure gas k at T, and that for $\mu^*(T, P)$ depends on the convention chosen for composition. Two common examples are:

$$a_k = \begin{cases} \gamma_n n_k & \text{mole fraction} \\ \gamma_c c_k & \text{molar concentration} \end{cases} \tag{D-64}$$

The standard state for solute could be taken to be the pure substance ($n_k = 1$) at T and P in a real or hypothetical ideal state for which $\gamma_n = 1$ and $\mu_k^* = \mu_k^0$, or a solution of k in a real or hypothetical ideal state ($\gamma_c = 1$) for which $c_k = 1$ and $\mu_k^* \neq \mu_k^0$. In any case, this procedure guarantees that equilibrium constants are independent of composition.

In general, the thermodynamic criteria for equilibria are expressed in terms of vanishing gradients of thermodynamic intensive properties as shown in the following chart.

Equilibrium		Criterion
Thermal		$\nabla T = 0$
Mechanical		$\nabla P = 0$
Chemical	Diffusion	$\nabla \mu_k = 0$ for all k
	Reaction	$\Delta G_j = 0$ for all j

For transport processes the gradient ∇ refers to spatial variation of an intensive variable (P), and for chemical reaction the gradient Δ refers to variation of a thermodynamic potential function (G) with respect to a reaction coordinate (ξ_j).

By applying the equations of Table D-2 to vapor-liquid equilibrium (VLE)

$$(\mu_k)_L = (\mu_k)_V \tag{D-65}$$

we obtain

$$K(T, P) = \frac{\exp(\mu_k^* - \mu_k^+)}{R_g T} \tag{D-66}$$

for the distribution coefficient of substance k between two phases

$$K = \frac{f_k}{a_k} \tag{D-67}$$

For reaction equilibrium

$$\Delta G_r = \sum_k \nu_k \mu_k = 0 \tag{D-68}$$

we obtain

$$K(T, P) = \exp\left(-\frac{\Delta G_r^*}{R_g T}\right) \tag{D-69}$$

for the equilibrium constant

$$K \equiv \prod_k a_k^{\nu_k} \tag{D-70}$$

Frequently solutions exhibit ideal behavior over limited regions of composition and nonideal behavior elsewhere (3). Common examples of equation D-67 for such ideal

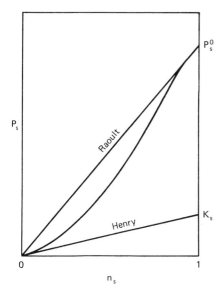

Figure D-1. VLE curve contrasting limiting ideal regions represented by Raoult's law ($n_s \to 1$) and Henry's law ($n_s \to 0$).

regions are Henry's law for solute s

$$P_s = K_s n_s \tag{D-71}$$

and Raoult's law for solvent s, where

$$K_s = P_s^0 \tag{D-72}$$

Both imply that partial pressure of s in the vapor phase P_s varies linearly with mole fraction in the liquid phase n_s at equilibrium. They are illustrated in Fig. D-1. In the nonideal liquid region, we have

$$P_s = K_s \gamma_n n_s \tag{D-73}$$

assuming that the vapor remains ideal. Note that K_s for solute may refer to a hypothetical standard state for pure solute which is obtained by extrapolating Henry's law from infinite dilution ($n_s \to 0$) where the solution is ideal ($\gamma_n = 1$) to $n_s = 1$.

Table D-1. Some Capacity Properties

Mechanical

  Fluid

$$\frac{1}{V}\left(\frac{\partial V}{\partial P}\right)_T \equiv -K_T \qquad = \text{ compressibility}$$

  Solid

$$\left(\frac{\partial \varepsilon}{\partial \tau}\right)_T \equiv J_T \qquad = \text{ compliance (reciprocal of modulus)}$$

Thermal

$$T\left(\frac{\partial S}{\partial T}\right)_P \equiv C_P \qquad = \text{ heat capacity at constant pressure}$$

$$T\left(\frac{\partial S}{\partial T}\right)_V \equiv C_V \qquad = \text{ heat capacity at constant volume}$$

Thermomechanical (Coupled)

$$\frac{1}{V}\left(\frac{\partial V}{\partial T}\right)_P \qquad = \text{ volumetric thermal expansion coefficient}$$

$$\left(\frac{\partial \varepsilon}{\partial T}\right)_\tau \qquad = \text{ linear thermal expansion coefficient}$$

When the solute is a liquid, it is sometimes convenient to write equation D-73 in terms of P_s^0, as in Chapter 6. Clearly, then, a different activity coefficient γ_n' must be defined

$$P_s = P_s^0 \gamma_n' n_s \tag{D-74}$$

Since γ_n' approaches unity as $n_s \rightarrow 1$, it must approach a value other than unity as $n_s \rightarrow 0$. In this case, the standard state for pure solute is real and Henry's constant in the region of ideality is $P_s^0 \gamma_n'$.

REFERENCES

1. H. B. Callen, *Thermodynamics*, Wiley, New York (1960).
2. I. Prigogine and R. Defay, *Chemical Thermodynamics*, Longmans, Green, London (1954).
3. K. G. Denbigh, *The Principles of Chemical Equilibria*, 3rd ed., Cambridge University Press, Cambridge, England (1971).

APPENDIX E

Chemical Kinetics

Following Boudart (1), we distinguish between a reaction rate function and a rate equation. The latter consists of a rate function together with a material balance equation. The rate function,

$$r = f(c_c, T) \tag{E-1}$$

may be interpreted as a scalar constitutive equation (Appendix F) for chemical reaction. It has dimensions of moles per unit volume per unit time and is an intensive quantity that depends on intensive variables only, such as molar concentrations of species $c_1, \ldots, c_N$ and reaction temperature. It is defined for each reaction as the rate of increase with time of the extent of that reaction ξ (Appendix D). Since $c_c(x_c, t)$ and $T(x_c, t)$, and therefore $r(x_c, t)$, are all time-varying point properties of a reacting system, which may in general be distributed nonuniformly in space within a reactor, the rate equation for a closed system (batch reactor) is

$$\frac{d\xi}{dt} = \int_{V(t)} r \, dV \equiv \langle r \rangle V \tag{E-2}$$

where $\langle \; \rangle$ signifies volume average (Appendix F) and only for spatially uniform reactors do we obtain the simple equation (2, 3)

$$\frac{1}{V} \frac{d\xi}{dt} = r \tag{E-3}$$

The corresponding rate equation for any species k participating in a stoichiometrically simple reaction (single independent equation) can be written as

$$\frac{dN_k}{dt} = \int_{V(t)} r_k \, dV \equiv \langle r_k \rangle V \tag{E-4}$$

714

by virtue of equation D-42, where the rate function for the formation of species k is

$$r_k = \nu_k r \tag{E-5}$$

For complex (multiple) reactions we have

$$r_k = \sum_j r_{jk} = \sum_j \nu_{jk} r_j \tag{E-6}$$

where r_{jk} represents the rate formation of substance k in reaction j and r_j is the rate of increase of the extent of the jth reaction ξ_j.

$$r_{jk} = \nu_{jk} r_j \tag{E-7}$$

The rate function for each reaction j

$$r_j = f_j(c_c, T) \tag{E-8}$$

could in turn consist of a quotient of summations,

$$r_j = \frac{\sum_i f_{ji}(c_c, T)}{\sum_l f_{ji}(c_c, T)} \tag{E-9}$$

each term of which is itself a rate function. In general, we see then that rate functions can be very complex.

When a reaction is an elementary step, representing events at the molecular level, and is irreversible the mass-action rate law takes on the simple form

$$r = k \prod_j c_j^{n_j} \tag{E-10}$$

where the reaction orders n_j are identical to the corresponding stoichiometric coefficients ν_j in the elementary equation. The more complex forms cited earlier, $f(c_c)$, generally represent the overall rate function for a composite reaction, consisting of a sequence of elementary steps and involving reaction intermediates (free radical, ionic, or catalytic).

The units of k are determined by the product of the reciprocal units of $\prod_j c_j^{n_j}$, which are arbitrary, and the desired units for r, which are moles per unit reaction volume per unit time for homogeneous reactions and moles per unit catalyst area per unit time for solid-catalyzed heterogeneous reactions. Thus the units of k might be (volume mol.$^{-1})^{n-1} t^{-1}$ for homogeneous reactions and (area mol.$^{-1})^{n-1} t^{-1}$ for heterogeneous reactions, where $n = \sum_j n_j$ is the reaction order. Since r must ultimately be expressed per unit volume of reactor occupied, one would additionally require the total surface area per unit weight of catalyst, the density of the catalyst, and the void fraction of the catalyst bed.

Frequently reactions, whether they are simple or complex, obey a rate function like the one in equation E-10. We shall call such reactions kinetically simple. Even when reactions are not kinetically simple, it is common engineering practice to fit rate data to this simple format whenever possible. Thus, for kinetically simple or pseudo simple reactions, we write

$$r = k_{ap}\prod_j c_j^{n_j} \tag{E-11}$$

where reaction orders n_j and apparent rate constants k_{ap} are determined empirically from initial rate $(r)_0$ measurements at various initial concentrations $(c_j)_0$ and temperature (isothermal) levels T_0. Graphical analysis is sometimes used in which the desired quantities appear as slopes and intercepts in logarithmic coordinates.

$$\ln(r)_0 = \sum_j n_j \ln(c_j)_0 + \ln k_{ap} \tag{E-12}$$

The temperature-dependence of rate constants is represented by the Arrhenius expression

$$k = A \exp\left(-\frac{E}{R_g T}\right) \tag{E-13}$$

where A is the preexponential factor and E is the activation energy. Arrhenius plots of $\ln(r_0)$ versus $1/T_0$ yield values for apparent overall activation energy E_{ap}.

When the reaction mechanism is known in detail, application of the mass action rate law to each elementary step leads to an overall composite rate function. If such functions are kinetically simple, then k_{ap} is a lumped rate constant.

$$k_{ap} = \prod_j k_j^{n_j} \tag{E-14}$$

which consists of products and quotients of rate constants for the individual elementary steps, and k_j is the constant for the jth step.

The heat of reaction j is given by (Appendix D)

$$\Delta H_j = \sum_j \nu_{jk} H_k \tag{D-52}$$

and the volume-specific heat generation rate for chemical reactions in general, r_G (energy per unit time per unit volume) is given by

$$r_G = -\sum_j \sum_k H_k r_{jk} = -\sum_j \sum_k H_k \nu_{jk} r_j = -\sum_j \Delta H_j r_j \tag{E-15}$$

Consequences of being stoichiometrically simple are that

$$\Delta H_r = \sum_k \nu_k H_k \tag{E-16}$$

and

$$r_G = -\sum_k H_k r_k = -\Delta H_r r \tag{E-17}$$

but r is not necessarily simple as defined by equation E-11. A consequence of being kinetically simple is that rate function r is separable, consisting of the product of a concentration-dependent term and a temperature-dependent term. Thus, by substituting equation E-13 into E-14 and the result into equation E-11, we obtain

$$r = A_{ap} \prod_k c_k^{n_k} \exp\left(-\frac{E_{ap}}{R_g T}\right) \tag{E-18}$$

By substituting this result into equation E-17, we conclude that the heat function r_G, too, is separable, assuming of course that ΔH_r is constant. To facilitate our analysis of thermal runaway in Chapter 4, we define all reactions whose functions r_G may be described by the single product $-\Delta H_r r$ as thermally simple. When a thermally simple reaction is kinetically simple (or pseudo simple) in addition, then r_G, of course, takes on the convenient form

$$r_G = -\Delta H_r A_{ap} \prod_k c_k^{n_k} \exp\left(-\frac{E_{ap}}{R_g T}\right) \tag{E-19}$$

In the text, frequent use is made of dimensionless rate functions, defined in general as

$$\hat{r} \equiv \frac{r}{r_r} \tag{E-20}$$

where subscript r denotes an arbitrary reference state. When rate function r is kinetically simple, this definition gives

$$\hat{r} = k_{ap} \prod_k \hat{c}_k^{n_k} \tag{E-21}$$

where $\hat{c}_k \equiv c_k/(c_k)_r$. The dimensionless rate constant $\hat{k}_{ap}$ may be written in two alternative forms: one in terms of dimensionless temperature $\hat{T} \equiv (T - T_r)/T_r$ and the other in terms of dimensionless temperature $\theta \equiv \hat{E}\hat{T}$, where $\hat{E} \equiv E_{ap}/R_g T_r$ is dimensionless activation energy. Thus,

$$\hat{k}_{ap} \equiv \frac{k_{ap}}{(k_{ap})_r} = \exp\frac{\hat{E}\hat{T}}{(1+\hat{T})} = \exp\frac{\theta}{(1+\varepsilon\theta)} \tag{E-22}$$

where $\varepsilon \equiv \hat{E}^{-1}$. Temperature θ is preferred for reasons that are discussed in Chapter 4, and additionally because for "hot" reactions (large $\hat{E}$, small ε) expression E-22 simplifies to

$$\hat{k}_{ap} \cong \exp\theta \tag{E-23}$$

REFERENCES

1. M. Boudart, *Kinetics of Chemical Processes*, Prentice-Hall, Englewood Cliffs, NJ (1968).
2. K. G. Denbigh and J. C. R. Turner, *Chemical Reactor Theory*, 2nd ed., Cambridge University Press, Cambridge, England (1971).
3. I. Prigogine and R. Defay, *Chemical Thermodynamics*, Longmans, Green, London (1954).

APPENDIX F

Transport Theory

Balance equations are mathematical statements of the physical laws of conservation of mass, momentum (Newton's second law of motion), and energy (first law of thermodynamics), and the nonconservation of entropy (second law of thermodynamics). They state that the rate of accumulation by a system of an extensive property Q (Appendix D) is equal to the net rate with which Q is transported into the system across its boundaries plus the generation rate of Q within it (if Q is not conserved). Thus we have

$$\frac{dQ}{dt} = \dot{Q}_{\text{net in}} + R_Q \tag{F-1}$$

where the dot signifies transport rate.

FLUXES AND FLOW RATES

When applying equation F-1 to specific systems it is usually more convenient to write it in terms of point functions $q'(x_c, t)$, $f_Q(x_c, t)$, and $r_Q(x_c, t)$

$$\frac{d}{dt} \int_V q' \, dV = - \int_S \bar{n} \cdot (f_Q) \, ds + \int_V r_Q \, dV \tag{F-2}$$

where $q' \equiv \rho q$ is volume-specific (intensive) property Q, f_Q is the flux (flow rate per unit surface area) of property Q, and r_Q is the generation rate of Q per unit volume. Since the unit normal vector n points away from the surface S surrounding the volume V of the system in question, the normal component of f_Q

$$f_n = n \cdot f_Q \tag{F-3}$$

is negative for inflow and positive for outflow. Thus the surface integral in equation

F-2 signifies net outflow and

$$\dot{Q}_{\text{net in}} = -\int_S \boldsymbol{n} \cdot \boldsymbol{f}_Q \, ds = -\int_S f_n \, ds \qquad (F\text{-}4)$$

Flux $\boldsymbol{f}_Q$ represents either $\boldsymbol{n}_Q$ or $\boldsymbol{j}_Q$, depending on whether transport across S is convective or molecular (conductive), respectively.

In the former case, $\boldsymbol{f}_Q$ is the total flux of property Q including convective and molecular transport (1)

$$\underset{\text{total}}{\boldsymbol{n}_Q} = \underset{\text{convective}}{q'\boldsymbol{v}} + \underset{\substack{\text{conductive} \\ \text{or molecular}}}{\boldsymbol{j}_Q} \qquad (F\text{-}5)$$

where $\boldsymbol{v}$ is the mass-average velocity. In the latter case, $\boldsymbol{f}_Q$ is simply $\boldsymbol{j}_Q$. Numerous examples of flux vectors $\boldsymbol{n}_Q$ and $\boldsymbol{j}_Q$ are listed in Table F-1. The inclusion of $\boldsymbol{T}$ and $\rho \boldsymbol{v v}$ requires further explanation. First, they are both tensors of rank two. Perhaps, therefore, it would have been more appropriate to list for Q the j-component of momentum, which is a scalar quantity. Properties q and q' would then be v_j and ρv_j, respectively, and the transport fluxes for component j would be vectors (columns) $\rho \boldsymbol{v} v_j$ and $\boldsymbol{T}_j$, shown below. Second, it is necessary in either case to give stress to the fluid mechanical interpretation of conductive (molecular) momentum flux. For example, while row matrix $[T_{j1} \, T_{j2} \, T_{j3}]$ conventionally represents the stress vector acting on component j of a differential surface element, we now interpret column matrix

$$\begin{bmatrix} T_{1j} \\ T_{2j} \\ T_{3j} \end{bmatrix} \equiv \boldsymbol{T}_j$$

as the molecular flux for component j of momentum, analogous to the interpretation

Table F-1. Transport Properties and Fluxes

Q	q	q'	$\overline{\overline{n}}_Q$	$\overline{\overline{j}}_Q$
m	1	ρ	$\rho \overline{v}$	0
m_k	w_k	ρ_k	$\rho_k \overline{v} + \overline{j}_k = \overline{n}_k$	$\overline{j}_k$
V	ρ^{-1}	1	0	$-\overline{v}$
$m\overline{v}$	$\overline{v}$	$\rho \overline{v}$	$\rho \overline{\overline{vv}} + \overline{\overline{T}}$	$\overline{\overline{T}} \equiv \overline{\overline{\tau}} + p\overline{\overline{I}}$
E	e	ρe	$\rho e \overline{v} + \overline{j}_e$	$\overline{j}_e \equiv \overline{q} + \overline{\overline{T}} \cdot \overline{v} + \sum_k H_k \overline{j}_k$

Double overbar = tensor of rank 2; single overbar = tensor of rank 1 (vector).

of column matrix

$$\begin{bmatrix} \rho v_1 v_j \\ \rho v_2 v_j \\ \rho v_3 v_j \end{bmatrix} \equiv \rho \boldsymbol{v} v_j$$

as the convective flux for component j of momentum. Thus vectors $\boldsymbol{n} \cdot \rho \boldsymbol{vv}$ and $\boldsymbol{n} \cdot \boldsymbol{T}$ represent the transport rates per unit area of all three components of momentum across a differential surface oriented in the direction of its unit normal vector $\boldsymbol{n}$. The more conventional interpretation of $\boldsymbol{n} \cdot \boldsymbol{T}$ is, of course, the stress vector acting on a differential surface with orientation $\boldsymbol{n}$.

In general, each j_Q is related to at least one driving force from among the gradients ∇P of general intensive property P (Appendix D) by a constitutive tensor equation, whose simplest form is linear

$$j_Q = \boldsymbol{k} \nabla P \tag{F-6}$$

For uncoupled systems, corresponding flux-force pairs $j_Q, \nabla P$ are related by a simple scalar equation

$$j_Q = -k \nabla P \tag{F-7}$$

where k is a conductivity property. To be thermodynamically consistent, such pairs must satisfy the equation for rate of entropy production (2)

$$r_S = -\sum_i j_{Qi} \cdot \nabla P_i \tag{F-8}$$

which follows from the Gibbs equation (Appendix D) after substitution of the appropriate balance equations for energy, material and entropy. The inequality

$$r_S \geqslant 0 \tag{F-9}$$

is a statement of the second law of thermodynamics.

Fourier's law and Fick's law are examples of equation F-7. Corresponding sets of transport fluxes, constitutive equations, and conductivity properties are listed in Table F-2. Fluxes and forces in Newton's law are represented by tensors of rank two instead of one, as noted earlier, because the extensive property transported, namely momentum, is a vector. It should be noted that only the dynamic part of the stress tensor, τ, appears in Table F-2, and that it is related only to the symmetric part of the deformation rate tensor, ∇v_s. The static part of the stress tensor represents hydrostatic pressure pI.

The symmetrical strain rate tensor ∇v_s may be obtained from the deformation rate tensor ∇v by Cauchy decomposition (3)

$$\nabla v = \nabla v_s + \nabla v_a \tag{F-10}$$

Table F-2.   Constitutive Equations

Diffusion	$\bar{j}_k = -\rho \nabla w_k$	Fick's law
Heat Transfer	$\bar{q} = -k \nabla T$	Fourier's law
Viscous flow	$\bar{\bar{\tau}} = -\eta \nabla \bar{v}_s$	Newton's law

Double overbar = tensor of rank 2; single overbar = tensor of rank 1 (vector).

where

$$\nabla v_s \equiv \frac{\nabla v + \nabla v^T}{2} \tag{F-11}$$

and ∇v_a is the antisymmetric rotation rate tensor.

Another example, not included, is the chemical rate law in which the fluxes and forces are scalar quantities (Appendix E).

All balances are classified as macroscopic or microscopic (infinitesimal), depending on the system to which they apply. A system is a region in space of arbitrary size and shape whose boundaries may be real (e.g., the walls and openings of a reaction vessel) or imaginary (e.g., a differential volume element or "shell" within a reactor). Differential regions may be further subdivided into two fundamentally distinct types, the control volume (CV) and the control mass (CM). They differ from one another in that the surface of the CV is fixed in space whereas that of the CM, $S(t)$, moves with the mass-average velocity v of the matter lying on it (1). The choice between them is dictated by convenience. We use the CV exclusively in the text.

MACROSCOPIC BALANCES

For any macroscopic CV of arbitrary size and shape, equation F-2 can be written in any one of several equivalent forms:

$$\frac{d}{dt}(\langle q' \rangle V) = -\sum_i \{n_{Qn}\}_i S_i - \{j_{Qn}\}_w S_w + \langle r_Q \rangle V \tag{F-12}$$

$$= -\sum_i \{q'v_n\}_i S_i - \sum_i \{j_{Qn}\}_i S_i - \{j_{Qn}\}_w S_w + \langle r_Q \rangle V \tag{F-13}$$

$$= -\Delta([q']\dot{V}) - \sum_i \{j_{Qn}\}_i S_i - \{j_{Qn}\}_w S_w + \langle r_Q \rangle V \tag{F-14}$$

$$= -\Delta([q]\dot{m}) - \sum_i \{j_{Qn}\}_i S_i - \{j_{Qn}\}_w S_w + \langle r_Q \rangle V \tag{F-15}$$

where subscript n signifies the flux vector component normal to the control surface and subscript w signifies control surfaces other than inlets and outlets (S_i). The symbols Σ and Δ represent the algebraic sum and net difference, respectively, of effluent and feed rates (out minus in). The brackets $\langle \rangle$ represent volume-average point variables. In general for any volume-specific variable ζ

$$\langle \zeta \rangle \equiv \frac{\int_V \zeta \, dV}{V} \tag{F-16}$$

Thus ζ includes, but is not limited to, the volume-specific quantity q'. It also includes such quantities as r_Q and c_k. We note that $\rho_k = \rho w_k$. The brackets $\{ \}$ represent area-average flux components normal to the control surface. In general for any flux f_Q

$$\{ f_n \} \equiv \frac{\int_S f_n \, ds}{S} \tag{F-17}$$

The brackets [ ] represent flow-average (cup-average) values of point variables over portions of the control surface across which bulk flow (convection) occurs, such as inlets and outlets. In general

$$[q'] \equiv \frac{\int_S q' v_n \, ds}{\dot{V}} \tag{F-18}$$

and

$$[q] \equiv \frac{\int_S q \rho v_n \, ds}{\dot{m}} \tag{F-19}$$

are volumetric flow average and mass flow average, respectively, where

$$\dot{V} \equiv \int_S v_n \, ds \tag{F-20}$$

and

$$\dot{m} \equiv \int_S \rho v_n \, ds \tag{F-21}$$

are volumetric and mass flow rates, respectively, across S. In particular, for incompressible fluids

$$[q'] = \rho[q] \tag{F-22}$$

Finally, it should be noted that the area-average term with subscript w includes such diverse quantities as work rate due to motion of control surfaces, heat rate due to conduction through control surface walls, and diffusion through a porous control surface.

MICROSCOPIC BALANCES

When the control region is microscopic, the fundamental differences between CV and CM are reflected in two distinct forms for the term on the LHS and the first term on the RHS of equations F-1 and F-2. With the aid of the Reynolds transport theorem (Leibnitz's rule for differentiating integrals) (3), we obtain for the term on the LHS

CV:
$$\frac{dQ}{dt} = \frac{d}{dt} \int_V q' \, dV = \int_V \frac{\partial q'}{\partial t} \, dV \tag{F-23}$$

CM:
$$\frac{dQ}{dt} = \frac{d}{dt} \int_{V(t)} q' \, dV = \int_{V(t)} \rho \frac{Dq}{Dt} \, dV \tag{F-24}$$

where the limit $V(t)$ reminds us that the control surface moves with the local mass average fluid velocity v and D/Dt is the substantial derivative operator (1),

$$\frac{D}{Dt} = \frac{\partial}{\partial t} + v \cdot \nabla \tag{F-25}$$

which expresses the time rate of change experienced by properties moving with the CM. On the RHS, we obtain

CV:
$$\dot{Q}_{\text{net in}} = - \int_S n \cdot n_Q \, ds \tag{F-26}$$

and

CM:
$$\dot{Q}_{\text{net in}} = - \int_{S(t)} n \cdot j_Q \, ds \tag{F-27}$$

In the latter, we need only flux j_Q because it represents transport relative to that of convecting surface (v) of the CM. If these respective balance statements are applied to arbitrary control regions of both types and all surface integrals are transformed via the divergence theorem

$$\int_S n \cdot f_Q \, ds = \int_V \nabla \cdot f_Q \, dV \tag{F-28}$$

we can deduce the following general differential balances (3)

CV:
$$\frac{\partial q'}{\partial t} = - \nabla \cdot n_Q + r_Q$$

CM:
$$\rho \frac{Dq}{Dt} = - \nabla \cdot j_Q + r_Q \tag{F-29}$$

MATERIAL BALANCES

For any macroscopic CV, the total and component material balances in the form of equation F-1 are, respectively,

$$\frac{dm}{dt} = \dot{m}_{\text{net in}} \tag{F-30}$$

and

$$\frac{dm_k}{dt} = (\dot{m}_k)_{\text{net in}} + M_k r_k \tag{F-31}$$

Alternatively, from equations F-12–F-15:

$$\frac{d}{dt}(\langle \rho \rangle V) = -\sum_i \{\rho v_n\}_i S_i = -\Delta([\rho]\dot{V}) \tag{F-32}$$

$$\frac{d}{dt}(\langle \rho_k \rangle V) = \sum_i \{n_{kn}\}_i S_i + \langle r_k \rangle V M_k \tag{F-33}$$

$$= -\Delta([\rho_k]\dot{V}) - \sum_i \{j_{kn}\}_i S_i + M_k \langle r_k \rangle V \tag{F-34}$$

$$= -\Delta([w_k]\dot{m}) - \sum_i \{j_{kn}\}_i S_i + M_k \langle r_k \rangle V \tag{F-35}$$

and for microscopic CV's in vector form (1)

$$\frac{\partial \rho}{\partial t} = -\nabla \cdot \rho v \tag{F-36}$$

and

$$\frac{\partial \rho_k}{\partial t} = -\nabla \cdot n_k + M_k r_k \tag{F-37}$$

ENERGY BALANCES

In most thermodynamics texts, the first law is written for internal energy (U) only and applied to closed systems. More generally, for total energy (E) and arbitrary CV's, it takes the form

$$\frac{dE}{dt} = \dot{E}_{\text{net in}} + \dot{Q} + \dot{W} \tag{F-38}$$

where $E = KE + PE + U$, $\dot{Q}$ represents the net rate of thermal energy (heat) transported in (not to be confused with the more general term $\dot{Q}_{\text{net in}}$ in equations F-1 and F-4) and $\dot{W} = -\int_S n \cdot T \cdot v \, ds$ represents the net rate of mechanical energy transported in (total work rate on system) including flow work (e.g., pV-type) at inlets and outlets as well as stress work due to moving control surfaces (shaft work, drag flow, etc.). Thus we see that $T \cdot v$ may be interpreted as the mechanical energy flux vector due to stresses. In terms of intensive properties we obtain, by neglecting conductive (molecular) transport across inlets and outlets (except p) and assuming that v is normal to these control surfaces

$$\frac{d}{dt}(\langle e' \rangle V) = -\sum_i \left\{ h' v_n + \frac{\rho v_n^3}{2} + \rho g y v_n \right\} S_i - \{q_n\}_w S_W + \dot{W}_s \quad \text{(F-39)}$$

where pV-type flow work has been included in the first term on the RHS, as is customary, so that $h' v_n = (u' + p) v_n$ and $\dot{W}_s$ signifies work rate due to moving control surfaces (walls) only. Distance y is measured from an arbitrary datum position in the gravitational field. For a microscopic control volume, including molecular flux contributions but neglecting coupled transport, we obtain

$$\frac{\delta e'}{\partial t} = -\nabla \cdot e' v - \nabla T \cdot v - \nabla \cdot q - \sum_k \nabla \cdot H'_k j_k \quad \text{(F-40)}$$

where q refers only to the thermal energy flux that is due to ∇T.

The familiar form of the first law of thermodynamics is actually a thermal energy balance (1). For an arbitrary macroscopic CV, this balance is

$$\frac{dU}{dt} = \dot{H}_{\text{net in}} + \dot{Q} + \dot{W}'_s \quad \text{(F-41)}$$

or

$$\frac{d}{dt}(\langle u' \rangle V) = -\sum_i \{h' v_n\}_i S_i - \{q_n\}_w S_w + \dot{W}'_s \quad \text{(F-42)}$$

where $\dot{W}'_s$ signifies that part of the total work rate ($\dot{W}$), which affects internal energy U only and excludes the remainder, which affects mechanical energy (kinetic and potential, KE and PE). Such a separation of work and energy is difficult to delineate for macroscopic systems in general. The thermal energy balance is virtually always the energy balance applied to chemical reactors. Since KE and PE changes are generally negligible, which negates the problem of separation and is tantamount to stating that $\dot{W}_s \cong \dot{W}'_s$. The remaining forms of the macroscopic balance follow by analogy with other balances.

For differential control systems, on the other hand, the work terms are readily distinguishable. Of the total specific work rate, $\nabla \cdot T \cdot v$, the "motion-work" part $v \cdot \nabla \cdot T$ contributes to mechanical energy (kinetic and potential) and the deformation-work part $T : \nabla v$ contributes to internal energy. Thus the microscopic thermal energy balance in control-mass form is

$$\rho \frac{Du}{Dt} = -\nabla \cdot q - \sum_k \nabla \cdot H'_k j_k - p \nabla \cdot v - \tau : \nabla v \quad \text{(F-43)}$$

where the last two terms on the RHS comprise $\dot{W}'_s$ per unit volume. The penultimate term represents pV-type deformation work. The last term represents viscous dissipation due to deformation and is a source term for thermal energy. When the thermal energy balance is written this way, its resemblance to the familiar form of the first law for systems of constant mass is apparent.

Application of thermal energy balances to chemical reactors requires that they be written in terms of temperature. To accomplish this for the microscopic balance, it is most convenient to begin with the Gibbs equation (Appendix D) in specific mass form

$$du = T ds - p d\rho^{-1} + \sum_k G'_k dw_k \tag{F-44}$$

where $ds = (c_p/T) dT - (\partial\rho^{-1}/\partial T)_{p,w_c} dp + \sum_k S'_k dw_k$ and s and S'_k represent specific entropy and partial entropy, respectively. Substitution of equation F-43 for du/dt and the k-component balance in control-mass form (transformed via equation F-25), $\rho(Dw_k/Dt) = -\nabla \cdot \boldsymbol{j}_k + r_k$, for dw_k/dt yields the following, commonly used microscopic thermal energy balance in control-volume form (transformed via equation F-25)

$$\rho c_p \frac{\partial T}{\partial t} = r_G - \rho c_p \boldsymbol{v} \cdot \nabla T - \nabla \cdot \boldsymbol{q} \tag{F-45}$$

if we neglect the thermal effects of interdiffusion ($\sum_k \boldsymbol{j}_k \cdot \nabla H'_k$), viscous dissipation ($\boldsymbol{\tau} : \nabla \boldsymbol{v}$) compared to reaction heat generation, $r_G = -\sum_k H_k r_k$ (Appendix E), and the term $(\partial \ln \rho^{-1}/\partial \ln T)\left(\dfrac{DP}{Dt}\right)$, which is zero when either density or pressure is constant.

Expressing the thermal energy balance in terms of temperature for macroscopic systems in general is difficult. It is preferable to write such balances for specific systems, such as the PFR and the MMIM, as required.

MOMENTUM BALANCE

The momentum balance for a differential CV in vector-tensor form is (1)

$$\frac{\partial \rho \boldsymbol{v}}{\partial t} = -\nabla \cdot \rho \boldsymbol{v} \boldsymbol{v} - \nabla p - \nabla \cdot \boldsymbol{\tau} + \rho \boldsymbol{g} \tag{F-46}$$

where $\rho \boldsymbol{v} \boldsymbol{v}$ is a tensor representing convective momentum transport rate and $\boldsymbol{g}$ represents a body force (gravitational) field.

SOME SPECIFIC SYSTEMS

For most reactors of practical interest molecular transport in and out, $\langle j_{kn}\rangle_i S_i$, is negligibly slow compared to convective transport. Furthermore, it is customary to use molar concentrations for reactive components. Thus, following division by molecular weight M_k, equation F-34 becomes

$$\frac{dN_k}{dt} = \frac{d}{dt}(\langle c_k\rangle V) = -\Delta([c_k]\dot{V}) + \langle r_k\rangle V \tag{F-47}$$

The batch reactor (BR) is a closed system. Consequently, the first term on the RHS is zero. Reaction density ρ is a variable in general but is frequently treated as a constant for liquid-phase reactions, even for polymerizations where material shrinkage may occur. Equation F-47 thus reduces to

$$\frac{d\langle c_k \rangle}{dt} = \langle r_k(c_c, T) \rangle \tag{F-48}$$

The ideal BR is assumed to be perfectly homogeneous due to complete microscopic mixing down to the molecular level. This renders all intensive variables such as concentrations, temperature, and pressure spatially uniform and reduces equation F-48 to the rate equation for component k commonly used in chemical kinetics

$$\frac{dc_k}{dt} = r_k(c_c, T) \tag{F-49}$$

This balance equation applies at once to the steady-state plug-flow reactor (PFR) as well, which is an idealized streamline reactor without residence time distribution or backmixing (Chapter 3), if we simply replace time t with age $\alpha = x/v$, where v is the uniform (plug) velocity and x is longitudinal position.

Continuous-flow reactors are open systems and are usually operated under steady-state conditions. Consequently, the transient term on the LHS is zero. By neglecting density changes, as before, $\Delta \dot{V} = 0$ and we obtain the equivalent of equation F-48 for continuous-flow vessels with a single inlet and outlet

$$\frac{\Delta[c_k]}{\lambda_v} = \langle r_k(c_c, T) \rangle \tag{F-50}$$

where $\lambda_v \equiv V/\dot{V}$ is the vessel space time.

The ideal mixer (IM) is a continuous-flow vessel which is assumed to be spatially uniform owing to complete microscopic mixing. Such uniformity eliminates the need for area-averaging and volume-averaging. Thus we obtain for component k

$$\frac{\Delta c_k}{\lambda_v} = r_k(c_c, T) \tag{F-51}$$

which is a rate equation like F-49, but is algebraic instead of differential. A common industrial reactor whose characteristics under ideal conditions are represented by the IM is the continuous stirred tank reactor (CSTR).

It should be noted that all inlet concentrations were assumed to be uniform due to the absence of reaction. Furthermore, all effluent concentrations in the IM are identical to their corresponding internal concentrations by the assumption of perfect mixing. The corresponding transient balance for constant reaction density is

$$\frac{dc_k}{dt} = \frac{(c_k)_0 - c_k}{\lambda_v} + r_k(c_c, T) \tag{F-52}$$

Writing a general macroscopic thermal energy balance in terms of temperature to parallel equation F-45 presents a problem, as previously noted. Since most texts in chemical reaction engineering deal with specific, idealized reactors only (PFR, dispersion model, CTR, CSTR), the need for such a balance does not arise. However, in keeping with our attempts to generalize the classification of reactors, it seems appropriate to give this matter some attention.

We begin by rewriting equation F-45, assuming constant ρ and c_p, as

$$\frac{\partial(\rho c_p T)}{\partial t} = r_G - \nabla \cdot \rho c_p T v - \nabla \cdot q \tag{F-53}$$

with the aid of the continuity equation (F-36). Then we integrate this result over an arbitrary CV, applying Leibnitz's rule and the divergence theorem, to obtain

$$\rho c_p \frac{d}{dt}(\langle T \rangle V) = \langle r_G \rangle V - \rho c_p \Delta([T]\dot{V}) - \{q_n\}_w S_w \tag{F-54}$$

where molecular conduction across the inlets and outlets has again been neglected. The appearance of flow-average temperature in the convection term is noteworthy. For the special case of an IM with constant volume V, equation F-54 reduces to

$$\rho c_p \frac{dT}{dt} = r_G + \frac{\rho c_p}{\lambda_v}(T_0 - T) - Us_v(T - T_R) \tag{F-55}$$

where U is an overall heat transfer coefficient. For a streamline vessel of constant cross-sectional area S, equation F-54, applied to the differential control volume $V = S\Delta x$, in the limit $\Delta x \to 0$ becomes

$$\rho c_p \frac{\partial}{\partial t}\{T\} = \{r_G\} - \rho c_p\{v\}\frac{\partial[T]}{\partial x} - \{q_n\}_w S_w \tag{F-56}$$

When spatial variations are absent, we may write this result simply as

$$\rho c_p \frac{dT}{d\alpha} = r_G - Us_v(T - T_R) \tag{F-57}$$

which, like equation F-49, applies at once to the ideal BR ($\alpha = t$) and the steady-state PFR ($\alpha = x/v$).

REFERENCES

1. R. B. Bird, W. E. Stewart, and E. N. Lightfoot, *Transport Phenomena*, Wiley, New York (1960).
2. S. R. DeGroot and P. Mazur, *Non-Equilibrium Thermodynamics*, North-Holland, Amsterdam (1962).
3. R. Aris, *Vectors, Tensors and the Basic Equations of Fluid Mechanics*, Prentice-Hall, Englewood Cliffs, NJ (1962).

APPENDIX G

Characteristic Times and Dimensionless Groups

A characteristic time (CT) or time constant is a measure of the time scale associated with a rate process. The CT is a powerful tool in both analysis and design of complex processes. By simply comparing CT's for various simultaneous rate processes, one can often make deductions about those processes that are surprisingly accurate. Most dimensionless groups used in engineering design and scaling may be expressed as ratios of CT's. Heavy use is made of this method in the text.

A characteristic time that gives at least an order-of-magnitude indication of process time (seconds, minutes, hours, etc.) may be defined intuitively for any rate process, even when no mathematical model is available. A more precise CT may be derived from a mathematical model. Relaxation time and half-life are examples of CT's for first-order (linear) and second-order processes, respectively. We use the symbol λ (with an appropriate subscript) throughout the text to represent CT. Numerous examples are listed in Table G-1.

CT's are used throughout the text to classify reactions (e.g., step versus chain) and reactors (e.g., microsegregated versus backmixed, runaway versus nonrunaway), to predict reaction phenomena (e.g., monomer conversion and direction and magnitude of polymer property drift), and effects of imperfect mixing and nonuniform temperatures on monomer conversion and polymer properties. The strategy used throughout consists of the development of criteria for the dominance among concurrent, competitive processes of one over the others. Such criteria involve the ratios of CT's in the form of dimensionless groups. Some well-known dimensionless groups are listed in Table G-2.

Examples are the Damköhler, Peclet, and Nusselt numbers: $\lambda_v/\lambda_r \equiv \mathrm{Da}$, $\lambda_D/\lambda_v \equiv \mathrm{Pe}_D$ and $\lambda_H/\lambda_R \equiv \mathrm{Nu}$. Clearly, then, Da compares convection with reaction, Pe_D compares diffusion with convection, and Nu compares internal with external heat transfer. In the absence of deeper knowledge, the dominant process may be identified by comparing the appropriate dimensionless group to unity. For instance, high conversions may be expected when $\mathrm{Da} \gg 1$, and diffusion should be relatively unimportant if $\mathrm{Pe}_D \gg 1$. The search for more precise numerical criteria requires detailed modeling and/or experimentation and occupies a significant portion of the text. An example is the criterion for averting thermal runaway in lumped-parameter polymerizers, $a \equiv \lambda_{ad}/\lambda_H > a_{cr}$ where, instead of $a_{cr} = 1$, we find more precisely that $2 \leqslant a_{cr} \leqslant e$.

As an aid to the intuitive approach to defining CT's, we offer the following generalization. It appears that the CT for any process is the ratio of some measure of the amount of extensive property Q (Appendix D) stored to the rate of change of that property. Examples are space time, $\lambda_v = V/\dot{V}$, and the CT for reaction of substance k, $\lambda_k = c_k/r_k$, where c_k and r_k are evaluated at some reference condition, such as feed.

Frequently CT is the quotient of a capacitance property C and a conductance property K

$$\lambda = \frac{C}{K} \tag{G-1}$$

Table G-1.   Some Characteristic Times

CLASSIFICATION	CT	DEFINITION	PROCESS
KINETIC	λ_k	$(c_k)_r/(r_k)_r$	reaction of substance k
	λ_G	$\rho c_p T_r/(r_G)_r$	reaction heat generation
TRANSPORT	λ_D	ℓ^2/D	diffusion
	λ_H	ℓ^2/α_T	heat conduction
	λ_M	ℓ^2/ν	momentum transport by shear
	λ_c	ℓ_L/v_r	convection; ℓ_L is longitudinal dimension
	λ_s	ℓ_T/v_r	shear deformation; ℓ_T is transverse dimension
	λ_R	$\rho c_p \ell/U$	convective heat transfer
	λ_F	$\rho v_r \ell_L / \lvert \Delta P \rvert$	momentum transport by pressure
	λ_v	$V/\dot{V}_r$	volumetric throughput
	λ_{MR}	η/G	mechanical relaxation
	λ_G'	$c_p T_r \ell^2/\nu v_r^2$	viscous heat generation
	λ_g	v_r/g	buoyancy

Table G-2. Some Dimensionless Groups

CLASSIFICATION	SYMBOL	DEFINITION		NAME
		IN TERMS OF CT's	IN TERMS OF VARIABLES	
KINETIC				
	Da_I or Da	λ_c/λ_r or λ_v/λ_r	$Vr_r/c_r\dot{V}$	Damköhler number
	Da_{II}	λ_D/λ_r	$r_r\ell^2/Dc_r$	Damköhler number
	Da_{IV}	λ_H/λ_G	$\ell^2(r_G)_r/kT_r$	Damköhler number
TRANSPORT				
	Br	λ_H/λ'_G	nv_r^2/kT_r	Brinkman number
	Re	λ_M/λ_s	$\ell_T v_r/\nu$	Reynolds number
	Po	λ_M/λ_F	$\ell_T^2|\Delta P|/nv_r\ell_L$	Poiseuille number
	Eu	λ_c/λ_F	$|\Delta P|/\rho v_r^2$	Euler number
	Deb	λ_{MR}/λ_s	$nv_r/G\ell$	Deborah number
	Sc	λ_D/λ_M	ν/D	Schmidt number
	$Pe_D = ReSc$	λ_D/λ_c	$\ell v_r/D$	Peclet number for diffusion
	$Le = Sc/Pr$	λ_D/λ_H	α_T/D	Lewis number
	Pr	λ_H/λ_M	ν/α_T	Prandl number
	$Pe_H = RePr$	λ_H/λ_c	$\ell v_r/\alpha_k$	Peclet number for heat transfer
	$Nu = PeSt$	λ_H/λ_R	$U\ell/k$	Nusselt number
	St	λ_c/λ_R	$U/\rho c_p v_r$	Stanton number
	Fo	λ_v/λ_H	$\lambda_v\alpha_T/\ell^2$	Fourier number

Capacitance $C = \rho c V$ arises in thermodynamic equations of state (Appendix D)

$$dQ = C\,dP \qquad\qquad (G\text{-}2)$$

and conductance $K = kS/l$ derives from transport constitutive equations (Appendix F)

$$j_Q = k\nabla P \qquad\qquad (G\text{-}3)$$

where c and k are specific capacity and conductivity, respectively, and V, S, and l and characteristic volume, area, and length. After substituting for C and K, we obtain (assuming $S/V \equiv s_v = l^{-1}$)

$$\lambda = l^2 \frac{\rho c}{k} \qquad\qquad (G\text{-}4)$$

where the quantity $k/\rho c$ is a generalized diffusivity. Some CT's are listed in Table G-1 and numerous others specific to polymerizations appear throughout the text. Subscript r refers to a reference (arbitrary) condition.

Examples of equation G-4 are the CT for heat conduction $\lambda_H = l^2/\alpha_T$, where α_T is thermal diffusivity, and the CT for diffusion $\lambda_D = l^2/D$. Another for momentum transport by shear $\lambda_M = l^2/\nu$ follows as well if we interpret m as the capacitance property for momentum and ρ as the specific capacity (1). Thus, consistent with equation G-2, $Q = mv$, $P = v$, and kinematic viscosity $\nu = \eta/\rho$ is the corresponding diffusivity. Kinetic energy $mv^2/2$ is the potential function (Appendix D) relative to conjugate variables P and Q. For mechanical relaxation of polymeric materials, the characteristic time λ_{MR} (Table G-1) may be interpreted as the quotient of compliance (capacity) G^{-1} and fluidity (conductivity) η^{-1}.

We shall now propose a formal procedure for obtaining CT's starting with the balance equations in terms of intensive variables, which can always be written in the following general form, whether they are microscopic (distributed parameter) or macroscopic (lumped parameter):

$$\left.\begin{array}{ll}\text{micro} & \dfrac{\partial p}{\partial t}\\[2ex] \text{macro} & \dfrac{dp}{dt}\end{array}\right\} = \sum_j \dot{p}_j \qquad\qquad (G\text{-}5)$$

Quantity P represents any intensive property (Appendix D) including specific properties q and q' (Appendix F) as well as specific momentum v, and $\dot{p}_j$ represents the rate with which process j causes p to increase. When quantities p and $\dot{p}_j$ are made dimensionless through division by their corresponding reference values, and the defining equations

$$\hat{p} \equiv \frac{p}{p_r} \qquad\qquad (G\text{-}6)$$

$$\hat{\dot{p}}_j \equiv \frac{\dot{p}_j}{(\dot{p}_j)_r} \qquad\qquad (G\text{-}7)$$

are used to replace the aforementioned quantities in balance equations G-5, the resulting balance equations take on the general form

$$\left. \begin{array}{ll} \text{micro} & \dfrac{\partial \hat{p}}{\partial t} \\[2em] \text{macro} & \dfrac{d\hat{p}}{dt} \end{array} \right\} = \sum_j \lambda_j^{-1} \hat{p}_j \tag{G-8}$$

having dimensions of reciprocal time only. These equations have been designated as "semidimensionless" in the text. Their coefficients provide a formal definition for CT. Thus, for process j,

$$\lambda_j \equiv \frac{p_r}{(\dot{p}_j)_r} \tag{G-9}$$

This definition should lead to the same characteristic times as definition G-4.

As an example, consider the microscopic material balance for substance k

$$\frac{\partial c_k}{\partial t} = r_k + D\left(\frac{\partial^2 c_k}{\partial x^2} + \frac{\partial^2 c_k}{\partial y^2} \right) - v_z \frac{\partial c_k}{\partial z} \tag{G-10}$$

After rendering x, y, and z dimensionless via characteristic lengths X, Y, and L, respectively, v via characteristic velocity v_r and r_k via $(r_k)_r$ by the procedure described in Appendix E, and dividing through by $(c_k)_r$, we obtain the semidimensionless equation

$$\frac{\partial \hat{c}_k}{\partial t} = \frac{(\hat{r}_k)_0}{(c_k)_0} \hat{r}_k + \frac{D}{X^2} \frac{\partial^2 \hat{c}_k}{\partial \hat{x}^2} + \frac{D}{Y^2} \frac{\partial^2 \hat{c}_k}{\partial \hat{y}^2} - \frac{v_z}{L} \frac{\partial \hat{c}_k}{\partial \hat{z}}$$

The reader will recognize the coefficients on the RHS as reciprocals of CT's λ_k, λ_{Dx}, λ_{Dy}, and λ_c, respectively. It should be noted that we have reduced each position variable by its own characteristic length. This leads to a separate CT for diffusion in each transverse direction and a separate one for longitudinal convection.

As noted earlier, virtually all dimensionless groups may be interpreted as ratios of CT's for two simultaneous processes. We can deduce this simply by using any CT, say λ_l (the choice is arbitrary), to reduce time t in equation G-8, thus rendering it completely dimensionless. The result is

$$\left. \begin{array}{ll} \text{micro} & \dfrac{\partial \hat{p}}{\partial \hat{t}} \\[2em] \text{macro} & \dfrac{d\hat{p}}{d\hat{t}} \end{array} \right\} = \sum_j D_{lj} \hat{p}_j \tag{G-11}$$

where $\hat{t} \equiv t/\lambda_l$ and D_{lj} are dimensionless groups

$$D_{lj} \equiv \frac{\lambda_l}{\lambda_j} \tag{G-12}$$

Numerous examples are listed in Table G-2.

REFERENCES

1. R. C. L. Bosworth, *Transport Processes in Applied Chemistry*, Wiley, New York (1956).

Index